Advanced Time Domain Modeling for Electrical Engineering

The ACES Series on Computational and Numerical Modelling in Electrical Engineering

Andrew F. Peterson, PhD – Series Editor

The volumes in this series will encompass the development and application of numerical techniques to electrical and electronic systems, including the modelling of electromagnetic phenomena over all frequency ranges and closely related techniques for acoustic and optical analysis. The scope includes the use of computation for engineering design and optimization, as well as the application of commercial modelling tools to practical problems. The series will include titles for senior undergraduate and postgraduate education, research monographs for reference, and practitioner guides and handbooks.

Titles in the Series

K. Warnick, **"Numerical Methods for Engineering,"** 2010.
W. Yu, X. Yang and W. Li, **"VALU, AVX and GPU Acceleration Techniques for Parallel FDTD Methods,"** 2014.
A.Z. Elsherbeni, P. Nayeri and C.J. Reddy, **"Antenna Analysis and Design Using FEKO Electromagnetic Simulation Software,"** 2014.
A.Z. Elsherbeni and V. Demir, **"The Finite-Difference Time-Domain Method in Electromagnetics with MATLAB Simulations 2nd Edition,"** 2015.
M. Bakr, A.Z. Elsherbeni and V. Demir, **"Adjoint Sensitivity Analysis of High Frequency Structures with MATLAB,"** 2017.
O. Ergul, **"New Trends in Computational Electromagnetics,"** 2019.
D. Werner, **"Nanoantennas and Plasmonics: Modelling, design and fabrication,"** 2020.
K. Kobayashi and P. D. Smith, **"Advances in Mathematical Methods for Electromagnetics,"** 2020.
V Lancellotti, **"Advanced Theoretical and Numerical Electromagnetics, Volume 1: Static, stationary and time-varying fields,"** 2021.
V Lancellotti, **"Advanced Theoretical and Numerical Electromagnetics, Volume 2: Field representations and the method of moments,"** 2021.
S. Roy, **"Uncertainty Quantification of Electromagnetic Devices, Circuits, and Systems,"** 2021.

Advanced Time Domain Modeling for Electrical Engineering

Edited by
Rodolfo Araneo

The Institution of Engineering and Technology

Published by SciTech Publishing, an imprint of The Institution of Engineering and Technology, London, United Kingdom

The Institution of Engineering and Technology is registered as a Charity in England & Wales (no. 211014) and Scotland (no. SC038698).

First published 2022

The Institution of Engineering and Technology
Futures Place
Kings Way, Stevenage
Hertfordshire, SG1 2UA, United Kingdom

www.theiet.org

British Library Cataloguing in Publication Data
A catalogue record for this product is available from the British Library

ISBN 978-1-83953-153-8 (hardback)
ISBN 978-1-83953-154-5 (PDF)

Typeset in India by MPS Ltd

Contents

About the editor

Rodolfo Araneo (S'99, M'03, SM'09) received the M.S. (summa cum laude) and Ph.D. degrees in electrical engineering from the University of Rome "La Sapienza", Rome, Italy, in 1999 and 2002, respectively.

In 1999, he was a Visiting Student at the National Institute of Standards and Technology (NIST), Boulder, CO, USA, where he was engaged in TEM cells and shielding. In the same year, he was recipient of the Past President's Memorial Award from the IEEE Electromagnetic Compatibility Society. In 2000, he was a Visiting Researcher at the Department of Electrical and Computer Engineering, University of Missouri-Rolla (UMR), Rolla, MO, USA, where he was engaged in printed circuit boards and finite-difference time-domain techniques.

He is currently a Full Professor with the DIAEE at the University of Rome La Sapienza.

He has authored more than 150 papers in international journals and conference proceedings.

He is the coauthor of the book "Electromagnetic Shielding" (IEEE Wiley 2008) and "Electrical Safety Engineering of Renewable Energy Systems" (IEEE Wiley 2021).

He serves as a Reviewer for several international journals.

In 2018, he has been nominated Fellow of the Applied Computational Electromagnetics Society (ACES).

Since 2015, he is general chair of the IEEE International Conference on Environment and Electrical Engineering.

His research interests include electromagnetic compatibility, energy harvesting and piezotronics based on piezoelectric ZnO nanostructures, and graphene electrodynamics, the development of numerical and analytical techniques for modeling high-speed printed circuit boards, shielding, transmission lines, periodic structures, devices based on graphene.

Preface

I conceived to propose to IET the book "Advanced Time Domain Modeling for Electrical Engineering" during the ACES Conference that was held in Miami in 2019. After 3 years of work, the first edition was published in the "ACES Series on Computational and Numerical Modelling in Electrical Engineering."

I started studying numerical methods during my master thesis in 1999. I began my study with the finite difference time-domain method. After that, during my PhD course and university career, I studied and programmed quite all the numerical methods with respect to time and frequency domain. In recent years, my interest focused more on modeling phenomena in the time domain, where many aspects still deserve an in-depth research.

In its way, editing this book is a dream come true. The book is composed of 17 chapters. I invited the most prestigious research groups in each field (on the books of many of whom I have studied), whom I thank warmly for having accepted my request for this collaboration. After Chapter 1, which introduces the basic electromagnetic theory in the time domain, the book is composed of six chapters about numerical methods (from 2 to 7) and ten chapters about specific applications (from 8 to 17). The book reviews the basis and applications of the most relevant numerical methods in the time domain, presents substantial novel research studies, and discusses the challenges in this research:

- In Chapter 2, Dr Gedney covers the finite difference method.
- In Chapter 3, Dr Clemens *et al.* cover the finite element method.
- In Chapter 4, Dr Weile and Shanker *et al.* cover the late-time stable integral equations.
- In Chapter 5, Dr Christopoulos covers the transmission line method.
- In Chapter 6, Dr Ruehli and Antonini *et al.* cover the partial element equivalent circuit method.
- In Chapter 7, Dr Teixeira *et al.* cover the unconditionally stable methods.

From Chapter 8, the book reviews the applications of time-domain techniques as a solution for relevant engineering problems:

- In Chapter 8, Dr Grivet-Talocia *et al.* present linear macromodeling.
- In Chapter 9, Dr Andreotti and Verolino present the fundamental problem of a charged particle traveling along the axis of a circular waveguide.
- In Chapter 10, Dr Burghignoli and Lovat present analytical techniques for transient analysis, especially focusing on the Cagniard-de Hoop method.

- In Chapter 11, Dr Rachidi and Rubinstein *et al.* present the application of the finite-difference time-domain technique to lightning studies.
- In Chapter 12, Dr Andreotti, Rakov, and Verolino present the modeling of lightening and its interaction with overhead conductors.
- In Chapter 13, Dr He and Zhang present the transient behavior of grounding systems.
- In Chapter 14, Dr Besnier presents the statistics of electromagnetic reverberation chambers.
- In Chapter 15, Dr Aroudi *et al.* present the analysis of a class of dynamical systems with applications to power conversion circuits.
- In Chapter 16, Dr Brandao Faria presents the pure time-domain multiconductor transmission line formalism.
- In Chapter 17, Dr Celozzi and myself present shielding in the time domain.

I hope the book will offer the reader a comprehensive support in understanding time-domain techniques and can also be a valid starting point for developing future research activities.

Finally, I would like to acknowledge the support provided by the IET staff, and, in particular, the patient and assistance given by Olivia Wilkins and Nicki Dennis of IET, and N. Srinivasan of MPS Limited.

Rome, July 4, 2022
Rodolfo Araneo

Chapter 1

Introduction to time-domain electromagnetics

Giampiero Lovat[1] and Paolo Burghignoli[2]

1.1 Differential form of Maxwell's equations

A complete description of the macroscopic electromagnetism is provided by Maxwell's equations whose validity is taken as a postulate. Maxwell's equations can be used either in a differential (local) form or in an integral (global) form, and there has been a long debate over which is the best representation. When stationary media are considered, the main difference between the two representations consists in how they account for discontinuities of materials and/or sources. Basically, if one adopts the differential form, some boundary conditions at surface discontinuities must be postulated; on the other hand, if the integral forms are chosen, one must postulate their validity across such discontinuities [1].

Maxwell's equations can be expressed in scalar, vector, or tensor form, and different vector fields can be considered as fundamental. A full description of all these details can be found, e.g., in [2]. In this chapter, we assume the following differential form of the Maxwell equations:

$$\begin{aligned} \nabla \times \mathbf{e}(\mathbf{r},t) &= -\frac{\partial}{\partial t}\mathbf{b}(\mathbf{r},t), \\ \nabla \times \mathbf{h}(\mathbf{r},t) &= \mathbf{j}(\mathbf{r},t) + \frac{\partial}{\partial t}\mathbf{d}(\mathbf{r},t). \end{aligned} \tag{1.1}$$

and

$$\begin{aligned} \nabla \cdot \mathbf{d}(\mathbf{r},t) &= \rho_{\mathrm{e}}(\mathbf{r},t), \\ \nabla \cdot \mathbf{b}(\mathbf{r},t) &= 0. \end{aligned} \tag{1.2}$$

In this framework, the electromagnetic field—described by vectors $\mathbf{e}$ (electric field, unit of measure V/m), $\mathbf{h}$ (magnetic field, unit of measure A/m), $\mathbf{d}$ (electric displacement, unit of measure C/m^2), and $\mathbf{b}$ (magnetic induction, unit of measure Wb/m^2 or T) —arises from sources $\mathbf{j}$ (electric current density, unit of measure A/m^2) and ρ_{e} (electric charge density, unit of measure C/m^3).

[1]Sapienza University of Rome, Department of Astronautical, Electrical and Energetic Engineering (DIAEE), Italy
[2]Sapienza University of Rome, Department of Information Engineering, Electronics and Telecommunications (DIET), Italy

Except for static fields, if a time can be found before which all the fields and sources are identically zero, then the divergence equations (1.2) follow from the curl equations (1.1).

From (1.1) and (1.2), the so-called continuity equation can also be derived as

$$\nabla \cdot \mathbf{j}(\mathbf{r},t) = -\frac{\partial}{\partial t}\rho_{\mathrm{e}}(\mathbf{r},t). \tag{1.3}$$

It is worth noting that if $\rho_{\mathrm{e}}(\mathbf{r},t) = 0$ at some initial time t_0, the charge density $\rho_{\mathrm{e}}(\mathbf{r},t)$ can be expressed in terms of the current density $\mathbf{j}(\mathbf{r},t)$. In fact, by integrating (1.3) over time we have

$$\rho_{\mathrm{e}}(\mathbf{r},t) = -\int_{t_0}^{t} \nabla \cdot \mathbf{j}(\mathbf{r},t')\,\mathrm{d}t' \tag{1.4}$$

and the charge density can in principle be eliminated from Maxwell's equations. Moreover, by integrating (1.3) over a finite volume V (with boundary surface S_V having normal unit vector $\mathbf{u}_n$ oriented outwards), we have

$$\frac{\partial}{\partial t}\int_V \rho_{\mathrm{e}}(\mathbf{r},t)\,\mathrm{d}V = -\int_V \nabla \cdot \mathbf{j}(\mathbf{r},t)\,\mathrm{d}V = -\oint_{S_V} \mathbf{j}(\mathbf{r},t)\cdot\mathbf{u}_n\,\mathrm{d}S, \tag{1.5}$$

so that the variation of the total charge in V is related to the flux of $\mathbf{j}$ through S_V; in particular, if this flux is zero, then the total charge in V remains constant for all time (conservation of charge).

It can be useful to make the Maxwell equations symmetric by introducing fictitious magnetic current and charge densities $\mathbf{m}$ and ρ_{m} (units of measure V/m^2 and Wb/m^3, respectively) which satisfy a continuity equation similar to (1.2) so that (1.1) and (1.2) can be rewritten as

$$\begin{aligned} \nabla \times \mathbf{e}(\mathbf{r},t) &= -\frac{\partial}{\partial t}\mathbf{b}(\mathbf{r},t) - \mathbf{m}(\mathbf{r},t), \\ \nabla \times \mathbf{h}(\mathbf{r},t) &= \mathbf{j}(\mathbf{r},t) + \frac{\partial}{\partial t}\mathbf{d}(\mathbf{r},t) \end{aligned} \tag{1.6}$$

and

$$\begin{aligned} \nabla \cdot \mathbf{d}(\mathbf{r},t) &= \rho_{\mathrm{e}}(\mathbf{r},t), \\ \nabla \cdot \mathbf{b}(\mathbf{r},t) &= \rho_{\mathrm{m}}(\mathbf{r},t). \end{aligned} \tag{1.7}$$

As it will be shown later, the equivalence principle indeed requires the introduction of such fictitious quantities.

It is also useful to identify in Maxwell's equations some "impressed" source terms, which are independent of the unknown fields and are instead due to other external sources (magnetic sources can be only of this type). Such "impressed" sources are considered as known terms in Maxwell's differential equations and indicated by the subscript "*i*". In this connection, (1.6) and (1.7) can be expressed as

$$\begin{aligned} \nabla \times \mathbf{e}(\mathbf{r},t) &= -\frac{\partial}{\partial t}\mathbf{b}(\mathbf{r},t) - \mathbf{m}_{\mathrm{i}}(\mathbf{r},t), \\ \nabla \times \mathbf{h}(\mathbf{r},t) &= \mathbf{j}(\mathbf{r},t) + \frac{\partial}{\partial t}\mathbf{d}(\mathbf{r},t) + \mathbf{j}_{\mathrm{i}}(\mathbf{r},t) \end{aligned} \tag{1.8}$$

and

$$\begin{aligned}\nabla\cdot\mathbf{d}(\mathbf{r},t)&=\rho_{\mathrm{e}}(\mathbf{r},t)+\rho_{\mathrm{ei}}(\mathbf{r},t),\\ \nabla\cdot\mathbf{b}(\mathbf{r},t)&=\rho_{\mathrm{mi}}(\mathbf{r},t).\end{aligned}\tag{1.9}$$

The impressed-source concept is well known in the circuit theory. For example, independent voltage sources are voltage excitations that are independent of possible loads.

Although both the sources and the fields cannot have true spatial discontinuities, from a modeling point of view, it is useful to consider additionally sources in one or two dimensions. In this connection, surface- and/or line-source densities can be introduced in terms of the Dirac delta distribution δ, as (singular) idealizations of actual continuous volume densities [3].

1.2 Integral form of Maxwell's equations

Let us consider an open surface S_ℓ with a normal unit vector $\mathbf{u}_n$ having a boundary described by a closed path ℓ with tangent unit vector $\mathbf{u}_l$ oriented positively (i.e., ℓ points counterclockwise when $\mathbf{u}_n$ points toward the viewer). For simplicity, let us consider non-moving surfaces. By integrating (1.1) over S_ℓ and using the Stokes theorem, the integral form of the Maxwell equations is obtained, i.e.,

$$\begin{aligned}\oint_\ell \mathbf{e}(\mathbf{r},t)\cdot\mathbf{u}_l\,\mathrm{d}\mathrm{d}\ell&=-\frac{\mathrm{d}}{\mathrm{d}t}\int_{S_\ell}\mathbf{b}(\mathbf{r},t)\cdot\mathbf{u}_n\,\mathrm{d}S,\\ \oint_\ell \mathbf{h}(\mathbf{r},t)\cdot\mathbf{u}_l\,\mathrm{d}\ell&=\int_{S_\ell}\mathbf{j}(\mathbf{r},t)\cdot\mathbf{u}_n\,\mathrm{d}S+\frac{\mathrm{d}}{\mathrm{d}t}\int_{S_\ell}\mathbf{d}(\mathbf{r},t)\cdot\mathbf{u}_n\,\mathrm{d}S.\end{aligned}\tag{1.10}$$

The first of (1.10) is also known as Faraday's law, while the second of (1.10) is the so-called Ampere's law. The circulation of $\mathbf{e}$ about a contour ℓ is also called electromotive force (or emf), while the circulation of $\mathbf{h}$ is also called magnetomotive force (or mmf).

The divergence equations (1.2) can be expressed in integral form as well. In particular, let us consider a volume region V surrounded by a closed surface S_V with normal unit vector $\mathbf{u}_n$. Integrating the first equation in (1.2) over V, using the divergence theorem, and recognizing that the integral of the charge density is the total charge, we obtain

$$\oint_{S_V}\mathbf{d}(\mathbf{r},t)\cdot\mathbf{u}_n\,\mathrm{d}S=\int_V\rho_{\mathrm{e}}(\mathbf{r},t)\,\mathrm{d}V=q(t),\tag{1.11}$$

where $q(t)$ is the total charge contained within V at time t. Equation (1.11) is also known as the Gauss' law and expresses the fact that the flux of $\mathbf{d}$ passing through a closed surface is identical to the electric charge q contained within the volume bounded by that surface. Similarly, from the second equation in (1.2), the equation

$$\oint_{S_V}\mathbf{b}(\mathbf{r},t)\cdot\mathbf{u}_n\,\mathrm{d}S=0\tag{1.12}$$

is obtained.

1.3 Constitutive relations

By direct inspection of Maxwell's curl equations in (1.1), it is immediately clear that they represent 6 scalar equations with 15 unknown quantities. With fewer equations than unknowns no unique solution can be identified (the problem is said to be indefinite). The additional equations required to make the problem definite are those that describe the relations among the field quantities $\mathbf{e}, \mathbf{h}, \mathbf{d}, \mathbf{b}$, and $\mathbf{j}$, enforced by the medium filling the region where the electromagnetic phenomena occur. Such relations are called *constitutive relations*, and they depend on the properties of the medium supporting the electromagnetic field.

In nonmoving media, with the exclusion of magnetoelectric materials, the **d** field depends only on the **e** field, **b** depends only on **h**, and **j** depends only on **e**. These dependences are expressed as constitutive relations, with the **e** and **h** fields regarded as causes and the **d**, **b**, and **j** fields as effects.

If a linear combination of causes (with given coefficients) produces a linear combination of effects (with the same coefficients), the medium is said to be *linear* (otherwise *nonlinear*). In general, the constitutive relations are described by a set of constitutive parameters and a set of constitutive operators that relate the above-mentioned fields inside a region of space. The constitutive parameters can be constants of proportionality between the fields (the medium is thus said *isotropic*), or they can be components in a tensor relationship (the medium is said anisotropic). If the constitutive parameters are constant within a certain region of space, the medium is said *homogeneous* in that region (otherwise, the medium is inhomogeneous). If the constitutive parameters are constant with time, the medium is said *stationary* (otherwise, the medium is nonstationary).

If the constitutive operators are expressed in terms of time integrals, the medium is said to be *temporally dispersive*. If these operators involve space integrals, the medium is said *spatially dispersive*. Finally, we note that the constitutive parameters may depend on other nonelectromagnetic properties of the material and external conditions (temperature, pressure, etc.).

The simplest medium is *vacuum*. In vacuum the following constitutive relations hold:

$$\begin{aligned} \mathbf{d}(\mathbf{r},t) &= \varepsilon_0 \mathbf{e}(\mathbf{r},t), \\ \mathbf{b}(\mathbf{r},t) &= \mu_0 \mathbf{h}(\mathbf{r},t), \\ \mathbf{j}(\mathbf{r},t) &= \mathbf{0}. \end{aligned} \tag{1.13}$$

The quantities $\mu_0 \simeq 4\pi \cdot 10^{-7}$ H/m and $\varepsilon_0 \simeq 10^{-9}/(36\pi) \simeq 8.854 \times 10^{-12}$ F/m are the free-space magnetic permeability and dielectric permittivity, respectively. Their values are related to the speed of light in free space c_0 through $c_0 = 1/\sqrt{\mu_0 \varepsilon_0} \simeq 2.998 \times 10^8$ m/s.

For a linear, homogeneous, isotropic, and nondispersive material the constitutive relations can be expressed as

$$\begin{aligned} \mathbf{d}(\mathbf{r},t) &= \varepsilon \mathbf{e}(\mathbf{r},t), \\ \mathbf{b}(\mathbf{r},t) &= \mu \mathbf{h}(\mathbf{r},t), \\ \mathbf{j}(\mathbf{r},t) &= \sigma \mathbf{e}(\mathbf{r},t), \end{aligned} \tag{1.14}$$

where μ and ε are the magnetic permeability and dielectric permittivity of the medium, respectively. These quantities can be related to the corresponding free-space quantities through the dimensionless relative permeability μ_r and relative permittivity ε_r, such that $\mu = \mu_r\mu_0$ and $\varepsilon = \varepsilon_r\varepsilon_0$. The dimensionless quantities $\chi_m = \mu_r - 1$ and $\chi_e = \varepsilon_r - 1$ (known as magnetic and electric susceptibilities, respectively) are also used. The third equation of (1.14) expresses Ohm's law in the local form, and σ is the conductivity of the medium (unit of measure S/m). If the medium is inhomogeneous, μ, ε, or σ are quantities that depend on the vector position **r**. If the medium is anisotropic (but still linear and nondispersive), the constitutive relations can be written as

$$\begin{aligned} \mathbf{d}(\mathbf{r},t) &= \underline{\boldsymbol{\varepsilon}} \cdot \mathbf{e}(\mathbf{r},t), \\ \mathbf{b}(\mathbf{r},t) &= \underline{\boldsymbol{\mu}} \cdot \mathbf{h}(\mathbf{r},t), \\ \mathbf{j}(\mathbf{r},t) &= \underline{\boldsymbol{\sigma}} \cdot \mathbf{e}(\mathbf{r},t), \end{aligned} \tag{1.15}$$

where $\underline{\boldsymbol{\varepsilon}}, \underline{\boldsymbol{\mu}}$, and $\underline{\boldsymbol{\sigma}}$ are called the permittivity tensor, the permeability tensor, and the conductivity tensor, respectively (they are space-dependent quantities for inhomogeneous media).

For linear, inhomogeneous, anisotropic, stationary, and temporally dispersive materials, the constitutive relation between **d** and **e** is expressed by a superposition integral as

$$\mathbf{d}(\mathbf{r},t) = \int_{-\infty}^{t} \underline{\boldsymbol{\varepsilon}}(\mathbf{r},t-t') \cdot \mathbf{e}(\mathbf{r},t')\,\mathrm{d}t', \tag{1.16}$$

where we maintained the symbol $\underline{\boldsymbol{\varepsilon}}$ with a little abuse of notation. The constitutive relations for other field quantities have similar expressions. Causality is implied by the upper limit t in the integrals (this means that the effect cannot depend on future values of the cause). If the medium is nonstationary, $\underline{\boldsymbol{\varepsilon}}(\mathbf{r},t,t')$ has to be used instead of $\underline{\boldsymbol{\varepsilon}}(\mathbf{r},t-t')$. The important concept expressed by (1.16) is that the behavior of **d** at the time t depends not only on the value of **e** at the same time t but also on its values at all past times, thus giving rise to a time-lag between cause and effect. The upper limit of the integral in (1.16) can be extended to $+\infty$ by assuming that $\underline{\boldsymbol{\varepsilon}}(\mathbf{r},t,t') = \underline{\mathbf{0}}$ whenever $t' > t$, thus obtaining a *convolution* integral as

$$\mathbf{d}(\mathbf{r},t) = \int_{-\infty}^{+\infty} \underline{\boldsymbol{\varepsilon}}(\mathbf{r},t-t') \cdot \mathbf{e}(\mathbf{r},t')\,\mathrm{d}t' = \int_{-\infty}^{+\infty} \underline{\boldsymbol{\varepsilon}}(\mathbf{r},t') \cdot \mathbf{e}(\mathbf{r},t-t')\,\mathrm{d}t'. \tag{1.17}$$

Finally, if the medium is also spatially dispersive (and nonstationary), the constitutive relation takes the form

$$\mathbf{d}(\mathbf{r},t) = \int_V \left[\int_{-\infty}^{t} \underline{\boldsymbol{\varepsilon}}(\mathbf{r},\mathbf{r}',t,t') \cdot \mathbf{e}(\mathbf{r},\mathbf{r}',t')\,\mathrm{d}t'\right] \mathrm{d}V', \tag{1.18}$$

where V indicates the whole three-dimensional space; as before, we maintained the symbol $\underline{\boldsymbol{\varepsilon}}$ with a little abuse of notation and similar expressions hold for the constitutive relations of other field quantities. The integral over the volume V in (1.18) expresses the physical phenomenon for which the effect at the point $\mathbf{r}$ depends on the value of the cause in all the neighboring points $\mathbf{r}'$. An important point is that if the medium is spatially dispersive but homogeneous, the constitutive relations involve a convolution integral in the space domain.

For a linear, homogeneous, isotropic, and nondispersive material, the curl Maxwell equations can thus be written as

$$\begin{aligned} \nabla \times \mathbf{e}(\mathbf{r},t) &= -\mu \frac{\partial}{\partial t}\mathbf{h}(\mathbf{r},t) - \mathbf{m}_{\mathrm{i}}(\mathbf{r},t), \\ \nabla \times \mathbf{h}(\mathbf{r},t) &= \sigma\mathbf{e}(\mathbf{r},t) + \varepsilon\frac{\partial}{\partial t}\mathbf{e}(\mathbf{r},t) + \mathbf{j}_{\mathrm{i}}(\mathbf{r},t), \end{aligned} \tag{1.19}$$

and the divergence equations as

$$\begin{aligned} \nabla \cdot \mathbf{e}(\mathbf{r},t) &= \frac{\rho_{\mathrm{e}}(\mathbf{r},t) + \rho_{\mathrm{ei}}(\mathbf{r},t)}{\varepsilon}, \\ \nabla \cdot \mathbf{h}(\mathbf{r},t) &= \frac{\rho_{\mathrm{mi}}(\mathbf{r},t)}{\mu}. \end{aligned} \tag{1.20}$$

1.4 Discontinuities and singularities

As mentioned in the previous section, in the absence of discontinuities, Maxwell's equations in differential form are valid everywhere in space; nevertheless, for modeling purposes, discontinuities of material parameters or singular sources are often considered. In such cases, other field relationships must be postulated (alternatively, they can be derived from Maxwell's equations in the integral form if such integral forms are postulated to be valid also across the discontinuities).

Let us consider the presence of singular sources in the form of electric and magnetic surface densities: electric $\mathbf{j}_S$ (A/m), $\rho_{\mathrm{e}S}$ (C/m^2) and magnetic $\mathbf{m}_S$ (V/m), $\rho_{\mathrm{m}S}$ (Wb/m^2), distributed over a surface S, which separates two regions (regions 1 and 2, respectively), or discontinuities in the material parameters across the surface S. Let $\mathbf{u}_n$ be the unit vector normal to the surface S directed from region 2 to region 1. In such conditions, the following jump conditions hold:

$$\begin{aligned} \mathbf{u}_n &\times \left[\mathbf{h}(\mathbf{r}^+,t) - \mathbf{h}(\mathbf{r}^-,t)\right] = \mathbf{j}_S(\mathbf{r},t), \\ \mathbf{u}_n &\times \left[\mathbf{e}(\mathbf{r}^+,t) - \mathbf{e}(\mathbf{r}^-,t)\right] = -\mathbf{m}_S(\mathbf{r},t), \\ \mathbf{u}_n &\cdot \left[\mathbf{d}(\mathbf{r}^+,t) - \mathbf{d}(\mathbf{r}^-,t)\right] = \rho_{\mathrm{e}S}(\mathbf{r},t), \\ \mathbf{u}_n &\cdot \left[\mathbf{b}(\mathbf{r}^+,t) - \mathbf{b}(\mathbf{r}^-,t)\right] = \rho_{\mathrm{m}S}(\mathbf{r},t), \end{aligned} \tag{1.21}$$

where $\mathbf{r}^+$ and $\mathbf{r}^-$ indicate a point infinitesimally inside the regions 1 and 2, respectively, while $\mathbf{r}$ is along the surface S. Moreover

$$\begin{aligned} \mathbf{u}_n \cdot \left[\mathbf{j}(\mathbf{r}^+,t) - \mathbf{j}(\mathbf{r}^-,t)\right] &= -\nabla_S \cdot \mathbf{j}_S(\mathbf{r},t) - \frac{\partial \rho_{\mathrm{e}S}(\mathbf{r},t)}{\partial t}, \\ \mathbf{u}_n \cdot \left[\mathbf{m}(\mathbf{r}^+,t) - \mathbf{m}(\mathbf{r}^-,t)\right] &= -\nabla_S \cdot \mathbf{m}_S(\mathbf{r},t) - \frac{\partial \rho_{\mathrm{m}S}(\mathbf{r},t)}{\partial t}, \end{aligned} \tag{1.22}$$

where $\nabla_S[\cdot] = \nabla[\cdot] - \mathbf{u}_n \partial[\cdot]/\partial n$. It is clear that when $\mathbf{j}_S$ and $\mathbf{m}_S$ are zero, the tangential components of both electric and magnetic fields are continuous across the surface S. In particular, if discontinuities in the material parameters are present, the electric surface current density $\mathbf{j}_S$ may be different from zero at the boundary of a perfect electric conductor (PEC, within which $\mathbf{e}_2 = \mathbf{0}$), and the magnetic surface current density $\mathbf{m}_S$ may be different from zero at the boundary of a perfect magnetic conductor (PMC, within which $\mathbf{h}_2 = \mathbf{0}$). Then the jump conditions at the interface between the conventional medium and the PEC are written as

$$\begin{aligned}
&\mathbf{u}_n \times \mathbf{h}(\mathbf{r},t) = \mathbf{j}_S(\mathbf{r},t)\,,\\
&\mathbf{u}_n \times \mathbf{e}(\mathbf{r},t) = \mathbf{0}\,,\\
&\mathbf{u}_n \cdot \mathbf{d}(\mathbf{r},t) = \rho_{eS}(\mathbf{r},t)\,,\\
&\mathbf{u}_n \cdot \mathbf{b}(\mathbf{r},t) = 0\,,\\
&\mathbf{u}_n \cdot \mathbf{j}(\mathbf{r},t) = -\nabla_S \cdot \mathbf{j}_S(\mathbf{r},t) - \frac{\partial \rho_{eS}(\mathbf{r},t)}{\partial t}\,,\\
&\mathbf{u}_n \cdot \mathbf{m}(\mathbf{r},t) = 0\,.
\end{aligned} \tag{1.23}$$

Likewise, at the interface between a conventional medium and a PMC, the results are

$$\begin{aligned}
&\mathbf{u}_n \times \mathbf{h}(\mathbf{r},t) = \mathbf{0}\,,\\
&\mathbf{u}_n \times \mathbf{e}(\mathbf{r},t) = -\mathbf{m}_S(\mathbf{r},t)\,,\\
&\mathbf{u}_n \cdot \mathbf{d}(\mathbf{r},t) = 0\,,\\
&\mathbf{u}_n \cdot \mathbf{b}(\mathbf{r},t) = \rho_{mS}(\mathbf{r},t)\,,\\
&\mathbf{u}_n \cdot \mathbf{j}(\mathbf{r},t) = 0\,,\\
&\mathbf{u}_n \cdot \mathbf{m}(\mathbf{r},t) = -\nabla_S \cdot \mathbf{m}_S(\mathbf{r},t) - \frac{\partial \rho_{mS}(\mathbf{r},t)}{\partial t}\,.
\end{aligned} \tag{1.24}$$

In these jump conditions, the $\mathbf{u}_n$ unit vector points outside the conductors.

Finally, it is worth mentioning some other singular behaviors of fields and currents that may occur in correspondence to the edge of a dielectric or conducting wedge and to the tip of a dielectric or conducting cone. The solution of the electromagnetic problem in such cases is not unique, unless the singular behavior is specified. The order of singularity can be determined by requiring that the energy stored in the proximity of the edge or of the tip remains finite. Further details can be found in [3,4].

1.5 Initial conditions, boundary conditions, and causality

As was noted earlier, Maxwell's equations together with the constitutive relations represent a set of partial differential equations. However, it is well known that in order to obtain a solution for this set of equations, both initial and boundary conditions must be specified. The initial conditions are represented by the constraints that the electromagnetic field must satisfy at a given time, while boundary conditions are, in general, constraints that the electromagnetic field must satisfy over certain surfaces of the three-dimensional space, usually surfaces that separate regions of space filled with different materials. In these cases, the boundary conditions coincide with the jump conditions illustrated in the previous section.

Physical grounds require that all the fields satisfy the law of causality. Two different causality conditions can be defined, i.e., a weak and a strong causality condition. The *weak causality* condition states that all fields have to be zero for $t \leq t_0$ if the sources are zero for $t \leq t_0$. In other words, an output does not exist until an input is applied. If such a weak causality condition is assumed for electromagnetic fields, Maxwell's equations imply a *strong causality* condition, i.e., that the fields are zero beyond a distance $c(t - t_0)$ (where $c = 1/\sqrt{\mu\varepsilon}$ is the speed of light in a medium with constitutive parameters μ and ε) from a point source that is zero for $t \leq t_0$. Since the electromagnetic field equations and weak causality imply strong causality, the latter will be referred to in the following.

1.6 Fundamental theorems

1.6.1 Uniqueness theorem

As in any other problem of mathematical physics, the uniqueness property is a fundamental condition for a problem to be well-posed in the sense of Hadamard (along with the existence property and the continuous dependence on data). Uniqueness theorems generally provide sufficient conditions for ensuring that a solution (provided it exists) is unique. In fact, it is of critical importance to know that the solution that one can obtain through different techniques is also unique. Furthermore, the uniqueness theorem is a fundamental tool for the development of other important theorems, such as the equivalence theorems.

The uniqueness theorem can be formulated in the time domain as follows: given a stationary region V in space, filled with a linear, stationary, non-dispersive medium, and hosting prescribed impressed currents $\mathbf{j}_i$, $\mathbf{j}_{mi}$, there exists a unique electromagnetic field $\mathbf{e}$, $\mathbf{h}$ that satisfies both the Maxwell equations and the constitutive relations in V for $t \geq t_0$, provided that the tangential component of $\mathbf{e}$ over the boundary, or the tangential component of $\mathbf{h}$ over the boundary, or the former over part of the boundary and the latter over the remaining part of the boundary, are specified for $t \geq t_0$ (*boundary conditions*), and the electromagnetic field is assigned everywhere in V at $t = t_0$ (*initial conditions*). The theorem can easily be proved by contradiction, by assuming that two distinct electromagnetic fields $\{\mathbf{e}_1, \mathbf{h}_1\}$ and $\{\mathbf{e}_2, \mathbf{h}_2\}$ are solutions of the Maxwell equations with the same source terms. The difference electromagnetic field $\{\mathbf{e}_d, \mathbf{h}_d\}$ (where $\mathbf{e}_d = \mathbf{e}_1 - \mathbf{e}_2$ and $\mathbf{h}_d = \mathbf{h}_1 - \mathbf{h}_2$) thus satisfies the homogeneous Maxwell equations with homogeneous boundary conditions and homogeneous initial conditions at $t = t_0$. By suitably manipulating Maxwell's equations, it can thus be shown that it must result $\mathbf{e}_d = \mathbf{0}$ and $\mathbf{h}_d = \mathbf{0}$ and the theorem is proved [5].

1.6.2 Equivalence theorem

The equivalence theorem is a consequence of Maxwell's equations and the uniqueness theorem. Basically, it allows the original electromagnetic problem to be replaced with an equivalent problem whose solution coincides with that of the original problem in a finite region of space. To be effective, the equivalent problem should be easier to solve than the original one. The principle can be formulated in different forms.

1.6.2.1 The Love equivalence principle

The first form of the equivalence principle (also known as the *Love equivalence principle* [4,6]) establishes that the electromagnetic field $\{\mathbf{e}, \mathbf{h}\}$ outside a region V bounded by a surface S enclosing the sources $\{\mathbf{j}_i, \mathbf{m}_i\}$ is equal to that produced by the equivalent sources (*Huygens sources*) $\{\mathbf{j}_S, \mathbf{m}_S\}$ distributed over the surface S and given by

$$\begin{aligned} \mathbf{j}_S(\mathbf{r}, t) &= \mathbf{u}_n \times \mathbf{h}_S(\mathbf{r}, t), \\ \mathbf{m}_S(\mathbf{r}, t) &= -\mathbf{u}_n \times \mathbf{e}_S(\mathbf{r}, t), \end{aligned} \tag{1.25}$$

where $\{\mathbf{e}_S, \mathbf{h}_S\}$ is the electromagnetic field $\{\mathbf{e}, \mathbf{h}\}$ in correspondence to the surface S and $\mathbf{u}_n$ is the unit vector normal to S pointing outside the region V. It can be shown that the field produced by the equivalent sources inside V is identically zero. The equivalent sources are considered as known terms in the formulation of the problem. However, they depend on the field $\{\mathbf{e}, \mathbf{h}\}$, which is unknown. In practice, there are many problems for which approximate expressions can be found for the equivalent currents, and, in any case, they are extensively used to formulate exact boundary integral equations for the considered problem.

1.6.2.2 The Schelkunoff equivalence principle

A second form of the equivalence principle is known as the *Schelkunoff equivalence principle* [4,7]. It is based on the fact that, according to the Love equivalence principle, the field produced by the equivalent sources inside V is identically zero. It differs from the Love equivalence principle since the medium filling the region V is replaced with a PEC (the boundary conditions on S are not changed). In this way, the equivalent magnetic current $\mathbf{m}_S$ is the only radiating source. Dually, the region V could be replaced with a PMC; in this case, $\mathbf{j}_S$ would be the only radiating source. However, it must be pointed out that the two situations considered by Love and Schelkunoff are different: in the latter, the presence of a PEC (or PMC) body must be explicitly taken into account.

1.7 Wave equations

The curl Maxwell equations (1.19) are coupled partial differential equations of the first order having the electric and magnetic fields as unknowns. Such coupled equations can easily be converted into two uncoupled partial differential equations of second order (called *wave equations*) having either the electric or the magnetic field as unknown. If suitable conditions are not met, the solutions of the wave equations are only a subset of solutions of Maxwell's equations. In other words, not all the electromagnetic fields are electromagnetic waves (i.e., satisfy the wave equation). In particular, Hansen and Yaghjian have shown that, in the absence of boundaries, Maxwell's equations and wave equations are equivalent if the field quantities are causal and sufficiently smooth functions and if the sources lie in a bounded region and are zero before a certain initial time t_0 [8] (it is worth noting that to derive Maxwell's equation from wave equations the continuity equation has to be assumed valid).

However, if boundaries are present (such as PEC surfaces), boundary field values required to guarantee uniqueness of the field and solutions of the wave equation and of Maxwell's equations may differ (since, e.g., wave equations need also divergence equations to derive the curl equations) and this may result in spurious solutions in numerical methods [9].

To illustrate how wave equations are derived from Maxwell's equations, let us consider for simplicity a linear, homogeneous, isotropic, and nondispersive medium with constitutive parameters ε, μ, and σ. Maxwell's equations are described by (1.19). Taking the curl of both sides of the first and second equations in (1.19), we have

$$\begin{aligned}
\nabla\times\nabla\times\mathbf{e}(\mathbf{r},t) &= -\mu\frac{\partial}{\partial t}\nabla\times\mathbf{h}(\mathbf{r},t)-\nabla\times\mathbf{m}_\mathrm{i}(\mathbf{r},t),\\
\nabla\times\nabla\times\mathbf{h}(\mathbf{r},t) &= \sigma\nabla\times\mathbf{e}(\mathbf{r},t)+\varepsilon\frac{\partial}{\partial t}\nabla\times\mathbf{e}(\mathbf{r},t)+\nabla\times\mathbf{j}_\mathrm{i}(\mathbf{r},t).
\end{aligned} \tag{1.26}$$

Substituting the second of (1.19) inside the first of (1.26) and using the vector identity $\nabla\times\nabla\times[\cdot]=\nabla\nabla\cdot[\cdot]-\nabla^2[\cdot]$, the first of (1.26) can be rewritten as

$$\begin{aligned}
\nabla[\nabla\cdot\mathbf{e}(\mathbf{r},t)]-\nabla^2\mathbf{e}(\mathbf{r},t) &= -\mu\frac{\partial}{\partial t}\left[\sigma\mathbf{e}(\mathbf{r},t)+\varepsilon\frac{\partial}{\partial t}\mathbf{e}(\mathbf{r},t)+\mathbf{j}_\mathrm{i}(\mathbf{r},t)\right]\\
&\quad -\nabla\times\mathbf{m}_\mathrm{i}(\mathbf{r},t).
\end{aligned} \tag{1.27}$$

By using the first equation of (1.20) and rearranging the terms, we thus obtain

$$\begin{aligned}
\nabla^2\mathbf{e}(\mathbf{r},t)-\frac{1}{c^2}\frac{\partial^2\mathbf{e}(\mathbf{r},t)}{\partial t^2}-\mu\sigma\frac{\partial\mathbf{e}(\mathbf{r},t)}{\partial t}\\
=\nabla\times\mathbf{m}_\mathrm{i}(\mathbf{r},t)+\mu\frac{\partial\mathbf{j}_\mathrm{i}(\mathbf{r},t)}{\partial t}+\frac{\nabla[\rho_\mathrm{e}(\mathbf{r},t)+\rho_\mathrm{ei}(\mathbf{r},t)]}{\varepsilon},
\end{aligned} \tag{1.28}$$

having used $c=1/\sqrt{\mu\varepsilon}$. Equation (1.28) is the sought second-order partial differential equation for **e** (i.e., the vector wave equation for **e**).

Similarly, an uncoupled second-order partial differential equation for **h** is obtained as

$$\begin{aligned}
\nabla^2\mathbf{h}(\mathbf{r},t)-\frac{1}{c^2}\frac{\partial^2\mathbf{h}(\mathbf{r},t)}{\partial t^2}-\mu\sigma\frac{\partial\mathbf{h}(\mathbf{r},t)}{\partial t}\\
=-\nabla\times\mathbf{j}_\mathrm{i}(\mathbf{r},t)+\varepsilon\frac{\partial\mathbf{m}_\mathrm{i}(\mathbf{r},t)}{\partial t}+\sigma\mathbf{m}_\mathrm{i}(\mathbf{r},t)+\frac{\nabla\rho_\mathrm{mi}(\mathbf{r},t)}{\mu},
\end{aligned} \tag{1.29}$$

which is thus the vector wave equation for **h**.

For lossless media ($\sigma = 0$), the wave equations (1.28) and (1.29) become

$$\begin{aligned}\nabla^2\mathbf{e}(\mathbf{r},t) - \frac{1}{c^2}\frac{\partial^2\mathbf{e}(\mathbf{r},t)}{\partial t^2} &= \nabla\times\mathbf{m}_{\mathrm{i}}(\mathbf{r},t) + \mu\frac{\partial\mathbf{j}_{\mathrm{i}}(\mathbf{r},t)}{\partial t} + \frac{\nabla\rho_{\mathrm{ei}}(\mathbf{r},t)}{\varepsilon},\\ \nabla^2\mathbf{h}(\mathbf{r},t) - \frac{1}{c^2}\frac{\partial^2\mathbf{h}(\mathbf{r},t)}{\partial t^2} &= -\nabla\times\mathbf{j}_{\mathrm{i}}(\mathbf{r},t) + \varepsilon\frac{\partial\mathbf{m}_{\mathrm{i}}(\mathbf{r},t)}{\partial t} + \frac{\nabla\rho_{\mathrm{mi}}(\mathbf{r},t)}{\mu}.\end{aligned} \tag{1.30}$$

1.8 Transient uniform plane waves

As in the corresponding frequency domain, transient uniform plane waves are solutions of the homogeneous wave equations (1.28) and (1.29) in which the fields are invariant over a certain planar surface.

Without loss of generality, we can assume the xy plane as the plane of field invariance so that $\mathbf{e}(\mathbf{r},t) = \mathbf{e}(z,t)$ and $\mathbf{h}(\mathbf{r},t) = \mathbf{h}(z,t)$. Under this assumption, the homogeneous wave equation for the electric field becomes

$$\frac{\partial^2\mathbf{e}}{\partial z^2} - \mu\sigma\frac{\partial\mathbf{e}}{\partial t} - \frac{1}{c^2}\frac{\partial^2\mathbf{e}}{\partial t^2} = 0 \tag{1.31}$$

and from the first of (1.19) with $\mathbf{m}_{\mathrm{i}} = \mathbf{0}$, it follows that

$$-\mathbf{u}_x\frac{\partial e_x}{\partial z} + \mathbf{u}_y\frac{\partial e_y}{\partial z} = -\mu\frac{\partial\mathbf{h}}{\partial t} \tag{1.32}$$

and therefore

$$\frac{\partial h_z}{\partial t} = 0. \tag{1.33}$$

This means that h_z has to be constant with respect to time and since it thus not exhibit a wave behavior, it can be assumed $h_z = 0$.

With a similar reasoning, from the second of (1.19), it also follows that

$$\sigma e_z(z,t) + \varepsilon\frac{\partial e_z}{\partial t} = 0, \tag{1.34}$$

i.e.,

$$e_z(z,t) = e_0(z)e^{-\frac{\sigma}{\varepsilon}t}. \tag{1.35}$$

Again, since (1.35) does not represent a wave behavior, it can be assumed $e_z = 0$.

Since $e_z = h_z = 0$, both $\mathbf{e}(\mathbf{r},t)$ and $\mathbf{h}(\mathbf{r},t)$ are orthogonal to the z direction, so that the transient uniform plane wave is a TEM$_z$ field. By assuming linearly polarized fields:

$$\begin{aligned} \mathbf{e}(\mathbf{r},t) &= \mathbf{u}_e e(z,t)\,, \\ \mathbf{h}(\mathbf{r},t) &= \mathbf{u}_h h(z,t)\,, \end{aligned} \tag{1.36}$$

it results

$$\mathbf{u}_e \times \mathbf{u}_h = \pm \mathbf{u}_z\,. \tag{1.37}$$

Usually, the direction $\mathbf{u}_e$ is called the polarization direction of the plane wave.

Now, by scalarly multiplying (1.28) and (1.29) by $\mathbf{u}_e$ and $\mathbf{u}_h$, respectively, it follows that in a source-free region, it results

$$\begin{aligned} \frac{\partial^2 e(z,t)}{\partial z^2} - \mu\sigma \frac{\partial e(z,t)}{\partial t} - \frac{1}{c^2}\frac{\partial^2 e(z,t)}{\partial t^2} &= 0\,, \\ \frac{\partial^2 h(z,t)}{\partial z^2} - \mu\sigma \frac{\partial h(z,t)}{\partial t} - \frac{1}{c^2}\frac{\partial^2 h(z,t)}{\partial t^2} &= 0\,, \end{aligned} \tag{1.38}$$

i.e., the electric and magnetic fields satisfy the same kind of equation. It should initially be noted that, in the lossless case (i.e., $\sigma = 0$), the homogeneous wave equation

$$\frac{\partial^2 \psi(z,t)}{\partial z^2} - \frac{1}{c^2}\frac{\partial^2 \psi(z,t)}{\partial t^2} = 0 \tag{1.39}$$

is satisfied by any function whose argument is in the form

$$\psi(z,t) = f(z - ct) = g(t - z/c)\,. \tag{1.40}$$

In any case, to uniquely determine the scalar function $\psi(z,t)$ solution of (1.38) (or (1.39)), suitable boundary or initial conditions have to be specified. In particular, by requiring

$$\begin{aligned} \psi(z,t)\Big|_{z=0} &= f_1(t)\,, \\ \frac{\partial}{\partial z}\psi(z,t)\Big|_{z=0} &= f_2(t)\,, \end{aligned} \tag{1.41}$$

it can be shown that [5]

$$\begin{aligned} \psi(z,t) = {} & \frac{1}{2} e^{-\frac{\alpha}{c}z} f_1\left(t - \frac{z}{c}\right) + \frac{1}{2} e^{\frac{\alpha}{c}z} f_1\left(t + \frac{z}{c}\right) \\ & + \frac{c}{2} e^{-\alpha t} \int_{t-\frac{z}{c}}^{t+\frac{z}{c}} f_1(\tau) e^{\alpha\tau} \frac{\partial}{\partial z} J_0\left[\frac{\alpha}{c}\sqrt{z^2 - c^2(t-\tau)^2}\right] \mathrm{d}\tau \\ & + \frac{c}{2} e^{-\alpha t} \int_{t-\frac{z}{c}}^{t+\frac{z}{c}} f_2(\tau) e^{\alpha\tau} J_0\left[\frac{\alpha}{c}\sqrt{z^2 - c^2(t-\tau)^2}\right] \mathrm{d}\tau\,, \end{aligned} \tag{1.42}$$

where $\alpha = \sigma/(2\varepsilon)$ and $J_0(\cdot)$ is the zeroth order Bessel function of first kind. It should be noted that in lossless media (i.e., $\alpha = 0$) (1.42) reduces to

$$\psi(z,t) = \frac{1}{2} f_1\left(t - \frac{z}{c}\right) + \frac{1}{2} f_1\left(t + \frac{z}{c}\right) + \frac{c}{2}\int_{t-\frac{z}{c}}^{t+\frac{z}{c}} f_2(\tau)\,\mathrm{d}\tau \,. \tag{1.43}$$

The physical interpretation of (1.42) and (1.43) is the following. The initial pulse $f_1(t)$ initially splits into two waves of half the initial amplitude (at $z = 0$ the integrals are zero). These partial waves propagate to positive and negative z with a velocity c and, in the lossy case, they are exponentially attenuated in the direction of propagation. In the lossless case, the partial waves propagate without any distortion and are superimposed to other two waves propagating without distortion since it can be shown that

$$\frac{c}{2}\int_{t-\frac{z}{c}}^{t+\frac{z}{c}} f_2(\tau)\,\mathrm{d}\tau = \frac{1}{2} f_3\left(t - \frac{z}{c}\right) - \frac{1}{2} f_3\left(t + \frac{z}{c}\right), \tag{1.44}$$

where

$$f_3(\tau) = -c\int_0^{\tau} f_2(t')\,\mathrm{d}t' \,. \tag{1.45}$$

On the other hand, in the presence of a finite conductivity, the initial function changes its form while propagating, since the integrals in (1.42) represent contributions that indefinitely remain nonzero at points which have been traversed by the wave front. Therefore, in the lossy case, the wave leaves a wake which decays exponentially with time.

By enforcing the initial conditions

$$\begin{aligned} &\psi(z,t)|_{t=0} = g_1(z)\,, \\ &\frac{\partial}{\partial z}\psi(z,t)\bigg|_{t=0} = g_2(z)\,, \end{aligned} \tag{1.46}$$

it can be shown that [5]

$$\begin{aligned} \psi(z,t) = {} & e^{-\alpha t}\left\{\frac{1}{2} g_1(z - ct) + \frac{1}{2} g_1(z + ct)\right. \\ & + \frac{1}{2c}\int_{z-ct}^{z+ct} g_1(\zeta)\frac{\partial}{\partial t} J_0\left[\frac{\alpha}{c}\sqrt{(z-\zeta)^2 - c^2t^2}\right]\mathrm{d}\zeta \\ & \left. + \frac{1}{2c}\int_{z-ct}^{z+ct} [\alpha g_1(\zeta) + g_2(\zeta)] J_0\left[\frac{\alpha}{c}\sqrt{(z-\zeta)^2 - c^2t^2}\right]\mathrm{d}\zeta\right\}, \end{aligned} \tag{1.47}$$

which for lossless media reduces to

$$\psi(z,t) = \frac{1}{2} g_1(z - ct) + \frac{1}{2} g_1(z + ct) + \frac{1}{2c}\int_{z-ct}^{z+ct} g_2(\zeta)\,\mathrm{d}\zeta \,. \tag{1.48}$$

It should be noted that (1.42) and (1.47) (and (1.43) and (1.48)) have the same essential characteristics. In particular, the initial field distribution along z initially splits into two partial waves which propagate for positive and negative z. In the lossy case, after the passage of the wave front, it remains a wake and the field exponentially decays with time.

All the above expressions can easily be generalized to a transient uniform plane wave whose plane of field invariance is arbitrary. For instance, the field of a transient uniform plane wave propagating along the direction $\mathbf{u}_k$ in a lossless medium can be expressed as

$$\begin{aligned} \mathbf{e}(\mathbf{r},t) &= \mathbf{u}_e e_0 f\left(t - \frac{\mathbf{u}_k \cdot \mathbf{r}}{c}\right), \\ \mathbf{h}(\mathbf{r},t) &= \mathbf{u}_h h_0 f\left(t - \frac{\mathbf{u}_k \cdot \mathbf{r}}{c}\right). \end{aligned} \tag{1.49}$$

With reference to (1.49), the fields are invariant over the planes orthogonal to $\mathbf{u}_k$. Moreover, it results $\mathbf{u}_e \times \mathbf{u}_h = \mathbf{u}_k$ and it is also easy to show that the ratio of the amplitudes of the electric and magnetic fields is independent of the time and observation point. In particular, it results

$$|\mathbf{e}(\mathbf{r},t)| = \eta|\mathbf{h}(\mathbf{r},t)|. \tag{1.50}$$

The quantity $\eta = \sqrt{\mu/\varepsilon}$ is called intrinsic impedance of the medium and carries units of ohms.

1.9 Electromagnetic potentials and gauge transformations

As is well known, a general vector field can be decomposed into the sum of an irrotational vector (i.e., a vector with a zero curl) and a solenoidal vector (i.e., a vector with a zero divergence). Such a decomposition is known as *Helmholtz decomposition* (or Helmholtz theorem) [2]. Moreover, the solenoidal part is expressed as the curl of a vector potential function and the irrotational part as the gradient of a scalar potential function. Usually, such a decomposition is initially used to express the electric and the magnetic-induction fields so that

$$\mathbf{e}(\mathbf{r},t) = \nabla \times \mathbf{f}(\mathbf{r},t) + \nabla v(\mathbf{r},t) \tag{1.51}$$

and

$$\mathbf{b}(\mathbf{r},t) = \nabla \times \mathbf{a}(\mathbf{r},t) + \nabla w(\mathbf{r},t), \tag{1.52}$$

where $\mathbf{f}$ and $\mathbf{a}$ are the vector electric and magnetic potentials, respectively, while v and w are the scalar electric and magnetic potentials, respectively. In what follows we refer to electromagnetic fields in a linear, homogeneous, isotropic, and time-invariant medium with constitutive parameters ε, μ, and σ, so that (1.52) can also be related to the magnetic field $\mathbf{h}$ by expressing

$$\mathbf{h}(\mathbf{r},t) = \frac{1}{\mu}\nabla \times \mathbf{a}(\mathbf{r},t) + \frac{1}{\mu}\nabla w(\mathbf{r},t). \tag{1.53}$$

Electric and magnetic fields in (1.51) and (1.53) are related by Maxwell's equations, so that a particular relationship must also exist for the potentials. It is convenient to investigate such a relationship by considering separately the cases for which $\mathbf{j}_\mathrm{i} = \mathbf{0}$ and $\mathbf{m}_\mathrm{i} = \mathbf{0}$.

In particular, in the absence of magnetic currents (i.e., $\mathbf{m}_\mathrm{i} = \mathbf{0}$), we have

$$\nabla \times \mathbf{e}(\mathbf{r}, t) = -\mu \frac{\partial}{\partial t} \mathbf{h}(\mathbf{r}, t). \tag{1.54}$$

From (1.54), it follows that $\mathbf{h}$ is solenoidal so that it must be $\nabla w = \mathbf{0}$ and therefore $w = 0$. From (1.51) and (1.54), it results

$$\nabla \times \left[\nabla \times \mathbf{f}(\mathbf{r}, t) + \frac{\partial \mathbf{a}(\mathbf{r}, t)}{\partial t}\right] = \mathbf{0}, \tag{1.55}$$

so that

$$\nabla \times \mathbf{f}(\mathbf{r}, t) = -\frac{\partial \mathbf{a}(\mathbf{r}, t)}{\partial t} + \nabla \phi(\mathbf{r}, t), \tag{1.56}$$

where ϕ is a suitable scalar function. From (1.51) and (1.55), by improperly redefining $v = -v - \phi$, we finally obtain

$$\begin{aligned} \mathbf{e}(\mathbf{r}, t) &= -\frac{\partial \mathbf{a}(\mathbf{r}, t)}{\partial t} - \nabla v(\mathbf{r}, t), \\ \mathbf{h}(\mathbf{r}, t) &= \nabla \times \mathbf{a}(\mathbf{r}, t). \end{aligned} \tag{1.57}$$

1.9.1 Coulomb gauge

It should be noted that, in general, the first equation in (1.57) could not be a decomposition in solenoidal and irrotational components since, while $\nabla \times \mathbf{a}$ is certainly solenoidal, the same cannot be said about the potential $\mathbf{a}$. On the other hand, the Helmholtz theorem establishes that a vector field is uniquely determined by its curl and divergence. The potential $\mathbf{a}$ can thus be chosen in such a way that

$$\nabla \cdot \mathbf{a}(\mathbf{r}, t) = 0, \tag{1.58}$$

so that the first of (1.57) really represents a solenoidal–irrotational decomposition. The condition (1.58), that together with the second of (1.57) uniquely defines the potential $\mathbf{a}_\mathrm{C}$, is called *Coulomb gauge.* It should be noted that the use of the potentials

$$\begin{aligned} \mathbf{a}'(\mathbf{r}, t) &= \mathbf{a}(\mathbf{r}, t) + \nabla \Psi(\mathbf{r}, t), \\ v'(\mathbf{r}, t) &= v(\mathbf{r}, t) - \frac{\partial \Psi(\mathbf{r}, t)}{\partial t}, \end{aligned} \tag{1.59}$$

does not change (1.57). The transformation (1.59) is called *gauge transformation* and by suitably choosing the scalar function Ψ, the divergence of $\mathbf{a}$ can be specified to any value without changing (1.57). Using (1.57) inside the second equation in (1.19), we obtain

$$\frac{1}{\mu}\nabla\times\nabla\times\mathbf{a}(\mathbf{r},t) = -\sigma\frac{\partial\mathbf{a}(\mathbf{r},t)}{\partial t} - \sigma\nabla v(\mathbf{r},t) - \varepsilon\frac{\partial^2\mathbf{a}(\mathbf{r},t)}{\partial t^2}$$
$$-\varepsilon\frac{\partial}{\partial t}\nabla v(\mathbf{r},t) + \mathbf{j}_{\mathrm{i}}(\mathbf{r},t) \tag{1.60}$$

and taking the divergence of (1.60), we have

$$-\sigma\frac{\partial}{\partial t}\nabla\cdot\mathbf{a}(\mathbf{r},t) - \sigma\nabla\cdot\nabla v(\mathbf{r},t) - \varepsilon\frac{\partial^2}{\partial t^2}\nabla\cdot\mathbf{a}(\mathbf{r},t)$$
$$-\varepsilon\frac{\partial}{\partial t}\nabla^2 v(\mathbf{r},t) + \nabla\cdot\mathbf{j}_{\mathrm{i}}(\mathbf{r},t) = 0. \tag{1.61}$$

Using the Coulomb gauge (1.58) and the continuity equation (1.3), a differential equation is obtained for the scalar electric potential (in the Coulomb gauge) v_{C}, i.e.,

$$\frac{\partial}{\partial t}\left[\rho_{\mathrm{ei}}(\mathbf{r},t) + \varepsilon\nabla^2 v_{\mathrm{C}}(\mathbf{r},t)\right] = -\sigma\nabla^2 v_{\mathrm{C}}(\mathbf{r},t), \tag{1.62}$$

which for a lossless medium ($\sigma = 0$) becomes

$$\nabla^2 v_{\mathrm{C}}(\mathbf{r},t) = -\frac{\rho_{\mathrm{ei}}(\mathbf{r},t)}{\varepsilon}. \tag{1.63}$$

The Coulomb gauge also allows to write (1.60) as

$$\nabla^2\mathbf{a}_{\mathrm{C}}(\mathbf{r},t) - \mu\varepsilon\frac{\partial^2\mathbf{a}_{\mathrm{C}}(\mathbf{r},t)}{\partial t^2} = -\mu\mathbf{j}_{\mathrm{i}}(\mathbf{r},t) + \mu\sigma\frac{\partial\mathbf{a}_{\mathrm{C}}(\mathbf{r},t)}{\partial t} + \mu\sigma\nabla v_{\mathrm{C}}(\mathbf{r},t)$$
$$+\frac{1}{c^2}\frac{\partial}{\partial t}\nabla v_{\mathrm{C}}(\mathbf{r},t), \tag{1.64}$$

that for a lossless medium becomes

$$\nabla^2\mathbf{a}_{\mathrm{C}}(\mathbf{r},t) - \frac{1}{c^2}\frac{\partial^2\mathbf{a}_{\mathrm{C}}(\mathbf{r},t)}{\partial t^2} = -\mu\mathbf{j}_{\mathrm{i}}(\mathbf{r},t) + \frac{1}{c^2}\frac{\partial}{\partial t}\nabla v_{\mathrm{C}}(\mathbf{r},t). \tag{1.65}$$

However, it can be shown that in (1.65), the ∇v_{C} term cancels with the irrotational component of the $\mathbf{j}_{\mathrm{i}}$ term [2] so that (1.65) can be more correctly written as

$$\nabla^2\mathbf{a}_{\mathrm{C}}(\mathbf{r},t) - \frac{1}{c^2}\frac{\partial^2\mathbf{a}_{\mathrm{C}}(\mathbf{r},t)}{\partial t^2} = -\mu\mathbf{j}_{\mathrm{i}}^{\mathrm{sol}}(\mathbf{r},t), \tag{1.66}$$

where $\mathbf{j}_\mathrm{i}^\mathrm{sol}$ indicates the solenoidal component of the impressed electric current density. In this way, (1.63) and (1.66) represent two uncoupled second-order differential equations for the potentials, i.e., a Poisson equation for the scalar potential v_C and a wave equation for the vector potential $\mathbf{a}_\mathrm{C}$.

1.9.2 Lorenz gauge

A different gauge which is widely used in radiation and guided-wave problems is the so-called *Lorenz gauge* in which the divergence of the vector potential $\mathbf{a}_\mathrm{L}$ and the scalar potential v_L are related through

$$\nabla \cdot \mathbf{a}_\mathrm{L}(\mathbf{r}, t) = -\frac{1}{c^2}\frac{\partial v_\mathrm{L}(\mathbf{r}, t)}{\partial t} - \mu\sigma v_\mathrm{L}(\mathbf{r}, t), \tag{1.67}$$

so that the equation for the vector potential $\mathbf{a}_\mathrm{L}$ in the Lorenz gauge is obtained as

$$\nabla^2 \mathbf{a}_\mathrm{L}(\mathbf{r}, t) - \mu\sigma\frac{\partial \mathbf{a}_\mathrm{L}(\mathbf{r}, t)}{\partial t} - \frac{1}{c^2}\frac{\partial^2 \mathbf{a}_\mathrm{L}(\mathbf{r}, t)}{\partial t^2} = -\mu \mathbf{j}_\mathrm{i}(\mathbf{r}, t), \tag{1.68}$$

while the scalar potential v_L in the Lorenz gauge has to be determined through (1.67). For lossless media, (1.68) becomes

$$\nabla^2 \mathbf{a}_\mathrm{L}(\mathbf{r}, t) - \frac{1}{c^2}\frac{\partial^2 \mathbf{a}_\mathrm{L}(\mathbf{r}, t)}{\partial t^2} = -\mu \mathbf{j}_\mathrm{i}(\mathbf{r}, t) \tag{1.69}$$

and the scalar potential v_L is found as a solution of the equation

$$\nabla^2 v_\mathrm{L}(\mathbf{r}, t) - \frac{1}{c^2}\frac{\partial^2 v_\mathrm{L}(\mathbf{r}, t)}{\partial t^2} = -\frac{\rho_\mathrm{ei}(\mathbf{r}, t)}{\varepsilon} \tag{1.70}$$

or as

$$v_\mathrm{L}(\mathbf{r}, t) = -c^2 \int_{-\infty}^{t} \nabla \cdot \mathbf{a}_\mathrm{L}(\mathbf{r}, t')\,\mathrm{d}t', \tag{1.71}$$

The electromagnetic field can thus expressed only in terms of the vector potential $\mathbf{a}_\mathrm{L}$ as

$$\begin{aligned} \mathbf{e}(\mathbf{r}, t) &= c^2 \nabla \int_{-\infty}^{t} \nabla \cdot \mathbf{a}_\mathrm{L}(\mathbf{r}, t')\,\mathrm{d}t' - \frac{\partial \mathbf{a}_\mathrm{L}(\mathbf{r}, t)}{\partial t}, \\ \mathbf{h}(\mathbf{r}, t) &= \nabla \times \mathbf{a}_\mathrm{L}(\mathbf{r}, t). \end{aligned} \tag{1.72}$$

It should be noted that, in a Lorenz gauge and for lossless media, both the scalar and the vector potentials satisfy a wave equation.

1.9.3 Hertzian potentials

Sometimes in electromagnetic problems, the electric Hertzian potential $\boldsymbol{\pi}_\mathrm{e}$ is used, defined as

$$\boldsymbol{\pi}_\mathrm{e}(\mathbf{r}, t) = -c^2 \int_{-\infty}^{t} \mathbf{a}_\mathrm{L}(\mathbf{r}, t')\,\mathrm{d}t' \tag{1.73}$$

or

$$\mathbf{a}_{\mathrm{L}}(\mathbf{r},t) = \frac{1}{c^2}\frac{\partial \boldsymbol{\pi}_{\mathrm{e}}(\mathbf{r},t)}{\partial t}. \tag{1.74}$$

In this way, the electromagnetic field in lossless media is expressed in terms of the electric Hertzian potential $\boldsymbol{\pi}_{\mathrm{e}}$ as

$$\begin{aligned} \mathbf{e}(\mathbf{r},t) &= \nabla\nabla\cdot\boldsymbol{\pi}_{\mathrm{e}}(\mathbf{r},t) - \frac{1}{c^2}\frac{\partial^2 \boldsymbol{\pi}_{\mathrm{e}}(\mathbf{r},t)}{\partial t^2}, \\ \mathbf{h}(\mathbf{r},t) &= \varepsilon\nabla\times\frac{\partial \boldsymbol{\pi}_{\mathrm{e}}(\mathbf{r},t)}{\partial t} \end{aligned} \tag{1.75}$$

and the Hertzian potential $\boldsymbol{\pi}_{\mathrm{e}}$ satisfies the wave equation

$$\nabla^2\boldsymbol{\pi}_{\mathrm{e}}(\mathbf{r},t) - \frac{1}{c^2}\frac{\partial^2 \boldsymbol{\pi}_{\mathrm{e}}(\mathbf{r},t)}{\partial t^2} = -\frac{1}{\varepsilon}\mathbf{p}_{\mathrm{i}}(\mathbf{r},t), \tag{1.76}$$

where the impressed polarization current $\mathbf{p}_{\mathrm{i}}$ is defined through

$$\mathbf{j}_{\mathrm{i}}(\mathbf{r},t) = \frac{\partial \mathbf{p}_{\mathrm{i}}(\mathbf{r},t)}{\partial t}. \tag{1.77}$$

By suitably expanding $\nabla\nabla\cdot\boldsymbol{\pi}_{\mathrm{e}}$ and using (1.76), the electric field can also be expressed as

$$\mathbf{e}(\mathbf{r},t) = \nabla\times\nabla\times\boldsymbol{\pi}_{\mathrm{e}}(\mathbf{r},t) - \frac{\mathbf{p}_{\mathrm{i}}(\mathbf{r},t)}{\varepsilon}. \tag{1.78}$$

By proceeding in an analogous way, it can be shown that in the presence of magnetic sources $\mathbf{m}_{\mathrm{i}}$ and ρ_{mi} and no electric sources, the electromagnetic field can be derived from the vector and scalar magnetic potentials $\mathbf{f}$ and w which, in a Coulomb gauge (i.e., $\nabla\cdot\mathbf{f}_{\mathrm{C}} = 0$) and for lossless media, satisfy the equations

$$\begin{aligned} \nabla^2\mathbf{f}_{\mathrm{C}}(\mathbf{r},t) - \frac{1}{c^2}\frac{\partial^2 \mathbf{f}_{\mathrm{C}}(\mathbf{r},t)}{\partial t^2} &= -\varepsilon\mathbf{m}_{\mathrm{i}}^{\mathrm{sol}}(\mathbf{r},t), \\ \nabla^2 w_{\mathrm{C}}(\mathbf{r},t) &= -\frac{\rho_{\mathrm{mi}}(\mathbf{r},t)}{\mu}, \end{aligned} \tag{1.79}$$

where $\mathbf{m}_{\mathrm{i}}^{\mathrm{sol}}$ indicates the solenoidal component of the impressed magnetic current density.

On the other hand, in a Lorenz gauge, the potentials satisfy the wave equations

$$\begin{aligned} \nabla^2\mathbf{f}_{\mathrm{L}}(\mathbf{r},t) - \frac{1}{c^2}\frac{\partial^2 \mathbf{f}_{\mathrm{L}}(\mathbf{r},t)}{\partial t^2} &= -\varepsilon\mathbf{m}_{\mathrm{i}}(\mathbf{r},t), \\ \nabla^2 w_{\mathrm{L}}(\mathbf{r},t) - \frac{1}{c^2}\frac{\partial^2 w_{\mathrm{L}}(\mathbf{r},t)}{\partial t^2} &= -\frac{\rho_{\mathrm{mi}}(\mathbf{r},t)}{\mu}. \end{aligned} \tag{1.80}$$

The Lorenz gauge is expressed by the condition

$$\nabla \cdot \mathbf{f}_{\mathrm{L}}(\mathbf{r}, t) = -\frac{1}{c^2} \frac{\partial w_{\mathrm{L}}(\mathbf{r}, t)}{\partial t} \tag{1.81}$$

and the electromagnetic field can thus be expressed in terms of the Lorenz potentials as

$$\begin{aligned} \mathbf{e}(\mathbf{r}, t) &= -\frac{1}{\varepsilon} \nabla \times \mathbf{f}_{\mathrm{L}}(\mathbf{r}, t), \\ \mathbf{h}(\mathbf{r}, t) &= -\frac{\partial \mathbf{f}_{\mathrm{L}}(\mathbf{r}, t)}{\partial t} - c^2 \nabla \int_{-\infty}^{t} \nabla \cdot \mathbf{f}_{\mathrm{L}}(\mathbf{r}, t')\, \mathrm{d}t'. \end{aligned} \tag{1.82}$$

The magnetic Hertzian potential $\boldsymbol{\pi}_{\mathrm{h}}$ can also be defined through

$$\mathbf{f}_{\mathrm{L}}(\mathbf{r}, t) = \frac{1}{c^2} \frac{\partial \boldsymbol{\pi}_{\mathrm{h}}(\mathbf{r}, t)}{\partial t}. \tag{1.83}$$

In this way, the electromagnetic field in lossless media is expressed in terms of the magnetic Hertzian potential $\boldsymbol{\pi}_{\mathrm{h}}$ as

$$\begin{aligned} \mathbf{e}(\mathbf{r}, t) &= -\mu \nabla \times \frac{\partial \boldsymbol{\pi}_{\mathrm{h}}(\mathbf{r}, t)}{\partial t}, \\ \mathbf{h}(\mathbf{r}, t) &= \nabla[\nabla \cdot \boldsymbol{\pi}_{\mathrm{h}}(\mathbf{r}, t)] - \frac{1}{c^2} \frac{\partial^2 \boldsymbol{\pi}_{\mathrm{h}}(\mathbf{r}, t)}{\partial t^2} \end{aligned} \tag{1.84}$$

and the Hertzian potential $\boldsymbol{\pi}_{\mathrm{h}}$ satisfies the wave equation

$$\nabla^2 \boldsymbol{\pi}_{\mathrm{h}}(\mathbf{r}, t) - \frac{1}{c^2} \frac{\partial^2 \boldsymbol{\pi}_{\mathrm{h}}(\mathbf{r}, t)}{\partial t^2} = -\frac{1}{\mu} \mathbf{m}_{\mathrm{mi}}(\mathbf{r}, t), \tag{1.85}$$

where the impressed magnetization current $\mathbf{m}_{\mathrm{mi}}$ is defined through

$$\mathbf{m}_{\mathrm{i}}(\mathbf{r}, t) = \frac{\partial \mathbf{m}_{\mathrm{mi}}(\mathbf{r}, t)}{\partial t}. \tag{1.86}$$

By suitably expanding $\nabla\nabla \cdot \boldsymbol{\pi}_{\mathrm{h}}$ and using (1.85), the magnetic field can also be expressed as

$$\mathbf{h}(\mathbf{r}, t) = \nabla \times \nabla \times \boldsymbol{\pi}_{\mathrm{h}}(\mathbf{r}, t) - \mathbf{m}_{\mathrm{mi}}(\mathbf{r}, t). \tag{1.87}$$

Finally, in the presence of both electric and magnetic sources, the electromagnetic field is expressed in terms of the Lorentz potentials (and omitting the subscript $\cdot_{\mathrm{L}}$) as

$$\begin{aligned} \mathbf{e}(\mathbf{r}, t) &= -\frac{\partial \mathbf{a}(\mathbf{r}, t)}{\partial t} + c^2 \int_{-\infty}^{t} \nabla\nabla \cdot \mathbf{a}(\mathbf{r}, t')\, \mathrm{d}t' - \frac{1}{\varepsilon} \nabla \times \mathbf{f}(\mathbf{r}, t), \\ \mathbf{h}(\mathbf{r}, t) &= \nabla \times \mathbf{a}(\mathbf{r}, t) - \frac{\partial \mathbf{f}(\mathbf{r}, t)}{\partial t} - c^2 \int_{-\infty}^{t} \nabla\nabla \cdot \mathbf{f}(\mathbf{r}, t')\, \mathrm{d}t'. \end{aligned} \tag{1.88}$$

and in terms of Hertzian potentials as

$$
\begin{aligned}
\mathbf{e}(\mathbf{r},t) &= \nabla\nabla\cdot\boldsymbol{\pi}_{\mathrm{e}}(\mathbf{r},t) - \frac{1}{c^2}\frac{\partial^2\boldsymbol{\pi}_{\mathrm{e}}(\mathbf{r},t)}{\partial t^2} - \mu\nabla\times\frac{\partial\boldsymbol{\pi}_{\mathrm{h}}(\mathbf{r},t)}{\partial t}, \\
\mathbf{h}(\mathbf{r},t) &= \varepsilon\nabla\times\frac{\partial\boldsymbol{\pi}_{\mathrm{e}}(\mathbf{r},t)}{\partial t} + \nabla\nabla\cdot\boldsymbol{\pi}_{\mathrm{h}}(\mathbf{r},t) - \frac{1}{c^2}\frac{\partial^2\boldsymbol{\pi}_{\mathrm{h}}(\mathbf{r},t)}{\partial t^2}.
\end{aligned}
\tag{1.89}
$$

1.10 Potentials in unbounded media: retarded potentials

Let us consider the time-domain potentials in the Lorenz gauge $\mathbf{a}$ and v (the subscript $\cdot_{\mathrm{L}}$ is suppressed throughout). In lossless media, such potentials satisfy the equations

$$
\begin{aligned}
\nabla^2\mathbf{a}(\mathbf{r},t) - \frac{1}{c^2}\frac{\partial^2\mathbf{a}(\mathbf{r},t)}{\partial t^2} &= -\mu\mathbf{j}_{\mathrm{i}}(\mathbf{r},t), \\
\nabla^2 v(\mathbf{r},t) - \frac{1}{c^2}\frac{\partial^2 v(\mathbf{r},t)}{\partial t^2} &= -\frac{\rho_{\mathrm{ei}}(\mathbf{r},t)}{\varepsilon}.
\end{aligned}
\tag{1.90}
$$

The potentials are linearly related to their sources, so that they can be expressed in terms of Green's functions. In an unbounded medium with constitutive parameters ε and μ, it thus results

$$
\begin{aligned}
\mathbf{a}(\mathbf{r},t) &= \mu\int_V g(\mathbf{r}-\mathbf{r}',t-t')\mathbf{j}_{\mathrm{i}}(\mathbf{r}',t')\,\mathrm{d}V'\,\mathrm{d}t', \\
v(\mathbf{r},t) &= \frac{1}{\varepsilon}\int_V g(\mathbf{r}-\mathbf{r}',t-t')\rho_{\mathrm{ei}}(\mathbf{r}',t')\,\mathrm{d}V'\,\mathrm{d}t'.
\end{aligned}
\tag{1.91}
$$

The time-domain Green function $g(\mathbf{r}-\mathbf{r}',t-t')$ is a function only of relative positions and times because of the translational invariance and satisfies the equation

$$
\nabla^2 g(\mathbf{r}-\mathbf{r}',t-t') - \frac{1}{c^2}\frac{\partial^2 g(\mathbf{r}-\mathbf{r}',t-t')}{\partial t^2} = -\delta(\mathbf{r}-\mathbf{r}')\delta(t-t'). \tag{1.92}
$$

To solve (1.92), we first Fourier transform it with respect to time in the frequency ω domain. By defining the Fourier transform

$$
G(\mathbf{r},\omega) = \int_{-\infty}^{\infty} g(\mathbf{r},t)e^{-j\omega t}\,\mathrm{d}t, \tag{1.93}
$$

(1.92) becomes

$$
-\left(\nabla^2 + \frac{\omega^2}{c^2}\right)G(\mathbf{r}-\mathbf{r}',\omega) = \delta(\mathbf{r}-\mathbf{r}'), \tag{1.94}
$$

which has the two fundamental solutions

$$
G(\mathbf{r}-\mathbf{r}',\omega) = \frac{e^{\pm j\omega|\mathbf{r}-\mathbf{r}'|/c}}{4\pi|\mathbf{r}-\mathbf{r}'|}. \tag{1.95}
$$

By inverse-Fourier transforming (1.95), it is immediate to obtain

$$g(\mathbf{r}-\mathbf{r}',t-t') = \frac{1}{4\pi|\mathbf{r}-\mathbf{r}'|}\delta\left[\pm\frac{|\mathbf{r}-\mathbf{r}'|}{c} - (t-t')\right]. \tag{1.96}$$

The choice of the sign + leads to the retarded Green function

$$g_{\mathrm{ret}}(\mathbf{r}-\mathbf{r}',t-t') = \frac{1}{4\pi|\mathbf{r}-\mathbf{r}'|}\delta\left[\frac{|\mathbf{r}-\mathbf{r}'|}{c} - (t-t')\right], \tag{1.97}$$

which is nonzero for

$$t = t' + \frac{|\mathbf{r}-\mathbf{r}'|}{c}. \tag{1.98}$$

This means that the signal propagates with the speed of light c from the source (at time t') to the observer (at time t): the effect occurs later than the cause. By taking the sign—in (1.96), the advanced Green function g_{adv} is obtained with the signal arriving at the observer before it is emitted by the source. Since the latter does not comply with causality, the retarded Green function is adopted as the solution which satisfies the correct temporal boundary condition. From (1.91) and (1.97), the explicit expressions for the time-domain potentials are therefore

$$\begin{aligned}\mathbf{a}(\mathbf{r},t) &= \frac{\mu}{4\pi}\int_{-\infty}^{\infty}\int_V \frac{1}{|\mathbf{r}-\mathbf{r}'|}\delta\left[\frac{|\mathbf{r}-\mathbf{r}'|}{c} - (t-t')\right]\mathbf{j}_{\mathrm{i}}(\mathbf{r}',t')\,\mathrm{d}V'\,\mathrm{d}t', \\ v(\mathbf{r},t) &= \frac{1}{4\pi\varepsilon}\int_{-\infty}^{\infty}\int_V \frac{1}{|\mathbf{r}-\mathbf{r}'|}\delta\left[\frac{|\mathbf{r}-\mathbf{r}'|}{c} - (t-t')\right]\rho_{\mathrm{ei}}(\mathbf{r}',t')\,\mathrm{d}V'\,\mathrm{d}t'.\end{aligned} \tag{1.99}$$

By integrating over t' and by defining

$$\mathbf{R} = R\mathbf{u}_R = |\mathbf{r}-\mathbf{r}'|\mathbf{u}_R, \tag{1.100}$$

one obtains

$$\begin{aligned}\mathbf{a}(\mathbf{r},t) &= \frac{\mu}{4\pi}\int_V \frac{1}{R}\mathbf{j}_{\mathrm{i}}(\mathbf{r}',t-R/c)\,\mathrm{d}V', \\ v(\mathbf{r},t) &= \frac{1}{4\pi\varepsilon}\int_V \frac{1}{R}\rho_{\mathrm{ei}}(\mathbf{r}',t-R/c)\,\mathrm{d}V',\end{aligned} \tag{1.101}$$

which are the so-called *retarded potentials*. The quantity $R = |\mathbf{r}-\mathbf{r}'|$ is the distance between the observation point $\mathbf{r}$ and the source point $\mathbf{r}'$ at which point, the volume element of integration $\mathrm{d}V'$ is located, the integration being carried out throughout the volume V that includes all charge and current sources that produce the electromagnetic field under consideration. In the above expressions the argument $t_{\mathrm{ret}} = t - R/c$ denotes that the pair of potentials $\mathbf{a}(\mathbf{r},t)$ and $v(\mathbf{r},t)$ in the Lorenz gauge are due to the charge and current distributions at the *retarded time* t_{ret}. Because of this retardation effect, the pair of potentials in (1.101) are called retarded potentials.

Similar considerations hold also for the Hertzian potentials (which have the same kind of Green's function). In particular, it is immediate to derive the solution of (1.76) as the retarded Hertzian potential

$$\boldsymbol{\pi}_{\mathrm{e}}(\mathbf{r},t) = \frac{1}{4\pi\varepsilon}\int_V \frac{1}{|R|}\mathbf{p}_{\mathrm{i}}(\mathbf{r}',t-R/c)\,\mathrm{d}V'. \tag{1.102}$$

1.11 Time-domain dipole fields in free space

The expressions of the retarded potentials derived in Section 1.10 can be used to derive the corresponding time-domain electric and magnetic field expressions radiating in free space ($\mu = \mu_0$ and $\varepsilon = \varepsilon_0$). In particular, let us consider an elemental electric dipole placed at $\mathbf{r}' = \mathbf{r}_0$ and oscillating along a direction specified by the unit vector $\mathbf{u}_d$. Such an elemental dipole is characterized through its electric moment, i.e.,

$$\mathbf{p}_{\mathrm{i}}(\mathbf{r},t) = p_{\mathrm{i}}(t)\delta(\mathbf{r}-\mathbf{r}_0)\mathbf{u}_d\,. \tag{1.103}$$

By defining

$$\mathbf{R}_0 = R_0\mathbf{u}_{R_0} = |\mathbf{r}-\mathbf{r}_0|\mathbf{u}_{R_0} \tag{1.104}$$

and using (1.102), the retarded Hertzian electric potentials is obtained as

$$\boldsymbol{\pi}_{\mathrm{e}}(\mathbf{r},t) = \frac{1}{4\pi\varepsilon_0 R_0}p_{\mathrm{i}}(t-R_0/c_0)\mathbf{u}_d\,, \tag{1.105}$$

where $c_0 = 1/\sqrt{\mu_0\varepsilon_0}$. The electric and magnetic fields can thus be derived from (1.75) and (1.105). After some algebraic manipulations, it results

$$\begin{aligned}\mathbf{e}(\mathbf{r},t) = {} & \frac{\eta_0}{4\pi}\left[\frac{3c_0}{R_0^3}p(t_{\mathrm{ret}}) + \frac{3}{R_0^2}\left.\frac{\mathrm{d}p}{\mathrm{d}t}\right|_{t_{\mathrm{ret}}} + \frac{1}{c_0R_0}\left.\frac{\mathrm{d}^2p}{\mathrm{d}t^2}\right|_{t_{\mathrm{ret}}}\right](\mathbf{u}_d\cdot\mathbf{u}_{R_0})\mathbf{u}_{R_0}\\ & -\frac{\eta_0}{4\pi}\left[\frac{c_0}{R_0^3}p(t_{\mathrm{ret}}) + \frac{1}{R_0^2}\left.\frac{\mathrm{d}p}{\mathrm{d}t}\right|_{t_{\mathrm{ret}}} + \frac{1}{c_0R_0}\left.\frac{\mathrm{d}^2p}{\mathrm{d}t^2}\right|_{t_{\mathrm{ret}}}\right]\mathbf{u}_d\end{aligned} \tag{1.106}$$

and

$$\mathbf{h}(\mathbf{r},t) = \frac{1}{4\pi}\left[\frac{1}{R_0^2}\left.\frac{\mathrm{d}p}{\mathrm{d}t}\right|_{t_{\mathrm{ret}}} + \frac{1}{c_0R_0}\left.\frac{\mathrm{d}^2p}{\mathrm{d}t^2}\right|_{t_{\mathrm{ret}}}\right](\mathbf{u}_d\times\mathbf{u}_{R_0}), \tag{1.107}$$

where $\eta_0 = \sqrt{\mu_0/\varepsilon_0}$ and $t_{\mathrm{ret}} = t - R_0/c_0$.

1.12 Volume integral representations of the fields

In order to derive a volume integral representation for the electromagnetic field, we start from a solution of Maxwell's equations in a medium with constitutive parameters ε and μ in the form

$$\begin{aligned} \mathbf{e}(\mathbf{r},t) &= c^2\nabla\times\nabla\times\int_{t_0}^{t}\mathbf{a}(\mathbf{r},t')\,\mathrm{d}t' - \frac{1}{\varepsilon}\int_{t_0}^{t}\mathbf{j}_\mathrm{i}(\mathbf{r},t')\,\mathrm{d}t', \\ \mathbf{h}(\mathbf{r},t) &= \frac{1}{\mu}\nabla\times\mathbf{a}(\mathbf{r},t), \end{aligned} \tag{1.108}$$

where

$$\mathbf{a}(\mathbf{r},t) = \frac{\mu}{4\pi}\int_V \frac{1}{|\mathbf{r}-\mathbf{r}'|}\mathbf{j}_\mathrm{i}\left(\mathbf{r}',t-\frac{|\mathbf{r}-\mathbf{r}'|}{c}\right)\mathrm{d}V' \tag{1.109}$$

and all the fields and sources have been assumed zero for $t \le t_0$ while the current sources are inside the volume V. The derivatives in (1.108) can be brought inside the integral of (1.110), but care has to be taken when the observation point lies inside the source region. In fact, as in the frequency-domain counterpart, the integrals have to be intended in a principal-value sense, i.e., by excluding a small volume V_δ (principal volume) around the observation point. Therefore, (1.109) has to be understood in the sense of

$$\mathbf{a}(\mathbf{r},t) = \frac{\mu}{4\pi}\lim_{\delta\to 0}\int_{V-V_\delta} \frac{1}{|\mathbf{r}-\mathbf{r}'|}\mathbf{j}_\mathrm{i}\left(\mathbf{r}',t-\frac{|\mathbf{r}-\mathbf{r}'|}{c}\right)\mathrm{d}V'. \tag{1.110}$$

It should be noted that the value of the integral is independent of the shape of V_δ. However, to correctly differentiate the vector potential in the source region the volume V_δ has to be correctly considered since the limits of integration do depend on the differentiation variable $\mathbf{r}$ [10].

Alternatively, the electric field can also be expressed as

$$\mathbf{e}(\mathbf{r},t) = -\frac{\partial}{\partial t}\mathbf{a}(\mathbf{r},t) - \nabla v(\mathbf{r},t), \tag{1.111}$$

with

$$v(\mathbf{r},t) = \frac{1}{4\pi\varepsilon}\lim_{\delta\to 0}\int_{V-V_\delta} \frac{1}{|\mathbf{r}-\mathbf{r}'|}\rho_\mathrm{ei}\left(\mathbf{r}',t-\frac{|\mathbf{r}-\mathbf{r}'|}{c}\right)\mathrm{d}V', \tag{1.112}$$

where, again, to include observation points inside the source region the principal-value integral has been considered.

From (1.108) and (1.110), the electromagnetic field can also be expressed as

$$\begin{aligned} \mathbf{e}(\mathbf{r},t) = \frac{1}{4\pi\varepsilon}\lim_{\delta\to 0}\int_{V-V_\delta^\mathrm{s}} &\left\{ \frac{1}{c^2 R}\frac{\partial}{\partial t}\mathbf{j}_\mathrm{i}\left(\mathbf{r}',t-\frac{R}{c}\right)\cdot(\mathbf{u}_R\mathbf{u}_R - \underline{\mathbf{I}}) \right. \\ &+ \frac{1}{R^2}\left[\frac{1}{c}\mathbf{j}_\mathrm{i}\left(\mathbf{r}',t-\frac{R}{c}\right) + \frac{1}{R}\int_{t_0}^{t}\mathbf{j}_\mathrm{i}\left(\mathbf{r}',t'-\frac{R}{c}\right)\mathrm{d}t'\right] \\ &\left.\cdot(3\mathbf{u}_R\mathbf{u}_R - \underline{\mathbf{I}})\right\}\mathrm{d}V' - \frac{1}{3\varepsilon}\int_{t_0}^{t}\mathbf{j}(\mathbf{r},t')\,\mathrm{d}t' \end{aligned} \tag{1.113}$$

and

$$\mathbf{h}(\mathbf{r},t) = \frac{1}{4\pi}\lim_{\delta\to 0}\int_{V-V_\delta}\left[\frac{1}{c}\frac{\partial}{\partial t}\mathbf{j}_\mathrm{i}\left(\mathbf{r}',t-\frac{R}{c}\right) + \frac{\mathbf{j}_\mathrm{i}(\mathbf{r}',t-R/c)}{R}\right] \times \frac{\mathbf{R}}{R^2}\,\mathrm{d}V', \tag{1.114}$$

where V_δ^s indicates a spherical principal volume and $\underline{\mathbf{I}}$ is the identity dyadic. For an arbitrarily shaped principal volume, instead of (1.113) the electric field can be expressed as

$$\begin{aligned}\mathbf{e}(\mathbf{r},t) = \frac{1}{4\pi\varepsilon}\lim_{\delta\to 0}\int_{V-V_\delta}&\left\{\frac{1}{c^2R}\frac{\partial}{\partial t}\mathbf{j}_\mathrm{i}\left(\mathbf{r}',t-\frac{R}{c}\right)\cdot(\mathbf{u}_R\mathbf{u}_R-\underline{\mathbf{I}})\right.\\ &+\frac{1}{R^2}\left[\frac{1}{c}\mathbf{j}_\mathrm{i}\left(\mathbf{r}',t-\frac{R}{c}\right)+\frac{1}{R}\int_{t_0}^{t}\mathbf{j}_\mathrm{i}\left(\mathbf{r}',t'-\frac{R}{c}\right)\mathrm{d}t'\right]\\ &\left.\cdot(3\mathbf{u}_R\mathbf{u}_R-\underline{\mathbf{I}})\right\}\mathrm{d}V' - \frac{1}{\varepsilon}\underline{\mathbf{L}}\cdot\int_{t_0}^{t}\mathbf{j}(\mathbf{r},t')\,\mathrm{d}t',\end{aligned} \tag{1.115}$$

where the depolarization dyadic $\underline{\mathbf{L}}$ is defined as

$$\underline{\mathbf{L}} = \frac{1}{4\pi}\int_{S_\delta}\frac{\mathbf{u}_n\mathbf{u}_r}{r}\,\mathrm{d}S, \tag{1.116}$$

where $\mathbf{u}_n$ is the outward unit normal vector to the surface S_δ of the principal volume V_δ. Obviously, for points $\mathbf{r}$ outside the current density distribution $\mathbf{j}_\mathrm{i}$, the integrand of the volume integral is not singular and the depolarization-dyadic term is zero [11,12].

Alternatively, from (1.111) to (1.112), the electric field can be expressed as

$$\begin{aligned}\mathbf{e}(\mathbf{r},t) = \frac{1}{4\pi\varepsilon}\lim_{\delta\to 0}\int_{V-V_\delta}&\left[\frac{1}{c^2R}\frac{\partial}{\partial t}\mathbf{j}_\mathrm{i}\left(\mathbf{r}',t-\frac{R}{c}\right)\cdot(\mathbf{u}_R\mathbf{u}_R-\underline{\mathbf{I}})\right.\\ &\left.+\frac{1}{c}R^2\mathbf{j}_\mathrm{i}\left(\mathbf{r}',t-\frac{R}{c}\right)\cdot(2\mathbf{u}_R\mathbf{u}_R-\underline{\mathbf{I}})+\rho_\mathrm{ei}\left(\mathbf{r}',t-\frac{R}{c}\right)\frac{\mathbf{u}_R}{R^2}\right]\mathrm{d}V'.\end{aligned} \tag{1.117}$$

It is worth noting that (1.113) is not valid for electrostatic fields, while (1.117) holds also for static currents and charge distributions [8].

If the observation point $\mathbf{r}$ lies outside the source region (i.e., $R \neq 0$), from (1.113)–(1.114) we explicitly have

$$\begin{aligned}\mathbf{e}(\mathbf{r},t) = \frac{1}{4\pi\varepsilon}\int_{V}&\left[\frac{1}{c^2R}\frac{\partial}{\partial t}\mathbf{j}_\mathrm{i}\left(\mathbf{r}',t-\frac{R}{c}\right)\cdot(\mathbf{u}_R\mathbf{u}_R-\underline{\mathbf{I}})\right.\\ &\left.+\frac{1}{c}R^2\mathbf{j}_\mathrm{i}\left(\mathbf{r}',t-\frac{R}{c}\right)\cdot(2\mathbf{u}_R\mathbf{u}_R-\underline{\mathbf{I}})+\rho_\mathrm{ei}\left(\mathbf{r}',t-\frac{R}{c}\right)\frac{\mathbf{u}_R}{R^2}\right]\mathrm{d}V'\end{aligned} \tag{1.118}$$

and

$$\mathbf{h}(\mathbf{r},t) = -\frac{1}{4\pi}\int_V \left[\frac{\mathbf{R}}{cR^2} \times \frac{\partial}{\partial t}\mathbf{j}_\mathrm{i}\left(\mathbf{r}',t-\frac{R}{c}\right)\right.$$

$$\left. + \frac{\mathbf{R}}{R^3} \times \frac{\mathbf{j}_\mathrm{i}(\mathbf{r}',t-R/c)}{R}\right] \mathrm{d}V'. \tag{1.119}$$

1.13 Surface integral representations of the fields

It is well known that in the frequency domain the electric and magnetic field outside the source region can be expressed in terms of the tangential electric and magnetic fields on a surface S which contains the field sources through the Stratton-Chu formula [13,14] or through the Franz formula [15,16]. The same can be obtained in the time domain, as shown in [8].

The starting point is the vector identity

$$\int_{V_\infty} \{\varphi(\mathbf{r})\nabla \times \nabla\boldsymbol{\Psi}(\mathbf{r}) + \boldsymbol{\Psi}(\mathbf{r})\nabla'2\varphi(\mathbf{r}) + [\nabla \cdot \boldsymbol{\Psi}(\mathbf{r})]\nabla\varphi(\mathbf{r})\}\mathrm{d}V$$

$$= \oint_S \{\varphi(\mathbf{r})\mathbf{u}_n \times \nabla \times \boldsymbol{\Psi}(\mathbf{r}) + [\mathbf{u}_n \times \boldsymbol{\Psi}(\mathbf{r})] \times \nabla\varphi(\mathbf{r}) + [\mathbf{u}_n \cdot \boldsymbol{\Psi}(\mathbf{r})]\nabla\varphi(\mathbf{r})\}\mathrm{d}S, \tag{1.120}$$

valid for a scalar function $\varphi(\mathbf{r})$ and a vector function $\boldsymbol{\Psi}(\mathbf{r})$, where V_∞ is the infinite volume outside the closed surface S and $\mathbf{u}_n$ is the unit normal to S pointing in V_∞. By choosing $\varphi(\mathbf{r}) = g_\mathrm{ret}(\mathbf{r}-\mathbf{r}',t-t')$ as in (1.97) and $\boldsymbol{\Psi}(\mathbf{r}) = \mathbf{e}(\mathbf{r},t)$, from (1.120), it can be obtained

$$\mathbf{e}(\mathbf{r},t')\delta(t-t') = -\frac{1}{c^2}\int_{V_\infty}\left[g_\mathrm{ret}(\mathbf{r}-\mathbf{r}',t-t')\frac{\partial^2\mathbf{e}(\mathbf{r}',t')}{\partial t'^2}\right.$$

$$\left. -\mathbf{e}(\mathbf{r}',t')\frac{\partial^2 g_\mathrm{ret}(\mathbf{r}-\mathbf{r}',t-t')}{\partial t'^2}\right]\mathrm{d}V'$$

$$+\oint_S\left\{-\mu g_\mathrm{ret}(\mathbf{r}-\mathbf{r}',t-t')\mathbf{u}_{n'} \times \frac{\partial\mathbf{h}(\mathbf{r}',t')}{\partial t'}\right.$$

$$+[\mathbf{u}_{n'} \times \mathbf{e}(\mathbf{r}',t')] \times \nabla' g_\mathrm{ret}(\mathbf{r}-\mathbf{r}',t-t')$$

$$\left. +[\mathbf{u}_{n'} \cdot \mathbf{e}(\mathbf{r}',t')]\nabla' g_\mathrm{ret}(\mathbf{r}-\mathbf{r}',t-t')\right\}\mathrm{d}S'. \tag{1.121}$$

By noting that $\nabla' g_{\text{ret}}(\mathbf{r}-\mathbf{r}',t-t') = -\nabla g_{\text{ret}}(\mathbf{r}-\mathbf{r}',t-t')$ and by expressing the volume integral as a time derivative, from (1.121) it follows

$$\begin{aligned}
\mathbf{e}(\mathbf{r},t')\delta(t-t') = &-\frac{1}{c^2}\frac{\partial}{\partial t'}\int_{V_\infty}\left[g_{\text{ret}}(\mathbf{r}-\mathbf{r}',t-t')\frac{\partial \mathbf{e}(\mathbf{r}',t')}{\partial t'}\right.\\
&\left.-\mathbf{e}(\mathbf{r}',t')\frac{\partial g_{\text{ret}}(\mathbf{r}-\mathbf{r}',t-t')}{\partial t'}\right]\mathrm{d}V'\\
&-\mu\oint_S\left[\mathbf{u}_{n'}\times\frac{\partial \mathbf{h}(\mathbf{r}',t')}{\partial t'}\right]g_{\text{ret}}(\mathbf{r}-\mathbf{r}',t-t')\,\mathrm{d}S'\\
&+\nabla\times\oint_S[\mathbf{u}_{n'}\times\mathbf{e}(\mathbf{r}',t')]g_{\text{ret}}(\mathbf{r}-\mathbf{r}',t-t')\,\mathrm{d}S'\\
&-\nabla\oint_S[\mathbf{u}_{n'}\cdot\mathbf{e}(\mathbf{r}',t')]g_{\text{ret}}(\mathbf{r}-\mathbf{r}',t-t')\,\mathrm{d}S'.
\end{aligned} \tag{1.122}$$

By integrating (1.122) over t', the volume-integral term vanishes and

$$\begin{aligned}
\mathbf{e}(\mathbf{r},t) = &-\mu\frac{\partial}{\partial t}\oint_s\frac{\mathbf{u}_{n'}\times\mathbf{h}(\mathbf{r}',t-R/c)}{4\pi R}\,\mathrm{d}S'\\
&+\nabla\times\oint_s\frac{\mathbf{u}_{n'}\times\mathbf{e}(\mathbf{r}',t-R/c)}{4\pi R}\,\mathrm{d}S' - \nabla\oint_s\frac{\mathbf{u}_{n'}\cdot\mathbf{e}(\mathbf{r}',t-R/c)}{4\pi R}\,\mathrm{d}S'.
\end{aligned} \tag{1.123}$$

In the same way, by choosing $\boldsymbol{\Psi} = \mathbf{h}$, it results

$$\begin{aligned}
\mathbf{h}(\mathbf{r},t) = &\,\varepsilon\frac{\partial}{\partial t}\oint_S\frac{\mathbf{u}_{n'}\times\mathbf{e}(\mathbf{r}',t-R/c)}{4\pi R}\,\mathrm{d}S'\\
&+\nabla\times\oint_s\frac{\mathbf{u}_{n'}\times\mathbf{h}(\mathbf{r}',t-R/c)}{4\pi R}\,\mathrm{d}S' - \nabla\oint_s\frac{\mathbf{u}_{n'}\cdot\mathbf{h}(\mathbf{r}',t-R/c)}{4\pi R}\,\mathrm{d}S'.
\end{aligned} \tag{1.124}$$

Equations (1.123) and (1.124) express the electric and magnetic fields outside a closed region in terms of the values of the electric and magnetic fields on a surface S. In this sense, this is the time-domain analog of the frequency-domain Stratton–Chu formula and involves both the tangential and the normal components of the field. However, it should be noted that the normal components can in turn be expressed in terms of the tangential components. In particular, by assuming that the fields are zero for $t \leq t_0$,

$$\begin{aligned}
\mathbf{u}_n\cdot\mathbf{e}(\mathbf{r},t) &= -\frac{1}{\varepsilon}\nabla_S\cdot\int_{t_0}^{t}\mathbf{u}_n\times\mathbf{h}(\mathbf{r},t')\,\mathrm{d}t', \qquad \mathbf{r}\in S,\\
\mathbf{u}_n\cdot\mathbf{h}(\mathbf{r},t) &= \frac{1}{\mu}\nabla_S\cdot\int_{t_0}^{t}\mathbf{u}_n\times\mathbf{e}(\mathbf{r},t')\,\mathrm{d}t', \qquad \mathbf{r}\in S,
\end{aligned} \tag{1.125}$$

where ∇_S is the surface nabla operator.

The gradient terms in (1.123) and (1.124) can be eliminated by integrating over time, taking the curl and using Maxwell's equations so that we have

$$\mathbf{e}(\mathbf{r},t') = \nabla \times \oint_S \frac{1}{4\pi R}\left[\mathbf{u}_{n'} \times \mathbf{h}\left(\mathbf{r}',t-\frac{R}{c}\right)\right] \mathrm{d}S'$$
$$-\nabla \times \nabla \times \oint_S \frac{1}{4\pi\mu R}\left[\mathbf{u}_{n'} \times \int_{t_0}^{t} \mathbf{e}\left(\mathbf{r}',t-\frac{R}{c}\right)\mathrm{d}t'\right]\mathrm{d}S' \qquad (1.126)$$

and

$$\mathbf{h}(\mathbf{r},t') = \nabla \times \oint_S \frac{1}{4\pi R}\left[\mathbf{u}_{n'} \times \mathbf{e}\left(\mathbf{r}',t-\frac{R}{c}\right)\right] \mathrm{d}S'$$
$$+\nabla \times \nabla \times \oint_S \frac{1}{4\pi\varepsilon R}\left[\mathbf{u}_{n'} \times \int_{t_0}^{t} \mathbf{h}\left(\mathbf{r}',t-\frac{R}{c}\right)\mathrm{d}t'\right]\mathrm{d}S'. \qquad (1.127)$$

Equations (1.126) and (1.127) are the time-domain analogs of the Franz formulation and the equivalent current densities

$$\begin{aligned}\mathbf{j}_S(\mathbf{r},t) &= \mathbf{u}_n \times \mathbf{h}(\mathbf{r},t), \\ \mathbf{m}_S(\mathbf{r},t) &= -\mathbf{u}_n \times \mathbf{e}(\mathbf{r},t)\end{aligned} \qquad (1.128)$$

are the *time-domain Huygens sources* of the electromagnetic field outside the closed surface S.

Through suitable vector identities, (1.126) and (1.127) can also be written as

$$\mathbf{e}(\mathbf{r},t') = \frac{1}{4\pi}\oint_S \left\{\left[\frac{1}{c}\mathbf{u}_{n'} \times \frac{\partial}{\partial t}\mathbf{e}\left(\mathbf{r}',t-\frac{R}{c}\right) + \frac{1}{R}\mathbf{u}_{n'} \times \mathbf{e}\left(\mathbf{r}',t-\frac{R}{c}\right)\right] \times \frac{\mathbf{u}_R}{R}\right.$$
$$+\frac{\mu}{R}\left[\mathbf{u}_{n'} \times \frac{\partial}{\partial t}\mathbf{h}\left(\mathbf{r}',t-\frac{R}{c}\right)\right]\cdot(\mathbf{u}_R\mathbf{u}_R - \underline{\mathbf{I}})$$
$$+\frac{1}{\varepsilon R^2}\left[\frac{1}{c}\mathbf{u}_{n'} \times \frac{\partial}{\partial t}\mathbf{h}\left(\mathbf{r}',t-\frac{R}{c}\right) + \frac{1}{R}\mathbf{u}_{n'} \times \int_{t_0}^{t-R/c} \mathbf{h}(\mathbf{r}',t')\,\mathrm{d}t'\right]$$
$$\left.\cdot(3\mathbf{u}_R\mathbf{u}_R - \underline{\mathbf{I}})\right\}\mathrm{d}S' \qquad (1.129)$$

and

$$\mathbf{h}(\mathbf{r},t') = \frac{1}{4\pi}\oint_S \left\{\left[\frac{1}{c}\mathbf{u}_{n'} \times \frac{\partial}{\partial t}\mathbf{h}\left(\mathbf{r}',t-\frac{R}{c}\right) + \frac{1}{R}\mathbf{u}_{n'} \times \mathbf{h}\left(\mathbf{r}',t-\frac{R}{c}\right)\right] \times \frac{\mathbf{u}_R}{R}\right.$$
$$-\frac{\varepsilon}{R}\left[\mathbf{u}_{n'} \times \frac{\partial}{\partial t}\mathbf{e}\left(\mathbf{r}',t-\frac{R}{c}\right)\right]\cdot(\mathbf{u}_R\mathbf{u}_R - \underline{\mathbf{I}})$$
$$-\frac{1}{\mu R^2}\left[\frac{1}{c}\mathbf{u}_{n'} \times \frac{\partial}{\partial t}\mathbf{e}\left(\mathbf{r}',t-\frac{R}{c}\right) + \frac{1}{R}\mathbf{u}_{n'} \times \int_{t_0}^{t-R/c} \mathbf{e}(\mathbf{r}',t')\,\mathrm{d}t'\right]$$
$$\left.\cdot(3\mathbf{u}_R\mathbf{u}_R - \underline{\mathbf{I}})\right\}\mathrm{d}S', \qquad (1.130)$$

i.e.,

$$\mathbf{e}(\mathbf{r},t') = \frac{1}{4\pi}\oint_S \left\{ \left[-\frac{1}{c}\frac{\partial}{\partial t}\mathbf{m}_S\left(\mathbf{r}',t-\frac{R}{c}\right) - \frac{1}{R}\mathbf{m}_S\left(\mathbf{r}',t-\frac{R}{c}\right)\right] \times \frac{\mathbf{u}_R}{R} \right.$$
$$+\frac{\mu}{R}\left[\frac{\partial}{\partial t}\mathbf{j}_S\left(\mathbf{r}',t-\frac{R}{c}\right)\right]\cdot(\mathbf{u}_R\mathbf{u}_R-\underline{\mathbf{I}})$$
$$+\frac{1}{\varepsilon R^2}\left[\frac{1}{c}\frac{\partial}{\partial t}\mathbf{j}_S\left(\mathbf{r}',t-\frac{R}{c}\right) + \frac{1}{R}\int_{t_0}^{t-R/c}\mathbf{j}_S(\mathbf{r}',t')\,\mathrm{d}t'\right]$$
$$\left.\cdot(3\mathbf{u}_R\mathbf{u}_R-\underline{\mathbf{I}})\right\}\mathrm{d}S' \tag{1.131}$$

and

$$\mathbf{h}(\mathbf{r},t') = \frac{1}{4\pi}\oint_S \left\{ \left[\frac{1}{c}\frac{\partial}{\partial t}\mathbf{j}_S\left(\mathbf{r}',t-\frac{R}{c}\right) + \frac{1}{R}\mathbf{j}_S\left(\mathbf{r}',t-\frac{R}{c}\right)\right] \times \frac{\mathbf{u}_R}{R} \right.$$
$$+\frac{\varepsilon}{R}\left[\frac{\partial}{\partial t}\mathbf{m}_S\left(\mathbf{r}',t-\frac{R}{c}\right)\right]\cdot(\mathbf{u}_R\mathbf{u}_R-\underline{\mathbf{I}})$$
$$+\frac{1}{\mu R^2}\left[\frac{1}{c}\frac{\partial}{\partial t}\mathbf{m}_S\left(\mathbf{r}',t-\frac{R}{c}\right) - \frac{1}{R}\times\int_{t_0}^{t-R/c}\mathbf{m}_S(\mathbf{r}',t')\,\mathrm{d}t'\right]$$
$$\left.\cdot(3\mathbf{u}_R\mathbf{u}_R-\underline{\mathbf{I}})\right\}\mathrm{d}S'. \tag{1.132}$$

It is worth noting that, contrarily to the time-domain Franz formulas (1.126) and (1.127) (or (1.129) and (1.130)), the time-domain Stratton–Chu representation (1.123) and (1.124) remains valid also for static fields.

1.14 Time-domain far fields

Starting from the volume integral representation of the fields outside the source region (1.118) and (1.119), it can easily be shown that the electric and magnetic fields decay as $1/r$ (or faster) as $r \to \infty$, provided that $\mathbf{j}_\mathrm{i}(\mathbf{r},t'-R/c)$ together with its time derivative and integral are integrable over V for all $(\mathbf{r},t)$, and the time derivative of $\mathbf{j}(\mathbf{r},t')$ is bounded by an integrable time-independent function [8]. Moreover, if the current and its first time derivative are continuous function of time, it can be shown that in free space

$$\mathbf{e}(\mathbf{r},t)=\frac{\eta_0\mathbf{u}_r}{4\pi c_0 r}\times\left[\mathbf{u}_r\times\int_V\frac{\partial}{\partial t}\mathbf{j}_\mathrm{i}\left(\mathbf{r}',t-\frac{r+\mathbf{u}_r\cdot\mathbf{r}'}{c_0}\right)\right]\mathrm{d}V'+O\left(\frac{1}{r^2}\right) \tag{1.133}$$

and

$$\mathbf{h}(\mathbf{r},t) = -\frac{1}{4\pi c_0 r}\mathbf{u}_r\times\left[\int_V\frac{\partial}{\partial t}\mathbf{j}_\mathrm{i}\left(\mathbf{r}',t-\frac{r+\mathbf{u}_r\cdot\mathbf{r}'}{c_0}\right)\right]\mathrm{d}V'+O\left(\frac{1}{r^2}\right). \tag{1.134}$$

It can be shown that (1.133) and (1.134) hold also under more general assumptions [8].

Analogously, starting from the surface integral representation of the fields (1.129) and (1.130) and under analogous assumptions on the Huygens sources $\mathbf{u}_n \times \mathbf{e}$ and $\mathbf{u}_n \times \mathbf{h}$ it can be shown that

$$\begin{aligned}\mathbf{e}(\mathbf{r},t) = &-\frac{1}{4\pi r}\mathbf{u}_r \times \oint_s \left\{ \frac{1}{c_0}\mathbf{u}_{n'} \times \frac{\partial}{\partial t}\mathbf{e}\left(\mathbf{r}', t - \frac{r + \mathbf{u}_r \cdot \mathbf{r}'}{c_0}\right) \right. \\ &\left. -\mu_0 \mathbf{u}_r \times \left[\mathbf{u}_{n'} \times \frac{\partial}{\partial t}\mathbf{h}\left(\mathbf{r}', t - \frac{r + \mathbf{u}_r \cdot \mathbf{r}'}{c_0}\right)\right]\right\} \mathrm{d}S' + O\left(\frac{1}{r^2}\right)\end{aligned} \tag{1.135}$$

and

$$\begin{aligned}\mathbf{h}(\mathbf{r},t) = &-\frac{1}{4\pi r}\mathbf{u}_r \times \oint_s \left\{ \frac{1}{c_0}\mathbf{u}_{n'} \times \frac{\partial}{\partial t}\mathbf{h}\left(\mathbf{r}', t - \frac{r + \mathbf{u}_r \cdot \mathbf{r}'}{c_0}\right) \right. \\ &\left. +\varepsilon_0 \mathbf{u}_r \times \left[\mathbf{u}_{n'} \times \frac{\partial}{\partial t}\mathbf{e}\left(\mathbf{r}', t - \frac{r + \mathbf{u}_r \cdot \mathbf{r}'}{c_0}\right)\right]\right\} \mathrm{d}S' + O\left(\frac{1}{r^2}\right).\end{aligned} \tag{1.136}$$

Therefore, under the above-mentioned conditions on the sources, the time-domain far fields decay as $1/r$ or faster as $r \to \infty$. However, when the above conditions are not satisfied, far-field decays slower than $1/r$ can occur (in particular, the so-called "electromagnetic missiles," i.e., far-field pulses with energy decay slower than $1/r^2$ can exist [17,18]).

When volume and Huygens' sources satisfy the conditions for which the fields decay as $1/r$ (or faster), it is possible to define the *time-domain far-field patterns* as

$$\begin{aligned}\mathbf{e}^{\mathrm{ff}}(\mathbf{u}_r, t) &= \lim_{r\to\infty} r\,\mathbf{e}(\mathbf{r}, t + r/c_0), \\ \mathbf{h}^{\mathrm{ff}}(\mathbf{u}_r, \varphi, t) &= \lim_{r\to\infty} r\,\mathbf{h}(\mathbf{r}, t + r/c_0),\end{aligned} \tag{1.137}$$

so that

$$\begin{aligned}\mathbf{e}(\mathbf{r},t) &\sim \frac{\mathbf{e}^{\mathrm{ff}}(\mathbf{u}_r, t - r/c_0)}{r}, \\ \mathbf{h}(\mathbf{r},t) &\sim \frac{\mathbf{h}^{\mathrm{ff}}(\mathbf{u}_r, t - r/c_0)}{r}\end{aligned} \tag{1.138}$$

and

$$\begin{aligned}\mathbf{e}^{\mathrm{ff}}(\mathbf{u}_r, t) &= -\eta_0 \mathbf{u}_r \times \mathbf{h}^{\mathrm{ff}}(\mathbf{u}_r, t), \\ \mathbf{h}^{\mathrm{ff}}(\mathbf{u}_r, t) &= \frac{1}{\eta_0}\mathbf{u}_r \times \mathbf{e}^{\mathrm{ff}}(\mathbf{u}_r, t).\end{aligned} \tag{1.139}$$

1.15 Time-domain reciprocity and energy theorems

Some fundamental theorems with applications to time-domain electromagnetic theory are illustrated in this section.

1.15.1 Reciprocity theorems

A reciprocity theorem is an explicit mathematical relationship between two different wavefields that can exist in the same space-time configuration. More specifically, a reciprocity theorem involves interaction between sets of fields and sources in different configurations.

The first statement of electromagnetic reciprocity is commonly credited to Lorentz [19], who dealt with time-harmonic (i.e., fixed-frequency) electromagnetic fields. The interaction quantity that occurs in Lorentz reciprocity theorem was indicated by Rumsey as the reaction between the sources and the fields in two different configurations (or states) [20]. However, while in time-harmonic reciprocity the interaction is only spatial, in time-domain reciprocity the interaction is spatio-temporal. In particular, two types of temporal interactions can be used, i.e., a *convolution-type* and a *correlation-type reciprocity*. Historically, in [21], the first time-domain reciprocity theorem (of correlation type) was derived for electromagnetic fields, while in [22], a convolution-type reciprocity theorem was developed. These theorems apply to homogeneous, isotropic and lossless [21] or lossy [22] media. In [23], the convolution-type reciprocity theorem was extended to dispersive and possibly nonhomogeneous or anisotropic media. In [24], both convolution-type and correlation-type reciprocity theorems have been presented for homogeneous, isotropic, and lossless media. Next, time-domain reciprocity theorems of both types have been developed for a large class of media by A. de Hoop [25,26].

Applications of reciprocity theorems range from the study of coupling between antennas to the study of both direct and inverse scattering problems, to integral-equation formulations of electromagnetic problems, and to the development of numerical methods [26–28].

In what follows, in order to present the time-domain reciprocity theorems, the proofs presented in [23,29] are illustrated.

We recall that the convolution integral between two functions $\mathbf{f}_1(\mathbf{r},t)$ and $\mathbf{f}_2(\mathbf{r},t)$ is defined as

$$\begin{aligned} C(\mathbf{f}_1,\mathbf{f}_2;\mathbf{r},t)=\mathbf{f}_1\otimes\mathbf{f}_2 &= \int_{-\infty}^{\infty}\mathbf{f}_1(\mathbf{r},\tau)\cdot\mathbf{f}_2(\mathbf{r},t-\tau)\,\mathrm{d}\tau \\ &= \int_{-\infty}^{\infty}\mathbf{f}_1(\mathbf{r},t-\tau)\cdot\mathbf{f}_2(\mathbf{r},\tau)\,\mathrm{d}\tau, \end{aligned} \tag{1.140}$$

while the correlation integral is

$$\begin{aligned} R(\mathbf{f}_1,\mathbf{f}_2;\mathbf{r},t)=\mathbf{f}_1\circledast\mathbf{f}_2 &= \int_{-\infty}^{\infty}\mathbf{f}_1(\mathbf{r},\tau)\cdot\mathbf{f}_2(\mathbf{r},\tau-t)\,\mathrm{d}\tau \\ &= \int_{-\infty}^{\infty}\mathbf{f}_1(\mathbf{r},t+\tau)\cdot\mathbf{f}_2(\mathbf{r},\tau)\,\mathrm{d}\tau. \end{aligned} \tag{1.141}$$

From (1.140) to (1.141), it can be observed that

$$\begin{aligned} &C(\mathbf{f}_1,\mathbf{f}_2;\mathbf{r},t)=C(\mathbf{f}_2,\mathbf{f}_1;\mathbf{r},t), \\ &R(\mathbf{f}_1,\mathbf{f}_2;\mathbf{r},t)=R(\mathbf{f}_2,\mathbf{f}_1;\mathbf{r},-t), \\ &C(\mathbf{f}_1,\mathbf{f}_2;\mathbf{r},t)=C(\mathbf{f}_1,\mathbf{f}_2;\mathbf{r},-t). \end{aligned} \tag{1.142}$$

1.15.1.1 Convolution-type reciprocity

A linear, time-independent medium is considered but, for simplicity, only the presence of electric-current sources is assumed (although the inclusion of magnetic currents is straightforward). Two independent electric-current sources are considered, i.e., $\mathbf{j}_{i1}(\mathbf{r}, t)$ and $\mathbf{j}_{i2}(\mathbf{r}, t)$ which are both confined in a certain volume, are identically zero for $t < t_0$, and radiate the electromagnetic fields $\{\mathbf{e}_1(\mathbf{r}, t), \mathbf{h}_1(\mathbf{r}, t)\}$ and $\{\mathbf{e}_2(\mathbf{r}, t), \mathbf{h}_2(\mathbf{r}, t)\}$, respectively. The convolution-type reciprocity theorem thus establishes that

$$\int_{V_0} \mathbf{e}_1(\mathbf{r}, t) \otimes \mathbf{j}_{i2}(\mathbf{r}, t)\, \mathrm{d}V = \int_{V_0} \mathbf{e}_2(\mathbf{r}, t) \otimes \mathbf{j}_{i1}(\mathbf{r}, t)\, \mathrm{d}V, \tag{1.143}$$

where V_0 is any volume containing both $\mathbf{j}_{i1}$ and $\mathbf{j}_{i2}$.

The proof of (1.143) proceeds as follows. Starting from Maxwell's equations (1.8) with $\mathbf{m}_i = \mathbf{0}$, written for $t = \tau$ and $t = t - \tau$ and combining them through the vector identity

$$\nabla \cdot (\mathbf{A} \times \mathbf{B}) = \mathbf{B} \cdot (\nabla \times \mathbf{A}) - \mathbf{A} \cdot (\nabla \times \mathbf{B}), \tag{1.144}$$

it is easy to obtain

$$\begin{aligned}
\nabla \cdot \quad & [\mathbf{e}_1(\mathbf{r}, t-\tau) \times \mathbf{h}_2(\mathbf{r}, \tau) - \mathbf{e}_2(\tau) \times \mathbf{h}_1(\mathbf{r}, t-\tau)] = \\
& \mathbf{h}_2(\mathbf{r}, \tau) \cdot \frac{\partial}{\partial \tau} \mathbf{b}_1(\mathbf{r}, t-\tau) + \mathbf{h}_1(\mathbf{r}, t-\tau) \cdot \frac{\partial}{\partial \tau} \mathbf{b}_2(\mathbf{r}, \tau) \\
& - \mathbf{e}_1(\mathbf{r}, t-\tau) \cdot \mathbf{j}_{i2}(\mathbf{r}, \tau) - \mathbf{e}_1(\mathbf{r}, t-\tau) \cdot \mathbf{j}_2(\mathbf{r}, \tau) \\
& - \mathbf{e}_1(\mathbf{r}, t-\tau) \cdot \frac{\partial}{\partial \tau} \mathbf{d}_2(\mathbf{r}, \tau) - \mathbf{e}_2(\mathbf{r}, \tau) \cdot \frac{\partial}{\partial \tau} \mathbf{d}_1(\mathbf{r}, t-\tau) \\
& + \mathbf{e}_2(\mathbf{r}, \tau) \cdot \mathbf{j}_{i1}(\mathbf{r}, t-\tau) + \mathbf{e}_2(\mathbf{r}, \tau) \cdot \mathbf{j}_1(\mathbf{r}, t-\tau).
\end{aligned} \tag{1.145}$$

For isotropic and dispersive media, the constitutive relations of the medium can be expressed as convolution integrals (the proof can be generalized for anisotropic but not gyrotropic media [23]). Using such constitutive relations, one simply obtains

$$\begin{aligned}
& \int_{-\infty}^{\infty} \mathbf{h}_1(\mathbf{r}, t-\tau) \cdot \frac{\partial}{\partial \tau} \mathbf{b}_2(\mathbf{r}, \tau) + \mathbf{h}_2(\mathbf{r}, \tau) \cdot \frac{\partial}{\partial \tau} \mathbf{b}_1(\mathbf{r}, t-\tau)\, \mathrm{d}\tau \\
& = \int_{-\infty}^{\infty} \mu(t') \int_{-\infty}^{\infty} \left[\mathbf{h}_1(\mathbf{r}, t-\tau) \cdot \frac{\partial}{\partial \tau} \mathbf{h}_2(\mathbf{r}, \tau - t') \right. \\
& \left. + \mathbf{h}_2(\mathbf{r}, \tau) \cdot \frac{\partial}{\partial \tau} \mathbf{h}_1(\mathbf{r}, t-\tau-t') \right] \mathrm{d}\tau\, \mathrm{d}t' \\
& = \int_{-\infty}^{\infty} \mu(t') \int_{-\infty}^{\infty} \left[\mathbf{h}_1(\mathbf{r}, t-\tau) \cdot \frac{\partial}{\partial \tau} \mathbf{h}_2(\mathbf{r}, \tau - t') \right. \\
& \left. + \mathbf{h}_2(\mathbf{r}, \tau - t') \cdot \frac{\partial}{\partial \tau} \mathbf{h}_1(\mathbf{r}, t-\tau) \right] \mathrm{d}\tau\, \mathrm{d}t' \\
& = \int_{-\infty}^{\infty} \mu(t') \left[\mathbf{h}_1(\mathbf{r}, t-\tau) \cdot \mathbf{h}_2(\mathbf{r}, \tau - t') \Big|_{\tau=-\infty}^{\tau=+\infty} \right] \mathrm{d}t' = 0,
\end{aligned} \tag{1.146}$$

since $\mathbf{h}_1(\mathbf{r}, -\infty) = \mathbf{h}_2(\mathbf{r}, -\infty) = \mathbf{0}$. In the same way, one can show that

$$\int_{-\infty}^{\infty} \left[\mathbf{e}_1(\mathbf{r}, t-\tau) \cdot \frac{\partial}{\partial \tau} \mathbf{d}_2(\mathbf{r}, \tau) + \mathbf{e}_2(\mathbf{r}, \tau) \cdot \frac{\partial}{\partial \tau} \mathbf{d}_1(\mathbf{r}, t-\tau) \right] \mathrm{d}\tau = 0. \tag{1.147}$$

Moreover, the term involving the conduction current is identically zero since

$$\begin{aligned}
&\int_{-\infty}^{\infty} [\mathbf{e}_1(\mathbf{r}, t-\tau) \cdot \mathbf{j}_2(\mathbf{r}, \tau) - \mathbf{e}_2(\mathbf{r}, \tau) \cdot \mathbf{j}_1(\mathbf{r}, t-\tau)] \mathrm{d}\tau \\
&= \int_{-\infty}^{\infty} \left[\mathbf{e}_1(\mathbf{r}, t-\tau) \cdot \int_{-\infty}^{\infty} \sigma(t') \mathbf{e}_2(\mathbf{r}, \tau - t') \mathrm{d}t' \right. \\
&\left. -\mathbf{e}_2(\mathbf{r}, \tau) \cdot \int_{-\infty}^{\infty} \sigma(t') \mathbf{e}_1(\mathbf{r}, t-\tau-t') \mathrm{d}t' \right] \mathrm{d}\tau \\
&= \int_{-\infty}^{\infty} \sigma(t') \int_{-\infty}^{\infty} [\mathbf{e}_1(\mathbf{r}, t-\tau) \cdot \mathbf{e}_2(\mathbf{r}, \tau - t') \\
&-\mathbf{e}_2(\mathbf{r}, \tau - t') \cdot \mathbf{e}_1(\mathbf{r}, t-\tau)] \mathrm{d}\tau \mathrm{d}t' = 0.
\end{aligned} \tag{1.148}$$

Therefore, by integrating (1.145) with respect to τ and using (1.146)–(1.148), one obtains

$$\begin{aligned}
&\int_{-\infty}^{\infty} \nabla \cdot [\mathbf{e}_1(\mathbf{r}, t-\tau) \times \mathbf{h}_2(\mathbf{r}, \tau) - \mathbf{e}_2(\mathbf{r}, \tau) \times \mathbf{h}_1(\mathbf{r}, t-\tau)] \mathrm{d}\tau \\
&= \int_{-\infty}^{\infty} [-\mathbf{e}_1(\mathbf{r}, t-\tau) \cdot \mathbf{j}_{i2}(\mathbf{r}, \tau) + \mathbf{e}_2(\mathbf{r}, \tau) \cdot \mathbf{j}_{i1}(\mathbf{r}, t-\tau)] \mathrm{d}\tau.
\end{aligned} \tag{1.149}$$

By integrating (1.149) over a volume V_0 containing both $\mathbf{j}_{i1}$ and $\mathbf{j}_{i2}$, interchanging the order of integrations, using the divergence theorem, and changing the order of integration again, there is obtained

$$\begin{aligned}
&\oint_{S_0} \int_{-\infty}^{\infty} [\mathbf{e}_1(\mathbf{r}, t-\tau) \times \mathbf{h}_2(\mathbf{r}, \tau) - \mathbf{e}_2(\mathbf{r}, \tau) \times \mathbf{h}_1(\mathbf{r}, t-\tau)] \cdot \mathbf{u}_n \mathrm{d}\tau \mathrm{d}S \\
&= \int_{V_0} \int_{-\infty}^{\infty} [\mathbf{e}_2(\mathbf{r}, \tau) \cdot \mathbf{j}_{i1}(\mathbf{r}, t-\tau) - \mathbf{e}_1(\mathbf{r}, t-\tau) \cdot \mathbf{j}_{i2}(\mathbf{r}, \tau)] \mathrm{d}\tau \mathrm{d}V,
\end{aligned} \tag{1.150}$$

where S_0 is the surface enclosing the volume V_0 and $\mathbf{u}_n$ is the unit vector normal to S_0. It should be noted that the surface integral in (1.150) is a constant over any surface S_0 which encloses both $\mathbf{j}_{i1}$ and $\mathbf{j}_{i2}$ (since the volume integral remains constant by increasing V_0 beyond a region that encloses both $\mathbf{j}_{i1}$ and $\mathbf{j}_{i2}$). Moreover, for any current $\mathbf{j}_i$ which is identically zero for $t < t_0$ and has a finite extent, for every finite t_1 there exists a finite sphere with radius r_1 on which the radiated field $\{\mathbf{e}(t), \mathbf{h}(t)\}$ is identically zero for $t \le t_1$. Thanks to the latter consideration, it can easily be shown that the surface integral in (1.150) can be made zero by choosing V_0 as a sphere with a sufficiently large radius. Therefore the surface integral in

(1.150) is identically zero over any surface S_0 which encloses both $\mathbf{j}_{i1}$ and $\mathbf{j}_{i2}$ (including a surface at infinity) and

$$\int_{V_0}\int_{-\infty}^{\infty}[\mathbf{e}_2(\mathbf{r},\tau)\cdot\mathbf{j}_{i1}(\mathbf{r},t-\tau)-\mathbf{e}_1(\mathbf{r},t-\tau)\cdot\mathbf{j}_{i2}(\mathbf{r},\tau)]\,\mathrm{d}\tau\,\mathrm{d}V=0, \tag{1.151}$$

which is identical to (1.143).

The time-domain convolution-type reciprocity theorem (1.143) is closely related to the Lorentz reciprocity theorem for time-harmonic fields [20] in the form

$$\int_{V_1}\mathbf{E}_2(\mathbf{r})\cdot\mathbf{J}_{i1}(\mathbf{r})\,\mathrm{d}V=\int_{V_2}\mathbf{E}_1(\mathbf{r})\cdot\mathbf{J}_{i2}(\mathbf{r})\,\mathrm{d}V, \tag{1.152}$$

where $\mathbf{E}_{1,2}$ and $\mathbf{J}_{i1,2}$ are the complex vectors corresponding to an harmonic time-dependence $e^{j\omega t}$ of the relevant time-domain quantities while V_1 and V_2 are volumes enclosing the impressed current densities $\mathbf{J}_{i1}$ and $\mathbf{J}_{i2}$, respectively.

1.15.1.2 Correlation-type reciprocity

In order to formulate the time-dependent correlation-type reciprocity theorem, two independent electric-current sources are considered, i.e., $\mathbf{j}_{i1}(\mathbf{r},t)$ and $\mathbf{j}_{i2}(\mathbf{r},t)$ which are both confined in a certain volume, are identically zero for $t_1<t<t_2$ (i.e., they have a finite temporal support), and radiate the retarded $\{\mathbf{e}_1(\mathbf{r},t),\mathbf{h}_1(\mathbf{r},t)\}$ and *advanced* $\{\mathbf{e}_2^{\mathrm{adv}}(\mathbf{r},t),\mathbf{h}_2^{\mathrm{adv}}(\mathbf{r},t)\}$ fields, respectively. The latter are obtained from an *advanced* Green function and are *noncausal*, i.e., $\mathbf{e}_2(\mathbf{r},t)$ and $\mathbf{h}_2(\mathbf{r},t)$ are zero for $t>t_2$. The correlation-type reciprocity theorem thus establishes that

$$\int_{V_0}\mathbf{e}_1(\mathbf{r},t)\circledast\mathbf{j}_{i2}(\mathbf{r},t)\,\mathrm{d}V=\int_{V_0}\mathbf{e}_2^{\mathrm{adv}}(\mathbf{r},t)\circledast\mathbf{j}_{i1}(\mathbf{r},t)\,\mathrm{d}V, \tag{1.153}$$

where V_0 is any volume containing both $\mathbf{j}_{i1}$ and $\mathbf{j}_{i2}$.

The first part of the proof of (1.153) proceeds analogously to the proof of (1.143), except that the combined Maxwell equations are written for $t=\tau$ and $t=\tau-t$ and integration in time is performed from T_1 to T_2 (where $T_1<t_1$ and $T_2>t_2$). In this way, it is simple to obtain the identity

$$\begin{aligned}
&\int_{T_1}^{T_2}\oint_{S_0}\mathbf{u}_n\cdot\left[\mathbf{e}_1(\mathbf{r},\tau-t)\times\mathbf{h}_2^{\mathrm{adv}}(\mathbf{r},\tau)-\mathbf{e}_2^{\mathrm{adv}}(\mathbf{r},\tau)\times\mathbf{h}_1(\mathbf{r},\tau-t)\right]\mathrm{d}S\,\mathrm{d}\tau\\
&\quad+\int_{T_1}^{T_2}\int_{V_0}\left[\mathbf{e}_2^{\mathrm{adv}}(\mathbf{r},\tau)\cdot\mathbf{j}_{i1}(\mathbf{r},\tau-t)+\mathbf{e}_1(\mathbf{r},\tau-t)\cdot\mathbf{j}_{i2}(\mathbf{r},\tau)\right]\mathrm{d}V\,\mathrm{d}\tau\\
&\quad=-\int_{-\infty}^{\infty}\mu(t')\int_{V_0}\left[\mathbf{h}_1(\mathbf{r},\tau-t)\cdot\mathbf{h}_2^{\mathrm{adv}}(\mathbf{r},\tau)\Big|_{\tau=T_1}^{\tau=T_2}\right]\mathrm{d}V\,\mathrm{d}t'\\
&\qquad-\int_{-\infty}^{\infty}\varepsilon(t')\int_{V_0}\left[\mathbf{e}_1(\mathbf{r},\tau-t)\cdot\mathbf{e}_2^{\mathrm{adv}}(\mathbf{r},\tau)\Big|_{\tau=T_1}^{\tau=T_2}\right]\mathrm{d}V\,\mathrm{d}t'\\
&-\int_{T_1}^{T_2}\int_{V_0}\left[\mathbf{e}_2^{\mathrm{adv}}(\mathbf{r},\tau)\cdot\mathbf{j}_{i1}(\mathbf{r},\tau-t)+\mathbf{e}_1(\mathbf{r},\tau-t)\cdot\mathbf{j}_{i2}(\mathbf{r},\tau)\right]\mathrm{d}V\,\mathrm{d}\tau.
\end{aligned} \tag{1.154}$$

The volume integrals in the right-hand side of (1.154) are zero since it results $\mathbf{e}_2^{\mathrm{adv}}(T_2)=\mathbf{h}_2^{\mathrm{adv}}(T_2)=\mathbf{0}$ for the advanced fields and $\mathbf{e}_1(T_1-t)=\mathbf{h}_2(T_1-t)=\mathbf{0}$

for the retarded fields for any $t > T_1 - t_1$. As for the proof of (1.143), the surface integral in (1.154) is a constant over any surface S_0 which encloses both $\mathbf{j}_{i1}$ and $\mathbf{j}_{i2}$. In particular, for every finite t there exists a finite sphere S_0 on which the retarded field $\{\mathbf{e}_1(\tau - t), \mathbf{h}_1(\tau - t)\}$ (with $T_1 < \tau < T_2$) is identically zero. Therefore, from (1.154) we have

$$\int_{-\infty}^{+\infty}\int_{V_0} \mathbf{j}_{i1}(\mathbf{r}, \tau - t) \cdot \mathbf{e}_2^{\text{adv}}(\mathbf{r}, \tau)\, \mathrm{d}V\, \mathrm{d}\tau = \int_{-\infty}^{+\infty}\int_{V_0} \mathbf{j}_{i2}(\mathbf{r}, \tau) \cdot \mathbf{e}_1(\mathbf{r}, \tau - t)\, \mathrm{d}V\, \mathrm{d}\tau, \tag{1.155}$$

which is identical to (1.153), once the integration is interchanged.

1.15.2 Poynting's theorem and electromagnetic momentum

A fundamental consequence of Maxwell's equations is *Poynting's theorem*, which provides a power balance for the electromagnetic field. Let us consider the divergence of the Poynting vector $\mathbf{s} = \mathbf{e} \times \mathbf{h}$, i.e.,

$$\nabla \cdot \mathbf{s}(\mathbf{r}, t) = \mathbf{h}(\mathbf{r}, t) \cdot \nabla \times \mathbf{e}(\mathbf{r}, t) - \mathbf{e}(\mathbf{r}, t) \cdot \nabla \times \mathbf{h}(\mathbf{r}, t). \tag{1.156}$$

By using the Maxwell equations, (1.156) becomes

$$\begin{aligned} &\nabla \cdot \mathbf{s}(\mathbf{r}, t) + \mathbf{h}(\mathbf{r}, t) \cdot \frac{\partial}{\partial t}\mathbf{b}(\mathbf{r}, t) + \mathbf{e}(\mathbf{r}, t) \cdot \mathbf{j}(\mathbf{r}, t) + \mathbf{e}(\mathbf{r}, t) \cdot \frac{\partial}{\partial t}\mathbf{d}(\mathbf{r}, t) \\ &\quad = -\mathbf{h}(\mathbf{r}, t) \cdot \mathbf{m}_{\mathrm{i}}(\mathbf{r}, t) - \mathbf{e}(\mathbf{r}, t) \cdot \mathbf{j}_{\mathrm{i}}(\mathbf{r}, t). \end{aligned} \tag{1.157}$$

By integrating (1.157) over a region V bounded by a surface S_V with unit normal vector $\mathbf{u}_n$ oriented towards the exterior of V and using the divergence theorem, we obtain

$$\begin{aligned} &\oint_S \mathbf{u}_n \cdot \mathbf{s}(\mathbf{r}, t)\, \mathrm{d}S + \int_V p_{\mathrm{d}}(\mathbf{r}, t)\, \mathrm{d}V + \int_V [p_h(\mathbf{r}, t) + p_e(\mathbf{r}, t)]\, \mathrm{d}V \\ &\quad = \int_V [p_{\mathrm{i}}(\mathbf{r}, t) + p_{\mathrm{mi}}(\mathbf{r}, t)]\, \mathrm{d}V, \end{aligned} \tag{1.158}$$

where

$$\begin{aligned} p_{\mathrm{d}}(\mathbf{r}, t) &= \mathbf{e}(\mathbf{r}, t) \cdot \mathbf{j}(\mathbf{r}, t), \\ p_h(\mathbf{r}, t) &= \mathbf{h}(\mathbf{r}, t) \cdot \frac{\partial}{\partial t}\mathbf{b}(\mathbf{r}, t), \\ p_e(\mathbf{r}, t) &= \mathbf{e}(\mathbf{r}, t) \cdot \frac{\partial}{\partial t}\mathbf{d}(\mathbf{r}, t), \\ p_{\mathrm{i}}(\mathbf{r}, t) &= -\mathbf{e}(\mathbf{r}, t) \cdot \mathbf{j}_{\mathrm{i}}(\mathbf{r}, t), \\ p_{\mathrm{mi}}(\mathbf{r}, t) &= -\mathbf{h}(\mathbf{r}, t) \cdot \mathbf{m}_{\mathrm{i}}(\mathbf{r}, t). \end{aligned} \tag{1.159}$$

The right-hand side of (1.158) expresses the power furnished by the impressed currents, i.e., by the sources of the electromagnetic field. The three addends at the left-hand side are the destinations of such power: the surface integral is the power that leaves the volume V by crossing its boundary S; the volume integral with p_{d} represents the power transferred from the field to the charges in V (and eventually dissipated into heat via Joule effect inside, e.g., a metal); finally, the volume integral with p_h and p_e represents the power exchanged with the electromagnetic field inside V in the form of stored magnetic and electric energy.

In addition to defining power and energy of the electromagnetic field, it is also possible to define an electromagnetic linear and angular momentum. Considering for simplicity a vacuum, let us introduce the tensor

$$\underline{\mathbf{t}}^{e}(\mathbf{r},t) = \varepsilon_0 \mathbf{e}(\mathbf{r},t)\mathbf{e}(\mathbf{r},t) - \frac{1}{2}\varepsilon_0 |\mathbf{e}(\mathbf{r},t)|^2 \underline{\mathbf{I}}\,, \tag{1.160}$$

where $\underline{\mathbf{I}}$ is the unity dyadic. From (1.161), it is easy to obtain

$$\nabla \cdot \underline{\mathbf{t}}^{e}(\mathbf{r},t) = \varepsilon_0 [\nabla \cdot \mathbf{e}(\mathbf{r},t)]\mathbf{e}(\mathbf{r},t) - \varepsilon_0 \nabla \times \mathbf{e}(\mathbf{r},t) \times \mathbf{e}(\mathbf{r},t) \tag{1.161}$$

and by using the (source-free) Maxwell equations in free space, we derive

$$\nabla \cdot \underline{\mathbf{t}}^{e}(\mathbf{r},t) = \varepsilon_0 \rho_{\mathrm{e}}(\mathbf{r},t)\mathbf{e}(\mathbf{r},t) + \varepsilon_0 \mu_0 \mathbf{e}(\mathbf{r},t) \times \frac{\partial}{\partial t}\mathbf{h}(\mathbf{r},t)\,. \tag{1.162}$$

In the same way, by defining

$$\underline{\mathbf{t}}^{h}(\mathbf{r},t) = \mu_0 \mathbf{h}(\mathbf{r},t)\mathbf{h}(\mathbf{r},t) - \frac{1}{2}\mu_0 |\mathbf{h}(\mathbf{r},t)|^2 \underline{\mathbf{I}}\,, \tag{1.163}$$

it is easy to derive the identity

$$\nabla \cdot \underline{\mathbf{t}}^{h}(\mathbf{r},t) = -\mu_0 \mathbf{h}(\mathbf{r},t) \times \mathbf{j}(\mathbf{r},t) - \varepsilon_0 \mu_0 \mathbf{h}(\mathbf{r},t) \times \frac{\partial}{\partial t}\mathbf{e}(\mathbf{r},t)\,. \tag{1.164}$$

By defining the *Maxwell stress tensor* $\underline{\mathbf{t}}^{\mathrm{em}}$ as

$$\underline{\mathbf{t}}^{\mathrm{em}}(\mathbf{r},t) = -\underline{\mathbf{t}}^{e}(\mathbf{r},t) - \underline{\mathbf{t}}^{h}(\mathbf{r},t)\,, \tag{1.165}$$

it results

$$\nabla \cdot \underline{\mathbf{t}}^{\mathrm{em}}(\mathbf{r},t) + \frac{1}{c_0^2}\frac{\partial}{\partial t}\mathbf{s}(\mathbf{r},t) = -[\varepsilon_0 \rho_{\mathrm{e}}(\mathbf{r},t)\mathbf{e}(\mathbf{r},t) + \mu_0 \mathbf{j}(\mathbf{r},t) \times \mathbf{h}(\mathbf{r},t)]\,. \tag{1.166}$$

Equation (1.166) is the point form of the electromagnetic conservation of momentum theorem [2]. By integrating (1.166) over a region V bounded by a surface S_V with unit normal vector $\mathbf{u}_n$ oriented toward the exterior of V and using the divergence theorem, we obtain

$$\int_V [\varepsilon_0 \rho_{\mathrm{e}}(\mathbf{r},t)\mathbf{e}(\mathbf{r},t) + \mu_0 \mathbf{j}(\mathbf{r},t) \times \mathbf{h}(\mathbf{r},t)]\,\mathrm{d}V + \oint_{S_V} \mathbf{u}_n \cdot \underline{\mathbf{t}}^{\mathrm{em}}(\mathbf{r},t)\,\mathrm{d}S + \frac{1}{c_0^2}\frac{\mathrm{d}}{\mathrm{d}t}\int_V \mathbf{s}(\mathbf{r},t)\,\mathrm{d}V = \mathbf{0}\,. \tag{1.167}$$

Equation (1.167) represents an equality between forces: the first term represents the Lorentz force acting on the charges and currents existing in the volume V, the second term represents the flow of electromagnetic momentum across S_V, and the third term represents the change in momentum within V. In particular, the vector $\mathbf{g}_{\mathrm{em}}$ defined as

$$\mathbf{g}_{\mathrm{em}}(\mathbf{r}, t) = \frac{1}{c_0^2}\mathbf{s}(\mathbf{r}, t) \tag{1.168}$$

is the mechanical momentum density of the electromagnetic field (*electromagnetic momentum*), while the vector $\mathbf{p} = \mathbf{u}_n \cdot \underline{\mathbf{t}}^{\mathrm{em}}$ can be considered as the *stress* transmitted through the surface element with unit normal $\mathbf{u}_n$ (whose component along $\mathbf{u}_n$ is also known as *radiation pressure*).

Finally, a balance equation can also be derived for the angular momentum, in the form

$$\begin{aligned}&\int_V \mathbf{r} \times [\varepsilon_0 \rho_e(\mathbf{r}, t)\mathbf{e}(\mathbf{r}, t) + \mu_0 \mathbf{j}(\mathbf{r}, t) \times \mathbf{h}(\mathbf{r}, t)]\,\mathrm{d}V + \oint_{S_V} \mathbf{u}_n \cdot \underline{\mathbf{m}}^{\mathrm{em}}(\mathbf{r}, t)\,\mathrm{d}S \\ &+ \frac{\mathrm{d}}{\mathrm{d}t}\int_V \mathbf{l}_{\mathrm{em}}(\mathbf{r}, t)\,\mathrm{d}V = \mathbf{0}.\end{aligned} \tag{1.169}$$

where $\mathbf{l}_{\mathrm{em}}(\mathbf{r}, t)$ is the *angular momentum density* of the electromagnetic field, defined by

$$\mathbf{l}_{\mathrm{em}}(\mathbf{r}, t) = \mathbf{r} \times \mathbf{g}_{\mathrm{em}}(\mathbf{r}, t) = \frac{1}{c_0^2}\mathbf{r} \times \mathbf{s}(\mathbf{r}, t), \tag{1.170}$$

whereas $\underline{\mathbf{m}}^{\mathrm{em}}(\mathbf{r}, t) = -\underline{\mathbf{t}}^{\mathrm{em}}(\mathbf{r}, t) \times \mathbf{r}$ is the tensor expressing the flux of angular momentum, so that the vector $\mathbf{b} = \mathbf{u}_n \cdot \underline{\mathbf{m}}^{\mathrm{em}}$ can be considered as the unit torque transmitted through the surface element with unit normal $\mathbf{u}_n$.

It should be noted that Maxwell's equations have been manipulated to obtain expressions that resemble those for mechanical momentum and energy in order to postulate the meaning of the individual terms appearing in the conservation theorems. However, an alternative approach is possible by postulating a single Lagrangian density function for the electromagnetic field, and then applying the stationary property of the action integral, but the results are precisely the same relations for energy and momentum as obtained from manipulating Maxwell's equations [30].

1.15.3 Time-domain optical theorem

The frequency-domain optical theorem (or extinction cross-section theorem) relates the power extinguished from a plane wave incident on an object to the scattering amplitude in the direction of the incident field and has a long history [31]. On the other hand, the time-domain version is much less known and has its roots in the work of de Hoop [32].

In order to formulate the scattering problem in the time domain, let us consider a scattering object with general constitutive parameters which occupies a finite volume V_s (in free space) bounded by the closed surface S_{V_s}; the infinite volume (in free space) external to V_s (but still bounded by the closed surface S_{V_s}) is denoted by V_0. The electromagnetic field in the configuration is $\{\mathbf{e}(\mathbf{r},t),\mathbf{h}(\mathbf{r},t)\}$ and in V_0 is written as the sum of the incident field $\{\mathbf{e}^{\mathrm{inc}}(\mathbf{r},t),\mathbf{h}^{\mathrm{inc}}(\mathbf{r},t)\}$ and the scattered field $\{\mathbf{e}_{scat}(\mathbf{r},t),\mathbf{h}_{scat}(\mathbf{r},t)\}$. In particular, the incident field is defined everywhere and inside V_s satisfies the homogeneous Maxwell equations, while the scattered field satisfies the homogeneous Maxwell equations in V_0. Moreover, at large distances from the scatterer, the scattered field can be expressed as

$$\begin{aligned}\mathbf{e}_{\mathrm{scat}}(\mathbf{r},t)&\sim\frac{\mathbf{e}^{\mathrm{ff}}_{\mathrm{scat}}(\mathbf{u}_r,t-r/c_0)}{r},\\ \mathbf{h}_{\mathrm{scat}}(\mathbf{r},t)&\sim\frac{\mathbf{h}^{\mathrm{ff}}_{\mathrm{scat}}(\mathbf{u}_r,t-r/c_0)}{r}\end{aligned}\tag{1.171}$$

and satisfies the corresponding relations (1.139). According to the Poynting theorem, the quantity

$$w_{\mathrm{inc}}(t)=-\oint_{S_{V_\mathrm{s}}}\mathbf{u}_n\cdot\left[\mathbf{e}^{\mathrm{inc}}(\mathbf{r},t)\times\mathbf{h}^{\mathrm{inc}}(\mathbf{r},t)\right]\mathrm{d}S\tag{1.172}$$

represents the instantaneous power flow of the incident wave across S_{V_s} toward V_s. On the other hand, the quantity

$$w_{\mathrm{scat}}(t)=\oint_{S_{V_\mathrm{s}}}\mathbf{u}_n\cdot\left[\mathbf{e}_{\mathrm{scat}}(\mathbf{r},t)\times\mathbf{h}_{\mathrm{scat}}(\mathbf{r},t)\right]\mathrm{d}S\tag{1.173}$$

represents the instantaneous power flow of the scattered wave across S_{V_s} toward V_0, while the quantity

$$w_{\mathrm{abs}}(t)=-\oint_{S_{V_\mathrm{s}}}\mathbf{u}_n\cdot\left[\mathbf{e}(\mathbf{r},t)\times\mathbf{h}(\mathbf{r},t)\right]\mathrm{d}S\tag{1.174}$$

represents the instantaneous flow of power which is absorbed by the scatterer.

In scattering problems, the incident field is often assumed as a uniform plane-wave field, i.e. (see (1.49))

$$\begin{aligned}\mathbf{e}^{\mathrm{inc}}(\mathbf{r},t)&=e^{\mathrm{inc}}\left(t-\frac{\mathbf{u}_k\cdot\mathbf{r}}{c_0}\right)\mathbf{u}_e,\\ \mathbf{h}^{\mathrm{inc}}(\mathbf{r},t)&=h^{\mathrm{inc}}\left(t-\frac{\mathbf{u}_k\cdot\mathbf{r}}{c_0}\right)\mathbf{u}_h,\end{aligned}\tag{1.175}$$

where $\mathbf{u}_e\times\mathbf{u}_h=\mathbf{u}_k$ and

$$\frac{\left|e^{\mathrm{inc}}(t-\mathbf{u}_k\cdot\mathbf{r}/c_0)\right|}{\left|h^{\mathrm{inc}}(t-\mathbf{u}_k\cdot\mathbf{r}/c_0)\right|}=\eta_0.\tag{1.176}$$

Now, by defining the instant t_0 at which the incident plane wave hits the scatterer, it is easy to show that

$$\int_{t_0}^{+\infty} w_{\text{inc}}(t)\,\mathrm{d}t = 0. \tag{1.177}$$

Moreover, the total energy absorbed by the obstacle W_{abs} and the total energy carried by the scattered field W_{scat} are

$$\begin{aligned} W_{\text{abs}} &= \int_{t_0}^{+\infty} w_{\text{abs}}(t)\,\mathrm{d}t, \\ W_{\text{scat}} &= \int_{t_0}^{+\infty} w_{\text{scat}}(t)\,\mathrm{d}t, \end{aligned} \tag{1.178}$$

so that using $\mathbf{e} = \mathbf{e}_{\text{inc}} + \mathbf{e}_{\text{scat}}$ and $\mathbf{h} = \mathbf{h}_{\text{inc}} + \mathbf{h}_{\text{scat}}$ we also have

$$\begin{aligned} W_{\text{abs}} + W_{\text{scat}} = -\int_{t_0}^{+\infty} \oint_{S_{V_s}} \mathbf{u}_n \\ \cdot \left[\mathbf{e}_{\text{inc}}(\mathbf{r},t) \times \mathbf{h}_{\text{scat}}(\mathbf{r},t) + \mathbf{e}_{\text{scat}}(\mathbf{r},t) \times \mathbf{h}_{\text{inc}}(\mathbf{r},t)\right] \mathrm{d}S\,\mathrm{d}t. \end{aligned} \tag{1.179}$$

Using the equivalent currents $\mathbf{j}_S$ and $\mathbf{m}_S$ introduced in the surface integral representation of the scattered field, i.e., $\mathbf{j}_S(\mathbf{r},t) = \mathbf{u}_n \times \mathbf{h}_{\text{scat}}(\mathbf{r},t)$ and $\mathbf{m}_S(\mathbf{r},t) = -\mathbf{u}_n \times \mathbf{e}_{\text{scat}}(\mathbf{r},t)$ on S_{V_s}, and the plane-wave incident field (1.175), we obtain

$$\begin{aligned} W_{\text{abs}} + W_{\text{scat}} = \int_{t_0}^{+\infty} \oint_{S_{V_s}} \Bigg[\mathbf{u}_e e^{\text{inc}}\left(t - \frac{\mathbf{u}_k \cdot \mathbf{r}}{c_0}\right) \cdot \mathbf{j}_S(\mathbf{r},t) \\ + \mathbf{u}_h h^{\text{inc}}\left(t - \frac{\mathbf{u}_k \cdot \mathbf{r}}{c_0}\right) \cdot \mathbf{m}_S(\mathbf{r},t)\Bigg] \mathrm{d}S\mathrm{d}t. \end{aligned} \tag{1.180}$$

By introducing the time t_{i} at which the incident field reaches the origin, it results $e^{\text{inc}}(t) = h^{\text{inc}}(t) = 0$ for $t < t_{\text{i}}$, so that, from (1.180), it results

$$\begin{aligned} W_{\text{abs}} + W_{\text{scat}} = \int_{t_{\text{i}}}^{+\infty} \oint_{S_{V_s}} \Bigg[e^{\text{inc}}(t)\mathbf{u}_e \cdot \mathbf{j}_S\left(\mathbf{r}, t + \frac{\mathbf{u}_k \cdot \mathbf{r}}{c_0}\right) \\ + h^{\text{inc}}(t)\mathbf{u}_h \cdot \mathbf{m}_S\left(\mathbf{r}, t + \frac{\mathbf{u}_k \cdot \mathbf{r}}{c_0}\right)\Bigg] \mathrm{d}S\,\mathrm{d}t. \end{aligned} \tag{1.181}$$

It is then possible to show that [32]

$$\begin{aligned} \mathbf{e}_{\text{scat}}^{\text{ff}} = -\mu_0 \frac{\partial}{\partial t}\left\{\oint_{S_{V_s}} \mathbf{j}_S\left(\mathbf{r}, t + \frac{\mathbf{u}_k \cdot \mathbf{r}}{c_0}\right) \mathrm{d}S - \mathbf{u}_k\left[\mathbf{u}_k \cdot \oint_{S_{V_s}} \mathbf{j}_S\left(\mathbf{r}, t + \frac{\mathbf{u}_k \cdot \mathbf{r}}{c_0}\right) \mathrm{d}S\right]\right\} \\ + \frac{1}{c_0}\mathbf{u}_k \times \frac{\partial}{\partial t}\oint_{S_{V_s}} \mathbf{m}_S\left(\mathbf{r}, t + \frac{\mathbf{u}_k \cdot \mathbf{r}}{c_0}\right) \mathrm{d}S \end{aligned} \tag{1.182}$$

and

$$\mathbf{h}_{\text{scat}}^{\text{ff}} = -\varepsilon_0 \frac{\partial}{\partial t}\left\{\oint_{S_{V_s}} \mathbf{m}_S\left(\mathbf{r}, t+\frac{\mathbf{u}_k \cdot \mathbf{r}}{c_0}\right)\mathrm{d}S - \mathbf{u}_k\left[\mathbf{u}_k \cdot \oint_{S_{V_s}} \mathbf{m}_S\left(\mathbf{r}, t+\frac{\mathbf{u}_k \cdot \mathbf{r}}{c_0}\right)\mathrm{d}S\right]\right\}$$
$$-\frac{1}{c_0}\mathbf{u}_k \times \frac{\partial}{\partial t}\oint_{S_{V_s}} \mathbf{j}_S\left(\mathbf{r}, t+\frac{\mathbf{u}_k \cdot \mathbf{r}}{c_0}\right)\mathrm{d}S. \tag{1.183}$$

By recalling that $\mathbf{u}_k \cdot \mathbf{u}_e = \mathbf{u}_k \cdot \mathbf{u}_h = 0$, from (1.181) to (1.183), we finally obtain

$$W_{\text{abs}} + W_{\text{scat}} = -\frac{1}{\mu_0}\int_{t_i}^{+\infty} \mathbf{u}_e e^{\text{inc}}(t) \cdot \left[\int_{t_i}^{t} \mathbf{e}_{\text{scat}}^{\text{ff}}(\mathbf{u}_k, t')\right]\mathrm{d}t \tag{1.184}$$

and

$$W_{\text{abs}} + W_{\text{scat}} = -\frac{1}{\varepsilon_0}\int_{t_i}^{+\infty} \mathbf{u}_h h^{\text{inc}}(t) \cdot \left[\int_{t_i}^{t} \mathbf{h}_{\text{scat}}^{\text{ff}}(\mathbf{u}_k, t')\right]\mathrm{d}t, \tag{1.185}$$

which represent the time-domain optical theorem. The time-domain optical theorem thus establishes that the sum of the scattered and absorbed energies is uniquely determined by the far-field amplitude of the scattered field observed in the direction of propagation of the incident plane wave during the time interval (t_i, ∞). Such a theorem, as suggested in [32], serves as a check on different approximations made in practical analysis of scattering problems. Other applications of the theorem are envisioned in [33,34], while in [35], a new generalization has recently been proposed.

References

[1] Tai CT. On the presentation of Maxwell's theory. *Proc IEEE.* 1972;60 (8):936–945.

[2] Rothwell EJ, Cloud MJ. *Electromagnetics*. 3rd ed. Boca Raton, NJ: CRC Press; 2017.

[3] Van Bladel J. *Singular Electromagnetic Fields and Sources*. New York, NY: IEEE Press; 1996.

[4] Collin RE. *Field Theory of Guided Waves.* New York, NY: IEEE Press; 1991.

[5] Stratton JA. *Electromagnetic Theory.* New York, NY: IEEE Press; 2007.

[6] Love AEH. The integration of the equations of propagation of electric waves. *Phil Trans Roy Soc London, Ser A.* 1901;197(287–299):1–45.

[7] Schelkunoff S. Some equivalence theorems of electromagnetics and their application to radiation problems. *Bell System Techn J.* 1936;15(1):92–112.

[8] Hansen T, Yaghjian AD. *Plane-Wave Theory of Time-Domain Fields: Near-Field Scanning Applications.* New York, NY: IEEE Press; 1999.

[9] Jiang BN, Wu J, Povinelli LA. The origin of spurious solutions in computational electromagnetics. *J Comput Phys.* 1996;125(1):104–123.

[10] Yaghjian AD. Maxwellian and cavity electromagnetic fields within continuous sources. *Am J Phys.* 1985;53(9):859–863.

[11] Yaghjian AD. Electric dyadic Green's functions in the source region. *Proc IEEE.* 1980;68(2):248–263.

[12] Yaghjian AD. A delta-distribution derivation of the electric field in the source region. *Electromagn.* 1982;2(2):161–167.

[13] Stratton JA, Chu L. Diffraction theory of electromagnetic waves. *Phys Rev.* 1939;56(1):99–107.

[14] Levine H, Schwinger J. On the theory of electromagnetic wave diffraction by an aperture in an infinite plane conducting screen. *Comm Pure Appl Math.* 1950;3(4):355–391.

[15] Franz W. Zur formulierung des huygensschen prinzips. *Z Naturforsch A.* 1948;3a(8–11):500–506.

[16] Tai CT. Kirchhoff theory: scalar, vector, or dyadic? *IEEE Trans Antennas Propag.* 1972;20(1):114–115.

[17] Wu TT. Electromagnetic missiles. *J Appl Phys.* 1985;57(7):2370–2373.

[18] Yaghjian AD, Hansen TB. Time-domain far fields. *J Appl Phys.* 1996;79 (6):2822–2830.

[19] Lorentz HA. The theorem of Poynting concerning the energy in the electromagnetic field and two general propositions concerning the propagation of light. *Versl K Akad Wet Amsterdam.* 1896;4:176–187.

[20] Rumsey V. Reaction concept in electromagnetic theory. *Phys Rev.* 1954;94 (6):1483–1491.

[21] Welch W. Reciprocity theorems for electromagnetic fields whose time dependence is arbitrary. *IRE Trans Antennas Propag.* 1960;AP-8(1): 68–73.

[22] Welch W. Comments on "Reciprocity theorems for electromagnetic fields whose time dependence is arbitrary". *IRE Trans Antennas Propag.* 1961; 9(1):114–115.

[23] Cheo B. A reciprocity theorem for electromagnetic fields with general time dependence. *IEEE Trans Antennas Propag.* 1965;13(2):278–284.

[24] Bojarski NN. Generalized reaction principles and reciprocity theorems for the wave equations, and the relationship between the time-advanced and time-retarded fields. *J Acoust Soc Amer.* 1983;74(1):281–285.

[25] de Hoop AT. Time-domain reciprocity theorems for electromagnetic fields in dispersive media. *Radio Sci.* 1987;22(7):1171–1178.

[26] de Hoop AT. *Handbook of Radiation and Scattering of Waves*. London, UK: Academic Press; 1995.

[27] Stumpf M. *Electromagnetic Reciprocity in Antenna Theory*. Piscataway, NJ: IEEE Press; 2017.

[28] Stumpf M. *Time-Domain Electromagnetic Reciprocity in Antenna Modeling*. Piscataway, NJ: IEEE Press; 2020.

[29] Shlivinski A. Time-domain transfer coupled response of antennas – Reciprocity theorem approach. *IEEE Trans Antennas Propag.* 2017; 65(4):1714–1727.
[30] Jackson JD. *Classical Electrodynamics.* 3rd ed. New York, NY: Wiley; 1998.
[31] Newton RG. Optical theorem and beyond. *Am J Phys.* 1976;44(7):639–642.
[32] De Hoop A. A time domain energy theorem for scattering of plane electromagnetic waves. *Radio Sci.* 1984;19(5):1179–1184.
[33] Karlsson A. Some results extracted from the time domain version of the optical theorem. *Radio Sci.* 2002;37(2):1–10.
[34] Štumpf M, Lager IE. The time-domain optical theorem in antenna theory. *IEEE Antennas Wireless Propag Lett.* 2014;14:895–897.
[35] Marengo EA, Tu J. Generalized optical theorem in the time domain. *Prog. Electromagn Res.* 2016;65:1–18.

Chapter 2
Finite-difference time-domain method

Stephen D. Gedney[1]

Abstract

The finite-difference time-domain (FDTD) method has been used to solve grand challenge problems in computational electromagnetics for the past five decades. This simple, yet, elegant method has O(N) computational efficiency and readily exploits modern computer architectures. The focus of this chapter is to provide a review of the FDTD method and several significant advances of the method, including the convolutional perfectly matched layer (CPML) mesh truncation algorithm, subcell algorithms, and more. Emphasis is given on the efficient implementation of the FDTD method on modern day multi-threaded and multi-processor computers, and the importance of establishing a common framework in the implementation. It is shown that several advanced concepts such as subcell algorithms can be reduced to effective local anisotropic materials and maintaining a highly efficient implementation.

2.1 Impact of the finite-difference time-domain method (FDTD)

The FDTD solution of Maxwell's equations was first introduced by Kane Yee in 1966 [1]. The original intent of Yee when developing this algorithm was to "teach himself FORTRAN" [2]. This seemingly naïve exercise has made a significant and continual impact within the field of computational electromagnetics (CEM) where for over five decades the FDTD method has been highly successful in efficiently solving large, complex engineering problems. In fact, within one decade of Yee's seminal paper, R. Holland used the FDTD method to compute the electromagnetic response of a full F-11 aircraft within an EMP simulator [3]. This was done in an era when the method of moments (MoM) was limited to solving small 2D and thin-wire problems and the finite-element method (FEM) was limited to scalar formulations. About a decade later, using the FDTD method,

[1]Department of Electrical Engineering, University of Colorado Denver, USA

Sullivan *et al.*, were the first ever to simulate the electromagnetic interaction with a detailed model of the full human body [4]. This was in the era of the newly introduced Rao–Wilton–Glisson (RWG) basis [5] and 3D MoM modeling was still in its infancy. The next decade saw an explosion of FDTD applications enabled by the crucial introduction of the Perfectly Matched Layer absorbing media by J. P. Berenger [6] and the wide availability of high-performance parallel computers [7–10]. More specifically, the FDTD method has been successfully applied to a plethora of applications such as antenna analysis [11–16], printed microwave circuits and devices [15,17–22], cancer detection [23–26], metamaterials [27–29], the modeling of multi-path effects of wireless signals in buildings and urban environments [30–32], electromagnetic cloaking [10,33], to name just a few.

The FDTD method has spanned far beyond traditional CEM applications and has been applied to a plethora of multi-physics problems. For example, the FDTD method has been used to model the non-linear Schrodinger equation [34,35]. More recently, the FDTD method was used to numerically solve the quantum Maxwell's equations for arbitrary inhomogeneous media based on the macroscopic theory on quantum electrodynamics [36]. Jamesina Simpson and Allen Taflove were the first to perform a 3D computation of pre-seismic electromagnetic phenomena propagated within the entire Earth-ionosphere cavity using the FDTD method [37]. The FDTD method has recently been combined with partial-wave microscopic spectroscopy for advanced cancer diagnosis [26]. Bio-electromagnetic studies involving the entire human body with millimeter scale detail have been shown to be possible with the FDTD method [38]. The FDTD method has also had a significant impact in photonics and nanotechnology [39].

The FDTD method is a powerful tool, however, it does have its limitations and is not necessarily the best method to use for all problems. For example, FDTD requires the discretization of an entire volume domain. Thus, simulations involving multiple objects separated by large distances are better suited for non-local methods such as the MoM. Next, the base FDTD method relies on a spatial discretization that consists of a uniform, orthogonal grid. Subsequently, problems that contain a lot of finely detailed and compact arbitrarily shaped geometries may be better suited for the FEM, which provides conformal, unstructured, curvilinear meshing [40]. Though, more recent methods have been introduced to more accurately model arbitrarily shaped bodies within the FDTD method such as local conformal grid algorithms for perfectly conductor boundary interfaces [41–44] or sub-pixel methods for lossless penetrable materials [45,46].

Another weakness of the FDTD method is that the intrinsic numerical error is dispersive and anisotropic, and the dispersion error accumulates over time and distance. Therefore, simulations over large domains or large time-spans can necessitate finer spatial meshing to maintain the desired level of accuracy. This can be potentially mitigated using higher-order methods such as (2,4) FDTD methods (second-order in time, fourth-order in space) [47–49], at the expense of additional computational effort per field update. Though, a challenge of higher-order difference methods is that high-order accuracy is lost at material boundary interfaces. While there is still on-going work on this topic, some recently proposed methods show some promise [50,51] in improving this error.

Finally, being a purely explicit time-marching scheme, the FDTD method has a stability criterion that restricts the time step based on the smallest discretization size. Significant effort has been invested in developing "unconditionally-stable" FDTD methods. The most popular methods have been the Alternating Direction Implicit (ADI-FDTD) method [52], the Locally-One-Dimensional (LOD-FDTD) method [53], and Crank Nicolson (CN-FDTD) methods [54–57]. Such methods rely on a splitting of Maxwell's equations and then deriving 1D implicit formulations embedded within the 3D explicit formulation. In general, these methods require the solution of a set of tri-diagonal matrices during each field update cycle. The benefit is that the time-step size is not restricted by geometric cell size, but rather by dispersion error. Ironically, the trade-off is that these algorithms generally lead to an increase in dispersion error as compared to the base FDTD scheme [58]. The most accurate of these methods appears to be the LOD-FDTD method [53]. An alternative means of achieving unconditional stability was proposed by Yan and Jiao [59]. The premise of this work is that the stability of the FDTD operator is impacted by a small number of eigenvalues that have magnitudes outside the unit circle if the stability limit is violated. The concept is to remove the eigenvectors corresponding to these unstable eigenvalues from the intrinsic system matrix, leading to stability. The contribution of this most recent work is that Yan and Jiao show that this can be done via solving local eigenvalue problems rather than a global eigenvalue problem, providing a more efficient algorithm. The proposed algorithm maintains the explicit nature of the FDTD method without the loss of accuracy. The trade-off is additional computational cost of the field updates (roughly an order of magnitude), which are approximated by a Taylor series representation of the matrix inverse.

Despite such limitations, the FDTD method has long served as a powerful tool for simulating multi-physics problems involving Maxwell's equations and beyond. The remainder of this chapter will focus on the fundamentals of the FDTD method including details of the CPML with complex frequency shifted (CFS) constitutive parameters. For the latter, distinct rules of thumb on how to properly choose the scaled constative parameters is detailed. The underlying framework for an efficient implementation of the FDTD method is given that will exploit modern day multi-threaded and distributed processing computers. The state-of-the-art conformal grid algorithms are provided and it is demonstrated that such algorithms can be simply represented as effective local anisotropic materials and fit directly within the base computational framework. The presentation of the material is intended to be at a tutorial level such that the reader can immediately implement an efficient FDTD algorithm with more advanced algorithms to further explore the vast potential of the FDTD method on their own.

2.2 The FDTD algorithm

2.2.1 Discretization of Maxwell's equations

Electromagnetic fields radiated by time-varying currents or charges are governed by Maxwell's equations. These can be expressed in differential form by the curl equations (Faraday's and Ampere's laws):

Integral form	Diff form
$\frac{\partial}{\partial t}\iint_S \vec{B}\cdot d\vec{s} = -\oint_C \vec{E}\cdot d\vec{\ell}$	$\frac{\partial \vec{B}}{\partial t} = -\nabla\times\vec{E}$
$\frac{\partial}{\partial t}\iint_S \vec{D}\cdot d\vec{s} = \oint_C \vec{H}\cdot d\vec{\ell} - \iint_S \vec{J}\cdot d\vec{s}$	$\frac{\partial \vec{D}}{\partial t} = \nabla\times\vec{H} - \vec{J}$

(2.1)

and Gauss's laws:

$$\oiint_S \vec{D}\cdot d\vec{s} = \iiint_V \rho dv, \quad \oiint_S \vec{B}\cdot d\vec{s} = 0 \tag{2.2}$$

where $\vec{B}$ and $\vec{D}$ are the magnetic and electric flux density vectors, $\vec{H}$ and $\vec{E}$ are the magnetic and electric field intensities, $\vec{J}$ and $\vec{M}$ are the electric and magnetic current densities, and ρ is the electric charge density (MKS units are assumed throughout the chapter). The flux densities and field intensities are related through the constitutive relations (expressed here for isotropic linear media):

$$\vec{D} = \varepsilon\vec{E} = \varepsilon_0\varepsilon_r\vec{E}, \quad \vec{B} = \mu\vec{H} = \mu_o\mu_r\vec{H}, \quad \vec{J} = \sigma\vec{E}, \tag{2.3}$$

where ε_o is the free-space permittivity, ε_r is the relative permittivity, μ_o is the free-space permeability, μ_r is the relative permeability, and σ is the electrical conductivity.

At a source-free interface boundary, *S*, separating two media $(\varepsilon_1, \sigma_1, \mu_1)$ and $(\varepsilon_2, \sigma_2, \mu_2)$, the fields and flux densities must satisfy the continuity relations:

$$\hat{n}\times\left(\vec{E}_1 - \vec{E}_2\right)\Big|_S = 0, \quad \hat{n}\times\left(\vec{H}_1 - \vec{H}_2\right)\Big|_S = 0, \tag{2.4}$$

$$\hat{n}\cdot\left(\vec{D}_1 - \vec{D}_2\right)\Big|_S = 0, \quad \hat{n}\cdot\left(\vec{B}_1 - \vec{B}_2\right)\Big|_S = 0, \tag{2.5}$$

where $\hat{n}$ is the unit vector normal to the boundary *S* directed into medium 1.

The discretization of Maxwell's equations as originally posed by Kane Yee is based on a staggering of the vector electric and magnetic fields in both space and time such that both the spatial and time derivatives of the curl equations can be expressed using second-order accurate central difference approximations [1]. This discretization is illustrated by the classic Yee-cell in Figure 2.1. The entire volume domain is discretized using a uniform rectangular grid. Each orthogonal cell has dimensions $(\Delta x, \Delta y, \Delta z)$. The electric field vectors are parallel to the grid edges and are assumed to be constant along the length of the edge. The magnetic flux vectors are normal to each face of the grid cell and are assumed to be constant over the grid face. A dual or secondary grid can be formulated with vertices at the cell centers of the primary grid, and edge centers are located at the face centers of the primary grid faces [60,61]. The magnetic field vectors are associated with the edges of the

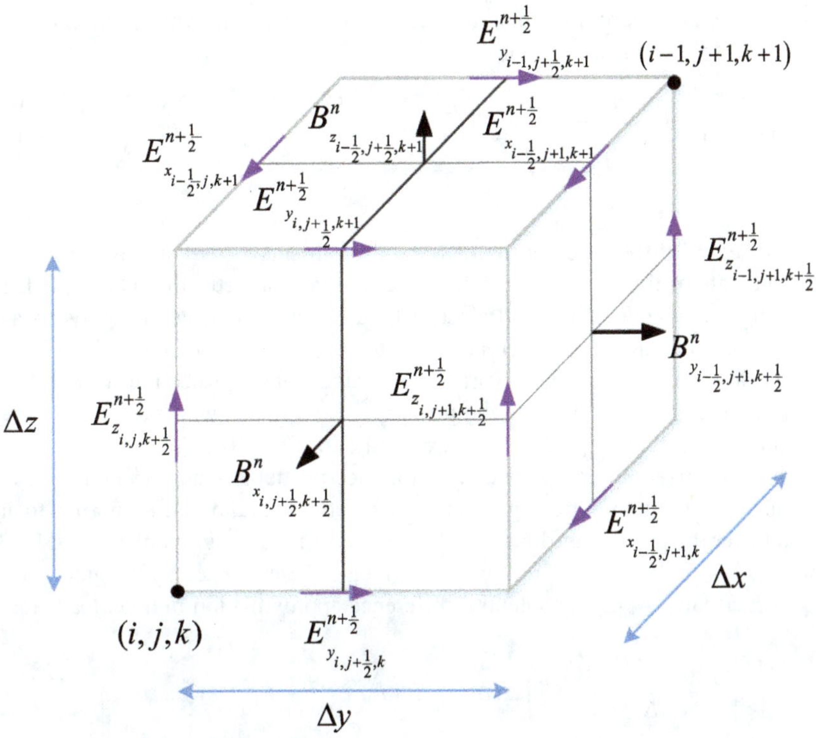

Figure 2.1 The primary grid cell (or Yee-cell) discretizing the fields

secondary grid, and the electric flux vector with the normal of the secondary grid faces, in a manner dual to the primary grid. By aligning electric material boundary interfaces with the primary grid, and magnetic material boundary interfaces with the secondary grid, this discretization also naturally satisfies the boundary conditions in (4) and (5), and hence will intrinsically satisfy Gauss' laws in (2). It is worth noting that this discretization naturally satisfies the de Rham diagram [62], and hence avoids spurious mode solutions.

The discrete coordinate within the uniform grid as illustrated in Figure 2.1 can be expressed as: $(x, y, z)_{i,j,k} = (i\Delta x, j\Delta y, k\Delta z)$. Similarly, time is uniformly discretized as $t = n\Delta t$. Consequently, a function $f(x,y,z,t)$ is expressed in discrete form as:

$$f^{n}_{i,j,k} = f(i\Delta x, j\Delta y, k\Delta z, n\Delta t) \tag{2.6}$$

Next, consider the approximation of the x-projection of Faraday's law at the center of the primary grid face (source free):

$$\mu \frac{\partial H_x}{\partial t} = \frac{\partial E_y}{\partial z} - \frac{\partial E_z}{\partial y} \tag{2.7}$$

at discrete coordinate $(i, j+1/2, k+1/2, n+1/2)$. An isotropic material is assumed and the constitutive relationship has been applied. This is naturally done using central difference approximations for the transverse space derivatives from the curl operator and for the time-derivative:

$$\mu\left(\frac{H_{x_{i,j+\frac{1}{2},k+\frac{1}{2}}}^{n+1}-H_{x_{i,j+\frac{1}{2},k+\frac{1}{2}}}^{n}}{\Delta t}\right)=\left(\frac{E_{y_{i,j+\frac{1}{2},k+1}}^{n+\frac{1}{2}}-E_{y_{i,j+\frac{1}{2},k}}^{n+\frac{1}{2}}}{\Delta z}\right)-\left(\frac{E_{z_{i,j+1,k+\frac{1}{2}}}^{n+\frac{1}{2}}-E_{z_{i,j,k+\frac{1}{2}}}^{n+\frac{1}{2}}}{\Delta y}\right). \tag{2.8}$$

The central difference approximations of both spatial derivatives are second-order accurate at the face center, where the normal magnetic flux is sampled. By staggering the magnetic flux and electric field in time, the time-derivative is also second-order accurate. The difference operator in (8) is thus second-order accurate in both space and time. Similar expressions for the y and z projections of Faraday's law are derived for each face of the primary grid. The same is true for Ampere's law using the secondary grid. This is explicitly done in [60,61].

The discretization also dictates that the electric field vector is parallel to and constant along a primary grid edge, and the magnetic flux (and field) is normal to and constant over the primary grid face. Following the "Finite Integration" method [103], Faraday's law in the integral form can also can also be evaluated by integrating over the x-normal face. Applying a central-difference approximation in time, leads to:

$$\begin{aligned}\mu\left(\frac{H_{x_{i,j+\frac{1}{2},k+\frac{1}{2}}}^{n+1}-H_{x_{i,j+\frac{1}{2},k+\frac{1}{2}}}^{n}}{\Delta t}\right)\Delta y\Delta z &= \Delta y\left(E_{y_{i,j+\frac{1}{2},k+1}}^{n+\frac{1}{2}}-E_{y_{i,j+\frac{1}{2},k}}^{n+\frac{1}{2}}\right)\\ &\quad -\Delta z\left(E_{z_{i,j+1,k+\frac{1}{2}}}^{n+\frac{1}{2}}-E_{z_{i,j,k+\frac{1}{2}}}^{n+\frac{1}{2}}\right)\end{aligned} \tag{2.9}$$

After dividing by the face area $\Delta y\Delta z$, this expression leads to exactly to (8) as derived using central difference approximations in time. However, it is found that using the integral form of Maxwell's equations in the discrete space can have several advantages when applying sub-cell model techniques. A further convenience is made by normalizing the fields by the edge length, and scaling the magnetic field by the free-space wave impedance, $\eta_o=\sqrt{\mu_o/\varepsilon_o}$:

$$e_{y_{i,j+\frac{1}{2},k+1}}^{n+\frac{1}{2}}=\Delta yE_{y_{i,j+\frac{1}{2},k+1}}^{n+\frac{1}{2}},\quad e_{z_{i,j+1,k+\frac{1}{2}}}^{n+\frac{1}{2}}=\Delta zE_{z_{i,j+1,k+\frac{1}{2}}}^{n+\frac{1}{2}},\quad h_{x_{i,j+\frac{1}{2},k+\frac{1}{2}}}^{n+1}=\Delta x\eta_oH_{x_{i,j+\frac{1}{2},k+\frac{1}{2}}}^{n+1} \tag{2.10}$$

Consequently, both (8) and (9) are rewritten as:

$$\frac{\mu_r}{c_o\Delta t}\left(h_{x_{i,j+\frac{1}{2},k+\frac{1}{2}}}^{n+1}-h_{x_{i,j+\frac{1}{2},k+\frac{1}{2}}}^{n}\right)=\frac{\Delta x}{\Delta y\Delta z}\left(e_{y_{i,j+\frac{1}{2},k+1}}^{n+\frac{1}{2}}-e_{y_{i,j+\frac{1}{2},k}}^{n+\frac{1}{2}}-e_{z_{i,j+1,k+\frac{1}{2}}}^{n+\frac{1}{2}}+e_{z_{i,j,k+\frac{1}{2}}}^{n+\frac{1}{2}}\right) \tag{2.11}$$

where $c_o=1/\sqrt{\varepsilon_o\mu_o}$ is the speed of light in vacuum. Similar normalizations are applied to the remaining fields, and all three projections of Faraday's law can be cast in a similar form.

Finally, if the initial conditions of all sampled magnetic and electric fields are known at discrete times n and $(n+1/2)$, respectively, then a recursive algorithm can be derived from (11) to compute the magnetic field at time step $(n+1)$:

$$h^{n+1}_{x_{i,j+\frac{1}{2},k+\frac{1}{2}}} = h^{n}_{x_{i,j+\frac{1}{2},k+\frac{1}{2}}} + \frac{c_o\Delta t}{\mu_r}\frac{\Delta x}{\Delta y\Delta z}\left(e^{n+\frac{1}{2}}_{y_{i,j+\frac{1}{2},k+1}} - e^{n+\frac{1}{2}}_{y_{i,j+\frac{1}{2},k}} - e^{n+\frac{1}{2}}_{z_{i,j+1,k+\frac{1}{2}}} + e^{n+\frac{1}{2}}_{z_{i,j,k+\frac{1}{2}}}\right). \tag{2.12}$$

Similar expressions can be derived for the y and z projections of the flux. By duality, from Ampere's law:

$$e^{n+1/2}_{x_{i+\frac{1}{2},j,k}} = e^{n-1/2}_{x_{i+\frac{1}{2},j,k}} + \frac{c_o\Delta t}{\varepsilon_r}\frac{\Delta x}{\Delta y\Delta z}\left(h^{n}_{z_{i+\frac{1}{2},j+\frac{1}{2},k}} - h^{n}_{z_{i+\frac{1}{2},j-\frac{1}{2},k}} - h^{n}_{y_{i+\frac{1}{2},j,k+\frac{1}{2}}} + h^{n}_{y_{i+\frac{1}{2},j,k-\frac{1}{2}}}\right). \tag{2.13}$$

Again, similar expressions are derived for the y and z projections of the electric flux density. It should be noted here that expressing the update equations in this form has several advantages. Most specifically, it reduces the number of floating point operations. For lossless materials, there is a single constant term. This term can be stored as a single array representative of 3D material parameters. This same array can be used to model non-uniform grids (e.g., $\Delta x(i)$), which naturally occur in perfectly matched layer regions. This same constant will also representative of deformations of the local grid when using sub-grid and conformal grid methods, as shown in Section 2.5. Consequently, the compact form of the recursive updates of (12) and (13) is representative of all field updates throughout the entire FDTD grid.

The recursive update equations (12) and (13) provide a recursive time-stepping solution procedure where the electric fields are solved at half time steps throughout all space followed by the magnetic fields. This solution is bounded (stable) provided that the time-step satisfies the Courant–Friedrichs–Lewy (CFL) stability limit [63]. This can be derived via an eigenvalue analysis of the combined recursive update operators [61], resulting in:

$$\Delta t < \min\frac{\sqrt{\mu_r\varepsilon_r}}{c_o}\bigg/\sqrt{\frac{1}{\Delta x^2}+\frac{1}{\Delta y^2}+\frac{1}{\Delta z^2}} \tag{2.14}$$

where the minimum is over the entire domain.

2.3 CPML

When modeling electromagnetic interactions in an unbounded domain, a challenge that any FEM or FDTD method faces is the need for a reflection-less truncation boundary at the extremities of the mesh. Historically, this was done with pseudo-differential absorbing boundary conditions, infinite elements, or a global integral operator. To date, the most effective means of doing so within the context of FDTD is the use of a *perfectly matched layer* (PML) absorbing media originally proposed by Jeanne-Pierre Bérenger [2,3]. The PML is said to be "perfectly matched" to a background medium in that an arbitrary polarized, broad-band electromagnetic wave impinging from an arbitrary angle on a PML half-space will penetrate into the medium without reflection. The background medium can be isotropic or anisotropic. It can also be inhomogeneous provided that the material is constant along the axis normal to the PML surface. A good example is a layered medium with material profile that is changing only in the direction tangential to the PML boundary.

Several PML formulations have been proposed. However, to date, the most robust and computational efficient method is the CPML with complex-frequency-shifted constitutive parameters proposed by Alan Roden [64,65]. The reader is referred to [64,65] or [61] for more details of the method. Of main focus here is the final formulation to study the efficient implementation of the method as well as the proper choice of scaled constitutive parameters.

Chew and Weedon [66] showed that the PML can also be posed in a coordinate frame with complex metric-tensor coefficients with gradient operator:

$$\tilde{\nabla} = \hat{x}\frac{1}{s_x}\frac{\partial}{\partial x} + \hat{y}\frac{1}{s_y}\frac{\partial}{\partial y} + \hat{z}\frac{1}{s_z}\frac{\partial}{\partial z}. \tag{2.15}$$

The CFS tensor coefficients are expressed in the frequency domain as:

$$s_x = \kappa_x + \frac{\sigma_x}{a_x + j\omega\varepsilon_o}, \tag{2.16}$$

where κ_x is the real tensor term that physically stretches the coordinate, σ_x is the axial conductivity, and a_x is the CFS term that shifts the pole off of the origin. The x-projection of the time-harmonic Faraday's law expressed in the complex stretched-coordinate space is expressed as:

$$-j\omega\mu H_x = \frac{1}{s_y}\frac{\partial}{\partial y}E_z - \frac{1}{s_z}\frac{\partial}{\partial z}E_y \tag{2.17}$$

where s_y and s_z are of the forms of (16) for a y-normal or z-normal PML region, respectively. Introducing the auxiliary variables

$$Q^E_{y,z} = -\frac{1}{B_y}\frac{\partial}{\partial y}E_z, Q^E_{z,y} = -\frac{1}{B_z}\frac{\partial}{\partial z}E_y \tag{2.18}$$

where

$$B_y = \frac{\sigma_y}{\kappa_y\left(\kappa_y\left(j\omega\varepsilon_o + a_y\right) + \sigma_y\right)} \tag{2.19}$$

Then, (17) can be expressed as:

$$-j\omega\mu H_x = \frac{1}{\kappa_y}\frac{\partial}{\partial y}E_z - \frac{1}{\kappa_z}\frac{\partial}{\partial z}E_y + Q^E_{y,z} - Q^E_{z,y} \tag{2.20}$$

This can be transformed into the time-domain. Then, applying the discrete approximations of the finite difference or finite-integration methods, this can be cast in a discrete form. Similar field normalizations are applied. However, what is realized is that the κ terms perform a real stretching of the discrete cells. Therefore, in the PML region, the normalizations used are $h_x = \kappa_x \eta_o \Delta x H_x$ $e_x = \kappa_x \Delta x E_x$, etc. Note that outside of the PML regions, $\kappa_x = 1$. The auxiliary variables are also normalized as $q^E_{y,z} = \kappa_z \Delta z \kappa_y \Delta y Q^E_{y,z}$ and $q^E_{z,y} = \kappa_z \Delta z \kappa_y \Delta y Q^E_{z,y}$. With this normalization, the recursive update equations within an arbitrary PML region for h_x is:

$$h^{n+1}_{x_{i,j+\frac{1}{2},k+\frac{1}{2}}} = h^{n}_{x_{i,j+\frac{1}{2},k+\frac{1}{2}}} + \frac{c_o \Delta t}{\mu_r} \frac{\kappa_{x_i} \Delta x}{\kappa_{z_{k+\frac{1}{2}}} \Delta z \kappa_{y_{j+\frac{1}{2}}} \Delta y} \left(\begin{array}{c} e^{n+\frac{1}{2}}_{y_{i,j+\frac{1}{2},k+1}} - e^{n+\frac{1}{2}}_{y_{i,j+\frac{1}{2},k}} - e^{n+\frac{1}{2}}_{z_{i,j+1,k+\frac{1}{2}}} + \\ e^{n+\frac{1}{2}}_{z_{i,j,k+\frac{1}{2}}} + q^{E^{n+\frac{1}{2}}}_{y,z_{i,j+\frac{1}{2},k+\frac{1}{2}}} - q^{E^{n+\frac{1}{2}}}_{z,y_{i,j+\frac{1}{2},k+\frac{1}{2}}} \end{array} \right) \tag{2.21}$$

where the auxiliary variables have are updated with the recursive relations:

$$q^{E^{n+\frac{1}{2}}}_{y,z_{i,j+\frac{1}{2},k+\frac{1}{2}}} = b_{y_{j+\frac{1}{2}}} q^{E^{n-\frac{1}{2}}}_{y,z_{i,j+\frac{1}{2},k+\frac{1}{2}}} - c_{y_{j+\frac{1}{2}}} \left(e^{n+\frac{1}{2}}_{z_{i,j+1,k+\frac{1}{2}}} - e^{n+\frac{1}{2}}_{z_{i,j,k+\frac{1}{2}}} \right) \tag{2.22}$$

with

$$b_{y_{j+\frac{1}{2}}} = e^{-\Delta t/\tau_{j+\frac{1}{2}}}, \quad c_{y_{j+\frac{1}{2}}} = \frac{\sigma_{y_{j+\frac{1}{2}}}}{\left(\kappa_{y_{j+\frac{1}{2}}} a_{y_{j+\frac{1}{2}}} + \sigma_{y_{j+\frac{1}{2}}} \right)} \left(1 - b_{y_{j+\frac{1}{2}}} \right), \tag{2.23}$$

and

$$\tau_{j+\frac{1}{2}} = \frac{\kappa_{y_{j+\frac{1}{2}}} \varepsilon_o}{\left(\kappa_{y_{j+\frac{1}{2}}} a_{y_{j+\frac{1}{2}}} + \sigma_{y_{j+\frac{1}{2}}} \right)} \tag{2.24}$$

Similar expressions are derived for the update equations involving h_y and h_z. Dual expressions are derived for the electric field.

The PML presents a discontinuity in both the electric and the magnetic material parameters at the PML interface. However, in the discrete FDTD space, electric material parameters boundaries are defined by the primary grid. Whereas, magnetic material boundaries are defined by the secondary grid. As a result, the PML is not "perfectly" matched in the discrete space, since the wave will encounter a non-PML material for one-half of a cell. If there is an abrupt change in material parameters, this will lead to a large reflection, nullifying the desired absorbing behavior of the PML. To mitigate this, Berenger brilliantly proposed to scale the material parameters such they present very small reflections at the PML interface, and larger reflections deeper in the PML once the wave has already decayed significantly. An effective approach is a simple 1D polynomial scaling of the PML constitutive parameters through the depth of the PML. In an x-normal PML:

$$\sigma_x(x) = \begin{cases} \dfrac{|x - x_o|^m}{d^m} \sigma_x^{\max}, & x_o \leq x \leq x_o + d \\ 0, & \text{else} \end{cases} \tag{2.25}$$

where d is the thickness of the PML layer, x_o is the location of the planar interface between the PML and the working volume of the FDTD space, and $\sigma_x^{\max}$ is the maximum value of σ_x at the back wall of the PML. Similarly,

$$\kappa_x(x) = \begin{cases} 1 + \dfrac{|x - x_o|^m}{d^m}\left(\kappa_x^{\max} - 1\right), & x_o \le x \le x_o + d \\ 1, & \text{else} \end{cases}, \tag{2.26}$$

The κ and σ terms are scaled such they are minimum at the front PML boundary and maximum at the back terminating wall. The CFS parameter is scaled oppositely such that it is maximum at the front wall, and trends to zero at the back wall:

$$a_x(x) = \begin{cases} a_x^{\max}\left|\dfrac{d + x_o - x}{d}\right|^{m_a}, & x_o \le x \le x_o + d \\ 0, & \text{else} \end{cases} \tag{2.27}$$

The reason for this is that the largest reflection of evanescent waves occurs at the front PML boundary and these reflections are reduced by the CFS term [65]. However, low-frequency traveling waves are poorly absorbed due to the CFS term. Therefore, it is scaled to zero to improve absorption of such waves deeper in the PML [61,67].

Extensive parameter studies have been conducted to optimize the choice of PML parameters [6,68–73]. Based on these studies, some general rules of thumb for choosing parameters with confidence for general problem sets is provided. A very robust definition for the optimal value of $\sigma^{\max}$ has been found to be computed as [74]

$$\sigma_x^{opt} = \frac{0.8(m+1)}{\eta_o \Delta x \sqrt{\varepsilon_{r,eff}\mu_{r,eff}}} \tag{2.28}$$

where Δx is the cell size along the normal. In general, it is recommended to choose cubic scaling, $m = 3$. $\varepsilon_{r,eff}$ represents and averaged or effective relative permittivity along the transverse direction. When terminating a medium that is inhomogeneous with the PML, the medium must extend through the PML to the back wall. Note that the medium must be held constant along the normal axis and should retain the inhomogeneous profile along the transverse axes. An illustrative example is a layered substrate of a printed circuit board, where the dielectrics and ground planes would extend all the way through the PML and maintain the same cross-sectional profile. In general, a good choice for $\varepsilon_{r,eff}$ is the averaged value of ε_r. If it is a quasi-TEM waveguide, then the $\varepsilon_{r,eff}$ of the wave is a good choice.

The appropriate choice of κ requires more knowledge of the problem, as this term accelerates the decay of evanescent waves. First, if multiple conductors passes through the PML region and a TEM wave is expected to be dominant, one would set $\kappa^{\max} = 1$. If near fields or strongly evanescent waves are anticipated, a good choice for $\kappa^{\max}$ is 10. Note that $\kappa^{\max}$ does not scale with Δx. Also, the polynomial scaling for κ should match that of σ ($m = 3$ is recommended).

Choosing the CFS parameter also requires knowledge of the types of fields to be absorbed. Again, if multiple conductors passes through the PML region and a quasi-TEM wave is likely to be dominant, one would set $a_{\max} = 0$, since low-frequency propagating waves are poorly absorbed if a is non-zero. When expecting near fields or evanescent waves, a recommend choice is linear scaling ($m_a = 1$) and

$$a_x^{\text{opt}} = 9 \times 10^{-3} \pi\, \sigma_x^{\text{opt}} \tag{2.29}$$

On a final note, when computing the scaled constitutive parameters, it is important to do so at the face center of the quantity computed (secondary faces for h-field updates, and primary faces for e-field updates). This is represented by the notation in (21)–(24) where the tensor coefficient subscripts correspond to the coordinate at which the scaled constitutive parameter is computed, which is at the face center for h_x.

2.4 Implementation

At the heart of an efficient implementation of the FDTD algorithm on a modern day computer are six triply nest loops, represented by (13) for electric field updates and (12) for magnetic field updates. These loops will generally consume 98+% of the overall computation time and should be implemented as efficiently as possible. On single, multi-threaded processors, these loops are easily parallelized using OpenMP. Though, to prevent cache misses, the looping over space should follow the contiguous memory arrangement of the field arrays. These loops can also be accelerated using tools such as the Intel instruction set extensions (SSE) and Intel Advanced Vector Extensions (AVX) on Intel processors. For distributed or cluster computing using a MIMD paradigm, a spatial decomposition of the grid should be done in a manner to minimize boundary interface surface area [7]. Typically, one would divide the grid along primary boundaries, such that tangential electric fields are on the partition boundaries. The electric field on the shared boundary is redundantly stored by all processors sharing that boundary. It is recommended to extend the magnetic field array one-half a cell beyond the boundary. At each time step, the magnetic fields are updated completely internally to all processors. The overlapped magnetic field at the shared boundary is then communicated to the local neighbor processors. Subsequently, the electric field is updated locally on processor using (13) [7]. Local, auxiliary updates, such as (22) within the PML region, are also done entirely locally and do not require inter-processor communication with this scheme (an exception could be fully anisotropic media). In this procedure, only one interprocessor communication per time step is required for field stepping.

Due to auxiliary updates in the PML region, processors containing PML will have a heavier load. Also, partitions containing plane-wave injection boundaries, near-field to far-field boundaries, or other auxiliary operators can further create a load imbalance. While the load can be estimated prior to the simulation and partition boundary dimensions determined a priori, an adaptive load balancing can also be performed, requiring re-partitioning during the simulation [9]. Typically, a balanced load can be realized in the first 10 time steps improving the efficiency of the distributed calculation for the remainder of the time-simulation.

One key to an efficient implementation is to eliminate any branches within the field update loops and minimize computations. As a result, the update equations in (12) and (13) will serve as a foot print for all field updates. The leading coefficients will be stored as a 3D array for each field. This increases memory to gain computational efficiency. A good illustration of this is showing that a field update in the PML region will be conducted in the same manner as an update outside the PML. To this end, (21) will be computed in two steps. In step one, the field update will be conducted as normal as:

$$\tilde{h}^{n+1}_{x_{i,j+\frac{1}{2},k+\frac{1}{2}}} = h^{n}_{x_{i,j+\frac{1}{2},k+\frac{1}{2}}} + coefhx_{i,j+\frac{1}{2},k+\frac{1}{2}}\left(e^{n+\frac{1}{2}}_{y_{i,j+\frac{1}{2},k+1}} - e^{n+\frac{1}{2}}_{y_{i,j+\frac{1}{2},k}} - e^{n+\frac{1}{2}}_{z_{i,j+1,k+\frac{1}{2}}} + e^{n+\frac{1}{2}}_{z_{i,j,k+\frac{1}{2}}}\right) \tag{2.30}$$

where, in the PML

$$coefhx_{i,j+\frac{1}{2},k+\frac{1}{2}} = \frac{c_o\Delta t}{\mu_{r_{i,j+\frac{1}{2},k+\frac{1}{2}}}} \frac{\kappa_{x_i}\Delta x}{\kappa_{z_{k+\frac{1}{2}}}\Delta z \kappa_{y_{j+\frac{1}{2}}}\Delta y} \tag{2.31}$$

Then, in an auxiliary update loop performed *only in the PML,* the auxiliary fields q^E are updated using (22) and these are superimposed into h as:

$$h^{n+1}_{x_{i,j+\frac{1}{2},k+\frac{1}{2}}} = \tilde{h}^{n+1}_{x_{i,j+\frac{1}{2},k+\frac{1}{2}}} + coefhx_{i,j+\frac{1}{2},k+\frac{1}{2}}\left(q^{E^{n+\frac{1}{2}}}_{yz_{i,j+\frac{1}{2},k+\frac{1}{2}}} - q^{E^{n+\frac{1}{2}}}_{zy_{i,j+\frac{1}{2},k+\frac{1}{2}}}\right) \tag{2.32}$$

A dual operation is done for the electric field updates with coefficient *coefex* for e_x. Consequently, the main field update loops are preserved, and the PML auxiliary parameters are only updated in the PML regions in a separate set of loops and then superimposed into the field as shown by using (32), thus completing the field update.

2.5 Subcell algorithms

2.5.1 Subcell models

As mentioned in the introductory section, one of the weaknesses of the FDTD algorithm is that it relies on a uniform rectangular grid. This presents a challenge when modeling arbitrarily shaped structures requiring a "staircase" approximation of surfaces that do not align to the grid. This can be mitigated by the use of "sub-cell" algorithms. Such algorithms make approximations by locally conforming the grid to the boundary of an arbitrary shaped device under test. This topic is by no means exhausted within this section. What is attempted instead is to provide current best practices for several commonly used sub-cell models and demonstrate how they can be implemented within the proposed computational framework presented in the previous section.

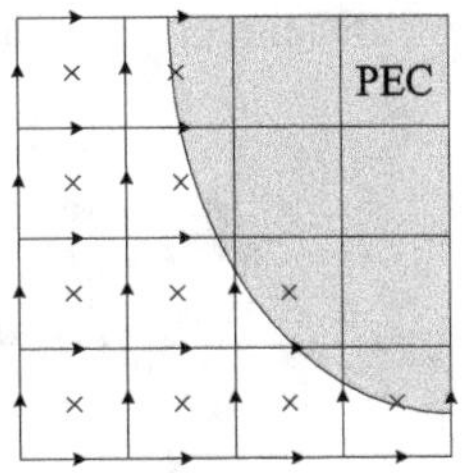

Figure 2.2 PEC object within an FDTD grid – a 2D slice of the primary Yee grid is shown

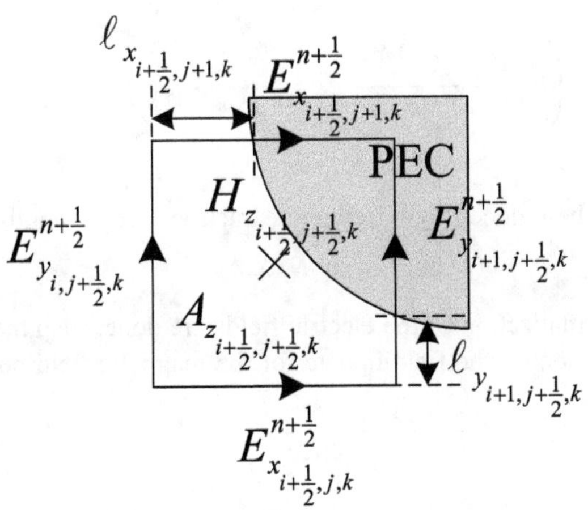

Figure 2.3 PEC object partially cutting through a primary cell face

2.5.2 Subcell for PEC objects

Consider an arbitrary shaped perfect electrical conducting (PEC) object placed within the FDTD grid, as illustrated in Figure 2.2. On the outer boundary of the PEC, the tangential electric field must be zero. Within the PEC, the electric (and magnetic) field will be zero. Forcing the fields within the PEC is easily enforced by setting the update coefficient *coefe_*= 0 for all electric field edges fully inside the PEC or fully on and tangential to the PEC surface. If the electric fields are initialized to zero, then the field in the PEC will remain 0 throughout the computation. H-fields associated with faces fully within the PEC will also remain zero.

Next, consider the fields near the boundary of the PEC – that is where the PEC cuts partially through a primary grid cell face, as illustrated by the z-normal face in Figure 2.3. Following the procedure of Dey and Mittra [42,75], let ℓ_x be the length of the edge *outside* the PEC (similarly for ℓ_y). Let A_z be the area of the face *outside* the PEC. Then, computing Faraday's law in the integral form only over the region outside the PEC and using central differencing on the time-derivative, leads to:

$$H^{n+1}_{z_{i+\frac{1}{2},j+\frac{1}{2},k}} = H^{n}_{z_{i+\frac{1}{2},j+\frac{1}{2},k}} - \frac{\Delta t}{\mu A_{z_{i+\frac{1}{2},j+\frac{1}{2},k}}} \cdot \left(E^{n+\frac{1}{2}}_{y_{i+1,j+\frac{1}{2},k}} \ell_{y_{i+1,j+\frac{1}{2},k}} - E^{n+\frac{1}{2}}_{y_{i,j+\frac{1}{2},k}} \ell_{y_{i,j+\frac{1}{2},k}} - E^{n+\frac{1}{2}}_{x_{i+\frac{1}{2},j+1,k}} \ell_{x_{i+\frac{1}{2},j+1,k}} + E^{n+\frac{1}{2}}_{x_{i+\frac{1}{2},j,k}} \ell_{x_{i+\frac{1}{2},j,k}} \right) \tag{2.33}$$

Note that if the edge is fully outside the PEC, then $\ell_x = \Delta x$. After applying the field normalization of (10), (33) is rewritten as:

$$h^{n+1}_{z_{i+\frac{1}{2},j+\frac{1}{2},k}} = h^{n}_{z_{i+\frac{1}{2},j+\frac{1}{2},k}} - \frac{c_o \Delta t}{\mu_r} \frac{\Delta x \Delta y}{A_{z_{i+\frac{1}{2},j+\frac{1}{2},k}}} \frac{\Delta z}{\Delta x \Delta y} \cdot \left(e^{n+\frac{1}{2}}_{y_{i+1,j+\frac{1}{2},k}} \frac{\ell_{y_{i+1,j+\frac{1}{2},k}}}{\Delta y} - e^{n+\frac{1}{2}}_{y_{i,j+\frac{1}{2},k}} \frac{\ell_{y_{i,j+\frac{1}{2},k}}}{\Delta y} - e^{n+\frac{1}{2}}_{x_{i+\frac{1}{2},j+1,k}} \frac{\ell_{x_{i+\frac{1}{2},j+1,k}}}{\Delta x} + e^{n+\frac{1}{2}}_{x_{i+\frac{1}{2},j,k}} \frac{\ell_{x_{i+\frac{1}{2},j,k}}}{\Delta x} \right) \tag{2.34}$$

The notation can be reduced by introducing fractional edge lengths and area as:

$$\delta_x = \ell_x/\Delta x, \quad \delta_y = \ell_y/\Delta y, \quad S_z = A_z/\Delta x \Delta y. \tag{2.35}$$

A second normalization of the electric fields are done, such that $\widetilde{e}_y = e_y \delta_y$ and $\widetilde{e}_z = e_z \delta_z$. Consequently, the field update for the magnetic field becomes:

$$h^{n+1}_{z_{i+\frac{1}{2},j+\frac{1}{2},k}} = h^{n}_{z_{i+\frac{1}{2},j+\frac{1}{2},k}} - \frac{c_o \Delta t}{\mu_r S_{z_{i+\frac{1}{2},j+\frac{1}{2},k}}} \frac{\Delta z}{\Delta x \Delta y} \cdot \left(\widetilde{e}^{n+\frac{1}{2}}_{y_{i+1,j+\frac{1}{2},k}} - \widetilde{e}^{n+\frac{1}{2}}_{y_{i,j+\frac{1}{2},k}} - \widetilde{e}^{n+\frac{1}{2}}_{x_{i+\frac{1}{2},j+1,k}} + \widetilde{e}^{n+\frac{1}{2}}_{x_{i+\frac{1}{2},j,k}} \right) \tag{2.36}$$

The face then has an effective permeability of $\mu_r S_z$, which is represented simply by the update coefficient.

Following the Dey-Mittra algorithm [42,75], the electric field partially outside the PEC is updated using a standard field update without deformation. With the proposed normalizations, from (13) this is expressed as:

$$\widetilde{e}^{n+1/2}_{x_{i+\frac{1}{2},j,k}} = \widetilde{e}^{n-1/2}_{x_{i+\frac{1}{2},j,k}} + c_o \Delta t \frac{\delta_{x_{i+\frac{1}{2},j,k}}}{\varepsilon_r} \frac{\Delta x}{\Delta y \Delta z} \left(h^{n}_{z_{i+\frac{1}{2},j+\frac{1}{2},k}} - h^{n}_{z_{i+\frac{1}{2},j-\frac{1}{2},k}} - h^{n}_{y_{i+\frac{1}{2},j,k+\frac{1}{2}}} - h^{n}_{y_{i+\frac{1}{2},j,k-\frac{1}{2}}} \right) \tag{2.37}$$

Consequently, this is interpreted as an effective local permittivity ε_r/δ_x.

Some comments need to be made at this point. The first is that the field updates near the surface of the PEC using the algorithm proposed by (36) and (37) exactly fit within the computational framework of (30) with the deformed contour being represented as an effective local anisotropic material. If any of the edges completely lie within the PEC, $\delta = 0$, and the field will simply be zero.

Next, since the effective relative material parameters can be <1, the stability criterion of (14) is impacted. More precisely, the stability criterion is:

$$\Delta t < \min \sqrt{\frac{S}{\delta}} \frac{\sqrt{\mu_r \varepsilon_r}}{c_o} \frac{1}{\sqrt{\frac{1}{\Delta x^2} + \frac{1}{\Delta y^2} + \frac{1}{\Delta z^2}}} \tag{2.38}$$

as determined by all faces in the grid (here μ_r and ε_r are the physical relative permeability and permittivity of the material outside the PEC). It is reasonable to expect situations where S/δ becomes very small, dramatically restricting the global time step. A means to mitigate this was proposed by Benkler *et al.* [41] by setting a minimum threshold for $\sqrt{S/\delta}$. For cells where $\sqrt{S/\delta}$ falls below some threshold, $\tau_{\min}$, instead set $S = \tau_{\min}^2 \delta$. This introduces local error at the expense of maintaining a reasonable time step.

Examples studied in [41] demonstrate that the proposed sub-cell conformal PEC model greatly improves accuracy without a significant sacrifice in the time step. It is further demonstrated here that this is simply a pre-processing step and can simply be modeled as a local material within a standard FDTD implementation.

2.5.3 Subcell for thin-wires

Next consider the situation when a perfectly conducting thin-wire with radius $a << \Delta x$ is aligned along the axis of the grid, as illustrated in Figure 2.4 (the proposed algorithm is actually valid if $a < \Delta x/2$). The previous sub-cell algorithm can certainly be applied in this case. However, for small radii, the tangential magnetic and normal electric fields near the surface of the wire have a $1/r$ type singularity. Approximating this singularity within the fields can significantly improve the accuracy of the simulation. While thin-wire sub-cell models are one of the earliest developed [76–79], it is reviewed here in order to put it in the context of fitting it to the foot print of the update equation in (30).

Consider the thin wire to be *z*-directed, and located on a set of primary grid edges at discrete *x,y* coordinates *i,j*, as illustrated in Figure 2.4. Ampere's law reveals that the

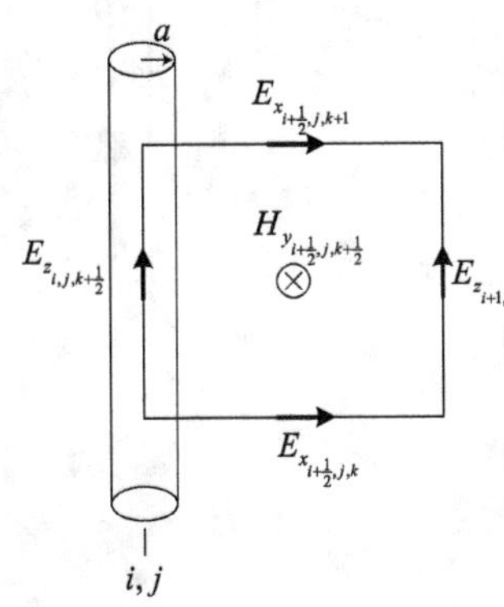

Figure 2.4 Thin PEC wire aligned along a z-axis of a primary grid cell

local tangential magnetic field will decay approximately as $1/r$ from the wire axis. Consider the local magnetic transverse magnetic field, H_y, shown in Figure 2.4. The $1/r$ variation in this face can be expressed as a function of x as:

$$H^n_{y_{i+\frac{1}{2},j,k+\frac{1}{2}}} = \widetilde{H}^n_{y_{i+\frac{1}{2},j,k+\frac{1}{2}}} \frac{\Delta x}{2} \frac{1}{|x - i\Delta x|} \tag{2.39}$$

where i is the x-index of the wire axis, and the $\Delta x/2$ scaling ensures the expression is unity at the cell center, or at $x = (i + 1/2)\Delta x$. The radial electric field can be also be scaled in this plane as:

$$E^{n+\frac{1}{2}}_{x_{i+\frac{1}{2},j,k}} = \widetilde{E}^{n+\frac{1}{2}}_{x_{i+\frac{1}{2},j,k}} \frac{\Delta x}{2} \frac{1}{|x - i\Delta x|} \tag{2.40}$$

Fields on the other three-quadrants of the wire are similarly scaled. Faraday's law in the integral form is next applied, where the surface is that of the face outside of the wire. After using a central-difference approximation of the time-derivative, this leads to the update equation:

$$\widetilde{H}^{n+1}_{y_{i+\frac{1}{2},j,k+\frac{1}{2}}} = \widetilde{H}^n_{y_{i+\frac{1}{2},j,k+\frac{1}{2}}} + \frac{\Delta t}{\mu} \left[\frac{E^{n+\frac{1}{2}}_{z_{i+1,j,k+\frac{1}{2}}}}{\frac{\Delta x}{2} \ln\left(\frac{\Delta x}{a}\right)} - \left(\frac{\widetilde{E}^{n+\frac{1}{2}}_{x_{i+\frac{1}{2},j,k+1}} - \widetilde{E}^{n+\frac{1}{2}}_{x_{i+\frac{1}{2},j,k}}}{\Delta z} \right) \right]. \tag{2.41}$$

Next, consider the integral for Ampere's law over the secondary face with the normal electric field, as illustrated in Figure 2.5. The issue is that the transverse

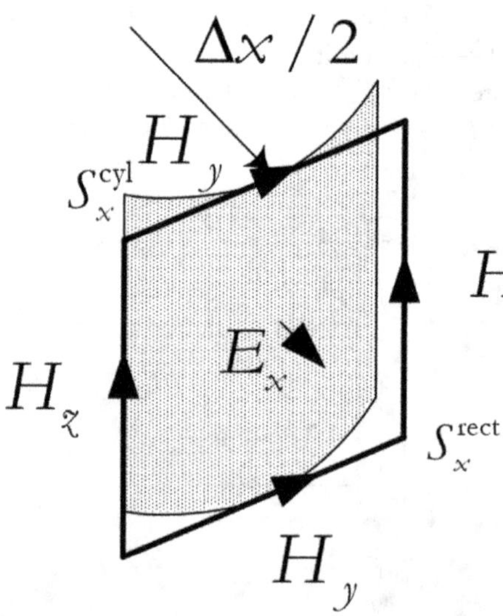

Figure 2.5 Curved secondary grid face near thin-wire

magnetic field, H_y and normal electric field are approximated by the radial functions in (39) and (40). However, the face integral spans y. The face integral of the electric field is then approximated as:

$$\iint_S \vec{E}\cdot d\vec{s} = \iint_{S_x^{\text{rect}}} \hat{r}\widetilde{E}_x \cdot \hat{x}ds = 2\widetilde{E}_x \int_0^{\Delta z}\int_0^{\frac{\Delta y}{2}} \cos\,\phi(y) \frac{\frac{\Delta x}{2}}{\sqrt{\left(\frac{\Delta x}{2}\right)^2 + y^2}} dydz$$

$$= \Delta z \Delta x \widetilde{E}_x \tan^{-1}\left(\frac{\Delta y}{\Delta x}\right) \tag{2.42}$$

A similar expression is derived for the line integral over the magnetic field. As a result, Ampere's law can be expressed in discrete form as:

$$\frac{\varepsilon \Delta z \Delta y \zeta_x}{\Delta t}\left(\widetilde{E}^{n+\frac{1}{2}}_{x_{i+\frac{1}{2},j,k}} - \widetilde{E}^{n-\frac{1}{2}}_{x_{i+\frac{1}{2},j,k}}\right) = \Delta y \zeta_x \left(\widetilde{H}_{y_{i+\frac{1}{2},j,k-\frac{1}{2}}} - \widetilde{H}_{y_{i+\frac{1}{2},j,k+\frac{1}{2}}}\right) + \Delta z\left(H_{z_{i+\frac{1}{2},j+\frac{1}{2},k}} - H_{z_{i+\frac{1}{2},j-\frac{1}{2},k}}\right) \tag{2.43}$$

where,

$$\zeta_x = \frac{\Delta x}{\Delta y}\tan^{-1}\left(\frac{\Delta y}{\Delta x}\right) \tag{2.44}$$

Then, applying the normalizations

$$\widetilde{e}^{n+\frac{1}{2}}_{x_{i+\frac{1}{2},j,k}} = \widetilde{E}^{n+\frac{1}{2}}_{x_{i+\frac{1}{2},j,k}} \frac{\Delta x}{2}\ln\left(\frac{\Delta x}{a}\right), \quad \widetilde{h}_{y_{i+\frac{1}{2},j,k+\frac{1}{2}}} = \Delta y \zeta_x \eta_o \widetilde{H}_{y_{i+\frac{1}{2},j,k+\frac{1}{2}}} \tag{2.45}$$

as well as the normalizations in (10) for the remaining fields, (41) and (43) become:

$$\widetilde{h}^{n+1}_{y_{i+\frac{1}{2},j,k+\frac{1}{2}}} = \widetilde{h}^{n}_{y_{i+\frac{1}{2},j,k+\frac{1}{2}}} + \frac{c\Delta t \Delta y \zeta_x}{\frac{\Delta x}{2}\ln\left(\frac{\Delta x}{a}\right)\Delta z}\left[\left(e^{n+\frac{1}{2}}_{z_{i+1,j,k+\frac{1}{2}}} - 0\right) - \left(\widetilde{e}^{n+\frac{1}{2}}_{x_{i+\frac{1}{2},j,k+1}} - \widetilde{e}^{n+\frac{1}{2}}_{x_{i+\frac{1}{2},j,k}}\right)\right] \tag{2.46}$$

$$\widetilde{e}^{n+\frac{1}{2}}_{x_{i+\frac{1}{2},j,k}} = \widetilde{e}^{n-\frac{1}{2}}_{x_{i+\frac{1}{2},j,k}} + \frac{c\Delta t \frac{\Delta x}{2}\ln\left(\frac{\Delta x}{a}\right)}{\varepsilon_{r_{i+\frac{1}{2},j,k}} \Delta z \Delta y \zeta_x}\left[\left(h_{z_{i+\frac{1}{2},j+\frac{1}{2},k}} - h_{z_{i+\frac{1}{2},j-\frac{1}{2},k}}\right) - \left(\widetilde{h}_{y_{i+\frac{1}{2},j,k+\frac{1}{2}}} - \widetilde{h}_{y_{i+\frac{1}{2},j,k-\frac{1}{2}}}\right)\right] \tag{2.47}$$

Similar expressions are derived for the radial electric and tangential magnetic field in the other three quadrants.

Some comments are warranted. The first is that again, the thin wire approximation can again be modeled as an effective material with

$$\mu_{eff} = \frac{1}{2\zeta_x} \ln\left(\frac{\Delta x}{a}\right), \quad \varepsilon_{eff} = \varepsilon_r \frac{2\zeta_x}{\ln(\Delta x/a)} \tag{2.48}$$

Very similar terms can be derived for all the remaining field updates.

The footprint of (46) and (47) preserves the desired base form. Due to the reciprocal nature of the effective material, there is clearly no impact on the stability condition. It is further noted that the zero'd electric field term in (46) would actually be the axial field $e_{z_{i,j,k+\frac{1}{2}}}^{n+\frac{1}{2}}$ rather than an explicit 0. During the pre-processing stage, the update coefficient *coefez* for electric fields on the wire axis would be set to 0, and hence the field will remain zero valued.

The presented methods can be extended to thin wires passing arbitrarily through the FDTD grid [80,81].

2.5.4 Conformal grid for dielectrics

Next consider a device under test that contains dielectric materials that are arbitrarily placed within the orthogonal FDTD grid as shown in Figure 2.6. Since the boundary does not conform to the grid, material boundary conditions are not properly enforced. A conformal algorithm to accurately model this within the FDTD method becomes necessary to avoid using a staircase approximation of the boundary. Early conformal algorithms, such as that proposed by Yu and Mittra [82] assumed an averaged permittivity value based on the percentage of the edge within each medium. A more accurate method was proposed by Wang, *et al.* [51], which approximated the local dielectric using a spatial averaging. As an illustration, consider the secondary grid face in Figure 2.6 centered at $i+\frac{1}{2}, j, k$. In this method, the electric field is assumed to be tangential to the interface at this cross section. The effective permittivity is assumed to be the

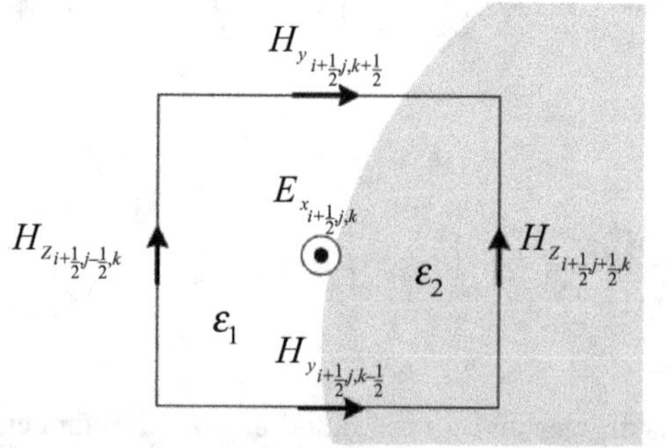

Figure 2.6 Dielectric object arbitrarily cutting through an FDTD grid – secondary grid face is shown

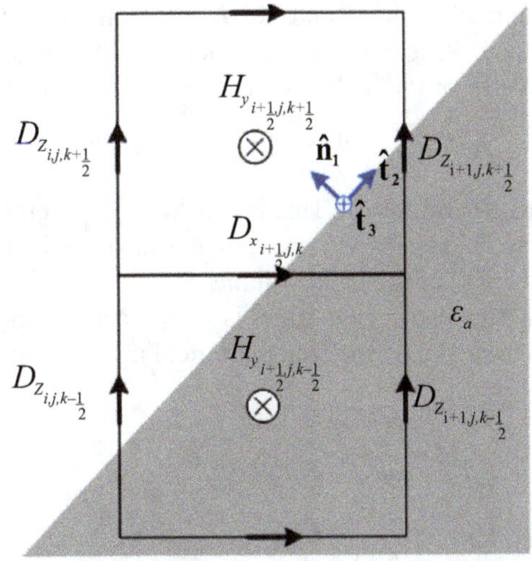

Figure 2.7 PEC object partially cutting through a primary cell face

average permittivity over the face:

$$\varepsilon_{r_{i+\frac{1}{2},j,k}} \approx \frac{1}{\Delta y \Delta z} \iint_S \varepsilon(y,z) ds \tag{2.49}$$

The field updates are then performed using (13) with this local permittivity. Wang *et al.*, showed that using this averaging scheme is more accurate than the Yu-Mittra formulation. It is also more accurate than using a volume averaging about the field point with the isotropic permittivity.

The Wang and Yu-Mittra methods both offer an improvement over a staircase approximation of the boundary, however, convergence of these methods is approximately first-order [45,83]. One of the drawbacks is that for an arbitrary surface, the appropriate field continuities across the dielectric boundary are somewhat obfuscated. An alternate conformal algorithm introduced by Farjadpour *et al.* [45] attempts to minimize the energy norm by treating the dielectric boundary interface as a local anisotropic material that separates the normal continuity of the displacement flux density and the tangential continuity of the electric fields. Their method is referred to as the subpixel smoothing algorithm.

The subpixel smoothing algorithm approximates the dielectric associated with an electric field edge as an averaged material over a volume centered about the edge center. However, a different averaging is performed for the normal electric field than for the tangential fields to compensate for the different boundary conditions. To illustrate this process, consider the primary grid illustrated in Figure 2.7 near a dielectric boundary. The subpixel smoothing for edge $i + \frac{1}{2}, j, k$ is defined via the mean of a quantity over a volume of dimension $\Delta x \times \Delta y \times \Delta z$ centered about the edge

$$\langle \chi \rangle_{i+\frac{1}{2},j,k} = \frac{1}{\Delta x \Delta y \Delta z} \int_{(j-\frac{1}{2})\Delta y}^{(j-\frac{1}{2})\Delta y} \int_{(j-\frac{1}{2})\Delta y}^{(j-\frac{1}{2})\Delta y} \int_{i\Delta x}^{(i+1)\Delta x} \chi dxdydz \tag{2.50}$$

The local unit normal to the boundary, $\hat{\mathbf{n}}_1$, is identified and a local projection matrix onto the normal, $\mathbf{P}$, is constructed, where the elements of $\mathbf{P}$ are given by $P_{i,j} = n_{1i}n_{1j}$. The inverse of the dielectric tensor is then defined as:

$$\bar{\bar{\varepsilon}}_r^{-1} = \mathbf{P}\langle\varepsilon_r^{-1}\rangle + (1 - \mathbf{P})\langle\varepsilon_r\rangle^{-1} \tag{2.51}$$

where ε_r is the relative permittivity. That is, for the normal projection, the inverse of the permittivity is averaged, whereas, for the tangential projections, just the permittivity is averaged over the local volume.

In this approach, the displacement flux is first updated using the discrete form of Ampere's law. That is, the normalized main field update equation (13) would be expressed as:

$$d_{x_{i+\frac{1}{2},j,k}}^{n+1/2} = d_{x_{i+\frac{1}{2},j,k}}^{n-1/2} + c_o\Delta t\frac{\Delta x}{\Delta y\Delta z}\left(h_{z_{i+\frac{1}{2},j+\frac{1}{2},k}}^n - h_{z_{i+\frac{1}{2},j-\frac{1}{2},k}}^n - h_{y_{i+\frac{1}{2},j,k+\frac{1}{2}}}^n - h_{y_{i+\frac{1}{2},j,k-\frac{1}{2}}}^n\right) \tag{2.52}$$

where $d_x = \Delta x D_x$ is the normalized flux density. The electric field is then computed using the inverse tensor of the relative permittivity, as posed by (51), times the full displacement flux vector. Since only d_x is known at this edge, the remaining terms are obtained by averaging the four adjacent vectors:

$$d_{z_{i+\frac{1}{2},j,k}}^{n+\frac{1}{2}} \approx \frac{1}{4}\left(d_{z_{i,j,k+\frac{1}{2}}}^{n+\frac{1}{2}} + d_{z_{i+1,j,k+\frac{1}{2}}}^{n+\frac{1}{2}} + d_{z_{i,j,k-\frac{1}{2}}}^{n+\frac{1}{2}} + d_{z_{i+1,j,k-\frac{1}{2}}}^{n+\frac{1}{2}}\right) \tag{2.53}$$

$$d_{y_{i+\frac{1}{2},j,k}}^{n+\frac{1}{2}} \approx \frac{1}{4}\left(d_{y_{i,j+\frac{1}{2},k}}^{n+\frac{1}{2}} + d_{y_{i+1,j+\frac{1}{2},k}}^{n+\frac{1}{2}} + d_{y_{i,j-\frac{1}{2},k}}^{n+\frac{1}{2}} + d_{y_{i+1,j-\frac{1}{2},k}}^{n+\frac{1}{2}}\right) \tag{2.54}$$

Then

$$e_{x_{i+\frac{1}{2},j,k}}^{n+1/2} = \left(\bar{\bar{\varepsilon}}_r^{-1}\right)_{1,1} d_{x_{i+\frac{1}{2},j,k}}^{n+1/2} + \left(\bar{\bar{\varepsilon}}_r^{-1}\right)_{1,2}\frac{\Delta x}{\Delta y} d_{y_{i+\frac{1}{2},j,k}}^{n+1/2} + \left(\bar{\bar{\varepsilon}}_r^{-1}\right)_{1,3}\frac{\Delta x}{\Delta z} d_{z_{i+\frac{1}{2},j,k}}^{n+1/2} \tag{2.55}$$

Similar expressions can be derived for e_y and e_z updates on y- and z-directed edges. A similar procedure could be derived for lossy media [84].

For the pixel smoothing algorithm, both the electric field and the flux density vectors would need to be stored. The main field update loops would be conducted globally using (52). However, it would not be necessary to perform (55) globally, but only over blocks where non-conforming dielectric materials exist.

2.6 Summary

The focus of this chapter has been on providing the fundamentals of the FDTD method and its efficient implementation. The important CPML algorithm with CFS constitutive parameters was also detailed as well as well-established rules of thumb for selecting values for the scaled constitutive parameters. Also detailed were several subcell models including thin-wire approximations, conformal FDTD for PEC, and dielectric objects. It was shown that these are conveniently represented as effective materials. With such a context, the stability of these methods is well understood. More importantly, using these models becomes a pre-processing step, and then such models preserve the base FDTD framework and can be executed in a very efficient manner.

The intent of this chapter was to be more tutorial, providing sufficient detail to enable one to fully implement these algorithms. However, in such a short space, many other topics have not been addressed, such as sources [85–87], near field to far field transformations [88–91], the treatment of non-linear and dispersive materials [10,33,84,92–96], the modeling of periodic structures [97–102], etc. Many such topics are available in the literature, as referenced, or in quality text books that cover such topics in detail [39,60,61].

References

[1] K. Yee, "Numerical solution of initial boundary value problems involving Maxwell's equations in isotropic media," *IEEE Transactions on Antennas and Propagation*, vol. 14, no. 3, pp. 302–307, 1966, doi: 10.1109/TAP.1966.1138693.

[2] K. Yee, "personal communication," ed, 1996.

[3] R. Holland, "THREDE – free-field EMP coupling and scattering code," *IEEE Transactions on Nuclear Science*, vol. 24, no. 6, pp. 2416–2421, 1977, doi: 10.1109/tns.1977.4329229.

[4] D. M. Sullivan, O. P. Gandhi, and A. Taflove, "Use of the finite-difference time-domain method for calculating EM absorption in man models," *IEEE Transactions on Biomedical Engineering*, vol. 35, no. 3, pp. 179–186, 1988, doi: 10.1109/10.1360.

[5] S. M. Rao, D. R. Wilton, and A. W. Glisson, "Electromagnetic scattering by surfaces of arbitrary shape," *IEEE Transactions on Antennas and Propagation*, vol. 30, no. 3, pp. 409–418, 1982, doi:10.1109/TAP.1982.1142818.

[6] J. P. Berenger, "Perfectly matched layer for the FDTD solution of wave-structure interaction problems," *IEEE Transactions on Antennas and Propagation*, vol. 44, no. 1, pp. 110–117, 1996.

[7] S. D. Gedney, "Finite-difference time-domain analysis of microwave circuit devices on high-performance vector/parallel computers," *IEEE Transactions on Microwave Theory and Techniques*, vol. 43, no. 10, pp. 2510–2514, 1995, doi:10.1109/22.466191.

[8] C. Guiffaut and K. Mahdjoubi, "A parallel FDTD algorithm using the MPI library," *IEEE Antennas and Propagation Magazine*, vol. 43, no. 2, pp. 94–103, 2001, doi:10.1109/74.924608.

[9] S. Liu, B. Zou, L. Zhang, and S. Ren, "Heterogeneous CPU+GPU-accelerated FDTD for scattering problems with dynamic load balancing," *IEEE Transactions on Antennas and Propagation*, vol. 68, no. 9, pp. 6734–6742, 2020, doi: 10.1109/TAP.2020.2990486.

[10] Y. Zhao and Y. Hao, "Full-wave parallel dispersive finite-difference time-domain modeling of three-dimensional electromagnetic cloaking structures," *Journal of Computational Physics*, vol. 228, no. 19, pp. 7300–7312, 2009, doi:10.1016/j.jcp.2009.06.026.

[11] J. G. Maloney, K. L. Shlager, and G. S. Smith, "A simple FDTD model for transient excitation of antennas by transmission-lines," *IEEE Transactions on Antennas and Propagation*, vol. 42, no. 2, pp. 289–292, 1994. [Online]. Available: <Go to ISI>://A1994NF48000023.

[12] W. V. Andrew, C. A. Balanis, P. A. Tirkas, J. Peng, and C. R. Birtcher, "Finite-difference time-domain analysis of HF antennas on helicopter airframes," *IEEE Transactions on Electromagnetic Compatibility*, vol. 39, no. 2, pp. 100–113, 1997, doi:10.1109/15.584932.

[13] C. R. Birtcher, C. A. Balanis, and D. DeCarlo, "Rotor-blade modulation on antenna amplitude pattern and polarization: predictions and measurements," *IEEE Transactions on Electromagnetic Compatibility*, vol. 41, no. 4, pp. 384–393, 1999, doi:10.1109/15.809828.

[14] M. Grimm and D. Manteuffel, "On-body antenna parameters," *IEEE Transactions on Antennas and Propagation*, vol. 63, no. 12, pp. 5812–5821, 2015, doi:10.1109/tap.2015.2482499.

[15] Y. Watanabe and H. Igarashi, "Accelerated FDTD analysis of antennas loaded by electric circuits," *IEEE Transactions on Antennas and Propagation*, vol. 60, no. 2, pp. 958–963, 2012, doi:10.1109/tap.2011.2173148.

[16] Y. Zhao, Y. Hao, and C. Parini, "FDTD characterization of UWB indoor radio channel including frequency dependent antenna directivities," *IEEE Antennas and Wireless Propagation Letters*, vol. 6, pp. 191–194, 2007, doi:10.1109/lawp.2007.891963.

[17] D. M. Sheen, S. M. Ali, M. D. Abouzahra, and J. A. Kong, "Application of the three-dimensional finite-difference time-domain method to the analysis of planar microstrip circuits," *IEEE Transactions on Microwave Theory and Techniques*, vol. 38, no. 7, pp. 849–857, 1990.

[18] A. C. M. Austin and C. D. Sarris, "Efficient analysis of geometrical uncertainty in the FDTD method using polynomial chaos with application to microwave circuits," *IEEE Transactions on Microwave Theory and Techniques*, vol. 61, no. 12, pp. 4293–4301, 2013, doi:10.1109/tmtt.2013.2281777.

[19] X. C. Li, J. F. Mao, and M. Swaminathan, "Transient analysis of CMOS-gate-driven RLGC interconnects based on FDTD," *IEEE Transactions on Computer-Aided Design of Integrated Circuits and Systems*, vol. 30, no. 4, pp. 574–583, 2011, doi:10.1109/tcad.2010.2095650.

[20] Z. G. Su and T. J. Brazil, "Direct incorporation of scattering parameters into the FDTD algorithm," *IEEE Transactions on Antennas and Propagation*, vol. 57, no. 4, pp. 1186–1190, 2009, doi:10.1109/tap.2009.2015830.

[21] D. C. Wittwer and R. W. Ziolkowski, "The effect of dielectric loss in FDTD simulations of microstrip structures," *IEEE Transactions on Microwave Theory and Techniques*, vol. 49, no. 2, pp. 250–262, 2001, doi:10.1109/22.903085.

[22] A. P. Zhao, J. Juntunen, and A. V. Raisanen, "An efficient FDTD algorithm for the analysis of microstrip patch antennas printed on a general

anisotropic dielectric substrate," *IEEE Transactions on Microwave Theory and Techniques*, vol. 47, no. 7, pp. 1142–1146 1999. [Online]. Available: <Go to ISI>://WOS:000081473000026.

[23] E. C. Fear, X. Li, S. C. Hagness, and M. A. Stuchly, "Confocal microwave imaging for breast cancer detection: localization of tumors in three dimensions," *IEEE Transactions on Biomedical Engineering*, vol. 49, no. 8, pp. 812–822, 2002, doi:10.1109/tbme.2002.800759.

[24] E. J. Bond, X. Li, S. C. Hagness, and B. D. Van Veen, "Microwave imaging via space–time beamforming for early detection of breast cancer," *IEEE Transactions on Antennas and Propagation*, vol. 51, no. 8, pp. 1690–1705, 2003 (in English), doi: 10.1109/tap.2003.815446.

[25] P. Kosmas and C. M. Rappaport, "Time reversal with the FDTD method for microwave breast cancer detection," *IEEE Transactions on Microwave Theory and Techniques*, vol. 53, no. 7, pp. 2317–2323, 2005, doi: 10.1109/tmtt.2005.850444.

[26] H. Subramanian, P. Pradhan, Y. Liu, *et al.*, "Partial-wave microscopic spectroscopy detects subwavelength refractive index fluctuations: an application to cancer diagnosis," *Optics Letters*, vol. 34, no. 4, pp. 518–520, 2009, doi:10.1364/ol.34.000518.

[27] S. L. Sun, K.-Y. Yang, C.-M. Wang, *et al.*, "High-efficiency broadband anomalous reflection by gradient meta-surfaces," *Nano Letters*, vol. 12, no. 12, pp. 6223–6229, 2012, doi:10.1021/nl3032668.

[28] F. Yang and Y. Rahmat-Samii, "Microstrip antennas integrated with electromagnetic band-gap (EBG) structures: a low mutual coupling design for array applications," *IEEE Transactions on Antennas and Propagation*, vol. 51, no. 10, pp. 2936–2946, 2003, doi:10.1109/tap.2003.817983.

[29] R. W. Ziolkowski, "Propagation in and scattering from a matched metamaterial having a zero index of refraction," *Physical Review E*, vol. 70, no. 4, 2004, Art no. 046608, doi:10.1103/PhysRevE.70.046608.

[30] A. C. M. Austin, M. J. Neve, and G. B. Rowe, "Modeling propagation in multifloor buildings using the FDTD method," *IEEE Transactions on Antennas and Propagation*, vol. 59, no. 11, pp. 4239–4246, 2011, doi:10.1109/tap.2011.2164181.

[31] T. Sekiguchi, T. Hikage, M. Yamamoto, *et al.*, "Numerical estimation of propagation path loss for wireless link design of WAIC systems installed on outside aircraft cabin based on large-scale FDTD simulation," *IEICE Communications Express*, vol. 8, no. 5, pp. 129–134, 2019, doi: 10.1587/comex.2018XBL0165.

[32] M. Thiel and K. Sarabandi, "3D-wave propagation analysis of indoor wireless channels utilizing hybrid methods," *IEEE Transactions on Antennas and Propagation*, vol. 57, no. 5, pp. 1539–1546, 2009, doi:10.1109/tap.2009.2016710.

[33] W. Yang, L. L. Liu, and Y. Q. Huang, "The FDTD simulation for the performance of dispersive cloak devices," *Applied Mathematics Letters*, vol. 88, pp. 171–178, 2019, doi:10.1016/j.aml.2018.08.020.

[34] A. Soriano, E. A. Navarro, J. A. Porti, and V. Such, "Analysis of the finite difference time domain technique to solve the Schrodinger equation for quantum devices," *Journal of Applied Physics*, vol. 95, no. 12, pp. 8011–8018, 2004, doi: 10.1063/1.1753661.

[35] F. I. Moxley, III, T. Byrnes, F. Fujiwara, and W. Dai, "A generalized finite-difference time-domain quantum method for the N-body interacting Hamiltonian," *Computer Physics Communications*, vol. 183, no. 11, pp. 2434–2440, 2012, doi:10.1016/j.cpc.2012.06.012.

[36] D.-Y. Na and W. C. Chew, "Time-domain quantum Maxwell solver," in *2020 IEEE International Symposium on Antennas and Propagation and North American Radio Science Meeting*, IEEE Antennas and Propagation Society International Symposium, 2020, pp. 1683–1684.

[37] J. J. Simpson and A. Taflove, "Electrokinetic effect of the Loma Prieta earthquake calculated by an entire-Earth FDTD solution of Maxwell's equations," *Geophysical Research Letters*, vol. 32, no. L09302, 2005.

[38] J. W. Baek, D. K. Kim, and K. Y. Jung, "Finite-difference time-domain modeling for electromagnetic wave analysis of human voxel model at millimeter-wave frequencies," *IEEE Access*, vol. 7, pp. 3635–3643, 2019, doi: 10.1109/access.2018.2888584.

[39] A. Taflove, A. F. Oskooi, and S. G. Johnson, *Advances in FDTD Computational Electrodynamics: Photonics and Nanotechnology*. Boston/ London: Artech House, 2013.

[40] J. M. Jin, *The Finite Element Method in Electromagnetics*. Holboken, NJ: John Wiley & Sons, Inc., 2014.

[41] S. Benkler, N. Chavannes, and N. Kuster, "A new 3-D conformal PEC FDTD scheme with user-defined geometric precision and derived stability criterion," *IEEE Transactions on Antennas and Propagation*, vol. 54, no. 6, pp. 1843–1849, 2006, doi: 10.1109/tap.2006.875909.

[42] S. Dey and R. Mittra, "A modified locally conformal finite-difference time-domain algorithm for modeling three-dimensional perfectly conducting objects," *Microwave and Optical Technology Letters*, vol. 17, no. 6, pp. 349–352, 1998. [Online]. Available: <Go to ISI>://000072488800004.

[43] W. H. Yu and R. Mittra, "A conformal FDTD software package modeling antennas and microstrip circuit components," *IEEE Antennas and Propagation Magazine*, vol. 42, no. 5, pp. 28–39, 2000. [Online]. Available: <Go to ISI>://000165180900005.

[44] W. H. Yu, R. Mittra, X. L. Yang, Y. J. Liu, Q. J. Rao, and A. Muto, "High-performance conformal FDTD techniques," *IEEE Microwave Magazine*, vol. 11, no. 4, pp. 42–55, 2010, doi: 10.1109/mmm.2010.936496.

[45] A. Farjadpour, D. Roundy, A. Rodriguez, *et al.*, "Improving accuracy by subpixel smoothing in the finite-difference time domain," *Optics Letters*, vol. 31, no. 20, pp. 2972–2974, 2006, doi:10.1364/ol.31.002972.

[46] A. F. Oskooi, C. Kottke, and S. G. Johnson, "Accurate finite-difference time-domain simulation of anisotropic media by subpixel smoothing," *Optics Letters*, vol. 34, no. 18, pp. 2778–2780, 2009, doi: 10.1364/ol.34.002778.

[47] G. Z. Chen, S. C. Yang, and D. L. Su, "An accurate three-dimensional FDTD(2,4) method on face-centered cubic grids with low numerical dispersion," *IEEE Antennas and Wireless Propagation Letters*, vol. 18, no. 9, pp. 1711–1715, 2019, doi: 10.1109/lawp.2019.2926423.

[48] S. V. Georgakopoulos, C. R. Birtcher, C. A. Balanis, and R. A. Renaut, "Higher-order finite-difference schemes for electromagnetic radiation, scattering, and penetration, Part I: Theory," *IEEE Antennas and Propagation Magazine*, vol. 44, no. 1, pp. 134–142, 2002, doi:10.1109/74.997945.

[49] M. F. Hadi and M. PiketMay, "A modified FDTD (2,4) scheme for modeling electrically large structures with high-phase accuracy," *IEEE Transactions on Antennas and Propagation*, vol. 45, no. 2, pp. 254–264, 1997, doi:10.1109/8.560344.

[50] M. E. Potter and M. D. Nauta, "FDTD on face-centered cubic (FCC) grids for Maxwell's equations," *IEEE Transactions on Antennas and Propagation*, vol. 61, no. 4, pp. 2116–2122, 2013, doi:10.1109/tap.2012.2234074.

[51] J. A. Wang, W. Y. Yin, P. G. Liu, and Q. H. Liu, "High-order interface treatment techniques for modeling curved dielectric objects," *IEEE Transactions on Antennas and Propagation*, vol. 58, no. 9, pp. 2946–2953, 2010, doi: 10.1109/tap.2010.2052562.

[52] T. Namiki, "3-D ADI-FDTD method – unconditionally stable time-domain algorithm for solving full vector Maxwell's equations," *IEEE Transactions on Microwave Theory and Techniques*, vol. 48, no. 10, pp. 1743–1748, 2000, doi:10.1109/22.873904.

[53] E. L. Tan, "Unconditionally stable LOD-FDTD method for 3-D Maxwell's equations," *IEEE Microwave and Wireless Components Letters*, vol. 17, no. 2, pp. 85–87, 2007, doi:10.1109/lmwc.2006.890166.

[54] M. Moradi, V. Nayyeri, and O. M. Ramahi, "An unconditionally stable single-field finite-difference time-domain method for the solution of Maxwell equations in three dimensions," *IEEE Transactions on Antennas and Propagation*, vol. 68, no. 5, pp. 3859–3868, 2020, doi: 10.1109/tap.2020.2975675.

[55] Y. S. Chung, T. K. Sarkar, B. H. Jung, and M. Salazar-Palma, "An unconditionally stable scheme for the finite-difference time-domain method," *IEEE Transactions on Microwave Theory and Techniques*, vol. 51, no. 3, pp. 697–704, 2003, doi: 10.1109/tmtt.2003.808732.

[56] G. Sun and C. W. Trueman, "Unconditionally stable Crank-Nicolson scheme for solving two-dimensional Maxwell's equations," *Electronics Letters*, vol. 39, no. 7, pp. 595–597, 2003, doi:10.1049/el:20030416.

[57] G. L. Sun and C. W. Trueman, "Efficient implementations of the Crank-Nicolson scheme for the finite-difference time-domain method," *IEEE Transactions on Microwave Theory and Techniques*, vol. 54, no. 5, pp. 2275–2284, 2006, doi:10.1109/tmtt.2006.873639.

[58] S. G. Garcia, T. W. Lee, and S. C. Hagness, "On the accuracy of the ADI-FDTD method," *IEEE Antennas and Wireless Propagation Letters*, vol. 1, pp. 31–34, 2002, doi: 10.1109/lawp.2002.802583.

[59] J. Yan and D. Jiao, "Fast explicit and unconditionally stable FDTD method for electromagnetic analysis," *IEEE Transactions on Microwave Theory and Techniques*, vol. 65, no. 8, pp. 2698–2710, 2017, doi:10.1109/tmtt.2017.2686862.

[60] A. Taflove and S. B. Hagness, Eds. *Computational Electrodynamics: The Finite-Difference Time-Domain*. Boston, MA: Artech House, 2000.

[61] S. D. Gedney, *Introduction to the Finite-Difference Time-Domain (FDTD) Method for Electromagnetics (Synthesis Lectures on Computational Electromagnetics)*. Fort Collins, CO: Morgan & Claypool, 2011.

[62] L. Demkowicz, P. Monk, L. Vardapetyan, and W. Rachowicz, "de Rham diagram for hp finite element spaces," *Computers & Mathematics with Applications*, vol. 39, no. 7–8, pp. 29–38, 2000, doi: 10.1016/s0898-1221(00)00062-6.

[63] R. Courant, K. O. Friedrichs, and H. Lewy, "On the partial differential equations of mathematical physics (translated)," *IBM Journal of Research and Development*, vol. 11, no. 2, pp. 215–234, 1967, doi:10.1147/rd.112.0215.

[64] J. A. Roden and S. D. Gedney, "Convolutional PML (CPML): an efficient FDTD implementation of the CFS-PML for arbitrary media," *Microwave and Optical Technology Letters*, vol. 27, pp. 334–339, 2000.

[65] S. D. Gedney and B. Zhao, "An auxiliary differential equation formulation for the complex-frequency shifted PML," *IEEE Transactions on Antennas and Propagation*, vol. 58, no. 3, pp. 838–847, 2010, doi:10.1109/TAP.2009.2037765.

[66] W. C. Chew and W. H. Weedon, "A 3D perfectly matched medium from modified Maxwells equations with stretched coordinates," *Microwave and Optical Technology Letters*, vol. 7, no. 13, pp. 599–604, 1994.

[67] J. A. Roden and S. D. Gedney, "An efficient FDTD implementation of the PML with CFS in general media," in 2000 *IEEE International Symposium on Antennas and Propagation,* 16–21 July, Salt Lake City, UT, 2000.

[68] J. P. Berenger, "Improved PML for the FDTD solution of wave-structure interaction problems," *IEEE Transactions on Antennas and Propagation*, vol. 45, no. 3, pp. 466–473, 1997.

[69] J. P. Berenger, "An effective PML for the absorption of evanescent waves in waveguides," *IEEE Microwave and Guided Wave Letters*, vol. 8, no. 5, pp. 188–190, 1998.

[70] J. P. Berenger, "Evanescent waves in PML's: origin of the numerical reflection in wave-structure interaction problems," *IEEE Transactions on Antennas and Propagation*, vol. 47, no. 10, pp. 1497–1503, 1999.

[71] J. P. Berenger, "Numerical reflection of evanescent waves by PMLs: origin and interpretation in the FDTD case. Expected consequences to other finite methods," *International Journal of Numerical Modelling-Electronic Networks Devices and Fields*, vol. 13, no. 2–3, pp. 103–114, 2000.

[72] J. P. Berenger, "Numerical reflection from FDTD-PMLs: a comparison of the split PML with the unsplit and CFSPMLs," *IEEE Transactions on Antennas and Propagation*, vol. 50, no. 3, pp. 258–265, 2002. [Online]. Available: <Go to ISI>://000175380200001.

[73] J. P. Berenger, "Application of the CFSPML to the absorption of evanescent waves in waveguides," *IEEE Microwave and Components Letters*, vol. 12, no. 6, pp. 218–220, 2002.

[74] S. D. Gedney, "An anisotropic perfectly matched layer-absorbing medium for the truncation of FDTD lattices," *IEEE Transactions on Antennas and Propagation*, vol. 44, no. 12, pp. 1630–1639, 1996.

[75] S. Dey and R. Mittra, "A locally conformal finite-difference time-domain (FDTD) algorithm for modeling three-dimensional perfectly conducting objects," *IEEE Microwave And Guided Wave Letters*, vol. 7, no. 9, pp. 273–275, 1997. [Online]. Available: <Go to ISI>://A1997XT11500005

[76] R. Holland and L. Simpson, "Finite-difference analysis of EMP coupling to thin struts and wires," *IEEE Transactions on Electromagnetic Compatibility*, vol. 23, no. 2, pp. 88–97, 1981.

[77] R. M. Makinen, J. S. Juntunen, and M. A. Kivikoski, "An improved thin-wire model for FDTD," *IEEE Transactions on Microwave Theory and Techniques*, vol. 50, no. 5, pp. 1245–1255, 2002. [Online]. Available: <Go to ISI>://000175334700001.

[78] R. M. Makinen and M. A. Kivikoski, "Stabilized resistive voltage source for FDTD thin-wire models," *IEEE Transactions on Antennas and Propagation*, vol. 51, no. 7, pp. 1615–1622, 2003, doi: 10.1109/tap.2003.813613.

[79] S. Watanabe and M. Taki, "An improved FDTD model for the feeding gap of a thin-wire antenna," *IEEE Microwave and Guided Wave Letters*, vol. 8, no. 4, pp. 152–154, 1998. [Online]. Available: <Go to ISI>://000072769100002.

[80] F. Edelvik, "A new technique for accurate and stable modeling of arbitrarily oriented thin wires in the FDTD method," *IEEE Transactions on Electromagnetic Compatibility*, vol. 45, no. 2, pp. 416–423, 2003, doi:10.1109/temc.2003.811294.

[81] G. Ledfelt, "A stable subcell model for arbitrarily oriented thin wires for the FDTD method," *International Journal of Numerical Modelling-Electronic Networks Devices and Fields*, vol. 15, nos. 5–6, pp. 503–515, 2002, doi: 10.1002/jnm.466.

[82] W. H. Yu and R. Mittra, "A conformal finite difference time domain technique for modeling curved dielectric surfaces," *IEEE Microwave and Wireless Components Letters*, vol. 11, no. 1, pp. 25–27, 2001, doi:10.1109/7260.905957.

[83] L. L. L. Fortes and S. T. M. Goncalves, "Wideband performance limitations of the C-FDTD in the discretization impoverishment of a curved surface," *Compel-the International Journal for Computation and Mathematics in Electrical and Electronic Engineering*, vol. 39, no. 5, pp. 1005–1015, 2020, doi: 10.1108/compel-01-2020-0048.

[84] A. Deinega and I. Valuev, "Subpixel smoothing for conductive and dispersive media in the finite-difference time-domain method," *Optics Letters*, vol. 32, no. 23, pp. 3429–3431, 2007, doi:10.1364/ol.32.003429.

[85] I. R. Capoglu and G. S. Smith, "A total-field/scattered-field plane-wave source for the FDTD analysis of layered media," *IEEE Transactions on Antennas and Propagation*, vol. 56, no. 1, pp. 158–169, 2008, doi:10.1109/tap.2007.913088.

[86] J. B. Schneider, "Plane waves in FDTD simulations and a nearly perfect total-field/scattered-field boundary," *IEEE Transactions on Antennas and Propagation*, vol. 52, no. 12, pp. 3280–3287, 2004, doi:10.1109/tap.2004.836403.

[87] T. M. Tan and M. Potter, "FDTD discrete planewave (FDTD-DPW) formulation for a perfectly matched source in TFSF simulations," *IEEE Transactions on Antennas and Propagation*, vol. 58, no. 8, pp. 2641–2648, 2010, doi:10.1109/tap.2010.2050446.

[88] X. Li, A. Taflove, and V. Backman, "Modified FDTD near-to-far-field transformation for improved backscattering calculation of strongly forward-scattering objects," *IEEE Antennas and Wireless Propagation Letters*, vol. 4, pp. 35–38, 2005, doi: 10.1109/lawp.2005.845038.

[89] R. J. Luebbers, K. S. Kunz, M. Schneider, and F. Hunsberger, "A finite-difference time-domain near zone to far zone transformation," *IEEE Transactions on Antennas and Propagation*, vol. 39, no. 4, pp. 429–433, 1991, doi:10.1109/8.81453.

[90] D. J. Robinson and J. B. Schneider, "On the use of the geometric mean in FDTD near-to-far-field transformations," *IEEE Transactions on Antennas and Propagation*, vol. 55, no. 11, pp. 3204–3211, 2007, doi:10.1109/tap.2007.908795.

[91] K. Umashankar and A. Taflove, "A novel method to analyze electromagnetic scattering of complex objects," *IEEE Transactions on Electromagnetic Compatibility*, vol. 24, no. 4, pp. 397–405, 1982, doi: 10.1109/temc.1982.304054.

[92] S. H. Chang and A. Taflove, "Finite-difference time-domain model of lasing action in a four-level two-electron atomic system," *Optics Express*, vol. 12, no. 16, pp. 3827–3833, 2004, doi:10.1364/opex.12.003827.

[93] G. M. Slavcheva, J. M. Arnold, and R. W. Ziolkowski, "FDTD simulation of the nonlinear gain dynamics in active optical waveguides and semiconductor microcavities," *IEEE Journal of Selected Topics in Quantum Electronics*, vol. 10, no. 5, pp. 1052–1062, 2004, doi: 10.1109/jstqe.2004.836023.

[94] S. D. Gedney, "An anisotropic PML absorbing media for the FDTD simulation of fields in lossy and dispersive media," *Electromagnetics*, vol. 16, no. 4, pp. 399–415, 1996.

[95] R. J. Luebbers and F. Hunsberger, "FDTD for Nth-order dispersive media," *IEEE Transactions on Antennas and Propagation*, vol. 40, no. 11, pp. 1297–1301, 1992.

[96] F. L. Teixeira, "Time-domain finite-difference and finite-element methods for Maxwell equations in complex media," *IEEE Transactions on Antennas and Propagation*, vol. 56, no. 8, pp. 2150–2166, 2008, doi:10.1109/tap.2008.926767.

[97] J. A. Roden, S. D. Gedney, M. P. Kesler, J. G. Maloney, and P. H. Harms, "Time-domain analysis of periodic structures at oblique incidence: orthogonal and nonorthogonal FDTD implementations," *IEEE Transactions on Microwave Theory and Techniques*, vol. 46, no. 4, pp. 420–427, 1998.

[98] K. ElMahgoub, F. Yang, A. Z. Elsherbeni, V. Demir, and J. Chen, "FDTD analysis of periodic structures with arbitrary skewed grid," *IEEE Transactions on Antennas and Propagation*, vol. 58, no. 8, pp. 2649–2657, 2010, doi:10.1109/tap.2010.2050445.

[99] B. Liang, M. Bai, H. Ma, N. M. Ou, and J. G. Miao, "Wideband analysis of periodic structures at oblique incidence by material independent FDTD algorithm," *IEEE Transactions on Antennas and Propagation*, vol. 62, no. 1, pp. 354–360, 2014, doi:10.1109/tap.2013.2287896.

[100] M. E. Veysoglu, R. T. Shin, and J. A. Kong, "A finite-difference time-domain analysis of wave scattering from periodic surfaces – oblique-incidence case," *Journal of Electromagnetic Waves and Applications*, vol. 7, no. 12, pp. 1595–1607, 1993, doi: 10.1163/156939393x00020.

[101] J. B. Wang, B. H. Zhou, B. Chen, and L. H. Shi, "Unconditionally stable FDTD method for solving oblique incident plane wave on periodic structures," *IEEE Microwave and Wireless Components Letters*, vol. 21, no. 12, pp. 637–639, 2011, doi:10.1109/lmwc.2011.2173181.

[102] F. Yang, J. Chen, R. Qiang, and A. Elsherbeni, "A simple and efficient FDTD/PBC algorithm for scattering analysis of periodic structures," *Radio Science*, vol. 42, no. 4, 2007, Art no. Rs4004, doi:10.1029/2006rs003526.

[103] T. Weiland, "A discretization method for the solution of Maxwell's equations for six-component fields," *Electronics and Communications, AEU*, vol. 31, no. 3, pp. 116–120, 1977.

Chapter 3

Finite-element methods in time domain with applications to low-frequency problems

Iryna Kulchytska-Ruchka[1], Idoia Cortes Garcia[2], Sebastian Schöps[1] and Markus Clemens[3]

The electromagnetic field theory is described by a set of partial differential equations (PDEs) known as the Maxwell equations [1,2]. They are given by

$$\mathrm{curl}\vec{E} = -\frac{\partial \vec{B}}{\partial t} \quad \text{(Faraday's law)}, \tag{3.1a}$$

$$\mathrm{curl}\vec{H} = \frac{\partial \vec{D}}{\partial t} + \vec{J} \quad \text{(Maxwell–Ampère's law)}, \tag{3.1b}$$

$$\mathrm{div}\vec{D} = \rho \quad \text{(electric Gauss' law)}, \tag{3.1c}$$

$$\mathrm{div}\vec{B} = 0 \quad \text{(magnetic Gauss' law)} \tag{3.1d}$$

with the *magnetic flux density* being $\vec{B}$, the *magnetic field strength* being $\vec{H}$, the *electric flux density* being $\vec{D}$, the *electric field strength* being $\vec{E}$, the *electric current density* being $\vec{J}$, and the *electric charge density* ρ. These quantities depend on the space variable $\vec{x} \in \Omega \subset \mathbb{R}^3$ and the time variable $t \in [0, T] \subset \mathbb{R}$, with $T > 0$. The vector fields are related through the constitutive relations [3]

$$\vec{D} = \varepsilon\vec{E} + \vec{P}, \qquad \vec{H} = \nu\vec{B} + \vec{M}, \qquad \vec{J} = \sigma\vec{E} + \vec{J}_{\mathrm{s}}. \tag{3.2}$$

The *electric permitivity* ε, the *magnetic reluctivity* ν, and the *electric conductivity* σ are in general rank-2 tensors; however, we assume them to be scalars, as we only consider isotropic materials. Besides, we assume the dependencies

$$\varepsilon = \varepsilon(\vec{x}, |\vec{E}|), \qquad \nu = \nu(\vec{x}, |\vec{B}|), \qquad \sigma = \sigma(\vec{x}, |\vec{E}|). \tag{3.3}$$

[1]Chair of Computational Electromagnetics, Technical University of Darmstadt, Germany
[2]Dynamics and Control Group, Eindhoven University of Technology, Netherlands
[3]Chair of Electromagnetic Theory, University of Wuppertal, Germany

The vector fields $\vec{P}$ and $\vec{M}$ are the *permanent electric polarization* and the *permanent magnetization*, which will be neglected in this chapter but can be easily incorporated as additional source terms. The vector field $\vec{J}_s$ is the *source current density* which commonly stems from an external current or voltage supply and can be, for example, expressed via

$$\vec{J}_s = \sum_{k=1}^{n_s} \vec{\chi}_k i_k \tag{3.4}$$

using the winding functions $\vec{\chi}_k(\vec{x}) \in \mathbb{R}^3$ [4] which spatially distribute the currents $i_k(t) \in \mathbb{R}$ flowing through n_s stranded conductors (coils).

When modeling electromagnetic fields in a finite computational domain Ω conditions on the boundary $\Gamma = \partial\Omega$ are imposed to define the fields uniquely and to model their behaviour at infinity. The most common types are the *electric boundary condition* (*EBC*) and the *magnetic boundary condition* (*MBC*) given by

$$\vec{n} \times \vec{E} = 0 \quad \text{on } \Gamma_{\mathrm{EBC}} \quad \text{and} \quad \vec{n} \times \vec{H} = 0 \quad \text{on } \Gamma_{\mathrm{MBC}}, \tag{3.5}$$

respectively [5, Section 1.3]. The vector $\vec{n}$ is the outward normal to the boundary $\Gamma = \Gamma_{\mathrm{EBC}} \cup \Gamma_{\mathrm{MBC}}$ where $\Gamma_{\mathrm{EBC}} \cap \Gamma_{\mathrm{MBC}} = \emptyset$. Other common boundary conditions (BCs) are periodic, anti-periodic, and open BCs. Besides, throughout this chapter we assume $\Omega \subset \mathbb{R}^3$ is open, bounded, Lipschitz, and contractible.

In general, electromagnetism includes phenomena attributed to electric and magnetic energies, as well as to Joule losses. However, for specific applications, some of these effects are negligible compared to the remaining ones. In those cases, simplifications of Maxwell's equations are considered [6], whose solution is easier compared to the full PDE set. Once an appropriate subset of PDEs is selected, a particular formulation is obtained through the choice of field quantities such as potentials and gauges. In this chapter, we focus on two settings: the magnetoquasistatic (MQS) and the electroquasistatic (EQS) ones, which neglect the electric and the magnetic effects, respectively. Another approximation used in the quasistatic regime, when inductive and capacitive effects are relevant, is the Darwin model [7].

This chapter is organised as follows. Section 3.1 provides several potential formulations, which are then discretised in space using the finite-element method (FEM) in Section 3.2. There, two problem settings in time domain are posed: an initial-value problem (IVP) and a time-periodic (TP) problem. The theory on implicit time-stepping schemes for IVPs is described and applied to the space-discrete system in Section 3.3. Section 3.4 deals with TP problems and presents a diagonalization approach for their efficient solution. A class of the parallel-in-time (PinT) methods described in Section 3.5 is used for a faster solution of IVPs and TP problems. The chapter is finalized with several examples from electrical engineering provided in Section 3.6 and conclusions in Section 3.7.

3.1 Quasistatic field formulations

Within the MQS setting, the electric flux density $\vec{D}$ is assumed to be independent from the time variable t, i.e.,

$$\frac{\partial \vec{D}}{\partial t} = 0. \tag{3.6}$$

Plugging (3.6) into the Maxwell–Ampère law (3.1b), we have

$$\mathrm{curl}\vec{H} = \vec{J}, \tag{3.7}$$

which together with the Faraday law (3.1a) and the magnetic Gauss law (3.4) describes the MQS approximation.

Within the EQS setting, the magnetic flux density $\vec{B}$ is assumed to be independent from the time variable t, i.e.,

$$\frac{\partial \vec{B}}{\partial t} = 0. \tag{3.8}$$

Plugging (3.8) into the Faraday law (3.1a), we have

$$\mathrm{curl}\vec{E} = 0, \tag{3.9}$$

which together with the Maxwell–Ampère law (3.1b) describes the EQS approximation.

3.1.1 *MQS: $\vec{A}$ - φ-formulation*

From the magnetic Gauss law (3.1d) and the Faraday law (3.1a), we conclude the existence of the *magnetic vector potential (MVP)* $\vec{A}$ and the *electric scalar potential* φ [8] such that

$$\vec{B} = \mathrm{curl}\vec{A}, \qquad \vec{E} = -\frac{\partial \vec{A}}{\partial t} - \mathrm{grad}\varphi, \tag{3.10}$$

since the divergence of the curl operator and the curl of the gradient operator are zero. The representations (3.10) and the constitutive relations (3.2) applied to the Maxwell–Ampère law for MQS (3.7) lead to the so called $\vec{A}$ - φ-formulation

$$\mathrm{curl}(\nu\mathrm{curl}\vec{A}) + \sigma\frac{\partial \vec{A}}{\partial t} + \sigma\mathrm{grad}\varphi = \vec{J}_{\mathrm{s}}. \tag{3.11}$$

To determine $\vec{A}$ and φ uniquely, we may include a *gauging condition*, for example the Coulomb gauge

$$\mathrm{div}\vec{A} = 0. \tag{3.12}$$

3.1.2 MQS: $\vec{A}^*$-formulation

To reduce the number of degrees of freedom in (3.11), we introduce

$$\vec{A}^* = \vec{A} + \int_0^t \mathrm{grad}\varphi \mathrm{d}s, \quad t \in [0, T], \quad (3.13)$$

and obtain the $\vec{A}^*$-formulation

$$\mathrm{curl}(\nu \mathrm{curl}\vec{A}^*) + \sigma \frac{\partial \vec{A}^*}{\partial t} = \vec{J}_{\mathrm{s}}. \tag{3.14}$$

This equation includes *eddy currents*, which lead to heat generation [5, Section 1.2, 9]. Heating of magnetic materials commonly deteriorates the performance of devices, as, e.g., electric machines. Therefore, it is essential to include the calculation of eddy currents and the corresponding heat and magnetic losses into the simulation of many electromagnetic devices. The BCs (3.5) are given in terms of the MVP by

$$\vec{n} \times \vec{A}^* = 0 \quad \text{on } \Gamma_{\mathrm{EBC}} \quad \text{and} \quad \vec{n} \times (\nu \mathrm{curl}\vec{A}^*) = 0 \quad \text{on } \Gamma_{\mathrm{MBC}}. \tag{3.15}$$

In the following, we will abuse the notations and use the symbol $\vec{A}$ for the MVP in the $\vec{A}^*$-formulation. In Section 3.2.2, we derive the weak formulation of (3.14) and (3.15) which is the first step toward the finite-element discretization in space.

3.1.3 EQS: φ-formulation

The Faraday law for EQS (3.9) and application of the divergence operator to the Maxwell-Ampère law (3.1b) yield

$$\vec{E} = -\mathrm{grad}\varphi, \qquad \mathrm{div}\frac{\partial \vec{D}}{\partial t} + \mathrm{div}\vec{J} = 0 \tag{3.16}$$

since (for sufficiently smooth field on a contractible domain) the curl of any gradient field and the divergence of any curl vanish. Here, φ is the electric scalar potential already introduced in Section 3.1.1. Using the material relations (3.2) and combining the two equations in (3.16), we obtain

$$\mathrm{div}\left(\varepsilon \frac{\partial}{\partial t}(\mathrm{grad}\varphi)\right) + \mathrm{div}(\sigma \mathrm{grad}\varphi) = 0 \tag{3.17}$$

under the assumption that $\mathrm{div}\vec{J}_{\mathrm{s}} = 0$. The BCs for φ are given by

$$\varphi = 0 \quad \text{on } \Gamma_{\mathrm{EBC}} \quad \text{and} \quad \mathrm{grad}\varphi \cdot \vec{n} = 0 \quad \text{on } \Gamma_{\mathrm{MBC}}. \tag{3.18}$$

The weak formulation in an appropriate function space necessary for the space discretization using the finite elements is derived in Section 3.2.3.

3.2 Space discretization using finite elements

FEM is a numerical discretization technique based on the weak solution of a considered PDE. It splits the computational domain into *elements* and approximates the weak solution by a linear combination of *shape functions* introduced on each element [10]. In this section, we first present the appropriate function spaces and their discrete counterparts which contain the sought solution. The weak formulations for EQS and MQS are then derived and approximated using the Ritz–Galerkin approach, which finally leads to space-discrete systems of differential and differential–algebraic equations (DAEs).

3.2.1 Function spaces

Let the spatial domain $\Omega \subset \mathbb{R}^3$ be open, bounded, Lipschitz, and contractible. We define the Hilbert spaces

$$H(\text{grad}; \Omega) = \{u \in L^2(\Omega) |\ \text{grad}u \in \mathbf{L}^2(\Omega)\}, \tag{3.19}$$

$$\mathbf{H}(\text{curl}; \Omega) = \{\vec{v} \in \mathbf{L}^2(\Omega) |\ \text{curl}\vec{v} \in \mathbf{L}^2(\Omega)\}, \tag{3.20}$$

$$\mathbf{H}(\text{div}; \Omega) = \{\vec{w} \in \mathbf{L}^2(\Omega) |\ \text{div}\vec{w} \in L^2(\Omega)\}, \tag{3.21}$$

where the grad, div, and curl operators are to be understood in the weak sense [11, Section 5.1.1], $L^2(\Omega)$ is the Lebesgue space of square integrable measurable functions [12, Chapter 2], and $\mathbf{L}^2(\Omega) = (L^2(\Omega))^3$ is its extension for vector fields. The Hilbert spaces (3.19)–(3.21) together with the grad, div, and curl operators define the de Rham sequence [11, Section 5.1.4]

$$\mathbb{R} \to H(\text{grad}; \Omega) \stackrel{grad}{} \mathbf{H}(\text{curl}; \Omega) \stackrel{curl}{} \mathbf{H}(\text{div}; \Omega) \stackrel{div}{} L^2(\Omega) \to \mathbb{R}. \tag{3.22}$$

When introducing a mesh on the domain Ω, e.g., by splitting it into N_v volumes (e.g., tetraedra), finite dimensional counterparts of the Hilbert spaces from (3.19) to (3.21) have to be considered. In particular, let $U_h \subseteq H(\text{grad}; \Omega)$ given by the N_n node basis functions u_i, $\mathbf{V}_h \subseteq \mathbf{H}(\text{curl}; \Omega)$ given by the N_e edge basis functions $\vec{v}_i$, and $\mathbf{W}_h \subseteq \mathbf{H}(\text{div}; \Omega)$ given by the N_f face basis functions $\vec{w}_i$ be the discrete analogues to the spaces in (3.19)–(3.21). Besides, let $S_h \subseteq L^2(\Omega)$ be the discrete variant of $L^2(\Omega)$. The subscript h is a mesh parameter denoting, e.g., the maximal size of the mesh elements. These finite dimensional spaces also form a de Rham sequence

$$\mathbb{R} \to U_h \stackrel{grad}{} \mathbf{V}_h \stackrel{curl}{} \mathbf{W}_h \stackrel{div}{} S_h \to \mathbb{R}, \tag{3.23}$$

also called the Whitney complex [11, Section 5.2]. We will derive the weak variant of the $\vec{A}^*$-formulation and approximate its weak solution by the edge basis functions from the space $\mathbf{V}_h \subseteq \mathbf{H}(\text{curl}; \Omega)$ in the following Section 3.2.2. On the other hand, the weak φ-formulation is discretised using the nodal basis functions from $U_h \subseteq H(\text{grad}; \Omega)$, as described in Section 3.2.3.

3.2.2 Weak formulation for MQS

Multiplying the $\vec{A}^*$-formulation (3.14) by a *test function* $\vec{v} \in \mathbf{H}(\mathrm{curl};\Omega)$ and integrating over the domain Ω, we obtain

$$\int_\Omega \sigma \frac{\partial \vec{A}}{\partial t} \cdot \vec{v}\,\mathrm{d}\Omega + \int_\Omega \mathrm{curl}(\nu \mathrm{curl}\vec{A}) \cdot \vec{v}\,\mathrm{d}\Omega = \int_\Omega \vec{J}_\mathrm{s} \cdot \vec{v}\,\mathrm{d}\Omega. \tag{3.24}$$

Application of a vector identity [11, Section 2.3.2] and the divergence theorem (also called the Gauss theorem), i.e., for arbitraty vector fields $\vec{F}$ and $\vec{G}$

$$\mathrm{div}(\vec{F} \times G) = (\mathrm{curl}\vec{F}) \cdot \vec{G} - \vec{F} \cdot (\mathrm{curl}\vec{G}) \quad \text{and} \quad \int_\Omega \mathrm{div}\vec{F}\,\mathrm{d}\Omega = \int_\Gamma \vec{F} \cdot \mathrm{d}\Gamma, \tag{3.25}$$

leads to the weak formulation for $\vec{A} \in \mathbf{H}(\mathrm{curl};\Omega)$

$$\int_\Omega \sigma \frac{\partial \vec{A}}{\partial t} \cdot \vec{v}\,\mathrm{d}\Omega + \int_\Omega \nu \mathrm{curl}\vec{A} \cdot \mathrm{curl}\vec{v}\,\mathrm{d}\Omega + \int_\Gamma (\nu \mathrm{curl}\vec{A} \times \vec{v}) \cdot \mathrm{d}\Gamma = \int_\Omega \vec{J}_\mathrm{s} \cdot \vec{v}\,\mathrm{d}\Omega. \tag{3.26}$$

Under the assumption that the MVP $\vec{A}$ and the test function $\vec{v}$ satisfy the BCs (3.15) the boundary integral in (3.26) vanishes, since

$$\begin{aligned}\int_\Gamma (\nu \mathrm{curl}\vec{A} \times \vec{v}) \cdot \mathrm{d}\Gamma &= \int_{\Gamma_\mathrm{EBC}} (\nu \mathrm{curl}\vec{A} \times \vec{v}) \cdot \mathrm{d}\Gamma + \int_{\Gamma_\mathrm{MBC}} (\nu \mathrm{curl}\vec{A} \times \vec{v}) \cdot \mathrm{d}\Gamma \\ &= \int_{\Gamma_\mathrm{EBC}} (\vec{v} \times \vec{n}) \cdot (\nu \mathrm{curl}\vec{A})\,\mathrm{d}\Gamma + \int_{\Gamma_\mathrm{MBC}} (\vec{n} \times (\nu \mathrm{curl}\vec{A})) \cdot \vec{v}\,\mathrm{d}\Gamma = 0.\end{aligned}$$

In order to find the solution $\vec{A}$, we represent it by the linear combination of *trial functions* $\vec{v}_i \in \mathbf{H}(\mathrm{curl};\Omega)$, i.e.,

$$\vec{A}(t,\vec{x}) = \sum_j a_j(t)\vec{v}_j(\vec{x}). \tag{3.27}$$

Plugging (3.27) into (3.26), restricting the solution to a finite-dimensional subspace $\mathbf{V}_h \subset \mathbf{H}(\mathrm{curl};\Omega)$, i.e., $\vec{v}_j \in \mathbf{V}_h$, $j = 1,\ldots,N_\mathrm{e}$, and choosing the test function $\vec{v} = \vec{v}_i \in \mathbf{V}_h$, $i = 1,\ldots,N_\mathrm{e}$, gives

$$\mathbf{M}_\sigma \frac{\mathrm{d}}{\mathrm{d}t}\mathbf{a}(t) + \mathbf{K}_\nu(\mathbf{a}(t))\mathbf{a}(t) = \mathbf{f}(t), \tag{3.28}$$

where the notations $\mathbf{a}(t) = \left[a_1(t),\ldots,a_{N_\mathrm{e}}(t)\right]^\mathrm{T} \in \mathbb{R}^{N_\mathrm{e}}$ and

$$2(\mathbf{M}_\sigma)_{ij} = \int_\Omega \sigma \vec{v}_j \cdot \vec{v}_i\,\mathrm{d}\Omega, \qquad (\mathbf{f}(t))_i = \int_\Omega \vec{J}_\mathrm{s} \cdot \vec{v}_i\,\mathrm{d}\Omega, \tag{3.29}$$

$$(\mathbf{K}_\nu(\mathbf{a}))_{ij} = \int_\Omega \nu \mathrm{curl}\vec{v}_j \cdot \mathrm{curl}\vec{v}_i\,\mathrm{d}\Omega, \qquad \nu = \nu(\vec{x}, |\sum_{k=1}^{N_\mathrm{e}} a_k \mathrm{curl}\vec{v}_k|), \tag{3.30}$$

with $\sigma = \sigma(\vec{x})$, $\vec{x} \in \Omega$, and $i,j = 1,\ldots,N_e$, are used. We see that (3.28) depends only on the time variable t and is called the semi-discrete or space-discrete eddy current equation. The fully discrete system is obtained by the numerical time-integration methods described in Section 3.3.

When the domain Ω contains only conductive materials (i.e., $\sigma > 0$ in Ω) the matrix $\mathbf{M}_\sigma$ from (3.29) is regular (i.e., $\det\mathbf{M}_\sigma \neq 0$) and (3.28) is a system of ordinary differential equations (ODEs). If there are non-conductive parts in Ω (i.e., $\sigma = 0$ in some parts of Ω), the matrix $\mathbf{M}_\sigma$ is singular (i.e., $\det\mathbf{M}_\sigma = 0$). In this case, (3.28) is a system of DAEs [13, Section VII.1] whose numerical solution might require a special treatment. To measure "how far away" a DAE system is from an ODE system and the possible numerical (or analytical) difficulties, it can convey, the notion of a *differentiation index* is used [13, Definition VII.1.2]. In this chapter, we will only deal with the DAEs of low index, i.e., index-1 DAEs. We note that ODEs can be classified as DAEs of index 0.

Finally, since there is the nonlinearity ν present in the eddy current formulation (3.28), a linearization method such as, e.g., the Newton method [14], has to be applied. For such algorithms where a Jacobian calculation is required, the *differential reluctivity* tensor (see [15]) given by

$$\boldsymbol{\nu}_{\mathrm{d}}(\vec{x},\vec{B}) = \nu(\vec{x},B)\mathbf{I} + 2\vec{B}\frac{\partial\nu(\vec{x},B)}{\partial B^2}\vec{B}^{\mathrm{T}} = \nu(\vec{x},B)\mathbf{I} + \frac{1}{B}\vec{B}\frac{\partial\nu(\vec{x},B)}{\partial B}\vec{B}^{\mathrm{T}} \tag{3.31}$$

is used, where $\mathbf{I}$ denotes the indentity (rank-2) tensor, $\vec{x} \in \Omega$, $B = |\vec{B}|$, and $\vec{B} = \mathrm{curl}\vec{A}$.

3.2.3 Weak formulation for EQS

The transient FEM solution was introduced within the EQS setting in [16]. To obtain the weak formulation for (3.17), we multiply it by a test function $u \in H(\mathrm{grad};\,\Omega)$ and integrate over the domain Ω, which yields

$$\int_\Omega div\left(\varepsilon\frac{\partial}{\partial t}(\mathrm{grad}\varphi)\right)u\,\mathrm{d}\Omega + \int_\Omega div(\sigma\mathrm{grad}\varphi)u\,\mathrm{d}\Omega = 0. \tag{3.32}$$

Application of the Green formula [17, Section 1.3]

$$\int_\Omega \mathrm{div}\vec{w}\,u\,\mathrm{d}\Omega = -\int_\Omega \vec{w}\cdot\mathrm{grad}u\,\mathrm{d}\Omega + \int_\Gamma(\vec{w}\cdot\vec{n})u\,\mathrm{d}\Gamma \tag{3.33}$$

leads to the weak formulation for $\varphi \in H(\mathrm{grad};\,\Omega)$

$$\begin{aligned}&\int_\Omega\left(\varepsilon\frac{\partial}{\partial t}(\mathrm{grad}\varphi)\right)\cdot\mathrm{grad}u\,\mathrm{d}\Omega + \int_\Omega \sigma\mathrm{grad}\varphi\cdot\mathrm{grad}u\,\mathrm{d}\Omega\\ &= \int_\Gamma\left(\left[\varepsilon\frac{\partial}{\partial t}(\mathrm{grad}\varphi) + \sigma\mathrm{grad}\varphi\right]\cdot\vec{n}\right)u\,\mathrm{d}\Gamma.\end{aligned} \tag{3.34}$$

Due to the BCs (3.18), the right-hand side (RHS) in (3.34) vanishes. For the representation of the solution φ, we exploit the trial functions $u_i \in H(\text{grad};\,\Omega)$, i.e.,

$$\varphi(t,\vec{x}) = \sum_j \phi_j(t)u_j(\vec{x}). \tag{3.35}$$

Plugging (3.35) into (3.34), restricting the solution to a finite-dimensional subspace $U_h \subset H(\text{grad};\,\Omega)$, i.e., $u_j \in U_h$, $j = 1,\ldots,N_\text{n}$, and choosing the test function $u = u_i \in U_h$, $i = 1,\ldots,N_\text{n}$, we derive the system of ODEs of the form (3.28), i.e.,

$$\mathbf{M}_\varepsilon \frac{\mathrm{d}}{\mathrm{d}t}\varphi(t) + \mathbf{K}_\sigma(\varphi(t))\varphi(t) = \mathbf{0}, \tag{3.36}$$

with the unknown vector $\mathbb{R}(t) = \left[\phi_1(t),\ldots,\phi_{N_\text{n}}(t)\right]^\text{T} \in \mathbb{R}^{N_\text{n}}$ and the matrices $\mathbf{M}_\varepsilon$ and $\mathbf{K}_\sigma$ given by

$$(\mathbf{M}_\varepsilon)_{ij} = \int_\Omega \varepsilon \text{grad} u_j \cdot \text{grad} u_i \,\mathrm{d}\Omega, \qquad \varepsilon = \varepsilon(\vec{x}), \tag{3.37}$$

$$(\mathbf{K}_\sigma(\varphi))_{ij} = \int_\Omega \sigma \text{grad} u_j \cdot \text{grad} u_i \,\mathrm{d}\Omega, \qquad \sigma = \sigma(\vec{x}, |\sum_{k=1}^{N_\text{n}} \phi_k \text{grad} u_k|), \tag{3.38}$$

with $\vec{x} \in \Omega$. The matrix $\mathbf{K}_\sigma$ is singular when non-conductive regions are present in the domain Ω. Nevertheless, the system (3.36) stays a system of ODEs also in this case, in contrast to the DAE system (3.28). In Section 3.2.4, we focus on two common time-domain formulations for the space-discrete eddy current equation (3.28). Nonetheless, this can be similarly formulated for the space-discrete EQS formulation (3.36).

3.2.4 *Initial-value and steady-state problems*

The input $\vec{J}_\text{s}$ of the eddy current system (3.14) modeling, e.g., an induction motor (IM), is often given by sinusoidal currents (also called *alternating currents* (AC))

$$i_k(t) = \hat{I} sin(2\pi f t - \varphi_k), \quad k = 1,\ldots,n_\text{s}, \tag{3.39}$$

with the amplitude $\hat{I}$, the frequency f, and the phase φ_k in (3.4). Solution of such an evolution problem excited with such periodic signals consists of a transient part, followed by a (periodic) steady state taking place once the transients are eventually damped out. Figure 3.1 illustrates the torque evolution of an induction motor used in an electric vehicle drive [18]. There, the steady state is reached at time instant $T_\text{st} = 0.26$.

For a unique solvability of the semi-discrete system (3.28) on a given time interval $(0, T]$, $T > 0$ an initial condition (IC)

$$\mathbf{a}(0) = \mathbf{a}_0 \tag{3.40}$$

has to be prescribed. Since (3.28) is commonly a DAE system, the chosen IC has to be consistent, i.e., such that the algebraic constraints in the system are satisfied

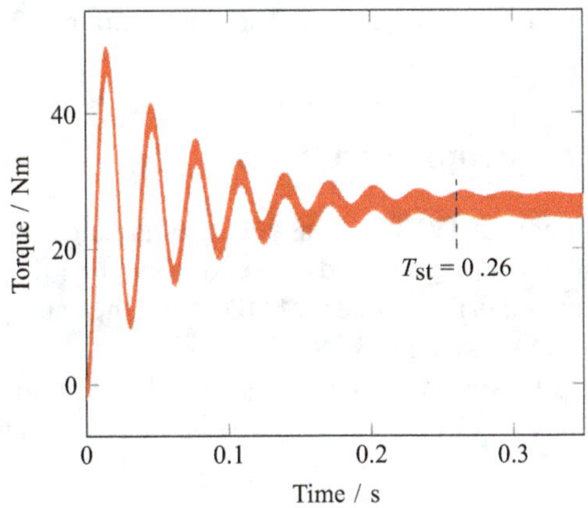

Figure 3.1 Evolution of an induction motor [18] © 2019 IEEE

[13, Section VII.1]. The classical approach to solve the IVP (3.28) and (3.40) is the numerical time stepping starting from the prescribed IC as it will be described in Section 3.3.

The steady-state solution can be obtained by solving an IVP numerically until the transient phenomenon fades away. However, the time stepping until the steady state is reached might require a long computational time. For example, to obtain the steady-state solution of the system depicted in Figure 3.1 one needs to solve 56,160 linear systems [18]. To speed up such calculations, various approaches have been applied such as the choice of a "good" IC [19] or the explicit error correction (EEC) [20]. They show a considerable reduction of the computational time for the electric motors simulations, but have several disadvantages. In particular, the algorithm in [19] is quite technical and developed for a specific motor type, i.e., a squirrel-cage IM. Similarly, the EEC method functions well only for particular operating points as shown in [18]. Such limitations can be overcome by using PinT methods, which are derived without being adjusted to a particular model at hand, as we describe in Section 3.5. There, we deal with one on such PinT algorithms called Parareal [21]. For systems with discontinuous inputs, e.g., modeled with pulse-width modulation (PWM) technique [22], a modification of the original Parareal algorithm was developed in Refs. [23,24], see Section 3.5.2.

Alternatively, the steady-state solution can be obtained by solving the TP problem, i.e., by searching for the solution to (3.28) on $(0, T)$, $T > 0$ which satisfies the TP condition

$$\mathbf{a}(0) = \mathbf{a}(T). \tag{3.41}$$

TP problems couple the initial and the end solutions and might lead to a higher computational effort compared to the solution of IVPs (see Section 3.4). In Section 3.4.1, we describe an algorithm which in the linear case allows to reduce the computational cost by diagonalization of the resulting block-cyclic system [25].

This approach can be also combined with the PinT solution [26], as presented in Section 3.5.3.

3.3 Implicit time-stepping methods

For a numerical solution of IVPs such as the ones arising from systems such as (3.28) and (3.40), time-stepping methods have to be applied [27]. We first provide the theoretical fundamentals for a system of ODEs and then apply them to the semi-discrete eddy current DAE system (3.28).

Consider an IVP for an ODE system with an unknown function $\mathbf{u} : [0, T] \rightarrow \mathbb{R}^d$ and the RHS $\mathbf{f} : \mathbb{R}^d \times [0, T] \rightarrow \mathbb{R}^d$, $d \geq 1$, such that

$$\frac{\mathrm{d}}{\mathrm{d}t}\mathbf{u}(t) = \mathbf{f}(\mathbf{u}(t), t), \quad t \in (0, T] \tag{3.42}$$

$$\mathbf{u}(0) = \mathbf{u}_0. \tag{3.43}$$

The idea of time discretization lies in the introduction of the partition of the time interval $[0, T]$

$$0 = t_0 < t_1 < \ldots < t_N = T \tag{3.44}$$

and approximation of the solution at $t = t_n$, i.e.,

$$\mathbf{u}_n \approx \mathbf{u}(t_n), \quad n = 1, \ldots, N. \tag{3.45}$$

The difference $l_n = \mathbf{u}(t_n) - \mathbf{u}_n$ is called the *local truncation error* (LTE) [28,27]. In the case of an equidistant partition of $[0, T]$ with the time step size $\Delta T = T/N$ and

$$t_n = t_{n-1} + \Delta T, \quad n = 1, \ldots, N, \tag{3.46}$$

the number $p \geq 1$ such that $l_n = \mathrm{O}(\Delta T^{p+1})$ is called the *order* of a numerical time-integration method.

There are several classifications of the numerical time integrators, e.g., one-step and multi-step, explicit and implicit, etc. In particular, a general formula of a *linear multi-step (k-step) method* [27, Chapter III] applied to (3.42) and (3.43) reads

$$\alpha_0 \mathbf{u}_n + \alpha_1 \mathbf{u}_{n+1} + \cdots + \alpha_k \mathbf{u}_{n+k} = \Delta T(\beta_0 \mathbf{f}_n + \beta_1 \mathbf{f}_{n+1} + \cdots + \beta_k \mathbf{f}_{n+k}), \tag{3.47}$$

where $\alpha_i, \beta_i \in \mathbb{R}$, $\mathbf{f}_{n+i} = \mathbf{f}(\mathbf{u}_{n+i}, t_{n+i})$, $i = 0, \ldots, k$, and $n = 0, \ldots, N - k$ for $k \geq 1$. We assume $\alpha_i \neq 0$ and $|\alpha_0| + |\beta_0| > 0$ [27, Section III.2]. For $k = 1$, we have a *one-step method.* If $\beta_k = 0$ the scheme (3.47) is *explicit*, while for $\beta_k \neq 0$, it is *implicit*.

In this section, we focus on implicit methods, since they are more suitable for *stiff* problems [13]. Although there is no rigorous definition of stiffness, we consider an ODE to be stiff if an applied explicit numerical scheme has stability problems, which limit time step size. Note that DAEs are sometimes called infinitely stiff. The defintions of stability are presented in Section 3.3.1.

3.3.1 Stability

Stability of a numerical time integrator is necessary to obtain a correct solution to a problem at hand. To discuss the stability properties of a method, we consider the Dahlquist test equation [13]

$$\frac{\mathrm{d}}{\mathrm{d}t}u(t) = \lambda u(t), \quad \lambda \in \mathbb{C}. \tag{3.48}$$

Assume a time-integration scheme applied to the ODE (3.48) on the equidistant grid (3.46) is given by

$$u_n = R(z)u_{n-1}, \quad z = \lambda \Delta T, \tag{3.49}$$

where $R(z)$ is the *stability function* of the numerical method. The set

$$S = \{z \in \mathbb{C} : \quad |R(z)| \leq 1\} \tag{3.50}$$

is called the *stability domain* [13, Section IV.2] of the method (3.49). We now name a couple of stability classifications in the following definition.

Definition 3.1 [13, Section IV.3] A time-integration scheme is called

- ***A-stable*** *if the stability domain S from (3.50) contains the entire complex left half-plane, i.e.,*

$$S \supseteq \mathbb{C}^- = \{z \in \mathbb{C} : \quad \mathrm{Re}(z) \leq 0\}.$$

- ***L-stable*** *if it is A-stable and*

$$\lim_{z \to \infty} R(z) = 0. \tag{3.51}$$

A-stability of a numerical method is very desirable for the solution of stiff problems. However, for A-stable methods, whose stability regions coincide with the negative half-plane, i.e., $S = \mathbb{C}^-$, the value of $|R(z)|$ can become very close to one [13]. In this case, stiff components of the numerical solution are damped out very slowly. L-stability ensures that $|R(z)| \ll 1$ for $z \to -\infty$, as one may expect from the solution.

As an example, we illustrate the stability properties of the two well-known one-step implicit time integrators: the implicit Euler (IE) method and the trapezoidal scheme [27, Section II.7]. The two methods applied to the Dahlquist equation (3.48) are given by (3.49) with the stability functions

$$R_1(z) = \frac{1}{1-z} \quad \text{and} \quad R_2(z) = \frac{1+0.5z}{1-0.5z}, \tag{3.52}$$

respectively. It is known that both the IE method and the trapezoidal rule are A-stable [13]. The IE method is also L-stable since for R_1 from (3.52), the condition (3.51) clearly holds. The trapezoidal rule is however not L-stable since (3.51) is not fulfilled for R_2. Further stability concepts can be found in [29, Section 6.1.3]. We now proceed to the class of the Runge–Kutta methods.

3.3.2 Runge–Kutta methods

On the equidistant partition (3.46), an *s-stage Runge–Kutta* (RK) method for (3.42) and (3.43) is given by the formula

$$\mathbf{u}_n = \mathbf{u}_{n-1} + \sum_{i=1}^{s} b_i k_i, \tag{3.53}$$

$$k_i = \mathbf{f}\left(t_{n-1} + c_i \Delta T, \mathbf{u}_n + \Delta T \sum_{j=1}^{s} a_{ij} k_j\right) \tag{3.54}$$

with $b_i, a_{ij} \in \mathbb{R}$ and

$$c_i = \sum_{j=1}^{s} a_{ij} \tag{3.55}$$

for $i = 1, \ldots, s$ and $n = 1, \ldots, N$. The RK methods are commonly represented by the Butcher tableau [27] as illustrated in Table 3.1. Besides, they can be further classified [27, Section II.7] as follows. The scheme (3.53) and (3.54) is

- an *explicit s-stage* RK method if $a_{ij} = 0$ for $i \leq j$;
- a *diagonal implicit* RK (DIRK) method if $a_{ij} = 0$ for $i < j$ and at least one $a_{ii} \neq 0$;
- a *singular* DIRK (SDIRK) method if it is a DIRK method and additionally all the diagonal elements are identical, i.e., $a_{ii} = \gamma$ for $i = 1, \ldots, s$;
- an *implicit* RK (IRK) method in all other cases.

The stability function of the s-stage IRK method (3.53) and (3.54) used in (3.49) is given by [13, Section IV.3]

$$R(z) = 1 + z\mathbf{b}^{\mathrm{T}}(\mathbf{I} - z\mathbf{A})^{-1}1 = \frac{det(\mathbf{I} - z\mathbf{A} + z1\mathbf{b}^{\mathrm{T}})}{det(\mathbf{I} - z\mathbf{A})}, \mathrm{T} \tag{3.56}$$

where $\mathbf{b}^{\mathrm{T}} = [b_1, \ldots, b_s]$, $\mathbf{A} \in \mathbb{R}^{s \times s}$ with $(\mathbf{A})_{ij} = a_{ij}$, $\mathbf{I} \in \mathbb{R}^{s \times s}$ is the identity matrix, and $1 = [1, \ldots, 1]^{\mathrm{T}}$. We note that the IE method and the trapezoidal scheme are the simplest one-stage IRK methods. In particular, the IE method has the coefficients $c_1 = a_{11} = b_1 = 1$, while the trapezoidal rule is defined by $c_1 = a_{11} = 0.5$ and $b_1 = 1$, see [27, Section II.7].

Table 3.1 The Butcher tableau for the s-stage RK method

c_1	a_{11}	a_{12}	$\cdots$	a_{1s}
c_2	a_{21}	a_{22}	$\cdots$	a_{2s}
$\vdots$	$\vdots$	$\vdots$	$\ddots$	$\vdots$
c_s	a_{s1}	a_{s2}	$\cdots$	a_{ss}
	b_1	b_2	$\cdots$	b_s

3.3.3 Application to the Eddy current problem

The time stepping methods of the form (3.47) or (3.53) and (3.54) can be applied to the time discretization of the IVP for the semi-discrete eddy current equation (3.28) and (3.40). In particular, it was solved with the IRK methods (3.53) and (3.54) in [30] and with their time-adaptive variants in Refs. [31,32]. We consider the IE method and the trapezoidal scheme in this section. Using the IE method on the equidistant partition (3.46), we can find $\mathbf{a}_n$ for $n = 1, \ldots, N$ from

$$\frac{1}{\Delta T}\mathbf{M}_\sigma[\mathbf{a}_n - \mathbf{a}_{n-1}] + \mathbf{K}_\nu(\mathbf{a}_n)\mathbf{a}_n = \mathbf{f}_n \tag{3.57}$$

with the RHS $\mathbf{f}_n = \mathbf{f}(t_n)$ starting from the IC $\mathbf{a}_0$ at t_0. As already mentioned in Section 3.2.4, the IC has to be consistent. Since (3.28) is an index-1 DAE system with linear mass matrix [33], after one step of the implicit Euler method, a consistent solution is obtained. For the computation of consistent ICs for higher-index (index-2) DAEs, we refer to [34,35].

Analogously, the trapezoidal rule applied to (3.28) and (3.40) on the equidistant partition (3.46) gives

$$\frac{1}{\Delta T}\mathbf{M}_\sigma[\mathbf{a}_n - \mathbf{a}_{n-1}] + 0.5[\mathbf{K}_\nu(\mathbf{a}_n)\mathbf{a}_n + \mathbf{K}_\nu(\mathbf{a}_{n-1})\mathbf{a}_{n-1}] = 0.5[\mathbf{f}_n + \mathbf{f}_{n-1}], \tag{3.58}$$

where $\mathbf{f}_n = \mathbf{f}(t_n)$ and $n = 1, \ldots, N$. The trapezoidal rule has (at similar computational effort) a slightly better convergence estimate but as already mentioned in Section 3.3.1, it does not damp high-frequency error components in the solution and is commonly not well-suited for systems with abrupt changes in the excitation $\mathbf{f}$ in (3.28) [36]. However, it is applicable to the φ-formulation (3.36) with a prescribed IC $\varphi(0) = \varphi_0$. For further research on time integration of EQS field models, we refer to [37].

Systems (3.57) and (3.58) are now fully discrete nonlinear algebraic equations which can be solved by applying a linearization approach at each time step. For this, the Newton method [14] can be used, i.e., for a given $\mathbf{x}^{(n,0)}$, e.g., $\mathbf{x}^{(n,0)} := \mathbf{a}_{n-1}$, calculate

$$\mathbf{J}_{\mathbf{F}_n}(\mathbf{x}^{(n,k)})\Delta\mathbf{x}^{(n,k)} = -\mathbf{F}_n(\mathbf{x}^{(n,k)}), \quad \mathbf{x}^{(n,k+1)} = \mathbf{x}^{(n,k)} + \Delta\mathbf{x}^{(n,k)} \tag{3.59}$$

for $k = 0, 1, \ldots, K_n$. The solution $\mathbf{a}_n$ at the time step n can be obtained via assigning $\mathbf{a}_n := \mathbf{x}^{(n,K_n)}$ once the algorithm (3.59) has converged at iteration K_n. For both considered time-stepping methods, the vector $\mathbf{F}_n(\mathbf{x})$ and its Jacobian matrix $\mathbf{J}_{\mathbf{F}_n}(\mathbf{x})$ are given by

$$\begin{aligned}\mathbf{F}_n(\mathbf{x}) = {} & \frac{1}{\Delta T}\mathbf{M}_\sigma[\mathbf{x} - \mathbf{a}_{n-1}] + \theta[\mathbf{K}_\nu(\mathbf{x})\mathbf{x} - \mathbf{f}_n] \\ & + (1-\theta)[\mathbf{K}_\nu(\mathbf{a}_{n-1})\mathbf{a}_{n-1} - \mathbf{f}_{n-1}] = \mathbf{0},\end{aligned} \tag{3.60}$$

$$\mathbf{J}_{\mathbf{F}_n}(\mathbf{x}) = \frac{1}{\Delta T}\mathbf{M}_\sigma + \theta\mathbf{K}_{\nu,\mathrm{d}}(\mathbf{x}), \tag{3.61}$$

respectively. We have $\theta = 1$ in the case of the IE method (3.57) and $\theta = 0.5$ in the case of the trapezoidal rule (61). For ν_{d} from (34) and vector $\mathbf{x} = \left[x_1, \ldots, x_{N_{\mathrm{e}}}\right]^{\mathrm{T}}$, the matrix $\mathbf{K}_{\nu,\mathrm{d}}(\mathbf{x})$ is defined as [38, Section A.3]

$$(\mathbf{K}_{\nu,\mathrm{d}}(\mathbf{x}))_{ij} = \int_{\Omega} (\mathrm{curl}\vec{v}_j)^{\mathrm{T}} \nu_{\mathrm{d}} \cdot \mathrm{curl}\vec{v}_i \,\mathrm{d}\Omega, \quad \nu_{\mathrm{d}} = \nu_{\mathrm{d}}\left(\vec{x}, \sum_{k=1}^{N_{\mathrm{e}}} x_k \mathrm{curl}\vec{v}_k\right), \tag{3.62}$$

$\vec{x} \in \Omega$. At each step, $n = 1, \ldots, N$ of the numerical time stepping K_n linear Newton systems (62) have to be solved. The total cost of the solution of (3.28) and (3.40) using an implicit one-step method (IE or trapezoidal scheme) then amounts to the solution of $K_1 + \cdots + K_N$ linear algebraic systems. For further results on space- and time-discretization of MQS and EQS formulations, we refer to Refs. [39–41].

In Section 3.4, we consider the TP condition (3.41) and apply the time-integration methods to calculate the TP solution to (3.28).

3.4 Solution of TP problems

We would now like to solve the TP eddy current problem (3.28) and (3.41). Considering the equidistant partition (3.46) of $[0, T]$, we have

$$\mathbf{a}_0 = \mathbf{a}_N, \tag{3.63}$$

which is the discrete variant of the TP condition (3.41). Using the IE discretization (3.57) and the TP condition (3.63), we can construct the nonlinear TP block system

$$\begin{bmatrix} \mathbf{Q}(\mathbf{a}_0) & & & -\mathbf{C} \\ -\mathbf{C} & \mathbf{Q}(\mathbf{a}_1) & & \\ & \ddots & \ddots & \\ & & -\mathbf{C} & \mathbf{Q}(\mathbf{a}_{N-1}) \end{bmatrix} \begin{bmatrix} \mathbf{a}_0 \\ \mathbf{a}_1 \\ \vdots \\ \mathbf{a}_{N-1} \end{bmatrix} = \begin{bmatrix} \mathbf{f}_N \\ \mathbf{f}_1 \\ \vdots \\ \mathbf{f}_{N-1} \end{bmatrix} \tag{3.64}$$

with $\mathbf{C} = 1/\Delta T \cdot \mathbf{M}_\sigma$ and $\mathbf{Q}(\mathbf{a}_n) = \mathbf{C} + \mathbf{K}_\nu(\mathbf{a}_n)$ for $n = 0, \ldots, N-1$. Analogously, in the case of the trapezoidal rule (3.58), we have the nonlinear TP block system

$$\begin{bmatrix} \mathbf{Q}(\mathbf{a}_0) & & & -\mathbf{C}(\mathbf{a}_{N-1}) \\ -\mathbf{C}(\mathbf{a}_0) & \mathbf{Q}(\mathbf{a}_1) & & \\ & \ddots & \ddots & \\ & & -\mathbf{C}(\mathbf{a}_{N-2}) & \mathbf{Q}(\mathbf{a}_{N-1}) \end{bmatrix} \begin{bmatrix} \mathbf{a}_0 \\ \mathbf{a}_1 \\ \vdots \\ \mathbf{a}_{N-1} \end{bmatrix}$$

$$= \begin{bmatrix} 0.5[\mathbf{f}_N + \mathbf{f}_{N-1}] \\ 0.5[\mathbf{f}_1 + \mathbf{f}_0] \\ \vdots \\ 0.5[\mathbf{f}_{N-1} + \mathbf{f}_{N-2}] \end{bmatrix} \tag{3.65}$$

with the matrices $\mathbf{C}(\mathbf{a}_n) = 1/\Delta T \cdot \mathbf{M}_\sigma - 0.5\mathbf{K}_\nu(\mathbf{a}_n)$ and $\mathbf{Q}(\mathbf{a}_n) = \mathbf{C}(\mathbf{a}_n) + \mathbf{K}_\nu(\mathbf{a}_n)$, for $n = 0, \ldots, N-1$. The systems (3.64) and (3.65) can then be solved by the Newton method. This approach is known in the engineering literature as TP-FEM [42].

In contrast to the solution of the IVP (3.28) and (3.40) through the classical time stepping, where N nonlinear systems of the dimension N_e are solved sequentially, the solution of the TP problem (3.28) and (3.41) requires the solution of the joint nonlinear system of dimension $N_\mathrm{e}N$. This might lead to a prohibitive computational effort.

Note that in the linear case, the matrices $\mathbf{C}$ and $\mathbf{Q}$ in (3.64) and (3.65) are constant and the system matrices possess a block-cyclic structure. In this case, it is possible to solve the block-cyclic linear systems efficiently using a diagonalization frequency-domain approach [25] which we discuss in Section 3.4.1.

3.4.1 Multi-harmonic diagonalization of cyclic systems

In this section, we describe a diagonalization method directly applicable to linear algebraic systems with a block-cyclic system matrix as introduced in [25]. It is based on the multi-harmonic (MH) transformation from time domain into frequency domain and backwards and is related to the well-known harmonic balance (HB) approach [44]. Let the TP system (3.64) be linear and given by

$$\mathbf{G}\,\mathbf{a}_{\mathrm{tot}} = \mathbf{f}_{\mathrm{tot}} \tag{3.66}$$

with the block-cyclic system matrix

$$\mathbf{G} = \begin{bmatrix} \mathbf{Q} & & & -\mathbf{C} \\ -\mathbf{C} & \mathbf{Q} & & \\ & \ddots & \ddots & \\ & & -\mathbf{C} & \mathbf{Q} \end{bmatrix}, \tag{3.67}$$

the unknown $\mathbf{a}_{\mathrm{tot}}$, and the RHS $\mathbf{f}_{\mathrm{tot}}$ given by

$$\mathbf{a}_{\mathrm{tot}} = \left[\mathbf{a}_0^{\mathrm{T}}, \mathbf{a}_1^{\mathrm{T}}, \ldots, \mathbf{a}_{N-1}^{\mathrm{T}}\right]^{\mathrm{T}}, \tag{3.68}$$

$$\mathbf{f}_{\mathrm{tot}} = \left[\mathbf{f}_N^{\mathrm{T}}, \mathbf{f}_1^{\mathrm{T}}, \ldots, \mathbf{f}_{N-1}^{\mathrm{T}}\right]^{\mathrm{T}}. \tag{3.69}$$

The matrices

$$\mathbf{C} = 1/\Delta T \cdot \mathbf{M}_\sigma, \quad \mathbf{Q} = \mathbf{C} + \mathbf{K}_\nu, \tag{3.70}$$

in (3.67) and $\mathbf{K}_\nu$ are constant (i.e., they do not depend on the solution). To overcome a possibly high computational effort required for the direct solution of the $(N_\mathrm{e}N)$-dimensional system (3.66), one can apply the diagonalization approach from [25]. The idea of the method is to transform the block-cyclic system in time domain into an equivalent block-diagonal system in frequency domain

$$\mathbf{G}\,\mathbf{u}_{\mathrm{tot}} = \mathbf{f}_{\mathrm{tot}} \tag{3.71}$$

with the block-diagonal system matrix

$$\mathbf{G} = \text{diag}[\mathbf{G}_0, \mathbf{G}_1, \ldots, \mathbf{G}_{N-1}], \tag{3.72}$$

the unknown $\mathbf{a}_{\text{tot}}$, and the RHS $\mathbf{f}_{\text{tot}}$ given by

$$\mathbf{a}_{\text{tot}} = [\mathbf{a}_0^{\text{T}}, \mathbf{a}_1^{\text{T}}, \ldots, \mathbf{a}_{N-1}^{\text{T}}]^{\text{T}}, \tag{3.73}$$

$$\mathbf{f}_{\text{tot}} = [\mathbf{f}_N^{\text{T}}, \mathbf{f}_1^{\text{T}}, \ldots, \mathbf{f}_{N-1}^{\text{T}}]^{\text{T}}. \tag{3.74}$$

This transformation is visualized in Figure 3.2. The systems (3.66) and (3.71) are related via

$$\mathbf{G} = (\mathbf{F} \otimes \mathbf{I})\mathbf{G}(\mathbf{F}^{\text{H}} \otimes \mathbf{I}), \quad \mathbf{u}_{\text{tot}} = (\mathbf{F} \otimes \mathbf{I})\mathbf{u}_{\text{tot}}, \quad \mathbf{f}_{\text{tot}} = (\mathbf{F} \otimes \mathbf{I})\mathbf{f}_{\text{tot}}. \tag{3.75}$$

The matrix $\mathbf{F}$ is the *discrete Fourier transform* (DFT) matrix whose elements are defined as

$$\mathbf{F}_{jn} = \frac{1}{\sqrt{N}} exp(-i\omega_{j-1}t_{n-1}), \qquad j, n = 1, \ldots, N \tag{3.76}$$

with the imaginary unit i, the angular frequencies

$$\omega_j = 2\pi j/T, \quad j = 0, \ldots, N-1, \tag{3.77}$$

and the time points $t_n = n\Delta T$ for $n = 0, \ldots, N-1$. The matrix $\mathbf{I} \in \mathbb{R}^{N_e \times N_e}$ is the identity matrix, $\otimes$ denotes the Kronecker product of two matrices, and $\mathbf{F}^{\text{H}}$ is the complex conjugate transpose matrix of $\mathbf{F}$. One can derive the expressions for the block matrices in (3.72) given by

$$\mathbf{G}_j = \mathbf{Q} - \mathbf{C}exp(-i\omega_j\Delta T), \quad j = 0, \ldots, N-1. \tag{3.78}$$

The RHS $\mathbf{f}_{\text{tot}}$ in (3.74) is obtained from $\mathbf{f}_{\text{tot}}$ in (3.69) through

$$\begin{aligned} \mathbf{f}_N &= \frac{1}{\sqrt{N}} \sum_{n=0}^{N-1} \mathbf{f}_n exp(-i\omega_0 t_n), \\ \mathbf{f}_j &= \frac{1}{\sqrt{N}} \sum_{n=0}^{N-1} \mathbf{f}_n exp(-i\omega_j t_n), \quad j = 1, \ldots, N-1. \end{aligned} \tag{3.79}$$

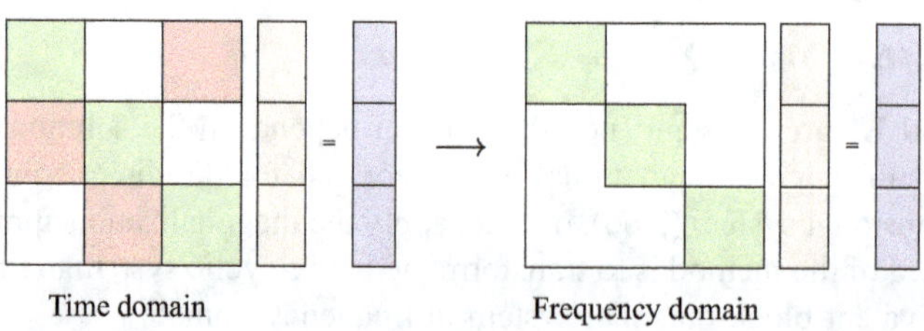

Figure 3.2 Sketch visualizing the DFT from time domain to frequency domain

Since the diagonal block matrices in the system (3.71) are decoupled, given the RHS components $\mathbf{f}_j$ for $j = 1, \ldots, N$ from (3.79), we can calculate the harmonic coefficients $\mathbf{u}_j$ separately by solving

$$\mathbf{G}_0\mathbf{u}_0 = \mathbf{f}_N \quad \text{and} \quad \mathbf{G}_j\mathbf{u}_j = \mathbf{f}_j, \quad j = 1, \ldots, N-1. \tag{3.80}$$

Clearly, the solution of N linear N_e-dimensional systems (3.80) as well as the construction of the harmonic components $\mathbf{f}_j$ can be calculated in parallel. Having calculated the solution $\mathbf{u}_{\text{tot}}$ in the frequency domain, we can obtain the solution $\mathbf{u}_{\text{tot}}$ in the time domain by applying the *inverse* DFT, i.e.,

$$\mathbf{u}_{\text{tot}} = (\mathbf{F}^{\text{H}} \otimes \mathbf{I})\mathbf{u}_{\text{tot}}, \qquad \mathbf{u}_n = \frac{1}{\sqrt{N}} \sum_{j=0}^{N-1} \mathbf{u}_j exp(i\omega_j t_n), \quad n = 0, \ldots, N-1. \tag{3.81}$$

The DFT and its inverse can be efficiently calculated using the *fast Fourier transform* (FFT) algorithm [43], thereby further reducing the complexity of the transformation.

Remark 3.1. We note that when $\Delta T \to 0$ the diagonalization of the TP linear system (3.66) using the DFT is equivalent to the application of the HB method [44] to the semi-discrete system (3.28) in case the matrix $\mathbf{K}_\nu$ is constant and the RHS $\mathbf{f}$ is periodic. Indeed, the HB method with N frequencies (3.77) reads

$$[i\omega_j\mathbf{M}_\sigma + \mathbf{K}_\nu]\mathbf{u}_j = \mathbf{f}_j, \quad j = 0, \ldots, N-1. \tag{3.82}$$

For the matrices $\mathbf{G}_j, j = 0, \ldots, N-1$ given in (3.78), we have

$$\lim_{\Delta T \to 0} \mathbf{G}_j = \lim_{\Delta T \to 0} [\mathbf{M}_\sigma(1 - exp(-i\omega_j \Delta T))/\Delta T + \mathbf{K}_\nu] = i\omega_j\mathbf{M}_\sigma + \mathbf{K}_\nu,$$

due to the definitions in (3.70). Thus, provided $\mathbf{f}_N = \mathbf{f}_0$ the system matrices in (3.80) head to the system matrices in (3.82) for all $j = 0, \ldots, N-1$ as $\Delta T \to 0$ (and $N \to \infty$).

3.5 PnT integration methods

As already mentioned in Section 3.2.4, the classical time stepping might require a long computational time, especially when aiming at the steady-state solution or in case the system possesses high-frequency dynamic behaviour (stemming, e.g., from a PWM excitation). In this section, we discuss a powerful tool for acceleration of the time stepping based on parallelization in time domain.

The introduction of the PinT methods dates back to the work of Nievergelt in 1964 [45]. An extensive overview of the existing approaches and their development can be found in [46]. Among various PinT methods such as, e.g., the multigrid reduction in time (MGRIT) [47] or ParaExp [48], we focus on the Parareal algorithm [21] in this chapter. It is a two-level method which as the other PinT approaches accelerates the standard time integration of IVPs through parallelization using available separate processing units. A major advantage of Parareal is its non-intrusiveness, i.e., existing black-box solvers can be used and integrated easily into the Parareal framework. The convergence of the Parareal method is well understood [49] and it has a property of converging in a finite number of iterations. We provide the Parareal theory in Section 3.5.1. In Section 3.5.2, a multirate Parareal method suitable for problems with discontinuous (PWM) excitations [23] is described.

Solution of TP problems such as (3.28) and (3.41) is also possible with PinT methods. Two extensions of the Parareal algorithm to TP problems were proposed in [50]. The application of the diagonalization approach from Section 3.4.1 to the Parareal-based methods has been proposed in [26,51,52]. We describe the existing Parareal-based approaches for the solution of TP problems in Section 3.5.3.

3.5.1 Parareal for IVPs

Consider the IVP for a system of ODEs in $\mathbb{R}^d$ (3.42)-(3.43). Let $\mathcal{F}(t, 0, \mathbf{u}_0)$ be the solution operator, which gives the solution to (3.42) at time $t \in (0, T]$ through the time integration starting from the IC $\mathbf{u}_0$ at $t = 0$. It can be, e.g., one of the RK methods (see Section 3.3.2) solving the IVP with step size δT.

The idea of the Parareal method is based on the partition of the time interval $[0, T]$ into N subintervals, e.g., one subinterval per processing unit, and the solution of the IVP on each subinterval in parallel. Thus, we consider the partition

$$0 = T_0 < T_1 < \cdots < T_N = T \tag{3.83}$$

and the corresponding IVP on the subinterval $[T_{n-1}, T_n]$

$$\frac{\mathrm{d}}{\mathrm{d}t}\mathbf{u}_n(t) = \mathbf{f}(\mathbf{u}_n(t), t), \quad t \in (T_{n-1}, T_n] \tag{3.84}$$

$$\mathbf{u}_n(T_{n-1}) = \mathbf{U}_{n-1}, \tag{3.85}$$

for an IC $\mathbf{U}_{n-1}$ and $n = 1, \ldots, N$. The ICs $\mathbf{U}_{n-1}$ are defined by the *matching conditions*

$$\mathbf{F}_0(\mathbf{U}_0) := \mathbf{u}_0 - \mathbf{U}_0 = \mathbf{0}, \tag{3.86}$$

$$\mathbf{F}_n(\mathbf{U}_{n-1}, \mathbf{U}_n) := \mathcal{F}(T_n, T_{n-1}, \mathbf{U}_{n-1}) - \mathbf{U}_n = \mathbf{0}, \quad n = 1, \ldots, N-1, \tag{3.87}$$

where $\mathbf{u}_0$ is given by (3.43) and $\mathcal{F}$ is the already mentioned solution operator of an IVP for the ODE system (3.42). The mismatches at the synchronization points T_n, $n = 1, \ldots, N-1$ for $N = 4$ are visualized in Figure 3.3. Our goal is to find the root of the mapping $\mathbf{F} : \mathbb{R}^{Nd} \to \mathbb{R}^{Nd}$, i.e., to solve

$$\mathbf{F}(\mathbf{U}) = \begin{bmatrix} \mathbf{F}_0(\mathbf{U}_0) \\ \mathbf{F}_1(\mathbf{U}_0, \mathbf{U}_1) \\ \vdots \\ \mathbf{F}_{N-1}(\mathbf{U}_{N-2}, \mathbf{U}_{N-1}) \end{bmatrix} = \mathbf{0}. \tag{3.88}$$

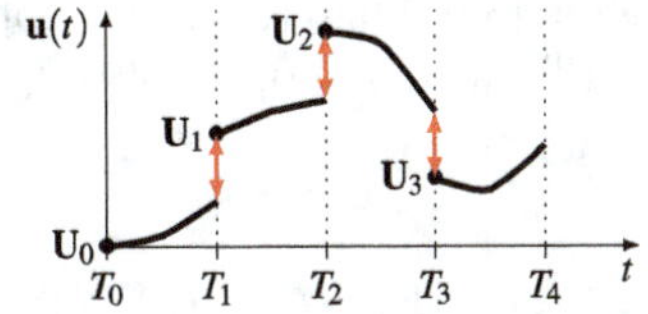

Figure 3.3 Mismatches at synchronization points T_n, $n = 1, 2, 3$

with $\mathbf{U} = [\mathbf{U}_0^{\mathrm{T}}, \mathbf{U}_1^{\mathrm{T}}, \ldots, \mathbf{U}_{N-1}^{\mathrm{T}}]^{\mathrm{T}}$. The system of nonlinear equations (3.88) can be solved using the Newton method, i.e., for a given $\mathbf{U}^{(k)}$ and $k = 0, 1, \ldots$ calculating

$$\mathbf{J}_{\mathbf{F}}(\mathbf{U}^{(k)})\Delta\mathbf{U}^{(k)} = -\mathbf{F}(\mathbf{U}^{(k)}), \quad \mathbf{U}^{(k+1)} = \mathbf{U}^{(k)} + \Delta\mathbf{U}^{(k)} \tag{3.89}$$

with the Jacobian matrix

$$\mathbf{J}_{\mathbf{F}}(\mathbf{U}) = 2.5pt \begin{bmatrix} -\mathbf{I} & & & \\ \dfrac{\partial \mathcal{F}_1(\mathbf{U}_0)}{\partial \mathbf{U}_0} & -\mathbf{I} & & \\ & \ddots & \ddots & \\ & & \dfrac{\partial \mathcal{F}_{N-1}(\mathbf{U}_{N-2})}{\partial \mathbf{U}_{N-2}} & -\mathbf{I} \end{bmatrix}, \tag{3.90}$$

where we have denoted $\mathcal{F}_n(\mathbf{U}_{n-1}) = \mathcal{F}(T_n, T_{n-1}, \mathbf{U}_{n-1})$ for $n = 1, \ldots, N-1$. The Newton iteration (3.89) can be written as

$$\mathbf{U}_0^{(k+1)} = \mathbf{u}_0, \tag{3.91}$$

$$\mathbf{U}_n^{(k+1)} = \mathcal{F}(T_n, T_{n-1}, \mathbf{U}_{n-1}^{(k)}) + \frac{\partial \mathcal{F}(T_n, T_{n-1}, \mathbf{U}_{n-1}^{(k)})}{\partial \mathbf{U}_{n-1}}[\mathbf{U}_{n-1}^{(k+1)} - \mathbf{U}_{n-1}^{(k)}] \tag{3.92}$$

for $k = 0, 1, \ldots$ and $n = 1, \ldots, N-1$. The described procedure for the solution of the IVP (3.42) and (3.43) is also known as the *multiple shooting method* [29, Section 8.2]. The Parareal algorithm can be derived from the multiple shooting iteration (3.91) and (3.92) via a finite-difference-based approximation of the derivative [53]

$$\begin{aligned} &\frac{\partial F(T_n, T_{n-1}, \mathbf{U}_{n-1}^{(k)})}{\partial \mathbf{U}_{n-1}}[\mathbf{U}_{n-1}^{(k+1)} - \mathbf{U}_{n-1}^{(k)}] \approx \\ &G(T_n, T_{n-1}, \mathbf{U}_{n-1}^{(k+1)}) - G(T_n, T_{n-1}, \mathbf{U}_{n-1}^{(k)}) \end{aligned} \tag{3.93}$$

for $n = 1, \ldots, N-1$. Here $\mathcal{G}$ is another solution operator applied to (3.84) and (3.85) which is less accurate and thus computationally cheaper to compute compared to $\mathcal{F}$. It can, e.g., be the same time-integration scheme as $\mathcal{F}$ but use a bigger step size $\Delta T \gg \delta T$. The operators $\mathcal{F}$ and $\mathcal{G}$ are called *the fine* and *the coarse propagators*, respectively.

Inserting the approximation (3.93) into (3.92), we obtain the Parareal iteration: given initial approximations $\mathbf{U}_n^{(0)}$ with $n = 0, \ldots, N - 2$, calculate

$$\mathbf{U}_0^{(k+1)} = \mathbf{u}_0, \tag{3.94}$$

$$\mathbf{U}_n^{(k+1)} = \mathcal{F}(T_n, T_{n-1}, \mathbf{U}_{n-1}^{(k)}) + \mathcal{G}(T_n, T_{n-1}, \mathbf{U}_{n-1}^{(k+1)}) - \mathcal{G}(T_n, T_{n-1}, \mathbf{U}_{n-1}^{(k)}) \tag{3.95}$$

for $k = 0, 1, \ldots$ and $n = 1, \ldots, N - 1$. The initial approximations $\mathbf{U}_n^{(0)}$ can be obtained, e.g., by applying the coarse solver sequentially, i.e., starting from $\mathbf{U}_0^{(0)} = \mathbf{u}_0$ calculate

$$\mathbf{U}_n^{(0)} = \mathcal{G}(T_n, T_{n-1}, \mathbf{U}_{n-1}^{(0)}), \quad n = 1, \ldots, N - 2. \tag{3.96}$$

We note from (3.95) that at iteration $k + 1$ the fine solution $\mathcal{F}(T_n, T_{n-1}, \mathbf{U}_{n-1}^{(k)})$ depends only on the solution at the previous iteration k. This allows to apply the fine propagator on each subinterval in parallel, thereby reducing the computational cost. On the other hand, the coarse solution $\mathcal{G}(T_n, T_{n-1}, \mathbf{U}_{n-1}^{(k+1)})$ depends on the solution at the same iteration $k + 1$ and therefore has to be calculated sequentially. However, the coarse propagator is cheaper to apply than the fine one. Therefore, it will commonly not affect the computational effort significantly. We now state the convergence result for the Parareal iteration (3.94) and (3.95) based on [49].

Theorem 3.1 [Convergence of Parareal [49]]. *Let RHS* $\mathbf{f}$ *in* (3.84) *be sufficiently smooth and let the partition* (3.83) *of the interval* $[0, T]$ *be equidistant with the step size* $\Delta T = T/N$. *We additionally assume*:

- $\mathcal{F}(T_n, T_{n-1}, \mathbf{U}_{n-1})$ *and* $\mathcal{F}(T_n, T_{n-1}, \mathbf{U}_{n-1})$ *are the exact and an approximate solution to* (3.84) and (3.85) *at* T_n, *respectively.*
- *The LTE is bounded by* $C_3 \Delta T^{p+1}$ *with* $p \geq 1$ *and can be expanded for small step size* ΔT *and an initial value* $\mathbf{U}$ *as*

$$\mathcal{F}(T_n, T_{n-1}, \mathbf{U}) - \mathcal{F}(T_n, T_{n-1}, \mathbf{U}) = c_{p+1}(\mathbf{U})\Delta T^{p+1} + c_{p+2}(\mathbf{U})\Delta T^{p+2} + \cdots$$

  where functions c_i, $i = p + 1, p + 2, \ldots$, *are continuously differentiable.*
- $\mathcal{F}$ *satisfies the Lipschitz condition*

$$\|\mathcal{G}(t + \Delta T, t, \mathbf{U}) - \mathcal{F}(t + \Delta T, t, \mathbf{V})\| \leq (1 + C_2 \Delta T)\|\mathbf{U} - \mathbf{V}\| \tag{3.97}$$

for $t \in [0, T]$, *for all* $\mathbf{U}, \mathbf{V}$, *and a constant* C_2.

Then at the Parareal iteration $k = 0, 1, \ldots, n$ from (97) to (98), we have for a constant C_1 the error bound

$$\|\mathbf{u}(T_n) - \mathbf{U}_n^{(k)}\| \leq \frac{C_3}{C_1} \frac{(C_1 \Delta T^{p+1})^{k+1}}{(k + 1)!} (1 + C_2 \Delta T)^{n-k-1} \prod_{j=0}^{k} (n - j). \tag{3.98}$$

We see from the estimate (3.98) that the error of the Parareal method at the synchronization point T_n vanishes at iteration $k = n$ for $n = 1, \ldots, N - 1$ due to the product on the RHS in (3.98). It means that at the Parareal iteration k, the exact time-stepping solution is obtained on all the subintervals $[T_{n-1}, T_n]$ for $n \leq k$. In practice, it can be expected that the method reaches a required tolerance much earlier in order to obtain a maximal possible speed up through the parallelization of the fine solutions.

3.5.2 Multirate Parareal for IVPs

As was shown in Theorem 3.1, the smoothness assumptions on the RHS were imposed. However, there are applications in electrical engineering such as, e.g., power converters, which include discontinuous signals. In particular, consider a PWM signal with the constant switching frequency f_s which can be constructed via

$$p(t) = \operatorname{sgn}[r(t) - c(t)], \tag{3.99}$$

where $r(t)$ is the *reference signal* and $c(t)$ is the *carrier signal*. An example of a PWM signal of $f_s = 500$ with $r(t) = sin(2\pi ft)$ and $f = 50$ is depicted in Figure 3.4. The carrier signal $c(t)$ is in this case the sawtooth carrier [22, Section 2.1]. We assume that the RHS in (3.84) can be expressed as a sum

$$\mathbf{f}(\mathbf{u}(t), t) = \overline{\mathbf{f}}(\mathbf{u}(t), t) + \widetilde{\mathbf{f}}(t), \tag{3.100}$$

where $\overline{\mathbf{f}}$ and $\widetilde{\mathbf{f}}$ satisfy the following assumption.

Assumption 3.2. *Assume, for functions $\overline{\mathbf{f}}$ and $\widetilde{\mathbf{f}}$ from (3.100) it holds that*

- $\overline{\mathbf{f}}$ *is bounded, sufficiently smooth in both arguments, and Lipschitz continuous in the first argument;*
- $\mathbf{f} \in \mathbf{L}^\gamma([0, T], \mathcal{F}^d)$, $\gamma \geq 1$ (*see* [54, Section 23.2]) *with a bounded norm*

$$C_\gamma = \|\mathbf{f}\|_{\mathbf{L}^\gamma([0,T],\mathbb{R}^d)} < \infty. \tag{3.101}$$

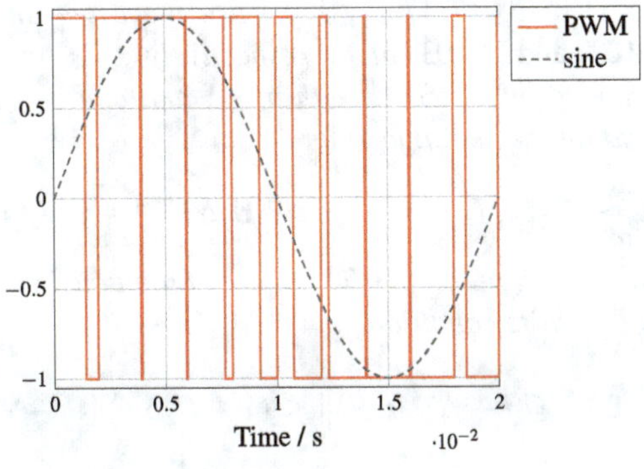

Figure 3.4 PWM signal of $f_s = 500$ Hz and reference signal of $f = 50$ Hz

Solvability and uniqueness of IVPs with discontinuous RHSs is assured by the Carathéodory theory [55].

The idea of the multirate Parareal algorithm for systems with discontinuous inputs presented in [23] is the introduction of the simplified coarse propagator $\bar{\mathcal{G}}$ which solves the IVP with a smooth RHS $\bar{\mathbf{f}}$ instead of the original RHS $\mathbf{f}$, i.e., $\bar{\mathcal{G}}(t, T_{n-1}, \mathbf{U}_{n-1})$ is the solution of

$$\frac{\mathrm{d}}{\mathrm{d}t}\bar{\mathbf{u}}_n(t) = \bar{\mathbf{f}}(\mathbf{u}_n(t), t), \quad t \in (T_{n-1}, T_n] \tag{3.102}$$

$$\bar{\mathbf{u}}_n(T_{n-1}) = \mathbf{U}_{n-1} \tag{3.103}$$

on the subinterval $[T_{n-1}, T_n]$. The fine propagator solves the IVP (3.84) and (3.85) with the discontinuous RHS $\mathbf{f}$ as in the standard Parareal method. The proposed multirate Parareal [23] differs from the classical Parareal by replacing $\mathcal{G}$ by the simplified coarse solver $\bar{\mathcal{G}}$, i.e.,

$$\mathbf{U}_0^{(k+1)} = \mathbf{u}_0, \tag{3.104}$$

$$\mathbf{U}_n^{(k+1)} = \mathcal{F}(T_n, T_{n-1}, \mathbf{U}_{n-1}^{(k)}) + \bar{\mathcal{G}}(T_n, T_{n-1}, \mathbf{U}_{n-1}^{(k+1)}) - \bar{\mathcal{G}}(T_n, T_{n-1}, \mathbf{U}_{n-1}^{(k)}) \tag{3.105}$$

for $k = 0, 1, \ldots$ and $n = 1, \ldots, N-1$. The smooth part of a PWM input can be obtained from the Fourier analysis, i.e., by extracting the principal frequency component(s). In case of the PWM signal from Figure 3.4, the natural choice is the reference sinusoidal signal $r(t)$. The convergence of the new Parareal method (3.104) and (3.105) is guaranteed according to Ref, [23] by the following theorem.

Theorem 3.3 (Convergence of the multirate Parareal [23]) *Let the RHS* $\mathbf{f}$ *in* (3.84) *be expressed by* (3.100) *fulfilling the Assumption 3.2 and let the partition* (3.83) *of the interval* $[0, T]$ *be equidistant with the step size* $\Delta T = T/N$. *We additionally assume:*

- $\mathcal{F}(T_n, T_{n-1}, \mathbf{U}_{n-1})$ *is the exact solution to* (3.84) and (3.85) at T_n.
- $\bar{\mathcal{F}}(T_n, T_{n-1}, \mathbf{U}_{n-1})$ *and* $\bar{\mathcal{G}}(T_n, T_{n-1}, \mathbf{U}_{n-1})$ *are the exact and an approximate solution to* (3.102) and (3.103) *at* T_n, *respectively.*
- *The LTE is bounded by* $\bar{C}_3 \Delta T^{p+1}$ *with* $p \geq 1$ *and can be expanded for small step size* ΔT *and an initial value* $\mathbf{U}$ *as*

$$\bar{\mathcal{F}}(T_n, T_{n-1}, \mathbf{U}) - \bar{\mathcal{G}}(T_n, T_{n-1}, \mathbf{U}) = \bar{c}_{p+1}(\mathbf{U})\Delta T^{p+1} + \bar{c}_{p+2}(\mathbf{U})\Delta T^{p+2} + \cdots$$

where functions $\bar{c}_i$, $i = p+1, p+2, \ldots$, *are continuously differentiable.*
- $\bar{\mathcal{G}}$ *satisfies the Lipschitz condition*

$$\|\mathcal{G}(t + \Delta T, t, \mathbf{U}) - \mathcal{G}(t + \Delta T, t, \mathbf{V})\| \leq (1 + C_2 \Delta T)\|\mathbf{U} - \mathbf{V}\| \tag{3.106}$$

for $t \in [0, T]$, *for all* $\mathbf{U}, \mathbf{V}$, *and a constant* C_2.

Then at the Parareal iteration $k = 0, 1, \ldots$ (3.104) and (3.105), we have for the constants $\bar{C}_1 > 0$ and $C_4 > 0$ the error bound

$$\begin{aligned}\|\mathbf{u}(T_n) - \mathbf{U}_n^{(k)}\| &\le \bar{C}_1^k \left[C_4 C_\gamma \Delta T^{(p+1)}k + 1/\zeta + \bar{C}_3 (\Delta T^{p+1})^{k+1}\right] \\ &\times \frac{(1 + C_2 \Delta T)^{n-k-1}}{(k+1)!} \prod_{j=0}^{k} (n-j)\end{aligned} \tag{3.107}$$

with the integer $\zeta \ge 1$ such that $1/\gamma + 1/\zeta = 1$ and constant C_γ defined in (3.101).

We see that the estimate (3.107) has an additional term compared to (3.98) which may lead to the reduction of the convergence order. Indeed, for $p \ge 1$ and $\zeta \ge 1$, we have for the powers in (3.107) the relation $(p+1)k + 1/\zeta < (p+1)k + (p+1)$. However, in practice, it is seen in [23] that there is no order reduction and the order $(p+1)(k+1)$ of the standard Parareal is maintained.

3.5.3 *Parareal for TP problems*

We now search for a function $\mathbf{u} : [0, T] \to \mathbb{R}^d$ such that

$$\frac{\mathrm{d}}{\mathrm{d}t}\mathbf{u}(t) = \mathbf{f}(\mathbf{u}(t), t), \quad t \in (0, T) \tag{3.108}$$

$$\mathbf{u}(0) = \mathbf{u}(T), \tag{3.109}$$

with the RHS $\mathbf{f} : \mathbb{R}^d \times [0, T] \to \mathbb{R}^d$, $d > 0$. Based on the Parareal method for IVPs (see Section 5.1), two PinT algorithms for TP problems were proposed in [50]. The first approach is called the *periodic Parareal algorithm with initial-value coarse problem* (PP-IC) which reads: for $k = 0, 1, \ldots$ and $n = 1, \ldots, N$ calculate

$$\mathbf{U}_0^{(k+1)} = \mathbf{U}_N^{(k)}, \tag{3.110}$$

$$\mathbf{U}_n^{(k+1)} = \mathcal{F}(T_n, T_{n-1}, \mathbf{U}_{n-1}^{(k)}) + \mathcal{G}(T_n, T_{n-1}, \mathbf{U}_{n-1}^{(k+1)}) - \mathcal{G}(T_n, T_{n-1}, \mathbf{U}_{n-1}^{(k)}). \tag{3.111}$$

We see that the only difference between Parareal and PP-IC is the update of the IC $\mathbf{U}_0^{(k+1})$ in (3.110). As Parareal, PP-IC is non-intrusive and therefore easy to implement. Note that the periodicity condition $\mathbf{u}_0 = \mathbf{u}_N$ is relaxed via (3.110) within PP-IC.

Remark 3.2. A natural extension of the multirate Parareal presented in Section 3.5.2 to TP problems was done based on the PP-IC algorithm in [24]. In particular, the mutlirate PP-IC iteration reads: for $k = 0, 1, \ldots$ and $n = 1, \ldots, N$ calculate

$$\mathbf{U}_0^{(k+1)} = \mathbf{U}_N^{(k)}, \tag{3.112}$$

$$\mathbf{U}_n^{(k+1)} = \mathcal{F}(T_n, T_{n-1}, \mathbf{U}_{n-1}^{(k)}) + \bar{\mathcal{G}}(T_n, T_{n-1}, \mathbf{U}_{n-1}^{(k+1)}) - \bar{\mathcal{G}}(T_n, T_{n-1}, \mathbf{U}_{n-1}^{(k)}). \tag{3.113}$$

As in the multirate Parareal for IVPs (3.104) and (3.105), the fine propagator $\mathcal{F}$ solves the original problem with the discontinuous (high-frequency) RHS (3.84) and (3.85), while the coarse propagator $\bar{\mathcal{G}}$ solves the problem with the smooth RHS (3.102) and (3.103) possessing a reduced dynamics. In [24], it is shown that the convergence estimate for the standard PP-IC algorithm (3.110) and (3.111) can be also used for the multirate PP-IC variant (3.112) and (3.113).

The second approach of [50] is the *periodic Parareal algorithm with periodic coarse problem* (PP-PC) which reads: for $k = 0, 1, \ldots$ and $n = 1, \ldots, N-1$ calculate

$$\mathbf{U}_0^{(k+1)} = \mathcal{F}(T_N, T_{N-1}, \mathbf{U}_{N-1}^{(k)}) + \mathcal{G}(T_N, T_{N-1}, \mathbf{U}_{N-1}^{(k+1)}) - \mathcal{G}(T_N, T_{N-1}, \mathbf{U}_{N-1}^{(k)}), \tag{3.114}$$

$$\mathbf{U}_n^{(k+1)} = \mathcal{F}(T_n, T_{n-1}, \mathbf{U}_{n-1}^{(k)}) + \mathcal{G}(T_n, T_{n-1}, \mathbf{U}_{n-1}^{(k+1)}) - \mathcal{G}(T_n, T_{n-1}, \mathbf{U}_{n-1}^{(k)}). \tag{3.115}$$

The PP-PC iteration can be derived from the multiple shooting method by searching for $\mathbf{U} = [\mathbf{U}_0^{\mathrm{T}}, \mathbf{U}_1^{\mathrm{T}}, \ldots, \mathbf{U}_{N-1}^{\mathrm{T}}]^{\mathrm{T}}$ fulfilling the periodicity condition

$$\mathbf{F}_0(\mathbf{U}_0, \mathbf{U}_{N-1}) := \mathcal{F}(T_N, T_{N-1}, \mathbf{U}_{N-1}) - \mathbf{U}_0 = \mathbf{0} \tag{3.116}$$

and the matching conditions (90), followed by the Parareal-based approximation of the derivative (3.93) [50]. In contrast to PP-IC, within PP-PC the direct periodic coupling is present on the coarse level. This situation is similar to the one described in Section 3.4, where the system of the form (3.64) has to be solved. Indeed, the PP-PC iteration (3.114) and (3.115) can be written in the matrix–vector operator form

$$\begin{bmatrix} -\mathbf{I} & & & G_N(\cdot) \\ G_1(\cdot) & -\mathbf{I} & & \\ & \ddots & \ddots & \\ & & G_{N-1}(\cdot) & -\mathbf{I} \end{bmatrix} \begin{bmatrix} \mathbf{U}_0^{(k+1)} \\ \mathbf{U}_1^{(k+1)} \\ \vdots \\ \mathbf{U}_{N-1}^{(k+1)} \end{bmatrix} = \begin{bmatrix} \mathbf{b}_N^{(k)} \\ \mathbf{b}_1^{(k)} \\ \vdots \\ \mathbf{b}_{N-1}^{(k)} \end{bmatrix} \tag{3.117}$$

with the notations $\mathcal{G}_n(\cdot) = \mathcal{G}(T_n, T_{n-1}, \cdot)$ for $n = 1, \ldots, N$ and

$$\mathbf{b}_n^{(k)} = \mathcal{G}(T_n, T_{n-1}, \mathbf{U}_{n-1}^{(k)}) - \mathcal{F}(T_n, T_{n-1}, \mathbf{U}_{n-1}^{(k)}), \quad n = 1, \ldots, N. \tag{3.118}$$

Since in general the system (3.108) can be nonlinear, an additional linearization of the PP-PC system (3.117), e.g., using the fixed point Jacobi-type iteration [24], has to be performed.

3.5.3.1 PP-PC with MH diagonalization

For the efficient solution of the PP-PC system, the MH diagonalization approach [25] described in Section 3.4.1 can be applied. The method can be easily used in the linear case as it was demonstrated in [26]. In particular, using the IE discretization of the linear TP eddy current problem on the equidistant partition (3.83) with the step size ΔT

$$\frac{1}{\Delta T}\mathbf{M}_\sigma[\mathbf{a}_n - \mathbf{a}_{n-1}^{(k+1)}] + \mathbf{K}_\nu\mathbf{a}_n = \mathbf{f}_n, \quad n = 1,\ldots,N \tag{3.119}$$

as the coarse solver $\mathcal{G}(T_n, T_{n-1}, \mathbf{a}_{n-1}^{(k+1)})$ with one IE step per subinterval $[T_{n-1}, T_n]$ [26], one derives

$$\mathbf{G}\,\mathbf{a}_{\text{tot}}^{(k+1)} = \mathbf{r}_{\text{tot}}^{(k)} \tag{3.120}$$

with the matrix $\mathbf{G}$ defined in (3.67) and the RHS

$$\mathbf{r}_{\text{tot}}^{(k)} = \left[(\mathbf{r}_N^{(k)})^{\mathrm{T}}, (\mathbf{r}_1^{(k)})^{\mathrm{T}}, \ldots, (\mathbf{r}_{N-1}^{(k)})^{\mathrm{T}}\right]^{\mathrm{T}}, \tag{3.121}$$

$$\mathbf{r}_n^{(k)} = -\mathbf{Q}\mathbf{b}_n^{(k)} + \mathbf{f}_n, \quad n = 1,\ldots,N \tag{3.122}$$

where $\mathbf{b}_n^{(k)}$ is defined in (3.118). In the nonlinear case, the MH diagonalization can be exploited within an applied linearization method, provided the system matrix at each iteration possesses the block-cyclic structure (3.67). For this purpose, the simplified Newton method was used in [52]. Similar ideas to combine the Parareal method with the Fourier-based diagonalization were also presented in Refs. [51,56].

In Section 3.6, we apply the described PinT time-integration methods to an induction motor (IM) and to a linear coaxial cable models.

3.6 Numerical examples

In this section, we apply the presented PinT methods to several examples from electrical engineering. In particular, the multirate Parareal and the multirate PP-IC algorithms are applied to an IM model supplied with PWM voltage source in Section 3.6.1. The performance of the PP-PC method with the MH correction is illustrated for a linear coaxial cable model in Section 3.6.2. The efficiency of the algorithms is evaluated in terms of the effective number of solved linear systems which is the total number of solution processes executed sequentially. In case of parallel computations, this corresponds to the maximum number of linear system solutions performed by a single core. Besides, for all the PinT methods, it is assumed that the number of subinterval N is equal to the number of available processing units, i.e., one processor governs exactly one subinterval.

3.6.1 Induction machine

We now apply the PinT time-integration methods to a three-phase (3ph) IM model. It represents an IM with a squirrel-cage rotor [57] with $n_{\text{rt}} = 32$ conductive bars in total. The 3ph current-carrying windings in stator are distributed among $n_{\text{st}} = 36$ slots. Such a configuration yields (due to the Maxwell–Ampère law (3.1b)) a magnetic flux distribution over 4 poles, which means that each pole covers a quarter of the motor. One pole of the two-dimensional (2D) IM model is depicted in Figure 3.5a.

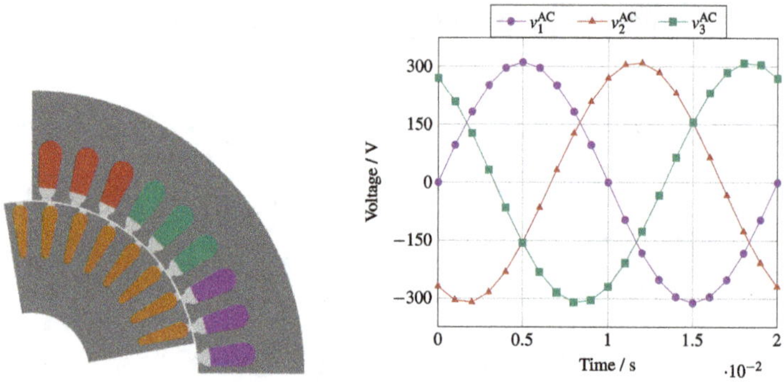

A 2D induction machine model [58]. (a)

3ph voltage of $f = 50$ Hz, (b)

Figure 3.5 One pole of a squirrel-cage induction motor and 3ph AC excitation

The electromagnetic phenomena in the IM can be described with sufficient accuracy by the MQS setting, in particular using the $\vec{A}^*$-formulation (see Section 3.1.2). We therefore consider for $\vec{x} \in \Omega$ and $t \in (0, T]$ the eddy current model

$$\mathrm{curl}(\nu(\vec{x}, |\mathrm{curl}\vec{A}|)\mathrm{curl}\vec{A}) + \sigma(\vec{x})\frac{\partial \vec{A}}{\partial t}(\vec{x}, t) = \sum_{k=1}^{3} \vec{\chi}_k(\vec{x}) i_k(t), \tag{3.123}$$

coupled to the 3ph voltage supply through

$$v_k(t) = R_k i_k(t) + \int_{\Omega} \vec{\chi}_k(\vec{x}) \cdot \frac{\partial \vec{A}}{\partial t}(\vec{x}, t)\mathrm{d}\Omega, \quad k = 1, 2, 3 \tag{3.124}$$

where R_k is the resistance of the stranded conductor k and $\vec{\chi}_k$ is the winding function already introduced in (3.4). Including the BC and the IC

$$\vec{n} \times \vec{A}(\vec{x}, t) = 0, \quad (\vec{x}, t) \in \partial\Omega \times (0, T], \tag{3.125}$$

$$\vec{A}(\vec{x}, 0) = 0, \quad \vec{x} \in \Omega \tag{3.126}$$

we discretize the planar problem in 2D using FEM in space with 4,400 degrees of freedom to obtain a time-dependent system of DAEs by Gmsh/GetDP [60].

3.6.1.1 Multirate Parareal for IVP

We consider the voltages (3.124) given by the 3ph PWM voltage source of 20 kHz, i.e.,

$$v_k^{\mathrm{PWM}}(t) = \mathrm{sgn}[m v_k^{\mathrm{AC}} - c(t)], \quad k = 1, 2, 3. \tag{3.127}$$

Here $m = 0.8$ is the modulation factor, v_k^{AC} is the 3ph sinusoidal reference signal given by

$$v_k^{\mathrm{AC}}(t) = \widehat{U} \sin(2\pi f t - (k-1) \cdot 2\pi/3) \quad \text{with } f = 50 \tag{3.128}$$

and depicted in Figure 3.5b, and

$$c(t) = \widehat{U}[2(f_s t - \lfloor f_s t \rfloor) - 1] \quad \text{with } f_s = 20\,\text{kHz} \tag{3.129}$$

is the sawtooth carrier obtained from a bipolar trailing-edge modulation [59]. The amplitude in (3.128) and (3.129) is $\widehat{U} = 220\sqrt{2} \approx 311.13$ V.

The multirate Parareal algorithm (3.104) and (3.105) (from Section 3.5.2) applied on one period $[0, T]$, $T = 1/f = 0.02$ uses the 3ph voltage inputs

$$v_k^{\mathcal{F}}(t) = f_{\text{relax}}(t) v_k^{\text{PWM}}(t), \qquad v_k^{\bar{\mathcal{G}}}(t) = f_{\text{relax}}(t) m v_k^{\text{AC}}(t), \qquad k = 1,2,3 \tag{3.130}$$

on the fine and on the coarse levels, respectively. The function

$$f_{\text{relax}}(t) = 0.5(1 - \cos(0.5\pi f t)) \quad \text{with } f = 50 \tag{3.131}$$

introduces a smooth relaxation of the applied voltages, used for reducing the transient part [60]. The fine propagator $\mathcal{F}$ uses the time step size $\delta T = 10^{-6}$ s, while the coarse propagator $\bar{\mathcal{G}}$ uses $\Delta T = 10^{-3}$ s. The considered time interval $[0, 0.02]$ s is partitioned into $N = 20$ subintervals, i.e., one coarse step is done per subinterval. Both fine and coarse solvers use the IE method for problems with the corresponding voltage inputs v_k from (3.130).

In [23], a comparison of the standard Parareal (3.94) and (3.95) where the PWM voltage source is used on both fine and coarse levels and the multirate Parareal method is shown. The multirate Parareal requires a quarter of the number of iterations fewer than the standard Parareal to obtain the solution up to the tolerance $< 1.5 \times 10^{-5}$ [23]. In terms of the effective number of solved linear systems, the multirate Parareal calculates 121,264 linear solutions (in 6 iterations), the standard Parareal requires the solution of 156,309 linear systems (in 8 iterations), while the classical sequential time-stepping on $[0, 0.02]$ s with the IE scheme and step size $\delta T = 10^{-6}$ s needs 244,104 linear systems solution [23].

3.6.1.2 Multirate Parareal for steady-state calculation

We now focus on the steady state calculation (as described in Section 3.2.4) for the IM model supplied with a 3ph PWM voltage input of $f_s = 20$ kHz. For this, three possible solution approaches are considered: the classical time stepping until the steady state is reached at the time instant T_{st}, application of the multirate Parareal for IVPs on the time interval $[0, T_{\text{st}}]$, and the multirate PP-IC method applied on one period $[0, T]$. As already mentioned in Remark 3.2, the multirate variant (3.112) and (3.113) of the PP-IC agorithm is introduced in [24]. The inputs of the fine and the coarse problems are given by v_k^{PWM} from (3.128) and $m v_k^{\text{AC}}$ from (3.128), respectively.

The sequential time stepping performed using the IE method with the time step size $\delta T = 10^{-6}$ s starting from the IC (3.126) attained the steady state after the calculation over 9 periods, i.e., at time $T_{\text{st}} = 9T = 9 \times 0.02\,\text{s} = 0.18$. The multirate Parareal was applied on the interval $[0, 9T] = [0, 0.18]$ s split into $N = 20$ subintervals with the coarse time step $\Delta T = 9T/N = 0.18/20\,\text{s} = 0.009$ and the fine time step size $\delta T = 10^{-6}$ s [24]. Finally, the multirate PP-IC method calculated the steady-state solution directly on the period $[0, T] = [0, 0.02]$ s with $N = 20$ subintervals, the coarse time step size $\Delta T = T/N = 10^{-3}$ s, and the fine time step $\delta T = 10^{-6}$ s [24]. Computational costs in terms of the effective number of linear systems solutions and the corresponding percentages are given in Table 3.2.

3.6.2 Coaxial cable

We now illustrate the performance of the considered approaches to obtain the steady-state solution for a linear 2D coaxial cable model used in Refs. [26,52]. It is described by the eddy current equation (3.14), where we assume that the reluctivity function ν is independent of the MVP $\vec{A}$. The excitation of the system is defined by AC $i(t) = 100sin(2\pi 50t)$ flowing through the coil. The 2D model is discretised in space with FEM using 2,269 degrees of freedom. The PP-IC method (3.110) and (3.111), the TP-FEM method with the MH diagonalization (from Sections 3.4 and 3.4.1), and the PP-PC algorithm with the MH diagonalization (from Section 3.5.3.1) are applied on the time interval $[0, T] = [0, 0.02]$ s splitted into N equal subintervals of the size $\Delta T = T/N$. The methods are compared to the classical time-stepping with IE and the time step $\delta T = 10^{-5}$ s starting from zero IC. The sequential solution required calculation over $k^* = 42$ periods, thereby solving 84,000 linear systems until the steady state [52].

Table 3.2 Effective number of linear system solutions for the IM model with the PWM voltage input [24]

Sequential	Multirate Parareal	Multirate PP-IC
2,176,179 (100%)	583,707 (26.82%)	194,038 (8.92%)

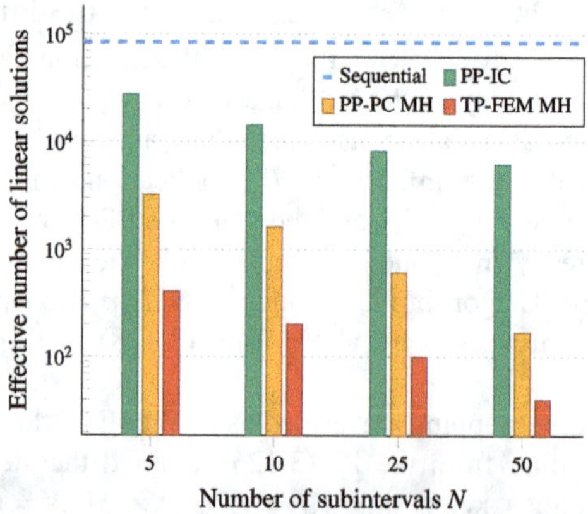

Figure 3.6 Computational costs for the linear coaxial cable model [52]

Table 3.3 Effective number of linear system solutions for the linear coaxial cable model [52]

Sequential	N	PP-IC	PP-PC MH	TP-FEM MH
84,000 (100%)	5	27,540 (32.8%)	3,216 (3.8%)	400 (0.47%)
	10	14,280 (17%)	1,616 (1.9%)	200 (0.24%)
	25	8,160 (9.7%)	612 (0.7%)	100 (0.12%)
	50	6,120 (7.3%)	168 (0.2%)	40 (0.05%)

Both PP-IC and PP-PC MH use the IE discretizations with the fine and the coarse step sizes equal to $\delta T = 10^{-5}$ s and $\Delta T = 10^{-3}$ s, respectively. TP-FEM MH discretizes the system only on the fine grid with $\delta T = 10^{-5}$ s, which requires solution of $N_f = 2,000$ independent linear systems. This can be parallelized among N available processors by assigning N_f/N system solutions to each processor, which corresponds to the optimal scaling. Figure 3.6 and Table 3.3 illustrate the costs in terms of the number of solved linear systems for different number of cores N keeping the precision on the fine level unchanged (strong scaling) [52]. As expected, in the linear case, the least computational effort was required by TP-FEM MH. However, it can be seen in [52] that for a nonlinear model the PP-PC MH approach with a simplified Newton linearization gave the best results.

3.7 Conclusions

In this chapter, several low-frequency formulations were presented and discretized with the FEM in the appropriate function spaces. We considered two settings to calculate the steady state of the space-discrete system: the initial value and the TP problems, whose solutions are commonly calculated using implicit time stepping methods. Since the classical sequential solution often leads to prohibitively long computation times, PinT integration methods were proposed. Additionally, for TP problems a multi-harmonic (MH) diagonalization was used to simplify the solution of large block systems. Incorporation of the MH approach into the PinT framework allowed to introduce parallelism on the coarse level. Finally, the considered parallelization approaches were applied to an induction machine model and to a coaxial cable models which showed a considerably smaller computational effort compared to the sequential time-stepping approach.

Acknowledgement

This work was supported by the European High-Performance Computing Joint Undertaking (JU) under the grant agreement No. 955701. The JU receives support from the European Union's Horizon 2020 research and innovation programme and Belgium, France, Germany, Switzerland.

References

[1] Maxwell JC. A dynamical theory of the electromagnetic field. Philosophical *Transactions of The Royal Society of London* 1865; CLV:459–512.
[2] Heaviside O. On the forces, stresses, and fluxes of energy in the electromagnetic field. *Proceedings of the Royal Society of London* 1892;50:126–129.
[3] Assous F, Ciarlet P, Labrunie S. *Mathematical Foundations of Computational Electromagnetism.* New York, NY: Springer.
[4] Schöps S, De Gersem H, Weiland T. Winding functions in transient magnetoquasistatic field-circuit coupled simulations. *Proceedings of the Royal Society of London* 2013;32(6):2063–2083.
[5] Alonso Rodríguez A, Valli A. Eddy current approximation of Maxwell equations. *Vol. 4 of Modeling, Simulation and Applications.* New York, NY: Springer.
[6] Dirks HK. Quasi-stationary fields for microelectronic applications. *Electrical Engineering* 1996;79(2):145–155.
[7] Koch S, Weiland T. Different types of quasistationary formulations for time domain simulations. *Radio Science* 2011;46(5).
[8] Griffiths DF. *Introduction to Electrodynamics*. London: Prentice-Hall.
[9] Ida N, Bastos JPA. *Electromagnetics and Calculation of Fields*. 2nd ed. New York, NY: Springer.
[10] Brenner SC, Scott LR. The mathematical theory of finite element methods. vol. 15 *of Texts in Applied Mathematics.* 3rd ed. New York, NY: Springer.
[11] Bossavit A. Computational Electromagnetism: Variational Formulations, Complementarity, *Edge Elements.* London: Academic Press.
[12] Adams RA. Sobolev Spaces. *Pure and Applied Mathematics.* London: Academic Press.
[13] Hairer E, NØrsett SP, Wanner G. Solving Ordinary Differential Equations II: Stiff and Differential-Algebraic Problems. 2nd ed. *Springer Series in Computational Mathematics.* New York, NY: Springer.
[14] Deuflhard P. *Newton Methods for Nonlinear Problems: Affine Invariance and Adaptive Algorithms.* New York, NY: Springer.
[15] Gyselinck J, Dular P, Geuzaine C, *et al.* Harmonic-balance finite-element modeling of electromagnetic devices: a novel approach. *IEEE Transactions on Magnetics* 2002;38(2):521–524.
[16] Egiziano L, Tucci V, Petrarca C, *et al.* A Galerkin model to study the field distribution in electrical components employing nonlinear stress grading materials. *IEEE Transactions on Dielectrics and Electrical Insulation* 1999;6(6):765–773.
[17] Quarteroni A, Valli A. Numerical Approximation of Partial Differential Equations. *Vol. 23 of Springer Series in Computational Mathematics.* New York, NY: Springer.

[18] Bast D, Kulchytska-Ruchka I, Schöps S, *et al.* Accelerated steady-state torque computation for induction machines using parallel-in-time algorithms. *IEEE Transactions on Magnetics* 2020;56(2):1–9.

[19] Bermúdez A, Gómez D, Piñeiro M, *et al.* A novel numerical method for accelerating the computation of the steady-state in induction machines. *Computers & Mathematics with Applications* 2020;79(2):274–292.

[20] Takahashi Y, Tokumasu T, Kameari A, *et al.* Convergence acceleration of time-periodic electromagnetic field analysis by singularity decomposition-explicit error correction method. *IEEE Transactions on Magnetics* 2010;46(8):2947–2950.

[21] Lions JL, Maday Y, Turinici G. A parareal in time discretization of PDEs. *Comptes Rendus de l'Académie des Sciences – Series* 2001;332(7):661–668.

[22] Vasca F, Iannelli L, editors. *Dynamics and Control of Switched Electronic Systems.* New York, NY: Springer.

[23] Gander MJ, Kulchytska-Ruchka I, Niyonzima I, *et al.* A new Parareal algorithm for problems with discontinuous sources. *SIAM Journal on Scientific Computing* 2018;41(2):B375–B395.

[24] Gander MJ, Kulchytska-Ruchka I, Schöps S. A new Parareal algorithm for time-periodic problems with discontinuous inputs. In: *Domain Decomposition Methods in Science and Engineering XXV*. Lecture Notes in Computational Science and Engineering, vol 138. Cham: Springer; 2020.

[25] Bíró O, Preis K. An efficient time domain method for nonlinear periodic eddy current problems. *IEEE Transactions on Magnetics* 2006;42(4):695–698.

[26] Kulchytska-Ruchka I, De Gersem H, Schöps S. An efficient steady-state analysis of the eddy current problem using a parallel-in-time algorithm. In: *The Tenth International Conference on Computational Electromagnetics* (CEM 2019).

[27] Hairer E, NØrsett SP, Wanner G. Solving Ordinary Differential Equations I: Nonstiff Problems. 2nd ed. *Springer Series in Computational Mathematics.* New York, NY: Springer.

[28] Burrage K. *Parallel and Sequential Methods for Ordinary Differential Equations.* Oxford: Oxford University Press.

[29] Deuflhard P, Deuflhard P, Bornemann F. *Scientific Computing with Ordinary Differential Equations.* 1st ed. New York, NY: Springer.

[30] Nicolet A, Delincé F. Implicit Runge-Kutta methods for transient magnetic field computation. *IEEE Transactions on Magnetics* 1996;32(3):1405–1408.

[31] Cameron F, Piche R, Forsman K. Variable step size time-integration methods for transient eddy current problems. *IEEE Transactions on Magnetics* 1998;34(5):3319–3322.

[32] Clemens M, Wilke M, Weiland T. 3-D transient eddy-current simulations using FI2TD schemes with variable time-step selection. *IEEE Transactions on Magnetics* 2002;38(2):605–608.

[33] Schöps S, Niyonzima I, Clemens M. Parallel-in-time simulation of Eddy current problems using Parareal. *IEEE Transactions on Magnetics* 2017;54(3):1–4.

[34] Estévez Schwarz D. Consistent initialization for index-2 differential algebraic equations and its application to circuit simulation [Dissertation].

[35] Estévez Schwarz D, Lamour R. The computation of consistent initial values for nonlinear index-2 differential–algebraic equations. *Numerical Algorithms* 2001;26(1):49–75.

[36] Clemens M, Weiland T. Numerical algorithms for the FDiTD and FDFD simulation of slowly varying electromagnetic fields. *International Journal of Numerical Modelling* 1999;12(1–2):3–22.

[37] Steinmetz T, Helias M, Wimmer G, *et al.* Electro-quasistatic field simulations based on a discrete electromagnetism formulation. *IEEE Transactions on Magnetics* 2006;42(4):755–758.

[38] Schöps S. Multiscale Modeling and Multirate Time-Integration of Field/Circuit Coupled Problems [Dissertation]. VDI Verlag. Fortschritt-Berichte VDI, *Reihe* 21. Available from: http://elpub.bib.uni-wuppertal.de/servlets/DocumentServlet?id=2132.

[39] Clemens M, Wilke M, Weiland T. Linear-implicit time-integration schemes for error-controlled transient nonlinear magnetic field simulations. *IEEE Transactions on Magnetics* 2003;39(3):1175–1178.

[40] Clemens M, Wilke M, Schuhmann R, *et al.* Subspace projection extrapolation scheme for transient field simulations. *IEEE Transactions on Magnetics* 2004;40(2):934–937.

[41] Clemens M. Large systems of equations in a discrete electromagnetism: formulations and numerical algorithms. *IEE Proceedings - Science, Measurement and Technology* 2005;152(2):50–72.

[42] Hara T, Naito T, Umoto J. Time-periodic finite element method for nonlinear diffusion equations. *IEEE Transactions on Magnetics* 1985;21(6):2261–2264.

[43] Trefethen LN. *Finite Difference and Spectral Methods for Ordinary and Partial Differential Equations.* Ithaca, NY: Cornell University.

[44] Urabe M. Galerkin's procedure for nonlinear periodic systems. *Archive for Rational Mechanics and Analysis* 1965;20(2):120–152.

[45] Nievergelt J. Parallel methods for integrating ordinary differential equations. *Communications of the ACM* 1964;7:731–733.

[46] Gander MJ. 50 years of time parallel time integration. In: *Multiple Shooting and Time Domain Decomposition Methods*, Contributions in Mathematical and Computational Sciences, vol 9, Cham: Springer; pp. 69–113.

[47] Falgout RD, Friedhoff S, Kolev TV, *et al.* Parallel time integration with multigrid. *SIAM Journal on Scientific Computing* 2014;36(6):C635–C661.

[48] Gander MJ, Güttel S. PARAEXP: a parallel integrator for linear initial-value problems. *SIAM Journal on Scientific Computing* 2013;35(2):C123–C142.

[49] Gander MJ, Hairer E. Nonlinear convergence analysis for the parareal algorithm. In: Langer U, Discacciati M, Keyes DE, *et al.*, editors. *Domain Decomposition Methods in Science and Engineering XVII.* New York, NY: Springer. pp. 45–56.

[50] Gander MJ, Jiang YL, Song B, *et al.* Analysis of two parareal algorithms for time-periodic problems. *SIAM Journal on Scientific Computing* 2013;35(5):A2393–A2415.

[51] Wu SL. Towards parallel coarse grid correction for the Parareal algorithm. *SIAM Journal on Scientific Computing* 2018;40(3):A1446–A1472.

[52] Kulchytska-Ruchka I, Schöps S. Efficient parallel-in-time solution of time-periodic problems using a multi-harmonic coarse grid correction. *SIAM Journal on Scientific Computing* 2021;43(1):C61–C88.

[53] Gander MJ, Vandewalle S. Analysis of the parareal time-parallel time-integration method. *SIAM Journal on Scientific Computing* 2007;29(2):556–578. Available from: https://doi.org/10.1137/05064607X.

[54] Zeidler E. *Nonlinear Functional Analysis and its Applications II/A. Linear Monotone Operators*. New York, NY: Springer-Verlag *Inc*.

[55] Filippov AF. Differential equations with discontinuous righthand sides: control systems. Vol. 18 *of Mathematics and Its Applications*. Dordrecht: Springer; 1988.

[56] Gander MJ, Liu J, Wu SL, *et al.* ParaDiag: Parallel-in-Time Algorithms Based on the Diagonalization Technique; 2020. ArXiv: 2005.09158.

[57] Hughes TJR, Cottrell JA, Bazilevs Y. Isogeometric analysis: CAD, finite elements, NURBS, exact geometry and mesh refinement. *Computer Methods in Applied Mechanics and Engineering* 2005;194:4135–4195.

[58] Gyselinck J, Vandevelde L, Melkebeek J. Multi-slice FE modeling of electrical machines with skewed slots-the skew discretization error. *IEEE Transactions on Magnetics* 2001;37(5):3233–3237.

[59] Sun J. *Pulse-Width Modulation*. London: Springer-Verlag; pp. 25–61.

[60] Geuzaine C. GetDP: a general finite-element solver for the de Rham complex. *PAMM* 2007;7:1010603–1010604.

Chapter 4

Transient analysis of scattering from composite objects using late-time stable TDIEs

A.J. Pray[1], N. V. Nair[1], D. S. Weile[2] and B. Shanker[1]

4.1 Introduction

Analysis of scattering and propagation of fields in the time domain offers a number of advantages as opposed to that in frequency domain. The obvious one is that it is possible to obtain behavior over the desired spectrum, including early time behavior, ease of inclusion of non-linearities, be it point non-linearities as in circuit elements or bulk non-linearities [1–4], it is a natural framework to understand light™matter interaction [5–7], and so on. Given these benefits, it is natural to develop a transient field solver. This was indeed the case; the finite difference time domain (FDTD) is perhaps the most popular transient analysis tool [1,2]. Its popularity, in large part, is thanks to the simplicity of both the formulation and the implementation. As a result, this approach has seen applications to a wide variety of problems. Along the same vein, there has been extensive body of work on transient differential equation-based solvers that alleviate some of the bottlenecks of FDTD [4,8,9]. Unfortunately, much as we would like to delve into this topic, this chapter is devoted to a promising alternative–transient boundary integral equation solvers.

In the Fourier domain, the use of boundary integral-based solution techniques is rampant, be it for analysis of radiation and scattering or to impose exact boundary conditions [10]. This popularity is in large part to overcoming the computational bottleneck associated with evaluation of matrix vector products (and storage of matrices); note, if there are N_s spatial degrees of freedom, the cost of computation of a matrix vector product as well as storage scales as $O(N_s^2)$. The development of the fast multipole method reduced this to $\mathcal{O}(N_s \log N_s)$ [11–13]. The landscape in the time domain is very different. While time domain integral equations have been an active research topic since the 1960s [14], they have yet to be as widely employed as their Fourier domain counterparts. This is despite an extensive body of work that has addressed two fundamental bottlenecks: (i) late time stability and (ii) computational complexity [15–20]. Methods that reduce computational

[1]Department of ECE, Michigan State University, USA
[2]Department of ECE, University of Delaware, USA

complexity, from $\mathcal{O}(N_t N_s^2)$ to $\mathcal{O}(N_t N_s \log^2 N_s)$ where N_t is the number of degrees of freedom in time, have been introduced by the authors [20,21]. However, what has remained reasonably elusive is late time stability.

The research in understanding and prescribing methods to overcoming stability is extensive. While a number of stabilization schemes have been developed over the last two decades [17,22–26], four techniques stand out; these are Lubich quadrature [27,28], smooth and band-limited temporal basis functions [24], quasi-exact integration [29,30], and the separable expansion method [31]. Lubich quadrature is based on using a Laplace transform to the appropriate TDIE and using a finite differencing scheme to map to the $\mathcal{Z}$-transform domain. This discretized equation is then solved by marching in time. These methods have yielded excellent results and have robust stability properties, but are characterized by interactions with infinite temporal tails which in turn lead to higher scaling in both computational cost and memory. Methods to overcome this cost scaling are current topics of research. Alternatively, smooth and band-limited temporal basis functions can be reasonably effective in stabilizing TDIE solutions to either the electric field integral equation (EFIE) for open regions or the combined field integral equation (CFIE) for closed regions when integrated together with methods to avoid the static null space. However, their stability has not been examined for late time and multiple transits across the domain for example in resonant cavities. The smoothness of these functions allows for accurate evaluation of retarded potential integrals using conventional quadrature rules. However, these functions are symmetric about their interpolation point and, therefore, produce marching systems that are non-causal, requiring extrapolation. The extrapolation produces additional errors.

The stabilization of time domain integral equations has been a topic of significant interest in the applied mathematics literature as well as evidenced by [32–37]. It has been shown that full space–time Galerkin methods are stable [36], but their implementation is challenging. It requires evaluating five-dimensional integrals (4 in space and 1 in time) and determining regions of continuity. Approximate evaluation of these integrals is performed in quasi-exact integration methods [29,30,36,38]. Various numerical experiments from different authors suggest that it suffices to only calculate a subset of the four spatial integrals analytically, while the remaining integrals can be approximated by a suitable quadrature rule. For instance, in [29,36], only the inner 2 integrals are evaluated analytically, while the outer 2 are evaluated approximately. Ref. [30] transforms the 2 surface integrals into 1 volume integral, which is evaluated analytically, and 1 line integral, which is approximated. The analytical evaluation of these surface or volume integrals is only possible on flat tessellations. Accurate approximation of these integrals on a higher order tessellation requires a different approach. Starting from the functional framework of [37,39] derived the appropriate functional framework for the differentiated EFIE and demonstrated that their results (marching on in time (MOT)-matrix elements) are identical to those derived using quasi-exact schemes for low-order tessellations, and late time stable solutions were obtained. However, the results depend critically on exact source integrals, time step size, etc. Unfortunately, to the best of our knowledge, extensive data are not available.

While the stability using quasi-exact schemes has been demonstrated for the EFIE, its use for stabilizing higher order geometries is almost impossible. Consider the method developed in [29] as well as papers that followed [30,38]. To extend this to higher order, one needs to find the intersection of curvilinear elements with spheres to demarcate regions of continuity.

A purely numerical approach was introduced in [31], i.e., it did not involve analytical integration, to stabilizing TDIE solutions. Instead, it invoked an interesting proposition. A high level description of the method was to develop an alternate representation of Green's function such that its convolution with the spatial basis can be effected to desired precision. In what follows, this approach will be referred to as the separable expansion method. Here, the temporal basis sets used to represent the current and measure the field are functions whose support spans $[-1, 1]$ and collocation in time, respectively. It can be shown that the MOT system that arises from this procedure is tantamount to those arising from using temporal basis that satisfy the properties defined by Terrasse [37]. The work presented in both [31] and [40] was restricted to scattering from perfect electrically conducting objects, [41] uses this to couple in TDIE and non-linear circuits, and [42–44] have used this approach for analysis of transient potential integral equations in a variety of settings. A related approach (a series representation of the retarded potential) was used to derive quasi-analytic (and robust) solutions to scattering from spherical objects [45–47].

This chapter presents an extension of the separable expansion technique to multiply connected domains, which are either impenetrable with a PEC boundary, or penetrable with constant permittivity and permeability. To demonstrate the stability and validity of the solution obtained via the separable expansion method, comparisons will be made to the quasi-exact integration method [29]. Note, the material here borrows from unpublished work in [48]; addition details can be found in this citation. This chapter is organized as follows. In Section 4.2, the scattering problem will be introduced. Next, Section 4.3 will detail the discretization of the scattering problem using the method of moments, the main source of late time instability, i.e., inaccuracy in the computation of matrix elements, quasi-exact schemes, bounds on error inherent in these methods, as well as integration with TDIEs. Numerical results will be presented in Sections 4.4 and 4.5 will draw conclusions.

4.2 Problem statement

Consider a body Ω composed of Q homogeneous regions, either penetrable $(\varepsilon_{r,q}, \mu_{r,q})$ or PEC, illuminated by a field $\{\mathbf{E}^i(\mathbf{r}, t), \mathbf{H}^i(\mathbf{r}, t)\}$ that is assumed band-limited and zero for $t \leq 0$. Let the medium exterior to the scatterer (assumed to be free space) be denoted by Ω_0. A generic description of the problem is depicted in Figure. 4.1. Within each region, Ω_q, assume there exist an impressed source (or sources), which give rise to fields $\{\mathbf{E}_q^{inc}, \mathbf{H}_q^{inc}\}$. The total fields in region in Ω_q are then given by

$$\mathbf{E}_q^{tot}(\mathbf{r}, t) = \mathbf{E}_q^{inc}(\mathbf{r}, t) + \mathbf{E}_q^{scat}(\mathbf{r}, t) \circ \{\mathbf{J}_q, \mathbf{M}_q\} \tag{4.1a}$$

$$\mathbf{H}_q^{tot}(\mathbf{r}, t) = \mathbf{H}_q^{inc}(\mathbf{r}, t) + \mathbf{H}_q^{scat}(\mathbf{r}, t) \circ \{\mathbf{J}_q, \mathbf{M}_q\} \tag{4.1b}$$

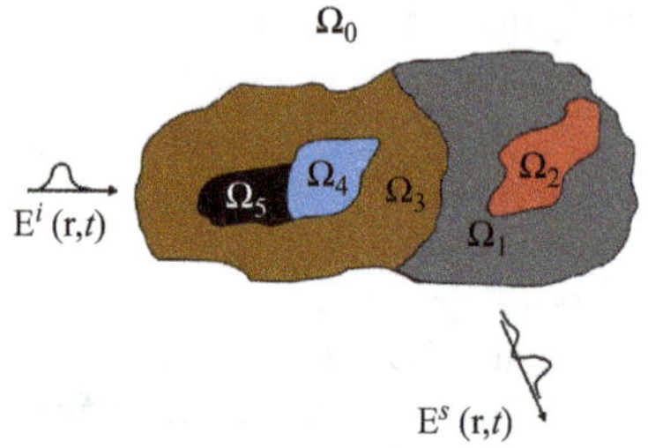

Figure 4.1 Generic description of scattering by a composite object

where $\{\mathbf{E}_q^{scat}, \mathbf{H}_q^{scat}\}$ are the fields scattered by equivalent sources $\{\mathbf{J}_q, \mathbf{M}_q\}$, residing on the boundary of Ω_q. In region q, the scattered fields $\{\mathbf{E}_q^{scat}, \mathbf{H}_q^{scat}\}$ are given as

$$\mathbf{E}_q^{scat} \circ \{\mathbf{J}_q, \mathbf{M}_q\} = -\mu_q \mathcal{L}_{e,q}(\mathbf{r}, t) \circ \{\mathbf{J}_q\} - \mathcal{L}_{h,q}(\mathbf{r}, t) \circ \{\mathbf{M}_q\} \tag{4.2a}$$

$$\mathbf{H}_q^{scat} \circ \{\mathbf{J}_q, \mathbf{M}_q\} = -\varepsilon_q \mathcal{L}_{e,q}(\mathbf{r}, t) \circ \{\mathbf{M}_q\} + \mathcal{L}_{h,q}(\mathbf{r}, t) \circ \{\mathbf{J}_q\} \tag{4.2b}$$

with the operators

$$L_{e,q} \circ \{X\} \doteq \int_0^t dt' \int_{\partial\Omega_q} dS \left(\partial_{t'}^2 \bar{\mathbf{I}} - c_q^2 \nabla\nabla\right) \cdot \frac{X(r', \tau_q)}{4\pi R} \tag{4.3a}$$

$$\mathcal{L}_{h,q} \circ \{X\} \doteq \int_{\partial\Omega_q} dS \nabla \times \frac{X(r', \tau_q)}{4\pi R} \tag{4.3b}$$

where c_q denotes the speed of light in medium q and $\tau_q = t - R/c_q$ is the retarded time.

Each region has an equivalent interior and exterior problem. Denoting the interior and exterior regions by $(+)$ and $(-)$, respectively, we have

$$\mathbf{J}_q^{\pm}(\mathbf{r}, t) = \pm\widehat{n}_q \times \mathbf{H}^{tot,\pm} \tag{4.4a}$$

$$\mathbf{M}_q^{\pm} = \mathbf{E}^{tot,\pm} \times \pm\widehat{n}_q \tag{4.4b}$$

where $\{\mathbf{E}^{tot,\pm}, \mathbf{H}^{tot,\pm}\}$ are the total fields for the equivalent interior and exterior problems and $\widehat{n}_q$ is the normal to surface $\partial\Omega_q$ pointing outward from region Ω_q. The boundary conditions can then be used to relate the interior and exterior currents. For dielectric interfaces, we have continuity in the tangential components of the fields at the boundary, $\partial\Omega_q$,

$$\widehat{n}_q \times \mathbf{E}^{tot,+}|_{\mathbf{r}\in\partial\Omega_q} = \widehat{n}_q \times \mathbf{E}^{tot,-}|_{\mathbf{r}\in\partial\Omega_q} \tag{4.5a}$$

$$\widehat{n}_q \times \mathbf{H}^{tot,+}|_{\mathbf{r}\in\partial\Omega_q} = \widehat{n}_q \times \mathbf{H}^{tot,-}|_{\mathbf{r}\in\partial\Omega_q} \tag{4.5b}$$

which lead to the conditions

$$\mathbf{J}_q(\mathbf{r},t) \doteq \mathbf{J}_q^+(\mathbf{r},t) = -\mathbf{J}_q^-(\mathbf{r},t) \tag{4.6a}$$

$$\mathbf{M}_q(\mathbf{r},t) \doteq \mathbf{M}_q^+(\mathbf{r},t) = -\mathbf{M}_q^-(\mathbf{r},t) \tag{4.6b}$$

For PEC boundaries, the internal fields are zero and the boundary conditions for the external fields are

$$\widehat{n}_q \times \mathbf{E}^{tot,+}\big|_{\mathbf{r}\in\partial\Omega_q} = 0 \quad (4.7a)$$

$$\widehat{n}_q \times \mathbf{H}^{tot,+}\big|_{\mathbf{r}\in\partial\Omega_q} = \mathbf{J}_q(r,t) \quad (4.7b)$$

This means that there are no equivalent magnetic currents, and only external electric currents as defined in (4.7b). Lastly, the tangential components of both $\mathbf{E}_q^{tot,\pm}$ and $\mathbf{H}_q^{tot,\pm}$ are equal to each other on either side of the boundary $\partial\Omega_q$ and may be combined to yield the PMCHWT formulation for dielectrics [49]

$$\begin{aligned}&\widehat{n}_q \times \left[\mathbf{E}_q^{inc,+}(\mathbf{r},t) - \mathbf{E}_q^{inc,-}(\mathbf{r},t)\right] = \\ &-\widehat{n}_q \times \left[\mathbf{E}_q^{scat,+}(\mathbf{r},t) \circ \left\{\mathbf{J}_q, \mathbf{M}_q\right\} + \mathbf{E}_q^{scat,-}(\mathbf{r},t) \circ \left\{\mathbf{J}_q, \mathbf{M}_q\right\}\right]\end{aligned} \tag{4.8a}$$

$$\begin{aligned}&\widehat{n}_q \times \left[\mathbf{H}_q^{inc,+}(\mathbf{r},t) - \mathbf{H}_q^{inc,-}(\mathbf{r},t)\right] = \\ &-\widehat{n}_q \times \left[\mathbf{H}_q^{scat,+}(\mathbf{r},t) \circ \left\{\mathbf{J}_q, \mathbf{M}_q\right\} + \mathbf{H}_q^{scat,-}(\mathbf{r},t) \circ \left\{\mathbf{J}_q, \mathbf{M}_q\right\}\right]\end{aligned} \tag{4.8b}$$

While the rest of the discussion will focus on the PMCHWT formulation, it should be noted that deficiencies of this formulation are well known in terms of low frequency and dense mesh breakdown [50,51]. Several other formulations have been developed in the recent years [52–54] specifically to mitigate this problem. While they have largely been examined in the Fourier domain, it is not difficult to conceive of a time domain version of these formulations. As most of these formulations exploit the same kernels described thus far, transitioning of these integral equations to time domain would not be an arduous tasks.

4.3 Discretization and evaluation of inner products

Given the requisite equations in the previous section, in what follows we discuss the discretization of these in both space and time. To do so, we follow the well-worn path on spatial representation and something slightly different on transient representation. Specifically, the equivalent magnetic and electric currents on $\partial\Omega_q$

are approximated as

$$\begin{aligned}\mathbf{J}_q(\mathbf{r},t) &= \sum_{n=1}^{N_q}\mathbf{S}_{n,q}(\mathbf{r})\left[\sum_{i=1}^{N_t}J_{q,n,i}T_i(t)\right]\\ \mathbf{M}_q(\mathbf{r},t) &= \sum_{n=1}^{N_q}\mathbf{S}_{n,q}(\mathbf{r})\left[\sum_{i=1}^{N_t}M_{q,n,i}T_i(t)\right]\end{aligned} \tag{4.9}$$

where the spatial basis functions, $\mathbf{S}_{n,q}(\mathbf{r})$, are RWG basis functions [55] and the temporal basis functions, $T_i(t) = T(t - i\Delta_t)$, are first-order shifted Lagrange polynomials (hat functions) [29]. Next, Galerkin testing is used in space and point testing is used in time to yield a marching system written as

$$\overline{\overline{Z}}_0\overline{I}_i = \overline{V}_i - \sum_{j=1}^{i-1}\overline{\overline{Z}}_{i-j}\overline{I}_j \tag{4.10a}$$

$$\overline{\overline{Z}}_{i-j} = \begin{bmatrix}\overline{\overline{Z}}_{i-j,0} & 0 & \dots & 0\\ 0 & \overline{\overline{Z}}_{i-j,1} & & \\ \cdot & & \cdot & \\ \cdot & & & \cdot \\ \cdot & & & \cdot \\ 0 & & & \overline{\overline{Z}}_{i-j,Q}\end{bmatrix} \tag{4.10b}$$

$$\overline{\overline{Z}}_{i-j,q} = \begin{bmatrix}\overline{\overline{Z}}^{\mathrm{ee}}_{i-j,q} & \overline{\overline{Z}}^{\mathrm{eh}}_{i-j,q}\\ \overline{\overline{Z}}^{\mathrm{he}}_{i-j,q} & \overline{\overline{Z}}^{\mathrm{hh}}_{i-j,q}\end{bmatrix} \tag{4.10c}$$

$$\left[\overline{\overline{Z}}^{\mathrm{ee}}_{i-j,q}\right]_{mn} = \mu_q\langle\delta(t-(i-j)\Delta_t)\mathbf{S}_{m,q}(\mathbf{r}),\{\mathcal{L}^+_{e,q}+\mathcal{L}^-_{e,q}\}\circ\{\mathbf{S}_{n,q}(\mathbf{r})T(t)\}\rangle \tag{4.11a}$$

$$\left[\overline{\overline{Z}}^{\mathrm{eh}}_{i-j,q}\right]_{mn} = \langle\delta(t-(i-j)\Delta_t)\mathbf{S}_{m,q}(\mathbf{r}),\{\mathcal{L}^+_{h,q}+\mathcal{L}^-_{h,q}\}\circ\{\mathbf{S}_{n,q}(\mathbf{r})T(t)\}\rangle \tag{4.11b}$$

$$\left[\overline{\overline{Z}}^{\mathrm{ee}}_{i-j,q}\right]_{mn} = -\langle\delta(t-(i-j)\Delta_t)\mathbf{S}_{m,q}(\mathbf{r}),\{\mathcal{L}^+_{h,q}+\mathcal{L}^-_{h,q}\}\circ\{\mathbf{S}_{n,q}(\mathbf{r})T(t)\}\rangle \tag{4.11c}$$

$$\left[\overline{\overline{Z}}^{\mathrm{ee}}_{i-j,q}\right]_{mn} = \varepsilon_q\langle\delta(t-(i-j)\Delta_t)\mathbf{S}_{m,q}(\mathbf{r}),\{\mathcal{L}^+_{e,q}+\mathcal{L}^-_{e,q}\}\circ\{\mathbf{S}_{n,q}(\mathbf{r})T(t)\}\rangle \tag{4.11d}$$

$$\overline{I}_j = \left[\overline{I}_{0,j}\,\overline{I}_{1,j}\dots\overline{I}_{Q,j}\right]^T \tag{4.12a}$$

$$\overline{I}_{q,j} = \left[J_{q,1,j}\, J_{q,2,j} \ldots J_{q,N_q,j}\, M_{q,1,j}\, M_{q,2,j} \ldots M_{q,N_q,j}\right]^T \tag{4.12b}$$

$$\overline{V}_i = \left[\overline{V}_{0,i}\, \overline{V}_{1,i} \ldots \overline{V}_{Q,i}\right]^T \tag{4.12c}$$

$$\begin{aligned}
\overline{V}_{q,i} = [&\langle \delta(t-(i-j)\Delta_t)\mathbf{S}_{1,q}(\mathbf{r}), \mathbf{E}_q^{inc,+}(\mathbf{r},t) - \mathbf{E}_q^{inc,-}(\mathbf{r},t)\rangle \\
&\langle \delta(t-(i-j)\Delta_t)\mathbf{S}_{2,q}(\mathbf{r}), \mathbf{E}_q^{inc,+}(\mathbf{r},t) - \mathbf{E}_q^{inc,-}(\mathbf{r},t)\rangle \ldots \\
&\langle \delta(t-(i-j)\Delta_t)\mathbf{S}_{N_q,q}(\mathbf{r}), \mathbf{E}_q^{inc,+}(\mathbf{r},t) - \mathbf{E}_q^{inc,-}(\mathbf{r},t)\rangle \\
&\langle \delta(t-(i-j)\Delta_t)\mathbf{S}_{1,q}(\mathbf{r}), \mathbf{H}_q^{inc,+}(\mathbf{r},t) - \mathbf{H}_q^{inc,-}(\mathbf{r},t)\rangle \ldots \\
&\langle \delta(t-(i-j)\Delta_t)\mathbf{S}_{2,q}(\mathbf{r}), \mathbf{H}_q^{inc,+}(\mathbf{r},t) - \mathbf{H}_q^{inc,-}(\mathbf{r},t)\rangle \\
&\langle \delta(t-(i-j)\Delta_t)\mathbf{S}_{N_q,q}(\mathbf{r}), \mathbf{H}_q^{inc,+}(\mathbf{r},t) - \mathbf{H}_q^{inc,-}(\mathbf{r},t)\rangle]^T
\end{aligned} \tag{4.12d}$$

4.3.1 Accurate evaluation of retarded potential integrals

As alluded to earlier in the chapter, the stability of time domain integral equation methods is tightly linked to basis set used for representing and testing, as well as consistent integration of the requisite inner products. Provided the correct basis sets in space and time are used, quasi-exact schemes provide robust and effective stabilization. This has been demonstrated for both scattering from perfect electrically conducting structures [30,38] and multiply connected structures [29]. As the name implies, this approach relies on both analytic and numerical techniques to evaluate the integrals in (4.11a)™(4.11d). In what follows, we will discuss this procedure briefly before proposing an alternate approach that only employs numerical methods. As is apparent, the latter is more advantageous as it can be readily extended to higher order basis sets i for both geometries and physics [40]. To elucidate the challenges faced in computing these matrix elements, consider the four-dimensional integral

$$\begin{aligned}
T_{mn}^{ij} &= \int_{T_m} d\mathbf{r}\nabla\cdot\mathbf{S}_m(\mathbf{r}) \int_{T_n} d\mathbf{r}'\nabla'\cdot\mathbf{S}_n(\mathbf{r}')\frac{\psi((i-j)\Delta_t - R/c)}{R} \\
&= C\int_{T_m} d\mathbf{r}\int_{T_n} d\mathbf{r}'\frac{\psi((i-j)\Delta_t - R/c)}{R}
\end{aligned} \tag{4.13}$$

where $\psi(t) = \int_{-\infty}^{t} dt' T(t')$ and $T_m \subset \mathrm{supp}\{\mathbf{S}_m(\mathbf{r})\}$ and $T_n \subset \mathrm{supp}\{\mathbf{S}_n(\mathbf{r})\}$ are assumed to be triangles. A constant C has been removed from the integrand as we assume $\mathbf{S}_n(\mathbf{r})$ to be an RWG basis function, whose divergence is constant on T_n. Such integrals appear in the scalar potential due to one spatio-temporal basis function with Galerkin testing applied in space and point testing applied in time. For clarity, we restrict our discussion to the scalar potential term. The procedure for applying these methods to the vector potential is straightforward. Note that RWG functions have support over *two* triangles, but we restrict our discussion to one test and one source triangle for simplicity. Typically, in MOT discretizations, $T(t)$ and $\int dt T(t)$ are only smooth to the zeroth (continuous) and the first order, respectively,

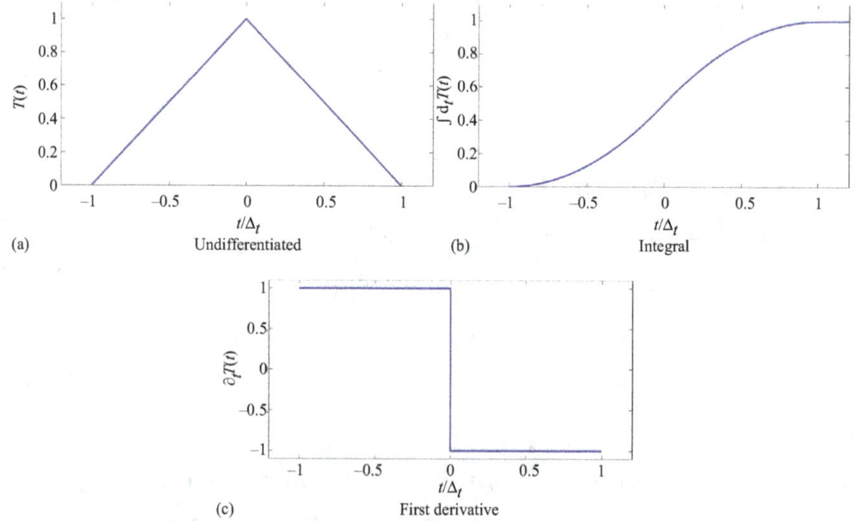

Figure 4.2 First-order shifted Lagrange polynomial: (a) undifferentiated, (b) integral, and (c) first derivative

and $\partial_t T(t)$ is discontinuous. This is shown for the oft-used shifted first-order Lagrange polynomials (hat or chapeau functions) in Figure 4.2. Therefore, integral (4.13) cannot be evaluated accurately using numerical quadrature. In the next section, we will review a class of methods for accurate evaluation (4.13).

4.3.2 Quasi-exact integration schemes

Although the exact evaluation of the four-dimensional integral (4.13) is impractical, schemes have been developed [29,30,56] to evaluate the integral to high precision by approximating a subset of the four dimensions and evaluating the remaining dimensions exactly on flat triangles with RWG functions for the spatial basis. These are the *quasi-exact integration* methods. In [29,56], analytical expressions are given for evaluating the two-dimensional source integral exactly. In contract, [29] approximates the testing integral using numerical quadrature to yield a set of observations points. Then, for a given test point $\mathbf{r}$, the procedure for accurately evaluating the source integral is as follows. The first step is to determine the sub-regions of T_n over which $T(k\Delta_t - R/c)$ is smooth. Let $\text{supp}\{T(t)\} = [-1, P\Delta_t]$, where P is an integer. The domains over which $T(k\Delta_t - R/c)$ is smooth are given by

$$\begin{aligned} k\Delta_t - \frac{R}{c} &\in (l, l+1)\Delta_t \\ \text{or } R &\in c\Delta_t(k-l-1, k-l) \quad l = -1, 0, \ldots, P-1 \end{aligned} \tag{4.14}$$

which describes the region between two spheres of radii $c\Delta_t(k-l-1)$ and $c\Delta_t(k-l)$ centered at $\mathbf{r}$. Denoting this region as $D_{k,l}$, we must find

$$C_{n,k,l} = D_{k,l} \cap T_n \tag{4.15}$$

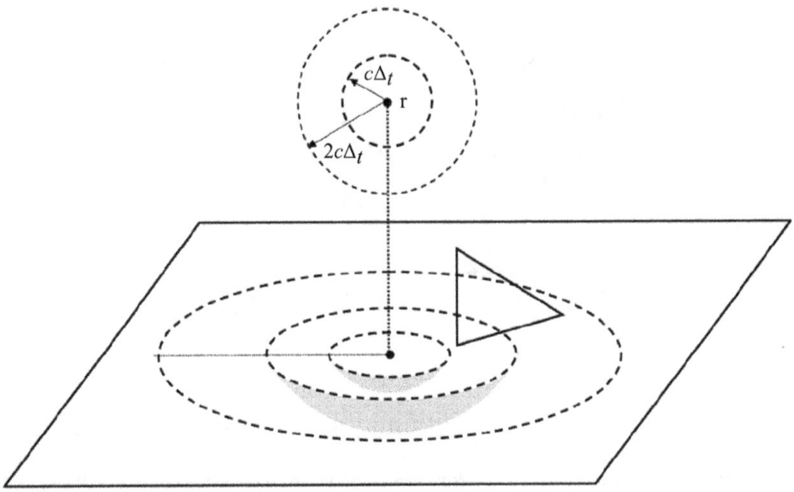

Figure 4.3 Regions of smoothness on source element (graphic from [29])

$C_{n,k,l}$ is the region on T_n over which $T(k\Delta_t - R/c)$ is smooth. Determining $C_{n,k,l}$ is tantamount to finding the arcs of intersection between spheres and source elements, T_n (see Figure 4.3). Analytical expressions exist for the integrals over these sub-regions. The procedure for finding the sub-regions, as described in [56], is to determine all the intersection points between these spheres and the edges of T_n. Next, these points need to be paired such that each pair corresponds to the end points of an arc of intersection (boundary between smooth sub-regions) on T_n. This is a cumbersome task as suggested in Figure 4.4. There are multitudinous ways in which a sphere can intersect a triangle and there can be up to six points of intersection between the sphere and triangle edges.

A second way of computing (4.13) is presented in [30], in which only one dimension of (4.13) is approximated. The integral $\mathcal{T}^{ij}_{mn}$ is rewritten as

$$\begin{aligned}\mathcal{T}^{ij}_{mn} &= C\int_{T_m} d\mathbf{r}\int_{T_n} d\mathbf{r}'\frac{\psi((i-j)\Delta_t - R/c)}{R}\\ &= C\int_{\mathbb{R}^3} d\mathbf{r}\Pi_{T_m}(\mathbf{r})\int_{\mathbb{R}^3} d\mathbf{r}'\Pi_{T_n}(\mathbf{r}')\frac{\psi((i-j)\Delta_t - R/c)}{R}\\ &= C\int_{\mathbb{R}^3} d\mathbf{r}\frac{\psi((i-j)\Delta_t - |\mathbf{r}|/c)}{|\mathbf{r}|}\Omega(\mathbf{r})\end{aligned} \tag{4.16}$$

The integrals have been transformed from surface to volume integrals by introducing the *indicator functions* $\Pi_{T_m}(\mathbf{r})$, defined as

$$\Pi_{T_m}(\mathbf{r}) = \begin{cases} 1 & \mathbf{r} \in T_m \\ 0 & \text{otherwise} \end{cases} \tag{4.17}$$

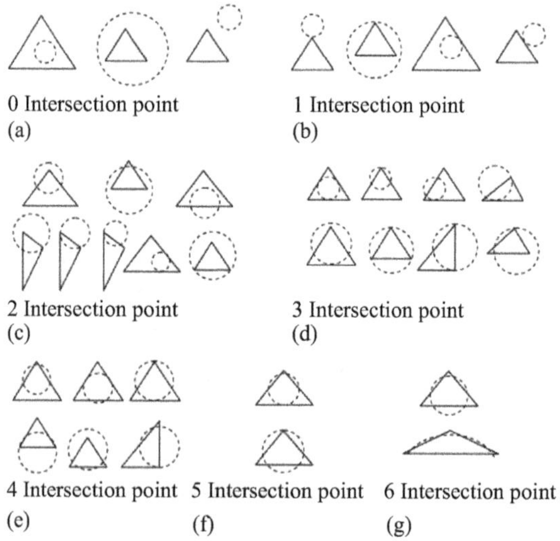

Figure 4.4 Various intersections between spheres and source triangles (graphic from [29])

$\Omega(r)$ is termed the *correlation function* and is given by

$$\Omega(\mathbf{r}) = \int_{\mathbb{R}^3} d\mathbf{r}' \Pi_{T_m}(\mathbf{r}')\Pi_{T_n}(\mathbf{r}' - \mathbf{r}) \tag{4.18}$$

Next, the last line of (4.16) is decomposed into a line and surface integral. The line integral is discretized and evaluated numerically. At each quadrature point on the line, the resulting surface integral is taken over a slice of the correlation function $\Omega(\mathbf{r})$. The support of a slice of $\Omega(\mathbf{r})$, being the correlation between two line segments, is composed of polygonal blocks. The form of $\Omega(\mathbf{r})$ on these surfaces is a summation of multi-nominals. The form of the integrand and the domain of integration lend themselves well to analytical evaluation. The line integral itself, however, is not smooth and it is therefore necessary to find critical points on the line, between which the integrand *is* smooth. The resulting line integrals are evaluated numerically over each smooth sub-domain.

4.3.3 Integration on curvilinear elements

One restriction is common to both quasi-exact integration methods [29] and [30], i.e. they apply only to flat domains. In [29], extension to curved elements leads to two complications. The first complication is in determining the sub-regions $C_{n,k,l}$. The intersection between a sphere and a curved triangle is no longer a simple circle, and the boundaries between sub-domains are not simply the arcs of a circle. Finding the intersection points on the edge of the curved triangle is made much more difficult, if not impossible. The second complication arises in deriving analytical expressions for the integrals over these sub-domains. For flat triangles, the integrals transform nicely into a polar coordinate system. For curved triangles, even if the boundaries of each sub-domain could be found, the integrals will not lend themselves to such a convenient transformation in general. The complications to [30] are similar and arise from the form of $\Omega(\mathbf{r})$ for indicator functions on curved elements.

An alternate, but more robust approach, is to develop methods that are not subjected to the vagaries of an analytical method, but can be resolved using well-tailored numerical methods. In what follows, we develop a method that can compute source integrals to arbitrary precision via standard quadrature rules. We will show that the spatial integrands become smooth. Additionally, we analyze the properties of these integrals in the Fourier domain.

4.3.4 Separable expansion

We begin by considering the two-dimensional source integral

$$\psi(\mathbf{r},t) = \int_{T_n} d\mathbf{r}' \frac{\delta(t - R/c) * T(t) S_n(\mathbf{r}')}{R} \tag{4.19}$$

This type of integral appears in the undifferentiated vector potential due to one spatio-temporal basis function. Next, we approximate this integral as

$$\begin{aligned}\psi(r,t) &= \delta(t-\zeta/c) * \int_{T_n} d\mathbf{r}' \frac{\sum_{l=0}^{\infty} a_l P_l(k_1(R-\zeta)/c + k_2) P_l(k_1 t + k_2) * T(t)\mathbf{S}_n(\mathbf{r}')}{R} \\ &\approx \delta(t-\zeta/c) * \sum_{l=0}^{M} a_l P_l(k_1 t + k_2) * T(t) \int_{T_n} d\mathbf{r}' \frac{P_l(k_1(R-\zeta)/c + k_2)\mathbf{S}_n(\mathbf{r}')}{R}\end{aligned} \tag{4.20}$$

where $a_l = k_1(2l+1)/2$, $k_1\alpha + k_2 = -1$, $k_1\beta + k_2 = 1$, $\beta > \alpha \geq 0$, and $P_l(k_1 x + k_2)$ are Legendre polynomials orthogonal over $x \in [\alpha, \beta]$. ζ represents the largest multiple of $c\Delta_t$ between $\mathbf{r}$ and T_n, as shown in Figure 4.5. This shift is used to minimize the spatial support of the Legendre polynomials. It is clear from (4.20) that, given proper choice of α and β, the spatial integrand is smooth over the entire domain (see Figure 4.5). This allows for the accurate numerical evaluation of (4.20) without having to determine shadow regions as in [29,30]. In this method, T_n must lie completely outside the sphere centered at $\mathbf{r}$ of radius ζ and must lie completely inside the sphere centered at $\mathbf{r}$ of radius $\zeta + (\beta - \alpha)c$. In other words, determining the support of the Legendre polynomials in (4.20) is equivalent to finding a spherical shell, denoted as $\mathbb{B}_{[c\alpha, c\beta]}(R - \zeta)$, that completely encloses T_n, where

$$\mathbb{B}_{[c\alpha,c\beta]}(|r|) = \begin{cases} 1 & |r| \in [c\alpha, c\beta] \\ 0 & \text{otherwise} \end{cases} \tag{4.21}$$

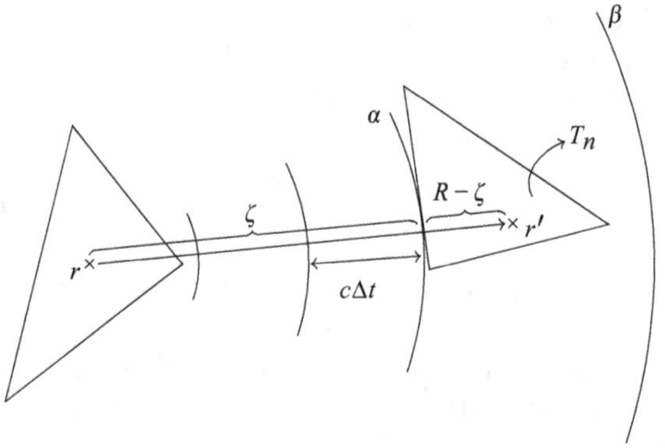

Figure 4.5 Spatial support of expansion (4.20)

This is a much simpler task than the determination of shadow regions. Finding the minimum and maximum distances on a flat triangle from **r** is a simple exercise. These same distances (the minimum and the maximum distance to the flat triangle formed by the curved triangle's vertices) can be used as an initial estimate when curvilinear elements are used. The inner and outer radii of the spherical shell can be decreased and increased based on the curvature of the element. The simplicity of this procedure enables the algorithm to be extended to higher order tessellations whereas [29] and [30] are restricted to flat tessellations.

4.3.5 Truncation error

While the integrand in (4.20) is clearly smooth over T_n, it remains to be seen whether the truncation in (4.20) is valid and its upper limit practical. In this section, we look at the error incurred through truncation of (4.20) to determine its practicality. To derive error bounds, we use Fourier analysis. Such analysis is facilitated through the use of indicator functions and pulse functions, which allow us to express the integrand in (4.20) in terms of entire domain functions. To begin we define

$$\Pi_T(t) = \begin{cases} 1 & t \in \text{supp}\{T(t)\} \\ 0 & \text{otherwise} \end{cases} \tag{4.22a}$$

which is an indicator function for $T(t)$, similar to $\Pi_{T_n}(\mathbf{r})$ from Section 4.3.2. Additionally, we define

$$\mathbb{H}_{[\alpha,\beta]}(t) = \begin{cases} 1 & t \in [\alpha, \beta] \\ 0 & \text{otherwise} \end{cases} \tag{4.22b}$$

which is a pulse function defined over the regions of orthogonality of $P_l(k_1 t + k_2)$, similar to the spherical shell of (4.21), which is displayed in Figure 4.6. In what follows, we will assume $\zeta = 0$. Using these definitions, (4.20) can be written as a

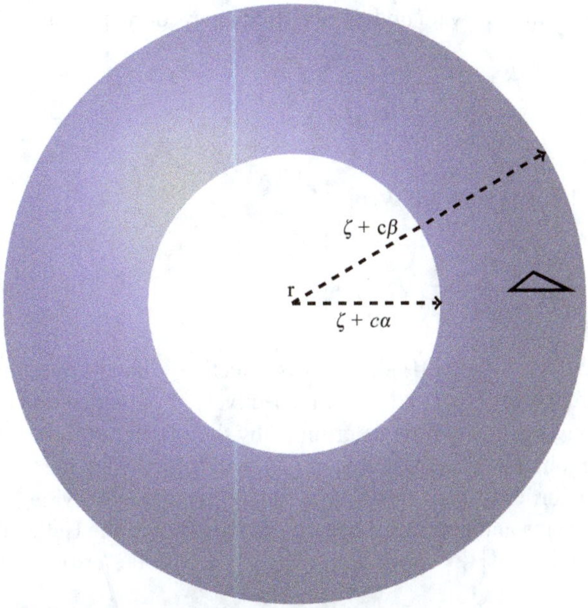

Figure 4.6 Domain of the spatial integrals

space–time convolution, which will transform to multiplications in the space–time frequency domain. Rewriting (4.20), we have

$$\begin{aligned}\psi(r,t) &= \int_{\mathbb{R}^3} dr' \Pi_{T_n}(r') S_n(r') \int_{-\infty}^{\infty} dt' \Pi_T(t') T(t') \frac{\delta\left(t - t' - \frac{R}{c}\right)}{4\pi R} \\ &\approx \sum_{l=0}^{M} a_l \left(\int_{\mathbb{R}^3} dr' \Pi_{T_n}(r') S_n(r') \mathbb{B}_{[c\alpha, c\beta]}(R) \frac{P_v(k_1 R/c + k_2)}{4\pi R} \right) \\ &\quad \left(\int_{-\infty}^{\infty} dt' \Pi_T(t') T(t') \mathbb{H}_{[\alpha,\beta]}(t - t') P_l(k_1(t - t') + k_2) \right)\end{aligned} \tag{4.23}$$

In order to determine the truncation error, we wish to evaluate

$$\begin{aligned}\varepsilon_M(r,t) \doteq \Bigg| &\int_{\mathbb{R}^3} dr' \Pi_{T_n}(r') S_n(r') \int_{-\infty}^{\infty} dt' \Pi_T(t') T(t') \frac{\delta\left(t - t' - \frac{R}{c}\right)}{4\pi R} \\ &- \sum_{l=0}^{M} a_l \left(\int_{\mathbb{R}^3} dr' \Pi_{T_n}(r') S_n(r') \mathbb{B}_{[c\alpha, c\beta]}(R) \frac{P_v(k_1 R/c + k_2)}{4\pi R} \right) \\ &\left(\int_{-\infty}^{\infty} dt' \Pi_T(t') T(t') \mathbb{H}_{[\alpha,\beta]}(t - t') P_l(k_1(t - t') + k_2) \right) \Bigg|\end{aligned} \tag{4.24}$$

This error will be evaluated in the frequency domain using the following definitions:

$$\widetilde{V} = \mathcal{F}_t\{V(t)\} = \int_{-\infty}^{\infty} dt V(t) e^{j\omega t} \tag{4.25a}$$

$$\overline{W} = \mathcal{F}_r\{\mathbf{W}(r)\} = \int_{\mathbb{R}^3} dr \mathbf{W}(r) e^{j\lambda \cdot r} \tag{4.25b}$$

$$\mathcal{F}_{r,t}\{X(\mathbf{r},t)\} = \mathcal{F}_r\{\mathcal{F}_t\{X(\mathbf{r},t)\}\} \tag{4.25c}$$

We assume $(\Pi_{T_n}(\mathbf{r})\mathbf{S}_n(\mathbf{r})\Pi_T(t)T(t))$ is effectively band-limited in spatial and temporal frequency to $|\lambda_{\max}|$ and $\omega_{\max}$, respectively. In practice, the upper limit M required for convergence will be determined by the smaller of these bandwidths. In a method of moments discretization, $|\lambda_{\max}|$ will depend on the size of a given element and the order of the spatial basis function. Analogously, $\omega_{\max}$ will depend on the time step size and order of the temporal basis function. The goal is to show a bounded error within $\{|\lambda| < |\lambda_{\max}|, \omega < \omega_{\max}\}$, where the error for a given truncation limit is

$$\begin{aligned} \widetilde{\varepsilon}_M(\lambda,\omega) = \Bigg| \Bigg(& F_{r,t}\left\{\frac{\delta(t-|r|/c)}{4\pi|r|}\right\} \\ & - F_{r,t}\left\{ \mathbb{B}_{[c\alpha,c\beta]}(r)\mathbb{H}_{[\alpha,\beta]}(t)\sum_{l=0}^{M} a_l \frac{P_l\left(k_1\frac{|r|}{c}+k_2\right)}{4\pi|r|} P_l(k_1t+k_2)\right\}\Bigg)\Bigg| \end{aligned} \tag{4.26}$$

A straightforward derivation shows that

$$\begin{aligned} & F_{r,t}\left\{ \mathbb{B}_{[c\alpha,c\beta]}(r)\mathbb{H}_{[\alpha,\beta]}(t)\sum_{l=0}^{M} a_l \frac{P_l\left(k_1\frac{|r|}{c}+k_2\right)}{4\pi|r|} P_l(k_1t+k_2)\right\} \\ & = \frac{c}{2\pi|\lambda|k_1} e^{-j\frac{\omega k_2}{k_1}} \sum_{l=0}^{M}(2l+1) j_l\left(\frac{\omega}{k_1}\right) j_l\left(\frac{|\lambda|c}{k_1}\right) \end{aligned} \tag{4.27}$$

where $j_l(z)$ is a spherical Bessel function, which satisfies the monotonicity relationship (see [57, Section 10.14.5])

$$|j_l(z)| \leq \sqrt{\frac{\pi}{2z}}\left(\frac{z}{\tilde{l}+\sqrt{\tilde{l}^2-z^2}}\right)^{\tilde{l}} e^{\sqrt{\tilde{l}^2-z^2}} \quad 0 < z \leq l+\frac{1}{2} \tag{4.28}$$

where $\widetilde{l} = l + \frac{1}{2}$. Substituting (4.27) into (4.26) and using the bound (4.3.5) yields

$$\widetilde{\varepsilon}_M(\lambda, \omega) \leq \frac{1}{4|\lambda|^{3/2}\omega^{1/2}} \sum_{l=M+1}^{\infty} \left[(2l+1) e^{\left(\frac{1}{k_1}\sqrt{\left(k_1 \widetilde{l}\right)^2 - (|\lambda| c)^2} + \frac{1}{k_1}\sqrt{\left(k_1 \widetilde{l}\right)^2 - \omega^2} \right)} \right.$$

$$\left. \left(\frac{\omega}{k_1 \widetilde{l} + \sqrt{\left(k_1 \widetilde{l}\right)^2 - \omega^2}} \right)^{\widetilde{l}} \left(\frac{|\lambda| c}{k_1 \widetilde{l} + \sqrt{\left(k_1 \widetilde{l}\right)^2 - (|\lambda| c)^2}} \right)^{\widetilde{l}} \right] \tag{4.29}$$

which can be relaxed to the simplified form

$$\widetilde{\varepsilon}_M(\lambda, \omega) < \frac{1}{4|\lambda|^{3/2}\omega^{1/2}} \sum_{l=M+1}^{\infty} (2l+1) \left(\frac{e^2 \omega |\lambda| c}{4(k_1 \widetilde{l})^2} \right)^{\widetilde{l}}$$

$$< \frac{1}{4|\lambda|^{3/2}\omega^{1/2}} \sum_{l=M+1}^{\infty} (2l+1) \left(\frac{e^2 \omega |\lambda| c}{4(k_1 \widetilde{M})^2} \right)^{\widetilde{l}} \tag{4.30}$$

where $\widetilde{M} = M + 3/2$. This is a standard arithmetico-geometric series, which converges to

$$\frac{1}{4|\lambda|^{3/2}\omega^{1/2}} \left(\frac{e^2 \omega |\lambda| c}{4k_1^2 \widetilde{M}^2 - e^2 \omega |\lambda| c} \right)^2 \left(\frac{e^2 \omega |\lambda| c^2}{4k_1^2 \widetilde{M}} \right)^{\widetilde{M}} \left[\frac{\left(2\widetilde{M} + 3\right) 4k_1^2 \widetilde{M}^2}{e^2 \omega |\lambda| c} - \left(2\widetilde{M} + 1\right) \right] \tag{4.31}$$

Setting $z_t = e\omega / \left(2k_1 \widetilde{M}\right)$ and $z_s = e|\lambda|c / \left(2k_1 \widetilde{M}\right)$ and using (4.31), it can be shown that

$$\widetilde{\varepsilon}_M(\lambda, \omega) < \frac{1}{4|\lambda|^{3/2}\omega^{1/2}} (z_t z_s)^M \left(\frac{z_t z_s}{1 - z_t z_s} \right)^{7/2} \left[\frac{(2M+6)}{z_t z_s} - (2M+4) \right] \tag{4.32}$$

for $(z_t z_s) < 1$, this series is rapidly convergent in M.

Next, we numerically test this bound. For simplicity, we only focus on the temporal portion of the expansion. In this test, the expansion

$$\widehat{f}(\omega) = \mathcal{F}_t \left\{ \sum_{l=0}^{M} a_l P_v(k_1 R/c + k_2) P_l(k_1 t + k_2) \right\} \tag{4.33}$$

is computed for a given value of R and compared against

$$\widetilde{f}(\omega) = \mathcal{F}_t\{\delta(t - R/c)\} = e^{-jkR} \tag{4.34}$$

over a sampling of N_f frequencies in the interval $[f_m, f_M]$, $f_M > f_m$. The error

$$\begin{aligned}\text{Error} &= \frac{\|\widehat{f}(\omega) - \widetilde{f}(\omega)\|}{\|\widetilde{f}(\omega)\|} \\ &= \frac{\left\| \sum_{l=M+1}^{\infty} a_l P_\nu(k_1 R/c + k_2)\mathcal{F}_t\{P_l(k_1 t + k_2)\} \right\|}{\|\widetilde{f}(\omega)\|}\end{aligned} \tag{4.35}$$

$$\|f(\omega)\| \doteq \sqrt{\sum_{i=1}^{N_f} |f(\omega_i)|^2} \tag{4.36}$$

is computed for various values of M with the width of the support of $\mathbb{H}$ set to $3\Delta_t$. The result is shown in Figure 4.7, while Figure 4.8 compares the analytical and approximated signals convolved with a band-limited function in the frequency domain for various M. As is evident from Figure 4.7, the expansion (4.20) quickly converges. Figure 4.7 suggests that the expansion (4.20) is practical.

A remark should also be made regarding the choice of M in practice. Given α (typically set to 0) and β for a given triangle, T_n, the value of M should be determined by (4.32) to achieve a given bound. Empirically, we have found that assuming a direct dependence $M = C\beta$ is sufficient to ensure convergence, with C typically set to either 2 or 3 for a typical simulation setup. Here, "typical" means edge lengths on the order of $\lambda_{\min}/10$ and $\Delta_t = 1/(20 f_{\max})$, where $f_{\max} = f_0 + B$ is the frequency at which the incident field power is 160 dB below its peak, $\lambda_{\min} = c/f_{\max}$, and f_0 is the center frequency and B is the effective bandwidth of the incident field described by (4.46).

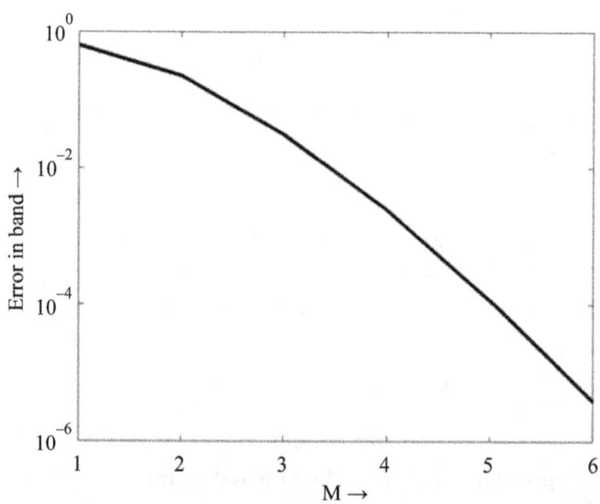

Figure 4.7 Convergence in M

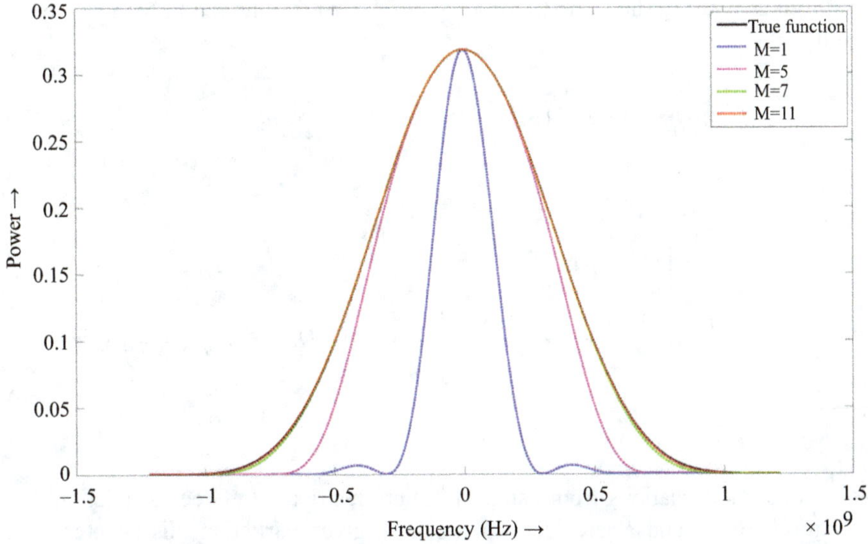

Figure 4.8 Spectra of approximating signals

4.3.6 Incorporation into method of moments scheme

In this section, we examine (4.10a) modified using the expansion in (4.20). We begin with the right hand side of (4.11a) for one spatio-temporal basis function and *one* region. Substituting in expansion (4.20), this becomes

$$\langle \delta(t - i\Delta_t)\mathbf{S}_m(\mathbf{r}), \mathcal{L}^+_{e,q}\{\mathbf{S}_n(\mathbf{r})T(t - j\Delta_t)\}\rangle = \sum_{\nu=0}^{M} a_\nu \left(\alpha_\nu^{m,n} \widetilde{T}^{\nu}_{i,j} + \varphi_\nu^{m,n} \widehat{T}^{\nu}_{i,j} \right) \tag{4.37a}$$

where

$$\alpha_\nu^{m,n} = \mu_q \int_{S_m} d\mathbf{r}\mathbf{S}_m(\mathbf{r}) \cdot \int_{S_n} d\mathbf{r}' \frac{\mathbf{S}_n(\mathbf{r}')P_\nu(k_1 R/c_q + k_2)}{4\pi R} \tag{4.37b}$$

$$\varphi_\nu^{m,n} = \frac{1}{\varepsilon_q} \int_{S_m} dr \nabla \cdot S_m(r) \int_{S_n} dr' \frac{\nabla \cdot S_n(r')P_\nu(k_1 R/c_q + k_2)}{4\pi R} \tag{4.37c}$$

$$\widetilde{T}^{\nu}_{i,j} = P_\nu(k_1 t + k_2) * T'(t - j\Delta_t)|_{t=i\Delta_t} \tag{4.37d}$$

$$\widehat{T}^{\nu}_{i,j} = P_\nu(k_1 t + k_2) * \int_{-\infty}^{t} d\tau T(\tau - j\Delta_t)|_{t=i\Delta_t} \tag{4.37e}$$

Similarly, using the same framework on (4.11b) for one spatio-temporal basis results in

$$\langle \delta(t - i\Delta_t)\mathbf{S}_m(\mathbf{r}), \mathcal{L}^{+}_{h,q}\{\mathbf{S}_n(\mathbf{r})T(t - j\Delta_t)\}\rangle \doteq \sum_{\nu=0}^{M} a_\nu \kappa_\nu^{m,n} T_{i,j}^{\nu} \tag{4.38a}$$

where

$$\begin{aligned}\kappa_\nu^{m,n} &= \frac{1}{4\pi}\int_{S_m} d\mathbf{r}\mathbf{S}_m(\mathbf{r})\cdot\widehat{n} \\ &\times \int_{S_n} d\mathbf{r}'\frac{1}{R}\left(\frac{P_\nu(k_1R/c_q + k_2)}{R} + \frac{k_1}{c_q}P_{\nu'}(k_1(R-\zeta)/c_q + k_2)\right)\mathbf{S}_n(\mathbf{r}')\times\widehat{R}\end{aligned} \tag{4.38b}$$

$$T_{i,j}^{\nu} = P_\nu(k_1t + k_2) * T(t - j\Delta_t)|_{t=i\Delta_t} \tag{4.38c}$$

As the modifications from using (4.20) are restricted to Green's function, the definitions of $\overline{V}_i$ and $\overline{I}_i$ are identical to those given earlier in this chapter. The spatial integrals (4.37b), (4.37c), and (4.38b) can be evaluated using standard quadrature rules. The upper limit M is used to determine the integration rule need for a given expansion. The temporal convolutions (4.37d), (4.37e), and (4.38c) can be evaluated analytically or via one-dimensional integration rules and are pre-computed for $l = 0, 1, ..., M$.

4.3.7 *Efficient evaluation of scalar potential*

A remark is in order for the temporal integration of the basis function in (4.37e). A naïve discretization of (4.10a) leads to a scaling of $O(N_t^2N_s^2)$. Instead, we make use of a recursive procedure which we outline here. We begin by noting that

$$\begin{aligned}\widehat{T}_{i,j}^{\nu} &= P_\nu(k_1t + k_2) * \int_{-\infty}^{t} d\tau T(\tau - j\Delta_t)\Big|_{t=i\Delta_t} \\ &= \int_{-\infty}^{i\Delta_t} dt P_l(k_1t' + k_2)\int_{-\infty}^{t-t'} d\tau T(\tau - j\Delta_t)\end{aligned} \tag{4.39a}$$

it can be shown that

$$\int_{-\infty}^{t-t'} d\tau T(\tau - j\Delta_t) = \begin{cases}\int_{-\infty}^{\infty} d\tau T(\tau) & t \geq \gamma\Delta_t \\ \int_{-\infty}^{t-t'} d\tau T(\tau - j\Delta_t) & t < \gamma\Delta_t\end{cases} \tag{4.39b}$$

where $\gamma = \beta - \alpha + i + 1$. Therefore, when $j \geq \gamma$, we have

$$\widehat{T}_{i,j}^{\nu} = \left(\int_{-\infty}^{\infty} dt T(t)\right)\left(\int_{-\infty}^{\infty} dt P_\nu(k_1t + k_2)\right) \tag{4.39c}$$

We introduce the charge term

$$\overline{C}_j = \overline{C}_{j-1} + \overline{I}_{j-1}\int_{-\infty}^{\infty} dt T(t),\ \overline{C}_1 = 0 \tag{4.40a}$$

Using the property that $\int_{-\infty}^{\infty} dt P_\nu(k_1 t + k_2) = \delta_{\nu 0}(\beta - \alpha)\Delta_t$ where δ_{ij} is the Kronecker delta function, we can rewrite (1.10a) (modified with (4.20)) as

$$\overline{\overline{Z}}_0^M \overline{I}_i = \overline{V}_i - \sum_{j=1}^{i-1} \overline{\overline{Z}}_{i-j}^M \overline{I}_j - \sum_{j=2}^{i-1} \overline{\overline{Z}}_{i-j}^{M,c} \overline{C}_j \tag{4.40b}$$

where

$$\left[\overline{\overline{Z}}_{i-j}^M\right]_{mn} = \langle \delta(t-(i-j)\Delta_t)\mathbf{S}_m(\mathbf{r}), \mathcal{L}_c^M\{\mathbf{S}_n(\mathbf{r})T(t)\}\rangle \tag{4.40c}$$

$$\mathcal{L}_c^M\{\cdot\} \doteq \alpha_c/\eta_0 \mathcal{L}_e^M\{\cdot\} + (1-\alpha_c)\mathcal{L}_h^M\{\cdot\} \tag{4.40d}$$

$$\left[\overline{\overline{Z}}_{i-j}^{M,c}\right]_{mn} = \left(\begin{matrix} (\beta-\alpha)\varphi_0^{m,n} & j \geq \gamma \\ 0 & \text{otherwise} \end{matrix}\right. \tag{4.40e}$$

The recursive computation of the charge terms $\overline{C}_j$ keeps the scaling of the solver at $O(N_t N_s^2)$.

4.3.8 Extension to higher order space–time Galerkin scheme

It is apparent that the procedure elucidated in the preceding subsections is rather readily extendable to higher order spatio-temporal basis sets. The main challenge is in determining appropriate integration rules to evaluate all integrals. For the quasi-exact integration methodology to be used within this framework, we would require the development of integration rules for $(p+1) \times (p+2)/2$ different integrals. As is evident, reduction to low order, in both space and time, is relatively seamless.

4.3.9 Interpolation properties

As shown in [40], higher order convergence of interpolation error can be achieved by expanding the unknown quantity using multiple basis functions within a single time step, and applying Galerkin testing to discretize the system. In what follows, we present a simpler convergence test to demonstrate and understand convergence. To that end, we represent a modulated Gaussian pulse using higher order temporal basis functions and then evaluate the error in convergence with delta function. This is done both analytically and using the truncated expansion (4.20). The control parameter is the number of harmonics M. The requisite expressions can be written as

$$\begin{aligned}\tilde{g}_\Delta^M(t) &= \sum_{j=1}^{N_t}\sum_{l=0}^{p} g^{j,l} T^l(t - j\Delta_t) * \left(\sum_{\nu=0}^{M} a_\nu P_\nu(k_1\Delta + k_2) P_\nu(k_1 t + k_2)\right) \\ &= \sum_{j=1}^{N_t}\sum_{l=0}^{p} g^{j,l} \sum_{\nu=0}^{M} a_\nu P_\nu(k_1\Delta + k_2) T^l(t - j\Delta_t) * P_\nu(k_1 t + k_2)\end{aligned} \tag{4.41}$$

Specifically, the modulated Gaussian pulse is defined as

$$g(t) = cos(2\pi f_0 t)e^{-(t-t_p)^2/2\sigma^2} \tag{4.42}$$

where, given some modulating frequency f_0 and maximum frequency $f_{\max}$, the values σ and t_p are defined as $\sigma = 3/(2\pi B)$ and $t_p = 6\sigma$, where $B = f_{\max} - f_0$. This pulse is shifted and approximated using the higher order basis as

$$g(t) \approx \tilde{g}(t) = \sum_{j=1}^{N_t} \sum_{l=0}^{p} g^{j,l} T^l(t - j\Delta_t) \tag{4.43}$$

The time step is defined as $\Delta_t = 1/(2k_{\text{samp}} f_{\max})$, where $k_{\text{samp}} > 0$ is the over-sampling factor. Let the interpolation error for a given p be defined as using the L_2 norm and define the error using the separable expansion with upper limit M as

$$\text{Error}^M = \frac{\|\overline{g}^M_{\text{approximate}} - \overline{g}_{\text{analytic}}\|}{\|\overline{g}_{\text{analytic}}\|} \tag{4.44a}$$

$$\overline{g}^M_{\text{approximate}} = \left[\tilde{g}^M_\Delta(\Delta_t)\, \tilde{g}^M_\Delta(2\Delta_t) \ldots \tilde{g}^M_\Delta(N_t\Delta_t)\right] \tag{4.44b}$$

The error (4.44a) is shown in Figure 4.9 as a function of the truncation limit for different p and with $k_{\text{samp}} = 10$. This is compared against the same error when the shift is achieved by convolution with a delta function. For each value of k_{samp}, one observes convergence with respect to M. It is expected that the error using (4.20) will converge to that using the delta shift and Figure 4.9 shows that the error indeed converges to this error (or the error is slightly better when (4.20) is used). The error rapidly improves as M is increased, to a value well below that of the delta-shifted interpolation.

Next, we show the error as a function of the oversampling factor, k_{samp}, in Figure 4.10. The support of the Legendre polynomials is chosen to reflect what would be expected in a typical method of moments discretization with edge lengths of approximately $\lambda/10$, i.e., the support is proportional to k_{samp}. It may be noticed

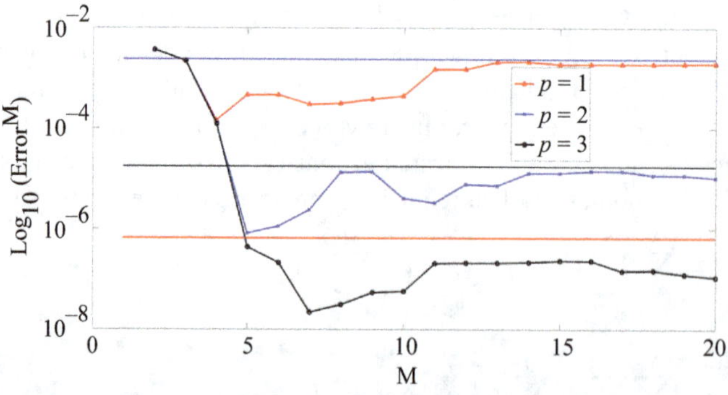

Figure 4.9 Interpolation errors for various upper limits of (4.20)

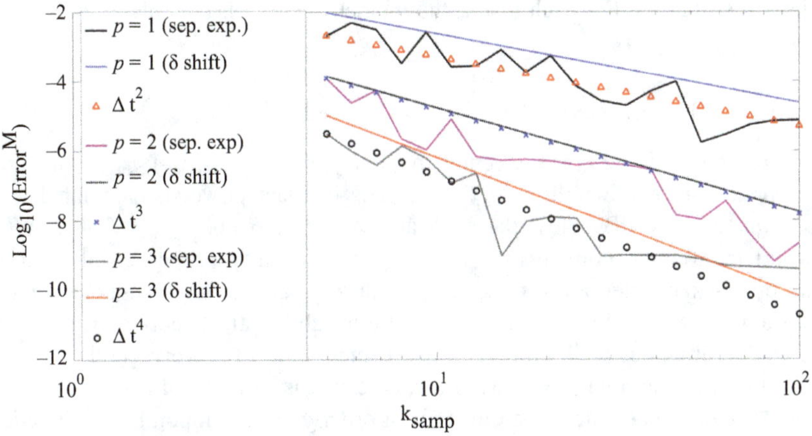

Figure 4.10 Interpolation errors for separable expansion

Table 4.1 Lower bound of M needed to reproduce interpolation accuracies (undifferentiated/first derivative/integral)

pk_{samp}	**5**	**10**	**20**	**40**
1	1/1/1	1/1/2	2/1/2	3/1/3
2	2/1/1	4/2/2	4/3/3	4/3/3
3	4/3/4	4/4/4	6/4/5	7/5/5

that the error for $p = 3$ does not converge at the same rate for large values of k_{samp}. This is due to the very large support of the Legendre polynomials and the corresponding number of harmonics increases faster than can be handled. Table 4.1 shows the number of harmonics required in (4.20) to reproduce the error produced using a delta shift. The upper limit needed to reproduce the interpolation error is defined as

$$M(\text{Error}) = \inf\{M \geq 0 : \text{Error}^M \leq \text{Error}\} \tag{4.45}$$

This is shown in Table 4.1 for the interpolation of the function, as well as for its derivative and integral, as required for the TD-EFIE. It can be seen that the number of harmonics required is numerically tractable.

4.4 Results

In this section, we present a number of results to serve two purposes; (a) demonstrate the utility of the method especially late time stability and (b) quantify the performance of the space-time separation method to quasi-exact integration for a variety of scatterers. For each result, the incident field will be a plane wave of the form

$$\mathbf{E}^{inc}(\mathbf{r},t) = \hat{u}E_0 \cos(2\pi f_0 \tau) e^{\frac{-(\tau-t_p)^2}{2\sigma^2}} \tag{4.46}$$

where $\tau = t - \mathbf{r}\cdot\hat{k}/c_0, f_0$ is the center frequency, $\sigma = 3/(2\pi f_{\max}), f_{\max} = f_0 + B$ is the approximate bandwidth, and $t_p = 6\sigma$. The incident power is calculated to be approximately 160 dB below the peak at $f_{\max} = f_0 + B$ and $f_{\min} = f_0 - B$. When appropriate, we will compare the accuracy and stability properties of solutions using quasi-exact integration to separable approximation. As quasi-exact integration is a well-validated method in terms of both stability and accuracy, we consider any result that agrees well with quasi-exact integration to be very good.

In the first example, we analyze a sphere that is discretized using curvilinear triangular elements (as described in [58]) of order $g = 3$. Temporal basis functions are the first-order Lagrange polynomials and the expansion of the spatio-temporal convolution is truncated using the criteria derived in this chapter, resulting in a maximum $M = 7$. The incident field is wide band with a center frequency of 150 MHz and bandwidth of 149.9MHz. It is seen that the current on the surface of the sphere is stable for 20, 000 time steps (Figure 4.11a) with $\alpha = 0.5$. Similarly, the far field observed due to this current, normalized to the incident field, can be used as a metric for comparison against an analytical frequency domain result. The excellent agreement with the analytical result at three different frequencies across the bandwidth validates the technique (Figure 4.11b).

Next, we present a result that has proven challenging to stabilize using traditional MOT-based schemes. We compute scattering from a nacelle geometry using the proposed scheme. The nacelle is constructed as a cylindrical shell with one end closed. The outer radius of the cylinder is 160 cm and the inner radius is 100 cm. The length of the cylinder is 300 cm. To be realistic, the edges of the cylinder are filleted at a 20° angle. The incident field is chosen to have the same functional dependence as in (4.46) with $f_0 = 1$ GHz and $B = 0.99$ GHz. The incident field is polarized along $\hat{\theta}$ and directed along $\theta = 90°$ and $\varphi = 5°$. The maximum M chosen across the geometry is 7 and the minimum is 2. The simulation is run for 100,000 time steps, corresponding to 753 transits. While the stability of the scheme is

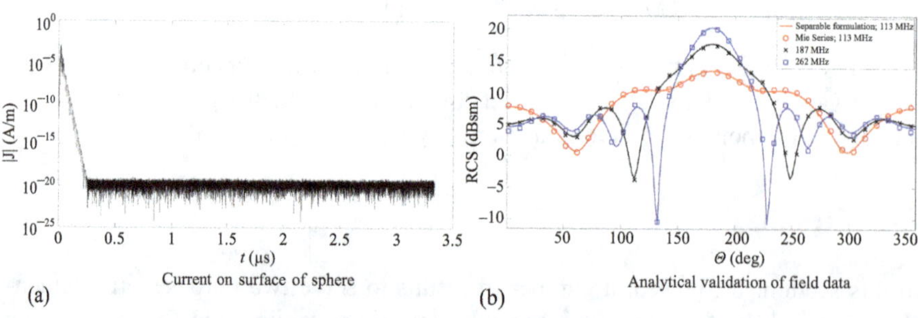

Figure 4.11 Validation results for scattering from a sphere

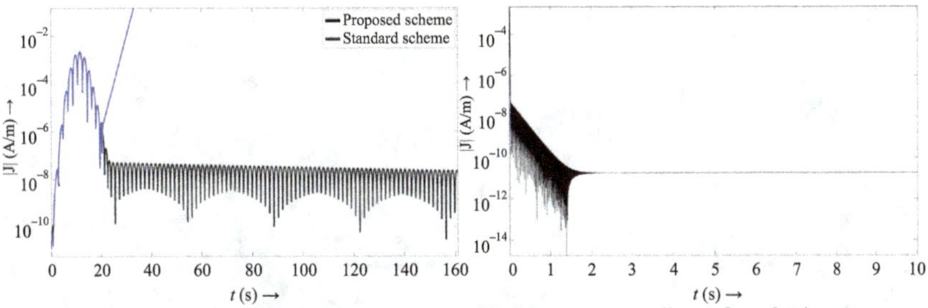

(a) Current on nacelle surface, for early time, comparison against standard MOT scheme

(b) Current on nacelle surface for long run time showing late time stability of proposed scheme

Figure 4.12 Construction of stable solutions on a nacelle geometry

proven for $\alpha > 0$, we have chosen $\alpha = 0$ in this case to illustrate the robustness of the approximation in terms of providing stable solutions for the more challenging case of the electric field integral equation. The early time data in Figure 4.12a show the current on the surface of a nacelle geometry computed using the proposed scheme compared against the scheme developed in [29]. The figure not only validates the method but clearly indicates the instability of the standard MOT schemes. Late time data in Figure 4.12b demonstrate the capability of the scheme developed in the paper to generate stable solutions for extremely long simulation times (753 transits) for problems of scattering from challenging geometries.

Next, we examine scattering from a dielectric sphere of radius 1 m with $\varepsilon_r = 2.0$. The geometry is discretized using 384 triangle elements. The excitation is incident in the $\widehat{k} = \widehat{z}$ direction with $\widehat{u} = \widehat{x}$ polarization. The frequency content of the excitation is f_0 =40 MHz and B =20 MHz. One current coefficient is computed for 4,800 time steps (424 transits across the geometry) and the result is shown for both methods in 4.13. The two results agree very well throughout the simulation. It can be seen that both results decay to a constant floor value and no instability is apparent. The agreement between the two currents is excellent, showing that the separable approximation can be used to yield accurate results.

For our next result, we examine a PEC sphere of radius 0.5 m, which is enclosed inside of a spherical dielectric region ($\varepsilon_r = 2.0$) of radius 1 m. Each boundary is discretized using 384 triangular elements. The incidence and the polarization of the plane wave are the same as the previous case, with f_0 =60 MHz and B =30 MHz. A current coefficient is plotted in Figure 4.14 for each solution for 4,500 time steps. Similar stability characteristics are seen as compared to the previous result. Each solution decays to a constant value and does not grow in magnitude. Again, the agreement between the two methods is quite good.

Next, we consider scattering from a thin dielectric box. This box has dimensions $0.5 \times 1 \times 0.1$ m and is discretized with 260 triangular elements. The incident field is given by $\widehat{k} = \widehat{z}, \widehat{u} = \widehat{x}, f_0$ =90 MHz and B =80 MHz. In Figure 4.15, 2,890 time samples of the current are shown for each solution. For this result, a slowly

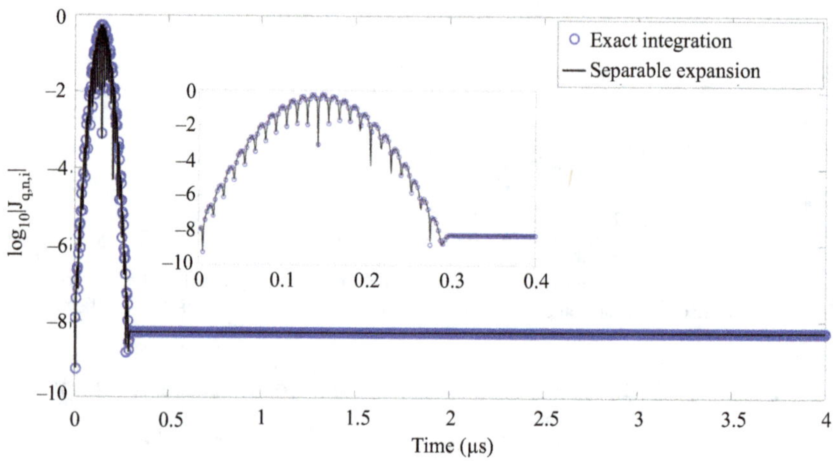

Figure 4.13 Current on sphere

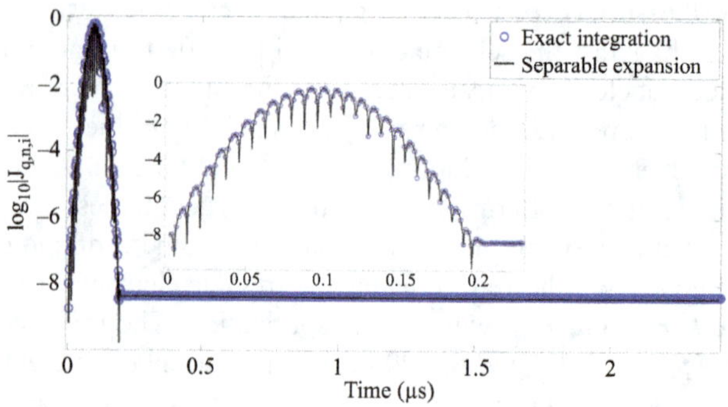

Figure 4.14 Current on composite sphere compared against exact integration

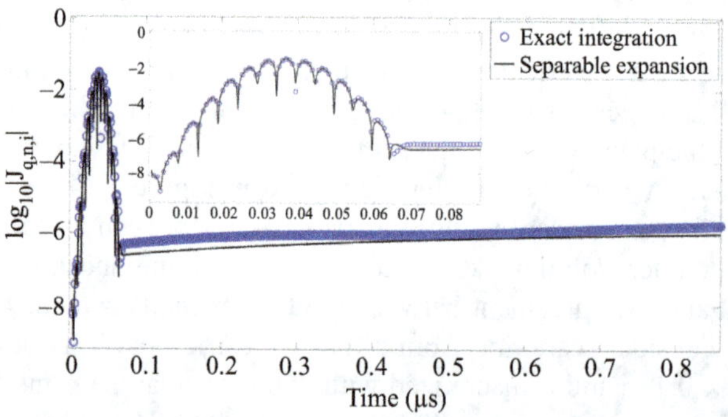

Figure 4.15 Current on thin box compared against exact integration

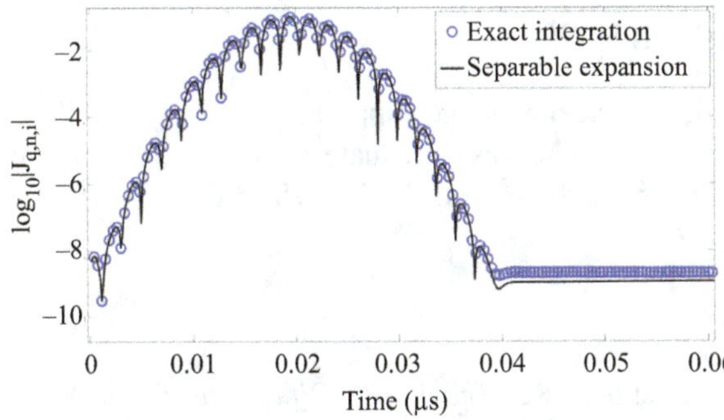

Figure 4.16 Current on conesphere compared against exact integration

growing low frequency term is seen for both solutions. We suspect that there is some inaccuracy in integration over the observation domains as higher integration rules were seen to have an effect on the level of instability seen. The tolerance of the iterative solution may also be a source of instability. The level of instability in the two solutions is roughly the same and the current coefficients agree well.

The last result is a dielectric cone-sphere ($\varepsilon_r == 1.5$) of 1,008 spatial unknowns, with the cone aligned in the $\widehat{x}$ direction. The radius of the sphere is 0.25m, and the height of the cone is 1 m. The incident wave is described by $f_0 == 250$ MHz, $B = 150$ MHz, $\widehat{k} = \widehat{y}$, and $\widehat{u} = \widehat{z}$. The resulting current is shown in Figure 4.16. As with the thin box, this is a more challenging problem than the spherical scatterers due to the sharp geometric features. Again, a slowly growing low-frequency instability is observed, but there is no evidence of high-frequency instability.

4.5 Summary

In this chapter, we have presented an extension of the method of separable expansion to the analysis of composite objects, i.e., objects comprising a mixture of piecewise homogeneous dielectric regions as well a perfectly conducting surfaces. This method is used to allow for accurate computation of impedance matrix elements using numerical quadrature, which is a necessary condition for obtaining stable solutions to TDIEs. The method, being based purely on numerical integration, is easily extended to higher order geometric surface discretizations. The stability of the method on these types of scatterers has been demonstrated for a variety of scatterers using quasi-exact integration solutions as a benchmark. In addition to these results, we have developed an analysis procedure for the errors incurred in the approximation. Altogether, this body of work indicates that using a separable approach to evaluating TDIE matrices is a viable approach. The slowly growing low-frequency instability we see in some of the results (for both the separable expansion and the quasi-exact method) has a well-known remedy and can be readily integrated into the system of equations presented here.

4.6 Acknowledgments

A.J. Pray would like to acknowledge support from SMART fellowship while this work was in progress when he was a graduate student, and B. Shanker would like to thank NSF for support via NSF-CMMI-1725278.

References

[1] Kunz KS, Luebbers RJ. *The Finite Difference Time Domain Method for Electromagnetics*. Boca Raton: CRC Press; 1993.
[2] Taflove A, Hagness S. *Computational Electrodynamics: The Finite Difference Time Domain Method*. Boston: Artech House; 2000.
[3] Vechinski DA, Rao SM. A stable procedure to calculate the transient scattering by conducting surfaces of arbitrary shape. *IEEE Transactions on Antennas and Propagation*. 1992;40(6):661–665.
[4] Jin J. *The Finite Element Method in Electromagnetics*. New York, NY: John Wiley & Sons; 2014.
[5] Baczewski AD. *Integral Equation and Discontinuous Galerkin Methods for the Analysis of Light–Matter Interaction*. East Lansing, MI: Michigan State University; 2013.
[6] Buschlinger R, Lorke M, Peschel U. Light-matter interaction and lasing in semiconductor nanowires: a combined finite-difference time-domain and semiconductor Bloch equation approach. *Physical Review B*. 2015;91:045203. Available from: https://link.aps.org/doi/10.1103/PhysRevB.91.045203.
[7] Glosser C, Shanker B, Piermarocchi C. Collective Rabi dynamics of electromagnetically coupled quantum-dot ensembles. *Physical Review A*. 2017;96(3):033816.
[8] Cohen G, Pernet S. *Finite Element and Discontinuous Galerkin Methods for Transient Wave Equations*. New York, NY: Springer; 2017.
[9] Hesthaven JS, Warburton T. *Nodal Discontinuous Galerkin Methods*. New York, NY: Springer; 2008.
[10] Peterson AF, Ray SL, Mittra R. *Computational Methods for Electromagnetics*. New York, NY: Wiley-IEEE Press; 1997.
[11] Coifman R, Rokhlin V, Wandzura S. The fast multipole method for the wave equation: a pedestrian prescription. *Antennas and Propagation Magazine, IEEE*. 1993;35(3):7–12.
[12] Song J, Lu CC, Chew WC. Multilevel fast multipole algorithm for electromagnetic scattering by large complex objects. *IEEE Transactions on Antennas and Propagation* 1997;45(10):1488–1493.
[13] Melapudi V, Shanker B, Seal S, *et al.* A scalable parallel wide-band MLFMA for efficient electromagnetic simulations on large scale clusters. *IEEE Transactions on Antennas and Propagation* 2011;59(7):2565–2577.
[14] Friedman M, Shaw R. Diffraction of pulses by cylindrical obstacles of arbitrary cross section. *Journal of Applied Mechanics* 1962;29(1):40–46.

[15] Rynne B. Stability and convergence of time marching methods in scattering problems. *IMA Journal of Applied Mathematics* 1985;35(3):297–310.

[16] Rynne B. Time domain scattering from arbitrary surfaces using the electric field integral equation. *Journal of Electromagnetic Waves and Applications* 1991;5(1):93–112.

[17] Sadigh A, Arvas E. Treating the instabilities in marching-on-in-time method from a different perspective [electromagnetic scattering]. *IEEE Transactions on Antennas and Propagation* 1993;41(12):1695–1702.

[18] Davies PJ. Numerical stability and convergence of approximations of retarded potential integral equations. *SIAM Journal on Numerical Analysis* 1994;31(3):856–920.

[19] Shanker B, Ergin AA, Aygun K, *et al.* Analysis of transient electromagnetic scattering from closed surfaces using a combined field integral equation. *IEEE Transactions on Antennas and Propagation* 2000;48(7):1064–1074.

[20] Shanker B, Ergin AA, Lu M, *et al.* Fast analysis of transient electromagnetic scattering phenomena using the multilevel plane wave time domain algorithm. *IEEE Transactions on Antennas and Propagation* 2003;51(3):628–641.

[21] Yilmaz AE, Jin JM, Michielssen E. Time domain adaptive integral method for surface integral equations. *IEEE Transactions on Antennas and Propagation* 2004;52(10):2692–2708.

[22] Rynne BP, Smith PD. Stability of time marching algorithms for the electric field integral equation. *Journal of Electromagnetic Waves and Applications* 1990;4(12):1181–1205.

[23] Hu JL, Chan CH, Xu Y. A new temporal basis function for the time-domain integral equation method. *Microwave and Wireless Components Letters, IEEE* 2001;11(11):465–466.

[24] Weile DS, Pisharody G, Chen NW, *et al.* A novel scheme for the solution of the time-domain integral equations of electromagnetics. *IEEE Transactions on Antennas and Propagation* 2004;52(1):283–295.

[25] Dodson SJ, Walker SP, Bluck MJ. Costs and cost scaling in time-domain integral-equation analysis of electromagnetic scattering. *IEEE Antennas and Propagation Magazine.* 1998;40(4):12–21.

[26] Sarkar TK, Lee W, Rao SM. Analysis of transient scattering from composite arbitrarily shaped complex structures. *IEEE Transactions on Antennas and Propagation.* 2000;48:1625 –1634.

[27] Lubich C. Convolution quadrature and discretized operational calculus. I. *Numerische Mathematik* 1988;52(2):129–145.

[28] Wang X, Weile DS, Member S. Implicit Runge-Kutta methods for the discretization of time domain integral equations. *IEEE Transactions on Antennas and Propagation* 2011;(c):1–13.

[29] Shanker B, Lu M, Yuan J, *et al.* Time domain integral equation analysis of scattering from composite bodies via exact evaluation of radiation fields. *IEEE Transactions on Antennas and Propagation* 2009;57(5): 1506–1520.

[30] Shi Y, Xia MY, Shan Chen R, *et al.* Stable electric field TDIE solvers via quasi-exact evaluation of MOT matrix elements. *IEEE Transactions on Antennas and Propagation* 2011;59(2):574–585.

[31] Pray AJ, Nair NV, Shanker B. Stability properties of the time domain electric field integral equation using a separable approximation for the convolution with the retarded potential. *IEEE Transactions on Antennas and Propagation* 2012;60(8):3772–3781.

[32] Abboud T, Sayah T. Potentiels retardes pour les equations de Maxwell avec condition d'impedence eneralisee; 1998. R. I. 382.

[33] Bachelot A, Lange V. Time dependent integral method for Maxwell's system. In: *Third International Conference on Mathematical and Numerical Aspects of Wave Propagation*; 1995.

[34] Bachelot A, Lange V. Time dependent integral method for Maxwell's system with impedance boundary condition. *Transactions on Modelling and Simulation* 1995;9:8.

[35] Ha-Duong T. On retarded potential boundary integral equations and their discretisation. In: *Topics in Computational Wave Propagation: Direct and Inverse Problems.* New York, NY: Springer-Verlag; 2003. p. 301–336.

[36] Ha-Duong T, Ludwig B, Terrasse I. A Galerkin BEM for transient acoustic scattering by an absorbing obstacle. *International Journal for Numerical Methods in Engineering* 2003;57:1845–1882.

[37] Terrasse I. Résolution mathématique et num'erique des équations de Maxwell instationnaires par une méthode de potentiels retardés. *Ecole Polytechnique*; 1993.

[38] Wang X, Shi Y, Lu M, *et al.* Stable and accurate marching-on-in-time solvers of time domain EFIE, MFIE, and CFIE based on quasi-exact integration technique. *IEEE Transactions on Antennas and Propagation* 2020;69(4):2218–2229.

[39] van't Wout E, van der Ven H, van der Heul DR, *et al.* A provably stable MoT scheme based on quadratic spline basis functions. In: *Antennas and Propagation Society International Symposium (APSURSI)*, 2012 IEEE. New York, NY: IEEE; 2012. p. 1–2.

[40] Pray AJ, Beghein Y, Nair NV, *et al.* A higher order space-time Galerkin scheme for time domain integral equations. *IEEE Transactions on Antennas and Propagation* 2014;62(12):6183–6191.

[41] O'Connor S, Hughey S, Dault D, *et al.* A novel port/network parameter extraction technique for coupling circuits with full-wave time-domain integral equation solvers. *IEEE Transactions on Microwave Theory and Techniques.* 2018;67(2):553–564.

[42] Roth TE, Chew WC. Development of stable **A-Φ** time-domain integral equations for multiscale electromagnetics. *IEEE Journal on Multiscale and Multiphysics Computational Techniques* 2018;3:255–265.

[43] Roth TE, Chew WC. Stability analysis and discretization of A-Φ time domain integral equations for multiscale electromagnetics. *Journal of Computational Physics* 2020;408:109102.

[44] Roth TR, Chew WC. Radiation gauge potential-based time domain integral equations for penetrable regions. *Progress In Electromagnetics Research* 2020;168:73–86.

[45] Li J, Dault D, Shanker B. A quasianalytical time domain solution for scattering from a homogeneous sphere. *Journal of the Acoustical Society of America* 2014;135(4):1676–1685.

[46] Li J, Shanker B. Time-dependent Debye–Mie series solutions for electromagnetic scattering. *IEEE Transactions on Antennas and Propagation.* 2015;63(8):3644–3653.

[47] Li J, Shanker B. Time-dependent Lorentz-Mie-Debye formulation for electromagnetic scattering from dielectric spheres. *Progress in Electromagnetics Research* 2015;154:195–208.

[48] Pray AJ. *A Robust Stabilization Methodology for Time Domain Integral Equations in Electromagnetics.* East Lansing, MI: Michigan State University; 2014.

[49] Medgyesi-Mitchang LN, Putnam JM, Gedera MB. Generalized method of moments for three-dimensional penetrable scatteres. *Journal of the Optical Society of America A* 1994;11:1383–1398.

[50] Cools K, Andriulli FP, Olyslager F, *et al.* Time domain Calderon identities and their application to the integral equation analysis of scattering by PEC objects. Part I: Preconditioning. *IEEE Transactions on Antennas and Propagation* 2009;57(8):2352–2364.

[51] Andriulli FP, Cools K, Bogaert I, *et al.* On a well-conditioned electric field integral operator for multiply connected geometries. *IEEE Transactions on Antennas and Propagation* 2013;61(4):2077–2087.

[52] Vico F, Ferrando M, Greengard L, *et al.* The decoupled potential integral equation for time-harmonic electromagnetic scattering. *Communications on Pure and Applied Mathematics* 2016;69(4):771–812.

[53] Vico F, Greengard L, Ferrando M. Decoupled field integral equations for electromagnetic scattering from homogeneous penetrable obstacles. *Communications in Partial Differential Equations* 2018;43(2):159–184.

[54] Li J, Fu X, Shanker B. Formulation and iso-geometric analysis of scalar integral equations for electromagnetic scattering. *IEEE Transactions on Antennas and Propagation* 2018;66(4):1957–1966.

[55] Rao S, Wilton D, Glisson A. Electromagnetic scattering by surfaces of arbitrary shape. *IEEE Transactions on Antennas and Propagation* 1982;30(3):409–418.

[56] Yucel AC, Ergin AA. Exact evaluation of retarded-time potential integrals for the RWG bases. *IEEE Transactions on Antennas and Propagation* 2006;54(5):1496–1502.

[57] Olver FWJ, Lozier DW, Boisvert RF, *et al.*, editors. *NIST Handbook of Mathematical Functions*. New York, NY: Cambridge University Press; 2010. Print companion to [59].

[58] Graglia RD, Wilton DR, Peterson AF. Higher order interpolatory vector bases for computational electromagnetics. *IEEE Transactions on Antennas and Propagation* 1997;45(3):329–342.
[59] NIST Digital Library of Mathematical Functions. Online companion to [57]. http://dlmf.nist.gov/, Release 1.0.5 of 2012-10-01. Available from: http://dlmf.nist.gov/.

Chapter 5
Transmission-line modeling method—TLM

Christos Christopoulos[1]

Numerical modeling for Computational ElectroMagnetics (CEM) is as much an art as a science. At all stages of the modeling processes, the modeler has to interpret the essential features and requirements of the problems at hand and decide on the best method to set up models which lead to accurate, elegant solution, which help the users of the model to engage in creative analysis and design. In this chapter we look at modeling from the point of view of the Transmission-Line Modelling or Matrix Method, in short, TLM.

TLM is put in the context of other approaches to CEM using generic classification schemes. Modeling in TLM proceeds by a repeated judicious use of a small number of basic building elements ("blocks" or "bricks"). The essential character and properties of these simple elements are described in detail and the way in which they may be combined to represent complexity in practical systems is outlined. In order to relate and contrast TLM to other similar methods these same basic elements are looked at from different perspectives.

The main task of this chapter is the description of the development of three-dimensional (3D) electromagnetic models based on TLM. An extensive discussion on the basic TLM algorithm is presented. Scatter and connect, excitation and output and how TLM relates to other methods are described.

The perennial problem of CEM is the issue of problem complexity. Whatever the basic modeling approach is, it eventually fails when confronted with challenging multi-scale problems. We describe here how TLM addresses multi-scaling through embedding procedures (macro-nodes, sub-cells).

A comprehensive list of references is given for readers to access more detailed or specialist work.

5.1 Introduction

In this chapter, the choices made by TLM in approximating the electromagnetic fields in time and space are described in broad terms. Approximation is inherent in all numerical schemes not just TLM. But, each numerical scheme, through the way

[1]Department of Electrical and Electronic Engineering, University of Nottingham, UK

approximations are introduced, results in errors which differ in magnitude, at different frequencies and in different direction of wave propagation. The basic building blocks of a TLM model are described and the way they may be combined to construct more complete models of entire complex systems is explained. Throughout, modeling errors are considered and the relationship of TLM to other techniques is highlighted. Before setting upon this task CEM is considered in general terms and a generic classification of numerical methods is presented so that TLM is set into context with other numerical methods. This highlights the strengths and weaknesses of TLM in relation to other generic modeling approaches. Two texts on TLM by the author of this chapter may also be consulted for a more detailed discussion of some of the issues addressed here [1,2].

5.1.1 Modeling for CEM

Modeling is the process of representing a physical system into another domain so that the study of the original physical system is more convenient. Modeling using analytical techniques is a mapping of a system into a set of integro-differential equations which then we attempt to solve. Unfortunately, only simple systems with a high degree of symmetry and low complexity are amenable to analytical solutions and hence in most cases we resort to numerical methods. Ultimately, a numerical model is a set of algorithms which are processed in a digital computer. These algorithms may be simply solving directly equations which represent an analytical mapping model of the system or may mimic, in some way, the behavior of the original physical system. The range of available numerical techniques reflects the different choices that modelers can make when preparing a particular problem for solution.

It goes without saying that a numerical model is not identical to the original system. It is simply an attempt to capture in software the essential behavior of the original system in as efficient and accurate manner as required by the nature of the problem at hand. The implication of this statement is that for the same physical system there may be more than one suitable model depending on exactly what we want to know about the system. Even if we limit ourselves to electrical behavior, there may be several models depending on whether we are interested in steady state response, transient response, response at very-high frequencies, etc. We should therefore be comfortable with the fact that different numerical methods are ideally suited for certain types of problem. No particular numerical modeling method can address efficiently and accurately all our modeling needs in electrical engineering. The important thing is to be sufficiently familiar with the main features of each modeling method so that we are able to select the one most suited to our particular needs. Much like in a workshop we have a tool box with a variety of tools and we choose the best for each task, similarly from our "numerical box," we should be able to choose the most appropriate numerical method to address each particular problem. However, the effort required in getting familiar with a numerical tool can be considerable. The present text, in attempting to sketch out the range of available methods, is intended to assist readers in this difficult task. In this particular chapter, we present the main features of TLM. It is not possible in a single chapter to cover in detail all aspects of TLM. We rather aim to present the features of TLM which distinguish it from other methods, and how these affect user experience, accuracy, and efficiency, in solving various classes of problems.

It is inevitable that when we seek to model something complex and unfamiliar, we try to connect it in some way with notions which are familiar to us. This has several advantages, since if we can establish such a connection (an analogy or correspondence) then understanding is considerably facilitated and we can use analysis techniques with which we are already familiar. In TLM, we do this very effectively by establishing a correspondence between electromagnetic field behavior and network behavior. As soon as we are able to do that, we can proceed with further analysis based on network theory. Electrical engineers have an intuitive understanding of the behavior and analysis of circuits using network theory. Moreover, there is a wealth of analysis techniques based on network theorems which may be readily used to understand fields. It has already been pointed out that all numerical models introduce errors. Errors can be reduced but cannot be entirely eliminated. In TLM, errors may be interpreted in terms of parasitic components (inductors, capacitors) which have a clear meaning to the non-professional modeler who is the user of numerical models. One of the advantages of TLM models is that there is a "physical feel" to their construction and errors. This is invaluable to the designer and analyst as it can support interpretation and creativity.

Having a model which is securely anchored in network theory means that unconditional stability is guaranteed. An additional advantage of TLM is that since its basic building blocks are simple circuit elements, it offers excellent scope for embedding objects into the spatial mesh whilst maintain numerical stability. This facility is crucial when tackling practical problems where multi-scaling is the prime challenge to all numerical modeling methods.

Some of the issue outlined above are clarified and addressed in the rest of this chapter.

5.1.2 Classification of CEM and TLM

Readers of this book and of the general scientific and commercial literature on CEM may be puzzled by the considerable number of methods and tools which are available. Some order can be put into this by attempting to place all available methods and tools into some general categories based on generic criteria. The advantage of this classification is that it offers intellectual economy and is also of practical significance. It will become evident that methods belonging to the different classes have broadly similar characteristics, strengths, and weaknesses. Placing TLM in such classification schemes brings out its main features and contrasts its capabilities with those of other available methods.

It is evident that several classification schemes are available depending on the selected classification criteria. Thus, we may select as criterion the domain of the variable (time or frequency) resulting in methods applied either in the time-domain (TD) or the frequency-domain (FD). Another criterion that may be selected is the domain of the operator used to formulate the system equations (differential or integral equations). We have thus methods based on solving differential equations (DE) or integral equations (IE). The manner in which space is tessellated in the preparation for numerical solution also results in another classification based on the nature of the cells (building blocks). These may be regular in shape or irregular, resulting in structured, unstructured, or multi-grid meshes.

For completeness, we point out that there are some methods which do not fall easily within these classifications e.g. very high-frequency ray optics methods, or, semi-analytical methods. These are specialist tools which are used under specific circumstances and are not generally applicable, versatile, numerical methods.

Thus, as an example, a method may be classed as time-domain TD, differential equation DE and with a structured mesh. In general, each method is best suited for a particular implementation e.g. DE and TD but the ingenuity of modelers has ensured that methods may be adapted to operate reasonably efficiently even outside their optimum classification category. The classification category confers to a method certain generic features some advantageous and some less so. To illustrate, a TD method is good for transient problems or problems where a broadband response is required but not so good when the steady-state response at a single frequency is required. An IE method is good when applied to problems where there are large volumes of free space but not so good in problems where there are parts of space containing complex shapes and different materials. In contrast, a DE method is good for problems with inherent complexity and material non-uniformity but less good when applied to open-boundary problems. A method employing a structured mesh is easy to apply but may result in a poor description of curved boundaries (stair-casing errors). An unstructured mesh gives a better approximation of curved boundaries but increases computational complexity and raises issues of mesh quality.

We see from the brief discussion above that whichever choice we make we will inherit some good and some less advantageous features. Therefore, no single method scores highly for all possible problem requirements. As already mentioned, methods can be adapted to operate outside their optimum domain, but it should be expected that this will be done at some cost in efficiency, accuracy, or ease in problem specification or data processing. It is natural to ask at this point whether it is possible to combine, in some kind of hybrid, more than one method each applied to model parts of the problem space best suited to it. As an example, a DE method may be applied to parts of the problem space where there is geometrical and material complexity and an IE method in the free-space between complex areas. This is possible and there is some work in the literature to show how this is done. There are however few mainstream general use tools based on hybrid methods due to the inherent interfacing complexity and the sophistication required from general users called upon to use such complex tools.

The preferred implementation of TLM is as a differential-equation DE, time-domain TD method much like FDTD. This is the implementation we focus on the present chapter. Frequency-domain implementations of TLM are also available [3]. There are also structured and unstructured mesh options and extensive developments in embedding techniques (sub-cell or macro-nodes) as discussed further in this chapter.

5.2 The basic TLM building elements

The description of all classical electrical phenomena relies on the availability of a small number of basic elements. A judicious combination of these elements can then be employed to describe the full complexity of all electrical applications. This approach has the advantage of elegance and simplicity and also supports an intuitive understanding of the vast range of electrical responses and applications. The basic elements which are described below in more detail are: two energy storage elements, a dissipation element and a source element. Two energy storage elements are required as by transfer of energy between them we can capture oscillatory and wave phenomena central to the electrical behavior of systems.

5.2.1 Lumped and distributed models of systems

An ideal lumped element has no spatial extent-all action is concentrated at a point. This is obviously an idealization. However, real behavior can be captured by the combination of several ideal elements.

A lumped capacitor C is an ideal element which stores energy in its associated electric field and if the potential difference between its terminals is v then the energy stored is,

$$w_C = \frac{1}{2}Cv^2 \tag{5.1}$$

Since energy cannot change instantaneously, if follows that that the voltage across the plates of a capacitor cannot change instantaneously. The terminal relationship between the voltage and the current across a capacitor may be described by the familiar differential equation:

$$i(t) = C\frac{dv(t)}{dt} \tag{5.2}$$

This expression indicates that the voltage and current across a capacitor are out of phase by 90°. This physical fact is captured in the frequency domain by shifting the relative position of the voltage and current phasors by 90°, by multiplying by $-j$,

$$\bar{V} = \bar{I}\frac{1}{j\omega C} \tag{5.3}$$

Since most numerical modeling and simulation take place in the time-domain using digital computers, it is inevitable that we seek to compute the behavior of elements such as the capacitor in the discrete time domain. In TLM, this is done by exploiting the fact that a segment of an ideal transmission line consists of distributed capacitance and inductance. We can thus approximately describe a capacitor by a segment of a transmission line where we have sought to minimize the influence of the distributed inductance. The impact of the inductance can be reduced, but not eliminated, and it constitutes the error in the model of the capacitor. Errors are inherent in all numerical models and not just in TLM. We may choose to represent the capacitor Figure 5.1a as a two-port device using a transmission line stub where the far end is an open-circuit as shown in Figure 5.1b. The open-circuit ensures that the voltage is maximized and the current minimized thus reducing the impact of the inductance as required. To fully, characterize this stub we need to specify its characteristic impedance Z_C and the round-trip transit time Δt. The round-trip transit time is the basic time-step of the numerical computation and it is selected for accuracy and adequate resolution of response details. The characteristic impedance is then chosen so that the stub represents as accurately as possible the required capacitance C. It can be shown that a capacitance C in a computation where the time-step is set at Δt is described by an open-circuit stub where [1],

$$Z_C = \frac{\Delta t}{2C} \tag{5.4}$$

Under these circumstances, the inductance which is inevitably associated with the stub is,

$$L_e = \frac{(\Delta t)^2}{4C} \tag{5.5}$$

This inductance represents the “error” in this model and can be reduced by a reduction in the time-step. This naturally results in a longer computation. A useful feature of TLM is that the error has a clear physical meaning i.e. a stray inductance which may in fact be a more realistic representation of a real capacitor. The error is in the model and not in the solution method and so its impact can be more easily assessed and controlled.

We may also choose to look at model accuracy in the frequency domain by calculating the input admittance of an open-circuit line and expanding it in a series,

$$\begin{aligned} Y_{in} &= \frac{j\tan\left(\omega\frac{\Delta t}{2}\right)}{Z_C} = \frac{j}{\Delta t/(2C)}\left[\frac{\omega\Delta t}{2} + \frac{1}{3}\left(\frac{\omega\Delta t}{2}\right)^3 + \cdots\right] \\ &= j\omega C + j\frac{C\omega^3(\Delta t)^2}{12} + \cdots \end{aligned}$$

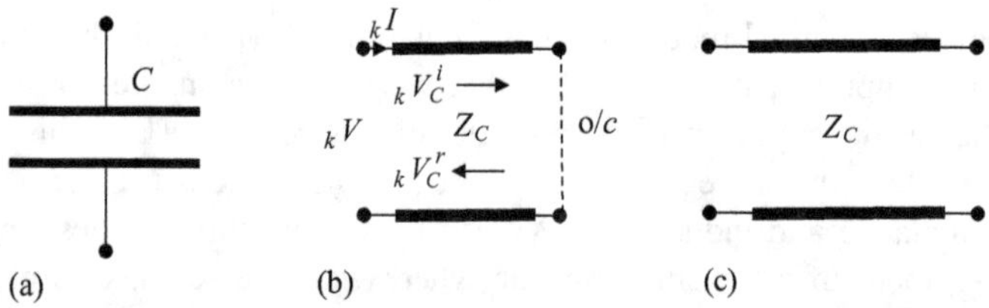

Figure 5.1 Capacitor models in TLM, (a) ideal capacitor, (b) its stub model, and (c) its link model

We see that the leading term represents the correct capacitive admittance but there are also higher order terms representing errors which may be reduced by decreasing the time-step.

Another approach is to represent a capacitor by a two-port link-line segment as shown in Figure 5.1c. Now the single transit time is Δt and the equivalent expressions corresponding to (5.4) and (5.5) are,

$$Z_C = \frac{\Delta t}{C}, \quad L_e = \frac{(\Delta t)^2}{C} \tag{5.6}$$

Using lumped components represents a discretization in space, essential for numerical computation, as otherwise we would need infinite memory even for the simplest problem. Modeling lumped components by transmission line stubs or link-lines represents discretization in time, essential for numerical computation, as otherwise computation even for the simplest problem would require an infinite time.

Taking as an example the stub model of the capacitor, computation in the discrete time domain proceeds by defining the following quantities: the total terminal voltage and current at time step k, ${}_kV$, ${}_kI$; the incident and reflected voltage pulses at time-step k, ${}_kV^i$,${}_kV^r$. The stub, looking into its terminal may be replaced by its Thevenin equivalent as shown in Figure 5.2.

In the equivalent, the factor two represents the fact that under open-circuit conditions at the terminals the incident voltage doubles. It follows that the terminal conditions at time-step k are represented by,

$${}_kV = 2{}_kV^i + {}_kIZ_C \tag{5.7}$$

Therefore, the reflected voltage at k is,

$${}_kV^r = {}_kV - {}_kV^i \tag{5.8}$$

This pulse travel toward the open circuit termination of the stub, it is reflected there with the same sign after time $\Delta t/2$ and then becomes incident after another delay $\Delta t/2$ at the stub terminal. This is in effect the incident voltage at time-step $k+1$,

$${}_{k+1}V^i = {}_kV^r = {}_kV - {}_kV^i \tag{5.9}$$

In TLM parlance, (5.8) represents the "scattering" process and (5.9) the "connection." It is evident that with the new incident voltage calculated as in (5.9)

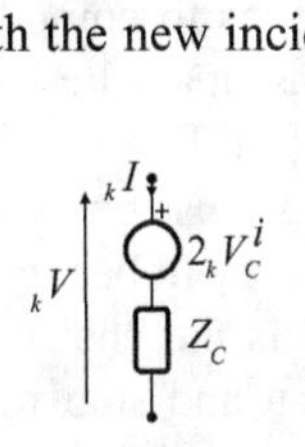

Figure 5.2 Thevenin equivalent of a stub capacitor model

the computation may be advanced by one time-step using (5.7) and (5.8). The TLM algorithm is thus a repetition of scattering and connect for as long as required. This routine is central to TLM whatever the complexity of the problem is e.g. multi-port systems, boundary conditions, etc.

An ideal inductor is a device storing energy in its associated magnetic field therefore requires a current flow. No instantaneous change in an inductor current is possible. The terminal relations in the time-domain for an inductor are,

$$v(t) = L\frac{di(t)}{dt} \tag{5.10}$$

The discrete modeling of an ideal inductor proceeds along similar lines as for a capacitor. The only difference here is that to minimize errors ("stray" capacitance) we employ a short-circuit stub since we want voltage minimization and hence energy stored in capacitance. The choice of stub impedance to model L and the associated capacitive error are,

$$Z_L = \frac{2L}{\Delta t}, \quad C_e = \frac{(\Delta t)^2}{4L} \tag{5.11}$$

The expressions for scattering are the same as for the capacitive stub, but connection is different for the inductive stub as reflection from a short circuit reverses the sign of the incident voltage and thus (6.9) is now,

$$_{k+1}V^i = -_kV^r = -_kV + {_kV^i} \tag{5.12}$$

For a "link" model of an inductor, the equivalent expressions to (5.11) are,

$$Z_L = \frac{L}{\Delta t}, \quad C_e = \frac{(\Delta t)^2}{L} \tag{5.13}$$

With the two energy storage elements (C and L), an energy dissipation element R, and energy sources (voltage or current source) we have a complete set of ideal elements to describe any linear electrical circuit and analyze it using the laws of network theory (Kirchhoff's voltage and current laws KVL and KCL, respectively). Further details may be found in [4–6]. The essence of TLM modeling is to capture all electrical phenomena by suitable lumped element models using the ideal elements above (space discretization), use "stub" or "link" transmission-line (TL) models for L and C (time discretization), and, implement an explicit routine of "scattering" and "connect" operations to compute responses for the time interval required. Other physical domains may be similarly addressed e.g. thermal, mechanical etc. as described in [7–9]. Naturally, the choice of the space discretization length and the timestep must be consistent with the physical requirements of each particular problem as it will become clearer in the following sections.

Another issue to be addressed is the choice between link and stub models. In general, link model result in explicit and stub models in implicit solution routines. Mixed stub and link models offer considerable flexibility in structuring complex circuits for efficient computational solution. Examples of this will be given in the next section.

It is recognized that the presence of nonlinear elements in complex circuits complicates and slows down computations. It is straightforward in TLM to isolate nonlinearities and deal with them locally thus greatly increasing computational efficiency [10–12]. A typical example is shown schematically in Figure 5.3a. Here we show schematically a large linear region A connected to a non-linear elements B represented for simplicity by a switch. The difficulty with this kind of circuit is that any iterative solution to deal with the non-linearity would involve manipulation of the entire network. In practical circuits such as typically found in power electronic drives with dozens of electronic switches at different states (open or closed) it would involve the setting up of a multitude of topologies to anticipate potential states. An elegant way of dealing with this in TLM is to recognize that in most practical arrangements some capacitance is associated with the non-linear element (e.g. stray capacitance) as shown in Figure 5.3b. The TLM model of this configuration focusing on region B is shown in Figure 5.3c where we employ a link model of the capacitor. At the interface between A and B, the Thevenin equivalent, looking both ways, is shown in Figure 5.3d. We see now that the large linear area A and the non-linear area B are decoupled. The iterative procedure need only be applied in B when calculating the reflected voltage. Several simultaneous non-linearities may be dealt with in a similar manner and only one circuit topology needs to be formulated irrespective of the switch(es) state. The state of the switch is communicated to the rest of the network as part of the "connection" process.

5.2.2 TLM models in one dimension

We address in this section the case of a distributed one-dimensional system as it permits, with the minimum of complexity, a clear presentation of TLM principles and the way it is applied to practical problems. Two- and three-dimensional problems are then a scaling up in complexity, but the modeling concepts are already firmly established in 1D where physical clarity is easier to achieve.

We first demonstrate the isomorphism between the equations describing guided wave propagation in a transmission line and free space one-dimensional field propagation. For a transmission line with parameters representing a segment of

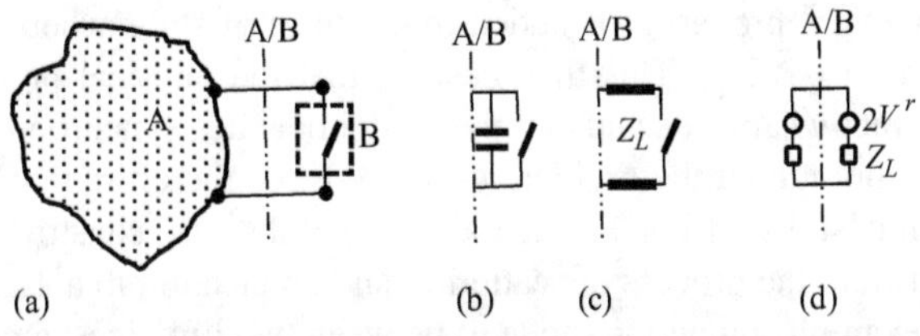

Figure 5.3 *(a) Decoupling non-linearity B from linear region A, (b) model of non-linear region including stray capacitance, (c) TLM equivalent, and (d) Thevenin equivalent*

length Δx, shunt resistance R, series inductance L and shunt capacitance C, the equation describing current evolution is,

$$\frac{\partial^2 i(x,t)}{\partial x^2} = \frac{L}{\Delta x}\frac{C}{\Delta x}\frac{\partial^2 i(x,t)}{\partial t^2} + \frac{L}{\Delta x}\frac{1}{R\Delta x}\frac{\partial i(x,t)}{\partial t} \tag{5.14}$$

where we have assumed that the segment length is very small, $\Delta x \to 0$. Similarly, for one-dimensional free-space field propagation the equation for the current density is

$$\frac{\partial^2 j(x,t)}{\partial x^2} = \mu\varepsilon\frac{\partial^2 j(x,t)}{\partial t^2} + \mu\sigma\frac{\partial j(x,t)}{\partial t} \tag{5.15}$$

The isomorphism between network and field equations (5.14) and (5.15) implies that we can solve field problems by reference to the solution of circuit problems using the network equations [13]. The network equations, KCL and KVL, are algebraic and easier to manipulate in comparison to Maxwell's equations which are differential equations. There are many techniques and theorems in network theory for efficient stable solutions of circuit problems which can be directly applied once the analogy between fields and circuits has been established. In one dimension, this analogy is directly apparent from (5.14) and (5.15), namely

$$i \leftrightarrow j \quad L/\Delta x \leftrightarrow \mu \quad C/\Delta x \leftrightarrow \varepsilon \quad 1/(R\Delta x) \leftrightarrow \sigma$$

Two steps are involved in model building: representation by lumped elements (space discretization-choice of Δx), and, representation of the lumped components by TL segments (stubs or links-time discretization Δt). For acceptable accuracy, the space discretization length must be much shorter than the shortest wavelength of interest. This gives an upper frequency limit to the accuracy of the model. Naturally, the space and time discretization lengths must accord with the speed of propagation of EM waves in the surrounding medium. It is known from TL theory that a L–C segment may be represented by a link TL of characteristic impedance and transit time given by, $Z = \sqrt{L/C}$, $\Delta t = \sqrt{LC}$ respectively. In complex problems (e.g. in non-uniform materials), we may end up with areas where a different transit time (timestep) applies. This complicates computation and, in such cases, we seek to impose a universal timestep applicable throughout the problem space. This is chosen to be the minimum value of $\sqrt{LC}$ and any additional L or C required to represent the material properties is added in the form of stubs where the roundtrip time is set at $\Delta t = \sqrt{LC}|_{\min}$.This then ensures that the physical properties of the medium are correctly captured and that the model operates synchronously (with the same timestep) throughout the problem space. We see therefore that the judicious use of links and stubs offers considerable flexibility in constructing efficient models. To illustrate the process of scattering and connection in a TLM model, we consider as an example a junction (node n) between two link lines, a resistor and an additional capacitor represented by a stub (Figure 5.4a). The incident and reflected pulses at node n at timestep k are also indicated in the figure. Subscripts L and R refer to pulse on the left and right of node n. The total nodal voltage at timestep k is

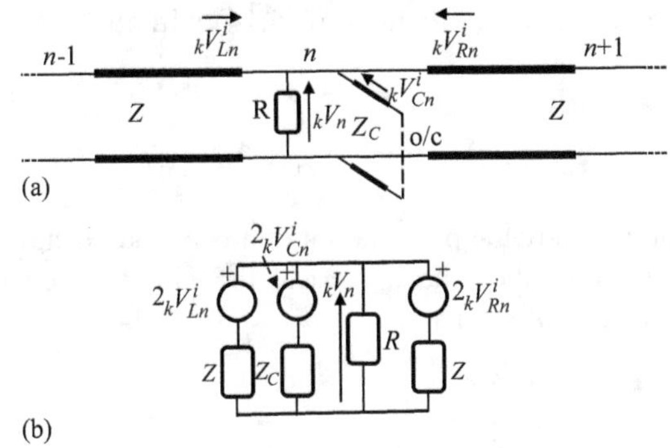

Figure 5.4 *(a) Node* n *representing the junction of two link lines, a capacitor represented by a stub and a resistor. (b) Thevenin equivalent circuit*

indicated by $_kV_n$. The Thevenin equivalent at node n and timestep k is shown in Figure 5.4b.

The total voltage may be directly obtained from the parallel generator theorem,

$$_kV_n = \frac{\frac{2V^i_{Ln}}{Z} + \frac{2V^i_{Cn}}{Z_C} + \frac{2V^i_{Rn}}{Z}}{\frac{2}{Z} + \frac{1}{Z_C} + \frac{1}{R}} \tag{5.16}$$

The scattered (reflected) voltages into the two link and one stub lines are,

$$_kV^r_{Ln} = {_kV_n} - {_kV^i_{Ln}}, \quad _kV^r_{Rn} = {_kV_n} - {_kV^i_{Rn}}, \quad _kV^r_{Cn} = {_kV_n} - {_kV^i_{Cn}} \tag{5.17}$$

Equations (5.16) and (5.17) describe the scattering process. Connection, the process where the incident voltage pulses at the next timestep $k+1$ are obtained, is also straight forward. For the capacitive stub we have,

$$_{k+1}V^i_{Cn} = {_kV^r_{Cn}} \tag{5.18}$$

The new incident voltage on node n from the left $_{k+1}V^i_{Ln}$, is simply the voltage reflected from the right of node n–1 at timestep k and so on. This type of calculation can be adapted to the presence of more specific lumped components including nonlinear ones to deal with very general configurations. It describes the essence of all TLM algorithms. Naturally, in more complex situations e.g. in three dimensions, scattering is more complicated but the essential principles described here remain. In Figure 5.4a, we could instead of a stub model for the capacitor employ a link model. This choice is up to the modeler although depending on the problem the distribution of modeling errors may change somewhat.

It is opportune at this stage to examine in more detail the impact of space discretization on model accuracy [2]. In deriving (5.14), we assumed that $\Delta x \to 0$. However, in numerical schemes, this condition is never achieved. For simplicity, we can derive the equivalent of (5.14) without losses and for a finite Δx to obtain,

$$LC\frac{\partial^2 v(x,t)}{\partial t^2} = v(x+\Delta x, t) + v(x-\Delta x, t) - 2v(x,t) \tag{5.19}$$

Assuming that we study the propagation of a wavelike disturbance (ω, k) with a phase velocity $u = \omega/k$ then the numerical model prediction based on (5.19) shows after some manipulation that,

$$\omega^2 = \frac{1}{LC}4\sin^2\left(\frac{k\Delta x}{2}\right) \tag{5.20}$$

It is clear from (5.20) that the velocity of propagation ω/k predicted by the numerical model is no longer constant. This aspect of the model is known as dispersion and it is the inevitable consequence of the discretization. If our choice of Δx is such that,

$$k\Delta x = 2\pi\frac{\Delta x}{\lambda} \ll 1 \rightarrow \quad \frac{\Delta x}{\lambda} \ll 1$$

then, we may use the small arguments approximation for the sin(.) in (5.20) to show that the velocity of propagation is approximately constant (no dispersion)

$$u = \frac{\omega}{k} = \frac{\Delta x}{\sqrt{LC}}$$

This is approximately true if the space discretization length is chosen to be much smaller that the shortest wavelength of interest. The rule of thumb is that a tenth of the wavelength is small enough for acceptable accuracy in most applications.

5.2.3 TLM models, integration algorithms, and wave digital filters

In this section, we consider the relationship between TLM models and other solution methods to highlight commonalities and differences.

It is straightforward to demonstrate that the stub model of a capacitor is equivalent to the discrete model of a capacitor obtained by using the trapezoidal rules [1]. Equation (5.2) may be written using the trapezoidal rule as,

$$\frac{{}_{k+1}I + {}_{k}I}{2} = C\frac{{}_{k+1}V - {}_{k}V}{\Delta t} \tag{5.21}$$

Re-arranging this equation, we obtain,

$${}_{k+1}I = \frac{{}_{k+1}V}{(\Delta t/2C)} - \left\{\frac{{}_{k}V}{(\Delta t/2C)} + {}_{k}I\right\} = \frac{{}_{k+1}V}{Z_C} - \left\{\frac{{}_{k}V}{Z_C} + {}_{k}I\right\} \tag{5.22}$$

Equation (5.22) is effectively the TLM stub capacitor model.

It is also clear that since the TLM models of lumped components describe differential (L) and integral (C) terms we may use them to convert directly into the discrete time-domain any arbitrary physical system (e.g. thermal, mechanical) described by integro-differential equations. Proportional terms are simply described by a resistor. Thus a differential term appears as,

$$V_L = L\frac{di}{dt} \Leftrightarrow V_L = i\left(\frac{2L}{\Delta t}\right) + 2V_L^i \tag{5.23}$$

And an integral term as,

$$V_C = \frac{1}{C}\int idt \Leftrightarrow V_C = i\left(\frac{\Delta t}{2C}\right) + 2V_C^i \tag{5.24}$$

Further details and the modeling of other terms may be found in [12].

TLM as a digital algorithm employing wave quantities may also be related to wave digital filters (WDF) [14–16]. WDFs are derived from analogue reference filters via a bilinear transformation using scattering parameters. This is shown schematically in Figure 5.5a which depicts a box containing an 1-port element of impedance Z described in the network domain by

$$V = IZ \tag{5.25}$$

In Figure 5.5b, the same element is described in the wave domain through the incident and reflected waves a and b (equivalent to V^i, V^r in TLM notation). We may define a port impedance R_0 with a value which can be set as to avoid a delay-free loop as shown below. The total port voltage and the current are related to the incident and reflected pulses as follows:

$$a = V + IR, \ \ b = V - IR \tag{5.26}$$

Combining (5.25) and (5.26), we obtain the reflection coefficient,

$$\Gamma = \frac{b}{a} = \frac{V - IR}{V + IR} = \frac{Z - R_0}{Z + R_0} \tag{5.27}$$

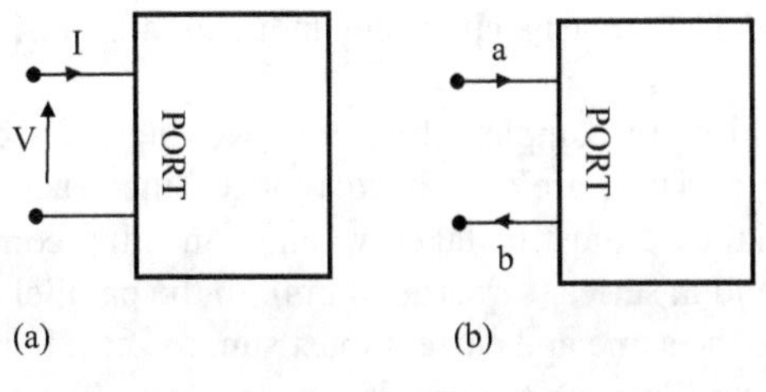

Figure 5.5 A one-port network in (a) network and (b) wave domains

We now illustrate the application of these ideas by considering the description of a capacitor C in the wave domain. For this component, the reflection coefficient in the s-domain is,

$$\Gamma_C(s) = \frac{1 - R_0 C s}{1 + R_0 C s} \tag{5.28}$$

We now apply the bilinear transformation

$$s = \frac{2}{\Delta t}\frac{1 - z^{-1}}{1 + z^{-1}} \tag{5.29}$$

to obtain the reflection coefficient in the z-domain relating incident and reflected waves at timestep k. In this expression, the operator z^{-1} represents a delay of one timestep Δt. Substituting (5.29) into (5.28), we obtain,

$$\frac{{}_k b}{{}_k a} = \frac{1 - R_0 C \frac{2}{\Delta t}\frac{1-z^{-1}}{1+z^{-1}}}{1 + R_0 C \frac{2}{\Delta t}\frac{1-z^{-1}}{1+z^{-1}}} \tag{5.30}$$

Multiplying out both sides of this equation, we obtain,

$$\begin{aligned} &{}_k b + {}_{k-1} b + R_0 \frac{2C}{\Delta t}\, {}_k b - R_0 \frac{2C}{\Delta t}\, {}_{k-1} b = \\ &{}_k a + {}_{k-1} a - R_0 \frac{2C}{\Delta t}\, {}_k a + R_0 \frac{2C}{\Delta t}\, {}_{k-1} a \end{aligned} \tag{5.31}$$

We see in (5.31) that ${}_k b$ depends on ${}_k a$. This instantaneous reflection requires iterative procedures and makes signal processing difficult in complex problems. For this reason, we may choose R_0 to cancel this dependence. Choosing $R_0 = \Delta t / 2C$ and substituting into (5.31), we obtain,

$${}_k b = {}_{k-1} a \tag{5.32}$$

This is precisely the stub model of a capacitor C in TLM. It should be noted that the TLM formulation has a clearer physical interpretation. A similar treatment of an inductor L gives,

$${}_k b = -{}_{k-1} a \tag{5.33}$$

The WDF models of a capacitor and inductor are shown schematically in Figure 5.6a and b.

In order to build more complex structures, we interconnect simple structures using what in WDF parlance are called “adaptors.” Since each port has a different impedance R_0 special care must be taken when connecting components in parallel (parallel adaptors) and in series (series adaptors). In the parallel case, all voltages at the junction must be the same and current must sum to zero. For the series case, the reverse holds true. We illustrate the use of adaptors for a 2D structure widely used in TLM—the “shunt” node shown in Figure 5.7a.

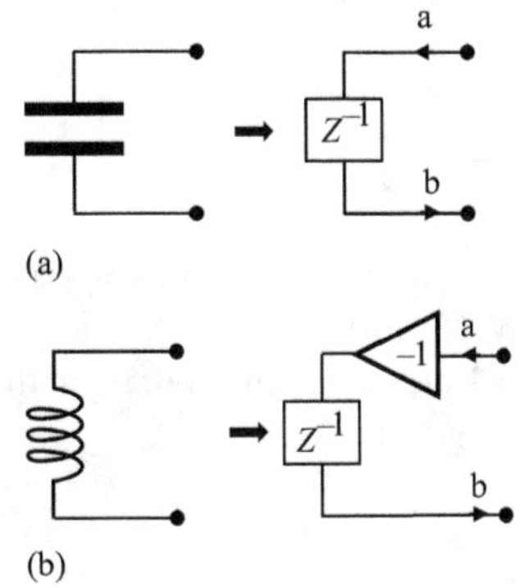

Figure 5.6 WDF models of (a) capacitor and (b) inductor

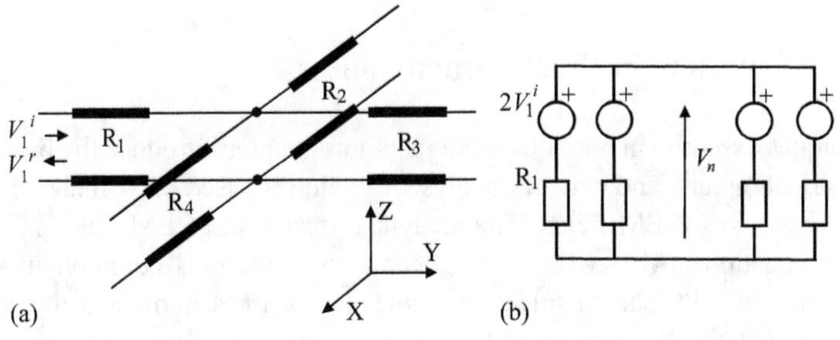

Figure 5.7 (a) A shunt TLM node and (b) its Thevenin equivalent

Using the normal notation for TLM for the incident voltage at each port n, V_n^i we can obtain the Thevenin equivalent shown in Figure 5.7b. The total voltage is then,

$$\begin{aligned} V_n &= \frac{2V_1^iG_1 + 2V_2^iG_2 + 2V_3^iG_3 + 2V_4^iG_4}{G_1 + G_2 + G_3 + G_4} \\ V_j^r &= V_n - V_j^i, \ j = 1 \cdots 4 \end{aligned} \tag{5.34}$$

For the WDF parallel adaptor eliminating voltage and current from the three equations demanding the same voltage across the junction and the one equation for zero total current at the junction and using (5.26), we obtain an expression relating the incident and reflected voltages at the junction,

$$\boldsymbol{V}^r = \{\boldsymbol{M} - \boldsymbol{1}\}\boldsymbol{V}^i \tag{5.35}$$

where $\boldsymbol{V}^r$ is a column matrix with four rows the elements of the four reflected voltages, and similarly $\boldsymbol{V}^i$ for the incident voltages. The matrix **1** is the identity matrix and $\boldsymbol{M}$ is a 4×4 matrix with the elements of column k,

$$m_k = \frac{2G_k}{\sum_{j=1}^{4} G_j} \tag{5.36}$$

For the common case where all four link-line impedances are equal, we obtain from both the TLM and WDF formulations the same scattering coefficients, namely, -0.5 for the diagonal elements and $+0.5$ for the off diagonal elements. The scattering matrix for the shunt node is thus,

$$S = 0.5 \begin{bmatrix} -1 & 1 & 1 & 1 \\ 1 & -1 & 1 & 1 \\ 1 & 1 & -1 & 1 \\ 1 & 1 & 1 & -1 \end{bmatrix} \tag{5.37}$$

Any structure described by WDF can be readily interfaced with TLM models [16].

5.3 TLM models in three dimensions

The techniques explored in previous sections of this chapter introduce the basic idea of using electrical circuits and network analysis techniques based on Kirchhoff's laws to model the behavior of EM fields. The analytical treatment of EM fields is based on Maxwell's equations (ME) which, for general configurations common in practical problems, are very difficult to solve. The objective of most numerical methods is to solve these equations by techniques such as approximating derivatives by finite differences (FDTD). TLM can be formulated by reference to ME as shown in Sections 5.3.6 and 5.3.7, but also directly from conservation and network laws as shown in [17,18]. For 3D fields, the application of TLM requires the construction of a 3D circuit of transmission lines such that the electric and magnetic field map directly to voltages and currents in this circuit. Naturally, for accuracy, a junction of interconnected transmission lines (the "node") can be developed to represent the fields in a volume of space (the "cell") provided the size of the cell in all dimensions is much smaller that the shortest wavelength of interest (typically $\leq \lambda_{min}/10$). The workhorse of 3D TLM is shown in Figure 5.8 and is known as the symmetrical condensed node (SCN).

This node models the EM fields in a cell of dimensions $\Delta x \times \Delta y \times \Delta z$. The three discretization lengths are chosen to be much smaller than the wavelength as already stated but otherwise can be chosen to fit problem requirements. The simplest and most common choice is to use a uniform cell (regular mesh), $\Delta x = \Delta y = \Delta z = \Delta \ell$ and this is our first choice here to introduce the basics of the SCN node. From Figure 5.8, we see that there are 12 transmission lines polarized in such a way as to have two orthogonal polarizations in each coordinate direction. As an example, Line 1 is polarized in the x-direction and represents signal propagating in the y-direction. Different alternative labeling schemes are used by authors to describe the various pulses propagating in this node. The availability of two orthogonal polarizations in each coordinate directions allows the description of any possible wave polarization. In this node, one

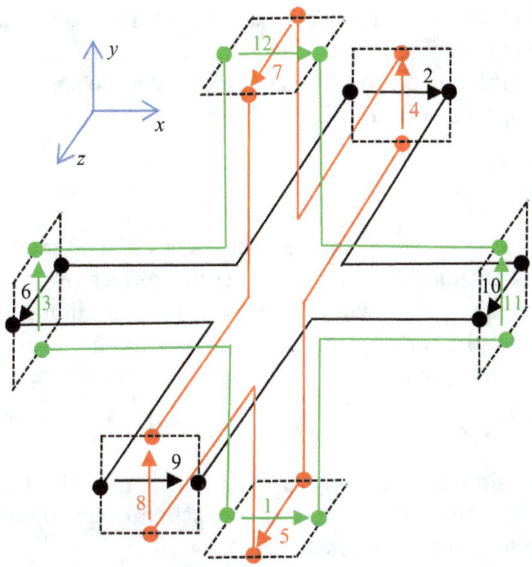

Figure 5.8 *The SCN node with twelve ports representing a cell of dimensions* $\Delta x\ \Delta y\ \Delta z$

can discern three structures of four series-connected lines, i.e. lines (1, 11, 12, 3), (5, 4, 7, 8), and (9, 10, 2, 6). Voltage pulses traveling toward the center of the node along Line 1 at time step k are termed as incident and those propagating away from the center toward the port as reflected, ${}_kV_1^i, {}_kV_1^r$, respectively. To calculate scattering in this node, we need to relate the vector of the 12 incident voltages to the vector of the 12 reflected voltages through an appropriate 12×12 scattering matrix,

$$\boldsymbol{V}^r = \boldsymbol{S}\boldsymbol{V}^i \tag{5.38}$$

The scattering matrix must satisfy network and conservation laws (it is a unitary matrix) and can be easily obtained for the uniform node [1,18],

$$S = 0.5\begin{bmatrix} 0 & 1 & 1 & 0 & 0 & 0 & 0 & 0 & 1 & 0 & -1 & 0 \\ 1 & 0 & 0 & 0 & 0 & 1 & 0 & 0 & 0 & -1 & 0 & 1 \\ 1 & 0 & 0 & 1 & 0 & 0 & 0 & 1 & 0 & 0 & 0 & -1 \\ 0 & 0 & 1 & 0 & 1 & 0 & -1 & 0 & 0 & 0 & 1 & 0 \\ 0 & 0 & 0 & 1 & 0 & 1 & 0 & -1 & 0 & 1 & 0 & 0 \\ 0 & 1 & 0 & 0 & 1 & 0 & 1 & 0 & -1 & 0 & 0 & 0 \\ 0 & 0 & 0 & -1 & 0 & 1 & 0 & 1 & 0 & 1 & 0 & 0 \\ 0 & 0 & 1 & 0 & -1 & 0 & 1 & 0 & 0 & 0 & 1 & 0 \\ 1 & 0 & 0 & 0 & 0 & -1 & 0 & 0 & 0 & 1 & 0 & 1 \\ 0 & -1 & 0 & 0 & 1 & 0 & 1 & 0 & 1 & 0 & 0 & 0 \\ -1 & 0 & 0 & 1 & 0 & 0 & 0 & 1 & 0 & 0 & 0 & 1 \\ 0 & 1 & -1 & 0 & 0 & 0 & 0 & 0 & 1 & 0 & 1 & 0 \end{bmatrix} \tag{5.39}$$

We see from (5.39) that to calculate each scattered (reflected) pulse we need to compute a weighted average of four incident pulses. It is also evident from this matrix that a pulse incident on port 1 takes two timesteps to appear in port 12. Thus, the velocity of propagation in the 3D TLM mesh is,

$$u_{TLM} = \frac{\Delta \ell}{2\Delta t} \tag{5.40}$$

Once $\Delta \ell$ is chosen, typically smaller that a tenth of the shortest wavelength of interest, then Δt is obtained from (5.40) to get the correct velocity of propagation for the modeled medium. Following scattering, the connection process is straightforward dictated by the topological details of the mesh. As in illustration, the connection in the x-direction for the two nodes shown in Figure 5.9 is,

$$_{k+1}V_6^i(x + \Delta x, y, z) = {}_kV_{10}^r(x, y, z) \tag{5.41}$$

Similar expressions hold for all the other ports. As in the 1D case, the TLM algorithm is a repetition of scattering and connection events throughout the mesh for as long as the response is required.

5.3.1 Mapping of fields to circuits in 3D

We now address the mapping of the field components to voltage pulses traveling in the TLs composing the SCN node. To map the x-component of the electric field, we identify ports polarized in the x-direction and calculate the average voltage to obtain,

$$E_x = -\frac{V_1^i + V_2^i + V_{12}^i + V_9^i}{2\Delta \ell} \tag{5.42}$$

Similarly, for the x-component of the magnetic field, we compute the current linked with H_x (current flowing in the y–z plane) to obtain,

$$H_x = \frac{V_4^i + V_7^i - V_5^i - V_8^i}{2Z\Delta \ell} \tag{5.43}$$

where Z is the characteristic impedance of each transmission line (e.g. 377 Ohm in free space). Equations (5.42) and (5.43) and their equivalents for the other field

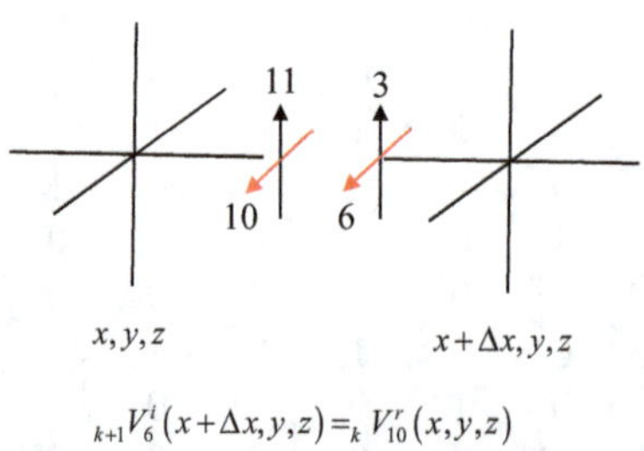

Figure 5.9 Schematic of "connection" in a TLM mesh

components map the field at the center of the node. An alternative is to map the fields at the cell boundaries (a one-to-one correspondence to voltage pulses). At the boundary between cells (x, y, z), $(x, y, z + \Delta\ell)$, we obtain,

$$E_x = -\frac{V_9^i(x,y,z) + V_2^i(x,y,z+\Delta\ell)}{\Delta\ell}$$
$$H_x = \frac{V_8^i(x,y,z) - V_4^i(x,y,z+\Delta\ell)}{Z\Delta\ell} \tag{5.44}$$

The expressions above also indicate how certain field components may be excited in the mesh by injecting suitable voltage pulses. For example, to inject an electric field component $E_x = E_0$ at a particular node centre, we choose,

$$V_1^i = V_2^i = V_9^i = V_{12}^i = -E_0\Delta\ell/2 \tag{5.45}$$

Then substituting (5.45) into (5.42) we obtain $E_x = E_0$. Similar relations are obtained for other field components.

5.3.2 Scatter and connect in TLM for irregular meshes

We have focused our attention so far on 3D TLM in a regular mesh where the basic cell is a cube of sides $\Delta\ell$. This has the advantage of simplicity and efficiency and allowed us to present the basics of field to circuit mapping, scattering, and connection. However, there are cases where we need to go beyond this simple case to include irregular cuboid cells and also unstructured meshes where the basic cell is an irregular tetrahedron. First, we discuss in general terms the problems faced by the modeler when choosing a mesh to fit a particular problem and the inherent computational difficulties that may arise.

The tessellation of space using basic cells may be attempted in several ways some of which are shown schematically in 2D in Figure 5.10.

Computational efficiency is best achieved if all cells are identical in size and propagation time (timestep) is the same throughout the mesh. This is easily achieved in the uniform mesh shown in Figure 5.10a. Here each cell has a single neighbor (one-to-one correspondence) and we operate with the same propagation delay (timestep) throughout the mesh (synchronism). From the computational point of view, this is ideal. However, in modeling terms, it creates difficulties in certain

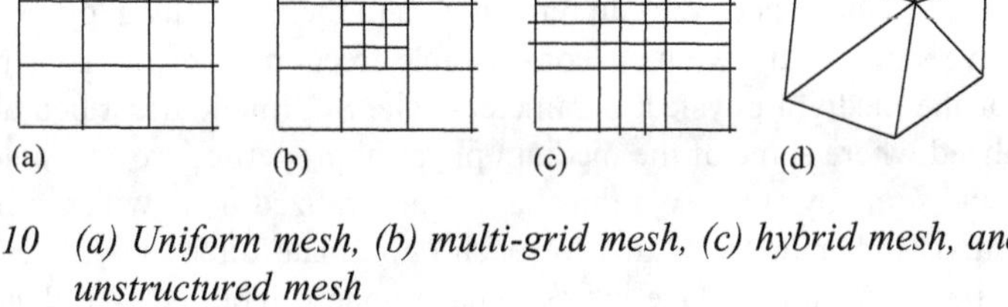

Figure 5.10 (a) Uniform mesh, (b) multi-grid mesh, (c) hybrid mesh, and (d) unstructured mesh

applications notably when we try to describe inclined boundaries (not coincident with coordinate directions). This results in "staircasing" errors [19]. Another, problem is the presence of fine (sub-wavelength) features which cannot be mapped accurately by the chosen space-step $\Delta\ell$. The option of decreasing the space-step is obviously available but at a great computational cost and it is rarely possible in practice. To address these problems, we have several options none of them without some computational cost. We may distort the mesh using irregular shaped cells, and/or, embed fine features into a coarse mesh using alternative techniques than distorting the mesh. We will address mesh distortion in this section, and embedding techniques in Section 5.4.

In the meshing shown in Figure 5.10b, we have fine- and coarse-mesh regions thus attempting to increase spatial resolution where needed and yet economize on computational resources elsewhere [20]. Two computational problems arise here. First, at the boundary between coarse and fine regions, there is no one-to-one correspondence—the coarse node has several fine neighbors, thus some kind of averaging (filtering) is required at the connection stage. Second, the fine and coarse regions operate with different time steps again implying some filtering process at the interface between the two. Filtering must be done carefully to abide by conservation laws and avoid instabilities. It is very difficult in general to avoid numerical artefacts (errors) at such an interface. The one-to-one correspondence problem can be addressed by abandoning the cubic cell and adopting a general cuboid shape (Figure 5.10c). But again, propagation delays in such a cell are not the same in each direction resulting in loss of synchronism and computational difficulties. Similar problems arise when modeling non-uniform materials where the velocity of propagation is not the same throughout the problem space. If we choose the same $\Delta\ell$ across the entire space, we are forced to have a different Δt thus losing synchronism. If we maintain the same Δt, then we are forced to have a different $\Delta\ell$, thus losing one-to-one correspondence. To overcome these difficulties in TLM, we maintain the same time step and the same space step by the judicious use of stubs. This is done by modifying the standard SCN described earlier, which consists of link lines only, by adding studs. To visualize the technique, consider that we model at the interface between two different media and to simplify we consider only one dimension. If we accept as a requirement that we have the same time step Δt in the two media, then if we use link lines only the corresponding space steps will be $\Delta x_1 = u_1\Delta t$, $\Delta x_2 = u_2\Delta t$ and since the velocity of propagation in the two media is different we end up with different space steps which in 3D causes complications. If we demand that $\Delta x_1 = \Delta x_2 = \Delta\ell$, then in the medium with the higher dielectric permittivity and magnetic permeability, we have underestimated in the model capacitance and inductance. We can remedy this in the model by adding capacitive and inductive stubs. Thus, we have considerable freedom in model construction to account for the medium physical parameters. The SCN node described above can be generalized where some of the medium physical properties are accounted for by link lines and some by stubs. We thus have a generalized node with six link lines, and capacitive and inductive stubs for each coordinate directions. Another three stubs may be added to introduce losses to each node if required. We thus obtain a node that can describe a very broad class of materials and cell shapes. In order for the node to capture accurately the medium properties, we need to define 6 link-line capacitances and 6 inductances, 3 capacitive stubs, and 3 inductive stubs—in total 18 parameters (eighteen unknowns). To represent correctly the total capacitance in x-polarization, we need to sum up the contributions of the two link lines directed along z and y and polarized in x and the x-polarized capacitive stub and demand that this sum is equal to the medium capacitance as shown below, where C_{yx} is the capacitance per unit length of the y-directed x-polarized link line and C_{ox} is the capacitance of the x-polarized open-circuit stub [21]:

$$C_{yx}\Delta y + C_{zx}\Delta z + C_{ox} = \varepsilon_x \frac{\Delta y \Delta z}{\Delta x} \tag{5.46}$$

Another two equations, similar to (5.46), are obtained for the y and z polarizations. For the inductance, we obtain another three equations e.g. for the x-direction,

$$L_{yz}\Delta y + L_{zy}\Delta z + L_{sx} = \mu_x \frac{\Delta y \Delta z}{\Delta x} \tag{5.47}$$

We now impose the synchronism requirement that the transit times on all six link lines are the same resulting in another six equations (stub roundtrip transit times are set equal to the link-line transit time). The 3 capacitance, 3 inductance, and 6 synchronism equations result in 12 equations in total. Since we have 18 unknowns, we have six degrees of freedom i.e. we can impose six additional conditions which may be convenient to us. This results in a number of nodes all modeling the same medium but with a different mix of links and stubs. One popular choice is to demand that all link lines have the same characteristic impedance set at the impedance of free-space $Z_0 = \sqrt{\mu_0/\varepsilon_0}$. For media with higher values of μ, ε we add stubs. This is known as the stubbed SCN node. Other choices are also possible as described in detail in [22,23]. The different node choices result in slightly different dispersion properties and the modeler may decide what is the best choice depending on the specifics of the problem at hand. A full scattering matrix may be obtained for each node, as we have shown in (5.39) for the stubless node, e.g. the hybrid HSCN [23]. However, a simpler and more elegant scattering formulation may be used which avoids an explicit derivation of a scattering matrix [24]. The basis of this formulation may be understood by reference to substructures of the 3D TLM node shown in Figure 5.11 where we depict a shunt substructure for E_y(a) and a series substructure for H_z(b). We can then calculate an equivalent y-directed voltage V_y and a current on the x–y plane I_z.

The reflected voltage for the case when the x-directed, y-polarized line have the same characteristic impedance Z throughout is then,

$$V_3^r = V_y + I_z Z - V_{11}^i \tag{5.48}$$

All other reflected voltages may be obtained in a similar way. For the most general case of a node of a general cuboid shape and different characteristic

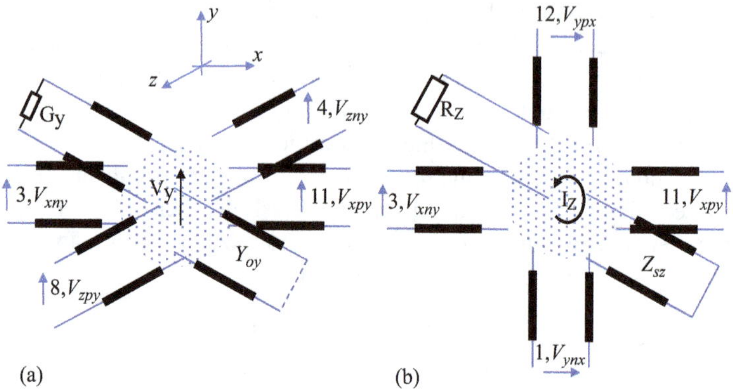

Figure 5.11 *(a) The "shunt" structure to determine* E_y *and (b) the "series" structure to determine* H_z

impedance for each line segment, the more intuitive notation of the original publications is used. Line 3 is x-directed on the negative side of the center and y-polarized hence voltages on this line and its characteristic admittance are labeled with the subscript *xny* and similarly for all other lines. The generalized scattering equations are then,

$$\begin{aligned} V^r_{inj} &= V_j \pm I_k Z_{inj} - V^i_{ipj} + h_{ij} \\ V^r_{ipj} &= V_j \mp I_k Z_{ipj} - V^i_{inj} + h_{ij} \end{aligned} \tag{5.49}$$

The upper and the lower sign apply, respectively to,

$$\begin{aligned} &(i,j,k) \in \{(x,y,z),(y,z,x),(z,x,y)\} \\ &(i,j,k) \in \{(x,z,y),(y,x,z),(z,y,x)\} \end{aligned}$$

The equivalent voltages and currents are given by,

$$\begin{aligned} V_i &= 2\frac{Y_{kni}V^i_{kni} + Y_{kpi}V^i_{kpi} + Y_{jni}V^i_{jni} + Y_{jpi}V^i_{jpi} + Y_{oi}V^i_{oi}}{Y_{kni} + Y_{kpi} + Y_{jni} + Y_{jpi} + Y_{oi} + G_i} \\ I_i &= 2\frac{V^i_{jpk} - V^i_{jnk} + V^i_{knj} - V^i_{kpj} - V^i_{si}}{Z_{jnk} + Z_{jpk} + Z_{knj} + Z_{kpj} + Z_{si} + R_i} \\ &(i,j,k) \in \{(x,y,z),(y,z,x),(z,x,y)\} \end{aligned} \tag{5.50}$$

The h-factors are obtained from,

$$h_{ij} = \frac{Z_{inj} - Z_{ipj}}{Z_{inj} + Z_{ipj}}\left(V^i_{inj} - V^i_{ipj}\right) \tag{5.51}$$

This is most general scattering formulation for general cuboid shaped nodes which can be used as building blocks to model a range of material and geometrical configurations. However, as already mentioned, staircasing errors persist and the modeling of fine irregularly shaped objects still presents a challenge. One of the limitations of a mesh consisting of identical cuboid cell is that it is structured, meaning that the local mesh environment around each feature is the same throughout the mesh. On the positive side, it is easy to design such a mesh and since all nodes are identical scattering and connection are straightforward. Mesh dispersion and anisotropy are the same throughout. On the negative side, this highly structured mesh is hard to deform in such a way as to conform with curved boundaries and fine features. We can escape from the restrictions of a structured mesh by using a more flexible basic cell in the form of a general tetrahedron shown schematically in Figure 5.10d. With such a structure the positive benefit is that we can increase the density of nodes near fine objects or where fields have high spatial gradients. There are, however, several negative aspects to an unstructured mesh based on tetrahedra. Since the shape of each tetrahedron may in general be different, it follows that the scattering properties are also different and a lot of scattering information has to be stored and processed for each node. More seriously, the shape of each cell cannot be arbitrarily chosen. Long thin tetrahedra shapes can in principle lead to negative propagation delays and cannot be allowed. Like with other modeling methods such as the finite element method (FEM) which routinely rely in the frequency-domain on unstructured tetrahedral meshes, the mesh must a Delaunay mesh [25]. Thus, an unstructured mesh for TD computations must obey certain restrictions and be of sufficient quality for stable and accurate computation. In contrast, the question of mesh quality hardly arises for a structured mesh. Practical system demands are such that even an unstructured mesh cannot service all modeling needs. In particular, fine intricate objects, embedded in systems cannot efficiently be mapped geometrically on a mesh. In such cases, modelers resort to embedding techniques where the outline of the embedded object is contained within a cell (structured or unstructured) and its impact on the surrounding space is accounted for by modifying the scattering properties of the containing cell.

We present the fundamentals of unstructured meshes in this section and of embedding techniques in the next section.

5.3.3 TLM scattering in the modal domain and the unstructured mesh

We start with scattering in TLM viewed from a different perspective to that hitherto adopted.

Every physical system responds to an arbitrary excitation in particular ways which are typical of its structure. These basic, elemental, fundamental, or characteristic responses, are generally known as its modes, or, using the German term, its eigensolutions. They tell us all there is to know about the natural response of the structure. The response to any given excitation is a combination of these elemental responses with different weight factors. It follows therefore that if we know the eigensolutions of a particular structure, then we can obtain responses to any input as a combination to these elemental responses. For the purposes of our study here, we are interested in the scattering from an TLM node representing a cell in space. The node may be cubic in shape, or a general tetrahedron. The cell may represent free space or another material or contain some other irregular conducting object. Can we study scattering in such circumstances by reference only to the modes (eigensolutions) of this cell? The advantage of such an option, if it is available, is that we characterize a cell by its modes and therefore no explicit mapping of its detailed structure to the mesh is required during the time-stepping procedure. The formal answer to this question is as follows:

To address the node scattering problem we start from the observation that on the surface of the cell the electric and magnetic field vectors are related through an admittance matrix $\boldsymbol{H} = \boldsymbol{YE}$ [26]. The electric and magnetic field vectors may be obtained from the incident and reflected voltage vectors and the admittance of the link lines of the node y_L, $\boldsymbol{E} \sim \boldsymbol{V}^i + \boldsymbol{V}^r$, $\boldsymbol{H} \sim y_L(\boldsymbol{V}^i - \boldsymbol{V}^r)$. Combining these expressions we obtain the scattering equation,

$$[\boldsymbol{Y} + y_L\boldsymbol{1}]\boldsymbol{V}^r = [-\boldsymbol{Y}+y_L\boldsymbol{1}]\boldsymbol{V}^i \tag{5.52}$$

$\boldsymbol{Y}$, $\boldsymbol{1}$ are the admittance matrix and the identity matrix, respectively. The formal solution to the scattering problem is embodied by (5.52) provided we know the admittance matrix and we are able to invert it. This not an easy task and the best approach is to find first the eigenvectors and eigenvalues of the admittance matrix, $\boldsymbol{U}_n$, γ_n, respectively, where, $\boldsymbol{YU}_n = \gamma_n\boldsymbol{U}_n$. Matrix $\boldsymbol{U}$, formed by the column eigenvectors $\boldsymbol{U}_n$, may be used to represent the admittance matrix (its spectral representation) $\boldsymbol{Y} = \boldsymbol{U\Lambda U}^{-1}$where $\boldsymbol{\Lambda}$ is a diagonal matrix with elements the eigenvalues of the admittance matrix. The set of eigenvectors is a complete orthogonal set which may be used to resolve any incident or reflected pulses into their modal components $\boldsymbol{X}$ i.e. $\boldsymbol{V}^{i,r} = \boldsymbol{UX}^{i,r}$. This is analogous in circuit analysis to decomposing a signal into its Fourier components and then evaluating the response to each component. Combining these expressions, we get the scattering in the modal domain

$$X_n^r = \frac{y_L - \gamma_n}{y_L + \gamma_n} X_n^r \tag{5.53}$$

The eigenvalue corresponding to each mode is the input admittance this mode encounters as it propagates along the modal transmission lines. In effect, we can view the scattering of a set of incident pulses on any TLM node as the decomposition into modal components (relative strength of each mode), the scattering of each mode according to (5.53), and the reconstitution of the total scattered voltage pulses from the scattered modal components for further processing (connection). This is shown schematically in Figure 5.12. Modal decomposition may be done as a signal

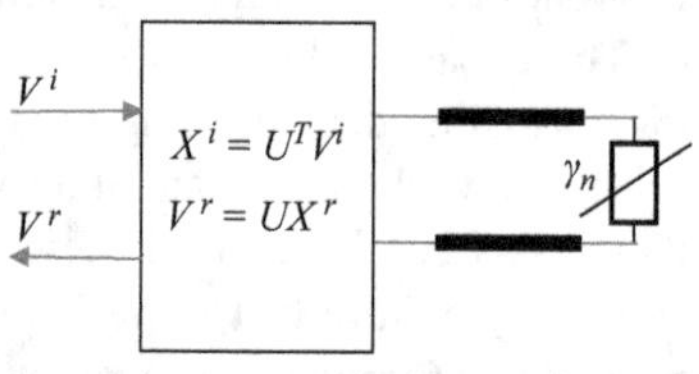

Figure 5.12 Decomposition of incident total voltages into modal components, scattering in the modal domain and reconstruction of total reflected voltages from the reflected modal components

processing task or inherently as part of the structure of the node. Obtaining the eigen solutions can be done either based on physical intuition, analytically or by numerical means. We will offer some examples of these various options.

By formulating scattering in this way, we can now see that embedding an object into a cell changes simply the eigen solutions. Interfacing with the rest of the mesh is essentially unchanged. To illustrate these ideas by a simple example, we examine modal decomposition in an asymmetrical 1D node shown in Figure 5.13a based on the approach outlined in [26,27].

Since we have two degrees of freedom, we choose two orthogonal modes to resolve fields and voltages,

$$E(x) = X_c \cos(kx) + X_s \sin(kx)$$

X_c, X_s are the strengths of each modal component of the field.

Applying this expression at the two extremities and adopting the small argument approximation for the cos(.) and sin(,) terms, we obtain,

$$\boldsymbol{E} = \begin{bmatrix} E_1 \\ E_2 \end{bmatrix} = \begin{bmatrix} 1 & -k\Delta x_1 \\ 1 & -k\Delta x_2 \end{bmatrix} \begin{bmatrix} X_c \\ X_s \end{bmatrix} = \boldsymbol{T}_e\boldsymbol{X} \rightarrow \boldsymbol{X} = \boldsymbol{T}_e^{-1}\boldsymbol{E} \tag{5.54}$$

In a similar manner using $j\omega\mu H = dE/dz$ we obtain the equivalent expression for the magnetic field,

$$j\omega\mu\boldsymbol{H} = \begin{bmatrix} k^2\Delta x_1 & k \\ -k^2\Delta x_2 & k \end{bmatrix} \begin{bmatrix} X_c \\ X_s \end{bmatrix} = \boldsymbol{T}_h\boldsymbol{X} \tag{5.55}$$

Combining (5.54) and (5.55) to eliminate $\boldsymbol{X}$, we obtain after some manipulation,

$$\begin{bmatrix} H_1 \\ H_2 \end{bmatrix} = \frac{1}{(\Delta x_1 + \Delta x_2)} \left\{ j\omega\varepsilon \begin{bmatrix} -\Delta x_1\Delta x_2 & -\Delta x_1^2 \\ \Delta x_2^2 & \Delta x_1\Delta x_2 \end{bmatrix} \begin{bmatrix} E_1 \\ E_2 \end{bmatrix} + \frac{1}{j\omega\mu} \begin{bmatrix} -1 & 1 \\ -1 & 1 \end{bmatrix} \begin{bmatrix} E_1 \\ E_2 \end{bmatrix} \right\} \tag{5.56}$$

We are now set to map this field expression into the nodal circuit parameters. We note that (5.56) represents the node admittance (matrix $\boldsymbol{Y}$ in the general case). The first term is capacitive in nature and the second inductive. Mapping requires

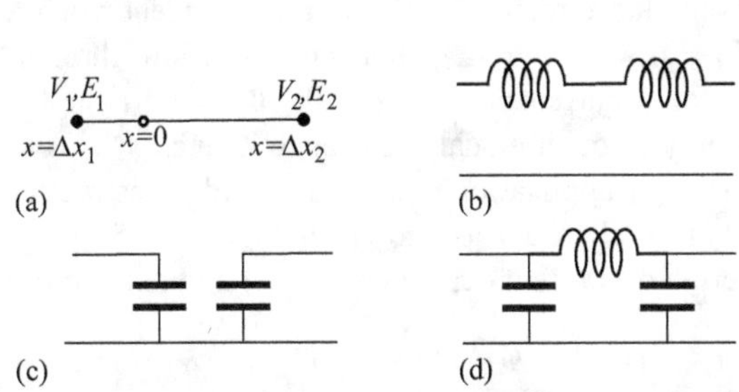

Figure 5.13 (a) Schematic of a one-dimensional asymmetric node, (b) and (c) inductive and capacitive components, respectively, and (d) combined circuit model at low-frequencies

that the resulting circuit must be reciprocal and therefore the capacitive and inductive matrices must be symmetric. The natural field-to-circuit mapping choice is,

$$\begin{bmatrix} E_1 \\ E_2 \end{bmatrix} \leftrightarrow \begin{bmatrix} V_1 \\ V_2 \end{bmatrix}, \quad \begin{bmatrix} H_1 \\ H_2 \end{bmatrix} \leftrightarrow \begin{bmatrix} I_1 \\ -I_2 \end{bmatrix}$$

The choice of negative sign for the mapping to current is such as to make the inductive matrix symmetric as required. However, the capacitive term remains non-reciprocal as shown below

$$\begin{bmatrix} I_1 \\ I_2 \end{bmatrix} = \frac{j\omega\varepsilon}{(\Delta x_1 + \Delta x_2)} \begin{bmatrix} -\Delta x_1 \Delta x_2 & -\Delta x_1^2 \\ \Delta x_2^2 & \Delta x_1 \Delta x_2 \end{bmatrix} \begin{bmatrix} V_1 \\ V_2 \end{bmatrix} + \frac{1}{j\omega\mu(\Delta x_1 + \Delta x_2)} \begin{bmatrix} -1 & 1 \\ 1 & -1 \end{bmatrix} \begin{bmatrix} V_1 \\ V_2 \end{bmatrix} \tag{5.57}$$

It is clear from (5.57) that the inductive terms dominates at LF where the model is valid and in principle we could neglect the capacitive term except in the case when $V_1 = V_2$. In this special case, the inductive term is zero and the capacitive term is then essential. The inductive term may be modeled as shown in Figure 5.13b and the capacitive term for $V_1 = V_2$ is as shown in Figure 5.13c. The complete circuit with this approximation is shown in Figure 5.13d. We see therefore that we may analyze responses in terms of fundamental modes and that we can construct physically realizable circuit models if we so wish. The same principles apply in more complex modeling situations in 2D and 3D.

We can re-consider the 2D shunt node shown in Figure 5.7 now seen from the modal point of view. A cylindrical expansion of the field around the center of the node has an infinite number of modes ($-\infty \leq n \leq \infty$) in a field expansion of the form, $A(r)\exp(j, n\phi)$ as shown in Figure 5.14a.

Since however we have four ports and thus only four degrees of freedom, we choose the lowest four modes with the following angle dependence: the common mode (response independent of the angle), $\sin\varphi$, $\cos\varphi$, $\cos 2\varphi$. A transformer network to analyze the four incident voltages into the four selected modes and for each mode to flow through the corresponding modal impedance is shown in Figure 5.14b. It can be readily confirmed from this circuit that if the four incident voltages contain common mode only ($V_1 = V_2 = V_3 = V_4$), then currents flow through Z_0 only. Similarly, if we inject mode 1 currents only, $V_1 = 0,\ V_2 = 1,\ V_3 = 0,\ V_4 = -1$ then these currents flow only through modal impedance Z_1 etc.

These ideas find particular application when we consider unstructured meshes based on a triangular shunt cell in 2D as shown in Figure 5.15a [27]. Here the modal expansion is in terms of angle independent, cos(.) and sin(.) modes,

$$E_z(r, \varphi) = A_0(r)X_0 + A_1(r)\cos(\varphi)X_1 + A_2(r)\sin(\varphi)X_2 \tag{5.58}$$

The analysis proceeds in exactly the same manner as for the 1D structure in Figure 5.13 But since here we have a node with three arms we eventually derive 3×3 capacitive and inductive matrices from which we can construct the TLM

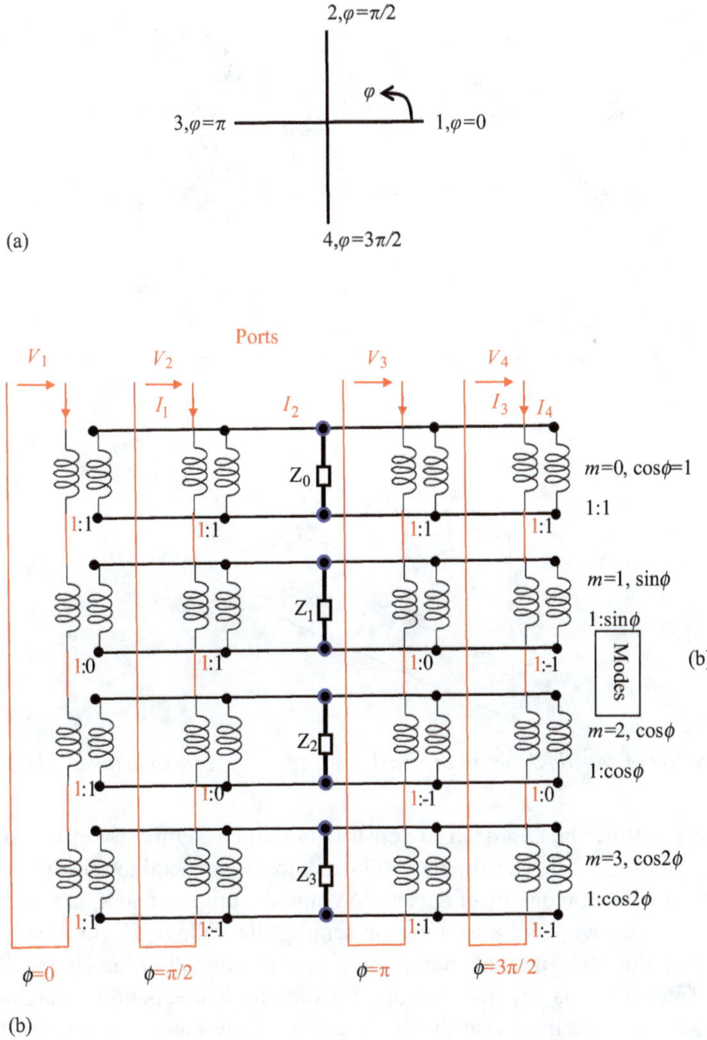

Figure 5.14 (a) A 2D 'shunt' node to illustrate cylindrical field expansion and (b) schematic of a transformer network to implement scattering in the modal domain

node shown in Figure 5.15b. The lumped components may be modeled in TLM by a mix of links and stubs using the appropriate time step. One issue apparent from the inductive and capacitive elements in Figure 5.15b is that they all depend on sin φ_i, hence, since for stability circuit components must be positive, it follows that these angles must be smaller than π (the Delaunay condition). To ensure continuity of the magnetic field on triangle boundaries, we may choose as the center of the node its circumcenter. If the circumradius is R, then we obtain factor $\alpha = 2R$.

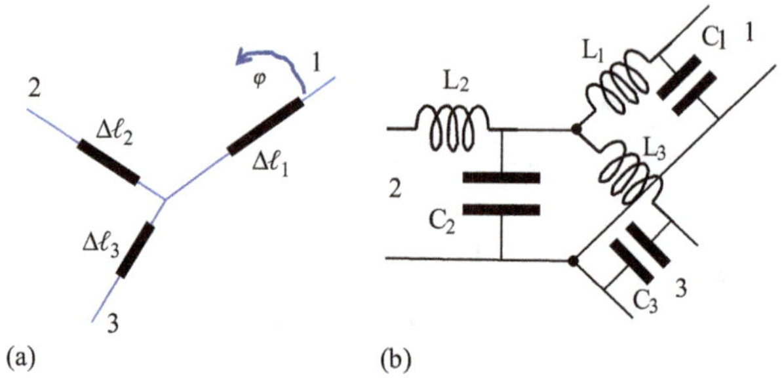

Figure 5.15 (a) Schematic of an unstructured 2D node and (b) its circuit equivalent. The component values are $L_i = \frac{\mu_0 \Delta\ell_i}{a \sin(\varphi_i)}$, $C_i = \frac{a\varepsilon_0 \sin(\varphi_i)\Delta\ell_i}{2}$

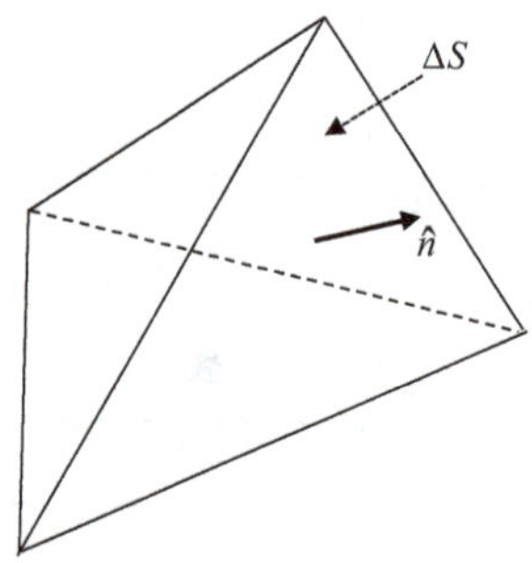

Figure 5.16 A tetrahedron as the basic cell of a TLM unstructured 3D mesh

We now outline the approach to constructing and a 3D mesh out of irregular tetrahedral nodes [28,29]. The simplest 3D structure is a general tetrahedron shown in Figure 5.16 where on one facet the area ΔS_i and the unit normal $\boldsymbol{n}_i$ are indicated.

On each facet, we have a vector representing the voltage $\boldsymbol{V}_i$ (i.e. two ports). The center of the node is the circumcenter of the tetrahedron and the link-line length for facet i, $\Delta\ell_i$, is half the distance between the corresponding neighboring circumcenters. It is assumed that the construction of the mesh is such that it conforms to the Delaunay conditions [25]. This may require extensive mesh optimization work in complex practical configurations. The starting point is to develop a modal expansion of the fields expressed as the weighted sum of spherical harmonics sampled at the ports of each node. Since we have eight degrees of freedom, we need eight terms in the modal expansion—three electric dipoles, three magnetic dipoles, and two quadrupoles. These may be expressed in terms of three normalized orthogonal vectors, $\boldsymbol{u}_0, \boldsymbol{u}_1, \boldsymbol{u}_2$. These vectors are the solution of a generalized eigenvalue problem,

$$\sum_{i=1}^{4} \Delta S_i \Delta\ell_i (\mathbf{1} - \boldsymbol{n}_i \boldsymbol{n}_i) \cdot \boldsymbol{u}_n = \lambda_n Z_0 \sum_{i=1}^{4} Y_i (\mathbf{1} - \boldsymbol{n}_i \boldsymbol{n}_i) \cdot \boldsymbol{u}_n \tag{5.59}$$

where Y_i is the admittance of the link lines modeling the inductance $L_i = \mu\Delta\ell_i/2$ and $\boldsymbol{n}_i\boldsymbol{n}_i$ is a dyadic product. The mapping of the fields to circuit quantities is shown below:

$$\boldsymbol{V}_i = \sum_{p=1}^{4}\left\{\frac{j\omega\mu\Delta\ell_i}{2}\delta_{ip} + \sum_{m=0}^{2}\sqrt{S_i}(\boldsymbol{1} - \boldsymbol{n}_i\boldsymbol{n}_i)\cdot\frac{2\boldsymbol{u}_m\boldsymbol{u}_m}{j\omega\varepsilon\lambda_m}\cdot(\boldsymbol{1} - \boldsymbol{n}_p\boldsymbol{n}_p)\sqrt{S_p}\right\}\cdot\boldsymbol{I}_p \tag{5.60}$$

The first term on the RHS represents inductive behavior and the second capacitive. This expression is represented by the circuit in Figure 5.17 where link lines represent the inductive terms and the open-circuit stubs the capacitive terms.

The transformer ratios are a schematic representation of the contribution of mode m to port voltage i,

$$\boldsymbol{V}_{im} = \left\{\sqrt{S_p}(\boldsymbol{1} - \boldsymbol{n}_p\boldsymbol{n}_p)\cdot\boldsymbol{u}_m\right\}V_m \tag{5.61}$$

Scattering is implemented in the usual way by expressing the total voltages and currents in (5.60) in terms of incident and reflected voltage to obtain an expression relating reflected to incident voltages.

A detailed study shows that the minimum eigenvalue (related to the minimum link line length) dictates the timestep. It follows therefore that very short line lengths result in very small timesteps and hence an inefficient computation. It is imperative that this problem is addressed to ensure that computation proceeds with a realistic timestep. In a series of publications, an approach of merging such tetrahedra, Figure 5.18, containing very short lines is described [29,30]. More than one such cell can be merged to form more complex shapes thus increasing the timestep by several orders of magnitude without significant loss in accuracy.

Although an unstructured mesh based on tetrahedra is a powerful modeling tool, it needs careful and intensive preprocessing design to improve mesh quality and maintain an acceptably long time-step. It should therefore be used when absolutely necessary e.g. in areas of intricate geometrical detail, high field gradient and for stitching together heterogenous mesh areas as shown schematically in Figure 5.19.

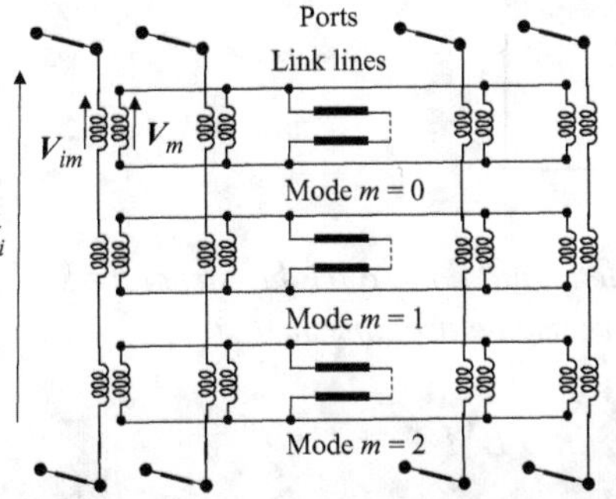

Figure 5.17 Schematic of a transformer network for implementing scattering in the modal domain for a tetrahedral cell (four incident vector voltages V_i)

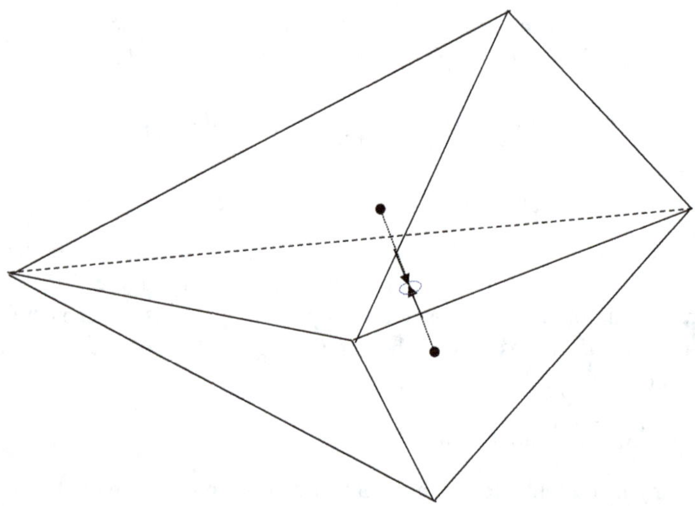

Figure 5.18 Schematic showing the merging of two adjacent tetrahedra when a link line is too short leading to unacceptably small timestep

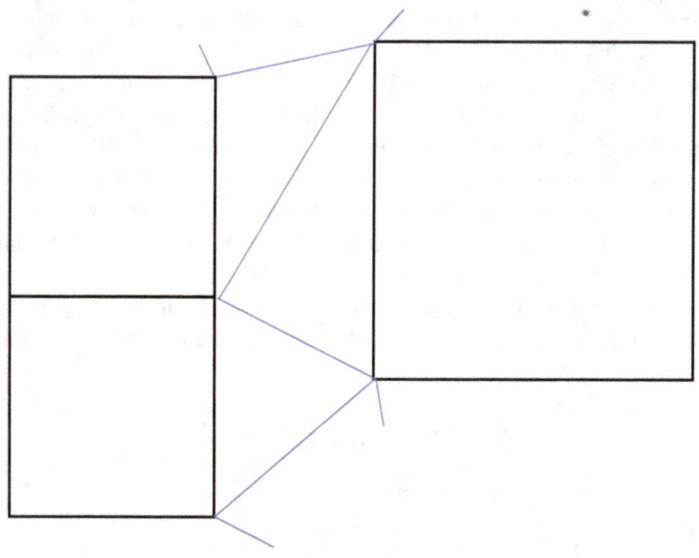

Figure 5.19 A fine (left) and coarse (right) structured mesh stitched together using an unstructured mesh

5.3.4 Boundaries in TLM

So far, we have addressed connection as a process where the neighbor is always another TLM node. However, in practical cases, we need to account for other types of boundary both internal and external to the mesh. A common case is a perfectly conducting boundary (PEC) which in TLM is simply dealt with by a short-circuit condition where the reflected voltage is minus the incident. In problems where we have symmetry, it is possible to save on simulation storage and time by introducing symmetry planes which in TLM imply that for any incident pulse on the symmetry plane from one half of the problem space, we have the same pulse incident on it from the other half i.e. an open-circuit boundary condition or, a perfect magnetic conductor (PMC). A more challenging problem is encountered when we model open-boundary problems i.e. problems which extend to infinity—there is no material boundary other than free-space. For modeling efficiency, we need to impose a numerical boundary since we cannot model infinite space by a volume discretization method. The simplest way to describe an open boundary in a TLM model is to demand zero reflection coefficient (matched boundary condition). This type of boundary is described as an absorbing boundary condition and work has been done in connection with TLM and FDTD to make it as perfect as possible. The matched boundary condition in TLM is very good for most problems. A better absorbing boundary condition based on the perfectly matched layer (PML) is also possible in TLM [31–33].

A further discussion of more complex boundaries will be presented in the next section of this chapter.

5.3.5 Dispersion in TLM

All numerical methods are to a certain degree inaccurate. It is clear that sampling in space and in time will result in errors which may be reduced by reducing the timestep and the space sampling step. However, in a numerical scheme, errors can never be zero. From the structure of the TLM node and mesh, it is evident that we are dealing with a low-pass network. Errors will get larger as the frequency rises (velocity errors). Also, waves propagating in different directions along the mesh encounter slightly different structures and therefore propagation is not perfectly isotropic. Setting a design standard of ten segments per wavelength at the highest frequency of interest means, in most cases, that these errors are kept low (typically a few percent) and thus acceptable in all but the most demanding applications. In critical cases, a more thorough assessment of errors is required and guidance may be obtained from specialist references [34–37].

5.3.6 TLM and other computational EM methods

In this subsection, we focus on the relation of TLM to other common CEM techniques to bring out generic and methodological aspects and other distinctive features. Relating TLM to other numerical methods is also of practical interest in understanding dispersion properties and facilitating interfacing (hybridization).

So far, we have presented TLM on the basis of the isomorphism between field and network equations and therefore were able to solve field problems by solving equivalent circuits of transmission lines using network laws. However, a more formal derivation is also possible. We may choose to start from a finite integration technique where fields are expressed in terms of incident and reflected voltage pulses. We then apply Ampere' and Faraday's laws along suitable integration paths on the TLM nodal structure to obtain a relationship between reflected and incident pulses, which turns out to be the same as the TLM scattering scheme. Details may be found in [38]. Another way is to relate TLM to the Method of Moments (MoM) [39]. In [40,41], it is shown that all finite-difference-based TD methods can be related to MoM with appropriate choices of basis and test functions.

An important feature of TLM is that fields are computed at the same point in space and in time (there is no offset) and this make it easier to embed features and interface with other solution regions. Also, as we show in the next section, the physical feel of the method and its unconditional stability make it ideal for embedding macro-nodes. The computational cost of scattering operations has also been reduced from its original count based on (5.39) by using the scheme described by (5.49). An example of a practical comparison of dispersion in TLM and FDTD for a particular problem is found in [37].

Hybridization of methods is desirable as it allows us to use the best suited method for each problem region. TLM had been combined with Integral-Equation TD methods to solve a particular class of problems where two problem regions with intricate geometrical detail and material non-uniformity are separated by large areas of free-space. In such cases, TLM may be used to model the two regions and the IETD method to model very efficiently the intervening space [42].

5.4 Dealing with complexity in TLM

Advances over the years have resulted in a body of analytical solutions to canonical problems. Several numerical techniques described in this book were developed to address a range of real-life problems which are not canonical in that they do not possess symmetries to facilitate the development of analytical solutions. We have seen in the previous section, in connection with TLM, that the mesh of grid points on which EM laws are enforced in a numerical solution may be suitably distorted (cuboid or tetrahedral shaped cells) to accommodate problem features and shapes commonly found in practical problems. Yet, such developments although important and necessary cannot accommodate the full complexity of the challenges encountered by the modeler [43–45].

We can illustrate this challenge by the example of an electrical cable-a common feature of most problems. The cable, typically a few mm in diameter, may be placed inside an aircraft. Here we have a feature which is electrically small in two dimensions (its diameter) and electrically long in the third (along its length). The aircraft is electrically large and in addition there will be several cables running close together in tight bundles. This is an example of a multi-scale problem where we have simultaneously electrically large and small features [45]. Increasing resolution (smaller $\Delta\ell$) to accommodate the electrically small features results is an unreasonably large number of cells to model the electrically large features. No amount of mesh distortion can address this problem satisfactorily. An additional difficulty is that problem definition (mapping features onto the mesh) becomes a lengthy and tedious task. It may be argued that small fine features are not important in the response and thus may be neglected. This may be true in some cases but clearly not for the diameter of a cable. The process of de-featuring, to simplify a problem for modeling, is a challenging one. Space limitations in a short chapter prevents addressing algorithmic methods for doing de-featuring based on model order reduction (MoR) techniques [46,47].

In this section, we focus on dealing with complexity by embedding into the mesh special cells which communicate with neighboring cells through scattered pulses representing the contents of the special cell (e.g. a cable, fine conducting object, unusual material, etc.). The geometrical structure of the mesh is therefore as if the entire problem space consists of free-space but some cells have special scattering properties to reflect their varying contents. In this way, we have economy in computation and problem definition and a satisfactory treatment of the fine features. Naturally, a combination of distorted mesh and embedded cells may be used in the same problem to deal with particularly challenging cases.

Examples of such embedding techniques are described briefly in the rest of this chapter with references to further study.

5.4.1 General principles of embedding fine objects in TLM

The principle underlying the design of special embedded nodes which embody in their scattering characteristics the properties of their embedded objects is based on the observation that very near the embedded object the fields are, to a good approximation (since $\Delta\ell \ll \lambda$), determined by the object and not by conditions prevailing elsewhere. Therefore, if by any means (based on intuition, analytics or numerical tests), we are able to obtain local solutions characteristic of the embedded object, then, in the manner we have done in Section 5.3.3 we can use this information to decompose incoming (incident pulses) to the special cell into a combination of local solutions (modes), present to each mode the correct impedance to obtain the reflected modal components, and then recombine to obtain the total reflected pulses for transmission to neighboring cells (connection). This process may be implemented in simple cases by suitably constructed nodes or by the process shown in Figure 5.12. The modal characterization of the special nodes is a pre-processing task and is not an overhead during the main computation. Some examples are shown in subsequent sections and more details may be found in the references given.

5.4.2 Embedded multi-conductor cables

The pioneering work in describing cables in numerical models is described in [48]. The essence of this type of model is to account for the local solution, simply by one static mode only, and interface the cable model to the fields at the boundaries of the cell. The TLM model based on this principle is shown in Figure 5.20 for a *z*-directed wire [49,50]. In this figure, Z_L, Z_C, and Z_{0W} represent the inductance and capacitance of the wire with respect to the cell boundary and *R*, *G* the series and shunt wire losses. Coupling to the field is represented by U_Z and $Z_0/4$ where the voltage source is the

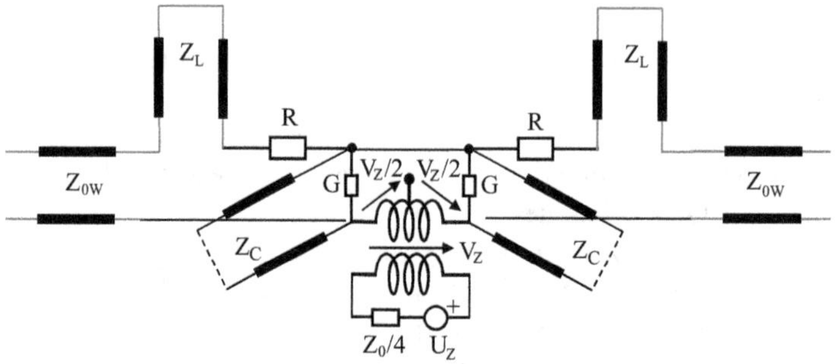

Figure 5.20 Embedding a wire into a TLM cell

average voltage of the four ports of the TLM node polarized in the z-direction. Coupling of the wire segment to segments adjacent to it is through the two link-lines Z_{OW}. The model has been generalized to account for multi-wire looms with bends and junctions embedded into TLM cells [51].

As already indicated in connection with our discussion on the modal view of TLM, it is possible to extract the modes of a cell containing one or more wires, and thus implement a scattering algorithm at the boundaries of the cell which accounts for the presence of a wire. In this way, we can include more modes than the single mode used above, we thus we are able to place wires off the node centre and generally enhance accuracy in multi-wire situations. As an example, in a 2D mesh of triangular cells with embedded wires, we can account for three modes and for square cells for four modes. Similar principles apply in 3D for cuboid or tetrahedral cells. Details may be found in [52–54]. Coaxial shielded cables may also be dealt in a similar way [55].

5.4.3 Fractional boundaries

A perennial problem of numerical modeling is the accurate description of inclined boundaries (not coinciding with coordinate directions) and of curved boundaries. Every numerical scheme that attempts to conform the mesh to the actual boundaries is necessarily imperfect. One approach is to use unstructured meshes as already indicate but this increases complexity and computational effort. A good compromise is to use a uniform structured mesh for most of the problem and restrict the unstructured mesh to areas near surfaces where a conformal mesh is required. An alternative is to use a uniform structured mesh everywhere but with a special facility to adjust the size of nodes near boundaries to achieve better conformity. This fractional adjustment is possible e.g. using models similar to those described in Section 5.4.2 [56,57].

5.4.4 Embedding complex materials

As we have already indicated, we may incorporate into TLM models lumped components such as L, R, C and thus model any frequency-dependent impedance by appropriate network synthesis. A parallel combination of RLC series branches may be used to approximate frequency resonances typical of various materials e.g. with several Debye or Lorenz-type resonances. Although this is possible in simple cases, it can be advantageous to have a general algorithmic procedure which can be applied to the most general cases. To illustrate the basics of such a procedure, we consider the modeling of a material with a relative dielectric constant of the form shown below,

$$\varepsilon_r(j\omega) = \frac{a_0 + a_1(j\omega) + a_2(j\omega)^2 + \cdots}{b_0 + b_1(j\omega) + b_2(j\omega)^2 + \cdots} - j\frac{\sigma_e}{\omega\varepsilon_0} \tag{5.62}$$

The response in (5.62) has a number of poles and zeros and a DC term on the RHS. The DC term may be modeled by a resistance $Z_{\sigma_e} = 1/\sigma_e\Delta\ell$. To model the first term on the RHS of (5.62), we introduce an extra capacitance (excess over free-space value) in the model $C_e = \varepsilon_0(\varepsilon_r - 1)\Delta\ell$. We thus have to model an impedance $Z_e(j\omega) = 1/j\omega C_e$. This impedance may be transformed in the z-domain by the application of the bilinear transform to obtain after some manipulation,

$$\begin{aligned} Z_e(z^{-1}) &= a'_0 + \frac{\alpha'_1 z^{-1} + \alpha'_2 z^{-2} + \cdots}{\beta'_0 + \beta'_1 z^{-1} + \beta'_2 z^{-2} + \cdots} = Z_{Cf} + Z_{Cs}(z^{-1}) \\ &= Z_{Cf} + V_{Cs}(z^{-1})/I(z^{-1}) \end{aligned} \tag{5.63}$$

On the RHS of (5.63), we have one instantaneous and one delayed part. The equivalent circuit representing (5.63) is an impedance Z_{Cf} in series with a voltage source V_{Cs} in which the current I flows. This equivalent is effectively connected to the respective TLM nodes to represent the material properties. The voltage at time k is obtained using an auxiliary quantity W,

$$\begin{aligned} {}_kW &= \frac{1}{\beta'_0}\left[{}_kI - \left(\beta'_1\,{}_{k-1}W + \beta'_2\,{}_{k-2}W + \cdots\right)\right] \\ {}_kV_{Cs} &= \alpha'_1\ {}_{k-1}W + \alpha'_2\ {}_{k-2}W + \cdots \end{aligned} \tag{5.64}$$

These expressions for modeling frequency-dependent materials are also suitable for modeling any impedance inserted into the model. Further details and elements of TLM source code to implement various TLM advanced algorithms, such as the one described above, may be found in [58]. Several approaches to material modeling in TLM may be found in [59–61].

5.4.5 Embedding arbitrarily shaped objects

In previous section, we presented the rudiments of embedding techniques for objects like cables using information on their modal response. In many cases, the modal response is available analytically and thus it is straightforward to employ the techniques outlined in Section 5.3.3. However, it is possible to extend this approach even in cases where the modal response cannot be obtained analytically e.g. in cases where the embedded object is irregular, lacking any symmetries, as shown in Figure 5.21.

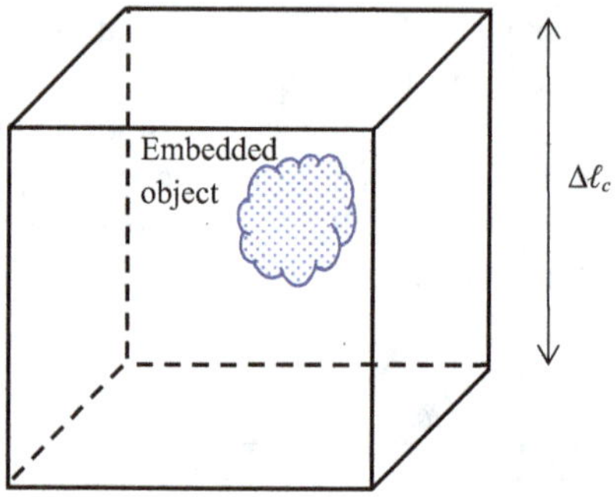

Figure 5.21 A TLM cell with an embedded arbitrary shaped object

In such cases, the first step is to obtain the local solutions numerically as a preprocessing task. Figure 5.21 shows a single cell with the embedded object with the global discretization length $\Delta\ell_c$ appropriate for the overall problem. To obtain local solutions numerically, we need to analyze numerically the responses of this cell using a finer discretization length $\Delta\ell_c/n$, where n is an integer suitable for describing the details of the embedded object. As a preprocessing task, we solve a problem with n^3 nodes several times, each corresponding to tangential fields incident on each cell face and for each of the two orthogonal polarizations. These numerical tests provide a number of local solutions which when averaged can be translated into voltages at the faces of the original coarse cell $\Delta\ell_c$. These are the local solutions needed for implementing the embedding technique. More details may be found in [62,63].

5.4.6 Behavioral models

We have already referred to the complexity of modern systems and the need to account as far as possible for the full spectrum of EM interactions across a wide bandwidth. The basic geometrical and material features may be modeled by distorting the mesh and/or embedding features into nodes to form macro-nodes embodying complex properties. In spite of these efforts, modelers are still challenged in trying to incorporate digital ICs (the main sources of EMI) into large-scale models of systems. This is essential from the EMC point of view and for signal integrity studies. However, the option of modeling fully the inner workings of ICs is rarely available for two reasons. First, the complexity of ICs with tens of thousands of components makes it unrealistic to even attempt this task. Second, the manufacturers of ICs have knowledge of the inner topology of ICs but limited knowledge of the inner geometry and, besides, for commercial reasons of confidentiality, are reluctant to provide such information even when it is available. We have therefore to compromise and replace components such as ICs by "behavioral" models. These are essentially voltage–current look-up tables at the IC ports which are then interfaced to field models such as TLM. A common industry standard is the so-called input/output buffer information specification (IBIS). IBIS models can be interfaced directly at the ports to TLM models as described in [64]. It has to be pointed out that IBIS models do not account for EM interactions inside ICs and that therefore their suitability for EMC and similar applications must be always carefully considered. More sophisticated models are being developed which attempt to account for a wider range of interactions [65,66].

5.4.7 Near-field scans

Another approach to handling complexity, especially in the case of ICs and complex PCBs, is to try and represent the EM radiation from such objects through a small number of equivalent radiating sources (electric and magnetic dipoles). This approach is akin to representing a complex circuit at its terminals by a Thevenin equivalent. This technique is depicted schematically in Figure 5.22a where a complex board is shown.

At a plane just above the board, the electric and/or magnetic fields are scanned with suitable probes (near-field scan). We then devise, through appropriate fitting routines, a small number of radiating dipoles which produce, as far as possible, the same radiating pattern as the original board, Figure 5.22b. It has to be appreciated that such measurement data is noisy and therefore special care must be taken to reduce noise prior to fitting (e.g. by Tikhonov regularization) [67,68]. Once a suitable set of equivalent dipoles has been obtained a reduced "equivalent" board consisting of ground plane (if present in the original), substrate and dipoles may be inserted in more complex structures, such as a cabinet with apertures, for further modeling using TLM. This approach substantially reduces complexity and can give acceptable accuracy.

5.4.8 TLM and stochastic models

We have so far focused on solving deterministic problems in CEM, whereby we start from precisely defined problem parameters and proceed to obtain the exact responses to specified excitations. Problems posed in this way are useful but in practice there is considerable uncertainty both in geometrical, material and

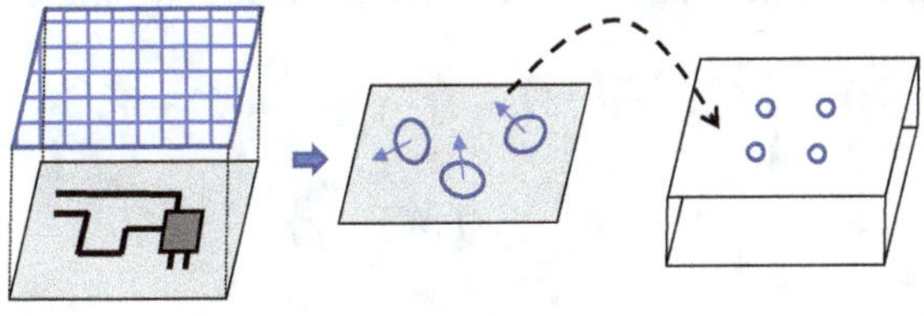

(a) (b) (c)

Figure 5.22 Near-field scanning procedures, (a) scanning on a fine grid of points on a plane above a PCB, (b) set of equivalent radiating dipoles, and (c) equivalent dipoles representing PCB radiation incorporated into more complex models excitation specifications. There is an element of randomness in many aspects of a problem which affects considerably actual responses. Therefore, it has to be accepted that responses are in practice random with a mean expected value and a standard deviation which characterize in a more meaningful way system behavior. It is necessary to go beyond deterministic models to encompass stochastic models. In this section, we give an indication of the tools available, progress made so far and future needs.

The conventional approach to dealing with uncertainty is to employ the Monte-Carlo method, where we run the same simulation a number of times using input parameters reflecting the potential spread of input vales and thus we obtain a number of responses from which meaningful statistical data may be extracted (mean and variance). The difficulty with this approach is that the number of numerical tests required is very large (tens of thousands) and this is impractical when we simulate large complex systems. What is needed is the development of techniques which allow us to extract meaningful statistical data from a small number of simulations. There are mathematical techniques based on principles similar to those employed in evaluating integrals from a finite number of values of the integrand (Gaussian Quadrature) which can be employed for this purpose [69]. The advantage of such techniques is that standard CEM tools such as TLM may be used without any modification. An example is the unscented transform (UT) technique which has been applied in connection with TLM [70,71]. We assume that in a complex system, the response y if a function of a random variable x. Because of the inherent complexity, the output is evaluated numerically, e.g. using TLM, for a particular value in the range of the random variable, e.g. a particular dimensional parameter of the system. Let us also assume that the random variable is, $x = \bar{X} + \widehat{x}$, where the first term on the right is its mean value and the second represents a zero mean Gaussian random variable. According to UT theory, to obtain the first two moments of the random variable representing the output we need to evaluate the $y = f(x)$ for the following three values of the input random variable, $\bar{X},\ \bar{X} + \sigma\sqrt{3},\ \bar{X} - \sigma\sqrt{3}$, where σ is the standard deviation of the input random variable. The output y is evaluated numerically using TLM to obtain, $f(\bar{X}),\ f(\bar{X} + \sigma\sqrt{3}),\ f(\bar{X} - \sigma\sqrt{3})$. From these three simulations, we can obtain, through weighted sums, estimates of the expected value and variance of the output,

$$\begin{aligned} \tilde{\bar{y}} &= E\{f(x)\} = \frac{1}{6}f\left(\bar{X} - \sigma\sqrt{3}\right) + \frac{2}{3}f(\bar{X}) + \frac{1}{6}f\left(\bar{X} + \sigma\sqrt{3}\right) \\ \tilde{\sigma}_y^2 &= \frac{1}{6}\left\{f\left(\bar{X} - \sigma\sqrt{3}\right) - \tilde{\bar{y}}\right\}^2 + \frac{2}{3}\{f(\bar{X}) - \tilde{\bar{y}}\}^2 + \frac{1}{6}\left\{f\left(\bar{X} + \sigma\sqrt{3}\right) - \tilde{\bar{y}}\right\}^2 \end{aligned} \tag{5.65}$$

An estimate of the statistical distribution of the output may also be made. Similar principles apply in the case of more than one random variable. However, the number of simulations required scales up rapidly with the number of random variables. If we have d independent RVs and we need N points per random variable, then the number of evaluations is N^d. Thus, for two RVs and three evaluation points, we need nine evaluations in total. In practical problems where many random variables are important, the number of simulations required reaches prohibitive numbers and, in such cases, dimensional reduction (sparcification) methods have been developed to reduce the computational effort [72].

To study EM emissions from complex boards where there is randomness in the signals generated by digital sources, field propagation may be traced through correlation functions (auto- and cross-correlations). More details of these emerging techniques may be found in [73,74].

5.5 Concluding remarks

We have presented the TLM method as it applies to solving electrical problems of some complexity. The basic techniques have been presented in some detail. Emphasis was placed in dealing with complexity by describing embedding techniques whereby advantage is taken of local solutions in the vicinity of complex objects to interface these solutions with surrounding fields obtained from the TLM method. This approach permits a largely uniform and structured mesh throughout the problem space thus enhancing efficiency and at the same time dealing with problematic features through the embedding of special nodes. This appears to be the optimum approach, however, each problem has its own characteristics and modelers must be prepared to adapt to particular requirements. In this respect, TLM, through its stability, efficiency, versatility, and physical feel, is an ideal tool at the hands of modelers confronted by increasingly complex practical problems.

References

[1] Christopoulos C. *The Transmission-Line Modeling Method TLM.* IEEE Press, New York, NY, 1995.

[2] Christopoulos C. '*The Transmission-Line Modeling (TLM) method in electromagnetics*', in *Synthesis Lectures in Computational Electromagnetics, Lecture 7*, Morgan & Claypool Publishers, San Rafael, CA, 2006.

[3] Johns P. D., Christopoulos C. 'A new frequency-domain TLM method for the solution of steady-state electromagnetic problems', *IEE Proc. Part A*, 1994, vol. 141, pp. 310–316.

[4] Johns P. B. 'Numerical modelling by the TLM method', in Wexler A. (ed). *Proceedings of The International Symposium on Large Engineering Systems*, Pergamon Press, New York, NY, 1977, pp. 139–151.

[5] Bandler J. W., Johns P. B., Rizk M. R. M. 'Transmission-line modelling and sensitivity evaluation for lumped network simulation and design in the time-domain', *J. Franklin Inst.*, 1977, vol. 304, no. 1, pp. 15–32.

[6] Johns P. B., O'Brien M. 'Use of the transmission line modelling (TLM) method to solve non-linear lumped networks', *Radio Electron. Eng.*, 1980, vol. 50, pp. 59–70.

[7] Johns P. B. 'A simple explicit and unconditionally stable numerical routine for the solution of diffusion equation', *Int. J. Numer. Model. Eng.*, 1977, vol. 11, pp. 1307–1328.

[8] Saleh A. H. M., Blanchfield P. 'Analysis of acoustic radiation pattens of array transducers using the TLM method', *Int. J. Numer. Model.*, 1990, vol. 3, pp. 39–56.

[9] Milan H. F. M., Gebremedhin K. G. 'General node for TLM method applied to bio-heat transfer', *Int. J. Numer. Model.*, 2018, vol. 31, no. 5, e2455, p. 13.

[10] Hui S. Y. R., Christopoulos C. 'A discrete approach to the modelling of power electronic switching networks', *Proceeding of the 20th Annual IEEE Power Electronics Specialist Conference*, 1989, vol. 1, pp. 130–137.

[11] Hui S. Y. R., Christopoulos C. 'The modelling of networks with frequently changing topology whilst maintaining a constant system matrix'. *Int. J. Numer. Model.*, 1989, vol. 3, pp. 11–21.

[12] Hui S. Y. R., Christopoulos C. 'Discrete transform technique for solving non-linear circuits and equations', *IEE Proc.-A, Sci. Measur. Technol.*, 1992, vol. 139, no. 6, pp. 321–328.

[13] Kron G. 'Equivalent circuit of the field equations of Maxwell', *Proc. IRE*, 1944, vol. 32, pp. 289–299.

[14] Fettweis A. 'Wave digital filters: theory and practice', *Proc. IEEE*, 1986, vol. 74, no. 2, pp. 270–327.

[15] Lawson S. S. 'Wave digital filters', in Chen W.-K. (ed.). *Passive, Active and Digital Filters*, CRC Press, London, 2006, pp. 19–33 to 19–55.

[16] Lorenz P., Russer P. 'Connection subnetworks for the TLM method', in Russer P., Siart U. (eds). *Time-Domain Methods in Electrodynamics*, Springer, Berlin Heidelberg, 2008, pp. 263–281.

[17] Akhtarzad S., Johns P. B. 'Solution of Maxwell's equations in three space dimensions and time by the TLM method of numerical analysis', *Proc. IEE*, 1975, vol. 122, pp. 1344–1348.

[18] Johns P. B. 'New symmetrical condensed node for three-dimensional solution of electromagnetic wave problems by TLM', *Electron. Lett.*, 1986, vol. 22, pp. 162–164.

[19] Cangellaris A. C., Wright D. B. 'Analysis of the numerical error caused by the stair-stepped approximation of conducting boundary in FDTD simulations of electromagnetic phenomena', *IEEE Trans. Antennas Propag.*, 1991, vol. 39, pp. 1518–1525.

[20] Wlodarczyk J. 'New multigrid interface for the TLM method', *Electron. Lett.*, 1996, vol. 32, no. 12, pp. 1111–1112.

[21] Trenkic V. The Development and Characterization of Advanced Nodes for the TLM Method. DPhil Thesis, The University of Nottingham, 1995.

[22] Trenkic V., Christopoulos C., Benson T. M. 'Advanced node formulations in TLM – The Adaptable Symmetrical Condensed Node (ASCN)', *IEEE Trans. Microwave Theory Techn.*, 1996, vol. 44, no. 12, pp. 2473–2478.

[23] Scaramuzza R. A., Lowery A. J. 'Hybrid condensed node for the TLM method', *Electron. Lett.*, 1990, vol. 26, pp. 1947–1949.

[24] Trenkic V., Christopoulos C., Benson T. M. 'Simple and elegant formulation of scattering in TLM nodes', *Electron. Lett.*, 1993, vol. 29, no. 18, pp. 1651–1652.

[25] Shewchuk J. R. 'Lecture Notes on Delaunay Mesh Generation', University of California at Berkeley, Febr 5 2012, 139 pages.

[26] Choong Y. K. Advanced Modal Expansion Techniques for the TLM Method. DPhil Thesis, The University of Nottingham, 2003.

[27] Sewell P., Wykes J., Benson T. M., Christopoulos C., Thomas T. M., Vukovic A. 'Transmission-line modelling using unstructured triangular meshes', *IEEE Trans. Microwave Theory Techn.*, 2004, vol. 52, no, 5, pp. 1490–1497.

[28] Sewell P., Benson T. M., Christopoulos C., Thomas D. W. P., Vukovic A. 'Transmission-line modelling (TLM) based upon unstructured tetrahedral meshes', *IEEE Trans. Microwave Theory Techn.*, 2005, vol. 53, no. 6, pp. 1919–1928.

[29] Sewell P., Benson T. M., Christopoulos C., Thomas D. W. P., Vukovic A., Wykes J. 'Implicit element clustering for tetrahedral transmission-line modelling (TLM). *IEEE Trans. Microwave Theory Techn.*, 2009, vol. 57, no. 6, pp. 2005–2014.

[30] Sewell P., Benson T. M., Vukovic A., Xuesong M. 'Complexity reduction of multiscale UTLM cell clusters', *IEEE J. Multiscale Multiphys. Comput. Techn.,* 2017, vol. 2, pp. 18–28.

[31] Pena N., Ney M. M. 'Absorbing boundary conditions using PML theory for 3-D TLM simulation', *IEEE Trans. Microwave Theory Techn.*, 1997, vol. 45, no. 10, pp. 1749–1755.

[32] Shao Z., Hong W., Wu H. 'A z-transform based absorbing boundary condition for 3-D TLM SCN method', *IEEE Trans. Microwave Theory Techn.*, 2002, vol. 50, no. 1, pp. 222–225.

[33] Odeyemi J., Vukovic A., Benson T. M., Sewell P. 'A complex domain mapping of the SCN for an effective PML implementation in TLM', *IEEE J. Antennas Propag.*, 2020, DOI: 10.1109/OJAP.2020.2986293.

[34] Nielsen J. S., Hoefer W. J. R. 'Generalized dispersion analysis and spurious modes in 2-D and 3-D TLM formulations', *IEEE Trans. Microwave Theory Techn.*, 1993, vol. 41, no. 8, pp. 1375–1384.

[35] Trenkic V., Christopoulos C., Benson T. M. 'Dispersion analysis of TLM symmetrical condensed node', *Electron. Lett.*, 1994, vol. 30, no. 25, pp. 2151–2153.

[36] Krumpholz M., Russer P. 'On dispersion in TLM and FDTD', *IEEE Trans. Microwave Theory Techn.*, 1994, vol. 42, no. 7, pp. 1274–1279.

[37] Trenkic V., Christopoulos C., Benson T. M. 'Analytical expansion of the dispersion relation for TLM condensed node', *IEEE Trans. Microwave Theory Techn.*, 1996, vol. 44, no. 12, pp. 2223–2230.

[38] Aidam M., Russer P. 'Derivation of the TLM method by finite integration'. *Int. J. Electron Commun. (AEU)*, 1997, vol. 51, pp. 35–39.

[39] Krumpholz M., Russer P. 'A field theoretical derivation of TLM', *IEEE Trans. Microwave Theory Techn.*, 1994, vol. 42, pp. 1660–1668.

[40] Chen Z., Ney M. M., Hoefer W. J. R. 'A new finite-difference formulation and its equivalence with the TLM SCN', *IEEE Trans. Microwave Theory Techn.*, 1991, vol. 39, pp. 2160–2169.
[41] Chen Z., Ney M. M. 'Recent progress in unifying time- and frequency-domain methods', in Russer P., Siart U. (eds). *Time Domain Methods in Electrodynamics*, Springer-Verlag, Berlin Heidelberg, 2008, pp. 31–39.
[42] Lindenmeier S., Pierantoni L., Russer P. 'Hybrid space discretization-integral equation methods for the numerical modelling of transients', *IEEE Trans. EMC*, 1999, vol. 41, no. 4, pp. 425–430.
[43] Christopoulos C. 'Modeling and simulation for EMC-Part I', **IEEE EMC Mag.*, 2015, vol. 4, no. 1, pp. 47–56.
[44] Christopoulos C. 'Modeling and simulation for EMC-Part II', **IEEE EMC Mag.*, 2015, vol. 4, no. 2, pp. 63–72.
[45] Christopoulos C. 'Multiscale modelling in time-domain electromagnetics', *Int. J. Electron. Commun. (AEU)*, 2003, vol. 57, no. 2, pp. 100–110.
[46] Cangellaris A. C., Pasha S., Prince J. L., Celik M. 'A new discrete TLM model for passive model order reduction and macromodeling in high-speed interconnections', *IEEE Trans. Adv. Packag*, 1999, vol. 22, no. 3, pp. 356–364.
[47] Lukashevich D., Cangellaris A. C., Russer P. 'Transmission line matrix method reduced order modeling', *MTT-S Int. Microwave Symp. Digest*, 2003, pp. 1125–1128.
[48] Holland R., Simpson L. 'Finite-difference analysis of EMP coupling in thin struts and wires', *IEEE Trans. EMC*, 1981, vol. 23, pp. 88–97.
[49] Paul J., Christopoulos C., Thomas D. W. P., Liu X. 'Time-domain modeling of electromagnetic wave interaction with thin wires using TLM', *IEEE Trans. EMC*, 2005, vol. 47, no. 3, pp. 447–455.
[50] Paul J., Christopoulos C., Thomas D. W. Correction to 'Time-domain modeling of electromagnetic wave interaction with thin wires using TLM', *IEEE Trans. EMC*, 2008, vol. 50, no. 2, pp. 450–451.
[51] Wlodarczyk A. J., Trenkic V., Scaramuzza R. A., Christopoulos C. 'A fully integrated multiconductor model for TLM', *IEEE Trans. Microwave Theory Techn.*, 1998, vol. 46, no. 12, pp. 2431–2437.
[52] Sewell P., Choong Y. K., Christopoulos C. 'An accurate thin-wire model for 3-D TLM simulations', *IEEE Trans. EMC*, 2003, vol. 45, no. 2, pp. 207–217.
[53] Wykes J. G., Sewell P., Christopoulos C., Benson T. M., Thomas D. W. P., Vukovic A. 'Embedded wire nodes in time-domain unstructured 2-D TLM', *Microwave Opt. Technol. Lett.*, 2008, vol. 50, no. 7, pp. 1753–1757.
[54] Meng X., Sewell P., Rahman N. H. A., Vukovic A., Benson T. M. 'Experimental benchmarking of unstructured transmission-line modeling (UTLM) method in modeling twisted wires', *ACES Exp. J.*, 2016, vol. 1, no. 3, pp. 101–104.
[55] Smartt C. J., Greedy S., Thomas D. W. P., Christopoulos C., Sewell P. 'Modeling and measurement of crosstalk between shielded cables'. *Proceedings of the EMC-Europe*; Wroclaw, Sept. 2010, pp. 86–90.

[56] Mueller U., Beyer A., Hoefer W. J. R. 'Moving boundaries in 2-D and 3-D TLM simulations realized by recursive formulas', *IEEE Trans. Microwave Theory Techn.*, 1992, vol. 40, no. 12, pp. 2267–2271.

[57] Panitz M., Paul J., Christopoulos C. 'A fractional boundary placement model using the transmission-line modeling (TLM) method', *IEEE Trans. MicrowaveTheory Techn.*, 2009, vol. 57, no. 3, pp. 637–646.

[58] GGIEMR web site; https://github.com/ggiemr/GGI_TLM.

[59] Paul J., Christopoulos C., Thomas D. W. P. 'Generalized material models in TLM-Part 1: materials with frequency-dependent properties', *IEEE Trans. Antennas Propag.*, 1999, vol. 47, no. 10, pp. 1528–1534.

[60] Paul J., Christopoulos C., Thomas D. W. P. 'Generalized material models in TLM-Part 2: materials with anisotropic properties', *IEEE Trans. Antennas Propag.*, 1999, vol. 47, no. 10, pp. 1535–1542.

[61] Paul J., Christopoulos C., Thomas D. W. P. 'Generalized material models in TLM-Part 3: materials with non-linear properties', *IEEE Trans. Antennas Propag.*, 2002, vol. 50, no. 7, pp. 997–1004.

[62] Biwojno K., Smartt C. J., Sewell P., Liu Y., Christopoulos C. 'General treatment of TLM node with embedded structures', *Proceedings of the CEM-TD Workshop*; Atlanta, USA, Sept. 2005, pp. 52–55.

[63] Biwojno K., Smartt C. J., Sewell P., Liu Y., Christopoulos C. 'TLM node with arbitrary embedded objects', *Int. J. Numer. Model.*, 2007, vol. 20, pp. 65–73.

[64] Scott I., Kumar V., Christopoulos C., Thomas D. W. P., Greedy S., Sewell P. 'Integration of behavioral models in the full-field TLM method', *IEEE Trans. EMC*, 2012, vol. 54, no. 2, pp. 359–366.

[65] Sicard E., Wu J., Shen R. J., *et al.* 'Recent advances in electromagnetic compatibility of 3-D ICs' Part I', *IEEE EMC Mag.*, 2015, vol. 4, no. 4, pp. 79–89.

[66] Sicard E., Wu J., Shen R. J., *et al.* 'Recent advances in electromagnetic compatibility of 3-D ICs' Part II', *IEEE EMC Mag.*, 2016, vol. 5, no. 1, pp. 66–75.

[67] Tong X., Thomas D. W. P., Nothofer A., Sewell P., Christopoulos C. 'Modelling electromagnetic emissions from printed circuit boards in closed environments using equivalent dipoles', *IEEE Trans. EMC*, 2010, vol. 52, no. 2, pp. 462–470.

[68] Tong X., Thomas D. W. P., Nothofer A., Christopoulos C., Sewell P. 'Reduction of sensitivity to measurement errors in the derivation of equivalent models of emission in numerical computation', *ACES J.*, 2010, vol. 26, no. 7, pp. 603–610.

[69] Julier S. J., Uhlman J. K. 'Unscented filtering and non-linear estimation', *Proc. IEEE*, 2004, vol. 92, no. 3, pp. 401–422.

[70] De Menezes L., Ajayi A., Christopoulos C., Sewell P., Borges G. A. 'Efficient computation of stochastic electromagnetic problems using unscented transforms', *IET Sci. Measur. Technol.*, 2008, vol. 2, no. 2, pp. 88–95.

[71] Thomas D. W. P., De Menezes L., Christopoulos C., Leferink F., Bergsma H. J. G. 'Estimation of probability distributions for cable coupling using unscented transforms', *Ann. Telecommun.*, 2011, vol. 66, pp. 475–482.

[72] Xu H., Rahman S. 'A generalized dimension reduction method for multidimensional integration in stochastic mechanics', *Int. J. Numer. Methods Eng.*, 2004, vol. 61, no. 12, pp. 1992–2019.

[73] Russer J. A., Cangellaris A., Russer P. 'Correlation transmission line matrix (CTLM) modeling of stochastic electromagnetic problems', *Int. Microwave Symp. MTT-S*, San Francisco, USA, May 2016, pp. 1–4.

[74] Haider M., Russer J. A. 'The correlation transmission line matrix (CTLM) method', *Proceedings of the ICEAA*, Verona, Italy, Sept 2017, pp. 1509–1512.

Chapter 6

Partial element equivalent circuit method in time-domain

Giulio Antonini[1], Daniele Romano[1], Luigi Lombardi[2] and Albert Ruehli[3]

The availability of very large-scale integrated circuits (VLSI) leads to the rapid growth in the design of electrical and electronic systems. This results in the ever increasing need for modeling and analysis of the design of such systems. Parts require the modeling of passive parasitic parts which may be the cause of electrical interferences or act as sources for electromagnetic compatibility (EMC) and signal integrity (SI) problems. The electromagnetic challenges of such effects along with the geometric complexity of electronic systems call for efficient electromagnetic methodologies and computer-aided design tools which allow a full-wave analysis of 3-D structures characterized by inhomogeneous materials with complex geometries.

Three fundamental popular computational electromagnetic (CEM) formulations which are used in such solvers are based on the finite element method (FEM) [1], the finite difference time-domain (FDTD) [2–4], and integral equation like the impedance moments methods (Z-MoM) [5]. It is known that the first two approaches are essentially partial differential equation (PDE)-based form of Maxwell's equations which can result in powerful techniques. Today, they are widely used for different classes of EM problems. The impedance Method of Moments (Z-MoM) solvers are based on an integral equation formulation of Maxwell's equations. The integral equation (IE) we focus on in this chapter is the partial element equivalent circuit (PEEC) method. Stemming originally from the works by Ruehli [6–8], in the following, the PEEC method is presented in this work with the aim to provide the reader with a step-by-step procedure to understand the formulation of a PEEC solver.

The main difference between the PEEC and the Z-MoM method is that the PEEC method leads to a general formulation in the descriptor form rather than an impedance result. The circuit interpretation of the electric field integral equation [9]

[1]UAq EMC Laboratory, Dipartimento di Ingegneria Industriale e dell'Informazione e di Economia, Università degli Studi dell'Aquila, Italy
[2]Micron Semiconductor, Avezzano, Italy
[3]UMRI/MST EMC Laboratory, Missouri University of Science and Technology, USA

is in terms of partial elements, namely resistances, partial inductances, and partial potential coefficients. Thus, the resulting equivalent circuit can be solved with SPICE-like circuit solvers [10] in both the time and the frequency domain. Furthermore, a systematic approach can be applied to reduce its complexity, taking into account the electrical size of the structure under analysis. For example, if the rise time of the excitation (i.e. of a pulse excitation *pr* the period of a time-harmonic excitation) is such that useful wavelengths of the highest frequencies are much larger than the size of the system. Solutions of this nature are not called *full-wave*.

IE methods are very effective for electromagnetic modeling of electromagnetic interference (EMI) and EMC problems. The first step of any integral equation-based method is the development of an integral equation formulation of Maxwell's equation. The most popular integral equation is the electric field integral equation which is obtained by enforcing the electric field at a point in the structure as the superposition of fields due to all electric currents as well as charges in the system [9,11].

Compared with differential equation (DE)-based methods, the matrices resulting from IE-based techniques solutions are smaller in size but are more dense. A key reason for the reduced number of unknowns is that no unknowns need to be assigned to the free space. Hence, the unknowns are representing the electric currents flowing through the conductors and charges on the surfaces of the dielectric parts . On the other hand, the density of matrices arising from IE solutions results from the fact that the element describing the electromagnetic interaction (electric and magnetic) between two discrete currents or charges in the structure must be taken into account.

This chapter is organized as follows: Section 6.1 presents the basic derivation of the PEEC method starting from the volume electric field integral equation (EFIE); the computation of the partial elements is briefly recalled in Section 6.2. Then, the time-domain formulation for both the quasi-static and the delayed PEEC formulation is revised in Section 6.3. Model order reduction techniques for quasi-static and delayed PEEC models are revised in Section 6.4. Section 6.5 reports numerical examples in EMC and SI areas.

6.1 Integral formulation of Maxwell's equations

Maxwell's equations allow to describe and relate the field vectors, current densities, and charge densities at any point in space at any time. In the differential form, Maxwell's equations can be written as [9]

$$\nabla \times \boldsymbol{H}(\boldsymbol{r},t) = \frac{\partial \boldsymbol{D}(\boldsymbol{r},t)}{\partial t} + \boldsymbol{J}(\boldsymbol{r},t) \tag{6.1a}$$

$$\nabla \times \boldsymbol{E}(\boldsymbol{r},t) = -\frac{\partial \boldsymbol{B}(\boldsymbol{r},t)}{\partial t} \tag{6.1b}$$

$$\nabla \cdot \boldsymbol{B}(\boldsymbol{r},t) = 0 \tag{6.1c}$$

$$\nabla \cdot \boldsymbol{D}(\boldsymbol{r},t) = q(\boldsymbol{r},t) \tag{6.1d}$$

where $q(\boldsymbol{r},t)$ is the charge density and $\boldsymbol{J}(\boldsymbol{r},t)$ is the current density. The magnetic field $\boldsymbol{H}(\boldsymbol{r},t)$, the magnetic flux density $\boldsymbol{B}(\boldsymbol{r},t)$, the electric field $\boldsymbol{E}(\boldsymbol{r},t)$, and the electric flux density $\boldsymbol{D}(\boldsymbol{r},t)$ satisfy the following constitutive relations

$$\boldsymbol{B}(\boldsymbol{r},t) = \mu_0\mu_r(\boldsymbol{r},t)\boldsymbol{H}(\boldsymbol{r},t) \tag{6.2a}$$

$$\boldsymbol{D}(\boldsymbol{r},t) = \varepsilon_0\varepsilon_r(\boldsymbol{r},t)\boldsymbol{E}(\boldsymbol{r},t) \tag{6.2b}$$

where $\mu_r(\boldsymbol{r},t)$ and $\varepsilon_r(\boldsymbol{r},t)$ are the time-varying relative permeability and the permittivity, respectively. Since the PEEC method is based on the volume equivalence principle, currents and charges are assumed to radiate in the background medium that is the vacuum. It is useful to express fields in terms of potential. From the solenoidality of $\boldsymbol{B}$, (6.1c), the magnetic vector potential is defined as

$$\boldsymbol{B}(\mathbf{r},t) = \nabla\times\boldsymbol{A}(\boldsymbol{r},t). \tag{6.3}$$

Substituting (6.3) into (6.1b), it follows that

$$\nabla\times\left(\boldsymbol{E}(\boldsymbol{r},t) + \frac{\partial\boldsymbol{A}(\boldsymbol{r},t)}{\partial t}\right) = \boldsymbol{0}. \tag{6.4}$$

The previous equation allows to define the electric scalar potential $\varphi(\boldsymbol{r},t)$ such that

$$\boldsymbol{E}(\boldsymbol{r},t) + \frac{\partial\boldsymbol{A}(\boldsymbol{r},t)}{\partial t} = -\nabla\phi(\boldsymbol{r},t). \tag{6.5}$$

Equation (6.5) relates the electric field $\boldsymbol{E}$ with the potentials $\boldsymbol{A}$ and ϕ. Then it becomes important to express the magnetic vector potential $\boldsymbol{A}$ and the electric scalar potential ϕ as a function of the current density $\boldsymbol{J}$ and the charge density q, respectively. To this aim, by substituting (6.3) and (6.5) into (6.1a), it follows that

$$\nabla\times\nabla\times\boldsymbol{A}(\boldsymbol{r},t) = \mu_0\varepsilon_0\frac{\partial}{\partial t}\left(-\frac{\partial\boldsymbol{A}(\boldsymbol{r},t)}{\partial t} - \nabla\phi(\boldsymbol{r},t)\right) - \mu_0\boldsymbol{J}(\boldsymbol{r},t). \tag{6.6}$$

The Laplacian identity

$$\nabla\times\nabla\times\boldsymbol{A}(\boldsymbol{r},t) = \nabla(\nabla\cdot\boldsymbol{A}(\boldsymbol{r},t)) - \nabla^2\boldsymbol{A}(\boldsymbol{r},t) \tag{6.7}$$

and the Lorenz gauge

$$\nabla\cdot\boldsymbol{A}(\boldsymbol{r},t) = -\mu_0\varepsilon_0\frac{\partial\phi(\boldsymbol{r},t)}{\partial t} \tag{6.8}$$

allows to obtain the Helmholtz equation for the magnetic vector potential

$$\nabla^2\boldsymbol{A}(\boldsymbol{r},t) - \mu_0\varepsilon_0\frac{\partial\boldsymbol{A}(\boldsymbol{r},t)}{\partial t^2} = -\mu_0\boldsymbol{J}(\boldsymbol{r},t). \tag{6.9}$$

Following an analogous approach, it is possible to express the potential $\varphi(\boldsymbol{r}, t)$ in terms of the charge density leading to the Helmholtz equation for the electric scalar potential

$$\nabla^2 \varphi(\boldsymbol{r}, t) - \mu_0 \varepsilon_0 \frac{\partial \phi(\boldsymbol{r}, t)}{\partial t^2} = -\frac{q(\boldsymbol{r}, t)}{\varepsilon_0}. \tag{6.10}$$

In an homogeneous medium, like the vacuum assumed as background medium, (6.9) has a closed-form solution for the magnetic vector potential $\mathbf{A}(\boldsymbol{r}, t)$ due to a current $\boldsymbol{J}(\boldsymbol{r}, t)$ in the volume V'; it is

$$\boldsymbol{A}(\boldsymbol{r}, t) = \frac{\mu_0}{4\pi} \int_{V'} \frac{\boldsymbol{J}(\mathbf{r}', t')}{|\boldsymbol{r} - \mathbf{r}'|} dV'. \tag{6.11}$$

Similarly, (6.10) has a closed-form solution for the electric scalar potential $\phi(\mathbf{r}, t)$ induced by the charge distribution $q(\mathbf{r}', t)$; assuming that the charge resides on the exterior surface of conductors and dielectrics, the solution of (6.10) in an homogenous medium is

$$\phi(\boldsymbol{r}, t) = \frac{1}{4\pi\varepsilon_0} \int_{S'} \frac{q(\mathbf{r}', t')}{|\boldsymbol{r} - \mathbf{r}'|} dS'. \tag{6.12}$$

It is to be remarked that, in (6.11) and (6.12), t' represents the time when the current and charge distributions, $\boldsymbol{J}$ and q, act as radiating sources of $\boldsymbol{A}$ and ϕ, respectively. It is different from t because of the finite value of the speed of light in the homogenous medium, $c_0 = 1/\sqrt{\mu_0 \varepsilon_0}$. They can be related by

$$t' = t - \frac{|\boldsymbol{r} - \mathbf{r}'|}{c}. \tag{6.13}$$

To derive (6.11) and (6.12), all the Maxwell's equations (6.1) have been used along with the Lorenz gauge (6.8). Equation (6.5) for the electric field has not been used yet.

We limit our analysis for now to the conductors; it will later be extended to dielectrics. The constitutive relation for a conductor reads

$$\boldsymbol{E}(\boldsymbol{r}, t) = \frac{\boldsymbol{J}(\boldsymbol{r}, t)}{\sigma} \tag{6.14}$$

where σ is the conductor conductivity. Substituting (6.14) into the electric field equation (6.5) and assuming that an external electric field $\boldsymbol{E}^{inc}(\boldsymbol{r}, t)$ can be impressed at point $\boldsymbol{r}$ at time t, the EFIE is obtained as

$$\boldsymbol{E}^{inc}(\boldsymbol{r}, t) = \frac{\boldsymbol{J}(\boldsymbol{r}, t)}{\sigma} + \frac{\partial}{\partial t} \frac{\mu_0}{4\pi} \int_{V'} \frac{\mathbf{J}(\mathbf{r}', t')}{|\boldsymbol{r} - \mathbf{r}'|} dV' + \nabla \phi(\boldsymbol{r}, t) \tag{6.15}$$

which holds at any point in a conductor and where the electric scalar potential is related to the charge distribution by (6.12), here repeated for clarity

$$\phi(\boldsymbol{r}, t) = \frac{1}{4\pi\varepsilon_0} \int_{S'} \frac{q(\mathbf{r}', t')}{|\boldsymbol{r} - \mathbf{r}'|} dS'. \tag{6.16}$$

Furthermore, the continuity equation guaranteeing the charge conservation must be enforced. Within volumes it reads

$$\nabla \cdot \boldsymbol{J}(\boldsymbol{r}, t) = -\frac{\partial q(\boldsymbol{r}, t)}{\partial t}. \tag{6.17}$$

By assuming that the charge is located only on the surface of conductors, within the volume of conductors, (6.17) reads

$$\nabla \cdot \boldsymbol{J}(\boldsymbol{r}, t) = 0 \tag{6.18}$$

while on the surface of conductors, using the surface divergence, it follows that

$$\widehat{\boldsymbol{n}} \cdot \boldsymbol{J}(\boldsymbol{r}, t) = \frac{\partial q(\boldsymbol{r}, t)}{\partial t} \tag{6.19}$$

where $\widehat{\boldsymbol{n}}$ is the outward normal to the surface S'.

Finally, the set of equations to be solved reads

$$\boldsymbol{E}^{inc}(\boldsymbol{r}, t) = \frac{\boldsymbol{J}(\boldsymbol{r}, t)}{\sigma} + \frac{\partial}{\partial t}\frac{\mu_0}{4\pi}\int_{V'} \frac{\boldsymbol{J}(\mathbf{r}', t')}{|\boldsymbol{r} - \mathbf{r}'|} dV' + \nabla \phi(\boldsymbol{r}, t) \tag{6.20a}$$

$$\phi(\boldsymbol{r}, t) = \frac{1}{4\pi\varepsilon_0}\int_{S'} \frac{q(\mathbf{r}', t')}{|\boldsymbol{r} - \mathbf{r}'|} dS' \qquad \boldsymbol{r} \in S' \tag{6.20b}$$

$$\nabla \cdot \boldsymbol{J}(\boldsymbol{r}, t) = 0 \qquad \boldsymbol{r} \in V' \tag{6.20c}$$

$$\widehat{\boldsymbol{n}} \cdot \boldsymbol{J}(\boldsymbol{r}, t) = \frac{\partial q(\boldsymbol{r}, t)}{\partial t} \qquad \boldsymbol{r} \in S'. \tag{6.20d}$$

The unknowns of such a problem are represented by the current density $\boldsymbol{J}(\boldsymbol{r}, t)$ flowing in the volumes of conductors and dielectrics, the charge density $q(\boldsymbol{r}, t)$ on the surface of conductors and dielectrics and the electric scalar potential $\phi(\boldsymbol{r}, t)$ which can be expressed as a function of the charge density for $\boldsymbol{r} \in S'$.

Equations (6.20) can be rewritten in the Laplace domain as

$$\boldsymbol{E}^{inc}(\boldsymbol{r}, s) = \frac{\boldsymbol{J}(\boldsymbol{r}, s)}{\sigma} + \frac{s\mu_0}{4\pi}\int_{V'} \frac{\boldsymbol{J}(\mathbf{r}', s)e^{-s\tau}}{|\boldsymbol{r} - \mathbf{r}'|} dV' + \nabla \phi(\boldsymbol{r}, s) \tag{6.21a}$$

$$\phi(\boldsymbol{r}, s) = \frac{1}{4\pi\varepsilon_0}\int_{S'} \frac{q(\mathbf{r}', s)e^{-s\tau}}{|\boldsymbol{r} - \mathbf{r}'|} dS' \qquad \boldsymbol{r} \in S' \tag{6.21b}$$

$$\nabla \cdot \boldsymbol{J}(\boldsymbol{r}, s) = 0 \qquad \boldsymbol{r} \in V' \tag{6.21c}$$

$$\widehat{\boldsymbol{n}} \cdot \boldsymbol{J}(\boldsymbol{r}, s) = sq(\boldsymbol{r}, s) \qquad \boldsymbol{r} \in S' \tag{6.21d}$$

where $\tau = |\boldsymbol{r} - \mathbf{r}'|/c_0$ and s is the Laplace variable.

The most popular method for the discretization of integral equations was called by Harrington the *method of moments* (MoM) [5] with different implementations [12–16]. Usually the solution is found in the frequency domain, assuming $s = j\omega$. Since we are interested into time-domain PEEC models, we will not make this assumption. Hence, in the following, we will remain in the Laplace-domain with s in the complex plane.

As a first step, the unknown quantities $\boldsymbol{J}(\boldsymbol{r}, s)$ and $q(\boldsymbol{r}, s)$ are approximated by a weighted sum of finite set of basis functions $\boldsymbol{b} \in R^3$ and $p \in R$:

$$\mathbf{J}(\boldsymbol{r}, s) \cong \sum_{n=1}^{N_v} \mathbf{b}_n(\boldsymbol{r}) I_n(s) \tag{6.22a}$$

$$q(\boldsymbol{r}, s) \cong \sum_{m=1}^{N_s} p_m(\boldsymbol{r}) Q_m(s) \tag{6.22b}$$

where $I_n(s)$ and $Q_m(s)$ are the basis function weights which must be determined at each angular frequency s, N_v, and N_s represent the number of volume and surface basis functions and the corresponding elementary volume and surface sub-regions, respectively. Expansions (6.22) are substituted into (6.21a) and (6.21b), yielding

$$\boldsymbol{E}^{inc}(\boldsymbol{r}, s) = \sum_{n=1}^{N_v} \frac{\boldsymbol{b}_n(\boldsymbol{r}) I_n(s)}{\sigma} + \frac{s\mu_0}{4\pi} \sum_{n=1}^{N_v} \int_{V_n} \boldsymbol{b}_n(\boldsymbol{r}_n) I_n(s) \frac{e^{-s\tau}}{|\boldsymbol{r} - \boldsymbol{r}_n|} dV_n + \nabla\phi(\boldsymbol{r}, s) \tag{6.23a}$$

$$\phi(\boldsymbol{r}, s) = \frac{1}{4\pi\varepsilon_0} \sum_{m=1}^{N_s} \int_{S_m} p_m(\boldsymbol{r}_m) Q_m(s) \frac{e^{-s\tau}}{|\boldsymbol{r} - \mathbf{r}_m|} dS_m. \tag{6.23b}$$

Next, the so-called Galerkin's testing or weighting process [15] is used to generate a system of equations for the unknowns weights $I_n(s), n = 1, \ldots, N_v$ and $Q_m(s), m = 1, \ldots, N_s$ by enforcing the residuals of (6.21a)–(6.21b) to be orthogonal to a set of weighting functions which are chosen to be coincident with the basis functions

$$\begin{aligned} &\langle -\boldsymbol{E}^{inc}(\boldsymbol{r}, s) + \frac{\sum_{n=1}^{N_v} \boldsymbol{b}_n(\boldsymbol{r}) I_n(s)}{\sigma} + \\ &+ \frac{s\mu_0}{4\pi} \left(\sum_{n=1}^{N_v} \int_{V_n} \boldsymbol{b}_n(\boldsymbol{r}_n) I_n(s) \frac{e^{-s\tau}}{|\boldsymbol{r} - \mathbf{r}_n|} \right) dV_n + \nabla\phi(\boldsymbol{r}, s), \boldsymbol{b}_i(\boldsymbol{r}) \rangle = 0 \end{aligned} \tag{6.24a}$$

$$\langle \phi(\boldsymbol{r}, s) - \frac{1}{4\pi\varepsilon_0} \sum_{m=1}^{N_s} \int_{S_m} p_m(\boldsymbol{r}_m) Q_m(s) \frac{e^{-s\tau}}{|\boldsymbol{r} - \mathbf{r}_m|} dS_m, p_j(\boldsymbol{r}) \rangle = 0 \tag{6.24b}$$

where the inner products are defined as

$$\langle \boldsymbol{f}(\boldsymbol{r}), \boldsymbol{b}_i(\boldsymbol{r}) \rangle = \int_{V_i} \boldsymbol{f}(\boldsymbol{r}) \cdot \boldsymbol{b}_i(\boldsymbol{r}) dV_i \qquad \textit{for } i = 1, \ldots, N_v \tag{6.25a}$$

$$\langle g(\boldsymbol{r}), p_j(\boldsymbol{r}) \rangle = \int_{S_j} g(\boldsymbol{r}) \cdot p_j(\boldsymbol{r}) dS_j \qquad \textit{for } j = 1, \ldots, N_s. \tag{6.25b}$$

6.1.1 Basis functions for the surfaces

A number of different typology of basis and weighting functions can be chosen to set (6.24a) and (6.24b). The most popular are the piecewise constant, piecewise linear, RWG [13] set of basis and/or weighting functions [17]. In the following, it will be assumed the piecewise constant set of functions which are more suited to model Manhattan type structures [17]. Thus, by assuming to deal with orthogonal conductors whose surface is discretized into N_s elementary rectangular patches which are electrically small compared with the wavelength of the highest frequency of interest. More specifically, the unknown electrical current and charge densities are taken to have constant values over each cell in the discrete model.

Under this assumption, the basis functions used to expand the charge density are chosen as

$$p_m(\boldsymbol{r}) = \begin{cases} \dfrac{1}{S_m} & \text{if } \boldsymbol{r} \in S_m \\ 0 & \text{otherwise.} \end{cases} \tag{6.26}$$

With such a choice of the basis function, the corresponding weight Q_m represents the charge on patch m. Finally, (6.23b) can be rewritten as

$$\phi(\boldsymbol{r}, s) = \sum_{m=1}^{N_s} \left[\frac{1}{4\pi\varepsilon_0} \frac{1}{S_m} \int_{S_m} \frac{e^{-s\tau}}{|\boldsymbol{r} - \boldsymbol{r}_m|} dS_m \right] Q_m(s) \tag{6.27}$$

which allows to evaluate the potential at point $\boldsymbol{r}$, at angular frequency s, due to the charge on the N_s patches on the interface surfaces between different media, including the background medium.

Applying Galerkin's scheme results in the evaluation of the average value of $\phi(\boldsymbol{r}, s)$ over the surface of each patch

$$\begin{aligned} \phi_\ell(s) &= \frac{1}{S_\ell} \int_{S_\ell} \phi(\boldsymbol{r}_\ell, s) dS_\ell = \sum_{m=1}^{N_s} \left[\frac{1}{4\pi\varepsilon_0} \frac{1}{S_\ell} \frac{1}{S_m} \int_{S_\ell} \int_{S_m} \frac{e^{-s\tau}}{|\boldsymbol{r}_\ell - \boldsymbol{r}_m|} dS_m dS_\ell \right] Q_m(s) = \\ &= P_{\ell,m}(s) Q_m(s) \quad \textit{for } \ell = 1, \ldots, N_s \end{aligned} \tag{6.28}$$

where coefficient of potential $P_{\ell,m}(s)$ is

$$P_{\ell,m}(s) = \frac{1}{4\pi\varepsilon_0} \frac{1}{S_\ell S_m} \int_{S_\ell} \int_{S_m} \frac{e^{-s\tau}}{|\boldsymbol{r}_\ell - \boldsymbol{r}_m|} dS_m dS_\ell. \tag{6.29}$$

Thus, the potential of the N_s patches can be related to the charges located on the same patches, at the angular frequency s, by

$$\boldsymbol{\Phi}(s) = \mathbf{P}(s)\mathbf{Q}(s) \tag{6.30}$$

where matrix $\mathbf{P}$ entries are known as coefficients of potential and are, in general, frequency dependent due to the full wave type of analysis. The displacement currents in the background medium are obtained as

$$\mathbf{I}_c(s) = s\mathbf{Q}(s) = s\mathbf{P}^{-1}(s)\mathbf{\Phi}(s). \tag{6.31}$$

6.1.2 Basis functions for volume currents

Conductor and dielectric volumes are discretized using a mesh of Manhattan type into N_v elementary orthogonal hexahedra (parallelepipeds) which are assumed electrically small compared with the wavelength of the highest frequency of interest. Let a_n the the cross section of volume V_n.

The basis functions used to expand the current density are chosen as

$$\boldsymbol{b}_n(\boldsymbol{r}) = \begin{cases} \dfrac{\widehat{\mathbf{u}}_n}{a_n} & \text{if } \boldsymbol{r} \in V_n \\ \mathbf{0} & \text{otherwise} \end{cases} \tag{6.32}$$

where $\widehat{\boldsymbol{u}}_i$ is the unit vector indicating the current orientation in volume V_n. With such a choice of the basis function, the corresponding weight represents the current flowing in the volume V_n with orientation $\widehat{\boldsymbol{u}}_i$. Equation (6.23a), after Galerkin's testing approach is applied, can be rewritten as

$$\frac{1}{a_i}\int_{a_i}\int_{\ell_i}\boldsymbol{E}^{inc}(\boldsymbol{r},s) = \frac{\ell_i I_i(s)}{\sigma a_i} + \frac{s\mu_0}{4\pi}\sum_{n=1}^{N_v}\frac{1}{a_i}\frac{1}{a_n}\int_{V_i}\int_{V_n}\widehat{\boldsymbol{u}}_i\cdot\widehat{\boldsymbol{u}}_n I_n(s)\frac{e^{-s\tau}}{|\boldsymbol{r}_i - \boldsymbol{r}_n|}dV_n dV_i + \\ + \phi_{2i}(s) - \phi_{1i}(s) \qquad \text{for } i = 1,\ldots,N_v \tag{6.33}$$

where ℓ_i denotes the length of the volume V_i. In deriving the previous equation, the external electric field $\boldsymbol{E}^{inc}(\boldsymbol{r},s)$ has been assumed uniform within the volume V_i. Applying Galerkin's testing approach to a volume cell m leads to an equivalent voltage source V_{S_i}, as described in [18]:

$$V_{S_i}(s) = -\frac{1}{a_i}\int_{a_i}\int_{\ell_i}\boldsymbol{E}^{inc}(\boldsymbol{r},s)dV \tag{6.34}$$

where

$$\boldsymbol{E}^{inc}(\boldsymbol{r},t) = E_x^{inc}(\boldsymbol{r},s)\widehat{x} + E_y^{inc}(\boldsymbol{r},s)\widehat{y} + E_z^{inc}(\boldsymbol{r},s)\widehat{z}. \tag{6.35}$$

The equivalent voltage source, $V_{S_i}(s)$, is placed in series with the inductive equivalent circuit of volume cell i.

Also, it has been considered that

$$\frac{1}{a_i}\int_{V_i}\widehat{\boldsymbol{u}}_i\cdot\nabla\phi(\boldsymbol{r},s)dV_i = \frac{1}{a_i}\int_{a_i}\left(\int_{\ell_i}\widehat{\boldsymbol{u}}_i\cdot\nabla\varphi(\boldsymbol{r},s)d\ell_i\right)da_i = \phi_{i_2}(s) - \phi_{i_1}(s) \tag{6.36}$$

where $\phi_{i_1}(s)$ and $\phi_{i_2}(s)$ represent the potential at the extremes of the volume V_i along the $\widehat{\boldsymbol{u}}_i$ direction.

Each term of equation (6.33) represents a voltage drop across volume V_i along the $\widehat{\boldsymbol{u}}_i$ direction and, thus, it can be rewritten as

$$\phi_{i_1}(s) - \phi_{i_2}(s) = V_{s_i}(s) + R_i I_i + s\sum_{n=1}^{N_v} L_{p_{i,n}} I_n(s) \tag{6.37}$$

where

$$R_i = \frac{\ell_i}{\sigma a_i} \tag{6.38}$$

is the resistance of the cell i where current flows along ℓ_i. More advanced models of dissipative phenomena within conductors, including the skin-effect are described in [19]. The quantity

$$L_{p_{i,n}}(s) = \frac{\mu_0}{4\pi}\frac{1}{a_i a_n}\int_{V_i}\int_{V_n} \widehat{\boldsymbol{u}}_i \cdot \widehat{\boldsymbol{u}}_n \frac{e^{-s\tau}}{|\boldsymbol{r}_i - \boldsymbol{r}_n|} dV_n dV_i \tag{6.39}$$

is the so-called partial inductance [20] between volume cells i and n. The product of partial inductance (6.39) with current $I_n(s)$ represents the flux $\Psi_i(s)$ whose interpretation can be found in [20,21]. Hence, the induced voltage on the N_v elementary volumes due to the magnetic field coupling is given by the following matrix expression:

$$\mathbf{E}(s) = s\boldsymbol{\Psi}(s) = s\mathbf{L}_p(s)\mathbf{I}(s). \tag{6.40}$$

Finally, the quantity

$$\phi_{i_1}(s) - \phi_{i_2}(s) \tag{6.41}$$

represents the difference of potential between nodes at the ends of volume V_i, along the $\widehat{\boldsymbol{u}}_i$ direction. An example of the equivalent circuit corresponding to (6.37) is depicted in Figure 6.1.

A different choice of the basis functions is presented in [22] where Whitney face and edge elements are adopted.

6.1.3 Dielectrics modeling

The PEEC-oriented modeling of dielectrics represents the displacement current due to the bound charges for dielectrics with $\varepsilon_r > 1$ separately from the conducting currents due to the free charges. In [23], it is shown that an elementary volume composed of a dielectric can be represented as the series of a partial inductance which is coupled with

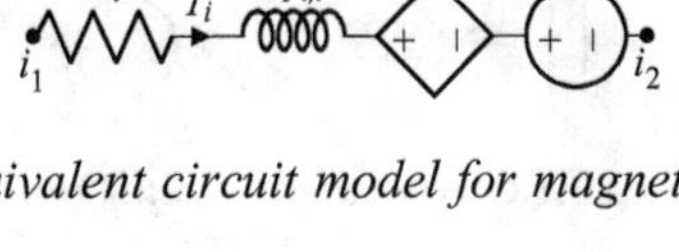

Figure 6.1 Equivalent circuit model for magnetic field coupling

all the other elementary volumes, exactly as it happens for conductors and an excess capacitance that takes the polarization current into account. Assuming again a mesh of Manhattan type, the excess capacitance of a volume V_i is written as

$$C_{e_i} = \frac{\varepsilon_0(\varepsilon_{ri} - 1)a_i}{\ell_i}. \tag{6.42}$$

Figure 6.2 shows the PEEC equivalent circuit of a dielectric volume.

More recently, PEEC models of dispersive and lossy dielectrics of Debye and Lorentz type have been proposed [24–27]. In time-domain, these models result in more complex circuits accounting for real and complex conjugate poles in the dielectric permittivity law. The modeling of anisotropic dielectrics in the PEEC framework has been presented in [28].

6.1.4 Enforcement of Kirchhoff's voltage and current laws

Once the equivalent circuit is generated, Kirchhoff's current and voltage laws can be enforced. The first set of equation can be obtained by enforcing the Kirchhoff's voltage law (KVL) to a mesh constituted by the resistive/dielctric-inductive branch connecting each couple of nodes and the capacitive branch connecting each node to infinity. Enforcing (6.37) to each elementary volume, a set of N_v equations is obtained that can be written in a compact matrix form as

$$-\mathbf{A}^T\mathbf{\Phi}(s) - \mathbf{R}\mathbf{I}(s) - s\mathbf{L}_p(s)\mathbf{I}(s) - \mathbf{V}_s(s) = \mathbf{0} \tag{6.43}$$

where vectors $\mathbf{\Phi}(s)$ and $\mathbf{I}(s)$ collect the potentials to infinity and the currents flowing through the longitudinal branches, respectively, and the matrix $\mathbf{A}$ is the connectivity matrix whose entries are

$$A_{n,k} = \begin{cases} +1 & \text{if current } I_k \text{ enters node } n \\ -1 & \text{if current } I_k \text{ leaves node } n \\ 0 & \text{otherwise} \end{cases} \tag{6.44}$$

and A^T denotes the transpose matrix A.

It is worth to notice that the discretization process described above has allowed to generate circuit topological elements such as branches, where currents I_i, $i = 1, \ldots, N_v$, flow and nodes, whose potential to infinity is ϕ_ℓ, $\ell = 1, \ldots, N$. It is to be remarked that, in the case of general 3D structures, $N > N_s$ because nodes interior to the conductors may occur.

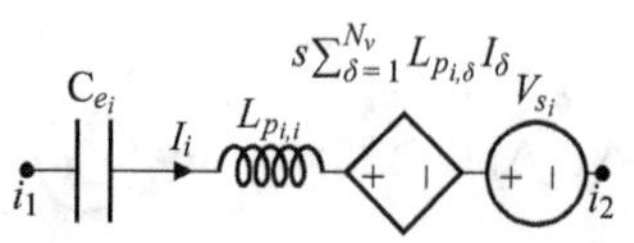

Figure 6.2 PEEC equivalent circuit of dielectrics

The PEEC method comprises the continuity equation in the form of Kirchhoff's current law (KCL). By considering that both $\mathbf{I}$ and $\mathbf{I}_c$ and external current sources $\mathbf{I}_s$ can be connected to each node, KCL can be written as

$$\mathbf{I}_c(s) - \mathbf{A}\mathbf{I}(s) = \mathbf{I}_s(s). \tag{6.45}$$

Considering that the displacement currents $\mathbf{I}_c$ can be expressed as a function of the potentials $\mathbf{\Phi}$ (6.31), it is possible to write

$$s\mathbf{P}^{-1}(s)\mathbf{\Phi}(s) + \mathbf{Y}_{le}(s)\mathbf{\Phi}(s) - \mathbf{A}\mathbf{I}(s) = \mathbf{I}_s(s) \tag{6.46}$$

where $\mathbf{Y}_{le}(s)$ denotes the admittance matrix of lumped elements connected to nodes.

In the Laplace-domain, matrix $\mathbf{P}(s)$ is to be regarded as a matrix of transfer functions and matrix inversion $\mathbf{P}^{-1}(s)$ at the symbolic level is unfeasible because of its complexity. The same happens in the Fourier-domain for large problems. Luckily, the previous equation can be re-written as

$$s\mathbf{\Phi}(s) + \mathbf{P}\mathbf{Y}_{le}\mathbf{\Phi}(s) - \mathbf{P}(s)\mathbf{A}\mathbf{I}(s) = \mathbf{P}(s)\mathbf{I}_s(s). \tag{6.47}$$

thus avoiding the inversion of matrix $\mathbf{P}(s)$.

6.1.5 DC solution

It is to be remarked the PEEC method admits a rigorous *DC* solution. In [29,30], a two-step process is proposed where first the magnetostatic problem is solved computing the electrical currents. Then, by using the solution of the magnetostatic problem as boundary conditions and enforcing the neutrality of disconnected objects, potentials to infinity and charges are computed.

6.2 Computation of partial elements

A crucial step in building a PEEC model is the computation of partial elements, namely partial inductances and coefficients of potential, accounting for the magnetic and electric couplings respectively, resistances which account for power dissipation in conductive materials, excess capacitances describing the polarization phenomena within dielectrics. Quite an extensive literature exists for the quasi-static case which assumes all the partial elements to be frequency independent and in the case of orthogonal meshes. Closed-form expressions of frequency-independent partial inductances for orthogonal volumes can be found in [17,20,31–34]. Closed-form expressions of frequency-independent coefficients of potential for rectangular patches are given in [17,35,36]. A multi-function approach has recently been proposed in [37] to compute partial elements of orthogonal meshes getting rid of the numerical errors due to the use of analytical formulas and finite machine precision. A more general approach to the computation of partial elements for non-orthogonal geometries can be found in [38–40]. Since their computation is out of the focus of this chapter, they will not be repeated here. The interested reader can refer to the existing literature.

6.3 Time-domain formulation

As seen before, the analysis of an electromagnetic system through the PEEC method is based on the EFIE, the continuity equation and the volume equivalence principle. It requires applying a volume and surface tessellation of the geometry in terms of elementary regions, hexahedra and patches, respectively, [39] over which the current and charge densities are expanded into a series of basis functions. Rectangular basis functions are adopted in the following, corresponding to assume constant current density and charge density over the elementary volume (inductive) and surface (capacitive) cells, respectively.

Conductors are modeled by their ohmic resistance (6.38), while dielectrics are modeled by the excess capacitance (6.42) modeling the excess charge due to the dielectric polarization [23]. Magnetic and electric field couplings are modeled by partial inductances $\mathbf{L}_p(s)$ in (6.40) and coefficients of potential $\mathbf{P}(s)$ in (6.30), respectively.

When (6.40) is inverse-Laplace transformed into the time-domain, it reads

$$\mathbf{e}(t) = \mathbf{L}_p(t)\frac{d}{dt}\mathbf{i}(t) \tag{6.48}$$

where the transient partial inductance between two volume cells i and n is given by

$$L_{p_{i,n}}(t) = \mathcal{L}^{-1}\left[L_{p_{i,n}}(s)\right] = \mathcal{L}^{-1}\left[\frac{\mu_0}{4\pi}\frac{1}{a_i a_n}\int_{V_i}\int_{V_n}\hat{\mathbf{u}}_i\cdot\hat{\mathbf{u}}_n\frac{e^{-s\tau}}{|\mathbf{r}_i - \mathbf{r}_n|}dV_n dV_i\right] \tag{6.49}$$

where $|\mathbf{r}_i - \mathbf{r}_n|$ is the distance between any two points in volumes V_i and V_n with a_i and a_n are their cross sections.

We have already assumed that charges are assumed to be located only on the surface of conductors and dielectrics. Surface node potentials and charges are related by the coefficients of potential matrix (6.30) which in the time-domain becomes

$$\boldsymbol{\phi}_s(t) = \mathbf{P}(t)\mathbf{q}_s(t). \tag{6.50}$$

Equation (6.50) states that node potentials are linear combination of charges residing on the surfaces of objects, either conductors or dielectrics. The coefficient of potential between two surface cells δ and γ is modeled by

$$P_{\ell,m}(t) = \mathcal{L}^{-1}\left[P_{\ell,m}(s)\right] = \mathcal{L}^{-1}\left[\frac{1}{4\pi\varepsilon_0}\frac{1}{S_\ell S_m}\int_{S_\ell}\int_{S_m}\frac{e^{-s\tau}}{|\mathbf{r}_\ell - \mathbf{r}_m|}dS_m dS_\ell\right] \tag{6.51}$$

where $|\mathbf{r}_\ell - \mathbf{r}_m|$ is the distance between any two points on surfaces ℓ and m, while S_ℓ and S_m denote the area of their respective surfaces.

If the electromagnetic problem is electrically small, e.g. $s\tau \cong 0$ and $e^{-s\tau} \cong 1$, (6.49) and (6.51) reduce to

$$L_{p_{i,n}}(t) = L_{p_{i,n}}^{(QS)}\delta(t) \tag{6.52a}$$

$$P_{\ell,m}(t) = P_{\ell,m}^{(QS)}\delta(t) \tag{6.52b}$$

where

$$L_{p_{i,n}}^{(QS)} = \frac{\mu_0}{4\pi} \frac{1}{a_i a_n} \int_{V_i} \int_{V_n} \widehat{\boldsymbol{u}}_i \cdot \widehat{\boldsymbol{u}}_n \frac{1}{|\boldsymbol{r}_i - \boldsymbol{r}_n|} dV_n dV_i \tag{6.53a}$$

$$P_{\ell,m}^{(QS)} = \frac{1}{4\pi\varepsilon_0} \frac{1}{S_\ell S_m} \int_{S_\ell} \int_{S_m} \frac{1}{|\boldsymbol{r}_\ell - \boldsymbol{r}_m|} dS_m dS_\ell. \tag{6.53b}$$

When the problem is electrically large, the interaction between far-apart volumes and surfaces is usually approximated by resorting to the so-called center-to-center approximation, assuming that $s\tau \cong sR_{cc}/c_0$. This assumption makes the exponential term constant over the integration domains. Hence, the time-domain mutual partial inductance (6.49) and mutual coefficient of potential (6.51) become

$$\begin{aligned} L_{p_{i,n}}(t) = L^{-1}\left[L_{p_{i,n}}(s)\right] &= L^{-1}\left[\frac{\mu_0}{4\pi} \frac{e^{-s\tau_{cc}}}{a_i a_n} \int_{V_i} \int_{V_n} \widehat{\boldsymbol{u}}_i \cdot \widehat{\boldsymbol{u}}_n \frac{1}{|\boldsymbol{r}_i - \boldsymbol{r}_n|} dV_n dV_i\right] \\ &= L_{p_{i,n}}^{(QS)} \delta(t - \tau_{cc}) \end{aligned} \tag{6.54a}$$

$$\begin{aligned} P_{\ell,m}(t) = L^{-1}\left[P_{\ell,m}(s)\right] &= L^{-1}\left[\frac{1}{4\pi\varepsilon_0} \frac{e^{-s\tau_{cc}}}{S_\ell S_m} \int_{S_\ell} \int_{S_m} \frac{1}{|\boldsymbol{r}_\ell - \boldsymbol{r}_m|} dS_m dS_\ell\right] \\ &= P_{\ell,m}^{(QS)} \delta(t - \tau_{cc}) \end{aligned} \tag{6.54b}$$

where $\tau_{cc} = R_{cc}/c_0$.

The voltage $e_i(t)$ induced on the volume cell i due to the time varying in the volume cell n, in the two cases, is

$$e_i^{QS}(t) = \int_{0-}^{t} L_{p_{i,n}}(\tau) \frac{di}{dt}(t - \tau) d\tau = L_{p_{i,n}}^{(QS)}) \frac{di}{dt}(t) \tag{6.55a}$$

$$e_i^{cc}(t) = \int_{0-}^{t} L_{p_{i,n}}(\tau) \frac{di}{dt}(t - \tau) d\tau = L_{p_{i,n}}^{(QS)}) \frac{di}{dt}(t - \tau_{cc}). \tag{6.55b}$$

Analogously, the potential $\phi_s(t)$ induced on the surface patch ℓ due to the charge on the patch m is

$$\phi_s^{QS}(t) = \int_{0-}^{t} P_{\ell,m}(\tau) q(t - \tau) d\tau = P_{\ell,m}^{(QS)} q(t) \tag{6.56a}$$

$$\phi_s^{cc}(t) = \int_{0-}^{t} P_{\ell,m}(\tau) q(t - \tau) d\tau = P_{\ell,m}^{(QS)} q(t - \tau_{cc}). \tag{6.56b}$$

It is to be remarked that an analytical formula for the full-wave partial inductances and coefficients of potential lying on the same plane are available and presented in [41]. Of course, it works for both the mutual and self cases.

A Taylor series-based computation of partial elements which is suited for time-domain analysis is presented in [42] and used in [43] to solve PEEC models.

6.3.1 Quasi-static PEEC formulation

In the general case of penetrable objects, it is worth partitioning the connectivity (incidence) matrix making a difference between the surface from the internal nodes. Let the discretized structure be characterized by n_b branches and n_n nodes. Among the n_b branches, let n_{bc} be the number of conductor branches and n_{bd} be the number of dielectric ones. Among the nodes, let n_{ns} be the number of surface nodes and n_{ni} that of the internal nodes, with the surface nodes numbered as first and the internal nodes appended to the surface ones. The connectivity matrix **A** is thus partitioned as

$$\underset{n_n \times n_b}{\mathbf{A}} = \begin{bmatrix} \underset{n_{ns} \times n_b}{\mathbf{A}_s} \\ \underset{n_{ni} \times n_b}{\mathbf{A}_i} \end{bmatrix}. \tag{6.57}$$

The enforcement of the KVL to the equivalent circuit obtained by the discretization of conductors [6] and dielectrics [44] and then applying the Galerkin procedure to the EFIE leads to

$$-\mathbf{A}^T\ \boldsymbol{\phi}(t) - \mathbf{L}_p^{(QS)} \frac{d\mathbf{i}(t)}{dt} - \mathbf{R}\mathbf{i}(t) - \Gamma \mathbf{v}_d(t) = \mathbf{v}_s(t) \tag{6.58}$$

where $\boldsymbol{\phi}(t)$ represents the node potentials of the reduced surface and internal nodes:

$$\boldsymbol{\phi}(t) = [\ \boldsymbol{\phi}_{sr}(t)\boldsymbol{\phi}_i(t)]^T \tag{6.59}$$

and the selection matrix Γ is

$$\Gamma = \begin{bmatrix} \underset{n_{bc} \times n_{bd}}{\mathbf{0}} \\ \underset{n_{bd} \times n_{bd}}{I} \end{bmatrix} \tag{6.60}$$

where I is the identity matrix. Matrix Γ is used to select the n_{bd} dielectric branches from the entire set of n_b branches. In addition to this, the displacement current through dielectric cells can be related to the corresponding voltage drop as

$$\Gamma^T \mathbf{i}(t) = \mathbf{C}_d \frac{d\mathbf{v}_d(t)}{dt}. \tag{6.61}$$

When two conductors and/or dielectrics are connected through some surface nodes, enforcement of equipotentiality and conservation of the charge need to be guaranteed. The equipotentiality of the coincident nodes need to be enforced but, at the same time, the conservation of the charge is to be preserved. The enforcement of equipotentiality of coincident nodes is equivalent to perform a node reduction and can be obtained by using a selection matrix **M**, defined as follows:

$$M_{ij} = \begin{cases} 1 & \text{if global node } i \text{ is coincident with reduced node } j \\ 0 & \text{otherwise.} \end{cases} \tag{6.62}$$

The reduction only attains to surface nodes. Hence, the node reduction is obtained as

$$\boldsymbol{\phi}_s(t) = \mathbf{M}\ \boldsymbol{\phi}_{sr}(t) \tag{6.63}$$

where $\boldsymbol{\phi}_s(t)$ and $\boldsymbol{\phi}_{sr}(t)$ are the vector of surface node potentials and that of reduced ones. Furthermore, the basis functions corresponding to touching panels are replaced by a new basis function relating the reduced node potential and the global charge on the touching panels. The conservation of charge can also be enforced through the matrix $\mathbf{M}$ as

$$\mathbf{q}_{sr}(t) = \mathbf{M}^T \mathbf{q}_s(t) \tag{6.64}$$

where $\mathbf{q}_s(t)$ and $\mathbf{q}_{sr}(t)$ are the vectors of the charges before and after connecting the objects. Upon substitution of (6.63) and (6.64) in (6.50), the reduced node potential and charges can be related as

$$\boldsymbol{\phi}_{sr}(t) = \underbrace{(\mathbf{M}^T \mathbf{P}^{(QS)^{-1}} \mathbf{M})^{-1}}_{\mathbf{P}_{sr}^{(QS)}} \mathbf{q}_{sr}(t) = \mathbf{P}_{sr}^{(QS)} \mathbf{q}_{sr}(t). \tag{6.65}$$

Hence, touching objects do not require to develop specialized basis functions, but only to properly enforce KCL.

The enforcement of the KCL to the n_{ns} surface nodes reads

$$\mathbf{P}_{sr}^{(QS)^{-1}} \frac{d\ \boldsymbol{\phi}_{sr}(t)}{dt} + \mathbf{G}_{le}\ \boldsymbol{\phi}_{sr}(t) - \mathbf{A}_s \mathbf{i}(t) = \mathbf{i}_s(t). \tag{6.66}$$

Since the net charge inside conductor or dielectric is zero and current sources are assumed to be connected to the surface of conductors, for the n_{bi} internal nodes, the KCL reads

$$-\mathbf{A}_i \mathbf{i}(t) = \mathbf{0}. \tag{6.67}$$

As consequence, (6.58)–(6.67) can be rewritten in the modified nodal analysis (MNA) form as

$$\begin{aligned}
&\underbrace{\begin{bmatrix} \mathbf{L}_p^{(QS)} & \mathbf{0} & \mathbf{0} & \mathbf{0} \\ \mathbf{0} & \mathbf{P}_{sr}^{(QS)^{-1}} & \mathbf{0} & \mathbf{0} \\ \mathbf{0} & \mathbf{0} & \mathbf{0} & \mathbf{0} \\ \mathbf{0} & \mathbf{0} & \mathbf{0} & \mathbf{C}_d \end{bmatrix}}_{\mathbf{C}} \frac{d\mathbf{x}(t)}{dt} = \\
&= -\underbrace{\begin{bmatrix} \mathbf{R} & \mathbf{A}_s^T & \mathbf{A}_i^T & \Gamma \\ -\mathbf{A}_s & \mathbf{G}_{le} & \mathbf{0} & \mathbf{0} \\ -\mathbf{A}_i & \mathbf{0} & \mathbf{0} & \mathbf{0} \\ -\Gamma^T & \mathbf{0} & \mathbf{0} & \mathbf{0} \end{bmatrix}}_{\mathbf{G}} \mathbf{x}(t) + \underbrace{\begin{bmatrix} I & \mathbf{0} \\ \mathbf{0} & I \\ \mathbf{0} & \mathbf{0} \\ \mathbf{0} & \mathbf{0} \\ \mathbf{0} & \mathbf{0} \end{bmatrix}}_{\mathbf{B}} \mathbf{u}(t)
\end{aligned} \tag{6.68}$$

where

$$\mathbf{x}(t) = [\mathbf{i}(t) \quad \boldsymbol{\phi}_{sr}(t) \quad \boldsymbol{\phi}_i(t) \quad \mathbf{v}_d(t)]^T \tag{6.69a}$$

$$\mathbf{u}(t) = [\,\mathbf{v}_s(t) \quad \mathbf{i}_s(t)\,]^T \tag{6.69b}$$

with $\mathbf{x}(t) \in \Re^{n_u \times 1}$, $n_u = n_b + n_n + n_{bd}$, and where $\mathbf{v}_s(t)$ and $\mathbf{i}_s(t)$ are the voltage and current sources which are applied to branches and nodes, respectively.

It is clear that this approach is unfeasible for large problems because the inversion and the explicit allocation of the full $\mathbf{P}^{(QS)}$ matrix are required.

Matrices **C** and **G** easily result to be very large when complex 3D structures are considered and the use of iterative solvers is in order [45]. Indeed, only matrix **C** has dense sub-blocks represented by the partial inductance matrix $\mathbf{L}_p^{(QS)}$ and the inverse of the reduced coefficient of potential matrix $\mathbf{P}_{sr}^{(QS)^{-1}}$, while matrix $\mathbf{C}_d$ is diagonal if pulse basis functions are used in Galerkin's procedure as expansion and testing functions. Matrix **G** can be stored as a sparse matrix since the connectivity matrices $\mathbf{A}, \mathbf{A}_s, \mathbf{A}_i$ are sparse matrices containing only $-1,+1$ coefficients; matrix **R** is diagonal if pulse basis functions are used in Galerkin's procedure as weighting and testing functions and, finally, also the selection matrix Γ is sparse by its definition (6.62). Compression techniques can be adopted for either reducing the memory storage of the dense sub-blocks and to accelerate the matrix vector products required by the iterative solution of (6.68), while avoiding to explicitly pre-computed matrix **C**. The adaptive cross approximation (ACA) algorithm [46] is well suited to compress the partial inductance matrix $\mathbf{L}_p^{(QS)}$ but cannot be efficiently used to compress the $\mathbf{P}_{sr}^{(QS)^{-1}}$ matrix [47,48]. In the next Section 6.3.2, a new formulation of the PEEC method is presented which allows to circumvent such a limitation making possible the use of the ACA algorithm to reduce the memory storage of the all dense sub-blocks and to accelerate the iterative solution of (6.68). The passivity of the quasi-static PEEC method is discussed in [49].

6.3.2 Improved MNA PEEC formulation

In order to avoid the inversion of $\mathbf{P}^{(QS)}$ and subsequently of $\mathbf{P}_{sr}^{(QS)}$, surface charges $\mathbf{q}_s$ are retained as additional unknowns. Hence, the enforcement of the KCL for the n_{ns} surface nodes reads

$$\mathbf{M}^T \frac{d\mathbf{q}_s(t)}{dt} + \mathbf{G}_{le} \; \boldsymbol{\phi}_{sr}(t) - \mathbf{A}_s \mathbf{i}(t) = \mathbf{i}_s(t). \tag{6.70}$$

The surface nodes potential to infinity $\boldsymbol{\phi}_{sr}(t)$ can be related to the charge located on the surface cells by combining (64) and (82):

$$\mathbf{M} \; \boldsymbol{\phi}_{sr}(t) = \mathbf{P}^{(QS)} \mathbf{q}_s(t). \tag{6.71}$$

By using (6.71), (6.68)–(6.69) can be rewritten in the MNA form as

$$\underbrace{\begin{bmatrix} \mathbf{L}_p^{(QS)} & \mathbf{0} & \mathbf{0} & \mathbf{0} & \mathbf{0} \\ \mathbf{0} & \mathbf{0} & \mathbf{0} & \mathbf{0} & \mathbf{M}^T \\ \mathbf{0} & \mathbf{0} & \mathbf{0} & \mathbf{0} & \mathbf{0} \\ \mathbf{0} & \mathbf{0} & \mathbf{0} & \mathbf{C}_d & \mathbf{0} \\ \mathbf{0} & \mathbf{0} & \mathbf{0} & \mathbf{0} & \mathbf{0} \end{bmatrix}}_{\mathbf{C}_{\text{NEW}}} \frac{d\mathbf{x}_{\text{NEW}}(t)}{dt} = \\ = -\underbrace{\begin{bmatrix} \mathbf{R} & \mathbf{A}_s^T & \mathbf{A}_i^T & \Gamma & \mathbf{0} \\ -\mathbf{A}_s & \mathbf{G}_{le} & \mathbf{0} & \mathbf{0} & \mathbf{0} \\ -\mathbf{A}_i & \mathbf{0} & \mathbf{0} & \mathbf{0} & \mathbf{0} \\ -\Gamma^T & \mathbf{0} & \mathbf{0} & \mathbf{0} & \mathbf{0} \\ \mathbf{0} & -\mathbf{M} & \mathbf{0} & \mathbf{0} & \mathbf{P}^{(QS)} \end{bmatrix}}_{\mathbf{G}_{\text{NEW}}} \mathbf{x}_{\text{NEW}}(t) + \underbrace{\begin{bmatrix} I & \mathbf{0} \\ \mathbf{0} & I \\ \mathbf{0} & \mathbf{0} \\ \mathbf{0} & \mathbf{0} \\ \mathbf{0} & \mathbf{0} \end{bmatrix}}_{\mathbf{B}_{\text{NEW}}} \mathbf{u}(t) \tag{6.72}$$

where $\mathbf{x}_{\text{NEW}}(t) = [\mathbf{i}(t) \quad \boldsymbol{\phi}_{sr}(t) \quad \boldsymbol{\phi}_i(t) \quad \mathbf{v}_d(t) \quad \mathbf{q}_s(t)]^T \in \Re^{n_u \times 1}$ with $n_u = n_b + n_n + n_p + n_{bd}$.

It is also worth to observe that the proposed PEEC formulation, based on separate charge and current basis functions, demonstrates favorable low-frequency behavior, as confirmed in [50,51].

Hence, in the proposed approach, the charges are maintained as unknowns and an additional set of equations relating potentials to charges through the coefficient of potential matrix $\mathbf{P}^{(QS)}$ is introduced. This allows to avoid the inversion of the $\mathbf{P}^{(QS)}$ matrix. Since $\mathbf{P}^{(QS)}$ no longer needs inverting, it can now be directly generated in compressed form. In this case also, matrices $\mathbf{C}_{\text{NEW}}$ and $\mathbf{G}_{\text{NEW}}$ are mostly sparse (matrix $\mathbf{M}$ is sparse by its definition (6.60)).

It is finally worth to mention that the MNA formulation (6.72) for the quasi-static PEEC method has been extended to consider lossy and dispersive media in [52,53], and to silicon and graphene in [54] and [55], respectively.

6.3.3 Delayed PEEC formulation

In this section, the introduction of propagation delays in the system equation (6.72) is presented. The coefficients in the matrices $\mathbf{L}_p^{(QS)}$ and $\mathbf{P}^{(QS)}$ are referred frequently as mutual (off-diagonal elements) or self (diagonal elements). Because of the finite value of the propagation speed, mutual elements are characterized by a certain delay. For each element, the propagation delay τ is computed as

$$\tau = \frac{d}{c} \tag{6.73}$$

where d is the distance between the cells and c is the propagation speed. In this context, distance d is the center to center separation between the cells involved in the mutual coupling phenomenon.

This means that, for the partial inductance $\mathbf{L}_{p,ij}^{(QS)}$, $d_{L_{ij}}$ is the distance between the centers of two volumes i and j, whereas for the coefficient of potential $\mathbf{P}_{ij}^{(QS)}$, $d_{P_{ij}}$ is the distance between the centers of surfaces i and j.

Based on what stated above, when delays are introduced in the system (6.72), the main effects are summarized as follows:

- $\mathbf{L}_p^{(QS)}$ has to be replaced with the instantaneous partial inductance matrix $\mathbf{L}_p^{(nd)}$ described as follows:

$$\begin{cases} \mathbf{L}_{p,ij}^{(nd)} = \mathbf{L}_{p,ij}^{(QS)} & \text{if } \dfrac{d_{L_{ij}}}{c} < \Delta t \\ \mathbf{L}_{p,ij}^{(nd)} = 0 & \text{if } \dfrac{d_{L_{ij}}}{c} \geq \Delta t \end{cases} \tag{6.74}$$

Δt being the time step. This is equivalent to neglect delays between two point when it is smaller than the time step. In Section 6.3.4, temporal basis functions are introduced which avoid or at least mitigate this effect.

- $\mathbf{P}^{(QS)}$ has to be replaced with the instantaneous potential matrix $\mathbf{P}_{nd}$ described as follows:

$$\begin{cases} \mathbf{P}_{ij}^{(nd)} = \mathbf{P}_{ij}^{(QS)} & \text{if } \dfrac{d_{P_{i,j}}}{c} < \Delta t \\ \mathbf{P}_{ij}^{(nd)} = 0 & \text{if } \dfrac{d_{P_{i,j}}}{c} \geq \Delta t. \end{cases} \tag{6.75}$$

- For each multiple of Δt, two matrices can be defined as

$$\mathbf{L}_p^{(d,k)} : \begin{cases} \mathbf{L}_{p,ij}^{(d,k)} = \mathbf{L}_{p,ij}^{(QS)} & \text{if } \dfrac{d_{L_{i,j}}}{c} \in [k \cdot \Delta t, (k+1)\Delta t) \\ \mathbf{L}_{p,ij}^{(nd)} = 0 & \text{otherwise} \end{cases} \tag{6.76a}$$

$$\mathbf{P}^{(d,k)} : \begin{cases} \mathbf{P}_{ij}^{(d,k)} = \mathbf{P}_{ij}^{(QS)} & \text{if } \dfrac{d_{P_{i,j}}}{c} \in [k \cdot \Delta t, (k+1)\Delta t) \\ \mathbf{P}_{ij}^{(d,k)} = 0 & \text{otherwise.} \end{cases} \tag{6.76b}$$

Those matrices, for $k = 1, \ldots, \lfloor \tau_{max}/\Delta t \rfloor$, relate the coupling effects in each point to the past values of the unknown in the others.

It is worth pointing out that for deeply disconnected geometries (as the dipoles in Section 6.5), the system delays resulting in the time-domain solver might be discontinuous. In those cases, there is a large amount of values of k for which the matrices $\mathbf{L}_p^{(d,k)}$ and $\mathbf{P}^{(d,k)}$ are null. For this reason, in practical implementations, it is a good choice to use two vectors τ_L and τ_P containing all the delays ordered from the smallest one to the highest one. It is important to know that at each delay can be associated to more than one off-diagonal element, due to the structure symmetries. It is straightforward to realize that the following identities always hold:

$$\mathbf{L}_p^{(QS)} = \mathbf{L}_p^{(nd)} + \sum_{k=1}^{\lfloor \tau_{L,max}/\Delta t \rfloor} \mathbf{L}_p^{(d,k)} \tag{6.77}$$

and

$$\mathbf{P}^{(QS)} = \mathbf{P}^{(nd)} + \sum_{k=1}^{\lfloor \tau_{P,max}/\Delta t \rfloor} \mathbf{P}^{(d,k)}. \tag{6.78}$$

The modifications introduced in this section affect directly only two equations in (6.72), the first and the last ones, sence they are the only involving partial inductances and coefficients of potential. In order to better clarify how they change when delays are taken into account, let us write them before the introduction of propagation delays:

$$\mathbf{L}_p^{(QS)} \frac{d\mathbf{i}(t)}{dt} + \mathbf{A}_s^T \ \boldsymbol{\phi}_{st}(t) + \mathbf{A}_i^T \ \boldsymbol{\phi}_i(t) + \mathbf{R}\mathbf{i}(t) + \Gamma \mathbf{v}_d(t) = \mathbf{v}_s(t) \tag{6.79a}$$

$$-\mathbf{M} \ \boldsymbol{\phi}_{st}(t) + \mathbf{P}^{(nd)} \mathbf{q}_s(t) = \mathbf{0}. \tag{6.79b}$$

For the sake of clarity, let us consider the *i*th row for each of them

$$\sum_{j=1}^{n_b} L_{p_{ij}}^{(QS)} \frac{d\mathbf{i}_j(t)}{dt} + \boldsymbol{\phi}_h(t) - \boldsymbol{\phi}_k(t) + \mathbf{R}_i \mathbf{i}_i(t) + \mathbf{v}_{d,i}(t) = \mathbf{v}_{s,i}(t) \tag{6.80a}$$

$$\sum_{j=1}^{n_p} P_{ij}^{(QS)} \mathbf{q}_{s,j}(t) = \boldsymbol{\phi}_{st,i}(t). \tag{6.80b}$$

If delays are included, (6.80) become

$$\sum_{j=1}^{n_b} L_{p_{ij}}^{(QS)} \frac{d\mathbf{i}_j\left(t - \frac{d_{L_{i,j}}}{c}\right)}{dt} + \varphi_h(t) - \varphi_k(t) + \mathbf{R}_i \mathbf{i}_i(t) + \mathbf{v}_{d,i}(t) = \mathbf{v}_{s,i}(t) \tag{6.81a}$$

$$\sum_{j=1}^{n_p} P_{ij}^{(QS)} \mathbf{q}_{s,j}\left(t - \frac{d_{P_{i,j}}}{c}\right) = \boldsymbol{\phi}_{st,i}(t). \tag{6.81b}$$

In the discrete time-domain and recovering to the matrix notation, we obtain

$$\begin{aligned} &\mathbf{L}_p^{(nd)} \frac{d\mathbf{i}(n \cdot \Delta t)}{dt} + \mathbf{A}_s^T \ \boldsymbol{\phi}_{st}(n \cdot \Delta t) + \mathbf{A}_i^T \ \boldsymbol{\phi}_i(n \cdot \Delta t) + \mathbf{R}\mathbf{i}(n \cdot \Delta t) + \Gamma \mathbf{v}_d(n \cdot \Delta t) = \\ &= \mathbf{v}_s(n \cdot \Delta t) - \sum_{k=1}^{\lfloor \tau_{L,max}/dt \rfloor} \mathbf{L}_p^{(d,k)} \frac{d\mathbf{i}((n-k) \cdot \Delta t)}{dt} \end{aligned} \tag{6.82a}$$

$$-\mathbf{M} \ \boldsymbol{\phi}_{st}(n \cdot \Delta t) + \mathbf{P}^{(nd)} \mathbf{q}_s(n \cdot \Delta t) = -\sum_{k=1}^{\lfloor \tau_{P,max}/\Delta t \rfloor} \mathbf{P}^{(d,k)} \mathbf{q}_s((n-k) \cdot \Delta t). \tag{6.82b}$$

From (6.82), it can be seen as the delayed portion of the system can be handled as a perturbation of the right-hand side (RHS) which is updated at each time step as the system evolves.

The time-domain sensitivity analysis of delayed PEEC models has been presented in [56].

6.3.4 Temporal basis functions

Coupling delays are achieved by dividing the distances between each couple of elementary inductive (volume) cells, or capacitive (surface) cells, by the speed of light [57]. Based on how they are computed, it is evident that delays are not related to the time step. More importantly it is not possible to pick a time step which is a sub-multiple of all the delays computed for the problem, as a result some sort of interpolation has to be applied. There are several ways to perform the interpolation, one consists in applying temporal basis functions which act as interpolators in time stepping. Time basis functions can be described as

$$T_n(t) = T(t - n\Delta t). \tag{6.83}$$

Independently from the expression used for the basis function, the following identity must hold

$$\sum_{n=-\infty}^{+\infty} T_n(t) = 1 \ \forall t \in \Re. \tag{6.84}$$

The basis function is used to expand the current time derivatives and charges, used in the system described in Section 6.3.3.

For the jth delayed term, it reads, respectively:

$$\mathbf{L}_p^{(d,j)} \frac{d\mathbf{i}((n-j)\cdot\Delta t)}{dt} \quad \rightarrow \quad \sum_{h=0}^{N_w-1} \mathbf{L}_p^{(d,j)} \frac{\delta\mathbf{i}((n-j-h)\cdot\Delta t)}{\delta t} T_j(\tau_L) \tag{6.85a}$$

$$\mathbf{P}^{(d,j)}\mathbf{q}_s((n-j)\cdot\Delta t) \quad \rightarrow \quad \sum_{h=0}^{N_w-1} \mathbf{P}^{(d,j)}\mathbf{q}_s((n-j-h)\cdot\Delta t) T_j(\tau_P) \tag{6.85b}$$

where N_w represents the width of the basis function and it is the number of Δt for which the temporal basis function is $\neq 0$. $\delta(\cdot)/\delta t$ is the discrete derivative function applied to the current. Equations (6.82) can be also seen as the ones resulting from a rectangular basis function with time support equal to Δt.

6.3.4.1 Triangular basis functions

The first basis function can be seen as linear interpolation in time-domain for the delays between the cells, the triangular basis functions [58]. It is very intuitive since it looks coherent with the physics which lies behind the propagation, each cells is traversed by the electromagnetic radiation progressively. The triangular shape has also been chosen because in general, it performs well considering the late time instability [59]. The support of such a function is two time steps; as a result the

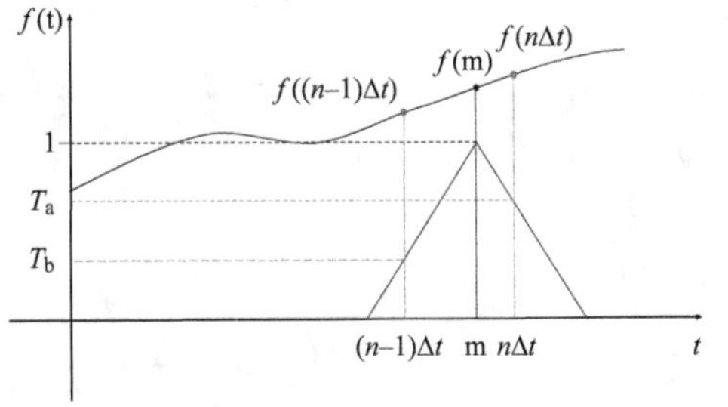

(a) Linear interpolation using triangular basis function of equation (1.86b).

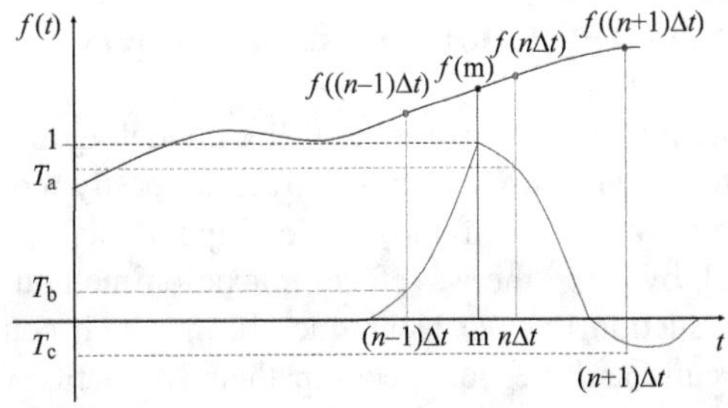

(b) Linear interpolation using polynomial basis function of equation (1.87b).

Figure 6.3 Linear interpolation using triangular and polynomial basis functions. © 2019 IET. Reprinted with permission from C. Gianfagna, L. Lombardi, G. Antonini, Marching-on-in-Time Solution of Delayed PEEC Models with Improved Temporal Basis Functions, IET-MAP, Vol. 13, no. 1, pp. 42–47, January 2019

interpolation will be performed between two consecutively stored samples for the variable [59] along the time axis (see Figure 6.3a).

$$T_n(t) = \begin{cases} 1 - \dfrac{|t - n\Delta t|}{\Delta t} & -\Delta t \le t - n\Delta t \le \Delta t \\ 0 & \text{otherwise} \end{cases} \tag{6.86a}$$

$$f(m) \simeq T_a f((n-1)\Delta t) + T_b f(n\Delta t). \tag{6.86b}$$

6.3.4.2 Polynomial basis functions

In order to provide additional alternatives, the polynomial basis function (see Figure 6.3b) adopted in [60] is also taken into account since it was proposed to improve the stability of delayed solvers and it seem to be well received in literature.

The function expression is

$$T_n(t) = \begin{cases} \dfrac{t^2}{2\Delta t^2} + \dfrac{3t}{2\Delta t} + 1 & -\Delta t \leq t - n\Delta t < 0 \\ -\dfrac{t^2}{\Delta t^2} + 1 & 0 \leq t - n\Delta t < \Delta t \\ \dfrac{t^2}{2\Delta t^2} - \dfrac{3t}{2\Delta t} + 1 & \Delta t \leq t - n\Delta t \leq 2\Delta t \\ 0 & \text{otherwise} \end{cases} \tag{6.87a}$$

$$f(m) \simeq T_a f((n-1)\Delta t) + T_b f(n\Delta t) + T_c f((n+1)\Delta t). \tag{6.87b}$$

6.4 Model order reduction of PEEC models

The equivalent circuits resulting from the PEEC modeling are typically very large and prohibitive storage and computing resources are easily required. Their analysis can be accelerated by means of the fast multipole method and multi-function techniques [61,62], by using the waveform relaxation method [63] or by resorting to model order reduction (MOR) techniques [64], which could reduce the complexity of the circuits and, hence, speed-up their simulation in the time-domain. Over the past three decades, Krylov subspace-based techniques have emerged as one of the most powerful tools for reduced-order modeling of large-scale systems [65]. Among the others, the PRIMA algorithm [66] has been found to be very effective to reduce linear time-invariant systems.

MOR techniques can be applied to both quasi-static and delayed PEEC models. The two cases are presented in the following sections.

6.4.1 MOR of quasi-static PEEC models

From Section 6.3.1, we have seen that the enforcement of KVL and KCL to a quasi-static PEEC circuit leads to the following set of first-order differential equations in a descriptor form:

$$\mathbf{C}\frac{d}{dt}\mathbf{x}(t) = -\mathbf{G}\mathbf{x}(t) + \mathbf{B}\mathbf{u}(t) \tag{6.88a}$$

$$\mathbf{y}(t) = \mathbf{L}^T\mathbf{x}(t) \tag{6.88b}$$

where $\mathbf{y}(t)$ represents the vector of the outputs that are assumed as a linear combination of the PEEC circuit quantities $\mathbf{x}(t)$. It is often the case we are interested into the currents and voltages at the same ports excited by the voltage and current sources, $\mathbf{v}_s(t)$ and $\mathbf{i}_s(t)$ respectively. In this case, it is easy to verify that $\mathbf{L} = \mathbf{B}$.

Basically, the reduced-order model is obtained by computing a projection matrix $\mathbf{K} \in \Re^{n_u \times n_r}$, of rank r such that

$$\mathbf{G}_r = \mathbf{K}^T\mathbf{G}\,\mathbf{K} \tag{6.89a}$$

```
Procedure: PRIMA(G, C, B, s0, q)
Output: K
1  solve (G+ s0C)U = B for U
2  (K0,T) = QR(U) //thin QR factorization of U
3  r = floor(q/np)
4  for i = 1 to r − 1 do
5      solve (G+ s0C)U = CKi−1 for U
6      for j = 1 to i do
7          U = U − Kj−1 Kj−1^T U
8      end for
9      (Ki,T) = QR(U) //thin QR factorization of U
10 end for
11 K = [K0, K1,..., Kr−1]
```

Figure 6.4 Pseudocode of the PRIMA algorithm to reduce quasi-static PEEC models

$$\mathbf{C}_r = \mathbf{K}^T\mathbf{C}\,\mathbf{K} \tag{6.89b}$$

$$\mathbf{B}_r = \mathbf{K}^T\mathbf{B}. \tag{6.89c}$$

Hence, the reduced-order model is obtained as a projection of the n_u-dimensional state space of the original system into the n_r-dimensional subspace spanned by the columns of the matrix **K**, provided that $n_r << n_u$.

The projection matrix **K** can be obtained by applying directly the PRIMA algorithm [66] or other MOR techniques to the PEEC equations (6.88) [67,68]. The pseudocode of such procedure is reported in Figure 6.4. The input parameters are represented by the matrices **G**, **C**, and **B** defined in Section 6.3.1, the expansion point s_0 and the order of the approximation q. As first step, the matrix **U** is determined as solution of the linear system $(\mathbf{G} + s_0\mathbf{C})\mathbf{U} = \mathbf{B}$ (line 1) and, then, the first n_p columns of the projection matrix **K** are computed by using the *thin QR factorization* of matrix **U** (line 2). Then, additional columns of the projection matrix are computed recursively (lines 3–10), leading to the full projection matrix **K** (the function floor (x), invoked at line 3, returns the first integer smaller than x).

More recently, a partitioned model order reduction technique, exploiting the block structure of matrices **C** and **G** and the low-rank behavior of magnetic and electric field couplings, has been presented in [69].

6.4.2 MOR of delayed PEEC models

If the center to center approximation (6.54) is adopted to compute the partial elements, the MNA form of PEEC models can be obtained by (6.81a), (6.70), and (6.81b). In a matrix form, it reads

$$\mathbf{C}(\tau)\frac{d\mathbf{x}(t)}{dt} = -\mathbf{G}(\tau)\mathbf{x}(t) + \mathbf{B}\mathbf{u}(t) \tag{6.90a}$$

$$\mathbf{y}(t) = \mathbf{L}^T\mathbf{x}(t) \tag{6.90b}$$

where $\tau \in \Re^{n_\tau \times 1}$ contains all delays τ_L, τ_P, $n_\tau = n_{\tau_L} + n_{\tau_P}$. These equations constitute a set of neutral delayed differential equations (NDDE), with delays not only in the state vector $\mathbf{x}(t)$ but also in its time derivative $d\mathbf{x}(t)/dt$. Each delayed entry of matrices $\mathbf{C}(\tau)$ and $\mathbf{G}(\tau)$ act as a delay operator for the corresponding entry of vector $\mathbf{x}(t)$. Hence, (6.90) can be re-written in the Laplace domain as:

$$s\mathbf{C}(s)\mathbf{X}(s) = -\mathbf{G}(s)\mathbf{X}(s) + \mathbf{B}\mathbf{U}(s) \tag{6.91a}$$

$$\mathbf{Y}(s) = \mathbf{B}^T\mathbf{X}(s) \tag{6.91b}$$

$$\mathbf{C}(s) = \mathbf{C}_0 + \sum_{k=1}^{n_\tau} \mathbf{C}_k e^{-s\tau_k} \tag{6.91c}$$

$$\mathbf{G}(s) = \mathbf{G}_0 + \sum_{k=1}^{n_\tau} \mathbf{G}_k e^{-s\tau_k}. \tag{6.91d}$$

In this case, a reduced order model can be obtained by performing a single-point expansion reduction. Denoting the orthogonal basis as $\mathbf{K} \in \Re^{n_u \times n_r}$, the reduced order system is given by

$$s\mathbf{C}_r(s)\mathbf{X}_r(s) = -\mathbf{G}_r(s)\mathbf{X}_r(s) + \mathbf{B}_r\mathbf{U}(s) \tag{6.92a}$$

$$\mathbf{Y}(s) = \mathbf{B}_r^T\mathbf{X}_r(s) \tag{6.92b}$$

$$\mathbf{C}_r(s) = \mathbf{C}_{r,0} + \sum_{k=1}^{n_\tau} \mathbf{C}_{r,k} e^{-s\tau_k} \tag{6.92c}$$

$$\mathbf{G}_r(s) = \mathbf{G}_{r,0} + \sum_{k=1}^{n_\tau} \mathbf{G}_{r,k} e^{-s\tau_k} \tag{6.92d}$$

where the following congruence transformations are used

$$\mathbf{C}_{r,i} = \mathbf{K}^T\mathbf{C}_i\mathbf{K}, \quad i = 0, \ldots, n_\tau \tag{6.93a}$$

$$\mathbf{G}_{r,i} = \mathbf{K}^T\mathbf{G}_i\mathbf{K}, \quad i = 0, \ldots, n_\tau \tag{6.93b}$$

$$\mathbf{B}_r = \mathbf{K}^T\mathbf{B} \tag{6.93c}$$

$$\mathbf{L}_r = \mathbf{K}^T\mathbf{L} \tag{6.93d}$$

and $\mathbf{X}_r(t)$ is a vector containing the state variables in the reduced domain. The crucial step in obtaining a compact and accurate reduced order model over a wide frequency range is represented by the construction of the orthogonal basis $\mathbf{K}$. The Arnoldi algorithm [66] can be adopted to this purpose, since it is numerically reliable and robust. The numerical robustness of the Arnoldi algorithm stems from its capability to obtain the orthogonal basis for the transfer function moments without computing these moments explicitly. Unfortunately, adapting the Arnoldi algorithm for NDDE systems is not a straightforward task, and therefore the original NDDE system has to be transformed in a suitable form to apply standard Arnoldi-based reduction. To address this issue, an equivalent first-order system is computed by expanding the exponential factors $e^{-s\tau_k}$ in a Taylor series form and using a companion form [70]. Then, the Arnoldi algorithm is applied to the first order equivalent system, allowing the computation of the corresponding orthogonal projection matrix **K** for the original NDDE system is extracted and a reduced NDDE system is obtained [70]. This single-point expansion-based MOR algorithm for NDDE systems was proposed assuming $s = 0$ as an expansion point. If any other expansion point $s = s_1, \ s_1 \neq 0$ is selected, setting $s = s_1 + \sigma$, where s_1 is a frequency shift and σ is the new Laplace variable, the NDDE system (6.91a)–(6.91b) becomes

$$\sigma\widehat{\mathbf{C}}(\sigma)\mathbf{X}(\sigma) = -\widehat{\mathbf{G}}(\sigma)\mathbf{X}(\sigma) + \mathbf{BU}(\sigma) \tag{6.94a}$$

$$\mathbf{Y}(\sigma) = \mathbf{B}^T\mathbf{X}(\sigma) \tag{6.94b}$$

$$\widehat{\mathbf{C}}(\sigma) = \widehat{\mathbf{C}}_0 + \sum_{k=1}^{n_\tau} \widehat{\mathbf{C}}_k e^{-\sigma\tau_k} \tag{6.94c}$$

$$\widehat{\mathbf{G}}(\sigma) = \widehat{\mathbf{G}}_0 + \sum_{k=1}^{n_\tau} \widehat{\mathbf{G}}_k e^{-\sigma\tau_k} \tag{6.94d}$$

where

$$\widehat{\mathbf{C}}_0 = \mathbf{C}_0 \tag{6.95a}$$

$$\widehat{\mathbf{G}}_0 = \mathbf{G}_0 + s_1\mathbf{C}_0 \tag{6.95b}$$

$$\widehat{\mathbf{C}}_k = \mathbf{C}_k e^{-s_1\tau_k}, \quad k = 1, \ldots, n_\tau \tag{6.95c}$$

$$\widehat{\mathbf{G}}_k = (\mathbf{G}_k + s_1\mathbf{C}_k)e^{-s_1\tau_k}, \quad k = 1, \ldots, n_\tau \tag{6.95d}$$

and the algorithm described in [70] can be applied. This single-point MOR approach [70] is able to preserve the NDDE formulation, but it may be not able to reduce NDDE systems with large delays, since the reduction of equivalent first-order systems becomes computationally expensive and sometimes unfeasible since exponential terms with large delays need many terms in the Taylor expansion to be accurately approximated. A valuable alternative is represented by the multipoint expansion [71] that is able to accurately reduce NDDE systems with large delays, since a small expansion Taylor order can be used for each expansion point and the accuracy of the reduced model is increased by adding new expansion points. Assuming that the order of the Taylor expansion is fixed for each expansion point, an adaptive algorithm is used to choose the expansion points. Following [70], the NDDE formulation is preserved in the reduced model. At each expansion point, the MOR algorithm described in [70] is applied and the corresponding projection matrix $\mathbf{K}_i, \ i = 1, \ldots, n_{points}$ is computed, where n_{points} denotes the number of expansion points. The final projection matrix **K** is based on the orthogonalization of the stack column collection of all single expansion point projection matrices. The computation of **K** in the case of multipoint expansion is explained through the

```
Procedure: MULTI-POINT(NDDEorig, {sk}_{k=1}^{n_points}, nr, q)
Output: K, L
K1 = ONE-POINT-DMOR(NDDEorig, s1, nr, q)
K = K1
for k = 2 to n_points do
K_k = ONE-POINT-DMOR(NDDEorig, s_k, n_r, q)
K = ([K, K_k])
end for
L = orth(K)
```

Figure 6.5 Pseudocode for computing the orthogonal basis **K** *in the multipoint expansion case*

algorithm given in Figure 6.5. The MOR algorithm described in [70] is called ONE-POINT-DMOR in Figure 6.5, where n_r represents the reduced order for each expansion point. Once the projection matrix **K** is computed, it is applied to the original NDDE system (6.91a) and (6.91b) and a reduced NDDE system (6.92a) and (6.92b) is obtained [72]. Parameterized model order reduction of both quasi-static and delayed PEEC models have been proposed in [73–76]. Parameterized reduced order models for sensitivity analysis of PEEC circuits have been presented in [77,78].

6.5 Examples

In this section, three examples of usage of the delayed PEEC formulation are presented. In the first one, Section 6.5.1, a stability comparison between different basis functions is shown. Then the proposed method and a commercial software have been used for both conduction based (Sections 6.5.2 and 6.5.3), as well as radiation based (Section 6.5.4) electromagnetic structures. All the PEEC time-domain simulations in the following examples have been carried out by backward difference scheme of the second order (BD2).

6.5.1 Stability analysis

In this example, the aim is to verify if any of the basis functions explored shows the ability to improve the stability while time stepping. In this test, the simulation of the dipoles shown in Figure 6.6 will be carried out by using different temporal basis functions: rectangular, polynomial and triangular. To address this issue, several time-domain simulations have been performed with different time steps.

Length, width and thickness of the dipoles are $\ell = 10$ cm $w = 1$ mm and $t = 1$ mm respectively; the gap between the dipole conductors is $g = 1$ mm and the distance from transmitter to receiver is $d = 30$ cm. Both the dipoles are terminated on a 73 Ω resistance. The first one is excited with the signal shown in Figure 6.7.

As it can be noticed from the results, the triangular basis function resulted to be always the most stable among the three, Figures 6.8. This result is confirmed even

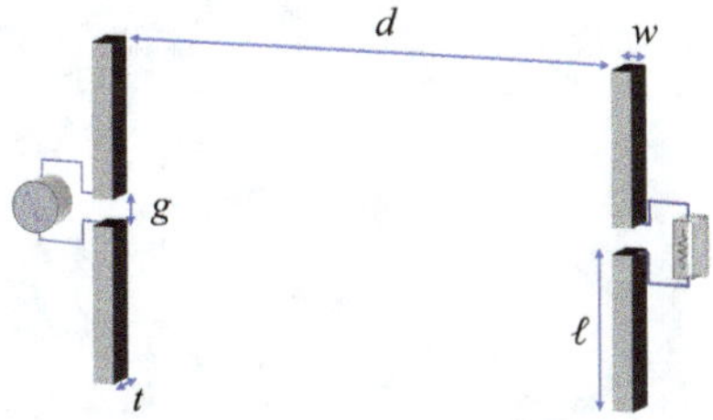

Figure 6.6 Dipole antennas. ©2019 IET. Reprinted with permission from C. Gianfagna, L. Lombardi, G. Antonini, Marching-on-in-Time Solution of Delayed PEEC Models with Improved Temporal Basis Functions, IET-MAP, Vol. 13, no. 1, pp. 42–47, January 2019

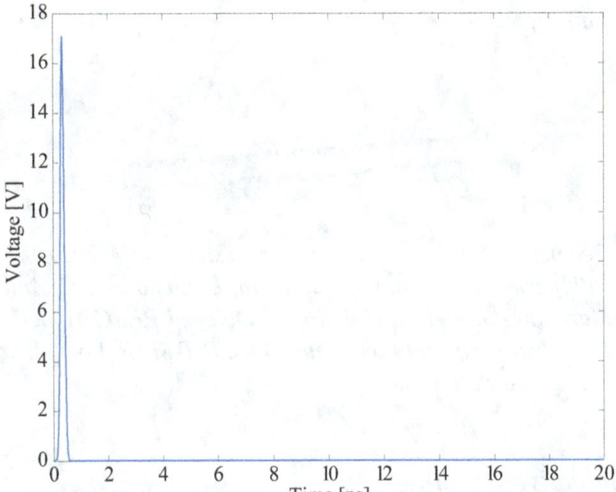

Figure 6.7 Excitation signal at the port of the transmitting dipole. ©2019 IET. Reprinted with permission from C. Gianfagna, L. Lombardi, G. Antonini, Marching-on-in-Time Solution of Delayed PEEC Models with Improved Temporal Basis Functions, IET-MAP, Vol. 13, no. 1, pp. 42–47, January 2019

for different geometrical setups. It can also be noticed that in this test the adoption of a polynomial basis functions provides improvements in the stability of the simulation if compared with the rectangular basis function, even if not as much as the triangular one.

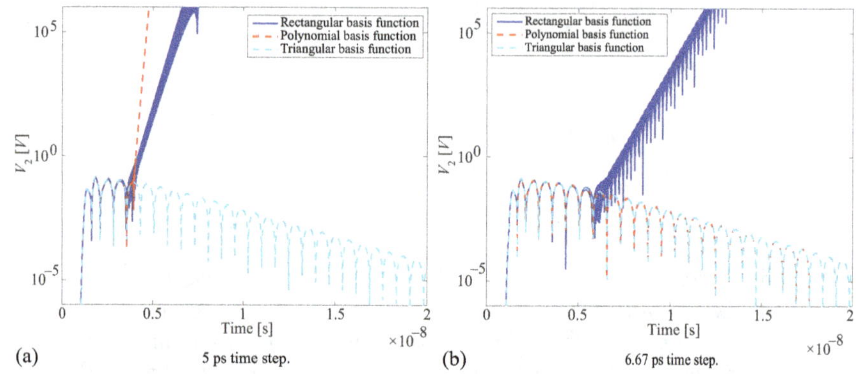

Figure 6.8 Stability test for two different time steps. ©2019 IET. Reprinted with permission from C. Gianfagna, L. Lombardi, G. Antonini, Marching-on-in-Time Solution of Delayed PEEC Models with Improved Temporal Basis Functions, IET-MAP, Vol. 13, no. 1, pp. 42–47, January 2019

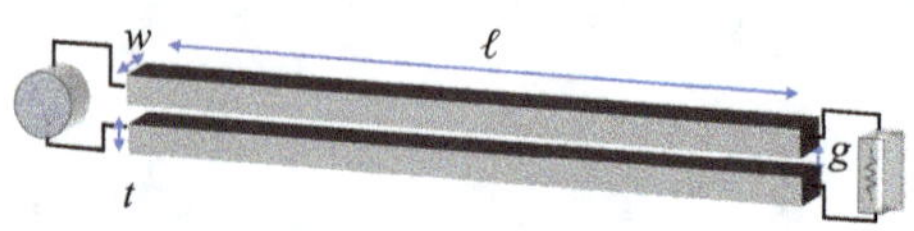

Figure 6.9 Geometry of two-conductor transmission line. ©2019 IET. Reprinted with permission from C. Gianfagna, L. Lombardi, G. Antonini, Marching-on-in-Time Solution of Delayed PEEC Models with Improved Temporal Basis Functions, IET-MAP, Vol. 13, no. 1, pp. 42–47, January 2019

6.5.2 Conductive coupling

Marching-on-in-time solvers are usually tested on scattering problems involving high frequencies and with no *DC* component. However, time-domain analysis are also very interesting in interconnect modeling with broadband excitations, from *DC* to up to the highest frequency in the bandwidth. In this test, the two-conductor transmission line shown in Figure 6.9 is considered. The input port of the transmission line is excited by a Gaussian pulse with a peak of 7 V, a rise time of 0.1 ns and a width of 0.5 ns.

The length, width, and thickness of the conductors are $\ell = 5$ cm, $w = 150$ μm, and $t = 100$ μm, respectively; the gap between the conductors is $g = 300$ μm. The structure under test is terminated on a 10 Ω resistance at the source end and on a 20 Ω resistance at the far end.

The results achieved by the proposed delayed PEEC method are shown in Figure 6.10. It is possible to see how the results of the proposed method match well the ones obtained with a full wave simulation based on the FDTD method.

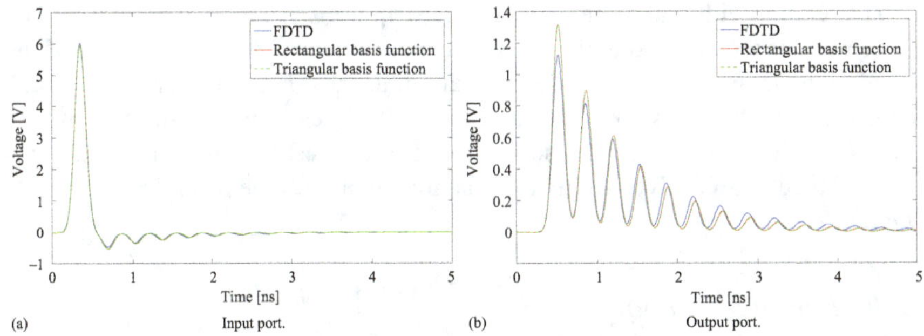

Figure 6.10 Transient port voltages of the transmission line (example 5.2). ©2019 IET. Reprinted with permission from C. Gianfagna, L. Lombardi, G. Antonini, Marching-on-in-Time Solution of Delayed PEEC Models with Improved Temporal Basis Functions, IET-MAP, Vol. 13, no. 1, pp. 42–47, January 2019

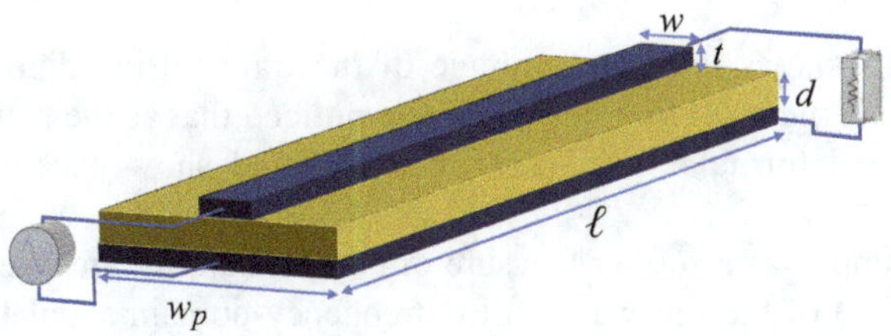

Figure 6.11 Geometry of the microstrip line. © 2019 IET. Reprinted with permission from C. Gianfagna, L. Lombardi, G. Antonini, Marching-on-in-Time Solution of Delayed PEEC Models with Improved Temporal Basis Functions, IET-MAP, Vol. 13, no. 1, pp. 42–47, January 2019

For the PEEC solver with the adopted Δt, the system has 101 delayed matrices for potential coefficients and 165 delayed coefficient matrices for partial inductance with a total of 17,380 unknowns.

Looking carefully at Figure 6.10, it can be appreciated how, in the case of conductive coupling, the use of delays does not affect significantly the results. This happens because, when terminations generate a *DC* path, the propagation is dominated by the conductive behavior and delays do not impact significantly the results.

6.5.3 Microstrip line

A second test for the modeling of conductive coupling makes use of a microstrip geometry. A dielectric substrate is placed between the microstrip and the ground, as shown in Figure 6.11. The input has form of a Gaussian pulse with a peak of 7 V, a rise time of 0.2 ns and a width of 1 ns.

The length, width, and thickness of the strip are $\ell = 3$ cm, $w = 0.6$ mm and $t = 0.1$ mm respectively; the dielectric thickness is $d = 0.8$ mm; the width of the ground plane is $w_p = 6.6$ mm. The structure under test is finally terminated on a 20 Ω resistance at the source end and on a 50 Ω resistance at the far end. In Figure 6.12, it can be seen as, also in this case, the results obtained by the proposed delayed PEEC solver are in a good agreement with the results of the FDTD solver.

6.5.4 Radiated coupling

The radiated coupling has been tested by using the two dipoles shown in Figure 6.6, already used for the stability test. Dipoles are terminated exactly as in Section 6.5.1. Even in this case, in order to perform the comparison, the input port of the first dipole is excited by a current source having a standard Gaussian pulse behavior with a pick of 17 V, a rise time of 0.1 ns and a width of 0.2 ns. In this case, the system related to the proposed PEEC solver presents 89 delayed coefficients for both potentials and partial inductances, the number of unknowns is 4,016.

Figure 6.13 shows the input voltage of the transmitting dipole and at the output voltage of the receiving one. It can be noticed that as the results achieved with the Delayed formulation of the PEEC method show excellent matching with FDTD.

In this example, as a further measure of comparison, the Inverse Fast Fourier Transform (IFFT) of the full-wave PEEC frequency domain simulation based on center to center delays has been included. By looking at the results, even the frequency domain simulation matches well with FDTD and PEEC results.

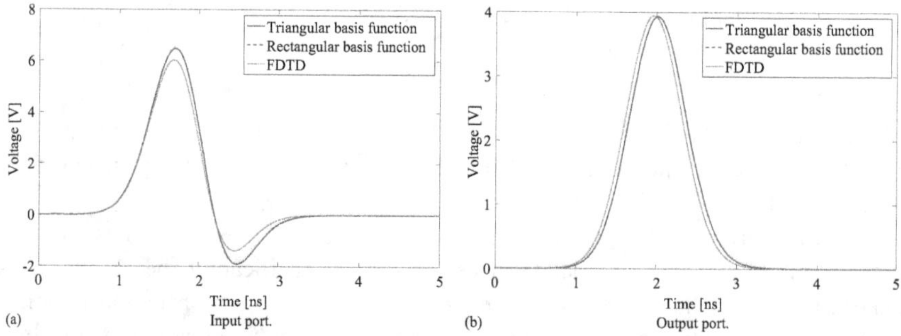

Figure 6.12 Transient port voltages of the microstrip (example 5.3). © 2019 IET. Reprinted with permission from C. Gianfagna, L. Lombardi, G. Antonini, Marching-on-in-Time Solution of Delayed PEEC Models with Improved Temporal Basis Functions, IET-MAP, Vol. 13, no. 1, pp. 42–47, January 2019

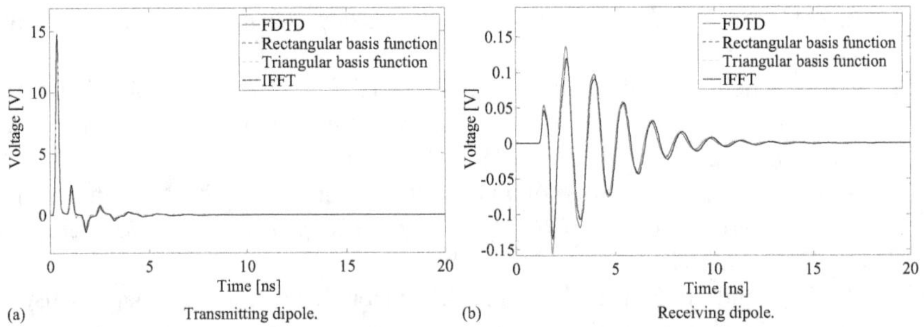

Figure 6.13 Transient port voltages of the dipole antennas (example 6.5.4). © 2019 IET. Reprinted with permission from C. Gianfagna, L. Lombardi, G. Antonini, Marching-on-in-Time Solution of Delayed PEEC Models with Improved Temporal Basis Functions, IET-MAP, Vol. 13, no. 1, pp. 42–47, January 2019

6.6 Conclusions

The partial element equivalent method is born in the 1970s and continues to evolve with various improvements and major breakthroughs. The request for more accurate modeling in electrical design has pushed forward the development of PEEC technique. This chapter gives an overview of the time-domain PEEC formulations. Among them, the quasi-static formulation is revised in which the propagation delays are totally neglected and, as such, is accurate for electrically small problems. Further, the delayed PEEC formulation is also revised which takes the center-to-center delays into account. Several temporal basis functions can be chosen to expand the circuit quantities. Since PEEC models are typically large, model order reduction techniques have been developed over the years for both quasi-static and delayed PEEC models and they are briefly presented in the chapter. The literature on time-domain analysis of PEEC models is quite large and the interested reader can refer to the references listed.

References

[1] J. M. Jin. *The Finite Element Method in Electromagnetics*. John Wiley and Sons, New York, 2nd edn., 2002.

[2] K. S. Yee. Numerical solution of initial boundary value problems involving Maxwell's equations in isotropic media. *IEEE Transactions on Antennas and Propagation*, 14(5):302–307, May 1966.

[3] A. Taflove. *Advances in Computational Electrodynamics*. Artech House, Norwood, MA, 1998.

[4] A. Taflove and S. C. Hagness. *Computational Electrodynamics*. Artech House, Norwood, MA, 2000.

[5] R. F. Harrington. *Field Computation by Moment Methods.* Macmillan, New York, NY, 1968.
[6] A. E. Ruehli. Equivalent circuit models for three-dimensional multiconductor systems. *IEEE Transactions on Microwave Theory and Techniques*, 22(3):216–221, 1974.
[7] A. E. Ruehli. Survey of computer-aided electrical analysis of integrated circuit interconnections. *IBM Journal of Research and Development*, 23 (6):626–639, 1979.
[8] H. Heeb and A. E. Ruehli. Three-dimensional interconnect analysis using partial element equivalent circuits. *IEEE Transactions on Circuits and Systems I: Fundamental Theory and Applications*, 39(11):974–982, 1992.
[9] C. A. Balanis. *Advanced Engineering Electromagnetics.* John Wiley and Sons, New York, NY, 1989.
[10] L. W. Nagel. SPICE: a computer program to simulate semiconductor circuits. *Electr. Res. Lab. Report ERL M520*, University of California, Berkeley, CA, May 1975.
[11] N. Morita, N. Kumagai, and J. R. Mautz. *Integral Equation Methods for Electromagnetics.* Artech House, Norwood, MA, 1990.
[12] A. Glisson and D. Wilton. Simple and efficient numerical methods for problems of electromagnetic radiation and scattering from surfaces. *IEEE Transactions on Antennas and Propagation*, 28(5):593–603, 1980.
[13] S. Rao, D. Wilton, and A. Glisson. Electromagnetic scattering by surfaces of arbitrary shape. *IEEE Transactions on Antennas and Propagation*, 30 (3):409–418, 1982.
[14] S. M. Rao and D. R. Wilton. Transient scattering by conducting surfaces of arbitrary shape. *IEEE Transactions on Antennas and Propagation*, 39(1):56–61, 1991.
[15] J. J. H. Wang. *Generalized Moment Method in Electromagnetics.* John Wiley and Sons, New York, NY, 1991.
[16] B. M. Kolundzija and B. D. Popovic. Entire-domain Galerkin method for analysis of metallic antennas and scatterers. *IEE Proceedings H – Microwaves, Antennas and Propagation*, 140(1):1–10, 1993.
[17] A. E. Ruehli, G. Antonini, and L. Jiang. *Circuit Oriented Electromagnetic Modeling Using the PEEC Techniques*. John Wiley & Sons, Inc., Hoboken, NJ, 2017.
[18] A. Ruehli, J. Garrett, and C. Paul. Circuit models for 3D structures with incident fields. In *1993 International Symposium on Electromagnetic Compatibility*, pp. 28–32, 1993.
[19] A. E. Ruehli, G. Antonini, and L. Jiang. Skin-effect loss models for time- and frequency-domain PEEC solver. *Proceedings of the IEEE*, 101(2):451–472, 2013.
[20] A. E. Ruehli. Inductance calculations in a complex integrated circuit environment. *IBM Journal of Research and Development*, 16(5):470–481, 1972.
[21] C. R. Paul. *Introduction to Electromagnetic Compatibility*. John Wiley and Sons, New York, NY, 1991.

[22] R. Torchio. A volume PEEC formulation based on the cell method for electromagnetic problems from low to high frequency. *IEEE Transactions on Antennas and Propagation*, 67(12):7452–7465, 2019.

[23] A. Ruehli and H. Heeb. Circuit models for three-dimensional geometries including dielectrics. *IEEE Transactions on Microwave Theory and Techniques*, 40(7):1507–1516, 1992.

[24] G. Antonini. PEEC modelling of Debye dispersive dielectrics. In WIT Press, C. A. Brebbia, D. Polyak editors, *Electrical Engineering and Electromagnetics*, pp. 126–133, WIT Press, Southampton, 2003.

[25] G. Antonini, A. E. Ruehli, and A. Haridass. Including dispersive dielectrics in PEEC models. In *Digest of Electr. Perf. Electronic Packaging*, pp. 349–352, Princeton, NJ, USA, 2003.

[26] G. Antonini, A. E. Ruehli, and A. Haridass. PEEC equivalent circuits for dispersive dielectrics. In *Proceedings of Piers-Progress in Electromagnetics Research Symposium*, Pisa, Italy, 2004.

[27] G. Antonini, A. E. Ruehli, and C. Yang. PEEC modeling of dispersive and lossy dielectrics. *IEEE Transactions on Advanced Packaging*, 31(4):768–782, 2008.

[28] A. Hartman, D. Romano, G. Antonini, and J. Ekman. Partial element equivalent circuit models of three-dimensional geometries including anisotropic dielectrics. *IEEE Transactions on Electromagnetic Compatibility*, 60 (3):696–704, 2018.

[29] D. Romano, G. Antonini, M. D'Emidio, D. Frigioni, A. Mori, and M. Bandinelli. Rigorous *DC* solution of partial element equivalent circuit models. *IEEE Transactions on Circuits and Systems I: Regular Papers*, 63 (9):1499–1510, 2016.

[30] D. Romano, I. Kovačević-Badstübner, M. Parise, U. Grossner, J. Ekman, and G. Antonini. Rigorous *dc* solution of partial element equivalent circuit models including conductive, dielectric, and magnetic materials. *IEEE Transactions on Electromagnetic Compatibility*, 62(3):870–879, 2020.

[31] F. W. Grover. *Inductance Calculations: Working Formulas and Tables*, Dover, New York, NY, 1962.

[32] C. Hoer and C. Love. Exact inductance equations for rectangular conductors with applications to more complicated geometries. *Journal of Research National Bureau of Standards-C. Engineering Instrumentation*, 69(C):127–137, 1965.

[33] P. A. Brennan, N. Raver, and A. E. Ruehli. Three-dimensional inductance computations with partial element equivalent circuits. *IBM Journal of Research and Development*, 23(6):661–668, 1979.

[34] C. R. Paul. *Inductance: Loop and Partial*. John Wiley & Sons, New York, NY, 2009.

[35] A. E. Ruehli and P. A. Brennan. Efficient capacitance calculations for three-dimensional multiconductor systems. *IEEE Transactions on Microwave Theory and Techniques*, 21(2):76–82, 1973.

[36] A. E. Ruehli and P. A. Brennan. Capacitance models for integrated circuit metallization wires. *IEEE Journal of Solid-State Circuits*, 10(6):530–536, 1975.

[37] I. Kovacevic-Badstuebner, D. Romano, L. Lombardi, U. Grossner, J. Ekman, and G. Antonini. Accurate calculation of partial inductances for the orthogonal PEEC formulation. *IEEE Transactions on Electromagnetic Compatibility*, 63(1):82–92, 2021.

[38] G. Antonini, A. Orlandi, and A. E. Ruehli. Analytical integration of quasi-static potential integrals on nonorthogonal coplanar quadrilaterals for the PEEC method. *IEEE Transactions on Electromagnetic Compatibility*, 44(2):399–403, 2002.

[39] A. E. Ruehli, G. Antonini, J. Esch, J. Ekman, A. Mayo, and A. Orlandi. Nonorthogonal PEEC formulation for time- and frequency-domain EM and circuit modeling. *IEEE Transactions on Electromagnetic Compatibility*, 45 (2):167–176, 2003.

[40] L. Di Angelo, D. Romano, G. Antonini, I. Kovacevic-Badstubner, and U. Grossner. Efficient computation of partial elements for non-orthogonal PEEC meshes. In *Proceeding. of the IEEE Int. Symp. on Electromagnetic Compatibility*, Glasgow, UK, August 2021.

[41] M. Štumpf, G. Antonini, and A. Ruehli. Cagniard-DeHoop technique-based computation of retarded partial coefficients: the coplanar case. *IEEE Access*, 8:148989–148996, 2020.

[42] L. Lombardi, G. Antonini, and A.E. Ruehli. Analytical evaluation of partial elements using a retarded taylor series expansion of the green's function. *IEEE Transactions on Microwave Theory and Techniques*, 66(5):2116–2127, 2018.

[43] F. Loreto, D. Romano, G. Antonini, A. Ruehli, L. Lombardi, and M. Parise. Taylor series expansion-based PEEC time domain solver for transient full-wave analysis. In *2020 International Symposium on Electromagnetic Compatibility – EMC EUROPE*, pp. 1–5, 2020.

[44] A. E. Ruehli and H. Heeb. Circuit models for three-dimensional geometries including dielectrics. *IEEE Transactions on Microwave Theory and Techniques*, 40(7):1507–1516, 1992.

[45] Y. Saad. *Iterative Methods for Sparse Linear Systems*. PWS Publishing Company, Boston, MA, 1996.

[46] Kezhong Zhao, M.N. Vouvakis, and J.-F. Lee. The adaptive cross approximation algorithm for accelerated method of moments computations of emc problems. *IEEE Transactions on Electromagnetic Compatibility*, 47(4):763–773, 2005.

[47] G. Antonini and D. Romano. Acceleration of time-domain nodal-analysis of partial element equivalent circuits through multiscale compressed decomposition. *IEEE Transactions on Electromagnetic Compatibility*, 56(4):970–979, 2014.

[48] G. Antonini and D. Romano. Adaptive-cross-approximation-based acceleration of transient analysis of quasi-static partial element equivalent circuits. *IET Microwaves, Antennas & Propagation*, 9(7):700–709, 2015.

[49] F. Ferranti, D. Romano, and G. Antonini. On the passivity of the quasi-static partial element equivalent circuit method. *International Journal of Circuit Theory and Applications*, 47(2):304–319, 2019.

[50] M. Taskinen and P. Yla-Oijala. Current and charge integral equation formulation. *IEEE Transactions on Antennas and Propagation*, 54(1):58–67, 2006.

[51] D. Gope, A. Ruehli, and V. Jandhyala. Solving low-frequency EM-CKT problems using the PEEC method. *IEEE Transactions on Advanced Packaging*, 30(2):313–320, 2007.

[52] D. Romano and G. Antonini. Augmented quasi-static partial element equivalent circuit models for transient analysis of lossy and dispersive magnetic materials. *IEEE Transactions on Magnetics*, 52(5):1–11, 2016.

[53] D. Romano and G. Antonini. Quasi-static partial element equivalent circuit models of magneto-dielectric materials. *IET Microwaves, Antennas and Propagation*, 11(6):915–922, 2017.

[54] Luigi Lombardi, Daniele Romano, and Giulio Antonini. Partial element equivalent circuit method modeling of silicon interconnects. *IEEE Transactions on Microwave Theory and Techniques*, 65(12):4794–4801, 2017.

[55] D. Romano and G. Antonini. Partial element equivalent circuit-based transient analysis of graphene-based interconnects. *IEEE Transactions on Electromagnetic Compatibility*, 58(3):801–810, 2016.

[56] L. Lombardi, M. S. Nakhla, F. Ferranti, and G. Antonini. Time-domain sensitivity analysis of delayed partial element equivalent circuits. *IEEE Transactions on Electromagnetic Compatibility*, 61(5):1465–1473, 2019.

[57] G. Antonini and P. Pepe. Input-to-state stability analysis of partial-element equivalent-circuit models. *IEEE Transactions on Circuits and Systems I: Regular Papers*, 56(3):673–684, 2009.

[58] D. Z. Ding L. P. Zha, Z. H. Fan, and R. S. Chen. Time domain analysis of electromagnetic scattering problems by using integral equation method with space-delayed temporal basis functions. *IEEE Transactions on Antennas and Propagation*, 62(11):5846–5851, 2014.

[59] E. Agastra, C. Dedeban, G. Guarnieri, *et al.* Space and time basis function design for the method of moments in time-domain analysis of wire and planar structures. *International Journal of RF and Microwave Computer-Aided Engineering*, 21(5):551–559, 2011.

[60] G. Manara, A. Monorchio, and R. Reggiannini. A space-time discretization criterion for a stable time-marching solution of the electric field integral equation. *IEEE Transactions on Antennas and Propagation*, 45(3):527–532, 1997.

[61] G. Antonini and A. E. Ruehli. Fast multipole and multifunction PEEC methods. *IEEE Transactions on Mobile Computing*, 2(4):288–298, 2003.

[62] G. Antonini. Fast multipole method for time domain PEEC analysis. *IEEE Transactions on Mobile Computing*, 2(4):275–287, 2003.

[63] G. Antonini and A. E. Ruehli. Waveform relaxation time domain solver for subsystem arrays. *IEEE Transactions on Advanced Packaging*, 33(4):760–768, 2010.

[64] S. X.-D. Tan and L. He. *Advanced Model Order Reduction Techniques in VLSI Design*. Cambridge University Press, Cambridge, *2007*.

[65] R. W. Freund. Krylov-subspace methods for reduced-order modeling in circuit simulation. *Journal of Computational and Applied Mathematics*, 123:395–421, 2000.

[66] A. Odabasioglu, M. Celik, and L. T. Pileggi. PRIMA: passive reduced-order interconnect macromodeling algorithm. *IEEE Transactions on Computer-Aided Design*, 17(8):645–654, 1998.
[67] J. Cullum, A. Ruehli, and T. Zhang. A method for reduced-order modeling and simulation of large interconnect circuits and its application to PEEC models with retardation. *IEEE Transactions on Circuits and Systems, II*, 47(4):261–373, 2000.
[68] M. Kamon, F. Wang, and J. White. Generating nearly optimally compact models from Krylov-subspace based reduced-order models. *IEEE Transactions on Circuits and Systems, II*, 47(4):239–248, 2000.
[69] D. Romano and G. Antonini. Partitioned model order reduction of partial element equivalent circuit models. *IEEE Transactions on Components, Packaging and Manufacturing Technology*, 4(9):1503–1514, 2014.
[70] W. Tseng, C. Chen, E. Gad, M. Nakhla, and R. Achar. Passive order reduction for RLC circuits with delay elements. *IEEE Transactions on Advanced Packaging*, 30(4):830–840, 2007.
[71] I. M. Elfadel and D. D. Ling. A block rational Arnoldi algorithm for multipoint passive model-order reduction of multiport RLC networks. In *IEEE/ACM International Conference on Computer-Aided Design*, pp. 66–71, Nov. 1997.
[72] F. Ferranti, M. S. Nakhla, G. Antonini, T. Dhaene, L. Knockaert, and A. E. Ruehli. Multipoint full-wave model order reduction for delayed PEEC models with large delays. *IEEE Transactions on Electromagnetic Compatibility*, 53(4):959–967, 2011.
[73] F. Ferranti, G. Antonini, T. Dhaene, and L. Knockaert. Guaranteed passive parameterized model order reduction of the partial element equivalent circuit (PEEC) method. *IEEE Transactions on Electromagnetic Compatibility*, 52 (4):974–984, 2010.
[74] F. Ferranti, G. Antonini, T. Dhaene, L. Knockaert, and A. E. Ruehli. Physics-based passivity-preserving parameterized model order reduction for PEEC circuit analysis. *IEEE Transactions on Components, Packaging and Manufacturing Technology*, 1(3):399–409, 2011.
[75] F. Ferranti, M. Nakhla, G. Antonini, T. Dhaene, L. Knockaert, and A. E. Ruehli. Interpolation-based parameterized model order reduction of delayed systems. *IEEE Transactions on Microwave Theory and Techniques*, 60 (3):431–440, 2012.
[76] L. Lombardi, Y. Tao, B. Nouri, F. Ferranti, G. Antonini, and M. S. Nakhla. Parameterized model order reduction of delayed PEEC circuits. *IEEE Transactions on Electromagnetic Compatibility*, 62(3):859–869, 2020.
[77] L. D. Camillis, F. Ferranti, G. Antonini, D. V. Ginste, and D. De Zutter. Parameterized partial element equivalent circuit method for sensitivity analysis of multiport systems. *IEEE Transactions on Components, Packaging and Manufacturing Technology*, 2(2):248–255, 2012.
[78] L. D. Camillis, F. Ferranti, and G. Antonini. Parameterized model order reduction for efficient time and frequency domain global sensitivity analysis of PEEC circuits. *ACES Journal*, 31(10):1170–1180, 2016.

Chapter 7
Unconditionally stable time-domain methods

Kyung-Young Jung[1] and Fernando L. Teixeira[2]

In this chapter, the fundamental aspects and several applications of unconditional stable time-domain methods for electromagnetic wave analysis are described. The update equations of alternating-direction-implicit (ADI) and locally-one-dimensional (LOD) finite-difference time-domain (FDTD) algorithms and their numerical errors are explained. The Perfectly Matched Layer (PML) implementation to truncate the computational domain is discussed. The iterative implementation to the ADI and LOD-FDTD algorithms is presented to reduce their numerical errors. In addition, the implementation of unconditionally stable FDTD algorithms in complex media is discussed. Finally, variant unconditionally stable FDTD methods are briefly discussed.

7.1 Introduction

The finite-difference time-domain (FDTD) method has been widely used to study electromagnetic problems due to its simplicity (including its matrix-free character), accuracy, and robustness [1–3]. However, the FDTD is a conditionally stable algorithm restricted by the Courant stability limit [4], which sets a maximum value for the time step size based on the spatial cell size. The Courant stability criterion hinders the applicability of the FDTD method especially for deep sub-wavelength problems or in problems requiring highly refined meshes to represent small geometrical features.

Unlike the conventional FDTD method, unconditionally stable FDTD methods have no limit on the time step size (except for accuracy considerations). In many unconditionally stable FDTD methods, the update equations are implicit and might necessitate the solution of a large sparse matrix system at each time step. Among unconditionally stable FDTD methods, the alternating-direction-implicit (ADI)-FDTD algorithm [5,6] and the locally-one-dimensional (LOD)-FDTD algorithm [7,8] have been popularly employed for a variety of electromagnetic wave problems. In these two representative unconditionally stable FDTD algorithms, a

[1]Department of Electronic Engineering, Hanyang University, Korea
[2]Electroscience Laboratory, The Ohio State University, USA

tridiagonal matrix system is obtained to update the electromagnetic field components which can be efficiently solved in $O(N)$ operations using the Thomas algorithm [9]. Therefore, ADI- or LOD-based FDTD algorithms have become attractive alternatives to the conventional FDTD method for use in problems in which the Courant limit would otherwise force the time step size to be extremely small. Despite these advantages, the ADI-FDTD and LOD-FDTD methods exhibit splitting error as well as dispersion error. In this chapter, the basic update equations of the ADI-FDTD and LOD-FDTD methods are derived and their numerical errors are analyzed. In order for any of these methods to model open-space problems, an absorbing boundary condition is required to properly truncate the computational domain. The perfectly matched layer (PML) is the current absorbing boundary condition of choice given that it can efficiently absorb electromagnetic waves even in complex media [10–12]. Accordingly, a PML implementation for unconditionally stable FDTD methods is presented here. The splitting error inherent to ADI-FDTD and LOD-FDTD methods can be significantly reduced by iterative approaches [13–15] which are also discussed in this chapter. The implementation of unconditionally stable methods to complex media with material dispersion is also discussed. Finally, some other variant unconditionally stable FDTD methods are briefly discussed.

7.2 ADI-FDTD algorithm

The source-free Maxwell's curl equations in simple (linear, isotropic, homogenous, non-dispersive) media write as

$$\varepsilon\frac{\partial \mathbf{E}}{\partial t} = \nabla \times \mathbf{H}, \tag{7.1a}$$

$$\mu\frac{\partial \mathbf{H}}{\partial t} = -\nabla \times \mathbf{E}. \tag{7.1b}$$

In Cartesian coordinates, the above relations can be expressed as

$$\varepsilon\frac{\partial E_x}{\partial t} = \frac{\partial H_z}{\partial y} - \frac{\partial H_y}{\partial z}, \tag{7.2a}$$

$$\varepsilon\frac{\partial E_y}{\partial t} = \frac{\partial H_x}{\partial z} - \frac{\partial H_z}{\partial x}, \tag{7.2b}$$

$$\varepsilon\frac{\partial E_z}{\partial t} = \frac{\partial H_y}{\partial x} - \frac{\partial H_x}{\partial y}, \tag{7.2c}$$

$$\mu\frac{\partial H_x}{\partial t} = \frac{\partial E_y}{\partial z} - \frac{\partial E_z}{\partial y}, \tag{7.2d}$$

$$\mu\frac{\partial H_y}{\partial t} = \frac{\partial E_z}{\partial x} - \frac{\partial E_x}{\partial z}, \tag{7.2e}$$

$$\mu\frac{\partial H_z}{\partial t} = \frac{\partial E_x}{\partial y} - \frac{\partial E_y}{\partial x}. \tag{7.2f}$$

Let us denote any function u at a discrete point in space and time as

$$u(i\Delta x, j\Delta y, k\Delta z, n\Delta t) = u^n_{i,j,k}, \tag{7.3}$$

where $\Delta\varsigma$ indicates the spatial cell size in the FDTD grid ($\varsigma = x,\ y,$ or z) and the subscript represents the space index. In addition, Δt indicates the time step size with the time index denoted by the superscript. In the conventional FDTD algorithm, Maxwell's curl equations are discretized as follows:

$$\varepsilon \frac{\partial E_x}{\partial t}\Big|^{n+1/2} = \frac{\partial H_z}{\partial y}\Big|^{n+1/2} - \frac{\partial H_y}{\partial z}\Big|^{n+1/2}, \tag{7.4a}$$

$$\varepsilon \frac{\partial E_y}{\partial t}\Big|^{n+1/2} = \frac{\partial H_x}{\partial z}\Big|^{n+1/2} - \frac{\partial H_z}{\partial x}\Big|^{n+1/2}, \tag{7.4b}$$

$$\varepsilon \frac{\partial E_z}{\partial t}\Big|^{n+1/2} = \frac{\partial H_y}{\partial x}\Big|^{n+1/2} - \frac{\partial H_x}{\partial y}\Big|^{n+1/2}, \tag{7.4c}$$

$$\mu \frac{\partial H_x}{\partial t}\Big|^{n} = \frac{\partial E_y}{\partial z}\Big|^{n} - \frac{\partial E_z}{\partial y}\Big|^{n}, \tag{7.4d}$$

$$\mu \frac{\partial H_y}{\partial t}\Big|^{n} = \frac{\partial E_z}{\partial x}\Big|^{n} - \frac{\partial E_x}{\partial z}\Big|^{n}, \tag{7.4e}$$

$$\mu \frac{\partial H_z}{\partial t}\Big|^{n} = \frac{\partial E_x}{\partial y}\Big|^{n} - \frac{\partial E_y}{\partial x}\Big|^{n}. \tag{7.4f}$$

The application of the central difference scheme (CDS) to discretize the partial derivatives in the above equations yields the following FDTD update equations:

$$E^{n+1}_{x_{i+1/2,j,k}} = E^{n}_{x_{i+1/2,j,k}} + \frac{\Delta t}{\varepsilon}\left(\frac{H^{n+1/2}_{z_{i+1/2,j+1/2,k}} - H^{n+1/2}_{z_{i+1/2,j-1/2,k}}}{\Delta y}\right) - \frac{\Delta t}{\varepsilon}\left(\frac{H^{n+1/2}_{y_{i+1/2,j,k+1/2}} - H^{n+1/2}_{y_{i+1/2,j,k-1/2}}}{\Delta z}\right), \tag{7.5}$$

$$E^{n+1}_{y_{i,j+1/2,k}} = E^{n}_{y_{i,j+1/2,k}} + \frac{\Delta t}{\varepsilon}\left(\frac{H^{n+1/2}_{x_{i,j+1/2,k+1/2}} - H^{n+1/2}_{x_{i,j+1/2,k-1/2}}}{\Delta z}\right) - \frac{\Delta t}{\varepsilon}\left(\frac{H^{n+1/2}_{z_{i+1/2,j+1/2,k}} - H^{n+1/2}_{z_{i-1/2,j+1/2,k}}}{\Delta x}\right), \tag{7.6}$$

$$E^{n+1}_{z_{i,j,k+1/2}} = E^{n}_{z_{i,j,k+1/2}} + \frac{\Delta t}{\varepsilon}\left(\frac{H^{n+1/2}_{y_{i+1/2,j,k+1/2}} - H^{n+1/2}_{y_{i-1/2,j,k+1/2}}}{\Delta x}\right) - \frac{\Delta t}{\varepsilon}\left(\frac{H^{n+1/2}_{x_{i,j+1/2,k+1/2}} - H^{n+1/2}_{x_{i,j-1/2,k+1/2}}}{\Delta y}\right), \tag{7.7}$$

$$H^{n+1/2}_{x_{i,j+1/2,k+1/2}} = H^{n-1/2}_{x_{i,j+1/2,k+1/2}} + \frac{\Delta t}{\mu}\left(\frac{E^n_{y_{i,j+1/2,k+1}} - E^n_{y_{i,j+1/2,k}}}{\Delta z}\right) - \frac{\Delta t}{\mu}\left(\frac{E^n_{z_{i,j+1,k+1/2}} - E^n_{z_{i,j,k+1/2}}}{\Delta y}\right), \tag{7.8}$$

$$H^{n+1/2}_{y_{i+1/2,j,k+1/2}} = H^{n-1/2}_{y_{i+1/2,j,k+1/2}} + \frac{\Delta t}{\mu}\left(\frac{E^n_{z_{i+1,j,k+1/2}} - E^n_{z_{i,j,k+1/2}}}{\Delta x}\right) - \frac{\Delta t}{\mu}\left(\frac{E^n_{x_{i+1/2,j,k+1}} - E^n_{x_{i+1/2,j,k}}}{\Delta z}\right), \tag{7.9}$$

$$H^{n+1/2}_{z_{i+1/2,j+1/2,k}} = H^{n-1/2}_{z_{i+1/2,j+1/2,k}} + \frac{\Delta t}{\mu}\left(\frac{E^n_{x_{i+1/2,j+1,k}} - E^n_{x_{i+1/2,j,k}}}{\Delta y}\right) - \frac{\Delta t}{\mu}\left(\frac{E^n_{y_{i+1,j+1/2,k}} - E^n_{y_{i,j+1/2,k}}}{\Delta x}\right). \tag{7.10}$$

In the above, the field components in the right-hand side (RHS) are obtained explicitly. In other words, the field components at the previous half time step are already known and thus no matrix calculation is involved in the conventional FDTD algorithm.

Now, let us consider the ADI-FDTD algorithm. In this case, the time marching from the nth time step to the $(n+1)$th time step is comprised of two sub-step time marching processes [16]. In the first sub-step, the Maxwell's curl equations are discretized as follows:

$$\varepsilon\frac{\partial E_x}{\partial t}\Big|^{n+1/4} = \frac{\partial H_z}{\partial y}\Big|^{n+1/2} - \frac{\partial H_y}{\partial z}\Big|^{n}, \tag{7.11a}$$

$$\varepsilon\frac{\partial E_y}{\partial t}\Big|^{n+1/4} = \frac{\partial H_x}{\partial z}\Big|^{n+1/2} - \frac{\partial H_z}{\partial x}\Big|^{n}, \tag{7.11b}$$

$$\varepsilon\frac{\partial E_z}{\partial t}\Big|^{n+1/4} = \frac{\partial H_y}{\partial x}\Big|^{n+1/2} - \frac{\partial H_x}{\partial y}\Big|^{n}, \tag{7.11c}$$

$$\mu\frac{\partial H_x}{\partial t}\Big|^{n+1/4} = \frac{\partial E_y}{\partial z}\Big|^{n+1/2} - \frac{\partial E_z}{\partial y}\Big|^{n}, \tag{7.11d}$$

$$\mu\frac{\partial H_y}{\partial t}\Big|^{n+1/4} = \frac{\partial E_z}{\partial x}\Big|^{n+1/2} - \frac{\partial E_x}{\partial z}\Big|^{n}, \tag{7.11e}$$

$$\mu\frac{\partial H_z}{\partial t}\Big|^{n+1/4} = \frac{\partial E_x}{\partial y}\Big|^{n+1/2} - \frac{\partial E_y}{\partial x}\Big|^{n}. \tag{7.11f}$$

In the second sub-step, the equations are discretized as follows:

$$\varepsilon\frac{\partial E_x}{\partial t}\Big|^{n+3/4} = \frac{\partial H_z}{\partial y}\Big|^{n+1/2} - \frac{\partial H_y}{\partial z}\Big|^{n+1}, \tag{7.12a}$$

$$\varepsilon \frac{\partial E_y}{\partial t}\Big|^{n+3/4} = \frac{\partial H_x}{\partial z}\Big|^{n+1/2} - \frac{\partial H_z}{\partial x}\Big|^{n+1}, \tag{7.12b}$$

$$\varepsilon \frac{\partial E_z}{\partial t}\Big|^{n+3/4} = \frac{\partial H_y}{\partial x}\Big|^{n+1/2} - \frac{\partial H_x}{\partial y}\Big|^{n+1}, \tag{7.12c}$$

$$\mu \frac{\partial H_x}{\partial t}\Big|^{n+3/4} = \frac{\partial E_y}{\partial z}\Big|^{n+1/2} - \frac{\partial E_z}{\partial y}\Big|^{n+1}, \tag{7.12d}$$

$$\mu \frac{\partial H_y}{\partial t}\Big|^{n+3/4} = \frac{\partial E_z}{\partial x}\Big|^{n+1/2} - \frac{\partial E_x}{\partial z}\Big|^{n+1}, \tag{7.12e}$$

$$\mu \frac{\partial H_z}{\partial t}\Big|^{n+3/4} = \frac{\partial E_x}{\partial y}\Big|^{n+1/2} - \frac{\partial E_y}{\partial x}\Big|^{n+1}. \tag{7.12f}$$

In the first sub-step, the first term in the RHS is expressed implicitly, whereas in the second sub-step, the second term in the RHS is expressed implicitly. This technique, known as the *alternating direction implicit* technique, is also available to express the implicit term in the opposite way. By using the CDS in the time and space domains, one obtains the following set of equations in the first sub-step:

$$E^{n+1/2}_{x_{i+1/2,j,k}} = E^{n}_{x_{i+1/2,j,k}} + \frac{\Delta t}{2\varepsilon}\left(\frac{H^{n+1/2}_{z_{i+1/2,j+1/2,k}} - H^{n+1/2}_{z_{i+1/2,j-1/2,k}}}{\Delta y}\right) - \frac{\Delta t}{2\varepsilon}\left(\frac{H^{n}_{y_{i+1/2,j,k+1/2}} - H^{n}_{y_{i+1/2,j,k-1/2}}}{\Delta z}\right), \tag{7.13}$$

$$E^{n+1/2}_{y_{i,j+1/2,k}} = E^{n}_{y_{i,j+1/2,k}} + \frac{\Delta t}{2\varepsilon}\left(\frac{H^{n+1/2}_{x_{i,j+1/2,k+1/2}} - H^{n+1/2}_{x_{i,j+1/2,k-1/2}}}{\Delta z}\right) - \frac{\Delta t}{2\varepsilon}\left(\frac{H^{n}_{z_{i+1/2,j+1/2,k}} - H^{n}_{z_{i-1/2,j+1/2,k}}}{\Delta x}\right), \tag{7.14}$$

$$E^{n+1/2}_{z_{i,j,k+1/2}} = E^{n}_{z_{i,j,k+1/2}} + \frac{\Delta t}{2\varepsilon}\left(\frac{H^{n+1/2}_{y_{i+1/2,j,k+1/2}} - H^{n+1/2}_{y_{i-1/2,j,k+1/2}}}{\Delta x}\right) - \frac{\Delta t}{2\varepsilon}\left(\frac{H^{n}_{x_{i,j+1/2,k+1/2}} - H^{n}_{x_{i,j-1/2,k+1/2}}}{\Delta y}\right), \tag{7.15}$$

$$H^{n+1/2}_{x_{i,j+1/2,k+1/2}} = H^{n}_{x_{i,j+1/2,k+1/2}} + \frac{\Delta t}{2\mu}\left(\frac{E^{n+1/2}_{y_{i,j+1/2,k+1}} - E^{n+1/2}_{y_{i,j+1/2,k}}}{\Delta z}\right) - \frac{\Delta t}{2\mu}\left(\frac{E^{n}_{z_{i,j+1,k+1/2}} - E^{n}_{z_{i,j,k+1/2}}}{\Delta y}\right), \tag{7.16}$$

$$H^{n+1/2}_{y_{i+1/2,j,k+1/2}} = H^{n}_{y_{i+1/2,j,k+1/2}} + \frac{\Delta t}{2\mu}\left(\frac{E^{n+1/2}_{z_{i+1,j,k+1/2}} - E^{n+1/2}_{z_{i,j,k+1/2}}}{\Delta x}\right) - \frac{\Delta t}{2\mu}\left(\frac{E^{n}_{x_{i+1/2,j,k+1}} - E^{n}_{x_{i+1/2,j,k}}}{\Delta z}\right), \tag{7.17}$$

$$H^{n+1/2}_{z_{i+1/2,j+1/2,k}} = H^{n}_{z_{i+1/2,j+1/2,k}} + \frac{\Delta t}{2\mu}\left(\frac{E^{n+1/2}_{x_{i+1/2,j+1,k}} - E^{n+1/2}_{x_{i+1/2,j,k}}}{\Delta y}\right) - \frac{\Delta t}{2\mu}\left(\frac{E^{n}_{y_{i+1,j+1/2,k}} - E^{n}_{y_{i,j+1/2,k}}}{\Delta x}\right). \tag{7.18}$$

Likewise, in the second sub-step, one obtains the following set of equations:

$$E^{n+1}_{x_{i+1/2,j,k}} = E^{n+1/2}_{x_{i+1/2,j,k}} + \frac{\Delta t}{2\varepsilon}\left(\frac{H^{n+1/2}_{z_{i+1/2,j+1/2,k}} - H^{n+1/2}_{z_{i+1/2,j-1/2,k}}}{\Delta y}\right) - \frac{\Delta t}{2\varepsilon}\left(\frac{H^{n+1}_{y_{i+1/2,j,k+1/2}} - H^{n+1}_{y_{i+1/2,j,k-1/2}}}{\Delta z}\right), \tag{7.19}$$

$$E^{n+1}_{y_{i,j+1/2,k}} = E^{n+1/2}_{y_{i,j+1/2,k}} + \frac{\Delta t}{2\varepsilon}\left(\frac{H^{n+1/2}_{x_{i,j+1/2,k+1/2}} - H^{n+1/2}_{x_{i,j+1/2,k-1/2}}}{\Delta z}\right) - \frac{\Delta t}{2\varepsilon}\left(\frac{H^{n+1}_{z_{i+1/2,j+1/2,k}} - H^{n+1}_{z_{i-1/2,j+1/2,k}}}{\Delta x}\right), \tag{7.20}$$

$$E^{n+1}_{z_{i,j,k+1/2}} = E^{n+1/2}_{z_{i,j,k+1/2}} + \frac{\Delta t}{2\varepsilon}\left(\frac{H^{n+1/2}_{y_{i+1/2,j,k+1/2}} - H^{n+1/2}_{y_{i-1/2,j,k+1/2}}}{\Delta x}\right) - \frac{\Delta t}{2\varepsilon}\left(\frac{H^{n+1}_{x_{i,j+1/2,k+1/2}} - H^{n+1}_{x_{i,j-1/2,k+1/2}}}{\Delta y}\right), \tag{7.21}$$

$$H^{n+1}_{x_{i,j+1/2,k+1/2}} = H^{n+1/2}_{x_{i,j+1/2,k+1/2}} + \frac{\Delta t}{2\mu}\left(\frac{E^{n+1/2}_{y_{i,j+1/2,k+1}} - E^{n+1/2}_{y_{i,j+1/2,k}}}{\Delta z}\right) - \frac{\Delta t}{2\mu}\left(\frac{E^{n+1}_{z_{i,j+1,k+1/2}} - E^{n+1}_{z_{i,j,k+1/2}}}{\Delta y}\right), \tag{7.22}$$

$$H^{n+1}_{y_{i+1/2,j,k+1/2}} = H^{n+1/2}_{y_{i+1/2,j,k+1/2}} + \frac{\Delta t}{2\mu}\left(\frac{E^{n+1/2}_{z_{i+1,j,k+1/2}} - E^{n+1/2}_{z_{i,j,k+1/2}}}{\Delta x}\right)$$
$$- \frac{\Delta t}{2\mu}\left(\frac{E^{n+1}_{x_{i+1/2,j,k+1}} - E^{n+1}_{x_{i+1/2,j,k}}}{\Delta z}\right), \tag{7.23}$$

$$H^{n+1}_{z_{i+1/2,j+1/2,k}} = H^{n+1/2}_{z_{i+1/2,j+1/2,k}} + \frac{\Delta t}{2\mu}\left(\frac{E^{n+1/2}_{x_{i+1/2,j+1,k}} - E^{n+1/2}_{x_{i+1/2,j,k}}}{\Delta y}\right)$$
$$- \frac{\Delta t}{2\mu}\left(\frac{E^{n+1}_{y_{i+1,j+1/2,k}} - E^{n+1}_{y_{i,j+1/2,k}}}{\Delta x}\right). \tag{7.24}$$

In the first sub-step, the field components at the $(n+1/2)$ time step in the left-hand side (LHS) cannot be directly updated since the field components inside the first parenthesis in the RHS are simultaneously required at the $(n+1/2)$ time step. Similarly, in the second sub-step, the field components at the $(n+1)$ time step in the LHS cannot be directly updated since the field components inside the second parenthesis in the RHS are simultaneously required at the $(n+1)$ time step. To resolve this issue, the update equations of magnetic field components are inserted in the corresponding update equations of the electric field components. For example, let us consider the E_x component: By substituting (7.18) into (7.13) and arranging the resulting equation, the E_x update equation in the first sub-step can be obtained as follows:

$$-\frac{C_bD_b}{(\Delta y)^2}E^{n+1/2}_{x_{i+1/2,j-1,k}} + \left(1+\frac{2C_bD_b}{(\Delta y)^2}\right)E^{n+1/2}_{x_{i+1/2,j,k}} - \frac{C_bD_b}{(\Delta y)^2}E^{n+1/2}_{x_{i+1/2,j+1,k}}$$
$$= E^{n}_{x_{i+1/2,j,k}} + \frac{C_b}{\Delta y}\left(H^{n}_{z_{i+1/2,j+1/2,k}} - H^{n}_{z_{i+1/2,j-1/2,k}}\right) - \frac{C_b}{\Delta z}\left(H^{n}_{y_{i+1/2,j,k+1/2}} - H^{n}_{y_{i+1/2,j,k-1/2}}\right)$$
$$-\frac{C_b D_b}{\Delta y\,\Delta x}\left(E^{n}_{y_{i+1,j+1/2,k}} - E^{n}_{y_{i,j+1/2,k}}\right) + \frac{C_b D_b}{\Delta y\,\Delta x}\left(E^{n}_{y_{i+1,j-1/2,k}} - E^{n}_{y_{i,j-1/2,k}}\right), \tag{7.25}$$

where $C_b = \frac{\Delta t}{2\varepsilon}$ and $D_b = \frac{\Delta t}{2\mu}$. In the above, all field components in the RHS are already known and the three adjacent E_x field components along the y direction in the LHS are updated implicitly. Importantly, this tridiagonal matrix system can be easily solved with $O(N)$ complexity using the Thomas algorithm. A similar procedure can be used to obtain the update equations of the other electric components. After updating the electric field components, the magnetic field components can be updated explicitly without difficulty.

Thomas algorithm

A tridiagonal matrix system can be efficiently solved by the Thomas algorithm. The tridagonal matrix can be formed as

$$\begin{bmatrix} b_1 & c_1 & 0 & \dots & 0 & 0 & 0 \\ a_2 & b_2 & c_2 & \dots & 0 & 0 & 0 \\ 0 & a_3 & b_3 & \dots & 0 & 0 & 0 \\ 0 & 0 & a_4 & \dots & c_{N-3} & 0 & 0 \\ 0 & 0 & 0 & \dots & b_{N-2} & c_{N-2} & 0 \\ 0 & 0 & 0 & \dots & a_{N-1} & b_{N-1} & c_{N-1} \\ 0 & 0 & 0 & \dots & 0 & a_N & b_N \end{bmatrix} \begin{bmatrix} u_1 \\ u_2 \\ u_3 \\ \dots \\ u_{N-2} \\ u_{N-1} \\ u_N \end{bmatrix} = \begin{bmatrix} r_1 \\ r_2 \\ r_3 \\ \dots \\ r_{N-2} \\ r_{N-1} \\ r_N \end{bmatrix},$$

where u_i indicates an unknown field component on the LHS and r_i represents a known value on the RHS. The Thomas algorithm proceeds as follows:

$i = 1:$

$u_i = c_i/b_i$

$g_i = r_i/b_i$

for $i = 2$ to $N - 1$

$u_i = c_i/(b_i - a_i u_{i-1})$

$g_i = (r_i - a_i g_{i-1})/(b_i - a_i u_{i-1})$

end for

$i = N:$

$u_i = (r_i - a_i g_{i-1})/(b_i - a_i u_{i-1})$

for $i = N - 1$ to 1

$u_i = g_i - u_i u_{i+1}$

end for

Let us consider next the second sub-step update equations. By substituting (7.23) into (7.19), the E_x update equation in the second sub-step is expressed as

$$\begin{aligned} &-\frac{C_b D_b}{(\Delta z)^2} E^{n+1}_{x_{i+1/2,j,k-1}} + \left(1 + \frac{2 C_b D_b}{(\Delta z)^2}\right) E^{n+1}_{x_{i+1/2,j,k}} - \frac{C_b D_b}{(\Delta z)^2} E^{n+1/2}_{x_{i+1/2,j,k+1}} \\ &= E^{n+1/2}_{x_{i+1/2,j,k}} + \frac{C_b}{\Delta y}\left(H^{n+1/2}_{z_{i+1/2,j+1/2,k}} - H^{n+1/2}_{z_{i+1/2,j-1/2,k}}\right) - \frac{C_b}{\Delta z}\left(H^{n+1/2}_{y_{i+1/2,j,k+1/2}} - H^{n+1/2}_{y_{i+1/2,j,k-1/2}}\right) \\ &- \frac{C_b D_b}{\Delta z\, \Delta x}\left(E^{n+1/2}_{z_{i+1,j,k+1/2}} - E^{n+1/2}_{z_{i,j,k+1/2}}\right) + \frac{C_b D_b}{\Delta z\, \Delta x}\left(E^{n+1/2}_{z_{i+1,j,k-1/2}} - E^{n+1/2}_{z_{i,j,k-1/2}}\right). \end{aligned} \tag{7.26}$$

A tridiagonal matrix system is obtained, which can again be solved using the Thomas algorithm. The update equations of E_y and E_z components in the second sub-step can be similarly derived. The magnetic field components are explicitly updated using (7.22)–(7.24) after updating the electric field components. The overall update procedure of the ADI-FDTD algorithm can be summarized as

Step 1. Update $\mathbf{E}^{n+1/2}$ implicitly (e.g. from (7.25)).

Step 2. Update $\mathbf{H}^{n+1/2}$ explicitly from (7.16) to (7.18).

Step 3. Update $\mathbf{E}^{n+1}$ implicitly (e.g. from (7.26)).

Step 4. Update $\mathbf{H}^{n+1}$ explicitly from (7.22) to (7.24).

Next, some important aspects of the ADI-FDTD algorithm accuracy are discussed in some detail. Of the several approaches available to examine its numerical accuracy and stability, we adopt the concise matrix-form approach [17] because it yields clearer insights regarding the differences between the ADI-FDTD and the LOD-FDTD algorithms. Maxwell's curl equations in matrix form can be written as

$$\frac{\partial \mathbf{u}}{\partial t} + [\mathcal{R}]\mathbf{u} = [0]_6, \tag{7.27}$$

with $\mathbf{u} = \left(E_x, E_y, E_z, H_x, H_y, H_z\right)^T$ and

$$[\mathcal{R}] = \begin{pmatrix} [0]_3 & -\dfrac{1}{\varepsilon}[\mathcal{C}] \\ \dfrac{1}{\mu}[\mathcal{C}] & [0]_3 \end{pmatrix}. \tag{7.28}$$

Here, $[0]_a$ is the $a \times a$ zero matrix and $[\mathcal{C}]$ is the curl matrix:

$$[\mathcal{C}] = \begin{bmatrix} 0 & -\partial_z & \partial_y \\ \partial_z & 0 & -\partial_x \\ -\partial_y & \partial_x & 0 \end{bmatrix}. \tag{7.29}$$

First, the Crank–Nicolson FDTD algorithm is applied as follows

$$\frac{\mathbf{u}_{i,j,k}^{n+1} - \mathbf{u}_{i,j,k}^{n}}{\Delta t} + [R]\frac{\mathbf{u}_{i,j,k}^{n+1} + \mathbf{u}_{i,j,k}^{n}}{2} = [0]_6, \tag{7.30}$$

where $[R]$ is the *numerical* (i.e. discrete) operator corresponding to $[\mathcal{R}]$. In other words, the space derivative ∂_ς is replaced by the central difference operator:

$$\delta_\varsigma f \equiv \frac{f_{\alpha+1/2} - f_{\alpha-1/2}}{\Delta\varsigma}, \tag{7.31}$$

where $\varsigma \in \{x, y, z\}$ and α indicate the grid point (i, j, k) in the ς-direction. In the Crank–Nicolson FDTD algorithm, all six field components are updated implicitly thus requiring a sparse (and non tridiagonal) linear solver for the solution. As such, the computational costs of the Crank–Nicolson FDTD algorithm are considerably larger. The ADI-FDTD algorithm is next derived as a perturbation on the Crank–Nicolson FDTD algorithm. To show this, let us rewrite the Crank–Nicolson FDTD as follows

$$\frac{\mathbf{u}_{i,j,k}^{n+1} - \mathbf{u}_{i,j,k}^{n}}{\Delta t} = [A]\frac{\mathbf{u}_{i,j,k}^{n+1} + \mathbf{u}_{i,j,k}^{n}}{2} + [B]\frac{\mathbf{u}_{i,j,k}^{n+1} + \mathbf{u}_{i,j,k}^{n}}{2}, \tag{7.32}$$

with

$$[A] = \begin{bmatrix} 0 & 0 & 0 & 0 & 0 & \delta_y/\varepsilon \\ 0 & 0 & 0 & \delta_z/\varepsilon & 0 & 0 \\ 0 & 0 & 0 & 0 & \delta_x/\varepsilon & 0 \\ 0 & \delta_z/\mu & 0 & 0 & 0 & 0 \\ 0 & 0 & \delta_x/\mu & 0 & 0 & 0 \\ \delta_y/\mu & 0 & 0 & 0 & 0 & 0 \end{bmatrix}, \tag{7.33}$$

$$[B] = -\begin{bmatrix} 0 & 0 & 0 & 0 & \delta_z/\varepsilon & 0 \\ 0 & 0 & 0 & 0 & 0 & \delta_x/\varepsilon \\ 0 & 0 & 0 & \delta_y/\varepsilon & 0 & 0 \\ 0 & 0 & \delta_y/\mu & 0 & 0 & 0 \\ \delta_z/\mu & 0 & 0 & 0 & 0 & 0 \\ 0 & \delta_x/\mu & 0 & 0 & 0 & 0 \end{bmatrix}. \tag{7.34}$$

Rearranging the above, the Crank–Nicolson FDTD algorithm can be represented as

$$\left([I] - \frac{\Delta t}{2}[A] - \frac{\Delta t}{2}[B]\right)\mathbf{u}^{n+1} = \left([I] + \frac{\Delta t}{2}[A] + \frac{\Delta t}{2}[B]\right)\mathbf{u}^{n}, \tag{7.35}$$

where $[I]$ is the identity matrix. The one-step ADI-FDTD algorithm proceeds by adding the splitting error

$$\left([I] - \frac{\Delta t}{2}[A]\right)\left([I] - \frac{\Delta t}{2}[B]\right)\mathbf{u}^{n+1} = \left([I] + \frac{\Delta t}{2}[A]\right)\left([I] + \frac{\Delta t}{2}[B]\right)\mathbf{u}^{n}. \tag{7.36}$$

The algorithm above is next factorized into two sub-step equations:

$$\left([I] - \frac{\Delta t}{2}[A]\right)\mathbf{u}^{n+1/2} = \left([I] + \frac{\Delta t}{2}[B]\right)\mathbf{u}^{n} \tag{7.37}$$

$$\left([I] - \frac{\Delta t}{2}[B]\right)\mathbf{u}^{n+1} = \left([I] + \frac{\Delta t}{2}[A]\right)\mathbf{u}^{n+1/2}. \tag{7.38}$$

It can be easily shown that the above ADI-FDTD algorithm is equivalent to the prior one-step equation by inserting (7.37) into (7.38), i.e.

$$\begin{aligned} \left([I] - \frac{\Delta t}{2}[B]\right)\mathbf{u}^{n+1} &= \left([I] + \frac{\Delta t}{2}[A]\right)\left([I] - \frac{\Delta t}{2}[A]\right)^{-1}\left([I] + \frac{\Delta t}{2}[B]\right)\mathbf{u}^{n} \\ \left([I] - \frac{\Delta t}{2}[B]\right)\mathbf{u}^{n+1} &= \left([I] - \frac{\Delta t}{2}[A]\right)^{-1}\left([I] + \frac{\Delta t}{2}[A]\right)\left([I] + \frac{\Delta t}{2}[B]\right)\mathbf{u}^{n} \\ \left([I] - \frac{\Delta t}{2}[A]\right)\left([I] - \frac{\Delta t}{2}[B]\right)\mathbf{u}^{n+1} &= \left([I] + \frac{\Delta t}{2}[A]\right)\left([I] + \frac{\Delta t}{2}[B]\right)\mathbf{u}^{n}. \end{aligned} \tag{7.39}$$

The dispersion relation of the ADI-FDTD algorithm can be derived based on a plane wave expansion. The numerical field vector on the FDTD grid is expressed as

$$\mathbf{u}^n_{i,j,k} = \mathbf{u}_0 e^{\mathrm{j}\omega n\Delta t} e^{-\mathrm{j}(i\Delta x\beta_x + j\Delta y\beta_y + k\Delta z\beta_z)}, \tag{7.40}$$

where β_ς is the real wavenumber, ω is the complex frequency, and $\mathrm{j} = \sqrt{-1}$. A series of mathematical manipulations are then carried out to obtain

$$\frac{2}{\Delta t}\tan\left(\omega\frac{\Delta t}{2}\right) = \sqrt{\frac{\beta_n^2 + \zeta^2}{\mu\varepsilon + \xi^2}}, \tag{7.41}$$

with

$$\beta_n^2 = \beta_{nx}^2 + \beta_{ny}^2 + \beta_{nz}^2 \tag{7.42}$$

$$\zeta^2 = \frac{\Delta t^2}{4\mu\varepsilon}\left(\beta_{nx}^2\beta_{ny}^2 + \beta_{ny}^2\beta_{nz}^2 + \beta_{nz}^2\beta_{nx}^2\right) \tag{7.43}$$

$$\xi^2 = \frac{\Delta t^6}{64\mu^2\varepsilon^2}\beta_{nx}^2\beta_{ny}^2\beta_{nz}^2. \tag{7.44}$$

Here, $\beta_{n\varsigma}$ is the numerical wavenumber, which is related to a real wavenumber by

$$\beta_{n\varsigma} = \frac{2}{\Delta\varsigma}\sin\left(\beta_\varsigma\frac{\Delta\varsigma}{2}\right). \tag{7.45}$$

The numerical (grid) dispersion relations of the Crank–Nicolson FDTD and the conventional FDTD formulations are summarized below for comparison.

- Crank–Nicolson FDTD

$$\frac{2}{\Delta t}\tan\left(\omega\frac{\Delta t}{2}\right) = \frac{\beta_n}{\sqrt{\mu\varepsilon}}. \tag{7.46}$$

- FDTD

$$\frac{2}{\Delta t}\sin\left(\omega\frac{\Delta t}{2}\right) = \frac{\beta_n}{\sqrt{\mu\varepsilon}}. \tag{7.47}$$

The above dispersion relations can be used to investigate the numerical stability. Although the ADI-FDTD and Crank–Nicolson FDTD algorithms are unconditionally stable, the conventional FDTD algorithm is conditionally stable [17]. In both ADI-FDTD and Crank–Nicolson FDTD, the inverse tangent function properties indicate that only real-valued frequencies exist for any real-valued wavenumbers. However, the inverse sine function is complex valued for (real-valued) arguments larger than one and thus the FDTD simulation can be unstable.

7.3 LOD-FDTD Algorithm

The LOD-FDTD algorithm is more computationally efficient than the ADI-FDTD algorithm, while maintaining the numerical unconditional stability [7,8,18]. To derive the LOD-FDTD algorithm, Maxwell's curl equations in the matrix form are expressed as

$$\frac{\partial \mathbf{u}}{\partial t} = [\mathcal{A}]\mathbf{u} + [\mathcal{B}]\mathbf{u}. \tag{7.48}$$

Here $[\mathcal{A}]$ and $[\mathcal{B}]$ are continuum operators and $[A]$ and $[B]$ are the corresponding discrete operators on the grid. In the LOD method, the above PDE is factorized into two sub-steps in a *locally one dimensional* fashion as follows:

First sub-step:

$$\frac{1}{2}\frac{\partial \mathbf{u}}{\partial t} = [\mathcal{A}]\mathbf{u}. \tag{7.49}$$

Second sub-step:

$$\frac{1}{2}\frac{\partial \mathbf{u}}{\partial t} = [\mathcal{B}]\mathbf{u}. \tag{7.50}$$

By discretizing the above equations at the $(n+1/4)$th time step and the $(n+3/4)$th time step respectively, and then utilizing CDS, we obtain

$$\left([I] - \frac{\Delta t}{2}[A]\right)\mathbf{u}^{n+1/2} = \left([I] + \frac{\Delta t}{2}[A]\right)\mathbf{u}^{n} \tag{7.51}$$

$$\left([I] - \frac{\Delta t}{2}[B]\right)\mathbf{u}^{n+1} = \left([I] + \frac{\Delta t}{2}[B]\right)\mathbf{u}^{n+1/2}. \tag{7.52}$$

The update equations of each field components are then expressed in the first sub-step as

$$E^{n+1/2}_{x_{i+1/2,j,k}} = E^{n}_{x_{i+1/2,j,k}} + \frac{\Delta t}{2\varepsilon}\left(\frac{H^{n+1/2}_{z_{i+1/2,j+1/2,k}} - H^{n+1/2}_{z_{i+1/2,j-1/2,k}}}{\Delta y}\right) + \frac{\Delta t}{2\varepsilon}\left(\frac{H^{n}_{z_{i+1/2,j+1/2,k}} - H^{n}_{z_{i+1/2,j-1/2,k}}}{\Delta y}\right), \tag{7.53}$$

$$E^{n+1/2}_{y_{i,j+1/2,k}} = E^{n}_{y_{i,j+1/2,k}} + \frac{\Delta t}{2\varepsilon}\left(\frac{H^{n+1/2}_{x_{i,j+1/2,k+1/2}} - H^{n+1/2}_{x_{i,j+1/2,k-1/2}}}{\Delta z}\right) + \frac{\Delta t}{2\varepsilon}\left(\frac{H^{n}_{x_{i,j+1/2,k+1/2}} - H^{n}_{x_{i,j+1/2,k-1/2}}}{\Delta z}\right), \tag{7.54}$$

$$E^{n+1/2}_{z_{i,j,k+1/2}} = E^{n}_{z_{i,j,k+1/2}} + \frac{\Delta t}{2\varepsilon}\left(\frac{H^{n+1/2}_{y_{i+1/2,j,k+1/2}} - H^{n+1/2}_{y_{i-1/2,j,k+1/2}}}{\Delta x}\right)$$

$$+ \frac{\Delta t}{2\varepsilon}\left(\frac{H^{n}_{y_{i+1/2,j,k+1/2}} - H^{n}_{y_{i-1/2,j,k+1/2}}}{\Delta x}\right), \tag{7.55}$$

$$H^{n+1/2}_{x_{i,j+1/2,k+1/2}} = H^{n}_{x_{i,j+1/2,k+1/2}} + \frac{\Delta t}{2\mu}\left(\frac{E^{n+1/2}_{y_{i,j+1/2,k+1}} - E^{n+1/2}_{y_{i,j+1/2,k}}}{\Delta z}\right)$$

$$+ \frac{\Delta t}{2\mu}\left(\frac{E^{n}_{y_{i,j+1/2,k+1}} - E^{n}_{y_{i,j+1/2,k}}}{\Delta z}\right), \tag{7.56}$$

$$H^{n+1/2}_{y_{i+1/2,j,k+1/2}} = H^{n}_{y_{i+1/2,j,k+1/2}} + \frac{\Delta t}{2\mu}\left(\frac{E^{n+1/2}_{z_{i+1,j,k+1/2}} - E^{n+1/2}_{z_{i,j,k+1/2}}}{\Delta x}\right)$$

$$+ \frac{\Delta t}{2\mu}\left(\frac{E^{n}_{z_{i+1,j,k+1/2}} - E^{n}_{z_{i,j,k+1/2}}}{\Delta x}\right), \tag{7.57}$$

$$H^{n+1/2}_{z_{i+1/2,j+1/2,k}} = H^{n}_{z_{i+1/2,j+1/2,k}} + \frac{\Delta t}{2\mu}\left(\frac{E^{n+1/2}_{x_{i+1/2,j+1,k}} - E^{n+1/2}_{x_{i+1/2,j,k}}}{\Delta y}\right)$$

$$+ \frac{\Delta t}{2\mu}\left(\frac{E^{n}_{x_{i+1/2,j+1,k}} - E^{n}_{x_{i+1/2,j,k}}}{\Delta y}\right), \tag{7.58}$$

and in the second sub-step as

$$E^{n+1}_{x_{i+1/2,j,k}} = E^{n+1/2}_{x_{i+1/2,j,k}} - \frac{\Delta t}{2\varepsilon}\left(\frac{H^{n+1}_{y_{i+1/2,j,k+1/2}} - H^{n+1}_{y_{i+1/2,j,k-1/2}}}{\Delta z}\right)$$

$$- \frac{\Delta t}{2\varepsilon}\left(\frac{H^{n+1/2}_{y_{i+1/2,j,k+1/2}} - H^{n+1/2}_{y_{i+1/2,j,k-1/2}}}{\Delta z}\right), \tag{7.59}$$

$$E^{n+1}_{y_{i,j+1/2,k}} = E^{n+1/2}_{y_{i,j+1/2,k}} - \frac{\Delta t}{2\varepsilon}\left(\frac{H^{n+1}_{z_{i+1/2,j+1/2,k}} - H^{n+1}_{z_{i-1/2,j+1/2,k}}}{\Delta x}\right)$$

$$- \frac{\Delta t}{2\varepsilon}\left(\frac{H^{n+1/2}_{z_{i+1/2,j+1/2,k}} - H^{n+1/2}_{z_{i-1/2,j+1/2,k}}}{\Delta x}\right), \tag{7.60}$$

$$E^{n+1}_{z_{i,j,k+1/2}} = E^{n+1/2}_{z_{i,j,k+1/2}} - \frac{\Delta t}{2\varepsilon}\left(\frac{H^{n+1}_{x_{i,j+1/2,k+1/2}} - H^{n+1}_{x_{i,j-1/2,k+1/2}}}{\Delta y}\right)$$
$$- \frac{\Delta t}{2\varepsilon}\left(\frac{H^{n+1/2}_{x_{i,j+1/2,k+1/2}} - H^{n+1/2}_{x_{i,j-1/2,k+1/2}}}{\Delta y}\right), \tag{7.61}$$

$$H^{n+1}_{x_{i,j+1/2,k+1/2}} = H^{n+1/2}_{x_{i,j+1/2,k+1/2}} - \frac{\Delta t}{2\mu}\left(\frac{E^{n+1}_{z_{i,j+1,k+1/2}} - E^{n+1}_{z_{i,j,k+1/2}}}{\Delta y}\right)$$
$$- \frac{\Delta t}{2\mu}\left(\frac{E^{n+1/2}_{z_{i,j+1,k+1/2}} - E^{n+1/2}_{z_{i,j,k+1/2}}}{\Delta y}\right), \tag{7.62}$$

$$H^{n+1}_{y_{i+1/2,j,k+1/2}} = H^{n+1/2}_{y_{i+1/2,j,k+1/2}} - \frac{\Delta t}{2\mu}\left(\frac{E^{n+1}_{x_{i+1/2,j,k+1}} - E^{n+1}_{x_{i+1/2,j,k}}}{\Delta z}\right)$$
$$- \frac{\Delta t}{2\mu}\left(\frac{E^{n+1/2}_{x_{i+1/2,j,k+1}} - E^{n+1/2}_{x_{i+1/2,j,k}}}{\Delta z}\right), \tag{7.63}$$

$$H^{n+1}_{z_{i+1/2,j+1/2,k}} = H^{n+1/2}_{z_{i+1/2,j+1/2,k}} - \frac{\Delta t}{2\mu}\left(\frac{E^{n+1}_{y_{i+1,j+1/2,k}} - E^{n+1}_{y_{i,j+1/2,k}}}{\Delta x}\right)$$
$$- \frac{\Delta t}{2\mu}\left(\frac{E^{n+1/2}_{y_{i+1,j+1/2,k}} - E^{n+1/2}_{y_{i,j+1/2,k}}}{\Delta x}\right). \tag{7.64}$$

Implicit update equations for the electric field components are derived by inserting the update equations of the magnetic field components into those of the electric field components. For example, in the first sub-step, the implicit update equation of the E_x field component is written as

$$-\frac{C_b D_b}{(\Delta y)^2} E^{n+1/2}_{x_{i+1/2,j-1,k}} + \left(1 + \frac{2C_b D_b}{(\Delta y)^2}\right) E^{n+1/2}_{x_{i+1/2,j,k}} - \frac{C_b D_b}{(\Delta y)^2} E^{n+1/2}_{x_{i+1/2,j+1,k}}$$
$$= \left(1 - \frac{2C_b D_b}{(\Delta y)^2}\right) E^{n}_{x_{i+1/2,j,k}} + \frac{2C_b}{\Delta y}\left(H^{n}_{z_{i+1/2,j+1/2,k}} - H^{n}_{z_{i+1/2,j-1/2,k}}\right)$$
$$+ \frac{C_b D_b}{(\Delta y)^2}\left(E^{n}_{x_{i+1/2,j-1,k}} + E^{n}_{x_{i+1/2,j+1,k}}\right). \tag{7.65}$$

The number of arithmetic operations on the RHS in the LOD-FDTD algorithm is less than in the ADI-FDTD algorithm, implying a better computational efficiency. Similar to the ADI-FDTD algorithm, the tridiagonal-matrix-form update equation of the electric field components can be efficiently solved by the Thomas algorithm in the LOD-FDTD algorithm. The overall update procedure of the LOD-FDTD algorithm can be summarized as follows:

Step 1. Update $\mathbf{E}^{n+1/2}$ implicitly (e.g. from (7.65)).
Step 2. Update $\mathbf{H}^{n+1/2}$ explicitly from (7.56) to (7.58).
Step 3. Update $\mathbf{E}^{n+1}$ implicitly.
Step 4. Update $\mathbf{H}^{n+1}$ explicitly from (7.62) to (7.64).

Note that the LOD-FDTD and ADI-FDTD dispersion relations are identical, which implies that they have the same dispersion error [8]. Similar to the ADI-FDTD algorithm, an inverse tangent function is involved in the LOD-FDTD dispersion relation so the unconditional stability of the algorithm is again evident.

7.4 PML implementation

When using PDE-based numerical methods such as finite difference and finite element methods, an absorbing boundary condition is necessary in open-boundary problems to minimize spurious reflections from the outer boundary of the (finite) computational domain. Of the absorbing boundary conditions available, the PML is the most popular due to its relative simplicity, efficiency, and robustness. The PML can be implemented using an anisotropic absorbing medium [19] or a stretched coordinate (SC) approach [11,12]. The anisotropic PML implementation is less straightforward in complex media. In the SC-based PML implementation, however, the introduction of a modified del operator is independent of the host medium properties, which greatly facilitates implementation.

In the SC-based PML formulation, a modified nabla operator is introduced as

$$\nabla \rightarrow \widetilde{\nabla} = \widehat{x}\frac{1}{s_x}\frac{\partial}{\partial x} + \widehat{y}\frac{1}{s_y}\frac{\partial}{\partial y} + \widehat{z}\frac{1}{s_z}\frac{\partial}{\partial z}, \tag{7.66}$$

where the complex stretching variables s_ς are given by

$$s_\varsigma = \kappa_\varsigma + \frac{\sigma_\varsigma}{\nu_\varsigma + j\omega\varepsilon_0} \tag{7.67}$$

and $\varsigma = x$, y, or z. In the PML implementation, $\sigma_\varsigma > 0$, $\kappa_\varsigma \geq 1$ and $\nu_\varsigma \geq 0$. The artificial conductivity σ_ς is used to absorb traveling electromagnetic waves along the ς, while the real stretching parameter κ_ς is used to increase attenuation of evanescent waves along that direction. When $\nu_\varsigma = 0$, the complex stretching variables have a pole at zero frequency, which leads to large reflections (or even instability) at low frequencies. The parameter $\nu_\varsigma > 0$ moves the pole into the complex plane to ensure late time stability for low-frequency waves. This PML implementation of the complex stretching variables is known as complex-frequency-shifted (CFS) PML [20,21]. The CFS-PML implementation also eliminates spurious late-time linear growth of axial fields [22]. Note that a graded profile should be employed for all PML parameters to reduce numerical reflections in the discrete FDTD grid [2].

The CFS-PML implementation to the ADI-FDTD algorithm [23] is next detailed. An analogous implementation can be adopted for the LOD-FDTD formulation without difficulty. Using the modified nabla operator, the field components in the CFS-PML can be written as

$$\varepsilon\frac{\partial E_x}{\partial t}=\frac{1}{\kappa_y}\frac{\partial H_z}{\partial y}-\frac{1}{\kappa_z}\frac{\partial H_y}{\partial z}+f_{xy}-f_{xz}, \tag{7.68a}$$

$$\varepsilon\frac{\partial E_y}{\partial t}=\frac{1}{\kappa_z}\frac{\partial H_x}{\partial z}-\frac{1}{\kappa_x}\frac{\partial H_z}{\partial x}+f_{yz}-f_{yx}, \tag{7.68b}$$

$$\varepsilon\frac{\partial E_z}{\partial t}=\frac{1}{\kappa_x}\frac{\partial H_y}{\partial x}-\frac{1}{\kappa_y}\frac{\partial H_x}{\partial y}+f_{zx}-f_{zy}, \tag{7.68c}$$

$$\mu\frac{\partial H_x}{\partial t}=\frac{1}{\kappa_z}\frac{\partial E_y}{\partial z}-\frac{1}{\kappa_y}\frac{\partial E_z}{\partial y}+g_{xz}-g_{xy}, \tag{7.68d}$$

$$\mu\frac{\partial H_y}{\partial t}=\frac{1}{\kappa_x}\frac{\partial E_z}{\partial x}-\frac{1}{\kappa_z}\frac{\partial E_x}{\partial z}+g_{yx}-g_{yz}, \tag{7.68e}$$

$$\mu\frac{\partial H_z}{\partial t}=\frac{1}{\kappa_y}\frac{\partial E_x}{\partial y}-\frac{1}{\kappa_x}\frac{\partial E_y}{\partial x}+g_{zy}-g_{zx}. \tag{7.68f}$$

The introduction of the 12 PML auxiliary variables seen in the RHS of the above equations will be addressed shortly. In the ADI method, the following update equations are then used:

In the first sub-step,

$$\begin{aligned}E^{n+1/2}_{x_{i+1/2,j,k}}&=E^{n}_{x_{i+1/2,j,k}}+\frac{\Delta t}{2\varepsilon\kappa_{y_j}}\left(\frac{H^{n+1/2}_{z_{i+1/2,j+1/2,k}}-H^{n+1/2}_{z_{i+1/2,j-1/2,k}}}{\Delta y}\right)\\&\quad-\frac{\Delta t}{2\varepsilon\kappa_{z_k}}\left(\frac{H^{n}_{y_{i+1/2,j,k+1/2}}-H^{n}_{y_{i+1/2,j,k-1/2}}}{\Delta z}\right)+\frac{\Delta t}{2\varepsilon}f^{n}_{xy_{i+1/2,j,k}}-\frac{\Delta t}{2\varepsilon}f^{n}_{xz_{i+1/2,j,k}},\end{aligned} \tag{7.69}$$

$$\begin{aligned}E^{n+1/2}_{y_{i,j+1/2,k}}&=E^{n}_{y_{i,j+1/2,k}}+\frac{\Delta t}{2\varepsilon\kappa_{z_k}}\left(\frac{H^{n+1/2}_{x_{i,j+1/2,k+1/2}}-H^{n+1/2}_{x_{i,j+1/2,k-1/2}}}{\Delta z}\right)\\&\quad-\frac{\Delta t}{2\varepsilon\kappa_{x_i}}\left(\frac{H^{n}_{z_{i+1/2,j+1/2,k}}-H^{n}_{z_{i-1/2,j+1/2,k}}}{\Delta x}\right)+\frac{\Delta t}{2\varepsilon}f^{n}_{yz_{i,j+1/2,k}}-\frac{\Delta t}{2\varepsilon}f^{n}_{yx_{i,j+1/2,k}},\end{aligned} \tag{7.70}$$

$$\begin{aligned}E^{n+1/2}_{z_{i,j,k+1/2}}&=E^{n}_{z_{i,j,k+1/2}}+\frac{\Delta t}{2\varepsilon\kappa_{x_i}}\left(\frac{H^{n+1/2}_{y_{i+1/2,j,k+1/2}}-H^{n+1/2}_{y_{i-1/2,j,k+1/2}}}{\Delta x}\right)\\&\quad-\frac{\Delta t}{2\varepsilon\kappa_{y_j}}\left(\frac{H^{n}_{x_{i,j+1/2,k+1/2}}-H^{n}_{x_{i,j-1/2,k+1/2}}}{\Delta y}\right)+\frac{\Delta t}{2\varepsilon}f^{n}_{zx_{i,j,k+1/2}}-\frac{\Delta t}{2\varepsilon}f^{n}_{zy_{i,j,k+1/2}},\end{aligned} \tag{7.71}$$

$$H^{n+1/2}_{x_{i,j+1/2,k+1/2}} = H^{n}_{x_{i,j+1/2,k+1/2}} + \frac{\Delta t}{2\mu\kappa_{z_{k+1/2}}}\left(\frac{E^{n+1/2}_{y_{i,j+1/2,k+1}} - E^{n+1/2}_{y_{i,j+1/2,k}}}{\Delta z}\right)$$
$$- \frac{\Delta t}{2\mu\kappa_{y_{j+1/2}}}\left(\frac{E^{n}_{z_{i,j+1,k+1/2}} - E^{n}_{z_{i,j,k+1/2}}}{\Delta y}\right) + \frac{\Delta t}{2\mu} g^{n}_{xz_{i,j+1/2,k+1/2}} - \frac{\Delta t}{2\mu} g^{n}_{xy_{i,j+1/2,k+1/2}}, \tag{7.72}$$

$$H^{n+1/2}_{y_{i+1/2,j,k+1/2}} = H^{n}_{y_{i+1/2,j,k+1/2}} + \frac{\Delta t}{2\mu\kappa_{x_{i+1/2}}}\left(\frac{E^{n+1/2}_{z_{i+1,j,k+1/2}} - E^{n+1/2}_{z_{i,j,k+1/2}}}{\Delta x}\right)$$
$$- \frac{\Delta t}{2\mu\kappa_{z_{k+1/2}}}\left(\frac{E^{n}_{x_{i+1/2,j,k+1}} - E^{n}_{x_{i+1/2,j,k}}}{\Delta z}\right) + \frac{\Delta t}{2\mu} g^{n}_{yx_{i+1/2,j,k+1/2}} - \frac{\Delta t}{2\mu} g^{n}_{yz_{i+1/2,j,k+1/2}}, \tag{7.73}$$

$$H^{n+1/2}_{z_{i+1/2,j+1/2,k}} = H^{n}_{z_{i+1/2,j+1/2,k}} + \frac{\Delta t}{2\mu\kappa_{y_{j+1/2}}}\left(\frac{E^{n+1/2}_{x_{i+1/2,j+1,k}} - E^{n+1/2}_{x_{i+1/2,j,k}}}{\Delta y}\right)$$
$$- \frac{\Delta t}{2\mu\kappa_{x_{i+1/2}}}\left(\frac{E^{n}_{y_{i+1,j+1/2,k}} - E^{n}_{y_{i,j+1/2,k}}}{\Delta x}\right) + \frac{\Delta t}{2\mu} g^{n}_{zy_{i+1/2,j+1/2,k}} - \frac{\Delta t}{2\mu} g^{n}_{zx_{i+1/2,j+1/2,k}}. \tag{7.74}$$

In the second sub-step,

$$E^{n+1}_{x_{i+1/2,j,k}} = E^{n+1/2}_{x_{i+1/2,j,k}} + \frac{\Delta t}{2\varepsilon\kappa_{y_j}}\left(\frac{H^{n+1/2}_{z_{i+1/2,j+1/2,k}} - H^{n+1/2}_{z_{i+1/2,j-1/2,k}}}{\Delta y}\right)$$
$$- \frac{\Delta t}{2\varepsilon\kappa_{z_k}}\left(\frac{H^{n+1}_{y_{i+1/2,j,k+1/2}} - H^{n+1}_{y_{i+1/2,j,k-1/2}}}{\Delta z}\right) + \frac{\Delta t}{2\varepsilon} f^{n+1/2}_{xy_{i+1/2,j,k}} - \frac{\Delta t}{2\varepsilon} f^{n+1/2}_{xz_{i+1/2,j,k}}, \tag{7.75}$$

$$E^{n+1}_{y_{i,j+1/2,k}} = E^{n+1/2}_{y_{i,j+1/2,k}} + \frac{\Delta t}{2\varepsilon\kappa_{z_k}}\left(\frac{H^{n+1/2}_{x_{i,j+1/2,k+1/2}} - H^{n+1/2}_{x_{i,j+1/2,k-1/2}}}{\Delta z}\right)$$
$$- \frac{\Delta t}{2\varepsilon\kappa_{x_i}}\left(\frac{H^{n+1}_{z_{i+1/2,j+1/2,k}} - H^{n+1}_{z_{i-1/2,j+1/2,k}}}{\Delta x}\right) + \frac{\Delta t}{2\varepsilon} f^{n+1/2}_{yz_{i,j+1/2,k}} - \frac{\Delta t}{2\varepsilon} f^{n+1/2}_{yx_{i,j+1/2,k}}, \tag{7.76}$$

$$E^{n+1}_{z_{i,j,k+1/2}} = E^{n+1/2}_{z_{i,j,k+1/2}} + \frac{\Delta t}{2\varepsilon\kappa_{x_i}}\left(\frac{H^{n+1/2}_{y_{i+1/2,j,k+1/2}} - H^{n+1/2}_{y_{i-1/2,j,k+1/2}}}{\Delta x}\right)$$
$$- \frac{\Delta t}{2\varepsilon\kappa_{y_j}}\left(\frac{H^{n+1}_{x_{i,j+1/2,k+1/2}} - H^{n+1}_{x_{i,j-1/2,k+1/2}}}{\Delta y}\right) + \frac{\Delta t}{2\varepsilon} f^{n+1/2}_{zx_{i,j,k+1/2}} - \frac{\Delta t}{2\varepsilon} f^{n+1/2}_{zy_{i,j,k+1/2}}, \tag{7.77}$$

$$H^{n+1}_{x_{i,j+1/2,k+1/2}} = H^{n+1/2}_{x_{i,j+1/2,k+1/2}} + \frac{\Delta t}{2\mu\kappa_{z_{k+1/2}}}\left(\frac{E^{n+1/2}_{y_{i,j+1/2,k+1}} - E^{n+1/2}_{y_{i,j+1/2,k}}}{\Delta z}\right)$$
$$-\frac{\Delta t}{2\mu\kappa_{y_{j+1/2}}}\left(\frac{E^{n+1}_{z_{i,j+1,k+1/2}} - E^{n+1}_{z_{i,j,k+1/2}}}{\Delta y}\right) + \frac{\Delta t}{2\mu} g^{n+1/2}_{xz_{i,j+1/2,k+1/2}} - \frac{\Delta t}{2\mu} g^{n+1/2}_{xy_{i,j+1/2,k+1/2}}, \tag{7.78}$$

$$H^{n+1}_{y_{i+1/2,j,k+1/2}} = H^{n+1/2}_{y_{i+1/2,j,k+1/2}} + \frac{\Delta t}{2\mu\kappa_{x_{i+1/2}}}\left(\frac{E^{n+1/2}_{z_{i+1,j,k+1/2}} - E^{n+1/2}_{z_{i,j,k+1/2}}}{\Delta x}\right)$$
$$-\frac{\Delta t}{2\mu\kappa_{z_{k+1/2}}}\left(\frac{E^{n+1}_{x_{i+1/2,j,k+1}} - E^{n+1}_{x_{i+1/2,j,k}}}{\Delta z}\right) + \frac{\Delta t}{2\mu} g^{n+1/2}_{yx_{i+1/2,j,k+1/2}} - \frac{\Delta t}{2\mu} g^{n+1/2}_{yz_{i+1/2,j,k+1/2}}, \tag{7.79}$$

$$H^{n+1}_{z_{i+1/2,j+1/2,k}} = H^{n+1/2}_{z_{i+1/2,j+1/2,k}} + \frac{\Delta t}{2\mu\kappa_{y_{j+1/2}}}\left(\frac{E^{n+1/2}_{x_{i+1/2,j+1,k}} - E^{n+1/2}_{x_{i+1/2,j,k}}}{\Delta y}\right)$$
$$-\frac{\Delta t}{2\mu\kappa_{x_{i+1/2}}}\left(\frac{E^{n+1}_{y_{i+1,j+1/2,k}} - E^{n+1}_{y_{i,j+1/2,k}}}{\Delta x}\right) + \frac{\Delta t}{2\mu} g^{n+1/2}_{zy_{i+1/2,j+1/2,k}} - \frac{\Delta t}{2\mu} g^{n+1/2}_{zx_{i+1/2,j+1/2,k}}. \tag{7.80}$$

In the above CFS-PML ADI-FDTD algorithm, the real part of the stretching parameters and extra auxiliary variables appear, which differs from the non-PML counterpart. Note that all auxiliary variables in the RHS are defined at a half-time step earlier than the field components in the LHS. The result is a simpler implementation and improved absorbing performance compared to the alternative PML implementation in which the time step for the one of auxiliary variables in the RHS is synchronous to a field component in the LHS [23]. The update equations of PML auxiliary variables can be implemented using recursive convolution. For example, one can write

$$f^{n+1/2}_{xy_{i+1/2,j,k}} = C_{fay_j} f^{n}_{xy_{i+1/2,j,k}} + C_{fby_j}\left(\frac{H^{n+1/2}_{z_{i+1/2,j+1/2,k}} - H^{n+1/2}_{z_{i+1/2,j-1/2,k}}}{\Delta y}\right), \tag{7.81}$$

$$g^{n+1/2}_{xy_{i,j+1/2,k+1/2}} = C_{gay_{j+1/2}} g^{n}_{xy_{i,j+1/2,k+1/2}} + C_{gby_{j+1/2}}\left(\frac{E^{n+1/2}_{z_{i,j+1,k+1/2}} - E^{n+1/2}_{z_{i,j,k+1/2}}}{\Delta y}\right), \tag{7.82}$$

$$f^{n+1}_{xy_{i+1/2,j,k}} = C_{fay_j} f^{n+1/2}_{xy_{i+1/2,j,k}} + C_{fby_j}\left(\frac{H^{n+1}_{z_{i+1/2,j+1/2,k}} - H^{n+1}_{z_{i+1/2,j-1/2,k}}}{\Delta y}\right), \tag{7.83}$$

$$g^{n+1}_{xy_{i,j+1/2,k+1/2}} = C_{gay_{j+1/2}} g^{n+1/2}_{xy_{i,j+1/2,k+1/2}} + C_{gby_{j+1/2}}\left(\frac{E^{n+1}_{z_{i,j+1,k+1/2}} - E^{n+1}_{z_{i,j,k+1/2}}}{\Delta y}\right), \tag{7.84}$$

with

$$C_{fay_j} = e^{-(\sigma_j/\kappa_j+\nu_j)\left(\frac{\Delta t}{2\varepsilon_0}\right)},$$

$$C_{fby_j} = \frac{\sigma_j}{\kappa_j(\sigma_j+\kappa_j\nu_j)}(C_{fay_j}-1),$$

$$C_{gay_{j+1/2}} = e^{-(\sigma_{j+1/2}/\kappa_{j+1/2}+\nu_{j+1/2})\left(\frac{\Delta t}{2\varepsilon_0}\right)},$$

$$C_{gby_{j+1/2}} = \frac{\sigma_{j+1/2}}{\kappa_{j+1/2}(\sigma_{j+1/2}+\kappa_{j+1/2}\nu_{j+1/2})}\left(C_{gay_{j+1/2}}-1\right).$$

In both sub-steps, substituting the update equations of the magnetic field components into the corresponding update equations of the electric field components yields the implicit (tridiagonal-matrix-form) update equations of the electric field components, which is again efficiently solved by the Thomas algorithm. The update steps of the CFS-PML ADI-FDTD algorithm is summarized as follows

Step 1. Update $\mathbf{E}^{n+1/2}$ implicitly.
Step 2. Update $\mathbf{H}^{n+1/2}$ explicitly.
Step 3. Update $\mathbf{f}^{n+1/2}$ explicitly.
Step 4. Update $\mathbf{g}^{n+1/2}$ explicitly.
Step 5. Update $\mathbf{E}^{n+1}$ implicitly.
Step 6. Update $\mathbf{H}^{n+1}$ explicitly.
Step 7. Update $\mathbf{f}^{n+1}$ explicitly.
Step 8. Update $\mathbf{g}^{n+1}$ explicitly.

Several PML implementations are available in the literature for both the ADI-FDTD [24–26] and the LOD-FDTD algorithms [27,28].

7.5 Iterative approach to reduce the splitting error

Both dispersion and splitting errors are present in the ADI-FDTD and LOD-FDTD algorithms. In this section, the origin of the splitting error and the use of iterative approaches to reduce it are discussed. Toward this purpose, the update equations of the ADI-FDTD and LOD-FDTD algorithms are first contrasted to those of the Crank–Nicolson FDTD algorithm.

7.5.1 Iterative ADI-FDTD algorithm

The update equations (7.35) of the Crank–Nicolson FDTD algorithm can be written as

$$\left([I]-\frac{\Delta t}{2}[A]\right)\left([I]-\frac{\Delta t}{2}[B]\right)\mathbf{u}^{n+1} = \left([I]+\frac{\Delta t}{2}[A]\right)\left([I]+\frac{\Delta t}{2}[B]\right)\mathbf{u}^{n} + \left(\frac{\Delta t}{2}\right)^2[A][B]\left(\mathbf{u}^{n+1}-\mathbf{u}^{n}\right). \tag{7.85}$$

The one time-step update equations of the ADI-FDTD algorithm is rewritten below for easy comparison

$$\left([I]-\frac{\Delta t}{2}[A]\right)\left([I]-\frac{\Delta t}{2}[B]\right)\mathbf{u}^{n+1}=\left([I]+\frac{\Delta t}{2}[A]\right)\left([I]+\frac{\Delta t}{2}[B]\right)\mathbf{u}^{n}. \tag{7.86}$$

A comparison of (7.86) with (7.85) indicates the following splitting error in the ADI-FDTD algorithm:

$$\left(\frac{\Delta t}{2}\right)^2[A][B]\left(\mathbf{u}^{n+1}-\mathbf{u}^{n}\right). \tag{7.87}$$

Note that this splitting error causes a deterioration of the ADI-FDTD solution with an increase in the time step size, the spatial derivatives of the field components, and/or the temporal variation of the field components. However, the following iterative approach can be used to improve the ADI-FDTD algorithm accuracy [13]

$$\begin{aligned}\left([I]-\frac{\Delta t}{2}[A]\right)\left([I]-\frac{\Delta t}{2}[B]\right)\mathbf{u}_{k+1}^{n+1}=&\left([I]+\frac{\Delta t}{2}[A]\right)\left([I]+\frac{\Delta t}{2}[B]\right)\mathbf{u}^{n}\\&+\left(\frac{\Delta t}{2}\right)^2[A][B]\left(\mathbf{u}_{k}^{n+1}-\mathbf{u}^{n}\right),\end{aligned} \tag{7.88}$$

where the subscript k denotes the kth iteration. The initial guess for the iterative term is usually set to be equal to the previous time step value, i.e., $\mathbf{u}_0^{n+1}=\mathbf{u}^n$. As mentioned previously, poor computational efficiency precludes the use of the one-step implementation of the ADI-FDTD algorithm. Therefore, the Peaceman–Rachford scheme is used to obtain the two-step implementation of the iterative ADI-FDTD algorithm as follows:

$$\left([I]-\frac{\Delta t}{2}[A]\right)\mathbf{u}^{n+1/2}=\left([I]+\frac{\Delta t}{2}[B]\right)\mathbf{u}^{n}+\frac{1}{2}\left(\frac{\Delta t}{2}\right)^2[A][B]\left(\mathbf{u}_{k}^{n+1}-\mathbf{u}^{n}\right), \tag{7.89}$$

$$\left([I]-\frac{\Delta t}{2}[B]\right)\mathbf{u}_{k+1}^{n+1}=\left([I]+\frac{\Delta t}{2}[A]\right)\mathbf{u}^{n+1/2}+\frac{1}{2}\left(\frac{\Delta t}{2}\right)^2[A][B]\left(\mathbf{u}_{k}^{n+1}-\mathbf{u}^{n}\right). \tag{7.90}$$

In this implementation, the splitting error correction term is decomposed into two halves and then considered equally in both sub-steps. Note that because the TE/TM coupling prevents proper function of the iterative ADI-FDTD [29], the iterative ADI-FDTD implementation is primarily applicable to 1-D or 2-D problems or to certain 3-D problems where such coupling is not present [30].

7.5.2 Iterative LOD-FDTD algorithm

Let us first investigate the splitting error of the LOD-FDTD algorithm and then apply the iterative approach to reduce this error. To determine the splitting error, a one-time step expression of the LOD-FDTD algorithm should be derived. By multiplying (7.52) with $\left([I] - \frac{\Delta t}{2}[A]\right)$, we have

$$\left([I] - \frac{\Delta t}{2}[A]\right)\left([I] - \frac{\Delta t}{2}[B]\right)\mathbf{u}^{n+1} = \left([I] - \frac{\Delta t}{2}[A]\right)\left([I] + \frac{\Delta t}{2}[B]\right)\mathbf{u}^{n+1/2}, \tag{7.91}$$

Rewriting the above, we have

$$\left([I]-\frac{\Delta t}{2}[A]\right)\left([I]-\frac{\Delta t}{2}[B]\right)\mathbf{u}^{n+1} = \left([I]+\frac{\Delta t}{2}[B]\right)\left([I]-\frac{\Delta t}{2}[A]\right)\mathbf{u}^{n+1/2} + \left(\frac{\Delta t}{2}\right)^2([B][A]-[A][B])\mathbf{u}^{n+1/2}, \tag{7.92}$$

Using (7.51), we have the following the one time-step expression of the LOD-FDTD algorithm:

$$\left([I] - \frac{\Delta t}{2}[A]\right)\left([I] - \frac{\Delta t}{2}[B]\right)\mathbf{u}^{n+1} = \left([I] + \frac{\Delta t}{2}[B]\right)\left([I] + \frac{\Delta t}{2}[A]\right)\mathbf{u}^{n} + \left(\frac{\Delta t}{2}\right)^2([B][A] - [A][B])\mathbf{u}^{n+1/2}, \tag{7.93}$$

The splitting error of the LOD-FDTD algorithm can then be obtained by comparing (7.93) with (7.35):

$$\left(\frac{\Delta t}{2}\right)^2[A][B]\left(\mathbf{u}^{n+1} + \mathbf{u}^{n+1/2}\right) - \left(\frac{\Delta t}{2}\right)^2[B][A]\left(\mathbf{u}^{n+1/2} + \mathbf{u}^{n}\right), \tag{7.94}$$

Note that the splitting error of the LOD-FDTD algorithm (7.94) has an extra non-commutativity term as compared to the ADI-FDTD algorithm (7.87). However, this extra noncommutativity error is negligible in many practical problems. Based on the iterative approach, the one time-step expression of the iterative LOD-FDTD algorithm may be written as follows [15]:

$$\begin{aligned}\left([I] - \frac{\Delta t}{2}[A]\right)\left([I] - \frac{\Delta t}{2}[B]\right)\mathbf{u}_{k+1}^{n+1} &= \left([I] + \frac{\Delta t}{2}[B]\right)\left([I] + \frac{\Delta t}{2}[A]\right)\mathbf{u}^{n}\\ &+\left(\frac{\Delta t}{2}\right)^2([B][A] - [A][B])\mathbf{u}^{n+1/2} + \left(\frac{\Delta t}{2}\right)^2[A][B](\mathbf{u}_k^{n+1} + \mathbf{u}^{n+1/2})\\ &-\left(\frac{\Delta t}{2}\right)^2[B][A]\left(\mathbf{u}^{n+1/2} + \mathbf{u}^{n}\right).\end{aligned} \tag{7.95}$$

The above one time-step implementation is not efficient and thus the two-step implementation of the iterative LOD-FDTD algorithm should be adopted instead. In general, however, this is not straightforward for 3-D problems and so a 2-D TE $_z$ problem is considered here. In this case, the field vector and the matrices are written as

$$\mathbf{u} = \left[E_x, E_y, H_z\right]^T, \tag{7.96}$$

$$[A] = \begin{bmatrix} 0 & 0 & \delta_y/\varepsilon \\ 0 & 0 & 0 \\ \delta_y/\mu & 0 & 0 \end{bmatrix}, \tag{7.97}$$

$$[B] = -\begin{bmatrix} 0 & 0 & 0 \\ 0 & 0 & \delta_x/\varepsilon \\ 0 & \delta_x/\mu & 0 \end{bmatrix}. \tag{7.98}$$

Two approaches are available for the two-step implementation depending on the sequence of the splitting error correction terms.

In the first approach,

$$\left([I] - \frac{\Delta t}{2}[A]\right)\mathbf{u}^{n+1/2} = \left([I] + \frac{\Delta t}{2}[A]\right)\mathbf{u}^n + \left(\frac{\Delta t}{2}\right)^2 [A][B]\left(\mathbf{u}_k^{n+1} + \mathbf{u}^{n+1/2}\right), \tag{7.99}$$

$$\left([I] - \frac{\Delta t}{2}[B]\right)\mathbf{u}_{k+1}^{n+1} = \left([I] + \frac{\Delta t}{2}[B]\right)\mathbf{u}^{n+1/2} - \left(\frac{\Delta t}{2}\right)^2 [B][A]\left(\mathbf{u}^{n+1/2} + \mathbf{u}^n\right). \tag{7.100}$$

Note the existence of the extra terms in the above two-step implementation (7.99) and (7.100) compared to the one time-step implementation (7.95):

$$\left(\frac{\Delta t}{2}\right)^3 [B][A][B]\left(\mathbf{u}_k^{n+1} + \mathbf{u}^{n+1/2}\right) + \left(\frac{\Delta t}{2}\right)^3 [A][B][A]\left(\mathbf{u}^{n+1/2} + \mathbf{u}^n\right). \tag{7.101}$$

In 2-D problems, the identities $[B][A][B] = [0]$ and $[A][B][A] = [0]$ always hold, making the first approach for the two-step iterative LOD-FDTD implementation equivalent to the one-step iterative LOD-FDTD implementation. The same iterative LOD-FDTD formulation can be applied to 2-D TM $_z$ problems with corresponding **u**, $[A]$, and $[B]$. However, $[B][A][B] \neq [0]$ and $[A][B][A] \neq [0]$ in general 3-D problems.

In the second approach,

$$\left([I] - \frac{\Delta t}{2}[A]\right)\mathbf{u}^{n+1/2} = \left([I] + \frac{\Delta t}{2}[A]\right)\mathbf{u}^n - \left(\frac{\Delta t}{2}\right)^2 [B][A]\left(\mathbf{u}^{n+1/2} + \mathbf{u}^n\right), \tag{7.102}$$

$$\left([I]-\frac{\Delta t}{2}[B]\right)\mathbf{u}_{k+1}^{n+1}=\left([I]+\frac{\Delta t}{2}[B]\right)\mathbf{u}^{n+1/2}+\left(\frac{\Delta t}{2}\right)^2[A][B]\left(\mathbf{u}_k^{n+1}+\mathbf{u}^{n+1/2}\right). \tag{7.103}$$

In this approach, the extra terms are as follows

$$-\left(\frac{\Delta t}{2}\right)^3[A][A][B]\left(\mathbf{u}_k^{n+1}+\mathbf{u}^{n+1/2}\right)-\left(\frac{\Delta t}{2}\right)^3[B][B][A]\left(\mathbf{u}^{n+1/2}+\mathbf{u}^{n}\right). \tag{7.104}$$

Note, however, that this alternative approach is not viable given that $[A][A][B]\neq[0]$ and $[B][B][A]\neq[0]$ in general.

Next, a 2-D TE $_z$ numerical example is employed to illustrate the effect of the iterative approach on reducing the splitting error in both the ADI-FDTD and LOD-FDTD algorithms [15,31]. We calculate the E_y field distribution at the center of a parallel plate waveguide (2 m width and 0.2 m separation) that is surrounded by perfect magnetic conductor (PMC) walls, as shown in Figure 7.1. The hard-source voltage excitation between the perfect electric conductors (PECs) plates is given by

$$v(t)=\begin{cases}\frac{1}{2}[1-\cos(2\pi f_0 t)] & 0\leq t\leq 0.66\ \mu\text{s}\\ 1 & t\geq 0.66\ \mu\text{s}\end{cases}$$

Here $f_0=750$ kHz is used. The space step size is $\Delta x=\Delta y=0.2$ m, corresponding to 2,000 points-per-wavelength at f_0. The Courant factor s is defined as $s=c_0\Delta t/\Delta x$, where c_0 is the speed of light in vacuum. Figure 7.2 shows E_y along the center plane of the parallel plate waveguide at $t=10$ μs, which is calculated using the iterative ADI-FDTD simulation with $s=5$ and the iterative LOD-FDTD simulation with $s=5$. The FDTD simulation with $s=0.5$ is used as reference solution. Note that E_y is plotted along $x=0-5$ m due to the symmetrical geometry. The effect of the iterative approach on reducing the splitting error is clearly

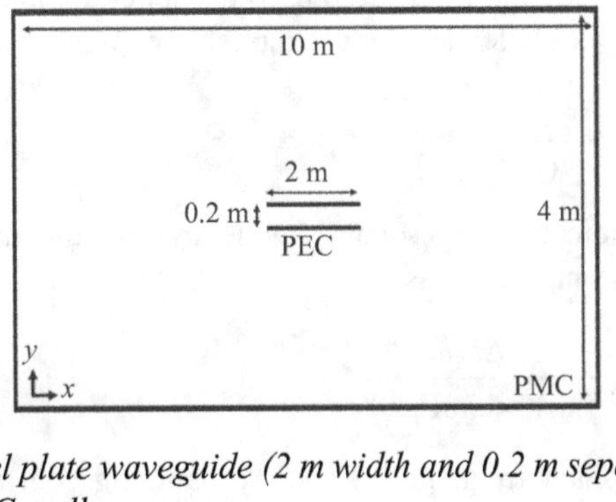

Figure 7.1 Parallel plate waveguide (2 m width and 0.2 m separation) surrounded by PMC walls

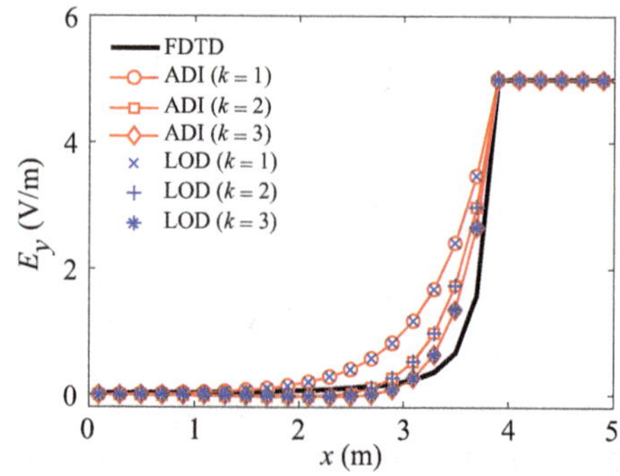

Figure 7.2 E_y along the x-axis calculated by the FDTD simulation, the iterative ADI-FDTD simulation, and the iterative LOD-FDTD simulation

observed in this figure. It is also observed that the extra noncommutativity error inherent to the LOD-FDTD algorithm is negligible in this example.

7.6 Extension to complex dispersive media

Material dispersion models should be employed to incorporate the frequency dependent behavior of complex media in wideband problems. Popular material dispersion models include Debye [32,33], Drude [34,35], and Lorentz models [32,36]. Of these, the Debye model is usually employed for lossy media such as soil, water, and certain human tissues. The relative permittivity in the Debye model is expressed as

$$\varepsilon_r(\omega) = \varepsilon_{r,\infty} + \frac{\varepsilon_s - \varepsilon_{r,\infty}}{1 + j\omega\tau}, \tag{7.105}$$

where $\varepsilon_{r,\infty}$ is the relative permittivity in the limit of very large frequencies (in the model range), ε_s is the static relative permittivity, and τ indicates the relaxation time constant. The Drude model is often used to model noble metals in the optical frequency range, for example. The Drude model is expressed as

$$\varepsilon_r(\omega) = \varepsilon_{r,\infty} + \frac{\omega_D^2}{j\omega(\Gamma_D + j\omega)}, \tag{7.106}$$

where ω_D and Γ indicate the resonant frequency and the damping coefficient, respectively. The Lorentz model is expressed as

$$\varepsilon_r(\omega) = \varepsilon_{r,\infty} + \frac{\Delta\varepsilon_L \omega_L^2}{(j\omega)^2 + j\omega\Gamma_L + \omega_L^2}, \tag{7.107}$$

where ω_L is the resonant frequency, Γ_L is the damping coefficient, and $\Delta\varepsilon_L$ is a weighting coefficient.

In this section, an extension of the ADI-FDTD algorithm to complex dispersive media is presented, specifically to the Drude–Lorentz multispecies dispersion model. ADI-FDTD and LOD-FDTD formulations for other dispersion models can be developed in a similar fashion. Although several time-domain approaches (e.g. piecewise linear recursive convolution [37], Z transform [38], and the auxiliary differential equation (ADE) [39]) are available in conjunction with such implementations, the ADE method is presented here due to its relative simplicity, efficiency, and easy extension to other complex (e.g. nonlinear) media [40,41].

Let us consider Au (gold) in the 500–1,000 nm wavelength range. It is well known that the simple Drude model cannot fit the Au dispersion in this entire wavelength range due to interband transitions below 700 nm. Alternatively, the Drude–Lorentz multispecies dispersion model should be employed. In this latter model, the relative permittivity is expressed as

$$\varepsilon_r(\omega) = \varepsilon_{r,\infty} + \frac{\omega_D^2}{j\omega(\Gamma_D + j\omega)} + \frac{\Delta\varepsilon_L\omega_L^2}{(j\omega)^2 + j\omega\Gamma_L + \omega_L^2}. \tag{7.108}$$

Inserting (7.108) into the frequency-domain Amperes' law, we have

$$\varepsilon_0\varepsilon_{r,\infty}j\omega\mathbf{E} + \mathbf{J}_D + j\omega\mathbf{Q}_L = \nabla \times \mathbf{H} \tag{7.109}$$

with

$$j\omega\mathbf{J}_D + \Gamma_D\mathbf{J}_D = \varepsilon_0\omega_D^2\mathbf{E}, \tag{7.110}$$

$$(j\omega)^2\mathbf{Q}_L + \Gamma_D j\omega\mathbf{Q}_L + \omega_L^2\mathbf{Q}_L = \varepsilon_0\Delta\varepsilon_L\omega_L^2\mathbf{E}. \tag{7.111}$$

By applying the inverse Fourier transform (IFT) into the above equations, we obtain

$$\varepsilon_0\varepsilon_{r,\infty}\frac{\partial\mathbf{E}}{\partial t} + \mathbf{J}_D + \frac{\partial\mathbf{Q}_L}{\partial t} = \nabla \times \mathbf{H}, \tag{7.112}$$

$$\frac{\partial\mathbf{J}_D}{\partial t} + \Gamma_D\mathbf{J}_D = \varepsilon_0\omega_D^2\mathbf{E}, \tag{7.113}$$

$$\frac{\partial^2\mathbf{Q}_L}{\partial t^2} + \Gamma_D\frac{\partial\mathbf{Q}_L}{\partial t} + \omega_L^2\mathbf{Q}_L = \varepsilon_0\Delta\varepsilon_L\omega_L^2\mathbf{E}. \tag{7.114}$$

In the above, two auxiliary variables are introduced: the current density term $\mathbf{J}_D$ for the Drude model and the polarization term $\mathbf{Q}_L$ for the Lorentz model. Although alternative choices are possible, the above choice uses the least amount of computer memory [42].

As presented previously, a two-step time marching approach is used in the ADI-FDTD algorithm. Applying the ADI scheme to the nabla operator in (7.112), the update equation for $E_x^{n+1/2}$ in the first sub-step is written as

$$
\begin{aligned}
E_{x_{i+1/2,j,k}}^{n+1/2} = E_{x_{i+1/2,j,k}}^{n} &+ \frac{\Delta t}{2\varepsilon_0\varepsilon_{r,\infty}}\left(\frac{H_{z_{i+1/2,j+1/2,k}}^{n+1/2} - H_{z_{i+1/2,j-1/2,k}}^{n+1/2}}{\Delta y}\right) \\
&- \frac{\Delta t}{2\varepsilon_0\varepsilon_{r,\infty}}\left(\frac{H_{y_{i+1/2,j,k+1/2}}^{n} - H_{y_{i+1/2,j,k-1/2}}^{n}}{\Delta z}\right) \\
&- \frac{\Delta t}{2\varepsilon_0\varepsilon_{r,\infty}}\left(\frac{J_{Dx_{i+1/2,j,k}}^{n+1/2} + J_{Dx_{i+1/2,j,k}}^{n}}{2}\right) \\
&- \frac{\Delta t}{2\varepsilon_0\varepsilon_{r,\infty}}\left(\frac{Q_{Lx_{i+1/2,j,k}}^{n+1/2} - Q_{Lx_{i+1/2,j,k}}^{n}}{\Delta t/2}\right).
\end{aligned}
\tag{7.115}
$$

Here, a central average scheme (CAS) and CDS are used for the Drude current term and the Lorentz polarization term, respectively. By applying the CDS and CAS to (7.113) and (7.114), the update equations for $J_{Dx}^{n+1/2}$ and $Q_{Lx}^{n+1/2}$ are derived as follows

$$
J_{Dx_{i+1/2,j,k}}^{n+1/2} = c_{D1} J_{Dx_{i+1/2,j,k}}^{n} + c_{D2}\left(E_{x_{i+1/2,j,k}}^{n+1/2} + E_{x_{i+1/2,j,k}}^{n}\right), \tag{7.116}
$$

$$
Q_{Lx_{i+1/2,j,k}}^{n+1/2} = c_{L1} Q_{Lx_{i+1/2,j,k}}^{n} + c_{L2} Q_{Lx_{i+1/2,j,k}}^{n-1/2} + c_{L3} E_{x_{i+1/2,j,k}}^{n}, \tag{7.117}
$$

with

$$
c_{D1} = \frac{1-0.5a_D}{1+0.5a_D}, \quad c_{D2} = \frac{0.25\Delta t \varepsilon_0 \omega_D^2}{1+0.5a_D},
$$

$$
c_{L1} = 2 - 0.25\left(\frac{\omega_L \Delta t)^2}{1+0.5a_L}\right), \quad c_{L2} = \frac{0.5a_L - 1}{1+0.5a_L}, \quad c_{L3} = \frac{0.25\varepsilon_0 \Delta\varepsilon_L (\omega_L \Delta t)^2}{1+0.5a_L}.
$$

Here, $a_M = 0.5\Gamma_M \Delta t$, $M = D$ (Drude) or L (Lorentz). The update equations for $\mathbf{H}^{n+1/2}$ remain the same as before unless a medium with frequency-dependent permeability is present.

As can be seen from (7.115), $E_x^{n+1/2}$ cannot be updated explicitly because $H_z^{n+1/2}$ from (7.18) and $J_{Dx}^{n+1/2}$ from (7.115) cannot be updated yet. Consequently, by substituting (7.18) and (7.116) into (7.115), we obtain the following tridiagonal-matrix update equation for $E_x^{n+1/2}$:

$$
\begin{aligned}
&-\frac{C_b D_b}{(\Delta y)^2} E_{x_{i+1/2,j-1,k}}^{n+1/2} + \left(1 + \frac{2C_b D_b}{(\Delta y)^2}\right) E_{x_{i+1/2,j,k}}^{n+1/2} - \frac{C_b D_b}{(\Delta y)^2} E_{x_{i+1/2,j+1,k}}^{n+1/2} \\
&= C_a E_{x_{i+1/2,j,k}}^n + \frac{C_b}{\Delta y}\left(H_{z_{i+1/2,j+1/2,k}}^n - H_{z_{i+1/2,j-1/2,k}}^n\right) - \frac{C_b}{\Delta z}\left(H_{y_{i+1/2,j,k+1/2}}^n - H_{y_{i+1/2,j,k-1/2}}^n\right) \\
&-\frac{C_b D_b}{\Delta y\,\Delta x}\left(E_{y_{i+1,j+1/2,k}}^n - E_{y_{i,j+1/2,k}}^n\right) + \frac{C_b D_b}{\Delta y\,\Delta x}\left(E_{y_{i+1,j-1/2,k}}^n - E_{y_{i,j-1/2,k}}^n\right) \\
&-C_b\left(\frac{Q_{Lx_{i+1/2,j,k}}^{n+1/2} - Q_{Lx_{i+1/2,j,k}}^n}{\Delta t/2}\right) + C_c J_{Dx_{i+1/2,j,k}}^n
\end{aligned}
\tag{7.118}
$$

with

$$
C_a = \frac{2\varepsilon_0\varepsilon_{r,\infty}/\Delta t - 0.5c_{D2}}{2\varepsilon_0\varepsilon_{r,\infty}/\Delta t + 0.5c_{D2}},
$$

$$
C_b = \frac{1}{2\varepsilon_0\varepsilon_{r,\infty}/\Delta t - 0.5c_{D2}},
$$

$$
C_c = -\frac{0.5(1 + c_{D2})}{2\varepsilon_0\varepsilon_{r,\infty}/\Delta t - 0.5c_{D2}},
$$

$$
D_b = \frac{\Delta t}{2\mu}.
$$

We can similarly derive the update equations for the other field components. The first sub-step update procedure of the ADI-FDTD algorithm for the Drude–Lorentz multispecies dispersion model is summarized below

Step 1. Update $\mathbf{Q}_L^{n+1/2}$ explicitly.
Step 2. Update $\mathbf{E}^{n+1/2}$ implicitly.
Step 3. Update $\mathbf{J}_D^{n+1/2}$ explicitly.
Step 4. Update $\mathbf{H}^{n+1/2}$ explicitly.

An analogous procedure can be employed for the update equations in the second sub-step.

Figure 7.3 shows the Au dispersion curve from experimental data [43], from the Drude model [44], and from the Drude–Lorentz model [45]. Here, the Drude parameters are $\varepsilon_{r,\infty} = 9.5$, $\omega_D = 2\pi \times 2165.03 \times 10^{12}$ (rad/s), and $\Gamma_D = 2\pi \times 33.43 \times 10^{12}$ (rad/s). In the Drude–Lorentz multispecies dispersion model, parameters are $\varepsilon_{r,\infty} = 5.9673$, $\omega_D = 2\pi \times 2113.6 \times 10^{12}$ (rad/s), $\Gamma_D = 2\pi \times 15.92 \times 10^{12}$ (rad/s), $\omega_L = 2\pi \times 650.07 \times 10^{12}$ (rad/s), $\Gamma_L = 2\pi\times 104.86 \times 10^{12}$ (rad/s), and $\Delta\varepsilon_L = 1.09$.

We analyze the localized surface plasmon resonance of a gold nanocube with 30 nm size in a SiO_2 host (with refractive index equal to 1.444) [31]. A z-polarized plane wave is excited and its time-domain signature is a sine wave modulated by a Gaussian pulse with half-power bandwidth (HPBW) between 500 and 1,000 nm. The time step size is defined as $\Delta t = CN\Delta s/\sqrt{3}c_0$, where CN indicates the Courant number, Δs (= 1 nm) is the grid size, and c_0 is the speed of light in vacuum. The computational domain comprises 90 × 90 × 90 cells. Figure 7.4 shows the spectra of the gold nanocube using both the standard FDTD and the ADI-FDTD algorithms. The spectral response is obtained by Fourier-transformation of the time-domain field data, followed by a normalization by the excitation pulse spectrum. Good agreement is observed between the ADI-FDTD and standard FDTD results. Table 7.1 summarizes the normalized central processing unit (CPU) time and the surface plasmon resonance wavelengths. For CN = 4, the CPU time of the ADI-FDTD simulation is reduced to approximately 58% of the FDTD simulation, while the resonance wavelength (relative) error is below 0.16%. For CN = 10, the CPU time required by the ADI-FDTD algorithm is approximately 24% of that by the FDTD algorithm.

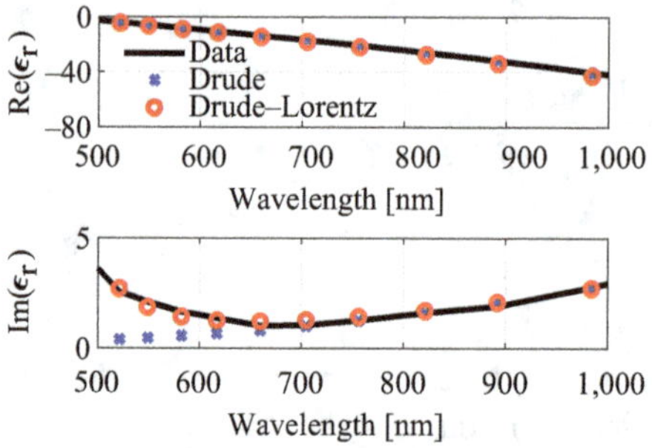

Figure 7.3 Dispersion curve of gold

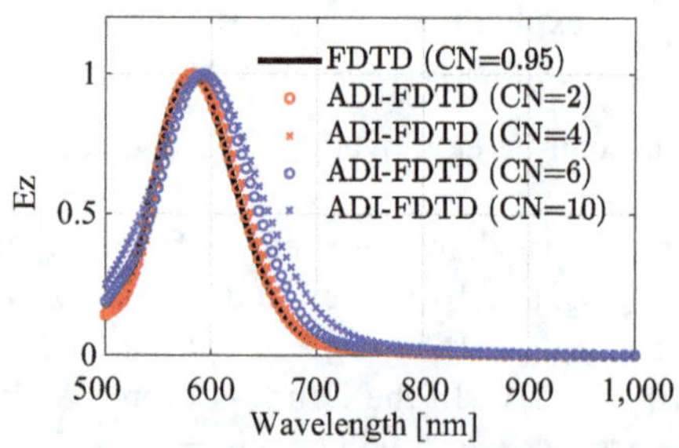

Figure 7.4 Spectra of a gold nanocube with the length of 30 nm surrounded by SiO_2

Table 7.1 Comparison between FDTD and ADI-FDTD results

Algorithm (CN)	CPU time (%)	Resonance wavelength (nm)
FDTD (CN = 0.95)	100	580.05
ADI-FDTD (CN = 2)	133	582.07
ADI-FDTD (CN = 4)	58	586.85
ADI-FDTD (CN = 6)	37	591.72
ADI-FDTD (CN = 10)	24	592.42

Several examples of unconditionally stable FDTD algorithms applied to frequency dispersive media are available in the literature, including isotropic plasmas [46,47], various metamaterials [48–50], plasmonic structures [51,52], and graphene [53–56]. In addition, unconditionally stable FDTD algorithms have been successfully applied to more complex dispersion models such as the modified Lorentz dispersion model [57] and the quadratic complex rational function dispersion model [58].

7.7 Split step, Leapfrog ADI, and hybrid-implicit-explicit FDTD methods

In addition to the ADI-FDTD and the LOD-FDTD methods, other types of unconditionally stable FDTD methods have been developed. Among those, perhaps the most popular are the Split-Step (SS) FDTD, the Leapfrog ADI-FDTD, and the Hybrid-Implicit-Explicit (HIE) FDTD methods.

In the SS-FDTD method, the time marching procedure from the nth time step to the $(n + 1)$th time step is decomposed into an arbitrary number of sub-steps [59–62] with the aim at increasing the overall accuracy of the solution. In particular, the so-called first-order SS-FDTD algorithm is equivalent to the LOD-FDTD algorithm.

In the Leapfrog ADI-FDTD method, one-step update equations are solved efficiently and thus the computational burden can be reduced to some extent, while preserving the unconditional stability property of the ADI-FDTD algorithm [63–66].

In the HIE-FDTD method [67–70], the maximum allowable time step is limited by the grid element size in only two (one) directions in 3-D (2-D) problems. As such, the Courant stability condition is still present but it is less stringent than in the standard FDTD algorithm. The HIE-FDTD algorithm is especially advantageous for electromagnetic problems with geometrical features that are small in some but not all the spatial directions, such as planar transmission lines and microstrip antennas.

7.8 Further unconditionally stable time-domain methods

It is also possible to develop unconditionally stable algorithms for the (slow-varying) time-domain *complex envelope* of the fields rather than for the time-domain field themselves [71–74]. In this case, the numerical dispersion is governed by the bandwidth to carrier frequency ratio rather than by the maximum frequency of the signal. Such unconditionally stable complex envelope methods incorporate an a priori, exact representation of the carrier frequency information through modified field update equations. Unconditionally stable complex envelope implementations are particularly useful for some narrowband problems.

Due to their popularity, we have focused on unconditionally stable FDTD methods in this chapter. Unconditionally stable methods are also routinely used in conjunction with finite element time-domain (FETD) simulations. However, since FETD simulations are most often based on unstructured (irregular) grids, ADI and LOD approaches are less applicable. Instead, unconditional stable algorithms such as the Crank–Nicolson [75] and the implicit Newmark-Beta [76] algorithms are most typically employed in FETD. These algorithms have the advantage of not producing any splitting errors but have the disadvantage of requiring an iterative sparse linear solver (matrix solver) at each time step. It is important to note the number of iterations required for convergence by any sparse linear solver increases with the condition number of the underlying matrix. For implicit FETD solvers, the condition number increases with the time step size. As a result, the consequent increase on the number of iterations required at each time step might partially negate the advantages of using larger time steps. For more details about this issue, the reader is referred to [77].

7.9 Summary and conclusions

In this chapter, we discussed the implementation of unconditionally stable FDTD algorithms, in particular the ADI-FDTD and LOD-FDTD algorithms. In both ADI- and LOD-FDTD algorithms, the time marching from the nth time step to the $(n+1)$th time step is decomposed into two sub-steps. In each sub-step, the electric field components are updated implicitly and the magnetic field components updated explicitly. The update equations of the electric field components are constructed as a tridiagonal matrix system that is efficiently solved by the Thomas algorithm with $O(N)$ complexity.

Numerical discretization artifacts were analyzed, including dispersion and splitting errors. The dispersion relations of ADI-FDTD and LOD-FDTD algorithms are identical to each other and expressed by the inverse tangent function term, which reflects the unconditional stability of these algorithms. The mitigation of the splitting error effected in iterative ADI-FDTD and LOD-FDTD algorithms was illustrated by means of 2-D numerical examples.

To enable the simulation of open-space problems in a finite computational domain, the derivation of a material-independent CFS-PML absorbing boundary condition was illustrated via the stretched coordinate approach. The CFS-PML implementation avoids late-time instabilities which may otherwise be present in the standard PML.

The implementation of the ADI-FDTD in dispersive media was illustrated by using the Drude–Lorentz multispecies dispersion model as an example. 3-D numerical simulations were used to demonstrate that the ADI-FDTD algorithm can reduce the CPU time considerably against the standard FDTD algorithm for certain dispersive-medium problems while retaining similar accuracy.

In addition to the ADI-FDTD and the LOD-FDTD methods, variant methods such as the SS-FDTD, the Leapfrog ADI-FDTD, and the HIE-FDTD methods were also briefly outlined.

Acknowledgment

The authors thank Dr J. Cho and Mr G. Kimball for their editorial assistance.

References

[1] Yee K. Numerical solution of initial boundary value problems involving Maxwell's equations in isotropic media. *IEEE Trans. Antennas Propagat.* 1966:14:302–7.

[2] Taflove A., Hagness S.C. *Computational Electrodynamics: The Finite-Difference Time-Domain Method*, 3rd ed. Norwood, MA, USA: Artech House; 2005.

[3] Gedney S.D. *Introduction to the Finite-Difference Time-Domain (FDTD) Method for Electromagnetics*, Lexington, KY, USA: Morgan & Claypool; 2010.

[4] Wang S., Teixeira, F.L. Some remarks on the stability of time-domain electromagnetic simulations. *IEEE Trans. Antennas Propag.* 2004:52(3): 895–8.

[5] Namiki T. A new FDTD algorithm based on alternating-direction implicit method. *IEEE Trans. Microw. Theory Tech.* 1999:47(10):2003–7.

[6] Zheng F., Chen Z., Zhang J. A finite-difference time-domain method without the Courant stability conditions. *IEEE Microw. Guided Wave Lett.* 1999: 9(11):441–3.

[7] Nascimento V.E., Teixeira F.L., Borges B.-H.V. Unconditionally stable finite-difference time-domain method based on the locally-one-dimensional technique. In *Proceeding of the 22nd Simpsio Brasileiro de Telecomun*, Campinas, Brazil, Sep 2005, pp. 288–91.

[8] Shibayama J., Muraki M., Yamauchi J., Nakano H. Efficient implicit FDTD algorithm based on locally one-dimensional scheme. *Electron. Lett.* 2005: 41(19):1046–7. See also *Errata, Electron. Lett.* 2006:42(8):497.

[9] Thomas J.W. *Numerical Partial Differential Equations: Finite Difference Methods*, Berlin, Germany: Springer-Verlag; 1995.

[10] Berenger J.P. A perfectly matched layer for the absorption of electromagnetic waves. *J. Comput. Phys.* 1994:114(2):185–200.

[11] Chew W.C., Weedon W.H. A 3D perfectly matched medium from modified Maxwell's equations with stretched coordinates. *Microw. Opt. Technol. Lett.* 1994:7(3):599–604.

[12] Teixeira F.L., Chew W.C. Systematic derivation of anisotropic PML absorbing media in cylindrical and spherical coordinates. *IEEE Microw. Guided Wave Lett.* 1997:7(11):371–3.

[13] Wang S., Teixeira F.L., Chen J. An iterative ADI-FDTD with reduced splitting error. *IEEE Microw. Wireless Compon. Lett.* 2005:15(2):92–4.

[14] Wang S, Chen J. Pre-iterative ADI-FDTD method for conductive medium. *IEEE Trans. Microw. Theory Tech.* 2005:53(6):1913–18.

[15] Jung K.-Y., Teixeira F.L. An iterative unconditionally stable LOD-FDTD method. *IEEE Microw. Wireless Compon. Lett.* 2008:18(2):76–8.

[16] Zheng F., Chen Z., Zhang J. Toward the development of a three-dimensional unconditionally stable finite-difference time-domain method. *IEEE Trans. Microw. Theory Tech.* 2000;53(6):1913–18.

[17] Garcia S.G., Rubio R.G., Bretones A.R., Martin R.G. On the dispersion relation of ADI-FDTD. *IEEE Microw. Wireless Compon. Lett.* 2006:16(6): 354–6.

[18] Tan E.L. Unconditionally stable LOD-FDTD method for 3-D Maxwell's equations. *IEEE Microw. Wireless Compon. Lett.* 2007:17(2):85–7.

[19] Zhao L., Cangellaris A.C. GT-PML: generalized theory of perfectly matched layers and its applications to the reflectionless truncation of finite-difference time-domain grids. *IEEE Trans. Microw. Theory Tech.* 1996:44(12): 2555–63.

[20] Kuzuoglu M., Mittra R. Frequency dependence of the constitutive parameters of causal perfectly matched anisotropic absorbers. *IEEE Microw. Guided Wave Lett.* 1996:6(12):447–9.

[21] Roden J.A., Gedney S.D. Convolution PML (CPML): an efficient FDTD implementation of the CFS-PML for arbitrary media. *Microw. Opt. Technol. Lett.* 2000:27(5):334–9.

[22] Becache E., Petropoulos P.G., Gedney S.D. On the long-time behavior of unsplit perfectly matched layers. *IEEE Trans. Antennas Propag.* 2004:52 (5):1335–42.

[23] Gedney S.D., Liu G., Roden J.A., Zhu A. Perfectly matched layer media with CFS for an unconditionally stable ADI-FDTD method. *IEEE Trans. Antennas Propag.* 2001:49(11):1554–9.

[24] Wang S., Teixeira F.L. An efficient PML implementation for the ADI-FDTD method. *IEEE Microw. Wireless Comp. Lett.* 2003:13(2):72–4. See also Author's Reply, *IEEE Microw. Wireless Compon. Lett.* 2004:14(5): 249–50.

[25] Rubio R.G., Garcia S.G., Bretones A.R., Martín R.G. Crank–Nicolson reformulation of ADI-FDTD PML extentions. *IEEE Antennas Wireless Propag. Lett.* 2006:5:357–60.

[26] Wang L., Liang C., Li L. Modification to convolution CFS-PML for the ADI-FDTD method.. *Microw. Opt. Technol. Lett.* 2006:48(2):261–5.

[27] Nascimento V.E., Borges B.-H., Teixeira F.L. Split-field PML implementation for the unconditionally stable LOD-FDTD method. *IEEE Microw. Wireless Compon. Lett.* 2006:16(7):398–400.

[28] Ramadan O. Efficient LOD-SC-PML formulations for electromagnetic fields in dispersive media. *IEEE Microw. Wireless Compon. Lett.* 2012:22 (6):297–9.

[29] Welfert B.D. Analysis of iterated ADI-FDTD schemes for Maxwell curl equations. *J. Comput. Phys.* 2007:222:9–27.

[30] Bhardwaj S., Teixeira F.L., Volakis J.L. Fast modeling of terahertz plasma-wave devices using unconditionally stable FDTD methods. *IEEE J. Multiscale Multiphys. Comp. Technol.* 2018:3:29–36.

[31] Jung K.-Y., Teixeira F.L. *Plasmonic and Dispersion-Engineered Metamaterials: Analysis and Algorithms*, Saarbrucken, Germany: VDM Verlag, 2009.

[32] Pereda J.A., Vielva L.A., Vegas Á., Prieto A. Analyzing the stability of the FDTD technique by combining the von Neumann method with the Routh-Hurwitz criterion. *IEEE Trans. Microw. Theory Techn.* 2001:49(2): 377–81.

[33] Lin Z., Qiu W., Pu J., Fang Y., Hu J. Accuracy and von Neumann stability of several highly accurate FDTD approaches for modelling Debye-type dielectric dispersion. *IET Microw. Antennas Propag.* 2018:12(2):211–6.

[34] Okoniewski M., Okoniewska E. Drude dispersion in ADE FDTD revisited. *Electron. Lett.* 2006:42(9):503–4.

[35] Jung K.-Y., Teixeira F.L., Reano R.M. Au/SiO nanoring plasmon waveguides at optical communication band. *J. Lightw. Technol.* 2007:25(9): 2757–65.

[36] Lin Z., Ou P., Jia Y., Zhang C. A highly accurate FDTD model for simulating Loretnz dielectric dispersion. *IEEE Photon. Technol. Lett.* 2009: 21(22):1692–4.

[37] Kelley D.F., Luebbers R.J. Piecewise linear recursive convolution for dispersive media using FDTD. *IEEE Trans. Antennas Propag.* 1996:44(6): 792–7.

[38] Sullivan D.M. Z-transform theory and the FDTD method. *IEEE Trans. Antennas Propag.* 1996:44(1):28–34.

[39] Okoniewski M., Mrozowski M., Stuchly M.A. Simple treatment of multi-term dispersion in FDTD. *IEEE Microw. Guided Wave Lett.* 1997:7(5): 121–3.

[40] Prokopidis K.P., Zografopoulos D.C. A unified FDTD/PML scheme based on critical points for accurate studies of plasmonic structures. *J. Lightw. Technol.* 2013:31(15):2467–76.

[41] Choi H., Baek J.-W., Jung K.-Y. Comprehensive study on numerical aspects of modified Lorentz model-based dispersive FDTD formulations. *IEEE Trans. Antennas Propag.* 2019:67(12):7643–8.

[42] Jung K.-Y., Teixeira F.L. Multispecies ADI-FDTD algorithm for nanoscale three-dimensional photonic metallic structures. *IEEE Photon. Technol. Lett.* 2007:19(8):586–8.

[43] Johnson P.B., Christy R.W. Optical constants of the noble metals. *Phys. Rev. B.* 1972:6(12):4370–9.

[44] Oubre C., Nordlander P. Optical properties of metallodielectric nanostructures calculated using the finite difference time domain method. *J. Phys. Chem. B.* 2004:108(46):17740–7.

[45] Vial A., Grimault A.-S., Macías D., Barchiesi D., Chapelle M.L. Improved analytical fit of gold dispersion: application to the modeling of extinction spectra with a finite-difference time-domain method. *Phys. Rev. B.* 2005:71:085416.

[46] Xu L., Yuan N. PLJERC-ADI-FDTD method for isotropic plasma. *IEEE Microw. Wireless Compon. Lett.* 2005:15(4):277–9.

[47] Wanjun S., Hou Z. RKADE-ADI FDTD applied to analysis of terahertz wave characteristics of plasma. *IET Microw. Antennas Propag.* 2017: 11(7):1071–7.

[48] Kantartzis N.V., Sounas D.L., Antonopoulos C.S., Tsiboukis T.D. A wideband ADI-FDTD algorithm for the design of double negative metamaterial-based waveguides and antenna substrates. *IEEE Trans. Magn.* 2007:43 (4):13297–32.

[49] Nascimento V.E., Jung K.-Y., Borges B.-H.V., Teixeira F.L. A study on unconditionally stable FDTD methods for the modeling of metamaterials. *J. Lightw. Techol.* 2009:27(19):4241–9.

[50] Heh D.Y., Tan E.L. Stable formulation of FADI-FDTD method for multiterm, doubly, second-order dispersive media. *IEEE Trans. Antennas Propag.* 2013:61(8):4167–75.

[51] Shibayama J., Takahashi R., Nomura A., Yamauchi J., Nakano H. Concise frequency-dependent formulation for LOD-FDTD method using Z transforms. *Electron. Lett.* 2008:44(16):949–50.

[52] Shibayama J., Ando R., Yamauchi J., Nakano H. Frequency-dependent 3-D LOD-FDTD method for the analysis of plasmonic devices. *IEEE Photon. Technol. Lett.* 2011:23(15):1070–2.

[53] Ahmed I., Khoo E.H., Li E. Efficient modeling and simulation of graphene devices with the LOD-FDTD method. *IEEE Microw. Wireless Compon. Lett.* 2013:23(6):306–8.

[54] Chen J., Hao G., Liu Q.-H. Using the ADI-FDTD method to simulate graphene-based FSS at terahertz frequency. *IEEE Trans. Electromagn. Compat.* 2017:59(4):1218–23.

[55] Gao J.-Y. , Wang X.-H., Teixeira F.L. Extended ADI-FDTD for anisotropic magnetized graphene sheets. *Electron. Lett.* 2019:55(15):833–5.

[56] Wang X.-H., Gao J.Y., Teixeira F.L. Unconditionally stable LOD-FDTD for modeling transmission properties of magnetized graphene sheet. *Opt. Commun.* 2019:442:90–4.

[57] Prokopidis K.P., Zografopoulos D.C. An ADI-FDTD formulation with modified Lorentz dispersion for the study of plasmonic structures. *IEEE Photon. Technol. Lett.* 2014:26(22):2267–70.

[58] Kim E.-K., Ha S.-G., Lee J., Park Y.B, Jung K.-Y. Three-dimensional efficient dispersive alternating-direction-implicit finite-difference time-domain

algorithm using a quadratic complex rational function. *Opt. Express.* 2015:23(2):873–81.

[59] Lee J., Fornberg B. A split step approach for the 3-D Maxwell's equations. *J. Comput. Appl. Math.* 2003:158:485–505.

[60] Lee J., Fornberg B. Some unconditional stable time stepping methods for the 3D Maxwell's equations. *J. Comput. Appl. Math.* 2004:166:497–523.

[61] Kusaf M., Oztoprak A.Y. Multi split-step unconditionally stable finite difference time domain methods. *Microw. Opt. Technol. Lett.* 2009:51(11): 2646–9.

[62] Singh G., Tan E.L., Chen Z.N. A split-step FDTD method for 3-D Maxwell's equations in general anisotropic media. *IEEE Trans. Antennas Propag.* 2010:58(11):3647–57.

[63] Cooke S.J., Botton M., Antonsen Jr. T.M., Levush B. A leapfrog formulation of the 3-D ADI-FDTD algorithm. *Int. J. Numer. Model.* 2009:22:187–200.

[64] Yang S.-C., Chen Z., Yu Y., Yin W.Y. An unconditionally stable one-step arbitrary-order leapfrog ADI-FDTD method and its numerical properties. *IEEE Trans. Antennas Propag.* 2012:60(4):1995–2003.

[65] Mao Y.-F., Chen B., Xia J.-L., Chen J., Tang J.-Z. Application of the leapfrog ADI FDTD method to periodic structures. *IEEE Antennas Wireless Propag. Lett.* 2013:12:599–602.

[66] Wang, X.-H., Gao J.Y., Chen Z., Teixeira, F.L. Unconditionally stable one-step leapfrog ADI-FDTD for dispersive media. *IEEE Trans. Antennas Propag.* 2019:67(4):2829–34.

[67] Huang B., Wang G., Jiang Y., Wang W. A hybrid implicit-explicit FDTD scheme with weakly conditional stability. *Microw. Opt. Technol. Lett.* 2003:39(2):97–101.

[68] Chen J., Wang J. 3-D FDTD method with weakly conditional stability. *Electron. Lett.* 2007:43(1):2–3.

[69] Unno M., Asai H. HIE-FDTD method for hybrid system with lumped elements and conductive media. *IEEE Microw. Wireless Compon. Lett.* 2011: 21(9):453–5.

[70] Wang, X.-H., Gao J.Y., Teixeira F.L. One-step leapfrog HIE-FDTD for Drude media. *IEEE Microw. Wireless Compon. Lett.* 2019:29(2):77–9

[71] Rao H., Scarmozzino R., Osgood Jr. R.M. An improved ADI-FDTD method and its application to photonic simulations. *IEEE Photon. Technol. Lett.* 2002:14:4779.

[72] Pinto D., Obayya S.S.A. Improved complex-envelope alternating-direction-implicit finite-difference time-domain method for photonic bandgap cavities. *J. Ligthw. Technol.* 2007:25(1):4407.

[73] Jung K.-Y., Teixeira F.L., Lee R. Complex envelope PML-ADI-FDTD method for lossy anisotropic dielectrics. *IEEE Antennas Wireless Propag. Lett.* 2007:6:643–6.

[74] Jung K.-Y., Teixeira F.L., Garcia S.G., Lee R. On numerical artifacts of the complex envelope ADI-FDTD method. *IEEE Trans. Antennas Propag.* 2009:57(2):491–8.

[75] Chen R.-S., Du L., Ye Z., Yang Y. An efficient algorithm for implementing the Crank–Nicolson scheme in the mixed finite-element time-domain method *IEEE Trans. Antennas Propag.* 2009:57(10):3216–22.

[76] Gedney S.D., Navsariwala U. An unconditionally stable finite element time-domain solution of the vector wave equation. *IEEE Microw. Guided Wave Lett.* 1995:5:332–4.

[77] Moon H., Teixeira F.L., Kim J., Omelchenko Y. Trade-offs for unconditional stability in the finite-element time-domain method. *IEEE Microw. Wireless Compon. Lett.* 2014:24(6):361–3.

Chapter 8
Time-domain linear macromodeling

Stefano Grivet-Talocia[1], *Tommaso Bradde*[1], *Marco De Stefano*[1] *and Alessandro Zanco*[1]

Linear macromodeling has rightly become a standard practice in computer-aided design (CAD) of electrical, electronic, and electromagnetic systems. We can safely state that all commercial tools and software environments include macromodeling algorithms to support design verification via numerical simulation at field or circuit level. In short, macromodels enable time- or frequency-domain simulations based on compact and reduced-order approximate models, which can be computed for virtually any type of linear and time-invariant device, component, subsystem, and even system.

What is a macromodel? This may be a very general term, which, however, assumes in this chapter a specific and well-defined meaning. Let us consider a generic structure that can be mathematically described through a LTI system of constitutive equations. The dynamic behavior of the structure relates its input and output signals through an operator that is best described in the frequency domain as a transfer function or frequency response. The physical structure can be extremely complex in terms of geometry, materials, electrical size, and number of inputs and outputs, and this complexity is inherited by the constitutive LTI equations. As an example, consider an electrically large linear electromagnetic subsystem described through space-discretized Maxwell equations, resulting in a system of coupled ordinary differential equations (ODEs) with possibly millions of unknowns. Yet, the input–output behavior is fully described by the frequency response. When this response can be described or approximated through a reduced number of degrees of freedom, linear macromodeling comes into play. A *macromodel* provides a compact set of ODEs with a much reduced number of unknowns (state variables) and with the same input–output signals of the original structure, and whose frequency response closely approximates that of the true system.

A macromodel must be not only accurate, but it should be obtained using fast and reliable algorithms, it should guarantee stable numerical results when used in system-level simulations, and it should be portable to any computing or simulation platform adopted by the application community. The reference scenario for linear macromodeling is a three-stage process. First, a characterization of the system response is

[1]Department of Electronics and Telecommunications, Politecnico di Torino, Italy

obtained, a typical case being the numerical solution of the underlying electromagnetic problem providing samples of the target frequency response within a given band of interest. Second, these samples are subjected to a macromodeling scheme that constructs a compact model that approximates this data. Third, the model is used (and reused) for fast system-level simulation, requiring computing resources that are negligible with respect to those required for the initial characterization.

Macromodels can be simulated both in time and in frequency-domain, even if the initial characterization is in the frequency domain. This is one particular aspect that boosted development of macromodeling schemes during the last two decades. In fact, the interaction of electrical signals or electromagnetic waves with non-ideal materials leads to losses, dispersion, couplings, and other effects that are best represented using first-principle models in the frequency domain. A direct simulation of such structures in the time domain is problematic. Macromodels provide an extremely efficient way to bridge the two domains, by translating a frequency-domain description into a closed-form set of ODEs, whose numerical solution in time domain is trivial. Hence, the title of this chapter, *Time-Domain Linear Macromodeling*.

This chapter is organized as follows. Section 8.1 provides some background and introduces the general framework that supports macromodeling developments and applications. Section 8.2 introduces well-known and basic algorithms for the construction of macromodels from frequency-domain sampled responses. We discuss the Vector Fitting algorithm in its basic and multiport declinations, and we provide guidelines for model realization either as state-space equations for inclusion in ODE solvers, and as equivalent circuits for inclusion in circuit solvers of the SPICE class. Section 8.3 presents the methods of choice for checking and enforcing model passivity, a fundamental property that guarantees robustness, reliability, and stability of transient simulations based on macromodels.

Section 8.4 presents more advanced topics by extending standard macromodeling algorithms to the multivariate case, where models embed in a closed form the dependence on external parameters, e.g. related to geometry or material properties. Such parameterized models are regarded as the main research direction nowadays, aiming at constructing scalable models that can be used not only in post-layout design verification but also during early design stages and for design centering, optimization, sensitivity, Monte-Carlo, and what-if analyses. We discuss the appropriate model structures for parameterization, model identification from sampled responses both in frequency and parameter spaces, and passivity enforcement. Finally, Section 8.5 presents a number of examples from various application scenarios.

Due to space limitation, this chapter provides only a concise overview of the main results. For a more complete introduction to the topic, we point the reader to the recent monograph [1], where one of the authors collected in the form of a textbook background, developments, results, algorithms, and applications in the broad field of passive macromodeling. We also mention the three-volume series *Model Order Reduction* [2–4], which provides a complete overview of state-of-the-art approaches for reducing complexity in numerical simulations, including passive macromodeling as one fundamental tool.

8.1 The macromodeling framework

8.1.1 The main objective

The general objective of most macromodeling flows is to represent the dynamic behaviour of a complex LTI system through a set of ODEs. Assuming a P-port structure, the most general representation can be formulated in the *descriptor* form [5–7]

$$\begin{cases} \mathbf{E}\dot{\mathbf{x}}(t) = \mathbf{A}\mathbf{x}(t) + \mathbf{B}\mathbf{u}(t) \\ \mathbf{y}(t) \quad = \mathbf{C}\mathbf{x}(t) + \mathbf{D}\mathbf{u}(t) \end{cases} \tag{8.1}$$

where t denotes time, $\mathbf{x} \in \mathbb{R}^N$ collects internal generalized state variables, and $\mathbf{u}, \mathbf{y} \in \mathbb{R}^P$ collect the inputs and the outputs of the model. The nature of these signals depends on the adopted electrical representation: for impedance and admittance (collectively immittance) systems, $\mathbf{u}$ and $\mathbf{y}$ are port currents (voltages) and voltages (currents). For scattering representations, $\mathbf{u}$ and $\mathbf{y}$ are, respectively, incident and reflected scattering waves at the ports, suitably defined through a set of real and positive reference impedances. The state-space (descriptor) matrices $\mathbf{E}, \mathbf{A} \in \mathbb{R}^{N\times N}$, $\mathbf{B} \in \mathbb{R}^{N\times P}$, $\mathbf{C} \in \mathbb{R}^{P\times N}$ and $\mathbf{D} \in \mathbb{R}^{P\times P}$ are real-valued and constant. Most often, $\mathbf{E} = \mathbf{I}$ so that (8.1) reduces to a regular *state-space* realization. In some applications, $\mathbf{E}$ may be singular, see e.g. Section 8.4. Macromodeling algorithms aim at deriving these matrices starting from some system characterization coming from first-principle discretization, simulation, or measurement process.

The frequency response associated to the model (8.1) is a rational function of the complex frequency s and can be expressed as

$$\mathbf{H}(s) = \mathbf{C}(s\mathbf{E} - \mathbf{A})^{-1}\mathbf{B} + \mathbf{D} = \sum_{n=1}^{\bar{n}} \frac{\mathbf{R}_n}{s - p_n} + \mathbf{R}_0 + s\mathbf{R}_{-1} \tag{8.2}$$

where the pole-residue expansion in terms of a common pole set p_n with associated $P \times P$ residue matrices $\mathbf{R}_n$ is introduced. These are either real or appear as complex conjugate pairs. The direct coupling $\mathbf{R}_0$ and the linear asymptotic term $\mathbf{R}_{-1}$ are real-valued, the latter missing in most cases of practical interest, in particular for scattering macromodels (whose frequency response must be bounded). The pole-residue expansion (8.2) is in fact the actual target of most macromodeling algorithms since uniquely defined, whereas the state-space matrices of (8.1) are not unique since depending on the particular coordinate system adopted to define the state variables.

8.1.2 Black-box modeling: intrusive vs non-intrusive

A macromodeling scheme generally aims at computing a *black-box* low-complexity representation or approximation of the system under modeling. The term "black-box" implies that the model dynamics (8.1) are not related to a physics-based description in terms of physically observable states, but rather provide a set of equations within a very broad class, whose frequency response (8.2) approximates the true system response with an acceptable accuracy.

Black-box macromodeling algorithms can be roughly divided in two classes. *Intrusive* approaches assume to have access to a fine-grained or detailed model, whose complexity is to be reduced. An example can be a system in form (8.1) arising from a space discretization of Maxwell's equations in both differential or integral form. The size of this initial system can be reduced by one of the several classical *Model Order Reduction* schemes [8–10], either through truncation [11–13] or projection [14]. Notable examples are Krylov subspace projection schemes [15–17], including the well-known PRIMA algorithm [18,19].

The second class of *non-intrusive* or *data-driven* macromodeling schemes do not assume an initial model, but rather construct a purely behavioural macromodel starting from a finite set of samples of the system response, either in frequency or time domain. These samples can come from direct measurement in case the hardware is available, but can also be obtained from the results of a first-principle field simulation, e.g. using a commercial solver. The latter scenario is by far the most common in electrical, electronic, and electromagnetic CAD applications: high-complexity designs are usually characterized by commercial solvers that do not provide access to the internal matrices arising from the discretization, but rather provide as output the samples of the (scattering) response of the system at well-defined interface ports.

In this chapter, we will only consider the latter data-driven macromodeling approaches, and we will provide an overview of the leading algorithms. A complete and detailed treatment of both approaches can be found in [1] and references therein. Throughout the following, we assume that the frequency response of the system is known at a finite set of frequencies $s_k = j2\pi f_k,\ \ k = 1, \ldots, \bar{k}$ over a given frequency band $\Omega = [f_{\min}, f_{\max}]$, with $f_1 = f_{\min}$ and $f_{\bar{k}} = f_{\max}$, as

$$\breve{\mathbf{H}}_k = \breve{\mathbf{H}}(s_k) \quad k = 1, \ldots, \bar{k} \tag{8.3}$$

where the accent $\breve{\mathbf{H}}_k$ labels the dataset that is fed as input to the macromodeling scheme. The objective is to find a macromodel in form (8.1) or (8.2) such that the following fitting condition:

$$\mathbf{H}(j2\pi f_k) \approx \breve{\mathbf{H}}_k, \quad k = 1, \ldots, \bar{k} \tag{8.4}$$

holds within an acceptable accuracy.

8.2 Vector Fitting

The frequency-domain macromodeling problem boils down to exploiting the fitting condition (8.4) to compute the parameters of a rational function of frequency. The latter can be expressed in several ways, including the pole-residue expansion (8.2), its equivalent pole-zero factorization, or a ratio of polynomials. None of these forms is however simple to handle, since minimization of the model-data error leads to a strongly non-convex optimization problem with possibly many local minima, which is impractical to be solved through standard descent methods. There is, however, a particular representation that lends itself to a straightforward linearization process, which is the foundation of the popular Vector Fitting (VF) scheme [20].

8.2.1 Rational barycentric forms

Let us start by assuming a one-port ($P = 1$) scalar example. We write the transfer function of the model as

$$H(s) = \frac{N(s)}{D(s)} = \frac{\sum_{n=0}^{\bar{n}} c_n \varphi_n(s)}{\sum_{n=0}^{\bar{n}} d_n \varphi_n(s)}, \tag{8.5}$$

where both the numerator and the denominator are expanded in terms of frequency-dependent basis functions $\varphi_n(s)$ with unknown coefficients c_n, d_n. Traditional approaches [21] have used monomials $\varphi_n(s) = s^n$, a choice that severely limits the order $\bar{n}$ of the expansion due to the intrinsically poor numerical conditioning of this basis. This difficulty is overcome by choosing $\varphi_n(s)$ to be the partial fractions based on a prescribed set of distinct $\bar{n}_r$ real poles $q_n \in \mathbb{R}^-$ and $\bar{n}_c$ complex pole pairs $p_n = p_n' \pm \mathrm{j}p_n'' \in \mathbb{C}^-$, where $\varphi_0(s) = 1$,

$$\varphi_n(s) = (s - q_n)^{-1}, \quad \text{for} \quad n = 1, \ldots, \bar{n}_r, \tag{8.6}$$

and

$$\begin{aligned} \varphi_{\bar{n}_r+2n-1}(s) &= (s - p_n)^{-1} + (s - p_n^*)^{-1} \\ \varphi_{\bar{n}_r+2n}(s) &= \mathrm{j}(s - p_n)^{-1} - \mathrm{j}(s - p_n^*)^{-1} \end{aligned} \tag{8.7}$$

for $n = 1, \ldots, \bar{n}_c$. In the following, the basis poles and associated basis functions for both real and complex conjugate poles will be collectively denoted as q_n and φ_n through (8.6) for notational compactness. The model structure (8.5) with these basis functions is usually denoted *rational barycentric form*. In order to guarantee uniqueness of this representation, we set $d_0 = 1$.

8.2.2 The generalized Sanathanan–Koerner iteration

Building on the well-known Sanathanan–Koerner (SK) iteration [21], we derive a numerically robust identification scheme for the coefficients of (8.5). We set up an iteration $\nu = 1, 2, \ldots$ and we update the iteration-dependent coefficients

$$\mathbf{x}_\nu = \left(c_0^\nu, \ldots, c_{\bar{n}}^\nu, d_1^\nu, \ldots, d_{\bar{n}}^\nu\right)^{\mathrm{T}} \tag{8.8}$$

by minimizing the euclidean norm $\|\mathbf{r}^\nu(\mathbf{x}_\nu)\|$ of the residual vector collecting the model-data error computed at all frequency points

$$r_k^\nu(\mathbf{x}_\nu) = \frac{D^\nu(s_k)\breve{H}_k - N^\nu(s_k)}{D^\nu(s_k)} \approx \frac{D^\nu(s_k)\breve{H}_k - N^\nu(s_k)}{D^{\nu-1}(s_k)} \tag{8.9}$$

where we introduced the explicit dependence of numerator and denominator on the unknowns $\mathbf{x}_\nu$ through superscript ν. Linearization of this residual is attained by using the estimate of the denominator available from previous iteration $\nu - 1$. Since all coefficients $\mathbf{x}_{\nu-1}$ are known at iteration ν, the residual (8.9) depends linearly on the coefficients (8.8), so that its minimization is achieved as a standard linear least squares problem. The scheme is initialized with $\mathbf{x}_0 = 0$ and stops when the coefficients stabilize.

At each iteration, minimization of $r_k^\nu(\mathbf{x}_\nu)$ amounts to solving

$$(\mathbf{W}_{\nu-1}\Psi)\mathbf{x}_\nu \approx \mathbf{W}_{\nu-1}\mathbf{b}, \tag{8.10}$$

in least squares sense, where

$$\mathbf{W}_{\nu-1} = \text{diag}_k\left\{D^{\nu-1}(s_k)^{-1}\right\}, \quad \mathbf{b} = \left(\breve{H}_1, \breve{H}_2, \ldots, \breve{H}_{\bar{k}}\right)^{\mathrm{T}}, \tag{8.11}$$

and

$$\Psi = (\,\Phi^\circ \quad -\breve{\mathbf{H}}\Phi\,), \qquad \Phi = \begin{pmatrix} \varphi_1(s_1) & \varphi_2(s_1) & \cdots & \varphi_{\bar{n}}(s_1) \\ \varphi_1(s_2) & \varphi_2(s_2) & \cdots & \varphi_{\bar{n}}(s_2) \\ \vdots & \vdots & & \vdots \\ \varphi_1(s_{\bar{k}}) & \varphi_2(s_{\bar{k}}) & \cdots & \varphi_{\bar{n}}(s_{\bar{k}}) \end{pmatrix}, \tag{8.12}$$

with $\Phi^\circ = (\,\mathbf{1} \quad \Phi\,)$ and $\mathbf{1} = (1, \ldots, 1)^{\mathrm{T}}$. This scheme is denoted as *Generalized Sanathanan–Koerner (GSK)* iteration, since it extends the SK iteration (based on the monomial basis) to more general choices of basis functions.

8.2.3 The basic VF scheme

The GSK iteration provides the main building block for Vector Fitting algorithm [20]. Whereas the GSK iteration retains the same "basis" poles q_n at each step, the VF scheme updates them after solving the least squares problem (8.10). This *pole relocation* process computes the zeros z_n^ν of the denominator

$$D^\nu(s) = 1 + \sum_{n=1}^{\bar{n}} \frac{d_n^\nu}{s - q_n^\nu} = \frac{\prod_{n=1}^{\bar{n}} (s - z_n^\nu)}{\prod_{n=1}^{\bar{n}} (s - q_n^\nu)} \tag{8.13}$$

which are in turn used to define the basis poles and the corresponding basis functions for the next iteration

$$q_n^{\nu+1} = z_n^\nu \quad \text{and} \quad \varphi_n^\nu(s) = \frac{1}{s - q_n^\nu} \qquad n = 1, \ldots, \bar{n}. \tag{8.14}$$

It is straightforward to see that the zeros z_n^ν can be computed as

$$\{z_n^\nu\} \equiv \text{eig}\{\mathbf{Q} - \mathbf{1}\cdot\mathbf{d}^{\mathrm{T}}\}, \quad \mathbf{Q} = \text{diag}\{q_1^\nu, \ldots, q_{\bar{n}}^\nu\}, \quad \mathbf{d}^{\mathrm{T}} = (d_1^\nu, \ldots, d_{\bar{n}}^\nu). \tag{8.15}$$

In summary, the basic VF scheme solves iteratively for $\nu = 1, 2, \ldots$ the following linear least squares problem

$$\left(1 + \sum_{n=1}^{\bar{n}} \frac{d_n^\nu}{s_k - q_n^\nu}\right)\breve{H}_k \approx c_0^\nu + \sum_{n=1}^{\bar{n}} \frac{c_n^\nu}{s_k - q_n^\nu} \qquad k = 1, \ldots, \bar{k} \tag{8.16}$$

which can be cast in a compact matrix form with a similar structure as (8.12),

$$\Psi_\nu \mathbf{x}_\nu \approx \mathbf{b} \quad \text{where} \quad \Psi_\nu = \left(\mathbf{\Phi}_\nu^\circ \quad -\breve{\mathbf{H}}\mathbf{\Phi}_\nu \right) \tag{8.17}$$

is now iteration-dependent since based on iteration-dependent poles.

When the iterations converge [1,22,23] and all unknowns stabilize, also the basis poles (8.14) converge to an invariant set $q_n^\nu \to p_n$, and the denominator estimate (8.13) converges identically to $D^\nu(s) \to 1$. Therefore, the pole-residue representation of the model is readily available. In practice, it is numerically preferable to solve an additional least squares problem based on the stablized poles p_n

$$\breve{H}_k \approx R_0 + \sum_{n=1}^{\bar{n}} \frac{R_n}{s_k - p_n}, \quad k = 1, \ldots, \bar{k} \tag{8.18}$$

to find the corresponding residues R_n.

8.2.4 Stability and realness

The basic VF scheme is easily modified to enforce model stability. It is sufficient to replace the pole relocation condition (8.14) with the following

$$q_n^{\nu+1} = -(z_n^\nu)^* \quad \text{when} \quad \mathrm{Re}\{z_n^\nu\} > 0, \tag{8.19}$$

to be used whenever a pole (real or conjugate pair) is detected unstable. This condition is usually denoted as *pole flipping,* since it corresponds to replace unstable poles with their symmetric image with respect to the imaginary axis.

An additional property that is essential for time-domain modeling is to ensure that the model has a real-valued impulse response, which in turn is guaranteed when $H(s^*) = H^*(s)$. This condition is guaranteed when complex conjugate pole pairs are treated using modified partial fractions as in (8.7), and the vector of unknowns (8.8) is constrained to be real-valued. The latter condition is easily enforced by replacing the complex-valued least squares system (8.17) with

$$\begin{pmatrix} \Psi_\nu' \\ \Psi_\nu'' \end{pmatrix} \mathbf{x}_\nu \approx \begin{pmatrix} \mathbf{b}' \\ \mathbf{b}'' \end{pmatrix} \tag{8.20}$$

where single- and double-prime symbols extract real and imaginary parts, respectively.

8.2.5 VF for multiport systems

When dealing with multiport systems, it is a common requirement to exploit a model structure with common poles for all transfer function elements. This is a general property of lumped circuits of great practical relevance, since it allows to represent system-level resonances, whose signature is present in all port responses. In fact, the term "vector" of Vector Fitting was historically adopted to denote the capability of this algorithm to fit concurrently multiple responses with common poles.

Let us denote a set of P frequency responses collected in vector $\breve{\mathbf{h}}(s) \in \mathbb{C}^P$, whose frequency samples are available as data vectors $\breve{\mathbf{h}}_k = \breve{\mathbf{h}}(s_k)$ for $s_k = \mathrm{j}2\pi f_k$, $k = 1,\ldots,\bar{k}$. All these samples are collected for each component $i = 1,\ldots,P$ as

$$\mathbf{b}^i = \left(\breve{h}_1^i,\ldots,\breve{h}_{\bar{k}}^i\right)^{\mathrm{T}} \quad \text{and} \quad \breve{\mathbf{H}}^i = \mathrm{diag}\{\breve{h}_1^i,\ldots,\breve{h}_{\bar{k}}^i\}. \tag{8.21}$$

We want to identify a vector $\mathbf{h}(s)$ of rational functions such that

$$\breve{\mathbf{h}}_k \approx \mathbf{h}(s_k) = \mathbf{r}_0 + \sum_{n=1}^{\bar{n}} \frac{\mathbf{r}_n}{s_k - p_n}, \quad k = 1,\ldots,\bar{k} \tag{8.22}$$

where the $\bar{n}$ poles p_n are common to all components, and where $\mathbf{r}_n$ are size-P vectors.

A model representation as (8.5) is assumed, where however $\mathbf{N}(s) \in \mathbb{C}^P$ and $\mathbf{c}_n \in \mathbb{C}^P$, and where the denominator $D(s)$ remains a scalar function of frequency. Following the same procedure of Section 8.2.3 and writing (8.16) for the i-th component of the model leads to a set of coupled least squares problems

$$\left(\mathbf{\Phi}_\nu^\circ - \breve{\mathbf{H}}^i\mathbf{\Phi}_\nu\right)\begin{pmatrix}\mathbf{a}^i\\ \mathbf{d}\end{pmatrix} \approx \mathbf{b}^i \quad i = 1,\ldots,P \tag{8.23}$$

where the unknown coefficients are collected in vectors

$$\mathbf{a}^i = \left(c_0^i,\ldots,c_n^i\right)^{\mathrm{T}}, \quad \mathbf{d} = (d_1,\ldots,d_n)^{\mathrm{T}}. \tag{8.24}$$

Systems (8.23) are coupled since they share a common set of unknowns $\mathbf{d}$. It turns out that the unknowns in each vector $\mathbf{a}^i$ belong only to the subproblem related to the i-th component and can be eliminated. This is achieved through the following "thin" QR decomposition [24]

$$\left(\mathbf{\Phi}_\nu^\circ - \breve{\mathbf{H}}^i\mathbf{\Phi}_\nu\right) = \mathbf{Q}_i\mathbf{R}_i = \mathbf{Q}_i\begin{pmatrix}\mathbf{R}_i^{11} & \mathbf{R}_i^{12}\\ \mathbf{0} & \mathbf{R}_i^{22}\end{pmatrix}, \tag{8.25}$$

where $\mathbf{Q}_i \in \mathbb{C}^{\bar{k}\times(2\bar{n}+1)}$ with $\mathbf{Q}_i^{\mathrm{H}}\mathbf{Q}_i = \mathbf{I}$, $\mathbf{R}_i \in \mathbb{C}^{(2\bar{n}+1)\times(2\bar{n}+1)}$, and where $\mathbf{R}_i^{11}, \mathbf{R}_i^{22}$ are square and upper triangular. Left-multiplication of both sides in (8.23) by $\mathbf{Q}_i^{\mathrm{H}}$ leads to

$$\begin{pmatrix}\mathbf{R}_i^{11} & \mathbf{R}_i^{12}\\ \mathbf{0} & \mathbf{R}_i^{22}\end{pmatrix}\begin{pmatrix}\mathbf{a}^i\\ \mathbf{d}\end{pmatrix} \approx \mathbf{Q}_i^{\mathrm{H}}\mathbf{b}^i = \begin{pmatrix}\mathbf{b}_1^i\\ \mathbf{b}_2^i\end{pmatrix}. \tag{8.26}$$

The second block row of this system depends only on the unknowns $\mathbf{d}$. Collecting all such block rows for $i = 1,\ldots,P$ leads to

$$\begin{pmatrix}\mathbf{R}_1^{22}\\ \vdots\\ \mathbf{R}_P^{22}\end{pmatrix}\mathbf{d} \approx \begin{pmatrix}\mathbf{b}_2^1\\ \vdots\\ \mathbf{b}_2^P\end{pmatrix} \tag{8.27}$$

whose least squares solution provides the denominator coefficients. The latter are in turn used to setup the pole relocation process as in (8.14) and (8.15), so that the next iteration can be carried out.

Once the basis poles stabilize, the model residues are computed as

$$\Phi^{\circ}\,\mathbf{r}^{i} \approx \mathbf{b}^{i} \quad i = 1,\ldots,P \tag{8.28}$$

where the system matrix Φ° is constructed based on the converged poles p_n and $\mathbf{r}^{i} = \left(r^{i}_{0},\ldots,r^{i}_{\bar{n}}\right)^{\mathrm{T}}$ collects the direct coupling constant and the residues for the i-th model component. This process of eliminating the numerator coefficients is the state-of-the-art VF implementation known as *Fast VF* [25]. Since the QR factorizations (8.25) can be computed independently, this scheme proves extremely efficient on parallel computing hardware, see [26,27].

The above derivations discussed the case of a vector of P frequency responses with common poles. A $P \times P$ matrix as in (8.2) is treated in the same way, provided that its P columns are stacked in a vector with P^2 components. Note that reciprocity of the model can be enforced without difficulty, by considering only the upper triangular part with $P(P+1)/2$ responses, which are fitted concurrently. As a last step, the lower triangular part is copied by enforcing symmetry.

8.2.6 State-space realizations

Exploitation of a macromodel within ODE solvers or circuit solvers of the SPICE class requires an intermediate realization step in form of state-space equations (8.1). Since these are not unique, there are many possible alternatives, as documented in [1]. We report here a particular choice that is suitable for a $P \times P$ model $\mathbf{H}(s)$ with common poles as in (8.2), and which can be obtained using the Fast VF algorithm.

Let us consider individually each j-th column of the transfer matrix

$$\mathbf{h}_j(s) = \mathbf{r}_{j,0} + \sum_{n=1}^{\bar{n}} \frac{\mathbf{r}_{j,n}}{s - p_n}, \quad j = 1,\ldots,P \quad \text{with} \quad \mathbf{r}_{j,n} \in \mathbb{C}^{P}. \tag{8.29}$$

This column can be interpreted as a single-input multiple-output (SIMO) submodel, for which a diagonal regular state-space realization is obtained as

$$\begin{aligned} \mathbf{A}_j &= \mathrm{diag}\{p_1,\ldots,p_{\bar{n}}\}, & \mathbf{b}_j &= (1,\ldots,1)^{\mathrm{T}}, \\ \mathbf{C}_j &= (\mathbf{r}_{j,1},\ldots,\mathbf{r}_{j,\bar{n}}), & \mathbf{d}_j &= \mathbf{r}_{j,0}\,. \end{aligned} \tag{8.30}$$

The above realization is diagonal but complex-valued in case some $p_n \in \mathbb{C}$. A purely real-valued realization can be derived through a similarity transformation. In practice, for each complex pole pair $p_{n,n+1} = \sigma_n \pm \mathrm{j}\omega_n$ with associated residues $\mathbf{r}'_{j,n} \pm \mathrm{j}\mathbf{r}''_{j,n}$, the corresponding size-2 blocks in $\mathbf{A}_j$, $\mathbf{b}_j$ and $\mathbf{C}_j$ are replaced with

$$\mathbf{A}_{j,n} = \begin{pmatrix} \sigma_n & \omega_n \\ -\omega_n & \sigma_n \end{pmatrix}, \quad \mathbf{b}_{j,n} = (2 \quad 0)^{\mathrm{T}}, \quad \mathbf{C}_{j,n} = \begin{pmatrix} \mathbf{r}'_{j,n} & \mathbf{r}''_{j,n} \end{pmatrix} \tag{8.31}$$

Putting together all columns leads to a state-space realization for the full model $\mathbf{H}(s)$, which is constructed with a multiple-SIMO structure, and whose state-space matrices read

$$\begin{aligned} \mathbf{A} &= \text{blkdiag}\{\mathbf{A}_1, \ldots, \mathbf{A}_P\}, & \mathbf{B} &= \text{blkdiag}\{\mathbf{b}_1, \ldots, \mathbf{b}_P\}, \\ \mathbf{C} &= (\mathbf{C}_1, \ldots, \mathbf{C}_P), & \mathbf{D} &= (\mathbf{d}_1, \ldots, \mathbf{d}_P). \end{aligned} \tag{8.32}$$

Note that although the number of common poles is $\bar{n}$, the number of state variables in (8.32) is $N = P\bar{n}$.

8.2.7 SPICE synthesis

A behavioral circuit that corresponds to the state-space realization (8.32) of the macromodel is readily obtained. Let us consider column j of the transfer function, which is excited only by input component u_j, with associated state-space matrices (8.30). Denoting with $x_{j,n}$ the state variable associated to a real pole $p_n \in \mathbb{R}$, the corresponding (uncoupled) state equation can be realized as in Figure 8.1. The coupled state-variables associated to a complex pole pair $p_{n,n+1} = \sigma_n \pm j\omega_n$ are described by the state equations (8.31), which are realized as in Figure 8.2.

The above equivalent circuits apply for each electrical representation of the multiport system, including impedance, admittance or scattering. The coupling of these cells for $j = 1, \ldots, P$ and $n = 1, \ldots, \bar{n}$ to input and output port variables

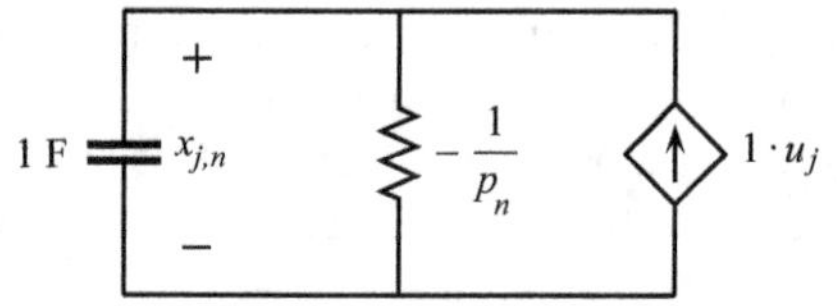

Figure 8.1 Direct circuit synthesis of the state equation corresponding to a real pole p_n of the j-th transfer function column

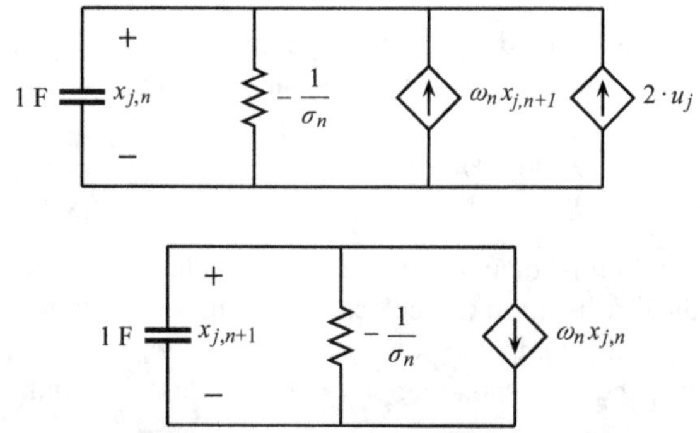

Figure 8.2 Direct circuit synthesis of the state equation corresponding to a complex pole pair $p_{n,n+1} = \sigma_n \pm j\omega_n$ of the j-th transfer function column

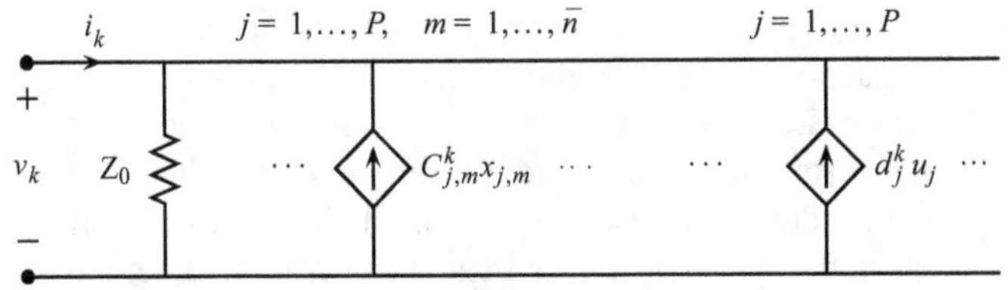

Figure 8.3 Circuit synthesis of the output state-space equation (8.1) at port k in the scattering representation, assuming a current normalization for the outgoing scattering wave $y_k = Y_0 v_k - i_k$, coherently with Figure 8.4. In case power normalization of scattering waves is used, all coefficients $C^k_{j,m}$ and d^k_j should be multiplied by $2\sqrt{Y_0}$

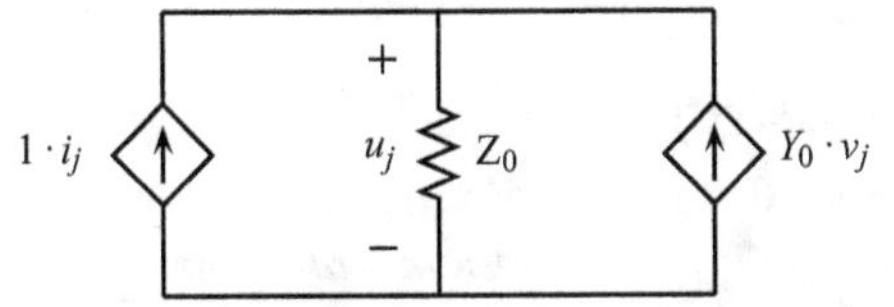

Figure 8.4 Synthesis of the incident (current-normalized) scattering wave $u_j = Y_0 v_j + i_j$ into port j as a single-cell voltage. In case power wave normalization is used, the gain of both controlled sources should be divided by $2\sqrt{Y_0}$, see [1]

depends instead on the adopted representation. We report in the following only the scattering case with port reference impedance $Z_0 \in \mathbb{R}$; all other representations are detailed in [1].

Denoting with $C^k_{j,m}$ element (k, m) of matrix $\mathbf{C}_j$ in (8.30) and (8.31) and as d^k_j the k-th component of vector $\mathbf{d}_j$ in (8.30), we see that the equivalent circuit of Figure 8.3 provides a direct synthesis of the output equation in (8.1), where $y_k = Y_0 v_k - i_k$ with $Y_0 = Z_0^{-1}$ is the (current-normalized) outgoing scattering wave at port k and $u_j = Y_0 v_j + i_j$ is the (current-normalized) incident scattering wave into port j. The signal u_j is constructed based on port voltages and currents through the elementary cell depicted in Figure 8.4.

8.3 Passivity

Passivity or *dissipativity* is a fundamental property of systems that are unable to generate energy on their own [28,29]. Electrical interconnects and basic components such as resistors, inductors, or capacitors are passive. The dynamics of such structures allow for redistribution, storage and dissipation of the energy received from the environment, but it is not possible that more energy is delivered from a passive structure than received from the external world. In our notation, this environment is represented by the termination network or system that is connected to the P ports of the structure for which we build the macromodel.

The importance of passivity becomes evident once we consider interconnected subsystems. It is guaranteed that the interconnection of passive subsystems is also passive. Since any passive system is also stable, using passive macromodels provides a guarantee of stability at the system level. No energy build up and blowing up solutions in time during a system-level simulation can occur if none of the individual models in the system is able to generate energy. In other words, a passive model cannot be the responsible for lack of stability in transient simulations. For the above reasons, passivity verification and enforcement in black-box macromodels have become a mandatory step of any CAD software, both under the standpoint of users and designers (who can exploit passive models with confidence) and of suppliers and CAD companies (who are guaranteed that their "product" models are reliable and robust). Macromodel passivity verification and enforcement are discussed in Sections 8.3.1 and 8.3.2, respectively.

8.3.1 Checking passivity

The following theorem [30–32] provides the main result.

Theorem 1: *A linear time-invariant system with transfer matrix* $\mathbf{H}(s)$ *is passive if and only if* $\mathbf{H}(s)$ *is positive real (PR, immittance case) or bounded real (BR, scattering case).*

A transfer function $\mathbf{H}(s)$ is positive (bounded) real when the following three conditions hold

1. each element of $\mathbf{H}(s)$ is defined and analytic in $\mathrm{Re}\{s\} > 0$;
2. $\mathbf{H}^*(s) = \mathbf{H}(s^*)$;
3. $\Upsilon(s) \geq 0$ for $\mathrm{Re}\{s\} > 0$, where

$$\Upsilon(s) = \begin{cases} \mathbf{H}(s) + \mathbf{H}(s)^{\mathrm{H}} & \text{Positive Real (immittance systems)} \\ \mathbf{I} - \mathbf{H}(s)^{\mathrm{H}}\mathbf{H}(s) & \text{Bounded Real (scattering systems)} \end{cases} \tag{8.33}$$

For the case under study in this chapter, where $\mathbf{H}(s)$ is a rational function of s, condition 1 states that all poles are stable, and condition 2 ensures a real-valued time response. The adopted model representation and the VF scheme ensure these conditions by construction. Condition 3 implies energy dissipativity and takes different forms (8.33) for immittance and scattering representations. Due to the regularity of $\mathbf{H}(s)$ for $\mathrm{Re}\{s\} > 0$ (with the additional requirement of asymptotic stability for immittance systems), this condition can be restricted to the imaginary axis $s = \mathrm{j}\omega$, providing the following *frequency-domain inequalities*

- Immittance models:

$$\mathbf{H}(\mathrm{j}\omega) + \mathbf{H}(\mathrm{j}\omega)^{\mathrm{H}} \geq 0 \;\Leftrightarrow\; \lambda_i\{\mathbf{H}(\mathrm{j}\omega) + \mathbf{H}(\mathrm{j}\omega)^{\mathrm{H}}\} \geq 0 \;\; \forall\omega \in \mathbb{R} \tag{8.34}$$

- Scattering models:

$$\mathbf{I} - \mathbf{H}(\mathrm{j}\omega)^{\mathrm{H}}\mathbf{H}(\mathrm{j}\omega) \geq 0 \;\Leftrightarrow\; \sigma_i\{\mathbf{H}(\mathrm{j}\omega)\} \leq 1 \;\; \forall\omega \in \mathbb{R} \tag{8.35}$$

where λ_i and σ_i denote the generic i-th eigenvalue and singular value, respectively.

Conditions (8.34) and (8.35) can be directly used to check a macromodel passivity based on a sampling process. Since, however, only a finite number of samples can be tested, purely algebraic tests that are equivalent to the PR and BR conditions (under technical assumptions, see [1]) have become very popular. The current method of choice for checking passivity is provided by the following *Hamiltonian matrix* test [33,34], which requires only a single eigenvalue calculation and is based on a state-space model representation. Macromodel (strict) passivity is implied by the following conditions

- Immittance models: assuming that $\mathbf{D}+\mathbf{D}^{\mathrm{T}}>0$, the Hamiltonian matrix

$$\mathbf{N}=\begin{pmatrix}\mathbf{A}-\mathbf{B}(\mathbf{D}+\mathbf{D}^{\mathrm{T}})^{-1}\mathbf{C} & -\mathbf{B}(\mathbf{D}+\mathbf{D}^{\mathrm{T}})^{-1}\mathbf{B}^{\mathrm{T}}\\ \mathbf{C}^{\mathrm{T}}(\mathbf{D}+\mathbf{D}^{\mathrm{T}})^{-1}\mathbf{C} & -\mathbf{A}^{\mathrm{T}}+\mathbf{C}^{\mathrm{T}}(\mathbf{D}+\mathbf{D}^{\mathrm{T}})^{-1}\mathbf{B}^{\mathrm{T}}\end{pmatrix} \tag{8.36}$$

  must not have purely imaginary eigenvalues $\mu_k=\mathrm{j}\omega_k$.
- Scattering models: assuming that $\|\mathbf{D}\|=\sigma_{\max}(\mathbf{D})<1$, the Hamiltonian matrix

$$\mathbf{M}=\begin{pmatrix}\mathbf{A}+\mathbf{B}(\mathbf{I}-\mathbf{D}^{\mathrm{T}}\mathbf{D})^{-1}\mathbf{D}^{\mathrm{T}}\mathbf{C} & \mathbf{B}(\mathbf{I}-\mathbf{D}^{\mathrm{T}}\mathbf{D})^{-1}\mathbf{B}^{\mathrm{T}}\\ -\mathbf{C}^{\mathrm{T}}(\mathbf{I}-\mathbf{D}\mathbf{D}^{\mathrm{T}})^{-1}\mathbf{C} & -\mathbf{A}^{\mathrm{T}}-\mathbf{C}^{\mathrm{T}}\mathbf{D}(\mathbf{I}-\mathbf{D}^{\mathrm{T}}\mathbf{D})^{-1}\mathbf{B}^{\mathrm{T}}\end{pmatrix} \tag{8.37}$$

must not have purely imaginary eigenvalues $\mu_k=\mathrm{j}\omega_k$.

The above Hamiltonian tests have two main advantages: they provide a pass/fail result in a predictable and finite time, and in addition they provide a precise localization of passivity violations, e.g., the frequency bands where the conditions (8.34) and (8.35) are not satisfied [1,34]. See Figure 8.5 for a graphical illustration. We will see in Section 8.3.2 that this information is essential for correcting a non-passive macromodel to recover its passivity. We remark that this information is not provided by a third class of tests based on the celebrated Kalman–Yakubovich–Popov (KYP) lemmas [28,32] also known as positive (bounded) real lemmas, which are not discussed here. See [1] for details and a complete literature review.

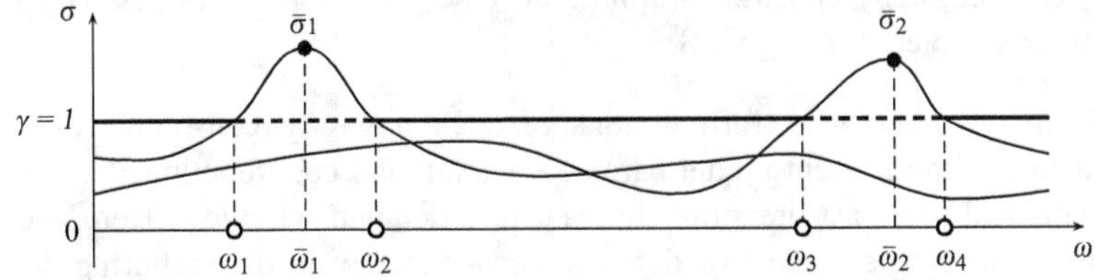

Figure 8.5 Graphical illustration of passivity conditions for a two-port scattering macromodel. Solid lines denote denote the singular values $\sigma_i(\omega)$ of $\mathbf{H}(s)$, and empty circles denote the frequencies ω_k corresponding to imaginary Hamiltonian eigenvalues $\mu_k=\mathrm{j}\omega_k$. These frequencies are the crossing points of the singular value trajectories with the passivity threshold $\gamma=1$ and delimit passive (thick solid line) and non-passive (thick dashed line) bands. Local singular value maxima $\bar{\sigma}_\ell$ occurring at frequencies $\bar{\omega}_\ell$ are depicted with black dots

8.3.2 Enforcing passivity

Macromodels are usually constrained to be passive through post-processing. Given an initial non-passive model $\mathbf{H}(s)$, a perturbation process is applied to its coefficients so that the resulting model $\widehat{\mathbf{H}}(s)$ is passive

$$\widehat{\mathbf{H}}(s) = \mathbf{H}(s) + \Delta\mathbf{H}(s). \tag{8.38}$$

The key ingredients for a successful passivity enforcement are

- A good choice of model coefficients to be perturbed. Considering (8.2), it is in principle feasible to perturb all state-space matrices or alternatively all pole-residue terms. In practice, the poles (equivalently, matrices **E** and **A**) are rarely perturbed since they provide an estimate of the system natural frequencies. It is quite common instead to perturb residues or output state matrices

$$\widehat{\mathbf{R}}_n = \mathbf{R}_n + \Delta\mathbf{R}_n \;\; \forall n \quad \text{or} \quad \widehat{\mathbf{C}} = \mathbf{C} + \Delta\mathbf{C}, \;\; \widehat{\mathbf{D}} = \mathbf{D} + \Delta\mathbf{D}. \tag{8.39}$$

  These two approaches are equivalent if the state-space realization of Section 8.2.6 is adopted.
- An appropriate cost function to be minimized, so that model accuracy is preserved. The two main approaches are model-based and data-based, respectively

$$\min\|\Delta\mathbf{H}(s)\|^2 \quad \text{and} \quad \min\sum_{k=1}^{\bar{k}} \left\|\widehat{\mathbf{H}}(s_k) - \breve{\mathbf{H}}_k\right\|_F^2 \tag{8.40}$$

  A model-based cost function minimizes the model perturbation in a suitable system norm, typically H_2 or H_∞. A data-based cost function minimizes the error of the passive model with respect to the initial data samples. A lot of flexibility can be exploited in the definition of these norms, including absolute or relative error definitions and adaptive, response-dependent or frequency-dependent weighting [35–38].
- A set of passivity constraints aiming at the elimination of the passivity violations, as depicted in Figure 8.5.

The many alternative formulations of such passivity constraints and their combination with different model parameterization and cost functions led to many algorithms and to a vast literature, see [1] for a detailed overview. Here, we concentrate on the single case of model-based cost function in the scattering domain, with perturbation based on local passivity constraints. The resulting scheme proves to be the most robust and reliable and can be considered as the state of the art when compared to competing approaches.

Cost function

We assume that the model is passive at $s = \infty$ with $\sigma_{\max}\{\mathbf{D}\} < 1$, and we perturb only the state-space matrix **C**. The model perturbation is written as

$$\Delta\mathbf{H}(s) = \Delta\mathbf{C}(s\mathbf{I} - \mathbf{A})^{-1}\mathbf{B} \quad \text{with} \quad \|\Delta\mathbf{H}\|_2^2 = \mathrm{tr}\left(\Delta\mathbf{C}\,\mathbf{W}_c\,\Delta\mathbf{C}^{\mathsf{T}}\right) \tag{8.41}$$

where tr is the matrix trace and $\mathbf{W}_c$ is the *controllability Gramian* [39] associated to the model, computed as the unique solution of the *Lyapunov* equation [8,35,40]

$$\mathbf{A}\mathbf{W}_c + \mathbf{W}_c\mathbf{A}^{\mathrm{T}} + \mathbf{B}\mathbf{B}^{\mathrm{T}} = \mathbf{0}. \tag{8.42}$$

An algebraic characterization of the model error norm is obtained as

$$\|\Delta\mathbf{H}\|_2^2 = \mathrm{tr}\left(\Delta\mathbf{C}\mathbf{Q}_c^{\mathrm{T}}\mathbf{Q}_c\Delta\mathbf{C}^{\mathrm{T}}\right) = \mathrm{tr}\left(\Xi\Xi^{\mathrm{T}}\right) = \|\Xi\|_F^2 = \|\boldsymbol{\xi}\|_2^2 \tag{8.43}$$

where the Cholesky factorization $\mathbf{W}_c = \mathbf{Q}_c^{\mathrm{T}}\mathbf{Q}_c$ is used to define a change of variable $\Xi = \Delta\mathbf{C}\mathbf{Q}_c^{\mathrm{T}}$ and matrix Ξ is reshaped by stacking its columns into the vector $\boldsymbol{\xi} = \mathrm{vec}(\Xi)$.

Local passivity constraints

We discuss only the scattering case, see [1] for generalizations. With reference to Figure 8.5, let us consider a singular value $\bar{\sigma}_\ell$ of $\mathbf{H}(j\omega)$ evaluated at the frequency $\bar{\omega}_\ell$, and let us assume that $\bar{\sigma}_\ell > 1$. This singular value should be perturbed so that

$$\widehat{\sigma}_\ell = \bar{\sigma}_\ell + \Delta\bar{\sigma}_\ell \leq 1 \quad \rightarrow \quad \Delta\bar{\sigma}_\ell \leq 1 - \bar{\sigma}_\ell \tag{8.44}$$

A local passivity constraint is obtained by writing the explicit form of the singular value perturbation $\Delta\bar{\sigma}_\ell$ in terms of the coefficients $\Delta\mathbf{C}$ or $\boldsymbol{\xi}$. This is possible by exploiting a result in matrix perturbation theory [41], which based on a first-order linearization states that

$$\Delta\bar{\sigma}_\ell \approx \mathrm{Re}\left\{\mathbf{u}_\ell^{\mathrm{H}}\Delta\mathbf{H}(j\bar{\omega}_\ell)\mathbf{v}_\ell\right\}, \tag{8.45}$$

where $\mathbf{u}_\ell, \mathbf{v}_\ell$ are the left and right singular vectors associated with $\bar{\sigma}_\ell$. Defining $\mathbf{Z}_\ell = (j\bar{\omega}_\ell\mathbf{I} - \mathbf{A})^{-1}\mathbf{B}$ we have

$$\Delta\bar{\sigma}_\ell \approx \mathbf{z}_\ell^{\mathrm{T}}\boldsymbol{\xi} \quad \text{with} \quad \mathbf{z}_\ell = \mathrm{Re}\left\{(\mathbf{Q}_c^{-\mathrm{T}}\mathbf{Z}_\ell\mathbf{v}_\ell) \otimes \mathbf{u}_\ell^*\right\} \tag{8.46}$$

where $\otimes$ is the Kronecker matrix product.

Iterative passivity enforcement

The above ingredients can be assembled in a passivity enforcement scheme that iteratively solves the following linearly-constrained minimum norm problem

$$\min_{\boldsymbol{\xi}} \|\boldsymbol{\xi}\| \quad s.t. \quad \mathbf{z}_\ell^{\mathrm{T}}\boldsymbol{\xi} \geq 1 - \bar{\sigma}_\ell \quad \ell = 1, \ldots, \bar{\ell} \tag{8.47}$$

where multiple local constraints are used to perturb concurrently multiple singular values exceeding $\gamma = 1$. These can be collected by running a Hamiltonian test to identify the bands with local passivity violations, and then performing a sampling within these bands to collect one or more constraints per band. The most effective constraints are placed at local maxima, as depicted in Figure 8.5. Since the local constraints (8.45) are not exact but based on a linear approximation, the solution of (8.47) needs to be iterated until all violations are removed.

8.3.3 An example

We illustrate the above-described macromodeling process on an illustrative test case coming from a signal integrity application, which is indeed one of the major fields where behavioral macromodeling played and is currently playing an important role. We consider an interconnect from a portion of a ball grid array (BGA) package formed by four adjacent traces, resulting in $P = 8$ ports. A full-wave solver was used to compute a total of $\bar{k} = 1{,}001$ frequency samples of the scattering responses, ranging from DC to 20 GHz. These samples were subjected to VF using a common set of $\bar{n} = 63$ poles. The resulting model was then checked for passivity, and three iterations of passivity enforcement were applied, in order to guarantee passivity. We used the constraints (8.47) derived from the associated Hamiltonian matrix, combined with a data-based cost function. All calculations were performed using a commercial implementation of these algorithms available in the tool IdEM [42].

Figure 8.6 plots one of the smallest (hence sensitive) scattering responses by comparing original data samples, the response of the non-passive model, and the corresponding response from the final passive model. All three curves are indistinguishable on this scale, as a confirmation of the excellent accuracy achieved by the model.

Figure 8.7 depicts the spectrum of all singular values of the model as it evolves through passivity enforcement iterations. At the initial iteration, a local passivity violation slightly outside the modeling bandwidth is detected through four imaginary Hamiltonian eigenvalues, with two local singular value maxima larger than the passivity threshold $\sigma = 1$. Both former and latter are iteratively removed through iterations, leading to a final passive model with no violations at iteration 3.

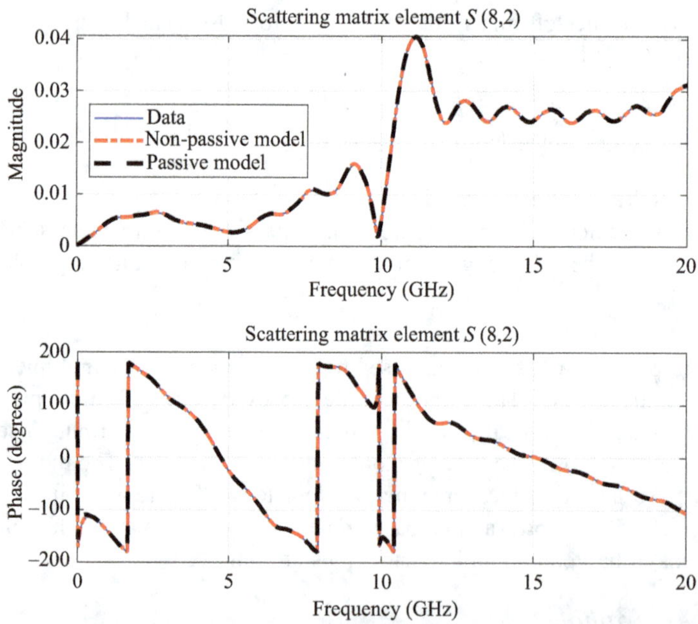

Figure 8.6 Validation of a package interconnect model before and after passivity enforcement for one of the scattering matrix elements

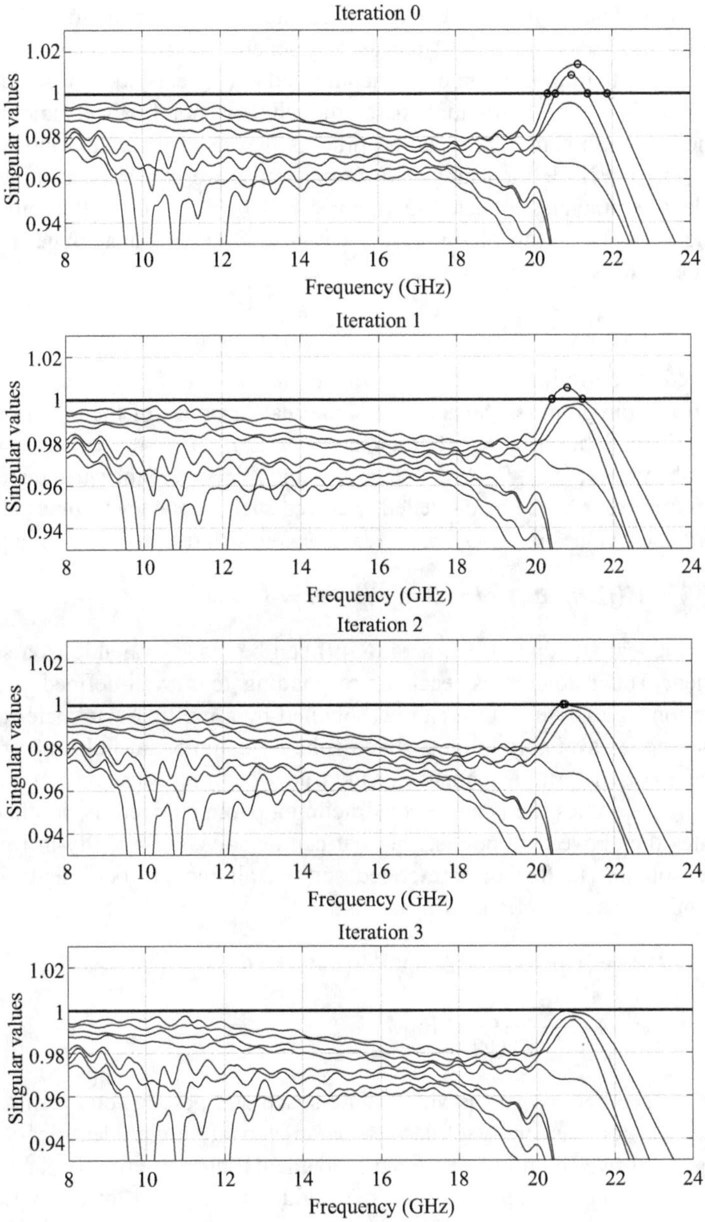

Figure 8.7 Iterative passivity enforcement for the package interconnect model

8.4 Parameterized macromodels

We now extend the macromodeling framework to the more general *multivariate* or *parameterized* setting. In this scenario, the system under investigation has a transfer matrix $\breve{\mathbf{H}}(s;\vartheta)$ that depends not only on frequency but also on some additional parameter $\vartheta \in \Theta = [\vartheta_{\min}, \vartheta_{\max}]$ such as a geometrical dimension or a material property. This situation is commonly encountered during early design stages, when the configuration of the structure under modeling is not finalized and must be determined through some optimization process that determines the optimal parameter ϑ_{opt} in a well-defined range. More generally, we can have a number ρ of independent parameters, collected in vector $\boldsymbol{\vartheta} = \left[\vartheta^1, \vartheta^2, \ldots, \vartheta^\rho\right]^{\mathrm{T}}$. We will assume that these parameters can attain values within a ρ-dimensional hyper-rectangle $\Theta \subseteq \mathbb{R}^\rho$ defined as

$$\Theta = [\vartheta^1_{\min}, \vartheta^1_{\max}] \times [\vartheta^2_{\min}, \vartheta^2_{\max}] \times \cdots \times [\vartheta^\rho_{\min}, \vartheta^\rho_{\max}] \tag{8.48}$$

and denoted in the following as the *parameter space*.

As in the foregoing sections, we assume that response of the structure under modeling is available at a set of frequency points f_k, $k = 1, \ldots, \bar{k}$ over a given frequency band $\Omega = [f_{\min}, f_{\max}]$, with $f_1 = f_{\min}$ and $f_{\bar{k}} = f_{\max}$. Since also dependence on the parameters $\boldsymbol{\vartheta}$ has to be modelled, we need such frequency samples also for a set of parameter values $\boldsymbol{\vartheta}_m \in \Theta$, $m = 1, \ldots, \bar{m}$. We will denote this training data as

$$\breve{\mathbf{H}}_{k,m} = \breve{\mathbf{H}}(\mathrm{j}2\pi f_k; \boldsymbol{\vartheta}_m) \quad k = 1, \ldots, \bar{k}, \quad m = 1, .., \bar{m}. \tag{8.49}$$

Under a practical standpoint, the data (8.49) can be made available as a set of $\bar{m}$ independent Touchstone files, each corresponding to a well-defined parameter configuration $\boldsymbol{\vartheta}_m$. These in turn can be obtained by a multiple parameter and frequency sweep, as available in most state of the art commercial field solvers.

Since a closed-form dependence of the true system response on s and $\boldsymbol{\vartheta}$ is not available, there comes the need of constructing a parameterized model that fits or interpolates the above data points, and that can be solved efficiently in frequency and time domain. In this parameterized setting, all model coefficients in (8.2) become multivariate functions of $\boldsymbol{\vartheta}$ as

$$\begin{aligned}\mathbf{H}(s;\boldsymbol{\vartheta}) &= \mathbf{C}(\boldsymbol{\vartheta})[s\mathbf{E}(\boldsymbol{\vartheta}) - \mathbf{A}(\boldsymbol{\vartheta})]^{-1}\mathbf{B}(\boldsymbol{\vartheta}) + \mathbf{D}(\boldsymbol{\vartheta}) \\ &= \sum_{n=1}^{\bar{n}} \frac{\mathbf{R}_n(\boldsymbol{\vartheta})}{s - p_n(\boldsymbol{\vartheta})} + \mathbf{R}_0(\boldsymbol{\vartheta}).\end{aligned} \tag{8.50}$$

We will see below that the above state-space and pole-residue forms are not appropriate for a robust and well-defined parameterization and identification from data. In general, we should preserve the fundamental property that $\mathbf{H}(s;\boldsymbol{\vartheta})$ for any fixed value of $\boldsymbol{\vartheta}$ is a rational function of s, which is required for a lumped macromodel structure in order to enable a SPICE netlist synthesis. More flexibility is allowed to represent dependence on $\boldsymbol{\vartheta}$, as appropriate for the application at hand. Assuming an expansion of all quantities that depend on $\boldsymbol{\vartheta}$ in terms of a set of basis functions with suitable coefficients, the latter can be determined through a fitting process

$$\mathbf{H}(\mathrm{j}2\pi f_k; \boldsymbol{\vartheta}_m) \approx \breve{\mathbf{H}}_{k,m}, \quad k = 1, \ldots, \bar{k}, \; m = 1, \ldots, \bar{m} \tag{8.51}$$

which extends (8.4) to the multivariate case. As in the standard non-parameterized case, the model obtained after fitting is subjected to stability and passivity enforcement. These two aspects are discussed in Sections 8.4.3 and 8.4.6, respectively. Some guidelines for SPICE synthesis of parameterized macromodels are available in [43].

8.4.1 Structure of parameterized models

The model structure should be tuned to the particular application or system under investigation, using the simplest format that is appropriate to represent the variations induced by the parameters.

8.4.1.1 Parameterized models with fixed poles

In some cases, it is sufficient to assume that the model poles do not depend on ϑ, which corresponds to a pole-residue representation

$$\mathbf{H}(s;\vartheta) = \sum_{n=1}^{\bar{n}} \frac{\mathbf{R}_n(\vartheta)}{s - p_n} + \mathbf{R}_0(\vartheta). \tag{8.52}$$

Model extraction is very simple in this case, following the steps below.

1. Compute the poles p_n through the Fast VF pole relocation iteration of Section 8.2.5. The constraint of common poles is achieved by processing together all $\bar{m} \cdot P^2$ responses in the training dataset.
2. Determine the set of residue matrices $\mathbf{R}_n^{(m)}$ for the m-th training parameter value ϑ_m in the data by solving the univariate fitting

$$\sum_{n=1}^{\bar{n}} \frac{\mathbf{R}_n^{(m)}}{j2\pi f_k - p_n} + \mathbf{R}_0^{(m)} \approx \breve{\mathbf{H}}_{k,m}, \quad k = 1, \ldots, \bar{k}, \quad \forall m = 1, \ldots, \bar{m} \tag{8.53}$$

   independently for each m.
3. Recover the continuous dependence of the model with respect to ϑ through some interpolation scheme applied to the direct coupling and residue matrices as

$$\mathbf{R}_n(\vartheta) = \sum_{m=1}^{\bar{m}} \mathbf{R}_n^{(m)} \xi_m(\vartheta) \quad \forall n = 0, \ldots, \bar{n} \tag{8.54}$$

where ξ_m are suitable basis functions (e.g. polynomials, trigonometric polynomials, splines, etc.).

The above approach has the advantage of reusing known and well-established software for VF, with minimal modifications. In addition, particular interpolation schemes exist (positive schemes for immittance or convex combinations for scattering) that preserve passivity if the individual models for each m are passive. Therefore, legacy passivity enforcement schemes can be applied to univariate (non-parameterized) *root* submodels (8.53), with a subsequent passivity-preserving interpolation. For more details on this approach, see [44–46].

8.4.1.2 Parameterizing model poles

Most structures that have some practical relevance in applications have poles that undergo significant variations with the parameters, making the invariant pole assumption (8.52) not feasible. From now on, we will concentrate on this scenario.

A model structure that is appropriate for the general parameterization case is based on a double expansion into frequency- and parameter-dependent basis functions of both numerator and denominator

$$\mathbf{H}(s;\boldsymbol{\vartheta}) = \frac{\mathsf{N}(s;\boldsymbol{\vartheta})}{\mathsf{D}(s;\boldsymbol{\vartheta})} = \frac{\sum_{n=0}^{\bar{n}} \mathbf{C}_n(\boldsymbol{\vartheta})\,\varphi_n(s)}{\sum_{n=0}^{\bar{n}} d_n(\boldsymbol{\vartheta})\,\varphi_n(s)}, \tag{8.55}$$

which extends (8.5) by introducing a parameter dependence in the numerator and denominator coefficients. Note that the denominator is scalar, whereas the numerator and its coefficients $\mathbf{C}_n(\boldsymbol{\vartheta})$ inherit the size of $\mathbf{H}$. The frequency-dependent basis functions $\varphi_n(s)$ are the same as in Vector Fitting, see (8.6) and (8.7). The (real-valued) numerator and denominator coefficients are further expressed as

$$\mathbf{C}_n(\boldsymbol{\vartheta}) = \sum_{\ell=1}^{\bar{\ell}} \mathbf{C}_{n,\ell}\,\xi_\ell(\boldsymbol{\vartheta}), \quad d_n(\boldsymbol{\vartheta}) = \sum_{\ell=1}^{\bar{\ell}} d_{n,\ell}\,\xi_\ell(\boldsymbol{\vartheta}) \tag{8.56}$$

where the (multivariate) basis functions $\xi_\ell(\boldsymbol{\vartheta})$ are responsible for reproducing the variations induced by the external parameters $\boldsymbol{\vartheta}$. We remark that the adopted model structure is the same as discussed in [1] and originally introduced in [47,48], where polynomials, piecewise polynomials, or trigonometric polynomials were used as basis functions ξ_ℓ. A common choice when parameters are separable is to consider each basis function $\xi_\ell(\boldsymbol{\vartheta})$ in (8.56) to be the product of univariate basis functions

$$\xi_\ell(\boldsymbol{\vartheta}) = \xi_{\ell_1}(\vartheta^1)\cdot\xi_{\ell_2}(\vartheta^2)\cdots\xi_{\ell_\rho}(\vartheta^\rho) \tag{8.57}$$

under a suitable mapping between the global linear index ℓ and the multi-index $(\ell_1, \ell_2, \ldots, \ell_\rho)$. With this choice, individual numerator and denominator expansion orders can be fine-tuned depending on the variability of the system behavior induced by each component of $\boldsymbol{\vartheta}$.

The model structure (8.55) provides implicitly parameterized poles and residues. Thanks to the assumed linear dependence of both model numerator $\mathsf{N}(s;\boldsymbol{\vartheta})$ and denominator $\mathsf{D}(s;\boldsymbol{\vartheta})$ on the respective coefficients $\mathbf{C}_{n,\ell}$ and $d_{n,\ell}$, the latter can be easily evaluated during a model identification step by enforcing the fitting condition (8.4). Details on this fitting follow in Section 8.4.2.

8.4.2 *The parameterized SK iteration*

Let us rewrite the general model structure (8.55) using (8.56) as

$$\mathbf{H}(s;\boldsymbol{\vartheta}) = \frac{\mathsf{N}(s;\boldsymbol{\vartheta})}{\mathsf{D}(s;\boldsymbol{\vartheta})} = \frac{\sum_{n=0}^{\bar{n}}\sum_{\ell=1}^{\bar{\ell}} \mathbf{C}_{n,\ell}\,\xi_\ell(\boldsymbol{\vartheta})\,\varphi_n(s)}{\sum_{n=0}^{\bar{n}}\sum_{\ell=1}^{\bar{\ell}} d_{n,\ell}\,\xi_\ell(\boldsymbol{\vartheta})\,\varphi_n(s)}. \tag{8.58}$$

Note that both numerator and denominator are linear combinations of basis function (products), and that the corresponding coefficients $\mathbf{C}_{n,\ell}$, $d_{n,\ell}$ enter linearly in both numerator and denominator. This situation is exactly the same as for Vector Fitting, which can be recovered as a particular case when $\bar{\ell} = 1$ and $\xi_1(\boldsymbol{\vartheta}) = 1$. Enforcing the fitting condition (8.4) leads to the following nonlinear and non-convex optimization problem

$$\min_{\mathbf{C}_{n,\ell}, d_{n,\ell}} \sum_{m,k} \left\| \frac{\mathsf{D}(\mathrm{j}2\pi f_k, \boldsymbol{\vartheta}_m)\, \breve{\mathbf{H}}_{k,m} - \mathsf{N}(\mathrm{j}2\pi f_k, \boldsymbol{\vartheta}_m)}{\mathsf{D}(\mathrm{j}2\pi f_k, \boldsymbol{\vartheta}_m)} \right\|_F^2. \tag{8.59}$$

A linearization of this problem is obtained through the GSK iteration as in Section 8.2.2. An iteration index $\nu = 1, 2, \ldots$ is introduced, and the following linear least squares problem is solved at each iteration

$$\min_{\mathbf{C}^{\nu}_{n,\ell}, d^{\nu}_{n,\ell}} \sum_{m,k} \left\| \frac{\mathsf{D}^{\nu}(\mathrm{j}2\pi f_k, \boldsymbol{\vartheta}_m)\, \breve{\mathbf{H}}_{k,m} - \mathsf{N}^{\nu}(\mathrm{j}2\pi f_k, \boldsymbol{\vartheta}_m)}{\mathsf{D}^{\nu-1}(\mathrm{j}2\pi f_k, \boldsymbol{\vartheta}_m)} \right\|_F^2. \tag{8.60}$$

where the unknowns $\mathbf{C}^{\nu}_{n,\ell}, d^{\nu}_{n,\ell}$ have now been equipped with an additional superscript since they are iteration-dependent. The denominator is "delayed" by one iteration, hence based on the coefficients $d^{\nu-1}_{n,\ell}$ which are known at iteration ν. Therefore, this denominator acts as a parameter- and frequency-dependent weight that is updated at each iteration, and (8.60) is a coupled linear least squares problem. The iterations are stopped when the estimates of these coefficients and the value of the cost function (8.60) stabilize, see [22,23] for a discussion on convergence. This scheme is usually denoted as *Parameterized Sanathanan-Koerner (PSK)* iteration.

Remark. The first iteration $\nu = 1$ is practically identical with the first Vector Fitting iteration. The only difference is that the regressor matrix, i.e., the matrix storing all basis functions (columns) at all data points (rows) is multivariate. The columns enclose each product combination of the parameter- and frequency-dependent basis functions, whereas each row corresponds to a combination of a frequency sample f_k at one specific parameter value $\boldsymbol{\vartheta}_m$.

Remark. After the first iteration (and after any subsequent iteration), the standard VF scheme evaluates the zeros z_n of the denominator and uses these zeros as new poles for the next iteration (the *pole relocation* step). This operation does not make sense in the parameterized case, since the zeros z^{ν}_n of the denominator

$$\mathsf{D}^{\nu}(s; \boldsymbol{\vartheta}) = \sum_{n=0}^{\bar{n}} \sum_{\ell=1}^{\bar{\ell}} d^{\nu}_{n,\ell}\, \xi_\ell(\boldsymbol{\vartheta})\, \varphi_n(s) \tag{8.61}$$

are parameter-dependent as $z^{\nu}_n = z^{\nu}_n(\boldsymbol{\vartheta})$. This is indeed the main objective of the entire parameterization, i.e., to obtain model poles that depend explicitly on the parameters $\boldsymbol{\vartheta}$. As a consequence, the VF pole relocation *cannot be performed* in the parameterized case, and the basis poles q_n *are kept constant through iterations*. Apart from the multivariate extension, this is the only major difference between parameterized and standard modeling in the proposed framework. The structure of the fitting equations for least-squares solution of (8.60) is instead very similar. The construction of a compact matrix form that generalizes (8.10) is left as an exercise for the Reader.

8.4.3 Enforcing stability

The fact that the poles are parameter-dependent and are not relocated through iterations has a major consequence related to stability. While in VF the model stability is trivially enforced by pole-flipping at any iteration, this procedure is not applicable in parameterized modeling. Therefore, other strategies are in order to make sure that the parameter-dependent poles are uniformly stable. We recall that these poles $p_n(\boldsymbol{\vartheta})$ are the zeros of the denominator

$$\mathsf{D}(s;\boldsymbol{\vartheta}) = \sum_{n=0}^{\bar{n}} \sum_{\ell=1}^{\bar{\ell}} d_{n,\ell}\, \xi_\ell(\boldsymbol{\vartheta})\, \varphi_n(s) \tag{8.62}$$

attained at the last iteration. These poles are multivariate (possibly non-smooth) functions of the parameters $\boldsymbol{\vartheta}$. A basic a-posteriori stability check can be performed through sampling in the parameter space Θ. However, such an approach is only able to perform a finite number of evaluations and may possibly miss stability violations. In addition, it is only a check and as such is unable to correct the model coefficients to enforce stability. It is preferable to embed in the model extraction process a set of stability constraints that enforce uniform stability by construction, so that

$$\mathrm{Re}\{p_n(\boldsymbol{\vartheta})\} < 0 \quad \forall \boldsymbol{\vartheta} \in \Theta. \tag{8.63}$$

A set of sufficient conditions for uniform stability are constructed as follows [43]. Considering the denominator $\mathsf{D}(s;\boldsymbol{\vartheta})$ as the frequency response of a scalar one-port immittance, we have that its poles coincide with the basis poles q_n, with corresponding residues

$$d_n(\boldsymbol{\vartheta}) = \sum_{\ell=1}^{\bar{\ell}} d_{n,\ell}\, \xi_\ell(\boldsymbol{\vartheta}). \tag{8.64}$$

Based on this structure, we conclude that

- $\mathsf{D}(s;\boldsymbol{\vartheta})$ is regular for $\mathrm{Re}\{s\} > 0$, since $\mathrm{Re}\{q_n\} < 0$ by assumption;
- $\mathsf{D}^*(s;\boldsymbol{\vartheta}) = \mathsf{D}(s^*;\boldsymbol{\vartheta})$;

Therefore, $\mathsf{D}(s,\boldsymbol{\vartheta})$ is a PR function (see Section 8.3.1) provided that the following condition is enforced

$$\mathrm{Re}\{\mathsf{D}(s;\boldsymbol{\vartheta})\} \geq 0 \quad \text{for} \quad \mathrm{Re}\{s\} > 0, \quad \forall \boldsymbol{\vartheta} \in \Theta \tag{8.65}$$

throughout the parameter space. Since [1,31]

- if $\mathsf{D}(s;\boldsymbol{\vartheta})$ is PR, then also $\mathsf{D}^{-1}(s;\boldsymbol{\vartheta})$ is PR (if it exists),
- any PR function is stable since regular for $\mathrm{Re}\{s\} > 0$,

we see that (8.65) is a sufficient condition to guarantee that the poles of $\mathsf{D}^{-1}(s;\boldsymbol{\vartheta})$, which are the zeros $p_n(\boldsymbol{\vartheta})$ of $\mathsf{D}(s;\boldsymbol{\vartheta})$, are stable as required in (8.63).

The above derivation showed that uniform stability of the parameterized macromodel (8.58) is guaranteed by the uniform nonnegativity constraint (8.65) on the denominator function. In turn, this is guaranteed by simpler conditions on the denominator coefficients, provided that the basis functions adopted in the parameterization are nonnegative

$$\xi_\ell(\boldsymbol{\vartheta}) \geq 0, \quad \forall \ell, \forall \boldsymbol{\vartheta} \in \Theta. \tag{8.66}$$

Under this assumption, let us rewrite the expansion of $\mathsf{D}(s;\boldsymbol{\vartheta})$ as

$$\mathsf{D}(s;\boldsymbol{\vartheta}) = \sum_{\ell=1}^{\bar{\ell}} \xi_\ell(\boldsymbol{\vartheta}) \left[\sum_{n=0}^{\bar{n}_r} \frac{d_{n,\ell}}{s-q_n} + \sum_{n=0}^{\bar{n}_c} \left(\frac{d'_{n,\ell} + \mathrm{j}d''_{n,\ell}}{s-p'_n - \mathrm{j}p''_n} + \frac{d'_{n,\ell} - \mathrm{j}d''_{n,\ell}}{s-p'_n + \mathrm{j}p''_n} \right) \right] \tag{8.67}$$

by splitting the contribution of the $\bar{n}_r$ real poles q_n with coefficients $d_{n,\ell}$ and $2 \cdot \bar{n}_c$ complex poles $p'_n \pm \mathrm{j}p''_n$ with coefficients $d'_{n,\ell} \pm \mathrm{j}d''_{n,\ell}$. It is straightforward to prove [49,50] that, when $q_n < 0$ and $p'_n < 0$, the following conditions:

$$\begin{cases} d_{n,\ell} > 0 \quad \forall n, \ell \\ -p'_n \cdot d'_{n,\ell} \pm p''_n \cdot d''_{n,\ell} > 0 \quad \forall n, \ell \end{cases} \tag{8.68}$$

imply (8.65) and are therefore sufficient for uniform stability. The main advantage in these conditions is that they are written as linear inequality constraints on the denominator coefficients, which in turn are part of the decision variables of the main PSK least squares problem (8.60).

Among the several choices of nonnegative basis functions, the Bernstein polynomials are particularly attractive. Given a maximum degree $\bar{\nu}$, these polynomials are defined in the univariate case as

$$b_{\nu,\bar{\nu}}(x) = \binom{\bar{\nu}}{\nu} x^\nu (1-x)^{\bar{\nu}-\nu}, \quad \nu = 0, \ldots, \bar{\nu}. \tag{8.69}$$

They are strictly positive in their normalized domain $x \in [0,1]$, except at the edges $x = 0$ and $x = 1$ where they have two zeros of order ν and $\bar{\nu} - \nu$, respectively. They provide uniform approximations for continuous functions [51], a fact that has been extensively used in computer graphics [52]. The positivity of a Bernstein polynomial approximation is implied by positivity of the corresponding coefficients

$$\sum_{\nu=0}^{\bar{\nu}} c_\nu b_{\nu,\bar{\nu}}(x) > 0, \ \forall x \quad \Leftarrow \quad c_\nu > 0, \ \forall \nu. \tag{8.70}$$

Therefore, a uniformly stable parameterized macromodeling scheme is readily obtained by enforcing conditions (8.68) at each PSK iteration after setting the basis functions $\xi_\ell(\boldsymbol{\vartheta})$ to be the Cartesian product of univariate Bernstein polynomials

$$\xi_\ell(\boldsymbol{\vartheta}) = \prod_{i=1}^{\rho} \xi_{\ell_i}(\vartheta_i), \quad \xi_{\ell_i}(\vartheta_i) = b_{\ell_i,\bar{\ell}_i}(\widehat{\vartheta}_i) \quad \widehat{\vartheta}_i = \frac{\vartheta_i - \vartheta_{\min}}{\vartheta_{\max} - \vartheta_{\min}}, \tag{8.71}$$

where index ℓ_i refers to the i-th coordinate in the parameter space, and ℓ is a global linear index spanning the multivariate basis of size $\bar{\ell} = \prod_i (1 + \bar{\ell}_i)$.

8.4.4 Descriptor realizations

A parameterized macromodel can be cast as a system of ODEs in several different ways. The one we present here is in descriptor form and provides the major advantage that the descriptor matrices inherit the smooth parameterization induced by model structure (8.55). We present here only the main result, leaving the details of the construction to the Reader (see also [47]).

The model (8.55) is rewritten and realized as

$$\mathbf{H}(s;\vartheta) = \mathbf{C}(\vartheta)(s\mathbf{E} - \mathbf{A}(\vartheta))^{-1}\mathbf{B} \quad \rightarrow \quad \begin{cases} \mathbf{E}\dot{\mathbf{x}}(t) = \mathbf{A}(\vartheta)\mathbf{x}(t) + \mathbf{B}\mathbf{u}(t) \\ \mathbf{y}(t) = \mathbf{C}(\vartheta)\mathbf{x}(t) \end{cases} \tag{8.72}$$

The descriptor matrices are defined as

$$\begin{aligned} \mathbf{E} &= \begin{bmatrix} \mathbf{I}_N & \mathbf{0}_{N,P} \\ \mathbf{0}_{P,N} & \mathbf{0}_{P,P} \end{bmatrix}, \qquad \mathbf{A}(\vartheta) = \begin{bmatrix} \mathbf{A}_0 & \mathbf{B}_0 \\ \mathbf{F}_2(\vartheta) & \mathbf{G}_2(\vartheta) \end{bmatrix}, \\ \mathbf{C}(\vartheta) &= [\,\mathbf{F}_1(\vartheta) \quad \mathbf{G}_1(\vartheta)\,], \quad \mathbf{B} = \begin{bmatrix} \mathbf{0}_{N,P} \\ -\mathbf{I}_P \end{bmatrix}, \end{aligned} \tag{8.73}$$

where $\mathbf{0}_{J,K}$ is a $J \times K$ all-zero matrix block, and where the generalized states are $\mathbf{x} \in \mathbb{R}^{\bar{N}}$ with $\bar{N} = N + P$ and $N = \bar{n}P$. We see that the descriptor matrix $\mathbf{E}$ is not parameterized and is singular, so that the last P rows provide purely algebraic relations. Matrices $\{\mathbf{A}_0, \mathbf{B}_0, \mathbf{F}_1(\vartheta), \mathbf{G}_1(\vartheta)\}$ and $\{\mathbf{A}_0, \mathbf{B}_0, \mathbf{F}_2(\vartheta), \mathbf{G}_2(\vartheta)\}$ are regular state-space realizations of the model numerator $\mathsf{N}(s;\vartheta)$ and the expanded denominator $\mathsf{D}(s;\vartheta)\mathbf{I}_P$, respectively. Each of these matrices is derived following the process of Section 8.2.6. Assuming q_n and p_n as $\bar{n}_r$ real and $\bar{n}_c$ complex-conjugate basis poles as in (8.6) and (8.7) and (8.67), we have constant dynamic matrices

$$\mathbf{A}_0 = \text{blkdiag}\{\mathbf{A}_{0r}, \mathbf{A}_{0c}\} \quad \mathbf{B}_0^{\mathrm{T}} = \left[\mathbf{B}_{0r}^{\mathrm{T}}, \mathbf{B}_{0c}^{\mathrm{T}}\right] \tag{8.74}$$

with individual blocks induced by real and complex pole pairs

$$\begin{aligned} \mathbf{A}_{0r} &= \text{blkdiag}\{q_n \mathbf{I}_P\}_{n=1}^{\bar{n}_r} & \mathbf{B}_{0r} &= [1, \ldots, 1]^{\mathrm{T}} \otimes \mathbf{I}_P \\ \mathbf{A}_{0c} &= \text{blkdiag}\left\{\begin{bmatrix} p_n' \mathbf{I}_P & p_n'' \mathbf{I}_P \\ -p_n'' \mathbf{I}_P & p_n' \mathbf{I}_P \end{bmatrix}\right\}_{n=1}^{\bar{n}_c} & \mathbf{B}_{0c} &= [2, 0, \ldots, 2, 0]^{\mathrm{T}} \otimes \mathbf{I}_P \end{aligned} \tag{8.75}$$

The other matrices in (8.73) are

$$\begin{aligned} \mathbf{F}_1(\vartheta) &= [\,\mathbf{C}_1(\vartheta) \quad \cdots \quad \mathbf{C}_{\bar{n}}(\vartheta)\,] & \mathbf{G}_1(\vartheta) &= \mathbf{C}_0(\vartheta) \\ \mathbf{F}_2(\vartheta) &= [\,\mathbf{I}_P d_1(\vartheta) \quad \cdots \quad \mathbf{I}_P d_{\bar{n}}(\vartheta)\,] & \mathbf{G}_2(\vartheta) &= \mathbf{I}_P d_0(\vartheta) \end{aligned} \tag{8.76}$$

The last definition confirms that a smooth parameterization (8.56) induces a smooth parameterization of matrices $\mathbf{F}_1(\boldsymbol{\vartheta})$, $\mathbf{G}_1(\boldsymbol{\vartheta})$, $\mathbf{F}_2(\boldsymbol{\vartheta})$, $\mathbf{G}_2(\boldsymbol{\vartheta})$ and consequently of the descriptor form (8.72) and (8.73).

8.4.5 Checking passivity

In this parameterized setting, the verification and enforcement of model passivity poses additional challenges with respect to the univariate case. In fact, the model must be compliant with passivity conditions *uniformly* in the parameter space, so that $\mathbf{H}(s;\boldsymbol{\vartheta})$ must fulfil the PR and BR conditions listed in Section 8.3.1 for all $\boldsymbol{\vartheta} \in \Theta$. Based on the adopted real and uniformly stable model representation, the constraints that need to be verified can be cast as

- Immittance models:

$$\lambda_i\{\mathbf{H}(\mathrm{j}\omega;\boldsymbol{\vartheta}) + \mathbf{H}(\mathrm{j}\omega;\boldsymbol{\vartheta})^{\mathrm{H}}\} \geq 0 \quad \forall\omega \in \mathbb{R} \quad \forall\boldsymbol{\vartheta} \in \Theta \tag{8.77}$$

- Scattering models:

$$\sigma_i\{\mathbf{H}(\mathrm{j}\omega;\boldsymbol{\vartheta})\} \leq 1 \quad \forall\omega \in \mathbb{R} \quad \forall\boldsymbol{\vartheta} \in \Theta \tag{8.78}$$

The univariate passivity verification presented in Section 8.3.1 applies only for fixed $\boldsymbol{\vartheta}$. It must thus be extended to scan the entire parameter space Θ, in order to find the sets $\Theta_\alpha \subset \Theta$ for $\alpha = 1, 2, \ldots$ where conditions (8.77) or (8.78) do not hold, and to define a set of constraints to be applied in order to perturb the model and enforce its passivity.

Also in the parameterized case we can benefit from Hamiltonian matrix tests, which avoid the need of sampling along the frequency axis to detect passivity violations. However, two generalizations are necessary. First, the macromodel realization in terms of a descriptor form (8.72) instead of a regular state-space form requires the eigensolution of a Skew-Hamiltonian/Hamiltonian (SHH) pencil [1,53–55] instead of a single Hamiltonian matrix. Second, this pencil is parameter-dependent. We report in the following the pencil $(\mathbf{M}(\boldsymbol{\vartheta}), \mathbf{K})$ that is appropriate for the scattering case (see [56] for the general case)

$$\mathbf{M}(\boldsymbol{\vartheta}) = \begin{pmatrix} \mathbf{A}(\boldsymbol{\vartheta}) & \mathbf{B}\mathbf{B}^{\mathrm{T}} \\ -\mathbf{C}^{\mathrm{T}}(\boldsymbol{\vartheta})\mathbf{C}(\boldsymbol{\vartheta}) & -\mathbf{A}^{\mathrm{T}}(\boldsymbol{\vartheta}) \end{pmatrix} \qquad \mathbf{K} = \begin{pmatrix} \mathbf{E}^{\mathrm{T}} & \mathbf{0} \\ \mathbf{0} & \mathbf{E} \end{pmatrix}. \tag{8.79}$$

whereas the main result for passivity characterization is stated as the following.

Theorem 2: *Let the matrix pencil* $(\mathbf{A}(\boldsymbol{\vartheta}), \mathbf{E})$ *have no purely imaginary eigenvalues. Then,* $\mathrm{j}\omega_k$ *is a generalized eigenvalue of* $(\mathbf{M}(\boldsymbol{\vartheta}), \mathbf{K})$ *if and only if* $\sigma = 1$ *is a singular value of* $\mathbf{H}(\mathrm{j}\omega_k, \boldsymbol{\vartheta})$.

As in the univariate case, the imaginary SHH pencil eigenvalues separate passive and non-passive frequency bands. The latter are then subjected to local frequency-domain sampling in order to determine local singular value maxima, as depicted in Figure 8.5. However, in this parameterized setting, both non-passive bands and corresponding local maxima are parameter-dependent. Therefore, the above result must be combined with a multivariate adaptive sampling in the parameter space, whose aim is to identify the regions $\Theta_\alpha \subset \Theta$ for $\alpha = 1, 2, \ldots$ where some passivity violation exist, and where local constraints for passivity enforcement will be placed. We omit the details on how this adaptive sampling can be carried out, see e.g. [56,57], but we report an application example in Figure 8.8, where a two-dimensional parameter space is adaptively sampled through a structured local grid refinement. At each grid point, the SHH pencil (8.79) is constructed and for each point where the SHH pencil has purely imaginary eigenvalues (red dots in the figure), the local singular value maxima and associated singular vectors are collected, together with their frequency and parameter space location, as $\{\bar{\omega}_\mu, \bar{\boldsymbol{\vartheta}}_\mu; \bar{\sigma}_\mu, \mathbf{u}_\mu, \mathbf{v}_\mu\}$ for $\mu = 1, \ldots, \bar{\mu}$.

8.4.6 Enforcing passivity

The passivity of a parameterized macromodel can be enforced by perturbation, as described in Section 8.3.2 for standard models. Here, we look for a perturbed model

$$\widehat{\mathbf{H}}(s; \boldsymbol{\vartheta}) = \mathbf{H}(s; \boldsymbol{\vartheta}) + \Delta\mathbf{H}(s; \boldsymbol{\vartheta}) \tag{8.80}$$

that is uniformly passive $\forall \boldsymbol{\vartheta} \in \Theta$. As in the non-parameterized case, we discuss below model parameterization, optimization goal for accuracy preservation, and passivity constriants.

Perturbation strategy

Also in the parameterized case, it is appropriate to preserve the (parameterized) model poles. This can be achieved by preserving the descriptor matrix $\mathbf{A}(\boldsymbol{\vartheta})$ of (8.73), which includes all coefficients related to the model denominator.

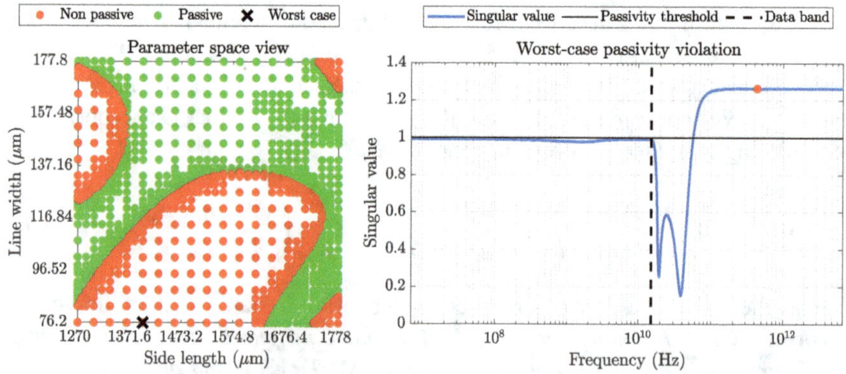

Figure 8.8 Modeling an integrated inductor, parameterized by its side length and line width. Left panel: adaptive sampling in a two-dimensional parameter space Θ, to detect the points where the parameterized model $\mathbf{H}(s; \boldsymbol{\vartheta})$ is passive (green dots) or non-passive (red dots). Right panel: largest singular value trajectory corresponding to the worst-case point (the cross in left panel) having the largest passivity violation (red dot)

Conversely, we perturb only the output descriptor matrix

$$\widehat{\mathbf{C}}(\boldsymbol{\vartheta}) = \mathbf{C}(\boldsymbol{\vartheta}) + \Delta\mathbf{C}(\boldsymbol{\vartheta}), \qquad \Delta\mathbf{C}(\boldsymbol{\vartheta}) = \sum_{\ell=1}^{\bar{\ell}} \Delta\mathbf{C}_\ell \xi_\ell(\boldsymbol{\vartheta}), \tag{8.81}$$

which stores all the numerator coefficients, see (8.73). The optimization variables are thus collected as $\mathbf{X} = (\Delta\mathbf{C}_1, \ldots, \Delta\mathbf{C}_{\bar{\ell}})$. The perturbation that is induced in the model can be compactly expressed as

$$\Delta\mathbf{H}(s;\boldsymbol{\vartheta}) = \Delta\mathbf{C}(\boldsymbol{\vartheta})(s\mathbf{E} - \mathbf{A}(\boldsymbol{\vartheta}))^{-1}\mathbf{B} = \mathbf{X}\cdot\mathbf{Z}(s;\boldsymbol{\vartheta}), \tag{8.82}$$

where

$$\mathbf{Z}(s;\boldsymbol{\vartheta}) = \{[\xi_1(\boldsymbol{\vartheta}), \ldots, \xi_{\bar{\ell}}(\boldsymbol{\vartheta})]^{\mathrm{T}} \otimes \mathbf{I}_{\bar{\mathrm{N}}}\}[s\mathbf{E} - \mathbf{A}(\boldsymbol{\vartheta})]^{-1}\mathbf{B} \tag{8.83}$$

Cost function
The above representation of the model perturbation provides a straightforward way to define a data-driven cost function for accuracy preservation. Considering the data samples (8.49) available for training the model, we evaluate the model perturbation at the same frequency and parameter points, and we enforce the corresponding cumulative norm to be minimal. The corresponding cost function to be minimized is

$$\mathrm{F}(\mathbf{x}) = \sum_{k=1}^{\bar{k}}\sum_{m=1}^{\bar{m}} \|\Delta\mathbf{H}(s_k, \boldsymbol{\vartheta}_m)\|^2 = \sum_{k=1}^{\bar{k}}\sum_{m=1}^{\bar{m}} \|\mathbf{X}\cdot\mathbf{Z}(s_k, \boldsymbol{\vartheta}_m)\|_F^2, \tag{8.84}$$

where $\mathbf{x} = \mathrm{vec}(\mathbf{X})$ stacks all decision variables in a column vector. Note that $\mathrm{F}(\mathbf{x})$ is a quadratic function of the decision variables $\mathbf{x}$.

Enforcing local passivity constraints
The formulation of local passivity constraints follows the same guidelines as in Section 8.3.2. Focusing on the scattering representation, we generalize the singular value perturbation (8.45) to include the parameter space location $\bar{\boldsymbol{\vartheta}}_\mu$ where each local singular value has been determined, and we write the following linear approximation

$$\widehat{\sigma}_\mu \approx \bar{\sigma}_\mu + \mathrm{Re}\left\{\mathbf{u}_\mu^{\mathrm{H}}\Delta\mathbf{H}(\mathrm{j}\bar{\omega}_\mu, \bar{\boldsymbol{\vartheta}}_\mu)\mathbf{v}_\mu\right\} = \bar{\sigma}_\mu + \mathbf{z}_\mu^{\mathrm{T}}\cdot\mathbf{x} \tag{8.85}$$

where

$$\mathbf{z}_\mu^{\mathrm{T}} = \mathrm{Re}\left\{[\mathbf{Z}(\mathrm{j}\bar{\omega}_\mu, \bar{\boldsymbol{\vartheta}}_\mu)\mathbf{v}_\mu]^{\mathrm{T}} \otimes \mathbf{u}_\mu^{\mathrm{H}}\right\}. \tag{8.86}$$

Setting the perturbed singular value $\widehat{\sigma}_\mu$ to be less than $\gamma = 1$ leads to the simple linear inequality constraint

$$\mathbf{z}_\mu^{\mathrm{T}}\cdot\mathbf{x} \le 1 - \bar{\sigma}_\mu \tag{8.87}$$

that must be enforced for all μ. Passivity enforcement is thus performed by iteratively solving the constrained minimization problem

$$\min_{\mathbf{x}} \mathrm{F}(\mathbf{x}), \quad \text{s.t.} \quad \mathbf{z}_{\mu}^{\mathrm{T}} \cdot \mathbf{x} \leq 1 - \bar{\sigma}_{\mu}, \quad \mu = 1, \ldots, \bar{\mu} \tag{8.88}$$

until no violations are detected.

8.5 Applications

We report here two application examples for which the discussed parameterized macromodeling framework provides clear advantages in CAD flows. The first example is an interconnect model relevant for signal integrity assessment of point-to-point links in Printed Circuit Boards (PCBs). The second example is an integrated inductor, for which a scalable simulation model is derived as an element of a component library.

8.5.1 A signal interconnect

We consider an electrical link that connects a driver and a receiver located on two different PCBs connected by a high-speed connector, see [58] for a complete description of the geometry. The overall length of the copper interconnect is 14 cm, which runs through an embedded layer with relative permittivity $\varepsilon_r = 3$ and loss tangent $\delta = 0.002$. The signal connection is provided by two stripline segments connected to the two sides of the connector and to the input/output ports by vertical vias of fixed antipad radius $b = 400$ μm. We consider as a free design parameter the via radius a, which is allowed to vary in the interval $[100, 300]$ μm. It is well-known that an optimal configuration of the via geometry exists that minimizes the impedance discontinuity, allowing to improve the signal integrity of the link. We want to investigate the influence of the via radius a on the transmission properties, by means of both frequency-domain and transient analyses. A parameterized macromodel is best suited for this task.

First, the frequency response of the link is characterized at $\bar{m} = 9$ different linearly spaced radius values, using a combination of a full wave solver and lossy transmission line models. This characterization is performed in the frequency domain, since loss mechanisms are best represented in the time-harmonic domain. For each fixed parameter value, $\bar{k} = 500$ frequency samples of the scattering parameters are computed in the interval $\Omega = [0, 10]$ GHz. These data are used to generate a rational macromodel of dynamic order $\bar{n} = 46$, in which poles and zeros are parameterized by means of Chebychev polynomials of degrees 2 and 4, respectively. The total runtime required to extract a passive model is 9 min on a standard laptop.

The worst-case RMS error of the resulting parameterized model is 1.04×10^{-2} if compared to additional validation samples not used for model extraction. Figure 8.9 compares the $S_{1,1}$ model responses for different values of a to the corresponding raw data samples. We see that the model is extremely accurate at all

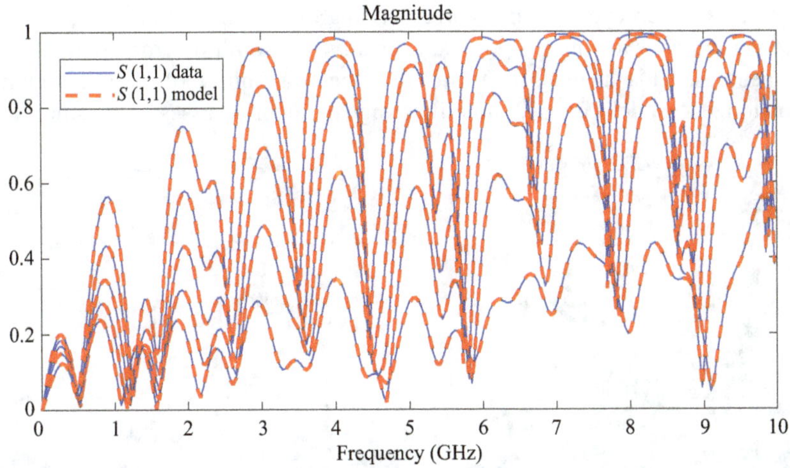

Figure 8.9 Comparing model responses (red dashed lines) to original frequency data used for model identification (blue solid lines) for a multiboard intertonnect. The various curves correspond to different values of the via radius a, used as a parameter for the model

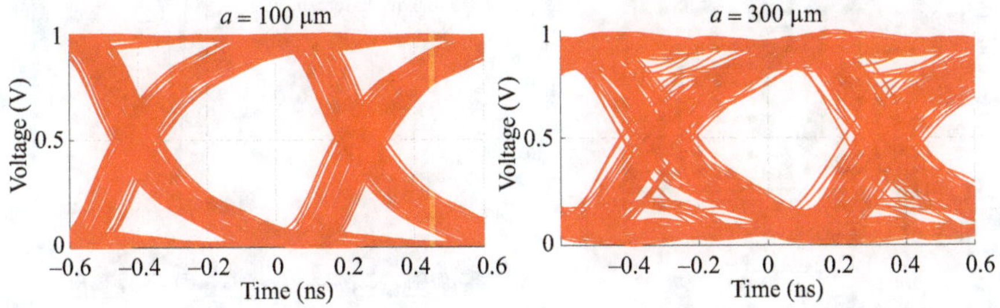

Figure 8.10 Eye-diagrams obtained through a SPICE time-domain simulation of the parameterized macromodel of the multiboard link, terminated by realistic (nonlinear) loads. The two diagrams correspond to two different values of the via radius a, used as a free parameter in the model

frequencies and for all parameter values. Note that this figure depicts the responses of the model after passivity verification and enforcement, so that the resulting synthesized parametric SPICE netlist can be safely used for transient simulation.

An assessment of how the via radius impacts the signal integrity of the link is performed by computing the time-domain eye diagram of the link. For this analysis, a pseudorandom bit sequence with 500 bits (bit time 650 ps, rise and fall times 100 ps) is applied on one end of the link, with a source impedance. At the far end, the link is terminated with a RC load ($R = 10\ \text{k}\Omega$, $C = 3$ pF). A transient simulation based on a freeware SPICE engine on a laptop requires about 20 s. Figure 8.10

compares two such eye diagrams corresponding to the smallest and largest values of the via radius, showing a significant sensitivity of both horizontal and vertical eye openings. This analysis confirms that the availability of SPICE-compliant parameterized models greatly simplifies signal integrity assessment since early

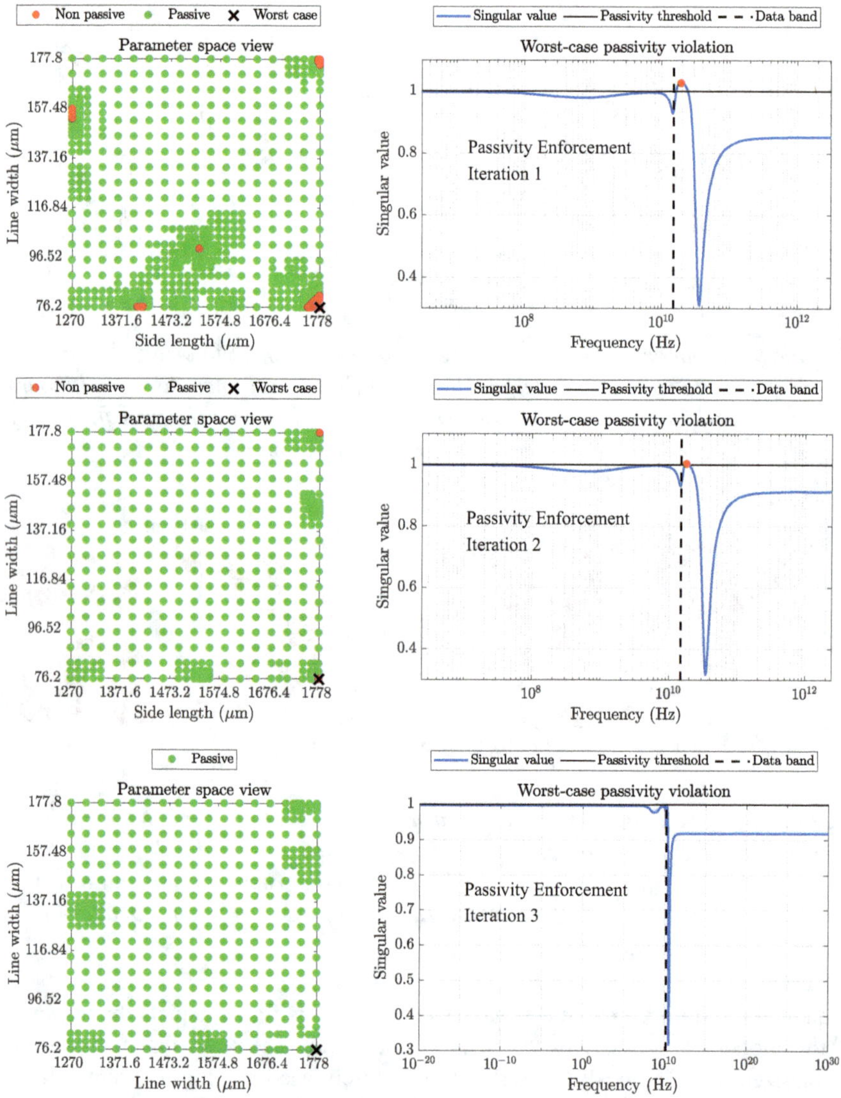

Figure 8.11 Evolution of the passivity violations of the integrated inductor model through passivity enforcement iterations. Left panels: location of passive (green) and non-passive (red) regions in the parameter space. Right panels: singular value trajectories with the worst-case violation (red dot), which corresponds to the parameter configuration highlighted with a cross in the left panels

design stages, where multiple design solutions need to be tested for their electrical performance through fast and accurate numerical simulations.

8.5.2 An integrated inductor

This second example is a 1-port integrated inductor with a square outline and 1.5 turns. The two free design parameters we consider are the side-length L of the structure and the width W of the traces. The two parameters are constrained to attain values within the intervals $L \in [1.27, 1.778]$ mm and $W \in [76, 177.8]$ μm. The main requirement is here to construct a parameterized model that reproduces with good accuracy the frequency behavior of the structure up to 15 GHz for any combination of the two parameters L and W, and that is available as a single SPICE netlist as part of a library.

The reference behaviour of the inductor is first computed via a full-wave frequency domain solver over a total of $\bar{m} = 25$ points in the parameter space, uniformly sampled over a grid of five fixed values for each parameter. For each of these parameter configurations, $\bar{k} = 590$ frequency samples of the scattering responses are evaluated over the frequency band of interest. This dataset is then used to extract a model with $\bar{n} = 7$ poles, in which the parameterized dependencies of the poles and the zeros are induced by means of bivariate linear and quadratic basis functions, respectively. Passivity enforcement is performed as part of model generation, to ensure its physical consistency and robustness/stability in transient

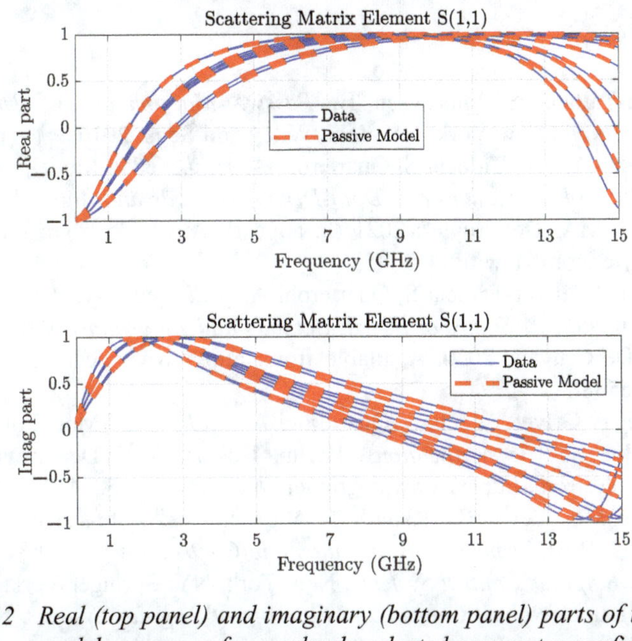

Figure 8.12 Real (top panel) and imaginary (bottom panel) parts of the inductor model responses for randomly selected parameter configurations. Solid blue and dashed red lines correspond to raw data and parameterized model responses, respectively

analyses. Evolution of the passivity violations through iterations is displayed in Figure 8.8 (first iteration) and in Figure 8.11 (successive iterations), which confirms that after the last (third) iteration the model is uniformly passive throughout the parameter space.

The quality of the resulting passive macromodel is shown in Figure 8.12, where real and imaginary parts of the frequency-domain reflection coefficient are displayed over a subset of randomly selected parameter configurations. Also for this example, there is no visual difference between model and data at this scale, as a confirmation of the excellent accuracy that can be provided by the proposed framework. The worst-case model-data error is below 0.5%.

8.6 Conclusions

In this chapter, we have presented both basic and advanced material on linear macromodeling, with emphasis on scalable parameterized models. We have discussed the various aspects that are relevant for a successful macromodel extraction, including choice of model structure, identification algorithms with stability and passivity constraints, and finally model realization in terms of differential equations or equivalent circuits. Although concise, the material in this chapter is reasonably self-contained and can be used as an introduction to the topic. We refer the reader to the list of references, in particular [1] for a complete treatment.

References

[1] Grivet-Talocia S, Gustavsen B. *Passive Macromodeling: Theory and Applications*. New York, NY: John Wiley and Sons; 2016.

[2] Benner P, Grivet-Talocia S, Quarteroni A, *et al.*, (eds). Model order reduction. In: *Vol. 1: System- and Data-Driven Methods and Algorithms*. Berlin, Boston, MA: De Gruyter; 2021 (to appear). Available from: https://www.degruyter.com/view/title/523453.

[3] Benner P, Grivet-Talocia S, Quarteroni A, *et al.*, (eds). Model order reduction. In: *Vol. 2: Snapshot-Based Methods and Algorithms*. Berlin, Boston, MA: De Gruyter; 2020. Available from: https://www.degruyter.com/view/title/565388.

[4] Benner P, Grivet-Talocia S, Quarteroni A, *et al.*, (eds). Model order reduction. In: *Vol. 3: Applications*. Berlin, Boston, MA: De Gruyter; 2020. Available from: https://www.degruyter.com/view/title/523454.

[5] Verghese G, Levy BC, Kailath T. A generalized state-space for singular systems. *IEEE Transactions on Automatic Control* 1981;26(4):811–831.

[6] Dai L. *Singular Control Systems*. New York, NY: Springer-Verlag; 1989.

[7] Xu S, Lam J. *Robust Control and Filtering of Singular Systems*. New York, NY: Springer; 2006.

[8] Antoulas AC. Approximation of large-scale dynamical systems. Houston, TX: *Society for Industrial and Applied Mathematics*; 2005.

[9] Benner P, Hinze M, ter Maten EJW. *Model Reduction for Circuit Simulation,* Vol. 74. New York, NY: Springer; 2010.
[10] Tan SXD, He L. *Advanced Model Order Reduction Techniques in VLSI Design*. New York, NY: Cambridge University Press; 2007.
[11] Moore B. Principal component analysis in linear systems: controllability, observability, and model reduction. *IEEE Transactions on Automatic Control* 1981;26(1):17–32.
[12] Glover K. All optimal Hankel-norm approximations of linear multivariable systems and their L_∞ error bounds. *International Journal of Control.* 1984;39(6):1115–1193.
[13] Enns DF. Model reduction with balanced realizations: an error bound and a frequency weighted generalization. In: *The 23rd IEEE Conference on Decision and Control*, 1984. p. 127–132.
[14] Schilders WHA, Van Der Vorst HA, Rommes *J. Model Order Reduction: Theory, Research Aspects and Applications*. New York, NY: Springer Verlag; 2008.
[15] Grimme EJ. Krylov Projection Methods For Model Reduction. Ph.D. thesis, Univ. Illinois Urbana Champaign; 1997.
[16] Boley DL. Krylov space methods on state-space control models. *Circuits, Systems, and Signal Processing* 1994;13(6):733–758.
[17] Elfadel IM, Ling DD. A block rational Arnoldi algorithm for multipoint passive model-order reduction of multiport RLC networks. In: *Proceedings of the 1997 IEEE/ACM International Conference on Computer-Aided Design*. IEEE Computer Society; 1997. pp. 66–71.
[18] Celik M, Pileggi L, Odabasioglu A. IC *Interconnect Analysis.* New York, NY: Springer; 2002.
[19] Odabasioglu A, Celik M, Pileggi LT. PRIMA: passive reduced-order interconnect macromodeling algorithm. *IEEE Transactions on Computer-Aided Design of Integrated Circuits and Systems* 1998;17(8):645–654.
[20] Gustavsen B, Semlyen A. Rational approximation of frequency domain responses by vector fitting. *IEEE Transactions on Power Delivery* 1999; 14(3):1052–1061.
[21] Sanathanan C, Koerner J. Transfer function synthesis as a ratio of two complex polynomials. *IEEE Transactions on Automatic Control* 1963;8(1): 56–58.
[22] Lefteriu S, Antoulas AC. On the convergence of the Vector-Fitting Algorithm. *IEEE Transactions on Microwave Theory and Techniques* 2013;61(4):1435–1443.
[23] Shi G. On the nonconvergence of the Vector Fitting Algorithm. *IEEE Transactions on Circuits and Systems II: Express Briefs* 2016;63(8):718–722.
[24] Golub GH, Van Loan CF. *Matrix Computations*, Vol. 3. Baltimore, MA: Johns Hopkins University Press; 1996.
[25] Deschrijver D, Mrozowski M, Dhaene T, *et al.* Macromodeling of multiport systems using a fast implementation of the Vector Fitting method. *Microwave and Wireless Components Letters, IEEE* 2008;18(6):383–385.

[26] Chinea A, Grivet-Talocia S. On the parallelization of Vector Fitting algorithms. *IEEE Transactions on Components, Packaging and Manufacturing Technology* 2011;1(11):1761–1773.

[27] Ganeshan S, Elumalai NK, Achar R, *et al.* GVF: GPU-based vector fitting for modeling of multiport tabulated data networks. *IEEE Transactions on Components, Packaging and Manufacturing Technology* 2020;10(8): 1375–1387.

[28] Scherer C, Weiland S. Linear matrix inequalities in control. *Lecture Notes, Dutch Institute for Systems and Control, Delft, The Netherlands;* 2000.

[29] Willems JC. Dissipative dynamical systems. Part I: General theory. *Archive for Rational Mechanics and Analysis* 1972;45(5):321–351. Available from: http://dx.doi.org/10.1007/BF00276493.

[30] Kuh ES, Rohrer RA. *Theory of Linear Active Networks.* UK: Holden-Day; 1967.

[31] Wohlers MR. *Lumped and Distributed Passive Networks.* London: Academic Press; 1969.

[32] Anderson BDO, Vongpanitlerd S. *Network Analysis and Synthesis.* Hoboken, NJ: Prentice-Hall; 1973.

[33] Boyd S, Balakrishnan V, Kabamba P. A bisection method for computing the H_∞ norm of a transfer matrix and related problems. *Mathematics of Control, Signals and Systems*. 1989;2(3):207–219.

[34] Grivet-Talocia S. Passivity enforcement via perturbation of Hamitonian matrices. *IEEE Trans Circuits and Systems I: Fundamental Theory and Applications* 2004;51(9):1755–1769.

[35] Zhou K, Doyle JC, Glover K. *Robust and Optimal Control.* Upper Saddle River, NJ: Prentice Hall; 1996.

[36] Grivet-Talocia S, Ubolli A. Passivity enforcement with relative error control. *IEEE Trans Microwave Theory and Techniques.* 2007;55(11): 2374–2383.

[37] Grivet-Talocia S, Ubolli A. A comparative study of passivity enforcement schemes for linear lumped macromodels. *IEEE Trans Advanced Packaging.* 2008;31(4):673–683.

[38] Coelho CP, Phillips J, Silveira LM. A convex programming approach for generating guaranteed passive approximations to tabulated frequency-data. *IEEE Transactions on Computer-Aided Design of Integrated Circuits and Systems* 2004;23(2):293–301.

[39] Kailath T. *Linear Systems.* Englewood Cliffs, NJ: Prentice-Hall; 1980.

[40] Laub AJ. *Matrix Analysis for Scientists and Engineers*. Houston, TX: Society for Industrial and Applied Mathematics; 2005.

[41] Wilkinson JH. *The Algebraic Eigenvalue Problem*. Oxford: Clarendon Press; 1965.

[42] IdEM 12, Dassault Systèmes [Online]. Available: www.3ds.com/products-services/simulia/products/idem/; 2018.

[43] Grivet-Talocia S, Trinchero R. Behavioral, parameterized, and broadband modeling of wired interconnects with internal discontinuities. *IEEE Transactions on Electromagnetic Compatibility* 2018;60(1):77–85.
[44] Ferranti F, Knockaert L, Dhaene T. Guaranteed passive parameterized admittance-based macromodeling. *IEEE Transactions on Advanced Packaging* 2010;33(3):623–629.
[45] Ferranti F, Knockaert L, Dhaene T. Parameterized S-parameter based macromodeling with guaranteed passivity. *Microwave and Wireless Components Letters, IEEE* 2009;19(10):608–610.
[46] Ferranti F, Dhaene T, Knockaert L. Compact and passive parametric macromodeling using reference macromodels and positive interpolation operators. *IEEE Transactions on Components, Packaging and Manufacturing Technology* 2012;2(12):2080–2088.
[47] Triverio P, Grivet-Talocia S, Nakhla MS. A parameterized macromodeling strategy with uniform stability test. *IEEE Transaction on Advanced Packaging* 2009;32(1):205–215.
[48] Grivet-Talocia S, Fevola E. Compact parameterized black-box modeling via Fourier-rational approximations. *IEEE Transactions on Electromagnetic Compatibility* 2017;59(4):1133–1142.
[49] Min SH, Swaminathan M. Construction of broadband passive macromodels from frequency data for simulation of distributed interconnect networks. *IEEE Transactions on Electromagnetic Compatibility* 2004;46(4):544–558.
[50] Zanco A, Grivet-Talocia S, Bradde T, *et al.* Uniformly stable parameterized macromodeling through positive definite basis functions. *IEEE Transactions on Components, Packaging and Manufacturing Technology* 2020;10(11): 1782–1794.
[51] Bernstein S. Démonstration du théorème de Weierstrass fondée sur le calcul des probabilités. *Communications de la Société mathématique de Kharkow 2-ée série* 1912;13(1):1–2.
[52] Farin G. *Curves and Surfaces for Computer-Aided Geometric Design: A Practical Guide.* London: Academic Press; 1988.
[53] Zhang Z, Wong N. Passivity check of S-parameter descriptor systems via S-parameter generalized Hamiltonian methods. *IEEE Transactions on Advanced Packaging* 2010;33(4):1034–1042.
[54] Zhang Z, Wong N. An efficient projector-based passivity test for descriptor systems. *IEEE Transactions on Computer-Aided Design of Integrated Circuits and Systems* 2010;29(8):1203–1214.
[55] Zhang Z, Wong N. Passivity test of immittance descriptor systems based on generalized hamiltonian methods. *IEEE Transactions on Circuits and Systems II: Express Briefs* 2010;57(1):61–65.
[56] Zanco A, Grivet-Talocia S, Bradde T, *et al.* Enforcing passivity of parameterized LTI macromodels via Hamiltonian-driven multivariate adaptive sampling. *IEEE Transactions on Computer-Aided Design of Integrated Circuits and Systems* 2018;39(1):225–238.

[57] Grivet-Talocia S. A perturbation scheme for passivity verification and enforcement of parameterized macromodels. *IEEE Transactions on Components, Packaging and Manufacturing Technology* 2017;7(11): 1869–1881.

[58] Preibisch J, Reuschel T, Scharff K, *et al.* Exploring efficient variability-aware analysis method for high-speed digital link design using PCE. *DesignCon*, 2017.

Chapter 9

A charged particle traveling along the axis of a circular waveguide

Amedeo Andreotti[1] *and Luigi Verolino*[1]

The aim of this chapter is to demonstrate that a satisfactory choice of the reference system and the use of Lorentz transformations can help determine the solution of electromagnetic problems, if moving sources are present in the system. This is a typical situation that occurs in particle accelerators, in particular, in linear accelerators, a functional scheme is shown in Figure 9.1. A linear particle accelerator, often shortened to *linac*, is a type of particle accelerator that accelerates charged subatomic particles or ions to a high speed by subjecting them to a series of oscillating electric potentials along a linear beamline. The design of a linac depends on the type of particle that is being accelerated: electrons, protons, or ions [1]. However, the length of the machine must be quite large in order to give sufficient thrust to accelerated particles, while in a circular machine, the packets can pass many times before being collided. The most powerful linear accelerator is located in Stanford, California at the Stanford linear accelerator centre (SLAC) laboratory and is about 3 km long. Electrons and positrons can be accelerated simultaneously and then separated and brought to collide. The maximum energy of each beam reached almost 50 GeV when, in the 1990s of the last century, the machine was used to study collisions e^+e^- at the Z_0 resonance, in

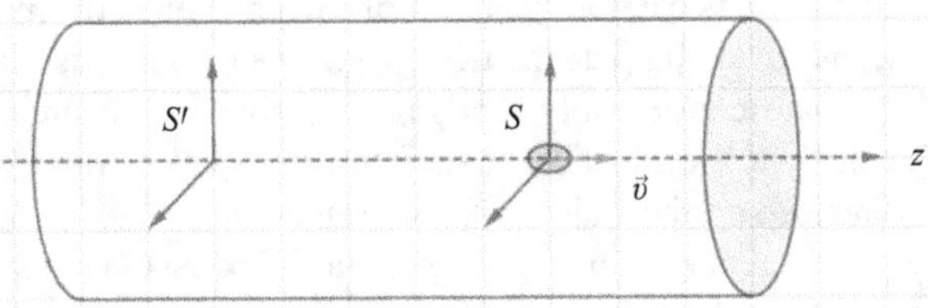

Figure 9.1 Particle in an infinite circular waveguide (pipe)

[1]Department of Electrical Engineering and Information Technology, University of Naples Federico II, Italy

competition with the LEP, circular accelerator, which in those years, was operating at CERN in Geneva. For the future, there are plans to build even longer and more powerful linear accelerators for fundamental research, such as International Linear Collider (ILC) for collisions e^+e^- at about 500–1,000 GeV in the center of mass or even Compact LInear Collider (CLIC) for energies up to 2,000–4,000 GeV, achieving such energies with circular machines would be practically impossible.

So, it is interesting to study the behaviour of the electromagnetic field supported in the time domain by a point charge, moving at a constant velocity along the axis of a circular waveguide: this field will be developed in a Fourier–Bessel series and the result will be a rapidly converging expansion in the complementary range of the classical integral representations, namely for large values of the argument. Quickly convergent expansions are nowadays necessary to test codes, useful to simulate beam dynamics in particle accelerators.

A charged particle accelerator is a complex machine that takes elementary charged particles and accelerates them to very high energies, and accelerator physics is a field of physics encompassing all the aspects required to design and operate the equipment and to understand the resulting dynamics of the charged particles. Tracking codes [2] are effective facilities in accelerator physics to simulate the dynamics of the interaction between the charged-particle beam and the surrounding structure in the time domain. In particle physics, tracking is the process of reconstructing the trajectory (or track) of electrically charged particles in a particle detector known as a tracker. The particles entering such a tracker leave a precise record of their passage through the device, by interaction with suitably constructed components and materials. The presence of a calibrated magnetic field, in all or part of the tracker, allows the local momentum of the charged particle to be directly determined from the reconstructed local curvature of the trajectory for known (or assumed) electric charge of the particle. The growing interest for these codes is testified by the effort made to obtain canonical solutions of classical electromagnetic problems directly into the time domain: a particle traveling at a constant velocity along the symmetry axis, the z-axis in this paper, of a perfectly conducting circular waveguide, which represents a pipe designing the vacuum chamber of an accelerating structure, is one of these problems. An efficient tracking code needs rapidly convergent expansions for the electromagnetic fields: unfortunately, all the classical formulations of Green's function in a pipe lead to integral formulae, useful to compute the fields when the integrand is a smooth function and does not fluctuate considerably, but impracticable in the presence of several oscillations. The analytical evaluation of such integrals will be discussed, in order to find the current induced on the wall of the waveguide.

However, what is reported in this chapter summarizes and completes what was published, years ago, in the papers [3–6].

9.1 Preliminary remarks

Let us consider an infinite circular waveguide, also indicated as pipe, of radius a as shown in Figure 9.1: a particle of charge q travels with uniform velocity $v = \beta c$, where c is the speed of light, along the axis of the structure, namely the z-axis of a cylindrical system. You can represent this particle as a current density $j(P,t)$, directed along the z-axis of the adopted reference, and you can write that

$$j(P,t) = j(x,y,z,t) = q\delta(x)\delta(y)\delta(z - vt).$$

We shall study the electromagnetic field created in the pipe, focusing our attention on the electric current flowing on the metallic cylindrical surface, and finding the general expansion of this current in the time domain.

An introductory description of the electromagnetic field supported by a particle, animated by a straight and uniform motion, can be found in the Edward Mills Purcell's classic textbook *Electricity and Magnetism*, Cambridge University Press, second edition (1985). It is well known that a charged particle traveling in the free space with uniform velocity does not radiate, and that the field produced has a TM structure and is given, in the frequency domain, by

$$\begin{aligned} \vec{H} &= \widehat{\varphi}\frac{q\kappa}{2\pi}K_1(\kappa r)\exp(-jkz/\beta), \\ &= \zeta_0\frac{q\kappa}{2\pi\beta\gamma}[\widehat{r}\gamma K_1(\kappa r) + \widehat{z}jK_0(\kappa r)]\exp(-jkz/\beta), \end{aligned} \tag{9.1}$$

where k is the wave number, $\zeta_0 = 120\pi\Omega$ is the characteristic impedance of vacuum, the Lorentz factor or Lorentz term, namely the factor by which time, length, and relativistic mass change for an object while that object is moving, is equal to

$$\begin{aligned} \gamma &= \frac{1}{\sqrt{1-\beta^2}} = 1 + \frac{1}{2}\beta^2 + \frac{3}{8}\beta^4 + \frac{5}{16}\beta^6 + \frac{35}{128}\beta^8 + \cdots \\ &= \sum_{n=0}^{\infty}\left(\prod_{k=1}^{n}\frac{2k-1}{2k}\right)\beta^{2n}, \end{aligned}$$

and we called for shortness $\kappa = k/(\beta\gamma) = \omega/(v\gamma)$ with ω being the angular frequency.

When the particle travels inside the tube, it is apparent that all electromagnetic quantifies can be imagined as the superposition of two terms: the first one represents the quantity due to the motion of the charge in free space without the pipe and the second one represents the same quantity supported by the current induced (always in frequency domain) on the tube wall ($r = a$),

$$\vec{J}(z,\omega) = \widehat{z}J(z,\omega), \tag{9.2}$$

directed all along the z-axis and sustaining a TM propagation, as in the case of the primary field in free space. The unknown of the problem is this current density $J(z,\omega)$, or its spatial Fourier transform, that is

$$F(w) = \frac{1}{2\pi}\int_{-\infty}^{\infty} J(z,\omega) \quad \exp(jwz)dw. \tag{9.3}$$

We have to find the relations linking all fields and/or potential to the selected unknown $F(w)$.

In this paper, we shall derive an expansion for the electromagnetic field of a charged particle travelling in a circular infinite waveguide whose rate of convergence is exponential; this aspect renders the expansion particularly attractive for large values of z. Besides, the proposed expansion is particularly convenient when the integral representations become impracticable. In other words, the proposed and the classical expansions are effective in two complementary ranges. A new formulation of the problem in the particle frame will transform the electrodynamics problem into a static one, and therefore a solution in cylindrical harmonics can be found.

The paper is organized in the following way. In Section 9.2, we present the general expansions of the electromagnetic field in the laboratory frame, where one has to solve a dynamic problem. Section 9.3 is dedicated to the study of a certain class of integrals, very useful for the physical applications described: the acceleration procedure is explained and a mathematical example and its asymptotic expansion is discussed. In Section 9.4, operating directly in the time domain, the calculation method used to obtain the solution in the traveling particle reference will be described and, using Lorentz transformations, the fields will be evaluated in the laboratory reference. Finally, Section 9.5 is dedicated to the conclusions and perspectives in order to try to generalize the proposed method. A particular attention will be put throughout the paper to the delicate case of the improper integral, when the integration domain becomes infinite, very interesting for many physical applications.

9.2 Solution in the laboratory frame

Starting with the Lorentz gauge of the vector potential, the azimuthal symmetry $\vec{A}(r,z,\omega) = \hat{z}A_z(r,z,\omega)$ enables us to write

$$\begin{aligned} A_z(r,z,\omega) &= \frac{\zeta_0 a}{4\pi c}\int_S J(z_0,\omega)\frac{\exp(-jkR)}{R}\, dS \\ &= \frac{\zeta_0 a}{4\pi c}\int_{-\infty}^{\infty} J(z_0,\omega)\left[\int_{-\infty}^{\infty} \exp(-jkR)\frac{d\varphi_0}{R}\right] dz_0, \end{aligned} \tag{9.4}$$

where S is the lateral surface of the pipe and R represents the distance between a generic observation point $P(r,\varphi,z)$ and the source point $P_0(r_0,\varphi_0,z_0)$. Analytically, it can be written that

$$R^2 = D^2 + (z - z_0)^2, \quad D^2 = r^2 + a^2 - 2ra\cos(\varphi - \varphi_0), \tag{9.5}$$

where we introduce the distance D in the transverse plane. Using the remarkable result [7]

$$\frac{\exp\left[-jk\sqrt{D^2+(z-z_0)^2}\right]}{\sqrt{D^2+(z-z_0)^2}} = \frac{1}{\pi}\int_{-\infty}^{\infty} I_0\left(D\sqrt{u^2-k^2}\right) K_0\left(D\sqrt{u^2-k^2}\right)\exp[-ju(z-z_0)]\,du, \tag{9.6}$$

where $I_0(x)$ and $K_0(x)$ are modified Bessel functions of order zero and $\mathrm{Im}\sqrt{k^2-u^2}$ is the branch cut, one can write that

$$\int_{-\infty}^{\infty} \exp(-jkR)\frac{d\varphi_0}{R} = 2\int_{-\infty}^{\infty} G(u,r)\exp[-ju(z-z_0)] \quad du, \tag{9.7}$$

where the function $G(u,r)$ can be found by applying the addition theorem for Bessel functions and is given by

$$G(u,r) = \begin{cases} I_0\left(a\sqrt{u^2-k^2}\right)K_0\left(r\sqrt{u^2-k^2}\right), & r \geq a, \\ I_0\left(r\sqrt{u^2-k^2}\right)K_0\left(a\sqrt{u^2-k^2}\right), & r \leq a. \end{cases} \tag{9.8}$$

The knowledge of the function $G(u,r)$ enables one to write simple integral relations linking the vector potential and the fields with the unknown $F(w)$ and, as we shall show, it will be quite easy to find an integral equation for it. In other words, the integral transform (3) renders algebraic the relation between the spectrum of the potentials and fields. Coming back to the vector potential and no longer reporting functional dependence on ω one can conclude

$$A_z(r,z,\omega) = \frac{\zeta_0 a}{c}\int_{-\infty}^{\infty} G(u,r)F(u)\exp(-juz)du, \tag{9.9}$$

from which the electromagnetic field follows, in particular the component

$$\begin{aligned} E_z(r,z) &= \frac{c}{jk}\left(\frac{\partial^2 A_z}{\partial z^2} + k^2 A_z\right) \\ &= \frac{\zeta_0 a}{c}\int_{-\infty}^{\infty} (k-u^2)G(u,r)F(u)\exp(-juz)du, \end{aligned}$$

the one we need to satisfy the boundary condition on the perfectly conducting tube. We are now ready to formulate the integral equation governing the phenomenon. The total electric field has to vanish on the lateral surface of the pipe ($r=a$), namely

$$\int_{-\infty}^{\infty} (k-u^2)G(u,r)F(u)\exp(-juz)du = \frac{q\kappa}{2\pi a}K_0(\kappa a)\exp\left(-\frac{jzk}{\beta}\right), \quad \forall z. \tag{9.10}$$

The integral equation (10) is easy to solve, because it can be interpreted as a Fourier transform. We have thus [8]

$$F(u) = -\frac{q}{2\pi a I_0(\kappa a)}\delta\left(u - \frac{k}{\beta}\right), \tag{9.11}$$

where $\delta(u)$ is the Dirac function, $F(u)$ is the key function to evaluate the electromagnetic field; but our attention will be pointed out on the induced current density which can be found taking the inverse Fourier transform of the definition (3)

$$\begin{aligned} J(z) &= \int_{-\infty}^{\infty} F(u) \quad exp(-juz) \quad du \\ &= -\frac{q}{2\pi a I_0(\kappa a)}\int_{-\infty}^{\infty} \delta\left(u - \frac{k}{\beta}\right) \quad exp(-juz) \quad du \\ &= -\frac{q\exp(-jkz/\beta)}{2\pi a I_0(\kappa a)}. \end{aligned} \tag{9.12}$$

Finally, the current flowing along the pipe is related to the flux of the current density: it is given by $I(z) = 2\pi a \quad J(z)$, namely

$$I(z) = -\frac{q\exp(-jkz/\beta)}{I_0(\kappa a)}. \tag{9.13}$$

The current $I(z)$ is a Fourier transform $I(z,\omega)$ because $k = \omega/c$; it represents the time Fourier transform of the current $i(z,t)$ flowing on the metallic tube. It is not so easy to come back to the time domain as we are going to prove. Anyway, taking the inverse Fourier transform, making a symbolic substitution, you can write that

$$\begin{aligned} i(z,t) &= \frac{1}{2\pi}\int_{-\infty}^{\infty} I(z,\omega)\exp(\quad j\omega t)d\omega \\ &= -\frac{q}{2\pi}\int_{-\infty}^{\infty} \frac{\exp[j\omega(t - z/v)]}{I_0[\omega a/(v\gamma)]}d\omega \\ &= -\frac{qv\gamma}{\pi a}g\left(\gamma\frac{z - vt}{a}\right), \end{aligned} \tag{9.14}$$

where all the problems of the inverse transformation are in the function $g(u)$, defined as

$$g(u) = \int_0^{\infty} \frac{\cos(ux)}{I_0(x)} dx. \tag{9.15}$$

The lower bound $\beta = 0$ is trivial, whereas the upper bound $\beta = 1$ gives

$$i(z,t) = -\frac{q}{2\pi}\int_{-\infty}^{\infty} \exp\left[j\omega\left(t - \frac{z}{c}\right)\right]d\omega = -q\delta\left(t - \frac{z}{c}\right).$$

The function $g(x)$, defined all over x, is a continuous function, whose evaluation ill be discussed in the next section. At the moment, it is important to stress only that it is a question of an even function and therefore it can be studied only in the restriction $x > 0$. The value for $x = 0$, numerically computed, is

$$g(0) = 2.083233277113131\cdots,$$

and it is also the maximum value of the function. This type of integrals does not exist in any table. However, by using the development technique that will be discussed, it is not difficult to show that

$$i(z,t) = -\frac{qv\gamma}{a}\sum_{k=1}^{\infty}\frac{\exp(-\gamma x_k|z-vt|/a)}{J_1(x_k)}, \tag{9.16}$$

where x_k represents a zero of the Bessel function, namely $J_0(x_k) = 0$ with k positive natural number. This series will be widely discussed in the next section, where the meaning of the different symbols used will also be explained; for the time being, it is sufficient to observe that the presence of an exponential term gives to the series excellent convergence properties, especially when it occurs that

$$\gamma|z - vt| \gg a,$$

since only a few terms are needed to describe it and obtain a reasonable approximate value. In the other case, a numerical treatment of the integral can be easily performed.

9.3 Evaluation of the integral

There are different classes of problems of mathematical physics, connected with cylindrical geometry, which can be investigated by means of the integral

$$\mathfrak{I} = \int_a^b \frac{f(x)}{I_0(x)}\,dx,$$

where $I_0(u)$ is the modified Bessel function of zeroth order, the function $f(x)$ is an arbitrary function that makes possible the integration and all the expansions proposed, and the values a and b are not negative integration limits, so that $a < b$. It is long time since this kind of integrals have been found in computational physics related to many physical applications, such as electrostatics, accelerator physics, but even very popular tables of integrals are lacking in this kind of integrals: there are not series expansions available and only numerical approximations are available in the literature [9,10]. The function $1/I_0(x)$ is represented in Figure 9.2: it is a continuous and smooth function and its asymptotic behaviour is given by

$$\frac{1}{I_0(x)} \cong \sqrt{2\pi x}\exp(-1)\text{for } x \gg 1.$$

It will be assumed in the following that the order of integration in repeated integrals, the orders of differentiation and integration, and the orders of summation and integration, can be reversed as necessary without explicit justification.

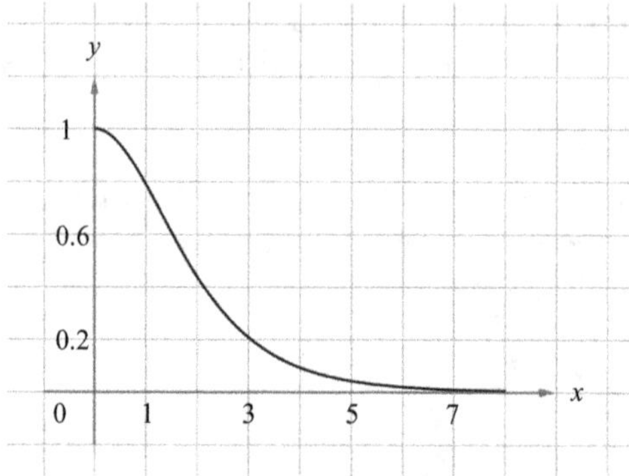

Figure 9.2 *The function* $1/I_0(x)$

9.3.1 The factorization of the Bessel function $J_0(x)$

We can start using the well-known factorization of the Bessel function [11]

$$J_0(x) = \prod_{s=1}^{\infty}\left(1 - \frac{x^2}{x_s^2}\right), \tag{9.17}$$

where x_s represents one of the zeros of the same function, namely

$$J_0(x_s) = 0, \quad \forall s \in \mathbb{N}_0.$$

The zeros of Bessel functions $J_\nu(x)$ are real and infinite, for any given real value of ν: they can be found on a lot of existing tables or can be evaluated by means of a computer program; in particular, we computed the zeros of the function $J_0(x)$ by means of special program library, such as the ones of the CERN. Nevertheless, only a few zeros have to be computed numerically, because large zeros are given by the expansion

$$x_k \cong \frac{\pi}{4}(4k-1) + \frac{1}{2\pi(4k-1)} - \frac{31}{6\pi^3(4k-1)^3} + \frac{3779}{15\pi^5(4k-1)^5} - \cdots.$$

This series is suitable for calculating all zeros, except the smallest ones, correctly to at least five digits. Thus, the numerical strategy has been the following: we computed the first 50 zeros solving the trascendental equation $J_0(x) = 0$, whereas the remaining ones can be found by means of the asymptotic expansion shown. The first five zeros are

$$x_k \cong \frac{\pi}{4}(4k-1) + \frac{1}{2\pi(4k-1)} - \frac{31}{6\pi^3(4k-1)^3} + \frac{3779}{15\pi^5(4k-1)^5} - \cdots.$$

This series is suitable for calculating all zeros, except the smallest ones, correctly to at least five digits. Thus, the numerical strategy has been the following: we computed the first 50 zeros solving the trascendental equation $J_0(x) = 0$, whereas the remaining ones can be found by means of the asymptotic expansion shown. The first five zeros are

$$x_1 \cong 2.4048256, \quad x_2 \cong 5.520078, \quad x_3 \cong 8.653728,$$

$$x_4 \cong 11.791534, \quad x_5 \cong 14.930918;$$

for larger zeros, the asymptotic formula gives a satisfying approximation.

9.3.2 The factorization of the modified Bessel function $I_0(x)$

Replacing the variable jx instead of the real x into equation (9.17), you get the $I_0(x)$ function, so that

$$I_0(x) = J_0(jx) = \prod_{s=1}^{\infty}\left(1 + \frac{x^2}{x_s^2}\right). \tag{9.18}$$

Taking the inverse of the previous infinite product, we can write the following partial fractions development, namely the Mittag–Leffler's theorem [12]:

$$\frac{1}{I_0(x)} = \prod_{s=1}^{\infty}\left(\frac{x_s^2}{x^2 + x_s^2}\right) = 2\sum_{k=1}^{\infty}\frac{x_k}{J_1(x_k)} \quad \frac{1}{x^2 + x_k^2}. \tag{9.19}$$

9.3.3 Evaluation of the zeros

There are many other ways to obtain expansion (9.19); the followed one seems the simpler and instructive manner.

It is worth noting that Euler, and independently Rayleigh, devised an ingenious method of calculating the smallest zeros of a function [11] by means of the knowledge of special series involving these zeros: they applied this method to determine the smallest zeros of $J_0(x)$. Starting from the formula (9.19), provided that $|x| < x_1$, one can find that

$$\frac{1}{x^2 + x_k^2} = \frac{1}{x_k^2}\sum_{n=0}^{\infty}(-1)^n\left(\frac{x}{x_k}\right)^{2n} \quad \rightarrow \quad \frac{1}{I_0(x)} = 2\sum_{k=1}^{\infty}(-1)^n S_n x^n,$$

where the expansion coefficients have to observe the following linear recurrence equation:

$$S_n = -\sum_{k=1}^{n}\frac{S_{n-k}}{4^k \quad k!^2}, \text{ with } S_0 = \frac{1}{2}.$$

Table 9.1 Expansion coefficients

n	S_n	$1/\sqrt[n]{S_n}$	S_n/S_{n+1}
0	0.5	∞	4
1	0.125	8	$5.\overline{3}$
2	0.0234375	6.531972647	5.684210527
3	$4.123263889 \cdot 10^{-3}$	6.236194087	5.763933177
4	$7.154676649 \cdot 10^{-4}$	6.114381003	5.779238567
5	$1.237996419 \cdot 10^{-4}$	6.045832659	5.782426502
6	$2.140963519 \cdot 10^{-5}$	6.001112783	5.783040993
7	$3.702141350 \cdot 10^{-6}$	5.969463296	5.783158382
8	$6.401590800 \cdot 10^{-7}$	5.945850856	5.783180723
9	$1.106932518 \cdot 10^{-7}$	5.927552760	...

The solution of this recurrence equation solution can be expressed in a closed form as a determinant, which is not useful for numerical calculations, namely

$$S_n = \frac{1}{2^{2n+1}} \begin{vmatrix} 1/1!^2 & 1 & 0 & \cdots & 0 \\ 1/2!^2 & 1/1!^2 & 1 & \cdots & 0 \\ 1/3!^2 & 1/2!^2 & 1/1!^2 & \cdots & 0 \\ \cdots & \cdots & \cdots & \cdots & \cdots \\ 1/(n-1)!^2 & 1/(n-2)!^2 & 1/(n-3)!^2 & \cdots & 1 \\ 1/n!^2 & 1/(n-1)!^2 & 1/(n-2)!^2 & \cdots & 1/1!^2 \end{vmatrix}$$

Numerical calculations performed by means of recurrence equation gave the results showed in Table 9.1.

Table 9.1 enables us to confirm the utility of the asymptotic formula

$$S_n \cong \frac{1}{x_1^{2n+1} \quad J_1(x_1)},$$

whence for $n > 2$ it gives at least five decimal correct digits. Besides, the same table could be used to compute the smallest zero of $J_0(x)$, because of the limit

$$\lim_{n\to\infty} \frac{S_n}{S_{n+1}} = \lim_{n\to\infty} \frac{1}{\sqrt[n]{S_n}} = x_1^2 \cong 5.78318596.$$

9.3.4 The expansion of the integral

In any case, returning to the problem under consideration, by means of the series representation (9.19), we can rewrite the class of integrals under test in the alternative form

$$\mathfrak{I} = \int_a^b \frac{f(x)}{I_0(x)} dx = \pi \sum_{k=1}^{\infty} \frac{A_k}{x_k J_1(x_k)}, \tag{9.20}$$

where the expansion coefficient A_k are given by

$$A_k = \frac{2}{\pi} x_k^2 \int_a^b \frac{f(x)}{x^2 + x_k^2} dx. \tag{9.21}$$

An analytical study of the properties of convergence of the series (9.20) is very complicated in the general case, depending on the properties of the function $f(x)$. However, an interesting feature of this expansion derives from the sign of $J_1(x_k)$ that is positive if k is odd, negative otherwise, thus we are dealing with an alternating series. It is well-known that if all the terms of this alternating series decrease monotonically in absolute value and approach zero, the series converges. Besides, the remainders of the series constitute a monotone sequence that approaches zero, in our case it is

$$\left| \sum_{k=n+1}^{\infty} \frac{A_k}{x_k J_1(x_k)} \right| < \left| \frac{A_{n+1}}{x_{n+1} J_1(x_{n+1})} \right|.$$

Unfortunately, expansion (9.20) exhibits a slow convergence, so that many terms are required to achieve a good approximation of integral: this implies that an acceleration procedure has to be considered.

9.3.5 *Acceleration of the convergence*

Let us consider the first term of the asymptotic behaviour of the expansion coefficients A_k, namely

$$A_\infty = \lim_{k\to\infty} A_k = \frac{2}{\pi} \int_a^b f(x) dx.$$

We can introduce the new coefficients defined as

$$A_k^{(0)} = A_k - A_\infty = -\frac{2}{\pi} \int_a^b \frac{x^2}{x^2 + x_k^2} f(x) \quad dx,$$

so that the series expansion (9.20) becomes

$$\mathfrak{I} = \int_a^b \frac{f(x)}{I_0(x)} dx = \frac{\pi}{2} A_\infty + \pi \sum_{k=1}^{\infty} \frac{A_k^{(0)}}{x_k J_1(x_k)}, \tag{9.22}$$

where the relevant result has been used

$$\frac{1}{I_0(x)} = 2 \sum_{k=1}^{\infty} \frac{x_k}{J_1(x_k)} \quad \frac{1}{x^2 + x_k^2} \quad \rightarrow$$

$$\frac{1}{I_0(0)} = 2 \sum_{k=1}^{\infty} \frac{1}{x_k J_1(x_k)} \quad \rightarrow \quad \sum_{k=1}^{\infty} \frac{1}{x_k J_1(x_k)} = \frac{1}{2}.$$

New expansion (9.22) converges more rapidly than (9.20), because the expansion coefficients $A_k^{(0)}$ approach zero, as k diverges, while the old coefficients (9.21) are approximated to a constant and not zero value.

This idea can be generalized, in order to increase the rate convergence of series (9.20). We start with the algebraic identity

$$\frac{x_k^2}{x^2+x_k^2} = \frac{(-1)^{n+1}}{x_k^{2n}}\frac{x^{2n+2}}{x^2+x_k^2} + \sum_{m=0}^{n}(-1)^m\left(\frac{x}{x_k}\right)^{2m} \quad (n = 0,\ 1,\ 2,\ 3,\ \cdots),$$

which, replaced in expression (21), provides the result

$$A_k = \frac{2}{\pi}\left[\frac{(-1)^{n+1}}{x_k^{2n}}\int_a^b \frac{x^{2n}f(x)}{x^2+x_k^2}\,dx + \sum_{m=0}^{n}\frac{(-1)^m}{x_k^{2m}}\int_a^b x^{2m}f(x)\,dx\right]$$
$$= A_k^{(n)} + \sum_{m=0}^{n}\frac{C_m}{x_k^{2m}}, \tag{9.23}$$

where, for the sake of shortness, it has been posed

$$A_k^{(n)} = \frac{2}{\pi}\frac{(-1)^{n+1}}{x_k^{2n}}\int_a^b \frac{x^{2n}f(x)}{x^2+x_k^2}\,dx, \quad C_m = \frac{2}{\pi}(-1)^m\int_a^b x^{2m}f(x)\,dx.$$

It is evident that $A_k^{(n)}$ approaches asymptotically zero, much faster when n increases. Coming back to (9.20), we have

$$\mathfrak{I} = \int_a^b \frac{f(x)}{I_0(x)}\,dx = \pi\sum_{k=1}^{\infty}\frac{A_k^{(n)}}{x_k J_1(x_k)} + \sum_{m=0}^{n} C_m S_m = \mathfrak{I}_n + \sum_{m=0}^{n} C_m S_m, \tag{9.24}$$

where the series defining $\mathfrak{I}_n$ converges faster than the one defining $\mathfrak{I}$ and the accelerating terms

$$S_m = \sum_{k=1}^{\infty}\frac{1}{x_k^{2m+1}J_1(x_k)}$$

can be put in a closed form as shown above.

9.3.6 An example

In order to check the effectiveness of the procedure described, consider an integral whose value is known, at least numerically. We consider now an integral that occurs frequently in applications, namely the one obtained with $f(x) = K_0(x)$, the modified Bessel function of the second kind, namely

$$\mathfrak{I} = \int_0^\infty \frac{K_0(x)}{I_0(x)} \quad dx.$$

In this case the expansion coefficient A_k becomes [7]

$$A_k = \frac{2}{\pi} x_k^2 \int_a^b \frac{f(x)}{x^2 + x_k^2} \quad dx = \frac{\pi}{2} x_k [\boldsymbol{H}_0(x_k) - Y_0(x_k)],$$

where $\boldsymbol{H}_0(x)$ is the Struve function, solutions of the non-homogeneous Bessel's differential equation, introduced by Hermann Struve in 1882, and $Y_0(x)$ is the Bessel function of the second kind. It is quite complicate to evaluate the Struve function but, nowadays, there are many good and popular codes, such as Wolfram, that can be used to evaluate it. The obtained series exhibits a slow convergence; summing 5,000 terms, we have

$$\int_0^\infty \frac{K_0(x)}{I_0(x)} dx \cong \pi \sum_{k=1}^{5,000} \frac{A_k}{x_k \quad J_1(x_k)} = 1.3519\cdots,$$

where only the first decimal digit is correct: it is therefore necessary to use the acceleration procedure, to obtain greater accuracy, to use fewer expansion terms. By virtue of asymptotic development [11]

$$\boldsymbol{H}_0(x) - Y_0(x) \cong \frac{2}{\pi x} - \frac{2}{\pi x^3} + \frac{18}{\pi x^5} + O\left(\frac{1}{x^6}\right), \quad as \quad x \to \infty,$$

For large values of x_k, we can write

$$A_k = \frac{\pi}{2} x_k [\boldsymbol{H}_0(x_k) - Y_0(x_k)] \cong 1 - \frac{1}{x_k^2} + \frac{9}{x_k^4},$$

so that by introducing the new coefficients B_k related to the old A_k according to the relations

$$B_k = A_k - 1 + \frac{1}{x_k^2} - \frac{9}{x_k^4},$$

the integral becomes

$$\int_0^\infty \frac{K_0(x)}{I_0(x)} dx = \pi \sum_{k=1}^\infty \frac{1}{J_1(x_k)} \left(\frac{1}{x_k} - \frac{1}{x_k^3} + \frac{9}{x_k^5}\right) + \pi \sum_{k=1}^\infty \frac{B_k}{x_k J_1(x_k)}.$$

The sum of the first series of this equation can be obtained by evaluating the function

$$F(x) = \frac{1}{I_0(x)} = 2 \sum_{k=1}^\infty \frac{x_k}{J_1(x_k)} \frac{1}{x^2 + x_k^2}$$

and its first four derivatives for $x = 0$. It turns out that

$$F(0) = 2\sum_{k=1}^{\infty}\frac{1}{x_k \ J_1(x_k)} = 1 \ \rightarrow \ \sum_{k=1}^{\infty}\frac{1}{x_k \ J_1(x_k)} = \frac{1}{2}, \quad F^{(1)}(0) = 0,$$

$$F^{(2)}(0) = -4\sum_{k=1}^{\infty}\frac{1}{x_k^3 \ J_1(x_k)} = -\frac{1}{2} \ \rightarrow \ \sum_{k=1}^{\infty}\frac{1}{x_k^3 \ J_1(x_k)} = \frac{1}{8}, \quad F^{(3)}(0) = 0,$$

$$F^{(4)}(0) = 48\sum_{k=1}^{\infty}\frac{1}{x_k^5 \ J_1(x_k)} = \frac{9}{8} \ \rightarrow \ \sum_{k=1}^{\infty}\frac{1}{x_k^5 \ J_1(x_k)} = \frac{3}{128}, \quad F^{(5)}(0) = 0,$$

so we may conclude that

$$\sum_{k=1}^{\infty}\frac{1}{J_1(x_k)}\left(\frac{1}{x_k} - \frac{1}{x_k^3} + \frac{9}{x_k^5}\right) = \frac{1}{2} - \frac{1}{8} + \frac{27}{128} = \frac{75}{128}.$$

Ultimately, you get the following value for the integral

$$\int_0^{\infty}\frac{K_0(x)}{I_0(x)}dx \cong \frac{75}{128}\pi + \pi\sum_{k=1}^{50}\frac{B_k}{x_k J_1(x_k)} = 1.36767686196\cdots,$$

where now only 50 terms give 11 decimal correct digits.

9.4 Solution in the particle frame

Now suppose you want to solve the problem in the particle reference system S. The electrostatic problem of a stationary particle on the axis of a conducting cylinder can be solved by integral transformations or by appropriate series developments: the second technique will be used in the following.

The potential for our configuration is axially symmetric, and in cylindrical coordinates has to satisfy the following Dirichlet problem for Poisson's equation

$$\begin{cases} \nabla^2\varphi(r,z) = \dfrac{1}{r}\dfrac{\partial}{\partial r}\left(r\dfrac{\partial\varphi}{\partial r}\right) + \dfrac{\partial^2\varphi}{\partial r^2} = -\dfrac{q}{\varepsilon_0} \ \delta(P), \\ \varphi(a,z) = 0 \quad \forall z, \end{cases} \tag{9.25}$$

and the condition that the potential is bounded when z tends to infinity inside the cylinder. Representing the function $\varphi(r,z)$, regular at all points inside the cylinder and satisfying the source-free equation, by means of the appropriate expansion in cylindrical coordinates, it results

$$\varphi(r,z) = \sum_{k=1}^{\infty} A_k \ J_0\left(x_k\frac{r}{a}\right)\exp\left(-x_k\frac{|z|}{a}\right), \tag{9.26}$$

where x_k are the zeros of the Bessel function $J_0(x)$. The evaluation of the unknown expansion coefficients A_k can be performed by using Gauss' theorem. It represents the surface of an infinite cylinder of radius $r < a$, whose symmetry axis coincides with the z-axis, we can state that

$$\oiint_{\Sigma} \nabla\varphi \cdot \hat{r} \quad dS = -\frac{q}{\varepsilon_0},$$

namely, using the series representation (9.22), we have

$$2\pi r \sum_{k=1}^{\infty} A_k \frac{d}{dr} \quad \left[J_0\left(x_k \frac{r}{a}\right)\right] \int_{-\infty}^{\infty} \exp\left(-x_k \frac{|z|}{a}\right) dz$$

$$\rightarrow \quad \sum_{k=1}^{\infty} \frac{A_k}{x_k} \quad \frac{d}{dr} \quad \left[J_0\left(x_k \frac{r}{a}\right)\right] = -\frac{q}{4\pi\varepsilon_0 r a}.$$

The last equation can be simplified by performing an integration over the interval (r, a), obtaining

$$\sum_{k=1}^{\infty} \frac{A_k}{x_k} \quad J_0\left(x_k \frac{r}{a}\right) = -\frac{q}{4\pi\varepsilon_0 a} \ln\frac{r}{a},$$

and then can be used to evaluate the unknown coefficients by making a projection on the complete set of Bessel functions

$$B = \left\{J_0\left(x_k \frac{r}{a}\right), \quad n \in \mathbb{N}\right\} \quad \rightarrow \quad \int_0^a r \quad J_0\left(x_k \frac{r}{a}\right) \quad J_0\left(x_k \frac{r}{a}\right) dr$$
$$= a^2 \begin{cases} 0, & n \neq k, \\ J_1^2(x_k)/2, & n = k, \end{cases}$$

that is

$$a^2 \frac{A_k}{2x_k} J_1^2(x_k) = -\frac{q}{4\pi\varepsilon_0 a} \int_0^a r \ln\frac{r}{a} \quad J_0\left(x_k \frac{r}{a}\right) dr$$
$$= -\frac{qa}{4\pi\varepsilon_0} \int_0^1 u \ln u \quad J_0(x_k u)\, du.$$

However, since it is well known that

$$\int_0^1 u \ln u \quad J_0(bu)\, du = \frac{J_0(b) - 1}{b^2} \quad \rightarrow \quad \int_0^1 u \ln u \quad J_0(x_k u) \quad du = -\frac{1}{x_k^2},$$

the expansion coefficient becomes

$$A_k = \frac{q}{2\pi\varepsilon_0 a} \frac{1}{x_k J_1^2(x_k)}.$$

By substituting the value of the coefficients A_k into expansion (22) of the potential, it is

$$\varphi(r, z) = \frac{q}{2\pi\varepsilon_0 a} \sum_{k=1}^{\infty} \frac{J_0(x_k r/a)}{x_k J_1^2(x_k)} \exp(-x_k |z|/a), \tag{9.27}$$

that can easily be put in the normalized form

$$\frac{2\pi\varepsilon_0 a}{q}\varphi(r,z) = \sum_{k=1}^{\infty} \frac{J_0(x_k r/a)}{x_k J_1^2(x_k)} \exp(-x_k|z|/a). \tag{9.28}$$

Figure 9.3 shows this normalized potential for different values of the dimensionless ratio r/a. It is apparent that expansion (9.28) exhibits an excellent rate of convergence for large values of z; besides, the exponential decrease of the generic term enables us to use this series also for small values of the longitudinal coordinate. But the convergence becomes problematic for values of z around zero, where the potential can be computed by means of formula obtained by Bernard W. Roos, in the book Analytical Functions and Distributions in Physics and Engineering (John Wiley & Sons, 1969)

$$\frac{2\pi\varepsilon_0 a}{q}\varphi(r,z) = \frac{a}{2\sqrt{r^2+z^2}} - \frac{a}{\pi}\int_0^{\infty} \frac{K_0(au)I_0(ru)}{I_0(au)} \cos(uz)\, du. \tag{9.29}$$

Therefore, we used formula (9.29) up to $|z| = a$, and continued with expansion (9.28) for $|z| > a$. The convergence of expansion (9.28) for different transverse positions and for $|z| = a$ is represented in Figure 9.4. It is worth noting that only three terms are necessary to stabilize the procedure of convergence.

From the potential (9.29), it is now easy to write an expansion for the normalized electric field

$$\frac{2\pi\varepsilon_0 a}{q}\overrightarrow{E(\vec{r},z)} = \sum_{k=1}^{\infty} \frac{\exp(-x_k|z|/a)}{J_1^2(x_k)}[\hat{r}J_1(x_k r/a) + \hat{z}\,\mathrm{sgn}(z)J_0(x_k r/a)], \tag{9.30}$$

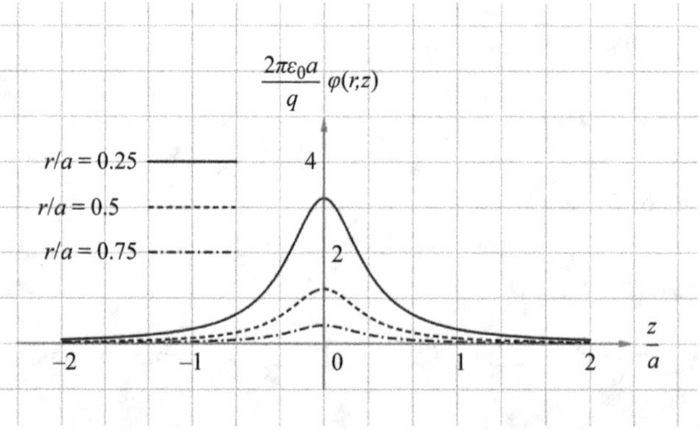

Figure 9.3 Normalized values of the potential in the particle frame for different values of r/a

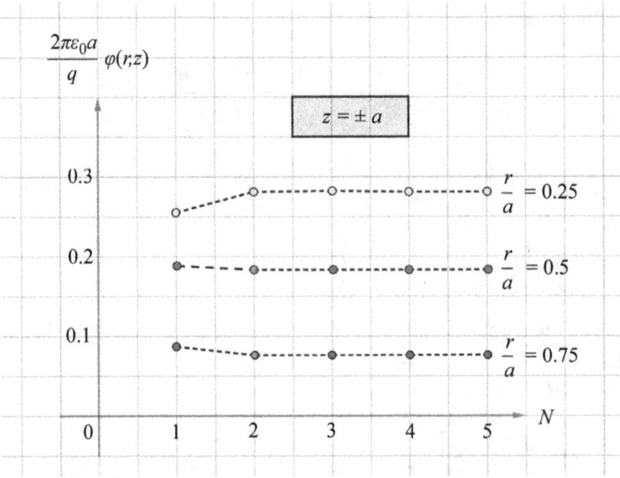

Figure 9.4 Convergence of the proposed expansion for the normalized potential

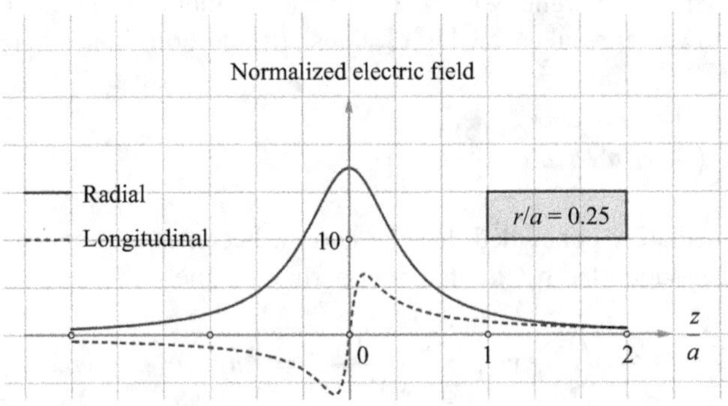

Figure 9.5 Normalized components of the electric field in the particle frame

which exhibits the same convergence properties discussed for the potential. The radial and longitudinal components of the electric field are represented in Figure 9.5 for a different given value of r/a.

9.5 Transformation in the laboratory frame

The field expansions in the laboratory frame S' can be obtained by means of the Lorentz transformation. The transformation formulae for the fields are given in a lot of classical textbook, such as *Electromagnetism* of Robert Stratman Elliott (McGraw-Hill, 1966)

$$E'_{\parallel} = E_{\parallel}, \quad E'_{\perp} = \gamma\left[E_{\perp} + \left(\vec{v} \times \vec{B}\right)_{\perp}\right],$$

with obvious meaning of the symbols used and gives the result

$$E_{r'}\left(r',z',t'\right) = \frac{q\gamma}{2\pi\varepsilon_0 a^2}\sum_{k=1}^{\infty}\frac{\exp\left(-x_k\gamma\left|z'-vt'\right|/a\right)}{J_1^2(x_k)}J_1\left(x_k r'/a\right),$$

$$E_{r'}\left(r',z',t'\right) = \frac{q\,\mathrm{sgn}\left(z'-vt'\right)}{2\pi\varepsilon_0 a^2}\sum_{k=1}^{\infty}\frac{\exp\left(-x_k\gamma\left|z'-vt'\right|/a\right)}{J_1^2(x_k)}J_0\left(x_k r'/a\right),$$

$$H_{\varphi'}\left(r',z',t'\right) = \beta\ \ E_{r'}\left(r',z',t'\right)/\zeta_0.$$

In particular, the knowledge of the magnetic field is useful to obtain the current flowing on the wall of the cylindrical pipe, already obtained as inverse cosine transform in [5]. We have

$$-\frac{a}{qv\gamma}i\left(z',t'\right) = \sum_{k=1}^{\infty}\frac{\exp\left(-\gamma x_k\left|z'-vt'\right|/a\right)}{J_1(x_k)}.$$

A plot of the normalized values of this current is given in Figure 9.6. If you approach the speed of light, you can see that the induced current on the tube becomes narrower and takes on high values. In addition, since it is worth the integral [7]

$$\int_{-\infty}^{\infty}\exp(-\gamma x_k|u|/a)\,du = \frac{2a}{\gamma x_k},$$

it can be immediately concluded that the current becomes an impulsive function, opposite to that supported by the inducing particle, namely

$$i\left(r',z'\right) = -\frac{qv}{2}\delta\left(z'-vt'\right)\sum_{k=1}^{\infty}\frac{1}{x_k J_1(x_k)} = -qv\ \ \delta\left(z'-vt'\right),$$

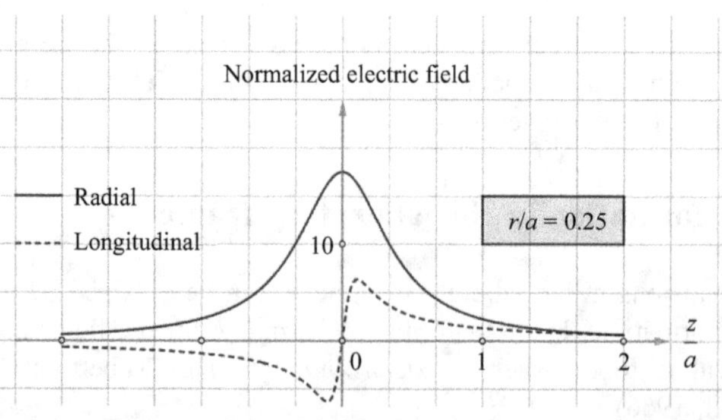

Figure 9.6 Normalized values of the surface current of the electric field in the laboratory frame

as was to be expected. In this way, direct solution was found in the time domain, using the solution of a stationary problem and making use of Lorentz relativistic transformations.

9.6 Concluding remarks

In this contribution, we directly described the electromagnetic field in the time domain sustained by a particle travelling in a perfectly conducting cylinder with uniform velocity along the symmetry axis of a circular waveguide by means of a Fourier–Bessel expansion whose general term is exponentially decaying. The classical solution is useful in the range of small values of $|z' - vt'|$; the proposed expansion, on the contrary, is helpful in the complementary range, namely for large values of $|z' - vt'|$. It would be desirable to carry out the same calculations with an off-axis particle. Finally, the possibility to manage rapidly convergent expansions can be favourable in the construction of tracking code, useful for numerical analysis of beam dynamics.

References

[1] S. Humphries, *Principles of Charged Particle Acceleration*, Mineola, NY: Dover Publications, 2013.

[2] L. J. Nevay, S. T. Boogert, J. Snuverink, *et al.* BDSIM: an accelerator tracking code with particle–matter interactions, *Computer Physics Communications*, 252, 107200, 2020.

[3] L. Verolino, Una certa classe di integrali, *Archimede*, 2, 527–533, 1993.

[4] L. De Menna, G. Miano, L. Verolino, A short note on some integrals involving Bessel functions, Il, *Nuovo Cimento* 109B(7), 805–808, 1994.

[5] G. Miano, L. Verolino, Some integrals involving Bessel functions, Il *Nuovo Cimento* 110B(4), 441–454, 1995.

[6] G. Miano, V.G. Vaccaro, L. Verolino, Time domain analysis of a charged particle travelling along the axis of a circular waveguide, *Il Nuovo Cimento* 111B(5), 659–664, 1996.

[7] I. S. Gradshteyn, I. M. Ryzhik, *Table of Integrals, Series, and Products*, London: Academic Press, 1980.

[8] B. W. Roos, *Analytical Functions and Distributions in Physics and Engineering*, New York, NY: John Wiley & Sons, 1969.

[9] W. R. Smythe, Charged spheroid in cylinder, *Journal of Mathematical Physics* 4, 833, 1963.

[10] Y. L. Luke, *Mathematical Functions and Their Approximations*, London: Academic Press, 1975.

[11] G. N. Watson, *A Treatise on the Theory of Bessel Functions*, Cambridge, MA: Cambridge University Press, 1944.

[12] E. G. Phillips, *Functions of a Complex Variable with Applications*, UK: Oliver and Boyd, 1954.

Chapter 10
Analytical techniques for transient analysis

Paolo Burghignoli[1] *and Giampiero Lovat*[2]

In this chapter, a variety of analytical techniques will be presented for the transient electromagnetic analysis of planar stratified structures made of linear, isotropic media. The field excited by simple canonical sources such as vertical or horizontal electric or magnetic dipoles will be considered, whence the solution of the electromagnetic problem for general extended sources can then be obtained by linear superposition.

The classical Cagniard-de Hoop (CDH) method will be addressed first. Originally developed for the transient analysis of a dipole source in the presence of a single interface between two semi-infinite non-dispersive media, it has been adapted over the years to considerably more general configurations. It will be exposed here in connection with the problem of a pulsed vertical electric dipole (VED) radiating in the presence of a thin metal screen. A modified version of the method will also be presented, considering a canonical configuration of importance for plasmonic applications, namely the excitation of a free-standing graphene sheet by a VED.

Alternative approaches will then be addressed that, in contrast with the CDH method, have a modal character, i.e., express the transient response of the multilayer structure in terms related to the relevant modal spectrum. These include the double-deformation method of Tsang and Kong, the Haddon leaky-mode method, and the Felsen–Niu unconventional spectral synthesis procedure. The first will be exposed for the case of a free-standing graphene sheet, whereas the latter two will be applied to a grounded dielectric slab.

Finally, an introduction to the time-domain version of the exact image theory developed by Nikoskinen and Lindell will conclude the chapter.

10.1 Cagniard-de Hoop method

10.1.1 Generalities

The CDH method is an ingenious analytical technique presented in the 1930s by Cagniard [1] (who however attributes the idea to Carson) to obtain an exact

[1]Sapienza University of Rome, Department of Information Engineering, Electronics and Telecommunications (DIET), Italy

[2]Sapienza University of Rome, Department of Astronautical, Electrical and Energetic Engineering (DIAEE), Italy

solution for lossless, non-dispersive half-spaces excited by transient *point* or *line* sources. Instead of inverse Fourier (or Laplace) transforming the relevant frequency-domain (FD) (or Laplace-domain, LD) solution, the CDH method manipulates the latter, typically expressed as an inverse Fourier transform of an analytical spectral-domain solution, in order to cast it in the form of a direct temporal Fourier (or Laplace) transform and thus obtains the time-domain (TD) solution by inspection. Depending on the considered problem, such a solution is expressed either analytically or in terms of single or double non-oscillating integrals over finite domains, with considerable computational advantages with respect to a canonical brute-force numerical inversion of the FD (or LD) solution.

After Cagniard, the CDH method was developed and generalized by various authors, in particular by de Hoop [2–5], who considerably simplified the original formulation and applied it to a number of important seismic, acoustic, and electromagnetic problems involving lossless non-dispersive media. The method was then extended to treat lossy half-spaces [6,7]. Various approximate CDH-like approaches have also been developed, in order to treat general dispersive media or planar structures with layers of finite thickness (see [5,8] and references therein). Most recently, it has been adopted to study planar configurations involving *thin sheets*, either with high permittivity or conductivity contrast [9–13] or made of graphene [14,15].

Let us illustrate more in detail the main steps of the method by considering, for simplicity, a generic two-dimensional (2-D) transient scalar field $a = a(x, z, t)$ in a planar medium stratified along the z-axis. Each layer is assumed to be uniform, lossless, nondispersive, and characterized by a wave slowness $s_m = 1/c_m$, where m is an integer index that labels the layer. The CDH method is then based on the following three steps:

1. The starting point is the representation of the LD scalar field $A(x, z, s)$ as an inverse Fourier transform of the corresponding spectral-domain field $\widetilde{A}(k_x, z, s)$:

$$A(x, z, s) = \frac{1}{2\pi} \int_{-\infty}^{+\infty} \widetilde{A}(k_x, z, s) e^{-jk_x x} \mathrm{d}k_x . \tag{10.1}$$

   Here the change of variable $k_x = -jss_x$ is performed, where s_x is the wave slowness in the x-direction, so that (10.1) becomes

$$A(x, z, s) = \int_{-j\infty}^{+j\infty} \frac{s}{2\pi j} \widetilde{A}(ss_x, z, s) e^{-ss_x x} \mathrm{d}s_x , \tag{10.2}$$

   where assuming a real value of s, the integration is now performed along the imaginary axis of the complex s_x-plane.
2. The integrand of the s_x-integral is then expressed as a sum of *generalized rays* A_i of the form

$$A_i(x, z, s) = \int_{-j\infty}^{+j\infty} s^n g_i(s_x) e^{-sw_i(s_x, x, z)} \mathrm{d}s_x , \tag{10.3}$$

where $w_i = s_x x + \sum_{m=1}^{M} h_{i,m}\sqrt{s_m^2 - s_x^2}$ and $h_{i,m}$ is the vertical travel distance of the ith ray inside the mth layer.

The crucial step of the method now aims at making the s_x-integral take the form of a forward Laplace transform in time. This is accomplished by letting $\tau = w_i(s_x, x, z)$ and deforming the integration path in (10.2) from the imaginary axis to the so-called *CDH path* C_{CDH}, defined as the image in the complex s_x-plane of the real interval $[\tau_i, +\infty)$, where τ_i is a suitable constant (the wavefront arrival time of the ith ray). Assuming that the deformation is possible without encountering singularities of the integrand, then (10.3) can be written as

$$\begin{aligned} A_i(x,z,s) &= s^n \int_{\tau_i}^{+\infty} [g_i(s_x)]_{s_x=s_x(\tau)} \frac{\mathrm{d}s_x}{\mathrm{d}\tau} e^{-s\tau}\mathrm{d}\tau \\ &= s^n \int_{0}^{+\infty} F_i(\tau)H(\tau-\tau_i)e^{-s\tau}\mathrm{d}\tau, \end{aligned} \tag{10.4}$$

where $F_i(\tau) = [g_i(s_x)]_{s_x=s_x(\tau)}\mathrm{d}s_x/\mathrm{d}\tau$. In (10.4) and in the rest of the chapter, $H(\cdot)$ indicates the Heaviside unit-step function defined as

$$H(x-x_0) = \begin{cases} 1 & x > x_0, \\ 0 & x < x_0. \end{cases} \tag{10.5}$$

3. The integral in (10.4) has now the form of a Laplace transform; since s was assumed to be an arbitrary real number, Lerch theorem ensures that the unique inverse Laplace transform of such an integral is equal to $F_i(\tau)H(\tau-\tau_i)$. By taking also into account the power of s in front of the integral, we finally have the sought TD waveform:

$$a_i(x,z,t) = \frac{\partial^n}{\partial t^n}[F_i(t)H(t-\tau_i)]. \tag{10.6}$$

When a single interface is present, i.e., the source radiates in the presence of a planar discontinuity between two uniform half-spaces, a single-ray (or body-wave) constituent is present. However, both in the single- and multiple-interface cases, depending on the relative position of the source and observation points, the path-deformation process to the CDH path may encounter branch-point singularities. In this case, it is necessary to detour around such singularities by suitably extending the CDH path and the relevant additional contribution to the TD waveform is the so called *lateral* (or head) wave.

The simplest configuration is perhaps the one constituted by a single planar sheet having negligible thickness (*thin sheet*). In this case, no lateral waves occur and a single-ray constituent is present. In the following sections, the application of the CDH method to specific thin-sheet configurations will be illustrated.

10.1.2 Thin metal sheet excited by a vertical dipole

Consider the canonical problem of excitation of a thin metal sheet by a VED (see Figure 10.1). The sheet, orthogonal to the vertical z axis, has thickness d, electric

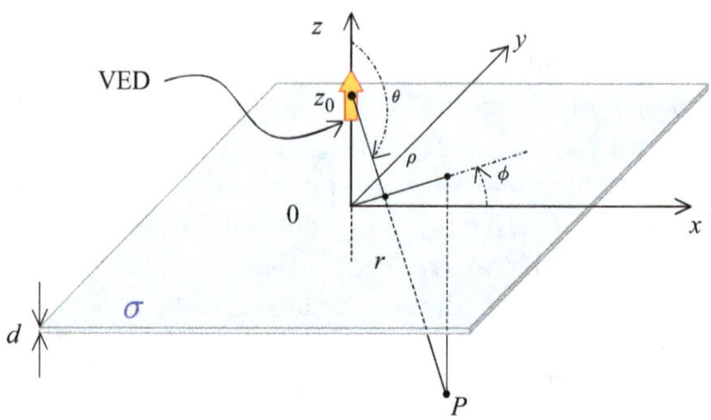

Figure 10.1 VED above a thin conductive sheet

permittivity $\varepsilon_s = \varepsilon_0\varepsilon_{sr}$, magnetic permeability μ_0, and bulk conductivity σ. The VED source has length ℓ, is located on the z axis at $z = z_0 > 0$, and carries a current moment $i(t)\ell = \delta(t)$, hence it can be represented in cylindrical coordinates (ρ, ϕ, z) by the transient impressed electric current density $\mathbf{j}(\mathbf{r}, t) = \delta(t)\delta(\rho)\delta(z - z_0)/(2\pi)\,\mathbf{u}_z$.

Such a current excites a purely TM^z field, which can be derived from a z-directed magnetic vector potential (in the Lorenz gauge) $\mathbf{a}(\mathbf{r}, t) = a_z(\mathbf{r}, t)\mathbf{u}_z$, where $a_z = g_a(\rho, z, t)$ is independent of the azimuthal angle ϕ (thanks to the rotational symmetry of the configuration) and is the relevant scalar *TD Green function* of the problem.

In the frequency domain the *incident*, free-space scalar Green function for the magnetic vector potential is

$$\begin{aligned} G_A^{\text{inc}}(\mathbf{r}, \omega) = \frac{e^{-jk_0 r}}{4\pi r} &= \frac{1}{(2\pi)^2}\int_{-\infty}^{+\infty}\int_{-\infty}^{+\infty} \frac{e^{-j\left(k_x x + k_y y + k_z|\Delta z|\right)}}{2jk_z}\,\mathrm{d}k_x \mathrm{d}k_y \\ &= -\frac{j}{4\pi}\int_0^{+\infty} \frac{k_\rho}{k_z} J_0\left(k_\rho \rho\right) e^{-jk_z|\Delta z|}\,\mathrm{d}k_\rho, \end{aligned} \tag{10.7}$$

where $k_0 = \omega/c_0$ is the free-space wavenumber, $c_0 = 1/\sqrt{\mu_0\varepsilon_0}$ is the speed of light in vacuum, $r = \sqrt{\rho^2 + \Delta z^2}$, $\Delta z = z - z_0$, $k_\rho = \sqrt{k_x^2 + k_y^2}$, $k_z = \sqrt{k_0^2 - k_\rho^2}$ (with $\mathrm{Im}(k_z) \leq 0$), and $J_n(\cdot)$ (here $n = 0$) is the Bessel function of the first kind and order n. As it can be seen, such a FD Green function can be expressed as a double inverse Fourier transform with respect to the spatial variables (x, y) or as a single inverse Hankel transform of order 0 with respect to the radial variable ρ [5].

The FD *transmitted* Green function $G_{A/F}(\mathbf{r}, \omega)$, i.e., the potential in the half space $z < 0$ below the screen, can be cast in the same form, i.e.,

$$\begin{aligned} G_A(\mathbf{r}, \omega) &= \frac{1}{(2\pi)^2}\int_{-\infty}^{\infty}\int_{-\infty}^{\infty} T\frac{e^{-j\left(k_x x + k_y y + k_z|\Delta z|\right)}}{2jk_z}\,\mathrm{d}k_x \mathrm{d}k_y \\ &= -\frac{j}{4\pi}\int_0^{\infty} \frac{k_\rho}{k_z} J_0\left(k_\rho \rho\right) T e^{-jk_z|\Delta z|}\,\mathrm{d}k_\rho, \end{aligned} \tag{10.8}$$

where T is the TM spectral transmission coefficient [5]:

$$T = T^{\mathrm{TM}}(k_\rho, \omega) = \frac{4Y_{c0}Y_{cs}e^{-jk_{zs}d}}{(Y_{c0}+Y_{cs})^2 - (Y_{c0}-Y_{cs})^2 e^{-2jk_{zs}d}}\,. \tag{10.9}$$

Here $k_{zs} = \sqrt{\omega^2\mu_0\varepsilon_{cs} - k_\rho^2}$ is the vertical wavenumber inside the sheet, with $\varepsilon_{cs} = \varepsilon_s - j\sigma/\omega$, and the spectral characteristic admittances are $Y_{c0} = \omega\varepsilon_0/k_z$ and $Y_{cs} = \omega\varepsilon_{cs}/k_{zs}$.

A direct inverse Fourier-transform of the second of (10.8) with respect to time would require the evaluation of double infinite integrals with oscillating integrands. The CDH method allows for alleviating the resulting computational burden by manipulating the first of (10.8) as described next.

Considering the rotational invariance of the structure around the z-axis, we may let the observation point belong to the plane $y = 0$, hence $x = \rho$. We then perform the crucial *change of variables* $k_x = \omega s_x$ and $k_y = \omega s_y$, from which $k_z = \omega\sqrt{s_0^2 - s_x^2 - s_y^2} = \omega s_z$, with $s_0 = 1/c_0$. Moving to the Laplace domain ($s = j\omega$), it then results

$$G_A(\mathbf{r}, s) = -\frac{s}{8\pi^2}\int_{-\infty}^{\infty}\int_{-\infty}^{\infty} T\frac{e^{-s(s_x\rho + s_z|\Delta z|)}}{s_z}\,\mathrm{d}s_x\mathrm{d}s_y\,. \tag{10.10}$$

The integration path in the s_y-plane is then deformed from the real axis to the imaginary axis (which is possible because no singularities of the integrand are encountered in the deformation) and let $s_y = -jq$, so that $s_z = \sqrt{s_0^2 + q^2 - s_x^2}$, thus obtaining

$$G_A(\mathbf{r}, s) = -\frac{js}{8\pi^2}\int_{-\infty}^{\infty}\int_{-\infty}^{\infty} T\frac{e^{-s(s_x\rho + s_z|\Delta z|)}}{s_z}\,\mathrm{d}s_x\mathrm{d}q\,. \tag{10.11}$$

Now we perform the key step of the procedure, by deforming the integration path in the s_x-plane of the inner integral in (10.11) from the real axis to the *hyperbolic CDH path* shown in Figure 10.2 (which again is possible because no singularities of the integrand are encountered in the deformation). Such a path is parameterized by the real variable $\tau > r\sqrt{s_0^2 + q^2}$ defined by

$$\tau = s_x\rho + s_z|\Delta z| \tag{10.12}$$

so that

$$\begin{aligned} s_x &= \frac{\tau}{r}\sin\theta \pm j\cos\theta\sqrt{\frac{\tau^2}{r^2} - (s_0^2 + q^2)}\,, \\ s_z &= \frac{\tau}{r}\cos\theta \mp j\sin\theta\sqrt{\frac{\tau^2}{r^2} - (s_0^2 + q^2)}\,, \end{aligned} \tag{10.13}$$

where $\sin\theta = \rho/r$, $\cos\theta = \Delta z/r$ (see Figure 10.1), and the upper/lower sign is referred to the upper/lower parts $P_{\mathrm{U/L}}$ of the CDH integration path, respectively.

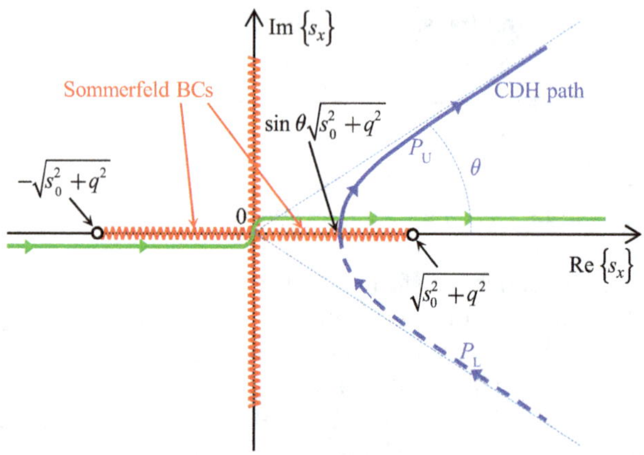

Figure 10.2 Complex plane of the s_x variable with the integration paths adopted in the CDH evaluation of the transmitted potential. Red, wiggly lines represent branch cuts; black circles represent the associated branch points; the green solid line is the original integration path; the blue line is the deformed CDH path, partially lying in the upper (proper) Riemann sheet (solid) and partially in the lower (improper) Riemann sheet (dashed)

Note that the integrand has branch-point singularities on the real axis at $s_x = \pm\sqrt{s_0^2+q^2}$ and that $P_{U/L}$ lie on the proper/improper Riemann sheets where $\mathrm{Im}\{s_z\}\leq 0$, respectively, defined by the Sommerfeld branch cuts shown in Figure 10.2.

Taking into account that $[T]_U = [T]_L^*$ (where the subscript U/L indicates that the quantity in square brackets is to be evaluated on $P_{U/L}$), (10.11) may be rewritten as

$$G_A(\mathbf{r},s) = \frac{s}{4\pi^2}\int_{-\infty}^{\infty}\int_{r\sqrt{s_0^2+q^2}}^{\infty}\frac{\mathrm{Re}\{[T]_U\}}{\sqrt{\tau^2-(s_0^2+q^2)r^2}}e^{-s\tau}\mathrm{d}\tau\mathrm{d}q. \tag{10.14}$$

By exchanging the order of integrations (and changing the integration limits accordingly), it results

$$G_A(\mathbf{r},s) = \frac{s}{4\pi^2}\int_{s_0r}^{\infty}\int_{-a(\tau)}^{+a(\tau)}\frac{\mathrm{Re}\{[T]_U\}}{\sqrt{\tau^2-(s_0^2+q^2)r^2}}\mathrm{d}q\,e^{-s\tau}\mathrm{d}\tau, \tag{10.15}$$

where $a(\tau) = \sqrt{\tau^2/r^2+s_0^2}$. Note that generally $\mathrm{Re}\{[T]_U\}$ depends on q, τ, *and* s, so that the τ-integral has not the form of a Laplace transform. However, by assuming that the sheet is *highly conductive* and *thin*, it can be modeled as a zero-thickness screen with surface (transition) admittance $Y_s = G_s$ with $G_s = \sigma d$ (cf. [12] for a derivation of this expression and a discussion of its limits of validity).

The transmission coefficient may then be written as

$$T = T(q,\tau) = \frac{2\varepsilon_0}{2\varepsilon_0 + s_z G_s}, \tag{10.16}$$

which is *independent* of s. The τ-integral in (10.15) has thus the form of a Laplace transform and one may deduce

$$g_a(\mathbf{r},t) = \frac{\partial}{\partial t} f(\mathbf{r},t) \tag{10.17}$$

with

$$f(\mathbf{r},t) = \frac{H(t - s_0 r)}{4\pi^2} \int_{-a(t)}^{+a(t)} \frac{C(q,r,\theta,t)}{\sqrt{t^2 - \left(s_0^2 + q^2\right) r^2}}\, \mathrm{d}q \tag{10.18}$$

and

$$C(q,r,\theta,t) = \mathrm{Re}\left\{[T]_{\mathrm{U}}\right\} \quad \text{with } \tau = t. \tag{10.19}$$

From (10.19) and (10.16), after lengthy manipulations, we obtain

$$C(q,r,\theta,t) = \frac{r(r + At\cos\theta)}{\gamma^2(r,\theta,t) - A^2 q^2 r^2 \sin^2\theta}, \tag{10.20}$$

where

$$\gamma^2(r,\theta,t) = (r + At\cos\theta)^2 + A^2 \sin^2\theta \left(t^2 - s_0^2 r^2\right) \tag{10.21}$$

and

$$A = \frac{G_s}{2\varepsilon_0}. \tag{10.22}$$

Remarkably, the integral in (10.18) can be evaluated in a *closed form*, resulting in

$$f(\mathbf{r},t) = \frac{H(t - s_0 r)}{4\pi\gamma(r,\theta,t)}. \tag{10.23}$$

Note that, in the limit $G_s \to 0$ (i.e., vanishing sheet), it results $f(\mathbf{r},t) = H(t - s_0 r)/(4\pi r)$ so that, upon derivation, the well-known TD Green function for the scalar wave equation in free space is recovered, i.e., $g_a^{\mathrm{inc}} = \delta(t - s_0 r)/(4\pi r)$.

Finally, the TD magnetic and electric potentials produced by a VED carrying an arbitrary current waveform $i(t)$ can be obtained by means of a time convolution between $g_a(\mathbf{r},t)$ and $\mu_0 i(t)\ell$, i.e.,

$$\begin{aligned} a_z(\rho,z,t) &= \mu_0 \int_{-\infty}^{+\infty} \frac{\partial f}{\partial \tau}\, i(t-\tau)\ell\,\mathrm{d}\tau \\ &= \mu_0 \int_{-\infty}^{t} f(\mathbf{r},\tau) \left.\frac{\mathrm{d}i}{\mathrm{d}t}\right|_{t-\tau} \ell\,\mathrm{d}\tau \end{aligned} \tag{10.24}$$

having integrated by parts in the latter expressions and having assumed $i(t) = 0$ for $t \leq 0$.

10.1.3 Free-standing graphene sheet

For a thin metal sheet, the CDH method leads to a direct determination of the TD Green function thanks to the fact that the sheet admittance is independent of frequency (or, equivalently, of the Laplace variable s). For thin sheets made of different materials, this may not be the case and hence the procedure described so far cannot be applied directly. A modified CDH approach is however sometimes feasible; it will be illustrated here considering a free-standing graphene sheet, again excited by a VED.

The electromagnetic properties of an infinite graphene sheet, in the absence of biasing magnetic fields and neglecting non-local effects that are unimportant below the high THz range, can be modeled in the FD through a scalar 2-D conductivity σ_{g} [S], which mainly depends on the radian frequency ω and a phenomenological scattering rate $\Gamma = 1/\tau_{\mathrm{s}}$ (where τ_{s} is the relaxation time), in addition to the chemical potential μ_{c} [16]. In particular, at room temperatures and up to the low THz regime, a Drude-like dispersion model holds for the 2-D conductivity [17], i.e.,

$$\sigma_{\mathrm{g}}(\omega) = \frac{\chi}{\Gamma + j\omega}, \tag{10.25}$$

where

$$\chi = \frac{q_e^2 k_{\mathrm{B}} T}{\pi \hbar^2} \ln\left\{2\left[1 + \cosh\left(\frac{\mu_{\mathrm{c}}}{k_{\mathrm{B}} T}\right)\right]\right\} \tag{10.26}$$

and where $-q_e$ is the electron charge, $\hbar$ is the reduced Planck constant, k_{B} is the Boltzmann constant, and T the absolute temperature.

The graphene sheet is assumed to lie on the $z = 0$ plane and to be excited by the same unit-amplitude transient VED considered in the case of a thin metal sheet, located on the z-axis at $z = z_0 > 0$. The excited field will then again be purely TM^z and our task is to derive an expression for the TD Green function g_a for the relevant z-directed magnetic vector potential.

Let us focus then on the reflected potential in $z > 0$. By proceeding as in the previous subsection, this can be expressed in the Laplace domain as (cf. (10.14))

$$G_A^{\mathrm{r}}(\mathbf{r}, s) = \frac{s}{4\pi^2} \int_{-\infty}^{\infty} \int_{r\sqrt{s_0^2+q^2}}^{\infty} \frac{\mathrm{Re}\{R|_{\mathrm{U}}\}}{\sqrt{\tau^2 - (s_0^2 + q^2) r^2}} e^{-s\tau} d\tau dq, \tag{10.27}$$

where the spectral reflection coefficient R is now dependent on s and is [5]

$$R(q, \tau, s) = \frac{a(q, \tau)}{b(q, \tau) + s} \tag{10.28}$$

with $a(q, \tau) = \chi_0 s_z(q, \tau)$, $b(q, \tau) = \Gamma + a(q, \tau)$, and $\chi_0 = \chi/(2\varepsilon_0)$.

From (10.14), it can easily be shown that

$$G_A^{\mathrm{r}}(\mathbf{r},s) = \frac{1}{4\pi^2}\int_{-\infty}^{+\infty}\int_{\gamma(q)}^{+\infty} \mathrm{Re}\{a(q,\tau)|_{\mathrm{U}}\}\frac{1}{\sqrt{\tau^2-\gamma^2(q)}}e^{-s\tau}\mathrm{d}q\mathrm{d}\tau$$
$$-\frac{1}{4\pi^2}\int_{-\infty}^{+\infty}\int_{\gamma(q)}^{+\infty}\mathrm{Re}\left\{\frac{c(q,\tau)}{b(q,\tau)+s}\bigg|_{\mathrm{U}}\right\}\frac{1}{\sqrt{\tau^2-\gamma^2(q)}}e^{-s\tau}\mathrm{d}q\mathrm{d}\tau$$
$$= G_A^{(1)}(\mathbf{r},s) - G_A^{(2)}(\mathbf{r},s), \tag{10.29}$$

where $c = c(q,\tau) = a(q,\tau)b(q,\tau)$.

The first term, $G_A^{(1)}(\mathbf{r},s)$, is analogous to (10.14) and can be treated similarly: by exchanging the order of integrations, a Laplace transform is obtained, whereby the relevant TD function $g_a^{(1)}(\mathbf{r},t) = L^{-1}\left\{G_A^{(1)}(\mathbf{r},s)\right\}$ can immediately be obtained by inspection (and in a closed form; cf. [15]).

The second term requires an additional step, namely, the s-dependent reflection coefficient is represented as a Laplace transform:

$$R(q,\tau,s) = L\left\{c(q,\tau)e^{-b(q,\tau)(t-\tau)}H(t-\tau)\right\} \tag{10.30}$$

whence

$$G_A^{(2)}(\mathbf{r},s) = \frac{1}{2\pi^2}$$
$$\cdot\int_0^{+\infty}\int_{\gamma(q)}^{+\infty}\int_0^{+\infty}\mathrm{Re}\left\{c(q,\tau)e^{-b(q,\tau)(t-\tau)}|_{\mathrm{U}}\right\}\frac{H(t-\tau)e^{-st}}{\sqrt{\tau^2-\gamma^2(q)}}\mathrm{d}q\mathrm{d}\tau\mathrm{d}t. \tag{10.31}$$

Exchanging the integrations and changing the integration limits accordingly [13], we have

$$G_A^{(2)}(\mathbf{r},s) = \frac{1}{2\pi^2 r}$$
$$\cdot\int_{s_0 r}^{+\infty}e^{-st}\int_{s_0 r}^{t}\int_0^{B(\tau)}\mathrm{d}q\frac{\mathrm{Re}\{c(q,\tau)e^{-b(q,\tau)(t-\tau)}|_{\mathrm{U}}\}}{\sqrt{B^2(\tau)-q^2}}\mathrm{d}t\mathrm{d}\tau \tag{10.32}$$

from which we can recognize

$$g_a^{(2)}(\mathbf{r},t) = L^{-1}\left\{G_A^{(2)}(\mathbf{r},s)\right\} = \frac{H(t-s_0 r)}{2\pi^2 r}\int_{s_0 r}^{t}\Psi(\tau)\mathrm{d}\tau, \tag{10.33}$$

where

$$\Psi(\tau) = \int_0^{B(\tau)}\mathrm{Re}\{c(q,\tau)e^{-b(q,\tau)(t-\tau)}|_{\mathrm{U}}\}\frac{1}{\sqrt{B^2(\tau)-q^2}}\mathrm{d}q \tag{10.34}$$

and the latter integral can be evaluated exactly in a closed form (for details, cf. [15]).

The TD Green function $g_a^{\mathrm{r}}(\mathbf{r},t) = L^{-1}\{G_A^{\mathrm{r}}(\mathbf{r},s)\} = g_a^{(1)}(\mathbf{r},t) + g_a^{(2)}(\mathbf{r},t)$ can thus finally be obtained. In contrast with the case of a thin metal sheet, such a Green function is not available in a closed form, but is expressed in terms of a *single integral* (the one in (10.33)); this is however straightforward to evaluate numerically, since its domain is finite and the integrand is well-behaved (i.e., regular and non-oscillatory).

10.2 Double-deformation method

The CDH method is an intrinsically *non-modal* technique, i.e., it provides a computationally convenient expression for the total transient field, but is not capable of sorting out the contribution of the pole singularities of the FD Green function that, as is well known, are associated with the propagation modes of the considered structure. For instance, in the case of a free-standing graphene sheet examined in the previous section, it does not allow for defining a TD counterpart of the surface-plasmon polariton (SPP) mode supported by the sheet, which has been intensely studied in recent years in view of its high-field confinement, long propagation distance, and tunability via electrostatic bias [18–21].

A modal alternative to the CDH method is provided by the so-called double-deformation technique (DDT), first proposed by Tsang and Kong to study the transient response of dielectric layers excited by line or point sources [22–24]. It will be illustrated here considering the TD magnetic induction produced by a VED in the presence of a free-standing graphene sheet modeled, as in the previous section, through a Drude-like conductivity [25].

In a nutshell, the starting point is the standard expression of the reflected induction (the incident induction is available in a closed form) as a double inverse-Fourier/inverse-Hankel transform, i.e.,

$$\begin{aligned} b_\varphi^{\mathrm{r}}(\rho,t) &= \frac{1}{\pi}\mathrm{Re}\left\{\int_0^\infty B_\varphi^{\mathrm{r}}(\rho,\omega)e^{j\omega t}\mathrm{d}\omega\right\} \\ &= \mathrm{Re}\left\{\int_0^\infty \int_{-\infty}^\infty \frac{k_\rho}{2\pi} H_1^{(2)}(k_\rho\rho)\widetilde{B}_\varphi^{\mathrm{r}}(k_\rho,\omega)\mathrm{d}k_\rho e^{j\omega t}\mathrm{d}\omega\right\}. \end{aligned} \tag{10.35}$$

where $b_\varphi^{\mathrm{r}}(\rho,t)$ is the (real) TD induction, $B_\varphi^{\mathrm{r}}(\rho,\omega)$ is the FD induction, $\widetilde{B}_\varphi^{\mathrm{r}}(k_\rho,\omega)$ is the spectral-domain induction, and $H_1^{(2)}(\cdot)$ indicates the Hankel function of second kind and order 1. This expression will be manipulated through a sequence of two deformations of the relevant complex integration paths (first in the radial-wavenumber plane and then in the radian-frequency plane) to suitable steepest-descent paths (SDPs). In this way, the total induction will be represented as a sum of residue contributions, arising from the poles captured in the deformation process, plus a double-integral SDP contribution.

The spectral-domain induction is available in a closed form [25], i.e.,

$$\widetilde{B}_\varphi^{\mathrm{r}}(k_\rho,\omega) = -\frac{j\mu_0}{4\pi} I(\omega)\ell \frac{k_\rho}{2\omega\varepsilon_0\bar{Z}_{\mathrm{g}}(\omega) + k_z(k_\rho,\omega)}, \tag{10.36}$$

where $k_z = \sqrt{k_0^2 - k_\rho^2}$ is the vertical wavenumber, $I(\omega)$ is the Fourier-transform of the source current waveform $i(t)$, and $\bar{Z}_g(\omega) = 1/\sigma_g(\omega)$ is the surface impedance of the graphene sheet. As a function of the radial wavenumber k_ρ, the spectral-domain induction has a single pair of pole singularities at $k_\rho = \pm k_{\rho\mathrm{SPP}}$, corresponding to the SPP mode supported by the graphene sheet, where

$$k_{\rho\mathrm{SPP}} = k_{\rho\mathrm{SPP}}(\omega) = \omega s_0 \sqrt{1 - 4\bar{Z}_g^2(\omega)} \tag{10.37}$$

is the frequency-dependent SPP wavenumber. Furthermore, algebraic branch-points located on the real axis at $k_\rho = \pm k_0$ are introduced by the square-root function that defines the vertical wavenumber k_z and a logarithmic branch point is also introduced at $k_\rho = 0$ by the Hankel function $H_1^{(2)}(\cdot)$. The integration path in the complex plane of the radial wavenumber k_ρ, the so-called (extended) Sommerfeld path, lies on the proper Riemann sheet where $\mathrm{Im}\{k_z\} \leq 0$, defined by the Sommerfeld branch-cuts $\mathrm{Im}\{k_z^2\} = 0$ and $\mathrm{Re}\{k_z^2\} \geq 0$. The situation is depicted in Figure 10.3.

The first step of the DDT consists in deforming the integration path of the inverse Hankel transform in (10.35) into the lower half of the k_ρ plane from the SIP to the vertical SDP $k_\rho = k_0 - jq$ ($q \in \Re_0^+$) through the branch-point $k_\rho = k_0 = \omega s_0$ (see Figure 10.4). Such an SDP is folded onto itself and lies partly on the proper Riemann sheet (solid line), partly on the improper sheet (dashed line).

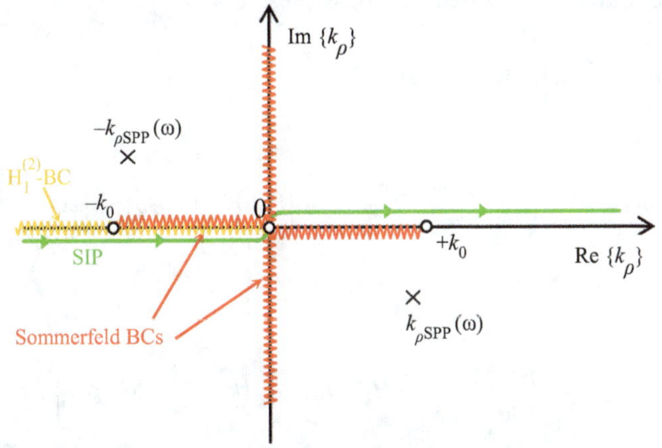

Figure 10.3 Singularities of the spectral-domain magnetic induction $\widetilde{B}_\varphi^{\mathrm{r}}(k_\rho, \omega)$ in the complex k_ρ plane: branch-points at $k_\rho = \pm k_0$ (circles), associated with the determination of the vertical wavenumber k_z, and the relevant Sommerfeld branch-cuts (red wiggly lines); branch-point at $k_\rho = 0$ (circle), associated with the Hankel function $H_1^{(2)}(\cdot)$, and the relevant branch-cut (orange wiggly line); poles at $k_\rho = \pm k_{\rho SPP}(\omega)$ (crosses), associated with FD SPP modes. The (extended) Sommerfeld integration path (SIP) for the calculation of the FD field B_φ as an inverse Hankel transform of order 1 is also indicated

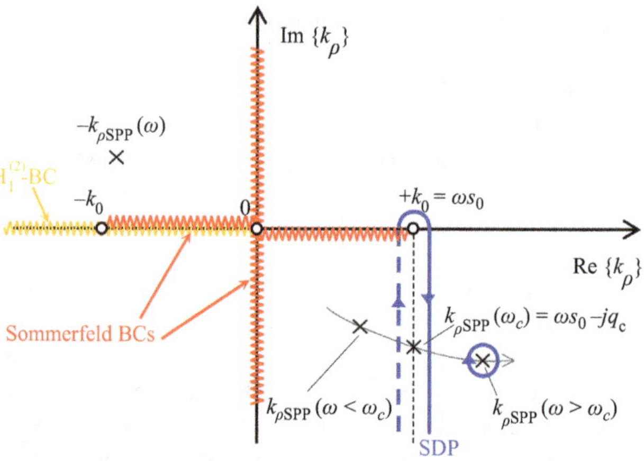

Figure 10.4 Path deformation in the k_ρ-plane: SDP through the branch point $k_\rho = k_0$ and evolution of the SPP pole $k_{\rho SPP}$ as a function of the radian frequency ω

By Cauchy and residue theorems, the original spectral integral can be represented as the sum of the pole contribution, $B_{\varphi SPP}(\rho, \omega)$, and the SDP contribution, $B_{\varphi SDP}(\rho, \omega)$, so that the TD field can be expressed as

$$b_\varphi^{\mathrm{r}}(\rho, t) = b'_{\varphi SPP}(\rho, t) + b_{\varphi SDP}(\rho, t), \tag{10.38}$$

where

$$b'_{\varphi SPP}(\rho, t) = \mathrm{Re}\left\{\int_{\omega_c}^{\infty} \frac{k_{\rho SPP}}{2\pi j} H_1^{(2)}\left[k_{\rho SPP}\rho\right] \mathrm{Res}\left[\widetilde{B}_\varphi^{\mathrm{r}}(k_\rho, \omega); k_{\rho SPP}\right] e^{j\omega t} \mathrm{d}\omega\right\} \tag{10.39}$$

and

$$b_{\varphi SDP}(\rho, t) = \mathrm{Re}\left\{\int_0^{\infty}\int_{\mathrm{SDP}} \frac{k_\rho}{2\pi} H_1^{(2)}(k_\rho\rho)\widetilde{B}_\varphi^{\mathrm{r}}(k_\rho, \omega)\mathrm{d}k_\rho e^{j\omega t}\mathrm{d}\omega\right\}. \tag{10.40}$$

Note that the lower limit of the ω-integral in (10.39) is not zero but is equal to ω_c, since the SPP pole $k_{\rho SPP}(\omega)$ is captured by the path deformation only for $\omega > \omega_c$. As a consequence, by the Paley–Wiener theorem, the relevant TD waveform $b'_{\varphi SPP}(\rho, t)$ is not causal [25].

The remedy to this is provided by the second step of the DDT. First, we exchange the order of integrations in (10.40). Then, we deform the integration path of the inner ω-integral from the positive real axis to the ω-SDP passing through the origin; this is the *positive* imaginary axis when $t > s_0\rho$, and the *negative* imaginary axis when $t < s_0\rho$ (see Figure 10.5).

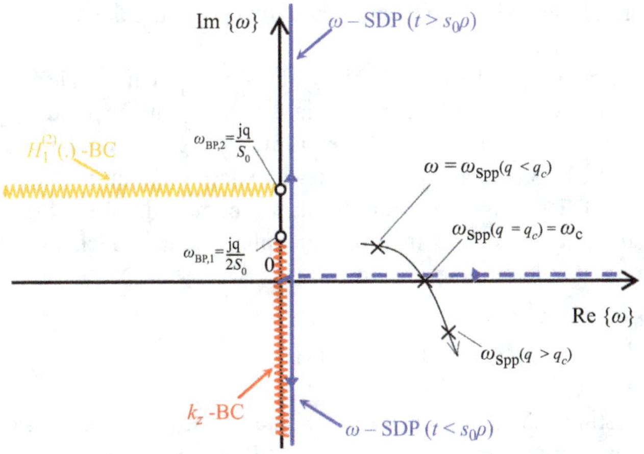

Figure 10.5 Path deformation in the ω-plane: SDPs through the origin for $t > s_0\rho$ (positive imaginary axis), $t < s_0\rho$ (negative imaginary axis), and evolution of the SPP pole $\omega_{\rho SPP}$ as a function of the imaginary part q of the radial wavenumber along the k_ρ-SDP ($k_\rho = k_0 - jq$). Two branch-point singularities are also shown with the relevant branch cuts, arising from the square-root function in the definition of the vertical wavenumber k_z and from the Hankel function $H_1^{(2)}(\cdot)$

In this last deformation, one has to take into account the various singularities of the integrand in the ω-plane and of their evolution as a function of the outer integration variable q (i.e., the imaginary part of the k_ρ wavenumber along the k_ρ-SDP). A detailed analysis [25] shows that the spectral Green function of a graphene sheet in free space has *four* pole singularities in the ω-plane. Of these, only one, denoted as $\omega_{\mathrm{SPP}}(q)$, crosses the positive real axis at $\omega = \omega_{\mathrm{c}}$ when $q = q_{\mathrm{c}}$ and can thus be associated with the FD SPP pole $k_{\rho\mathrm{SPP}}$; its evolution as a function of q is also reported in Figure 10.5.

The resulting representation for $b_{\varphi\mathrm{SDP}}$ can then be cast in the following form:

$$b_{\varphi\mathrm{SDP}}(\rho,t) = b''_{\varphi\mathrm{SPP}}(\rho,t) + b_{\varphi\mathrm{SDP}^2}(\rho,t) + b_{\varphi\mathrm{other}}(\rho,t)\,, \tag{10.41}$$

where $b''_{\varphi\mathrm{SPP}}(\rho,t)$ arises from the residue contribution of the pole $\omega_{\mathrm{SPP}}(q)$, $b_{\varphi\mathrm{SDP}^2}(\rho,t)$ is a double integral SDP contribution, and $b_{\varphi\mathrm{other}}(\rho,t)$ arises from the contribution of other Green function poles as well as of the ω-singularities of the source spectrum $I(\omega)$ (if any).

As shown in [25], the total field, sum of the incident field and of the reflected field (10.38) with (10.41), is identically zero before the first wavefront arrival, i.e., for $t < s_0\rho$, as required by causality. Furthermore, although the two SPP-related terms $b'_{\varphi\mathrm{SPP}}$ and $b''_{\varphi\mathrm{SPP}}$ are both different from zero for $t < s_0\rho$, their sum is instead identically zero for $t < s_0\rho$. This allows for defining a *causal* TD SPP field as

$$b_{\varphi\mathrm{SPP}}(\rho,t) = b'_{\varphi\mathrm{SPP}}(\rho,t) + b''_{\varphi\mathrm{SPP}}(\rho,t)\,, \tag{10.42}$$

i.e., as the sum of the residue contributions of *two* pole singularities (not one, as for the usual FD SPP field).

The modal nature of the DDT provides useful physical insight into the nature of the transient waveforms. In the present example, the method is particularly simple to implement, since the modal spectrum of the graphene sheet is constituted by a single mode. In other cases, e.g., finite-thickness slabs or general multilayered structures, the DDT requires to carefully track a certain number of the (most relevant) pole singularities in the involved complex plane, which may represent a drawback in its practical implementation.

10.3 Haddon leaky-mode method

Another interesting modal approach for the determination of the transient response of planar layered structures, based on the so-called *leaky modes*, was proposed by Haddon in the 1980s for geophysical problems involving elastic waves [26–29] and was subsequently applied to electromagnetic problems in [30,31]. Leaky modes are a class of partially guided waves supported by open structures, which radiate power in the outer infinite media while propagating, hence their wavenumber is complex, i.e., they attenuate, also in the absence of material losses. A vast literature exist on the relevant theory and applications, most notably to antennas of the traveling-wave type (see, e.g., [32] and references therein).

Similar to guided (or bound, or surface) modes, leaky (or radiating) modes are associated with pole singularities of the relevant spectral-domain Green function. However, whereas guided-mode poles reside on the *proper* Riemann sheet of the square-root functions that define the vertical wavenumbers inside the outer media (i.e., on the sheet where the radiation condition at infinity is satisfied), leaky-mode poles in uniform (i.e., non-periodic) structures made of ordinary materials reside on *improper* Riemann sheets, where such a condition is violated.

To illustrate, consider a grounded dielectric slab excited by a VED as shown in Figure 10.6.

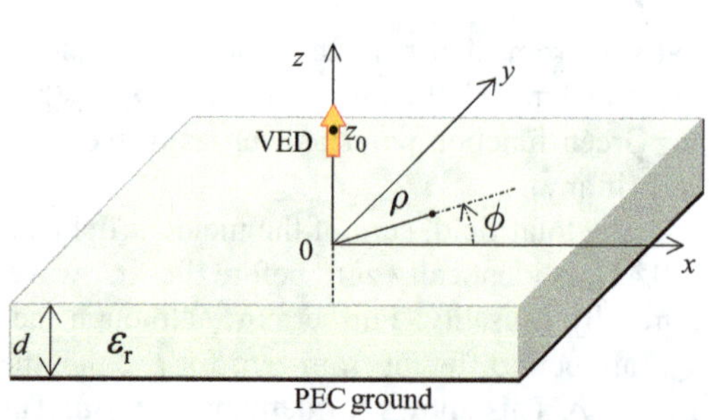

Figure 10.6 Grounded dielectric slab of thickness d and relative permittivity ε_r *excited by a VED*

The transient magnetic potential $a_z(\rho, z, t)$ for the TM^z field produced by the VED can be written as an inverse Fourier–Hankel transform of the relevant spectral-domain potential $\widetilde{A}_z(k_\rho, z, \omega)$ as

$$a_z(\rho, z, t) = \mathrm{Re}\left\{\frac{1}{\pi}\int_0^{+\infty} e^{j\omega t}\int_{\Gamma_1} \widetilde{A}_z(k_\rho, z, \omega) H_0^{(2)}(k_\rho \rho) k_\rho \mathrm{d}k_\rho \mathrm{d}\omega\right\}, \tag{10.43}$$

where the wavenumber (Hankel) integral is evaluated along the so-called extended Sommerfeld integration path Γ_1, equivalent to the entire real k_ρ-axis, and the frequency (Fourier) integral has been restricted to the positive real axis as the transient potential is real. It is worth noting that, by causality, the integration path in the ω-plane could be displaced below the real axis, where no ω-singularities occur.

Let us now consider the k_ρ-singularities of the integrand in the k_ρ-integral for a generic value of ω along such a displaced path (see Figure 10.7).

Since we have a single outer medium (air or vacuum above the slab), a single pair of branch points at $k_\rho = \pm k_0$ are present, associated with the square-root function that defines the vertical wavenumber $k_{z0} = \sqrt{k_0^2 - k_\rho^2}$ in the air region.

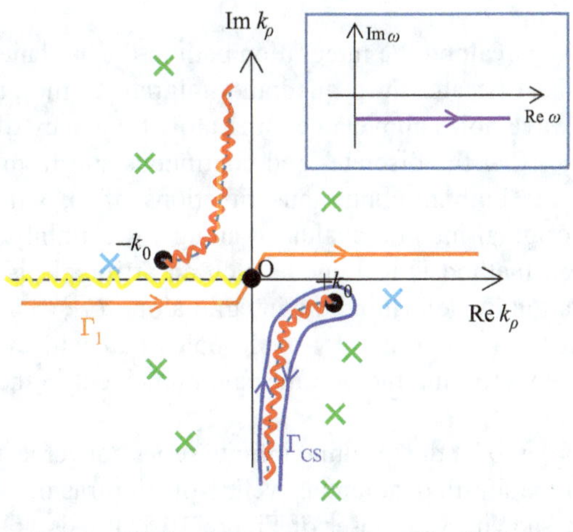

Figure 10.7 Singularities of the spectral integrand in the complex k_ρ-plane for a generic radian frequency ω belonging to the integration path $\omega = \omega_r - j\delta$, $\delta > 0$, $\omega_r \geq 0$ shown in the inset, and integration paths for the evaluation of the FD potential as an inverse Hankel transform. Black dots: branch points at $k_\rho = \pm k_0$ and $k_\rho = 0$; wiggly lines: branch cuts (red: Sommerfeld branch cuts associated with the square-root function $\sqrt{k_0^2 - k_\rho^2}$, yellow: standard branch cut associated with the function $H_0^{(2)}(k_\rho \rho)$); cyan crosses: bound-mode (proper) poles; green crosses: leaky-mode (improper) poles; Γ_1: extended Sommerfeld integration path; Γ_{CS}: continuous-spectrum integration path

The standard hyperbolic Sommerfeld branch cuts, with equations $\mathrm{Re}\{k_{z0}^2\} \geq 0$, $\mathrm{Im}\{k_{z0}^2\} = 0$, define the proper and improper Riemann sheets characterized by $\mathrm{Im}\{k_{x0}\} < 0$ and $\mathrm{Im}\{k_{x0}\} > 0$, respectively. At any (real) radian frequency ω, a finite number of bound-wave poles exists on the second and fourth quadrants of the proper sheet, whereas on the improper sheet an infinite number of leaky poles exists, located in all four quadrants. Therefore, it is possible to deform the integration path from Γ_1 to a path Γ_{CS} that runs along the two rims of the Sommerfeld branch cut in the fourth quadrant and thus we can obtain, by virtue of the Cauchy and residue theorems, the so-called *spectral representation* of the FD potential, i.e.,

$$A_z(\rho, z, \omega) = \sum_{n=1}^{N} A_{zn}(\rho, z, \omega) + \int_{\Gamma_{\mathrm{CS}}} \widetilde{A}_{\mathrm{SD}}(k_\rho, z) H_0^{(2)}(k_\rho \rho) k_\rho \mathrm{d}k_\rho, \qquad (10.44)$$

where $A_{zn}(\rho, z, \omega) = -2\pi j \mathrm{Res}\left[\widetilde{A}_{\mathrm{SD}}(k_\rho, z); k_{\rho n}\right] H_0^{(2)}(k_{\rho n}\rho) k_{\rho n}$ are the residue contributions of the proper bound-mode poles at $k_\rho = k_{\rho n}$. The finite sum of such contributions constitutes the *discrete-spectrum* contribution to the FD field, whereas the integral along the Sommerfeld branch cut constitutes the *continuous-spectrum* contribution.

Now, by varying ω along the integration path in the ω-plane, the leaky-wave poles located in the first and third quadrants migrate to the proper sheet, thus becoming bound-wave poles above a certain cutoff frequency (different for each mode). This implies that the discrete- and continuous-spectrum contributions to the FD potential are both discontinuous functions of ω, although their sum, equivalent to the original integral evaluated along Γ_1, certainly is not. The basic idea of the Haddon method is to remove such an artificial discontinuity of the spectral representation by deforming the integral along Γ_{CS} in order to enclose at all frequencies all the (infinite) leaky-mode poles that will eventually become bound-mode poles by crossing the Sommerfeld branch cut in the fourth quadrant of the k_ρ-plane.

However, in order to let the resulting infinite series converge, those leaky poles must be located in the fourth quadrant as well. Now, it turns out that, by using the ω-integration path shown in the inset of Figure 10.7, this is not the case, as the leaky-wave poles that would be involved belong to the first quadrant. However, by deforming the ω-integration path to the one lying above the real axis shown in the inset of Figure 10.8, the situation changes.

To maintain continuity, the Γ_1 contour should detour around the branch points at $\pm k_0$, which have now moved into the first and third quadrants; the associated branch cuts, however, still go to infinity in the fourth and second quadrants, respectively, so that Γ_1 partly belongs to the improper Riemann sheet (as indicated by its dashed-line portion in Figure 10.8). But, most importantly, now the leaky-mode poles that eventually become proper are those on the fourth quadrant, which produce a convergent infinite residue series. The Haddon method is thus based on evaluating the k_ρ-integral along a deformed k_ρ-path Γ_{H}, as the one shown in Figure 10.8, that detours around *all* such leaky-mode poles.

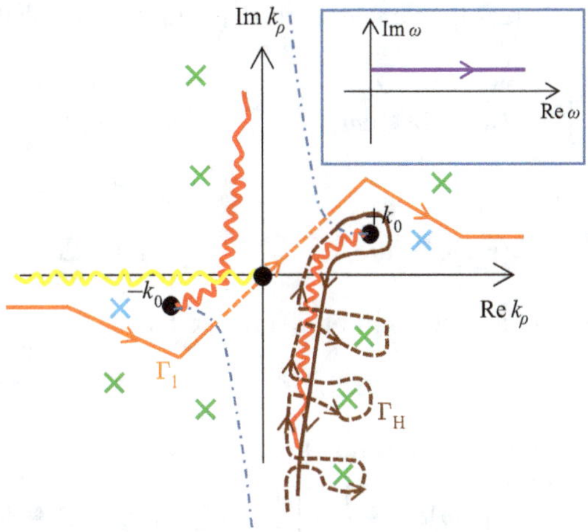

Figure 10.8 The same as in Figure 10.7, for the integration path $\omega = \omega_r + j\delta$, $\delta > 0$, $\omega_r \geq 0$ shown in the inset

Most notably, it can be shown that the ω-integral of such a deformed k_ρ-integral is *identically zero*. This implies that the transient potential can be represented *exactly* as an infinite summation of ω-integrals of leaky-mode residue contributions, i.e.,

$$a_z(\rho, z, t) = \mathrm{Re}\left\{\frac{1}{\pi}\int_0^{+\infty} e^{j\omega t}\sum_{n=1}^{+\infty} A_{zn}(\rho, z, \omega)\mathrm{d}\omega\right\}. \tag{10.45}$$

This is the main advantage of the Haddon method, i.e., it obviates the need of evaluating numerically any double integral. On the other hand, as for the double-deformation method, it requires an extensive preliminary dispersion analysis of the involved leaky-mode singularities.

10.4 Felsen–Niu unconventional spectral synthesis

In a series of papers, L. Felsen and F. Niu proposed various analysis and synthesis options in the wavenumber–frequency spectral domains in order to investigate pulse radiation in planarly stratified media [33–35]. In particular, an unconventional synthesis strategy has been proposed in which the frequency inversion is performed *before* the spatial wavenumber inversion, thus adopting TD leaky modes and TD surface modes with complex frequency and real spatial wavenumber as basis fields.

We consider the grounded dielectric slab of Figure 10.6 with relative permittivity ε_r and thickness d excited by a VED with a pulsed profile $f(t)$.

We start with the (x, y) spectral decomposition of $a_z(\mathbf{r}, t)$ at $\mathbf{r} = (x, y, z)$. For the time–frequency decomposition, it is useful to employ the analytic signal

a_z^+ defined for positive real frequencies ω throughout the lower half of the complex t plane, i.e.,

$$a_z^+(\mathbf{r}, t) = \frac{1}{\pi} \int_0^{+\infty} A_z(\mathbf{r}, \omega) e^{j\omega t} \mathrm{d}\omega, \tag{10.46}$$

where

$$A_z(\mathbf{r}, \omega) = \int_{-\infty}^{+\infty} a_z(\mathbf{r}, t) e^{-j\omega t} \mathrm{d}t \tag{10.47}$$

is the conventional Fourier spectrum of $a_z(\mathbf{r}, t)$, so that

$$a_z(\mathbf{r}, t) = \mathrm{Re}\{a_z^+(\mathbf{r}, t)\}, \qquad \mathrm{Im}\{t\} \leq 0. \tag{10.48}$$

The (x, y, t) spectral decomposition gives rise to

$$a_z^+(\rho, z, t) = \frac{1}{\pi} \int_0^{+\infty} F(\omega) e^{j\omega t} \frac{1}{(2\pi)^2} \iint_{-\infty}^{\infty} \widetilde{G}_a(k_\rho, z, \omega) e^{-j\mathbf{k}_\rho \cdot \boldsymbol{\rho}} \mathrm{d}k_x \mathrm{d}k_y \mathrm{d}\omega, \tag{10.49}$$

where $\mathbf{k}_\rho = (k_x, k_y)$ is the vector spectral wavenumber for the transverse $\boldsymbol{\rho} = (x, y)$ domain, $F(\omega)$ is the Fourier transform of the real input signal $f(t)$, i.e.,

$$F(\omega) = \int_{-\infty}^{+\infty} f(t) e^{-j\omega t} \mathrm{d}t, \tag{10.50}$$

and $\widetilde{G}_a$ is the zz-component of the spectral-domain dyadic potential Green function of the problem, i.e.,

$$\widetilde{G}_a(\mathbf{k}_\rho, z, \omega) = \iiint_{-\infty}^{\infty} g_a(\rho, z, t) e^{-j(\omega t - \mathbf{k}_\rho \cdot \boldsymbol{\rho})} \mathrm{d}\omega \mathrm{d}k_x \mathrm{d}k_y.$$

Equation (10.49) can be cast in the form

$$a_z^+(\rho, z, t) = \int_0^{+\infty} \iint_{-\infty}^{\infty} \frac{N(k_\rho, z, \omega)}{D(\omega, k_\rho)} e^{-j(\omega t - \mathbf{k}_\rho \cdot \boldsymbol{\rho})} \mathrm{d}k_x \mathrm{d}k_y \mathrm{d}\omega,$$

where $k_\rho = \sqrt{k_x^2 + k_y^2}$ and

$$D(\omega, k_\rho) = 1 - \Gamma(\omega, k_\rho) \tag{10.53}$$

with

$$\Gamma(\omega, k_\rho) = -R_{12}(\omega, k_\rho) e^{-jk_1(\omega, k_\rho) d}, \tag{10.54}$$

where R_{12} is the reflection coefficient looking upward from the dielectric medium at the interface $z = d$, i.e.,

$$R_{12}(\omega, k_\rho) = \frac{\varepsilon_r \sqrt{s_0^2 \omega^2 - k_\rho^2} - \sqrt{s_0^2 \omega^2 \varepsilon_r - k_\rho^2}}{\varepsilon_r \sqrt{s_0^2 \omega^2 - k_\rho^2} + \sqrt{s_0^2 \omega^2 \varepsilon_r - k_\rho^2}} \tag{10.55}$$

and

$$k_1(\omega, k_\rho) = \sqrt{s_0^2\omega^2\varepsilon_r - k_\rho^2}\,. \tag{10.56}$$

By using cylindrical coordinates (ρ, ϕ) in the spatial domain and (k_ρ, ψ) in the spectral-wavenumber domain and next introducing the spectral parameter $\xi = k_\rho/(s_0\omega)$, (10.52) can be expressed as

$$a_z^+(\rho, z, t) = \int_0^{+\infty}\int_0^{2\pi}\int_0^{\infty} \frac{I(\xi, \psi, z, \omega)}{D(\omega, \xi)}\,\mathrm{d}\xi\mathrm{d}\psi\mathrm{d}\omega\,, \tag{10.57}$$

where

$$I(\xi, \psi, z, \omega) = P(\xi, \psi)k_0^2 F(\omega)e^{j\omega[t-\tau(\xi,\psi)]} \tag{10.58}$$

where $P(\xi, \psi)$ is a function free of singularities and

$$\tau(\xi, \psi) = s_0\xi\rho\cos(\psi - \varphi) + \tau_z(\xi, z, z')\,, \tag{10.59}$$

where τ_z depends also on the slab thickness d. It is worth noting that, contrarily to the conventional Fourier–Hankel transform approach, the double spectral-wavenumber integration in the (ξ, ψ) domain has been retained, thus leaving the exponential function intact. This has the advantage of a simpler dependence of the integrand on ω when an unconventional reduction, where the ω integration is performed first, is adopted.

Since $|\Gamma| < 1$, the resonant denominator in (10.57) can be expanded in the power series

$$\frac{1}{D(\omega, \xi)} = \frac{1}{1 - \Gamma(\omega, \xi)} = \sum_{n=0}^{\infty}\Gamma^n(\omega, \xi)\,. \tag{10.60}$$

By a partial expansion of the denominator with a closed-form retention of the remainder, one has

$$\frac{1}{D(\omega, \xi)} = \sum_{n=0}^{N-1}\Gamma^n(\omega, \xi) + R(\omega, \xi) \tag{10.61}$$

with

$$R(\omega, \xi) = \frac{\Gamma^N(\omega, \xi)}{1 - \Gamma(\omega, \xi)} = \frac{\Gamma^{N-1}(\omega, \xi)}{\Gamma^{-1}(\omega, \xi) - 1}\,. \tag{10.62}$$

When (10.61) is substituted in (10.57), the finite series in (10.61) gives rise to a series of ray integrals that can be reduced explicitly by direct ω integration as in the spectral theory of transients [36,37]. On the other hand, the remaining integral originated from the remainder of the series R can be suitably modified to derive useful asymptotic expressions by deforming the path integral in the complex ω plane through the possible saddle points ω_s. In particular, from (10.57) and (10.61), one obtains

$$a_z^+(\mathbf{r},t) = \sum_{n=0}^{N-1} a_{z_n}^+(\mathbf{r},t) + \sum_{m=0}^{\infty} a_{zN_m}^+(\mathbf{r},t) + a_{zDN}^+(\mathbf{r},t), \tag{10.63}$$

where

$$a_{z_n}^+(\mathbf{r},t) = \int_0^{+\infty}\int_0^{2\pi}\int_0^{\infty} I(\xi,\psi,z,\omega)\Gamma^n(\omega,\xi)\mathrm{d}\xi\mathrm{d}\psi\mathrm{d}\omega \tag{10.64}$$

is the n-times reflected-ray (wave-front) integral. It can be shown that for $f(t) = \delta(t)$ each wave-front integral can be reduced to a single finite integral [36,37]. The term

$$a_{zN_m}^+(\mathbf{r},t) = 2\pi j \int_0^{2\pi}\int_0^{\sqrt{\varepsilon_r}} \left[\frac{I(\xi,\psi,z,\omega)}{\frac{\partial}{\partial\omega}D(\omega,\xi)}\right]_{\omega=\omega_m} H\left(\mathrm{Im}\{\mathbf{\Phi}_{N,m}\}\right)\mathrm{d}\xi\mathrm{d}\psi, \tag{10.65}$$

where

$$\mathbf{\Phi}_{N,m} = \omega_m\left[t - \tau(\xi,\psi) - 2s_0Nd\sqrt{\varepsilon_r - \xi^2}\right], \tag{10.66}$$

is the contribution from a pole $\omega_m(\xi)$ captured during the deformation of the real axis path into the SDP through the saddle point ω_s (transient mode) and defined by the dispersion equation

$$D(\omega_m,\xi) = 0. \tag{10.67}$$

The turn-on time (first appearance) of the transient mode parameterized by ξ is thus defined as the instant when $\omega_m(\xi)$ is captured by the SDP. Such a turn-on time depends on m, N, ξ, and ψ and is obtained as the solution of

$$\mathrm{Re}\{\mathbf{\Phi}_{N,m}\} = 0. \tag{10.68}$$

The higher-order modes (larger m) are turned on earlier than those with lower m.

The modes with $\mathrm{Im}\{\omega_m\} < 0$ are *transient leaky modes*, while those with $\mathrm{Im}\{\omega_m\} = 0$ are transient surface-wave modes [34]. Such transient modal fields in the integrand of (10.65) have the phase

$$\exp\left[-jk_{0m}\rho\cos(\psi-\phi) - jk_{0m}\sqrt{1-\xi^2}(z-d) + j\omega_m t\right], \tag{10.69}$$

where $k_{0m} = s_0\omega_m$. These fields decay in time but grow in the spatial domain (both along and transverse to z) for leaky modes, whereas they oscillate in time, decay along z, and propagate transverse to z for surface modes.

Finally, the term

$$a_{zDN}^+(\mathbf{r},t) = \int_0^{2\pi}\int_0^{\infty}\int_{\mathrm{SDP}} \frac{I(\xi,\psi,z,\omega)\Gamma^N(\omega,\xi)}{D(\omega,\xi)}\mathrm{d}\xi\mathrm{d}\psi\mathrm{d}\omega \tag{10.70}$$

is the contribution along the SDP through the saddle point ω_{sN}. In particular, the saddle point ω_{sN} is determined by [35]

$$\frac{\partial W}{\partial \omega}\Big|_{\omega=\omega_{sN}} = 0, \tag{10.71}$$

where

$$W(\omega,\xi,\psi) = \ln\left[\omega^2 F(\omega)\right] + j\omega\left[t - \tau(\xi,\psi) - 2s_0 N d\sqrt{\varepsilon_{\rm r} - \xi^2}\right] \tag{10.72}$$

and the corresponding SDP is defined by [35]

$$\begin{cases} \mathrm{Re}\{W(\omega,\xi,\psi) - W(\omega_{sN},\xi,\psi)\} \le 0, \\ \mathrm{Im}\{W(\omega,\xi,\psi) - W(\omega_{sN},\xi,\psi)\} = 0. \end{cases} \tag{10.73}$$

For $F(\omega)$ expressible as a finite-order polynomial in ω, the SDP is specified by

$$\begin{cases} \mathrm{Re}\left\{\omega\left[t - \tau(\xi,\psi) - 2s_0 N d\sqrt{\varepsilon_{\rm r} - \xi^2}\right]\right\} = 0, \\ \mathrm{Re}\{\omega\} \ge 0, \end{cases} \tag{10.74}$$

which defines a straight line from $\omega = 0$ in the right half of the complex ω-plane. When t is the complex and $\mathrm{Im}\{t\}$ is taken as a fixed parameter, for each ξ and ψ the SDP coincides with the positive imaginary axis at $\mathrm{Re}\{t\} = -\infty$, sweeps clockwise as $\mathrm{Re}\{t\}$ increases and coincides with the negative imaginary axis at $\mathrm{Re}\{t\} = \infty$. Since the poles ω_m considered before lies in the fourth quadrant, as $\mathrm{Re}\{t\}$ increases from $-\infty$ to ∞, ω_m will lie initially below, then on, and finally above the SDP thus defining the turn-on time in (10.68).

It is worth noting that the hybrid ray-mode decomposition can be easily reduced to either a pure ray- or a pure mode-fields representation. The ray-field representation is derived by letting $N = \infty$ in (10.61) (thus eliminating the remainder term) and obtaining

$$a_z^+(\mathbf{r},t) = \sum_{n=0}^{\infty} a_{z_n}^+(\mathbf{r},t). \tag{10.75}$$

The mode-field representation is instead derived by letting $N = 0$ in (10.61) (thus eliminating the series in (10.61) and maintaining $D(\omega,\xi)$ in its original form) and obtaining

$$a_z^+(\mathbf{r},t) = \sum_{m=0}^{\infty} a_{z_m}^+(\mathbf{r},t) + a_{zD}^+(\mathbf{r},t). \tag{10.76}$$

The SDP integral provides the transmitted fields along the direct ray path from source to receiver, while the residues at the captured poles give rise to the series of surface and leaky modes.

As mentioned above, the proposed unconventional formulation allows to derive asymptotic solutions which give physical and quantitative insight into the involved transient phenomena [35]. Especially, interesting is the behavior of the unconventional TD TM leaky modes which exhibit anomalies not encountered in their conventional time-harmonic counterpart when their angular spectra are near the Brewster angle for the air-dielectric interface [34].

10.5 Time-domain exact image theory

We finish this chapter by presenting the TD version of the exact image theory (EIT). The EIT was developed in the 1980s by Lindell and Alanen to extend the classical image theorems of electrostatics to the electrodynamic case of a time-harmonic source above a planar interface [38–40], and subsequently extended to other planar structures (for a review, cf. [41, Chapter 7]). A TD version of the EIT was developed by Nikoskinen and Lindell for a dipole source above a planar interface among two isotropic non-dispersive dielectric media [42]. It will be illustrated here for the case of a vertical magnetic dipole (VMD) above a thin metal sheet.

Consider a VMD of length ℓ, directed along the z-axis and carrying a voltage $v(t)$, located in air at a distance h above a metal plate of conductivity σ and thickness d (see Figure 10.9). The VMD can be represented by an impressed magnetic current $\mathbf{j}_{\mathrm{mi}}(\mathbf{r}, t) = v(t)\ell\delta(x)\delta(y)\delta(z-h)\mathbf{u}_z$. As concerns the metal plate, provided its thickness is negligible with respect to the skin depth at all frequencies in the spectrum of $v(t)$, it can be represented by an infinitesimally thin sheet with surface (transition) admittance $G_s = \sigma d$.

The source excites an azimuthally symmetric TE^z field derivable from a z-directed electric potential $f_z(\rho, z, t)$. The relevant spectral-domain potential is available in a simple closed form, i.e.,

$$\widetilde{F}_z(k_\rho, z, \omega) = \frac{1}{2jk_z}\left[e^{-jk_z|z-h|} + Re^{-jk_z(z+h)}\right], \tag{10.77}$$

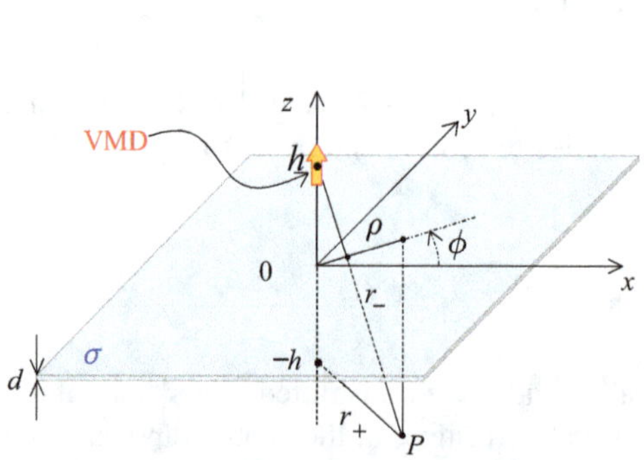

Figure 10.9 VMD placed at a distance h above an infinite thin metal sheet of conductivity σ and thickness d

where $k_z = \sqrt{k_0^2 - k_\rho^2}$, with $k_0 = \omega\sqrt{\mu_0\varepsilon_0}$, and

$$R = R(k_\rho, \omega) = -\frac{1}{1 + \frac{2k_z}{k_0 \widehat{G}_s}}, \tag{10.78}$$

with $\widehat{G}_s = \eta_0 G_s$, where $\eta_0 = \sqrt{\mu_0/\varepsilon_0}$.

The first addend in (10.77) is the free-space potential

$$f_z^{\rm inc}(\mathbf{r}, t) = c_0 \frac{\delta(r_- - c_0 t)}{4\pi r_-}, \tag{10.79}$$

where $r_- = \sqrt{\rho^2 + (z-h)^2}$ is the distance of the observation point from the source (see Figure 10.9). The EIT is based on representing R as a Laplace transform, so that also the reflected part of $\widetilde{F}_z$ can be expressed in terms of point sources (actually, an integral superposition of point sources placed at complex locations). Explicitly, by representing the reflection coefficient as

$$R(k_\rho, \omega) = -\int_0^{+\infty} e^{-p} e^{-\frac{2k_z}{k_0 \widehat{G}_s} p} \,\mathrm{d}p, \tag{10.80}$$

from (10.77), the reflected potential can be expressed as

$$\widetilde{F}_z^{\rm r}(k_\rho, z, \omega) = -\int_0^{+\infty} e^{-p} \frac{1}{2jk_z} e^{-jk_z\left(z+h-j\frac{2p}{k_0 \widehat{G}_s}\right)} \mathrm{d}p. \tag{10.81}$$

Now, the integrand beyond e^{-p} is the spectral-domain potential produced by a point source placed at the complex location $-h + j\zeta_{\rm TE}$, where $\zeta_{\rm TE} = 2p/\left(k_0 \widehat{G}_s\right)$. Therefore, the FD potential can be expressed as an exponentially convergent integral superposition of point-source contributions, i.e.,

$$F_z^{\rm r}(\rho, z, \omega) = -\int_0^{+\infty} e^{-p} \frac{e^{-jk_0 D}}{4\pi D} \mathrm{d}p, \tag{10.82}$$

where $D = D(p, \mathbf{r}, \omega) = \sqrt{\rho^2 + (z + h - j\zeta_{\rm TE})^2}$. By inverse Fourier-transforming (10.82), we have the sought TD potential

$$f_z^{\rm r}(\rho, z, t) = -\int_0^{+\infty} e^{-p} G_c(p, t) \mathrm{d}p, \tag{10.83}$$

where $G_c(p, t)$ is the TD potential produced by a point source at the complex location $-h + j\zeta_{\rm TE}$, whose expression can be calculated in a closed form:

$$\begin{aligned} G_c(p, t) = \frac{e^{-\alpha_+ c_0 t p}}{4\pi r_+} c_0 \{ \delta(r_+ - c_0 t) \\ - \left[\alpha_+ p J_0(\beta_+ p) + \frac{c_0 t \beta_+ p}{(c_0 t)^2 - r_+^2} J_1(\beta_+ p) \right] H(c_0 t - r_+) \}, \end{aligned} \tag{10.84}$$

where $r_+ = \sqrt{\rho^2 + (z+h)^2}$ is the distance of the observation point from the geometrical image of the source (see Figure 10.9) and the coefficients α_+, β_+ are given by $\alpha_+ = 2(z+h)/\left(\widehat{G}_s r_+^2\right)$ and $\beta_+ = 2\rho\sqrt{(c_0 t)^2 - r_+^2}/\left(\widehat{G}_s r_+^2\right)$.

In general, the TD response obtained with the EIT must be obtained by evaluating numerically the resulting p-integral. However, on the one hand, such an integral is rapidly (exponentially) convergent; on the other hand, in some cases, it can even be evaluated analytically; this is indeed the case of the configuration examined here (details are omitted for brevity).

References

[1] Cagniard L. *Réflexion et réfraction des ondes séismiques progressives*. Paris: Gauthiers-Villars; 1939. Available from: http://eudml.org/doc/192900.

[2] Pekeris CL. The seismic surface pulse. *Proc Natl Acad Sci.* 1955;41(7):469–480.

[3] de Hoop AT, Frankena HJ. Radiation of pulses generated by a vertical electric dipole above a plane, non-conducting, earth. *Appl Sci Res, Sec B.* 1960;8(1):369–377.

[4] de Hoop AT. Pulsed electromagnetic radiation from a line source in a two-media configuration. *Radio Sci.* 1979;14(2):253–268.

[5] Chew WC. *Waves and Fields in Inhomogenous Media*. Piscataway, NJ: IEEE Press; 1999.

[6] Kuester E. The transient electromagnetic field of a pulsed line source located above a dispersively reflecting surface. *IEEE Trans Antennas Propag.* 1984;32(11):1154–1162.

[7] Kooij BJ. The transient electromagnetic field of an electric line source above a plane conducting earth. *IEEE Trans Electromagn Compat.* 1991;33(1): 19–24.

[8] Štumpf M, Hoop ATD, Vandenbosch GAE. Generalized ray theory for time-domain electromagnetic fields in horizontally layered media. *IEEE Trans Antennas Propag.* 2013;61(5):2676–2687.

[9] de Hoop AT, Meng LL, Jiang LJ. Pulsed line source response of a thin sheet with high-contrast dielectric and conductive properties – A time-domain analysis. *IEEE Trans Antennas Propag.* 2013;61(11):5649–5657.

[10] Lovat G, Araneo R, Celozzi S. Time-domain shielding effectiveness of planar conductive nanoscreens. In: *Proceeding of the 2014 IEEE EMC-S International Symposium*; 2014. pp. 181–185.

[11] Lovat G, Araneo R, Celozzi S. Near-field time-domain shielding effectiveness of thin conductive screens. *Prog. Electromagn Res.* 2014;146:47–56.

[12] Burghignoli P, Lovat G, Araneo R, *et al.* Time-domain shielding of a thin conductive sheet in the presence of pulsed vertical dipoles. *IEEE Trans Electromagn Compat.* 2017;60(1):157–165.

[13] Burghignoli P, Lovat G, Araneo R, *et al.* Pulsed vertical dipole response of a thin sheet with high-contrast dielectric and conductive properties. *IEEE Trans Antennas Propag.* 2017;66(1):217–225.
[14] Lovat G, Araneo R. Semi-analytical representation of the two-dimensional time-domain Green's function of a graphene sheet in the intraband regime. *IEEE Trans Nanotechnol.* 2015;14(4):681–688.
[15] Lovat G, Burghignoli P, Araneo R, *et al.* Time-domain Green's function for a vertical dipole above a graphene sheet. *IEEE Trans Nanotechnol.* 2018; 17(4):841–851.
[16] Gusynin VP, Sharapov SG, Carbotte JP. Magneto-optical conductivity in graphene. *J. Condens. Matter Phys.* 2007;19(2):026222.
[17] Hanson GW. Dyadic Green's functions for an anisotropic non-local model of biased graphene. *IEEE Trans Antennas Propag.* 2008;56(3):747–757.
[18] Koppens FHL, Chang DE, Garcia de Abajo FJ. Graphene plasmonics: a platform for strong light–matter interactions. *Nano Lett.* 2011;11(8): 3370–3377.
[19] Batrakov KG, Saroka VA, Maksimenko SA, *et al.* Plasmon polariton deceleration in graphene structures. *J Nanophot.* 2012;6(1):061719.
[20] Bludov YV, Ferreira A, Peres NMR, *et al.* A primer on surface plasmon polaritons in graphene. *Int J Mod Phys B.* 2013;27(10):1341001.
[21] Garcia de Abajo FJ. Graphene plasmonics: challenges and opportunities. *ACS Phot.* 2014;1(3):135–152.
[22] Tsang L, Kong JA. Modified modal theory of transient response in layered media. *J Math Phys.* 1979;20(6):1170–1182.
[23] Ezzeddine A, Kong JA, Tsang L. Time response of a vertical electric dipole over a two-layer medium by the double deformation technique. *J Appl Phys.* 1982;53(2):813–822.
[24] Poh SY, Kong JA. Transient response of a vertical electric dipole (VED) on a two-layer medium. *J Electromagn Waves Appl.* 1987;1(2):133–158.
[25] Burghignoli P, Lovat G, Araneo R, *et al.* Time-domain surface plasmon polaritons on a graphene sheet. *Phys Rev B.* 2018;97(24):245418.
[26] Haddon R. Exact evaluation of the response of a layered elastic medium to an explosive point source using leaking modes. *Bull Seism Soc Am.* 1986; 76(6):1755–1775.
[27] Haddon R. Response of an oceanic wave guide to an explosive point source using leaking modes. *Bull Seism Soc Am.* 1987;77(5):1804–1822.
[28] Haddon R. Numerical evaluation of Green's functions for axisymmetric boreholes using leaking modes. Geophysics. 1987;52(8):1099–1105.
[29] Haddon R. Exact Green's functions using leaking modes for axisymmetric boreholes in solid elastic media. *Geophysics.* 1989;54(5):609–620.
[30] Duffy DG. Response of a grounded dielectric slab to an impulse line source using leaky modes. *IEEE Trans Antennas Propag.* 1994;42(3):340–346.
[31] Hanson GW, Yakovlev AB, Hao J. Leaky-wave analysis of transient fields due to sources in planarly layered media. *IEEE Trans Antennas Propag.* 2003;51(2):146–159.

[32] Monticone F, Alu A. Leaky-wave theory, techniques, and applications: from microwaves to visible frequencies. *Proc IEEE.* 2015;103(5):793–821.

[33] Felsen LB, Niu F. Spectral analysis and synthesis options for short pulse radiation from a point dipole in a grounded dielectric layer. *IEEE Trans Antennas Propag.* 1993;41(6):747–754.

[34] Niu F, Felsen LB. Time-domain leaky modes on layered media: dispersion characteristics and synthesis of pulsed radiation. *IEEE Trans Antennas Propag.* 1993;41(6):755–761.

[35] Niu F, Felsen LB. Asymptotic analysis and numerical evaluation of short pulse radiation from a point dipole in a grounded dielectric layer. *IEEE Trans Antennas Propag.* 1993;41(6):762–769.

[36] Heyman E, Felsen L. Weakly dispersive spectral theory of transients. Part I: formulation and interpretation. *IEEE Trans Antennas Propag.* 1987;35(1): 80–86.

[37] Heyman E, Felsen L. Weakly dispersive spectral theory of transients. Part II: evaluation of the spectral integral. *IEEE Trans Antennas Propag.* 1987;35 (5):574–580.

[38] Lindell IV, Alanen E. Exact image theory for the Sommerfeld half-space problem, Part I: vertical magnetic dipole. *IEEE Trans Antennas Propag.* 1984;32(2):126–133.

[39] Lindell IV, Alanen E. Exact image theory for the Sommerfeld half-space problem, Part II: vertical electrical dipole. *IEEE Trans Antennas Propag.* 1984;32(8):841–847.

[40] Lindell IV, Alanen E. Exact image theory for the Sommerfeld half-space problem, Part III: general formulation. *IEEE Trans Antennas Propag.* 1984;32(10):1027–1032.

[41] Lindell IV. *Methods for Electromagnetic Field Analysis*. Piscataway, NJ: IEEE Press; 1992.

[42] Nikoskinen KI, Lindell IV. Time-domain analysis of the Sommerfeld VMD problem based on the exact image theory. *IEEE Trans Antennas Propag.* 1990;38(2):241–250.

Chapter 11

The application of the finite-difference time-domain (FDTD) technique to lightning studies

Dongshuai Li[1], *Alejandro Luque*[1], *Farhad Rachidi*[2] *and Marcos Rubinstein*[3]

The finite-difference time-domain (FDTD) method, introduced by Yee in 1966 [1], is one of the most popular numerical methods for the solution of electromagnetic problems. As a powerful numerical technique, the FDTD method has a number of advantages with respect to other methods, such as,

- The ability to simulate complex geometries including non-linearity, inhomogeneity, and anisotropy.
- The ability to simulate large domains without having to solve matrix equations.
- The possibility of using arbitrary time-domain sources without concern for the time to frequency transform.
- The high data locality that allows codes to be parallelized easily to speed up the computation.

These advantages make FDTD especially well-suited for lightning electromagnetic pulse (LEMP) simulations (e.g., [2–5]). In this chapter, we first introduce the basic methodology of the FDTD technique, and then present a review of the applications of the FDTD method to lightning studies including (1) lightning electromagnetic fields at close, medium, and far ranges; (2) lightning electromagnetic fields propagation over mountainous terrain, and (3) lightning electromagnetic fields propagation in the earth-ionosphere waveguide (EIWG) and its interaction with the lower D-region of the ionosphere.

[1]Instituto de Astrofísica de Andalucía (IAA), CSIC, Spain
[2]Swiss Federal Institute of Technology (EPFL), Switzerland
[3]University of Applied Sciences and Arts Western Switzerland, Switzerland

11.1 Fundamentals of the FDTD method

11.1.1 FDTD basic equations in different coordinate systems

Based on the time-dependent Maxwell's equations, the partial differential equations (PDEs) are discretized using central-difference approximations to the space and time partial derivatives to generate FDTD's basic equations [6]. The equations can be written in various ways according to different coordinate systems, not only Cartesian coordinates but also curvilinear coordinates, such as cylindrical and spherical coordinates. Figure 11.1 shows the electromagnetic field components in a unit cube of the Yee algorithm for the FDTD method with spatial indices i, j, and k. In Cartesian coordinates, Maxwell's equations can be decomposed into a TE_z and a TM_z mode. For the TE_z mode, the H_z, E_x, E_y components are located in each lattice cut plane k, $k+1$, and so forth, while for the TM_z mode the E_z, H_x, H_y components are located in each lattice cut plane $k+\frac{1}{2}$, $k+\frac{3}{2}$, and so forth. The TE_z and TM_z modes are completely decoupled from each other to implement a three-dimensional (3D) grid.

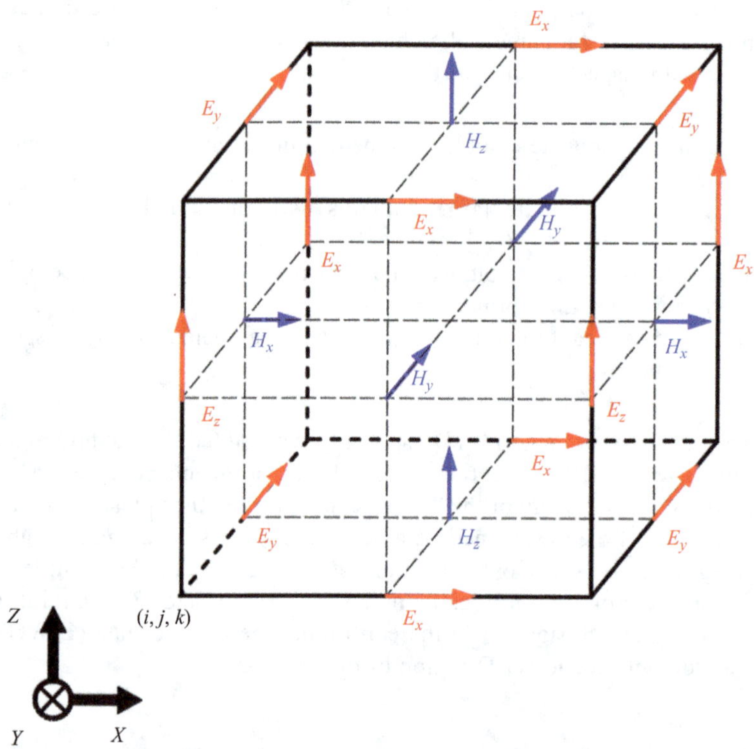

Figure 11.1 Positions of the E_x, E_y, and E_z components (red) of the electric field and the H_x, H_y, and H_z components (blue) of the magnetic field in a Yee cell for 3D Cartesian coordinates

In the following subsections, we will derive the update equations for 3D Cartesian coordinates, as well as 3D and two-dimensional (2D) cylindrical and spherical (polar) coordinates.

11.1.1.1 3D Cartesian coordinates

The FDTD discretization of Faraday's and Ampere's laws in Maxwell's equations leads to a total of six update equations in 3D Cartesian coordinates [6]. The update equations of the components of the electric field are:

$$\begin{aligned} E_x\big|_{i+1/2,j,k}^{n+1} &= C_a E_x\big|_{i+1/2,j,k}^{n} \\ &+ C_b\left[\frac{H_z\big|_{i+1/2,j+1/2,k}^{n+1/2} - H_z\big|_{i+1/2,j-1/2,k}^{n+1/2}}{\Delta_y}\right. \\ &\left. - \frac{H_y\big|_{i+1/2,j,k+1/2}^{n+1/2} - H_y\big|_{i+1/2,j,k-1/2}^{n+1/2}}{\Delta_z}\right], \end{aligned} \tag{11.1}$$

$$\begin{aligned} E_y\big|_{i,j+1/2,k}^{n+1} &= C_a E_y\big|_{i,j+1/2,k}^{n} \\ &+ C_b\left[\frac{H_x\big|_{i,j+1/2,k+1/2}^{n+1/2} - H_x\big|_{i,j+1/2,k-1/2}^{n+1/2}}{\Delta_z}\right. \\ &\left. - \frac{H_z\big|_{i+1/2,j+1/2,k}^{n+1/2} - H_z\big|_{i-1/2,j+1/2,k}^{n+1/2}}{\Delta_x}\right], \end{aligned} \tag{11.2}$$

$$\begin{aligned} E_z\big|_{i,j,k+1/2}^{n+1} &= C_a E_z\big|_{i,j,k+1/2}^{n} \\ &+ C_b\left[\frac{H_y\big|_{i+1/2,j,k+1/2}^{n+1/2} - H_y\big|_{i-1/2,j,k+1/2}^{n+1/2}}{\Delta_x}\right. \\ &\left. - \frac{H_x\big|_{i,j+1/2,k+1/2}^{n+1/2} - H_x\big|_{i,j-1/2,k+1/2}^{n+1/2}}{\Delta_y}\right], \end{aligned} \tag{11.3}$$

where the coefficients at the general E-field component location (i, j, k) are:

$$C_a\big|_{i,j,k} = \frac{2\varepsilon_{i,j,k} - \sigma_{i,j,k}\Delta_t}{2\varepsilon_{i,j,k} + \sigma_{i,j,k}\Delta_t}, \tag{11.4}$$

$$C_b\big|_{i,j,k} = \frac{2\Delta_t}{2\varepsilon_{i,j,k} + \sigma_{i,j,k}\Delta_t}, \tag{11.5}$$

where σ is the electric conductivity, $\varepsilon = \varepsilon_0\varepsilon_r$ is the permittivity of the medium, ε_0 is the free-space permittivity, and ε_r is the relative permittivity.

Following the same procedure, one can derive similar magnetic field expressions:

$$H_x\big|_{i,j+1/2,k+1/2}^{n+1/2} = D_a H_x\big|_{i,j+1/2,k+1/2}^{n-1/2} + D_b\left[\frac{E_y\big|_{i,j+1/2,k+1}^{n} - E_y\big|_{i,j+1/2,k}^{n}}{\Delta_z} - \frac{E_z\big|_{i,j+1,k+1/2}^{n} - E_z\big|_{i,j,k+1/2}^{n}}{\Delta_y}\right], \tag{11.6}$$

$$H_y\big|_{i+1/2,j,k+1/2}^{n+1/2} = D_a H_y\big|_{i+1/2,j,k+1/2}^{n-1/2} + D_b\left[\frac{E_z\big|_{i+1,j,k+1/2}^{n} - E_z\big|_{i,j,k+1/2}^{n}}{\Delta_x} - \frac{E_x\big|_{i+1/2,j,k+1}^{n} - E_x\big|_{i+1/2,j,k}^{n}}{\Delta_z}\right], \tag{11.7}$$

$$H_z\big|_{i+1/2,j+1/2,k}^{n+1/2} = D_a H_z\big|_{i+1/2,j+1/2,k}^{n-1/2} + D_b\left[\frac{E_x\big|_{i+1/2,j+1,k}^{n} - E_x\big|_{i+1/2,j,k}^{n}}{\Delta_y} - \frac{E_y\big|_{i+1,j+1/2,k}^{n} - E_y\big|_{i,j+1/2,k}^{n}}{\Delta_x}\right], \tag{11.8}$$

where the coefficients at the general H-field component location (i,j,k) are:

$$D_a\big|_{i,j,k} = \frac{2\mu_{i,j,k} - \sigma^*_{i,j,k}\Delta_t}{2\mu_{i,j,k} + \sigma^*_{i,j,k}\Delta_t}, \tag{11.9}$$

$$D_b\big|_{i,j,k} = \frac{2\Delta_t}{2\mu_{i,j,k} + \sigma^*_{i,j,k}\Delta_t}, \tag{11.10}$$

In (11.9) and (11.10), σ^* is the magnetic conductivity, $\mu = \mu_0\mu_r$, μ_0 is the free-space permeability, and μ_r is the relative permeability. Note that the coefficients in the above update equations are assumed to be constant in time. For a problem involving complex media, such as dispersive materials, the material coefficients ε and μ can be approximated as frequency-independent over a frequency band of interest [7].

11.1.1.2 3D cylindrical and spherical coordinates

In this subsection, we derive the update equations for 3D cylindrical and spherical coordinate systems. The positions of the different field components for the 3D cylindrical coordinate system are shown in Figure 11.2. Considering (i,j,k) corresponding to (r, φ, z), the update equations for 3D cylindrical coordinates can be

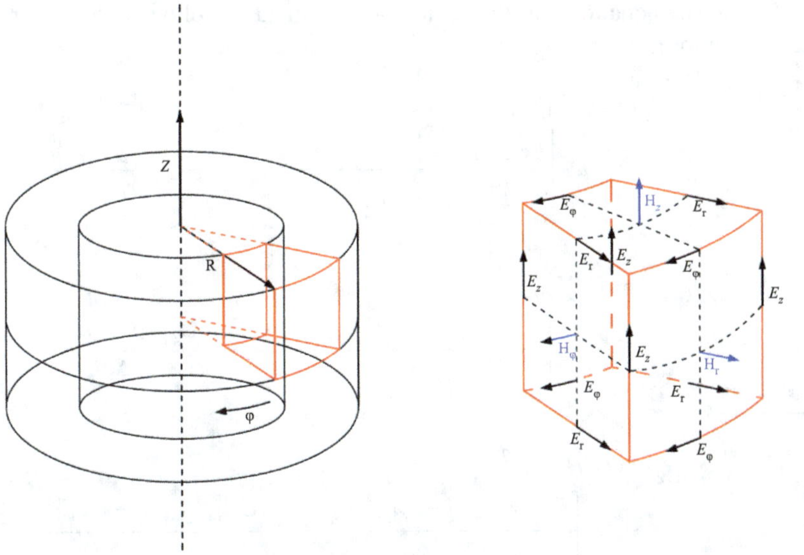

Figure 11.2 Positions of the FDTD field components in 3D cylindrical coordinates

obtained by applying the same discretization rules of 3D Cartesian coordinates [6,7]:

$$\begin{aligned} E_r|_{i+1/2,j,k}^{n+1} &= C_a E_r|_{i+1/2,j,k}^{n} \\ &+ C_b\left[\frac{H_z|_{i+1/2,j+1/2,k}^{n+1/2} - H_z|_{i+1/2,j-1/2,k}^{n+1/2}}{r_{i+1/2}\Delta_\varphi}\right. \\ &\left. - \frac{H_\varphi|_{i+1/2,j,k+1/2}^{n+1/2} - H_\varphi|_{i+1/2,j,k-1/2}^{n+1/2}}{\Delta_z}\right], \end{aligned} \tag{11.11}$$

$$\begin{aligned} E_z|_{i,j,k+1/2}^{n+1} &= C_a E_z|_{i,j,k+1/2}^{n} \\ &+ C_b\left[\frac{r_{i+1/2} H_\varphi|_{i+1/2,j,k+1/2}^{n+1/2} - r_{i-1/2} H_\varphi|_{i-1/2,j,k+1/2}^{n+1/2}}{r_i\Delta_r}\right. \\ &\left. - \frac{H_r|_{i,j+1/2,k+1/2}^{n+1/2} - H_r|_{i,j-1/2,k+1/2}^{n+1/2}}{r_i\Delta_\varphi}\right], \end{aligned} \tag{11.12}$$

$$\begin{aligned} E_\varphi|_{i,j+1/2,k}^{n+1} &= C_a E_\varphi|_{i,j+1/2,k}^{n} \\ &+ C_b\left[\frac{H_r|_{i,j+1/2,k+1/2}^{n+1/2} - H_r|_{i,j+1/2,k-1/2}^{n+1/2}}{\Delta_z}\right. \\ &\left. - \frac{H_z|_{i+1/2,j+1/2,k}^{n+1/2} - H_z|_{i-1/2,j+1/2,k}^{n+1/2}}{\Delta_r}\right], \end{aligned} \tag{11.13}$$

The update equations for the H-field are obtained by following a similar discretization process:

$$H_r\big|_{i,j+1/2,k+1/2}^{n+1/2} = D_a H_r\big|_{i,j+1/2,k+1/2}^{n-1/2} + D_b\left[\frac{E_\varphi\big|_{i,j+1/2,k+1}^{n} - E_\varphi\big|_{i,j+1/2,k}^{n}}{\Delta_z} - \frac{E_z\big|_{i,j+1,k+1/2}^{n} - E_z\big|_{i,j,k+1/2}^{n}}{r_i\Delta_\varphi}\right], \tag{11.14}$$

$$H_\varphi\big|_{i+1/2,j,k+1/2}^{n+1/2} = D_a H_\varphi\big|_{i+1/2,j,k+1/2}^{n-1/2} + D_b\left[\frac{E_z\big|_{i+1,j,k+1/2}^{n} - E_z\big|_{i,j,k+1/2}^{n}}{\Delta_r} - \frac{E_r\big|_{i+1/2,j,k+1}^{n} - E_r\big|_{i+1/2,j,k}^{n}}{\Delta_z}\right], \tag{11.15}$$

$$H_z\big|_{i+1/2,j+1/2,k}^{n+1/2} = D_a H_z\big|_{i+1/2,j+1/2,k}^{n-1/2} + D_b\left[\frac{E_r\big|_{i+1/2,j+1,k}^{n} - E_r\big|_{i+1/2,j,k}^{n}}{r_{i+1/2}\Delta_\varphi} - \frac{r_{i+1}E_\varphi\big|_{i+1,j+1/2,k}^{n} - r_i E_\varphi\big|_{i,j+1/2,k}^{n}}{r_{i+1/2}\Delta_r}\right], \tag{11.16}$$

where the coefficients C_a, C_b, D_a, and D_b can be defined in (11.4), (11.5), (11.9), and (11.10), respectively.

Note that a problem arises at $r = 0$: the update equations of the fields with a $1/r$ term will not work due to the division by zero. In this case, to update E_z on the z-axis, we apply Ampère's law to the field H_φ by taking its integral around a small circular region enclosing the axis at a radius $r = \Delta_r/2$ at the location $H_\varphi|_{1/2,j,k+1/2}$. As shown in Figure 11.3, we integrate H_φ along the curve C in a small circle of radius $r = \Delta_r/2$ in the φ direction:

$$E_z\big|_{0,j,k+1/2}^{n+1} = E_z\big|_{0,j,k+1/2}^{n} + \frac{4\Delta_t}{N_\varphi \varepsilon_0 \Delta_r}\sum_{k=1}^{N_\varphi} H_\varphi\big|_{1/2,j,k+1/2}^{n+1/2}. \tag{11.17}$$

where N_φ is the number of grid points in the φ direction.

By following the same principle shown in Figure 11.4, one can also find the FDTD update equations by considering (i, j, k) corresponding to (r, θ, φ) in 3D spherical coordinates [7–9]:

$$E_r\big|_{i+1/2,j,k}^{n+1} = C_a E_r\big|_{i+1/2,j,k}^{n} + C_b\left[\frac{\sin\theta_{j+1/2} H_\varphi\big|_{i+1/2,j+1/2,k}^{n+1/2} - \sin\theta_{j-1/2} H_\varphi\big|_{i+1/2,j-1/2,k}^{n+1/2}}{r_{i+1/2}\sin\theta_j \Delta_\theta} - \frac{H_\theta\big|_{i+1/2,j,k+1/2}^{n+1/2} - H_\theta\big|_{i+1/2,j,k-1/2}^{n+1/2}}{r_{i+1/2}\sin\theta_j \Delta_\varphi}\right], \tag{11.18}$$

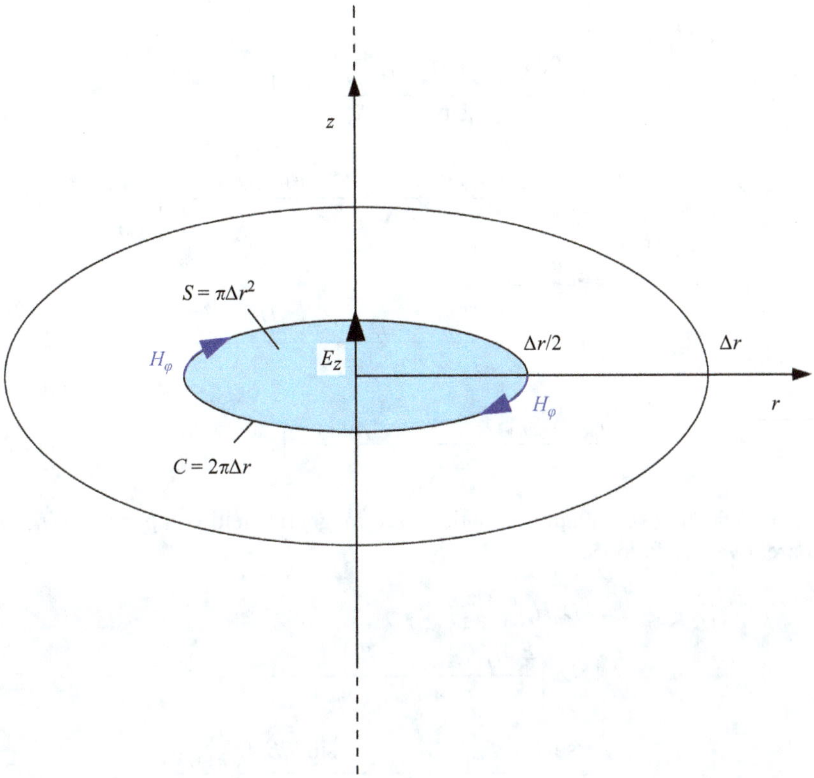

Figure 11.3 *Update E_z on the z-axis by applying Ampère's law. The marked blue circular surface enclosed by the curve C has a radius $r = \Delta_r/2$ and an area S*

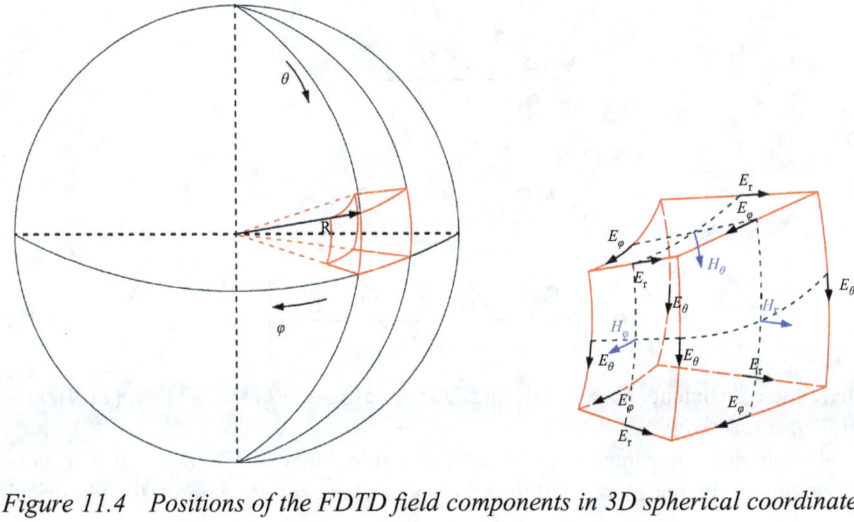

Figure 11.4 *Positions of the FDTD field components in 3D spherical coordinates*

$$
\begin{aligned}
E_\theta\big|_{i,j+1/2,k}^{n+1} &= C_a E_\theta\big|_{i,j+1/2,k}^{n} \\
&+ C_b\left[\frac{H_r\big|_{i,j+1/2,k+1/2}^{n+1/2} - H_r\big|_{i,j+1/2,k-1/2}^{n+1/2}}{r_i \sin\theta_{j+1/2}\Delta_\varphi}\right. \\
&\left.- \frac{r_{i+1/2}H_\varphi\big|_{i+1/2,j+1/2,k}^{n+1/2} - r_{i-1/2}H_\varphi\big|_{i-1/2,j+1/2,k}^{n+1/2}}{r_i\Delta_r}\right],
\end{aligned}
\tag{11.19}
$$

$$
\begin{aligned}
E_\varphi\big|_{i,j,k+1/2}^{n+1} &= C_a E_\varphi\big|_{i,j,k+1/2}^{n} \\
&+ C_b\left[\frac{r_{i+1/2}H_\theta\big|_{i+1/2,j,k+1/2}^{n+1/2} - r_{i-1/2}H_\theta\big|_{i-1/2,j,k+1/2}^{n+1/2}}{r_i\Delta_r}\right. \\
&\left.- \frac{H_r\big|_{i,j+1/2,k+1/2}^{n+1/2} - H_r\big|_{i,j-1/2,k+1/2}^{n+1/2}}{r_i\Delta_\theta}\right],
\end{aligned}
\tag{11.20}
$$

The field update equations for the three magnetic field components (H_r, H_θ, H_φ) are given as follows:

$$
\begin{aligned}
H_r\big|_{i,j+1/2,k+1/2}^{n+1/2} &= D_a H_r\big|_{i,j+1/2,k+1/2}^{n-1/2} \\
&+ D_b\left[\frac{E_\theta\big|_{i,j+1/2,k+1}^{n} - E_\theta\big|_{i,j+1/2,k}^{n}}{r_i \sin\theta_{j+1/2}\Delta_\varphi}\right. \\
&\left.- \frac{\sin\theta_{j+1}E_\varphi\big|_{i,j+1,k+1/2}^{n} - \sin\theta_j E_\varphi\big|_{i,j,k+1/2}^{n}}{r_i \sin\theta_{j+1/2}\Delta_\theta}\right],
\end{aligned}
\tag{11.21}
$$

$$
\begin{aligned}
H_\theta\big|_{i+1/2,j,k+1/2}^{n+1/2} &= D_a H_\theta\big|_{i+1/2,j,k+1/2}^{n-1/2} \\
&+ D_b\left[\frac{r_{i+1}E_\varphi\big|_{i+1,j,k+1/2}^{n} - r_i E_\varphi\big|_{i,j,k+1/2}^{n}}{r_{i+1/2}\Delta_r}\right. \\
&\left.- \frac{E_r\big|_{i+1/2,j,k+1}^{n} - E_r\big|_{i+1/2,j,k}^{n}}{r_{i+1/2}\sin\theta_j\Delta_\varphi}\right],
\end{aligned}
\tag{11.22}
$$

$$
\begin{aligned}
H_\varphi\big|_{i+1/2,j+1/2,k}^{n+1/2} &= D_a H_\varphi\big|_{i+1/2,j+1/2,k}^{n-1/2} \\
&+ D_b\left[\frac{E_r\big|_{i+1/2,j+1,k}^{n} - E_r\big|_{i+1/2,j,k}^{n}}{r_{i+1/2}\Delta_\theta}\right. \\
&\left.- \frac{r_{i+1}E_\theta\big|_{i+1,j+1/2,k}^{n} - r_i E_\theta\big|_{i,j+1/2,k}^{n}}{r_{i+1/2}\Delta_r}\right],
\end{aligned}
\tag{11.23}
$$

where the coefficients C_a, C_b, D_a, and D_b are defined in (11.4), (11.5), (11.9), and (11.10), respectively.

In spherical coordinates, we also have a singularity at $\theta = 0$ and $\theta = \pi$, corresponding to the north pole and the south pole of the sphere. Similar to cylindrical coordinates, the corrected update equations for the E_r components at $\theta = 0$ ($j = 1$ at the north pole) and $\theta = \pi$ ($j = N_\theta$ at the south pole, N_θ being the number of grid points in the θ direction) can be obtained by applying the Ampère's law contour integral around a small contour C enclosing the z-axis [7,9]:

$$\begin{aligned} E_r|_{i+1/2,j=1,k}^{n+1} &= E_r|_{i+1/2,j=1,k}^{n} \\ &+ \frac{\Delta_t \sin\theta_{1/2}}{2\pi\varepsilon_0 r_{i+1/2}(1-\cos\theta_{1/2})}\sum_{k=1}^{N_\varphi} H_\varphi|_{i+1/2,j=1,k}^{n+1/2}\Delta_\varphi, \end{aligned} \tag{11.24}$$

$$\begin{aligned} E_r|_{i+1/2,j=N_\theta,k}^{n+1} &= E_r|_{i+1/2,j=N_\theta,k}^{n} \\ &+ \frac{\Delta_t \sin\theta_{N_\theta-1/2}}{2\pi\varepsilon_0 r_{i+1/2}(1+\cos\theta_{N_\theta-1/2})}\sum_{k=1}^{N_\varphi} H_\varphi|_{i+1/2,j=N_\theta,k}^{n+1/2}\Delta_\varphi, \end{aligned} \tag{11.25}$$

where $\theta_{1/2} = \Delta_\theta/2$ and $\theta_{N_\theta-1/2} = \pi - \Delta_\theta/2$.

11.1.1.3 2D cylindrical and polar coordinates

In most LEMP simulations involving cloud-to-ground lightning flashes, the problem is symmetric around the vertical axis containing the electromagnetic source (a straight vertical channel), conventionally denoted as the z-axis. This allows us to reduce our problem to a 2D problem by means of cylindrical or polar symmetry, which vastly reduces the computation and storage required for the FDTD simulation. Generally, one can restrict the simulation to the TE_z mode and consider only the field components H_z, E_x, and E_y in 3D Cartesian coordinate systems and make use of the axial symmetry either in cylindrical or polar coordinates. Figure 11.5 shows the positions of E_r, E_z, and H_φ field components for 2D cylindrical coordinates. We begin with 2D cylindrical coordinates by assuming that the H_φ components are symmetric along the azimuthal φ direction ($\partial/\partial\varphi = 0$). Then, the TE mode of the previous 3D cylindrical update equations can be reduced and given by:

$$\begin{aligned} E_r|_{i+1/2,j}^{n+1} &= C_a E_r|_{i+1/2,j}^{n} \\ &- C_b\left[\frac{H_\varphi|_{i+1/2,j+1/2}^{n+1/2} - H_\varphi|_{i+1/2,j-1/2}^{n+1/2}}{\Delta_z}\right] \end{aligned} \tag{11.26}$$

$$\begin{aligned} E_z|_{i,j+1/2}^{n+1} &= C_a E_z|_{i,j+1/2}^{n} \\ &+ C_b\left[\frac{r_{i+1/2}H_\varphi|_{i+1/2,j+1/2}^{n+1/2} - r_{i-1/2}H_\varphi|_{i-1/2,j+1/2}^{n+1/2}}{r_i\Delta_r}\right] \end{aligned} \tag{11.27}$$

$$\begin{aligned} H_\varphi|_{i+1/2,j+1/2}^{n+1/2} &= D_a H_\varphi|_{i+1/2,j+1/2}^{n-1/2} \\ &+ D_b\left[\frac{E_z|_{i+1,j+1/2}^{n} - E_z|_{i,j+1/2}^{n}}{\Delta_r} - \frac{E_r|_{i+1/2,j+1}^{n} - E_r|_{i+1/2,j}^{n}}{\Delta_z}\right] \end{aligned} \tag{11.28}$$

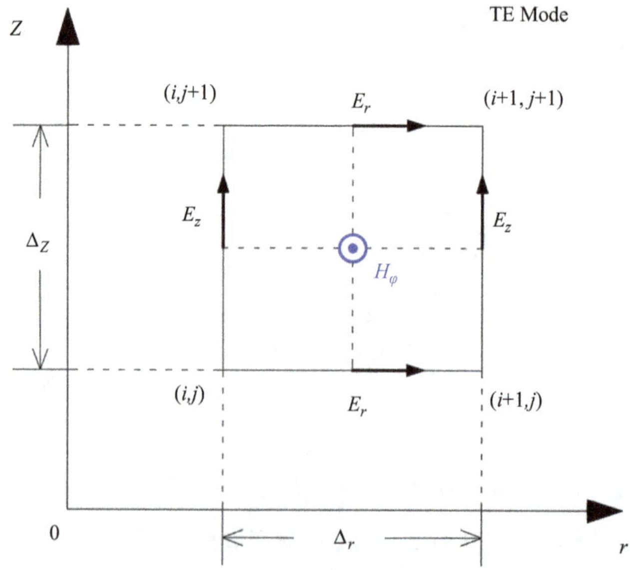

Figure 11.5 The positions of the E_r, E_z, and H_φ field components for 2D cylindrical coordinates. The H_φ components are symmetric along the azimuthal φ direction

where the coefficients C_a, C_b, D_a, and D_b are given by (11.4), (11.5), (11.9), and (11.10), respectively.

In order to deal with the singularity problems at $r = 0$ for 2D cylindrical coordinates, we assume that H_φ does not vary along a small circle around the z-axis with a radius $\Delta_r/2$. Thus, we have:

$$E_r|_{0,j+1/2}^{n+1} = E_r|_{0,j+1/2}^{n} + \frac{4\Delta_t}{\varepsilon_0 \Delta_r} H_\varphi|_{1/2,j+1/2}^{n+1/2} \tag{11.29}$$

Similarly, the update equations for 2D polar coordinates can be obtained considering symmetry along the azimuthal angle φ ($\partial/\partial\varphi = 0$) based on those in 3D spherical coordinates in Section 11.1.2. The positions of the E_r, E_θ, and H_φ field components for 2D polar coordinates are shown in Figure 11.6. The corresponding update equations are as below:

$$\begin{aligned} E_\theta|_{i+1/2,j}^{n+1} &= C_a E_\theta|_{i+1/2,j}^{n} \\ &- C_b \frac{r_{j+1/2} H_\varphi|_{i+1/2,j+1/2}^{n+1/2} - r_{j-1/2} H_\varphi|_{i+1/2,j-1/2}^{n+1/2}}{r_j \Delta_r} \end{aligned} \tag{11.30}$$

$$\begin{aligned} E_r|_{i,j+1/2}^{n+1} &= C_a E_r|_{i,j+1/2}^{n} \\ &+ C_b \frac{\sin\theta_{i+1/2} H_\varphi|_{i+1/2,j+1/2}^{n+1/2} - \sin\theta_{i-1/2} H_\varphi|_{i-1/2,j+1/2}^{n+1/2}}{r_{j+1/2} \sin\theta_i \Delta_\theta} \end{aligned} \tag{11.31}$$

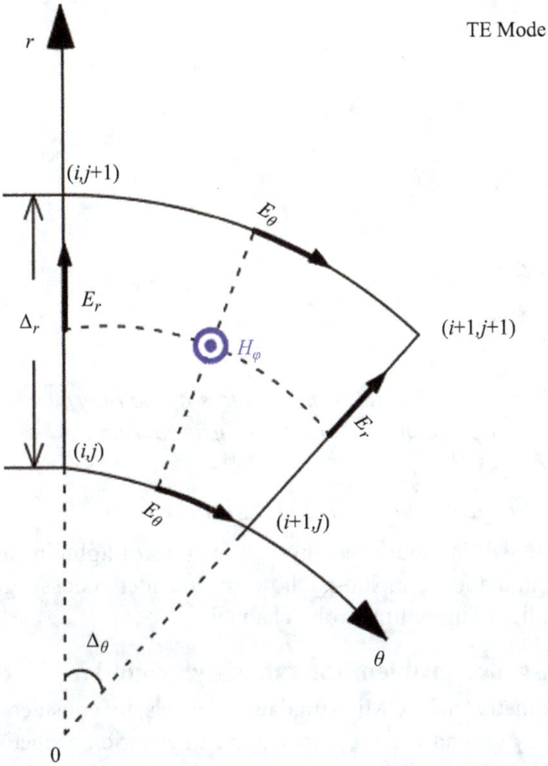

Figure 11.6 The positions of the E_r, E_θ, and H_φ field components for 2D polar coordinates. The H_φ component is symmetric along the azimuthal φ direction

$$H_\varphi|_{i+1/2,j+1/2}^{n+1/2} = D_a H_\varphi|_{i+1/2,j+1/2}^{n-1/2} + D_b \frac{E_r|_{i+1,j+1/2}^{n} - E_r|_{i,j+1/2}^{n}}{r_{j+1/2}\Delta_\theta} \tag{11.32}$$

where the coefficients C_a, C_b, D_a, and D_b are given by (11.4), (11.5), (11.9), and (11.10), respectively.

For the singularity at $\theta = 0$ in 2D polar coordinates, we assume that H_φ does not vary along a small circle around the r direction with radius $(r_{j+1/2}\Delta_\theta)/2$. Thus, we have:

$$E_r|_{0,j+1/2}^{n+1} = E_r|_{0,j+1/2}^{n} + \frac{4\Delta_t}{\varepsilon_0(r_{j+1/2}\Delta_\theta)} H_\varphi|_{1/2,j+1/2}^{n+1/2} \tag{11.33}$$

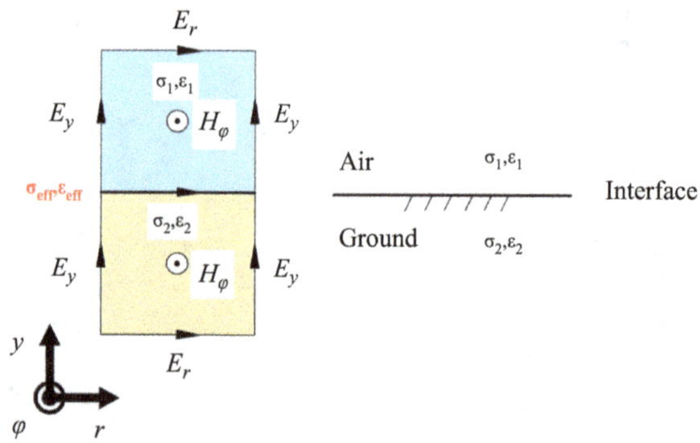

Figure 11.7 Illustration of material parameters at the interface between the air and lossy ground for 2D cylindrical coordinates

11.1.2 FDTD algorithm for LEMP simulations

In this section, we will introduce the physical features of lightning electromagnetic pulse (LEMP) simulations, including how to consider a lossy ground and the excitation of the lightning return stroke channel.

11.1.2.1 Half-space problem for the lossy ground

The general geometry for LEMP simulations needs to consider the half-space problem for a lossy ground with permittivity ε and electric conductivity σ, or more complex cases of the ground geometry such as those involving horizontal or vertical layers, and mountainous terrain profiles. In order to deal with the material at the interface between the air and the ground in FDTD modeling, as shown in Figure 11.7, the mean value of the permittivities and conductivities in the surrounding cells is taken as the effective permittivity ε_{eff} and effective electric conductivity σ_{eff} of the interface [6,10]:

$$\varepsilon_{eff} = \frac{1}{2}(\varepsilon_1 + \varepsilon_2) \tag{11.34}$$

$$\sigma_{eff} = \frac{1}{2}(\sigma_1 + \sigma_2) \tag{11.35}$$

Note that, in a single FDTD simulation, we can construct the material coefficients array according to the ground geometry and store it before the FDTD time-marching iteration starts, and then use it at the respective position while updating the E-field and H-field components.

11.1.2.2 Method of excitation of the lightning return stroke channel

The transmission-line (TL) return stroke model has been widely used for LEMP simulations. There are three different TL models proposed in the literature: (i) the TL model (TL) [11]; (ii) the modified TL model with linear current decay with height (MTLL) [12]; and (iii) the modified TL model with exponential current decay with height (MTLE) [13,14]. In the TL model, the temporal and spatial variation of the return stroke current as a function of the height z along the lightning channel $I(z,t)$, is given by $I(z,t) = I(0, t - z/v)$, $t \geq z/v$, where $I(0,t)$ is the current at ground level and v is the return stroke speed. According to the MTLL model, the return stroke current at any level along the channel is given by $I(z,t) = (1 - z/H)I(0, t - z/v)$, $t \geq z/v$, where H is the height of the return stroke channel. In the MTLE model, the spatial and temporal variation of the return stroke current is given by $I(z,t) = I(0, t - z/v)\exp(-z/\lambda)$, $t \geq z/v$, where $\lambda = 2$ km is a constant describing the current decay with height.

In order to include the TL model into the FDTD model, Baba and Rakov [15] proposed a method that uses a vertical phased current source array, which is valid for any return stroke speed v and does not require any unrealistic assumptions on the channel radius or on the source size. As shown in Figure 11.8, the entire lightning return stroke channel is represented by a vertical phased array of current sources that are progressively activated by an upward-moving return-stroke front with a speed v.

In the FDTD analysis, each current source of the lightning return stroke channel is located at the position of the electric field E_z surrounded by the magnetic

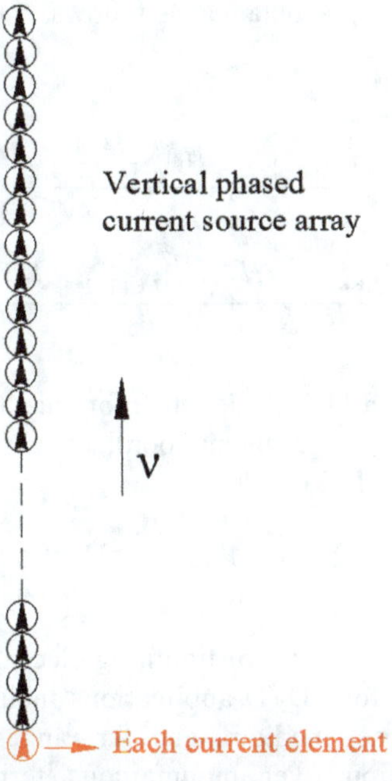

Figure 11.8 Vertical phased array of current sources that are progressively activated by an upward-moving return-stroke front with a speed v

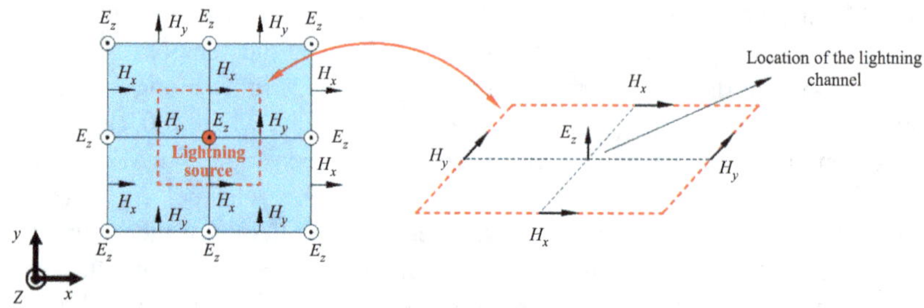

Figure 11.9 Each current source of the lightning return stroke channel in the FDTD procedure

fields H_y and H_x based on the integration path of Ampère's law, as shown in Figure 11.9. The equation based on Ampère's law can be written as:

$$\varepsilon_0 \frac{\partial \vec{E}}{\partial t} = \nabla \times \vec{H} - J_z, \tag{11.36}$$

where $J_z = I_z(0,t)/(\Delta_x \Delta_y)$ is the current density in the cell at which the current source is located. Then, the FDTD update equation with current source $I_z(0,t)$ in 3D Cartesian coordinates can be obtained as follows:

$$\begin{aligned} E_z\big|_{i,j,k+1/2}^{n+1} = {} & E_z\big|_{i,j,k+1/2}^{n} \\ & + \frac{\Delta_t}{\varepsilon_0}\left[\frac{H_y\big|_{i+1/2,j,k+1/2}^{n+1/2} - H_y\big|_{i-1/2,j,k+1/2}^{n+1/2}}{\Delta_x}\right. \\ & \left. - \frac{H_x\big|_{i,j+1/2,k+1/2}^{n+1/2} - H_x\big|_{i,j-1/2,k+1/2}^{n+1/2}}{\Delta_y} - \frac{I_z(0,t)\big|_{i,j,k+1/2}^{n+1/2}}{\Delta_x \Delta_y}\right]. \end{aligned} \tag{11.37}$$

As mentioned in Section 11.1, the iterative formula of the update equation near a current source is different for different coordinate systems in the FDTD model (see (11.17), (11.29), and (11.33)).

11.2 Application

In this section, we cover a number of lightning electromagnetic pulse (LEMP) topics of general interest for FDTD applications including (1) lightning electromagnetic fields at close, medium, and far ranges, (2) lightning electromagnetic field propagation over mountainous terrain, and (3) lightning electromagnetic field propagation in the earth-ionosphere waveguide (EIWG) and its interaction with the lower ionosphere.

11.2.1 Lightning electromagnetic fields at close, medium, and far ranges

In this subsection, we validate the FDTD model presented in Section 11.2 to calculate the electromagnetic fields from lightning at close, medium and far ranges and comparing the results against the well-known theoretical method proposed by Uman *et al.* [16]. In the calculation, we adopt the 2D cylindrical coordinate system in the FDTD model. The working space in the r and z directions is $20\ \text{km} \times 20\ \text{km}$, which is divided into square cells $\Delta_r \times \Delta_z = 2\ \text{m} \times 2\ \text{m}$. To avoid artificial reflections at the outer boundaries of the computational domain of FDTD model, we used Convolutional Perfectly Matched Layers (CPML) [17]. Note that the CPML absorbing boundary is used for the rest of FDTD simulations in this section (more details on absorbing boundary conditions (ABCs) for FDTD can be found in the box region below). The lightning channel is assumed to be straight and vertical above the ground with a height of $H = 8$ km. The current distribution along the lightning channel was specified according to the TL type model with exponential current decay (MTLE). As already mentioned in subsection 11.2.2, in the MTLE model, the temporal and spatial variation of the return stroke current $I(z,t)$ is given by $I(z,t) = I(0, t - z/v)\exp(-z/\lambda)$, $t \geq z/v$, where $I(0,t)$ is the current at ground level, $v = 1.5 \times 10^8$ m/s is the return stroke speed, and $\lambda = 2$ km is a constant describing the current decay with height [13]. Since Uman's equations adopted the method of images to solve the effect of the ground, for comparison, in our case, the ground is assumed to be perfectly conducting for the FDTD model as well.

The Heidler's functions are used to represent the first and subsequent return strokes:

$$\begin{aligned}
I(0,t) &= \frac{I_1}{\eta_1} \frac{(t/\tau_{11})^{n_1}}{\left[(t/\tau_{11})^{n_1} + 1\right]} e^{-t/\tau_{12}} + \frac{I_2}{\eta_2} \frac{(t/\tau_{21})^{n_2}}{\left[(t/\tau_{21})^{n_2} + 1\right]} e^{-t/\tau_{22}}, \\
\eta_1 &= \exp\left[-\frac{\tau_{11}}{\tau_{12}}\left(n_1 \frac{\tau_{12}}{\tau_{11}}\right)\right], \\
\eta_2 &= \exp\left[-\frac{\tau_{21}}{\tau_{22}}\left(n_2 \frac{\tau_{22}}{\tau_{11}}\right)\right].
\end{aligned} \tag{11.38}$$

As shown in Figure 11.10, two lightning current waveforms are considered corresponding to the typical first and subsequent return strokes with the parameters given in Table 11.1 [18]. Figure 11.11 gives the comparison results for the vertical electric fields between the FDTD method and Uman's theoretical method at distances ranging from 1 km to 10 km. It is noted that the results of the FDTD model match perfectly with those from Uman's equation at different distances for both first and subsequent return strokes.

Absorbing boundary conditions (ABCs)

A problem arises at the outer boundaries of an FDTD model due to the finite nature of any computation domain. The goal of ABCs is to approximately estimate the fields outside the finite simulation space to avoid unphysical reflections. The implementation of ABCs plays a very important role in the performance and

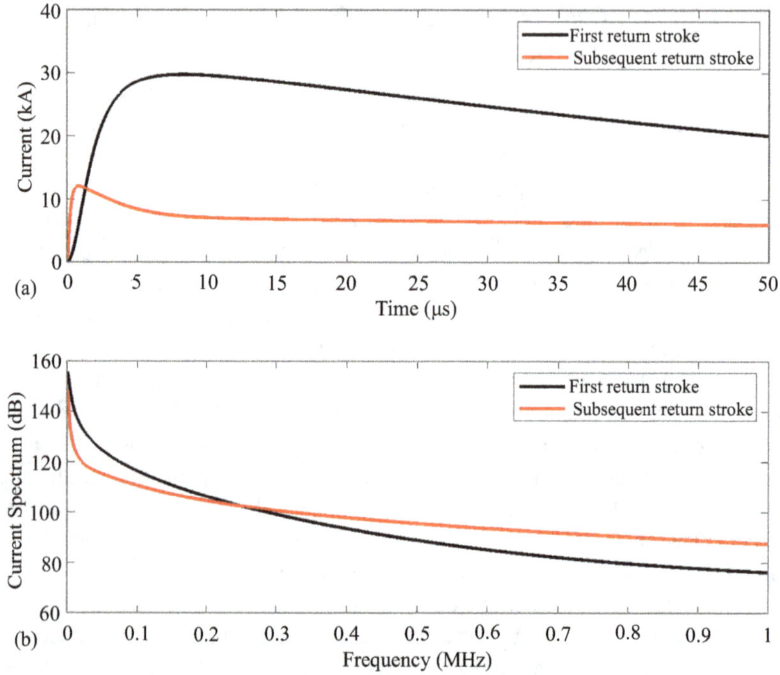

Figure 11.10 The current waveforms of the typical first and subsequent return strokes (top panel) and their frequency spectrum (bottom panel)

Table 11.1 Parameters for the typical first and subsequent return stroke currents

	I_1 (kA)	τ_{11} (μs)	τ_{12} (μs)	n_1	I_2 (kA)	τ_{21} (μs)	τ_{22} (μs)	n_2
First	28	1.8	95	2.0	–	–	–	–
Subsequent	10.7	0.25	2.5	2.0	6.5	2.0	230	2.0

accuracy of FDTD methods. There are different types of ABCs proposed in the literature: (i) ABCs based on the radiation wave equation, such as the first-order/second-order Mur boundary, or higher-order Mur boundary; (ii) ABCs based on an absorbing material/medium, such as perfectly matched layer (PML), uniaxial perfectly matched layer (UPML) and the CPML. Further details about ABCs for the FDTD method can be found in FDTD textbooks, such as [6,7].

Compared with Uman's equation, the FDTD method is more flexible to simulate complex geometries of the ground, including non-linearity, inhomogeneity and anisotropy, as in the case of a homogeneous lossy ground, horizontally stratified ground and vertically stratified ground, a mountainous terrain profile and the effect of the

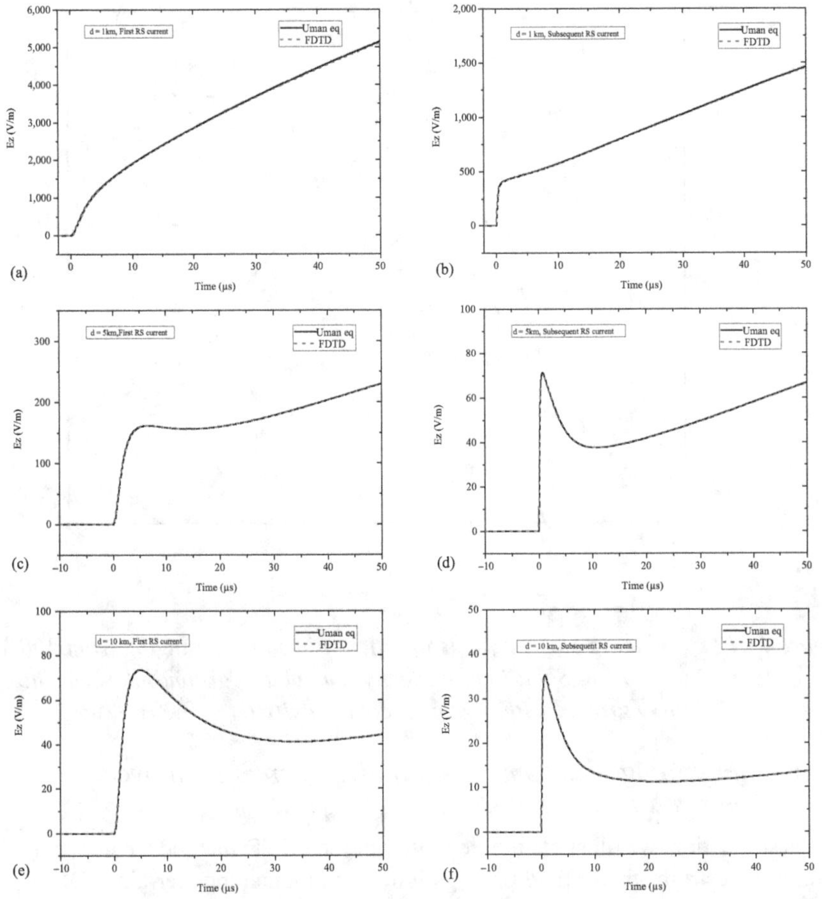

Figure 11.11 Comparison of FDTD modeling results with the theoretical Uman's method at different distances

Earth's curvature. Figure 11.12 shows an example of the vertical electric field from lightning at distances ranging from 100 km to 500 km by considering the effect of the ground conductivity and the earth curvature. In this case, 2D polar coordinates are used to include the effect of the Earth's curvature. The lightning current is simulated by using a bi-Gaussian function $I(t) = I_0(e^{-t^2/\tau_1^2} - e^{-t^2/\tau_2^2})$, where $I_0 = 100$ kA, the rise time $\tau_1 = 10$ μs and the fall time $\tau_2 = 100$ μs. The current distribution along the channel is specified according to the MTLE model. The ground is assumed to be a homogeneous lossy ground with a ground conductivity $\sigma_g = 0.001$ S/m and a permittivity $\varepsilon_{rg} = 10$. For comparison, the results corresponding to a perfectly conducting ground ($\sigma_g = \infty$) are also shown in the same figure.

The finite ground conductivity affects essentially the early time response of lightning-radiated fields by attenuating its peak and increasing its risetime. More details on more complex cases, such as a horizontally layered ground and a mixed ocean-land path can be found in [19–22] and the references quoted there.

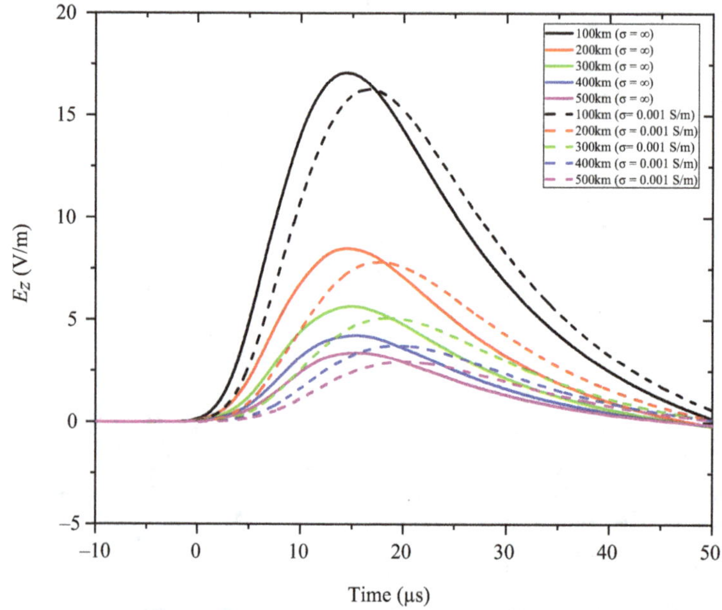

Figure 11.12 Lightning electric fields at different distances ranging from 100 km to 500 km. Solid line: perfectly conducting ground; dashed line: lossy ground with a ground conductivity $\sigma_g = 0.001$ S/m

11.2.2 Lightning electromagnetic fields propagation over mountainous terrain

In this subsection, we illustrate the flexibility of the FDTD method by analyzing the lightning electromagnetic field propagation over mountainous terrain. 2D and 3D FDTD models are developed that include irregular terrain based on the global digital elevation model version 2 (GDEM V2) data*. As shown in Figure 11.13, we focus on the region around the Säntis Tower located in the Swiss Alps. The 124-m Säntis Tower has been instrumented since 2010 and it serves as an experimental station for the direct measurement of lightning currents [23–25]. The electric fields will be calculated at a total of six measurement stations, including four stations from European Cooperation for Lightning Detection (EUCLID) network (labeled S_1 to S_4), two stations named Herisau and Neudorf in Figure 11.13.

11.2.2.1 2D symmetric cylindrical FDTD model

In order to save computational cost with sufficient accuracy, we use a 2D symmetric cylindrical FDTD model by considering the 2D cross-section of the

*The Advanced Spaceborne Thermal Emission and Reflection Radiometer (ASTER) Global Digital Elevation Model Version 2 (GDEM V2) product is available for any user pursuant to an agreement with the Ministry of Economy, Trade and Industry of Japan (METI) and the National Aeronautics and Space Administration (NASA), which is available at https://asterweb.jpl.nasa.gov/gdem.asp

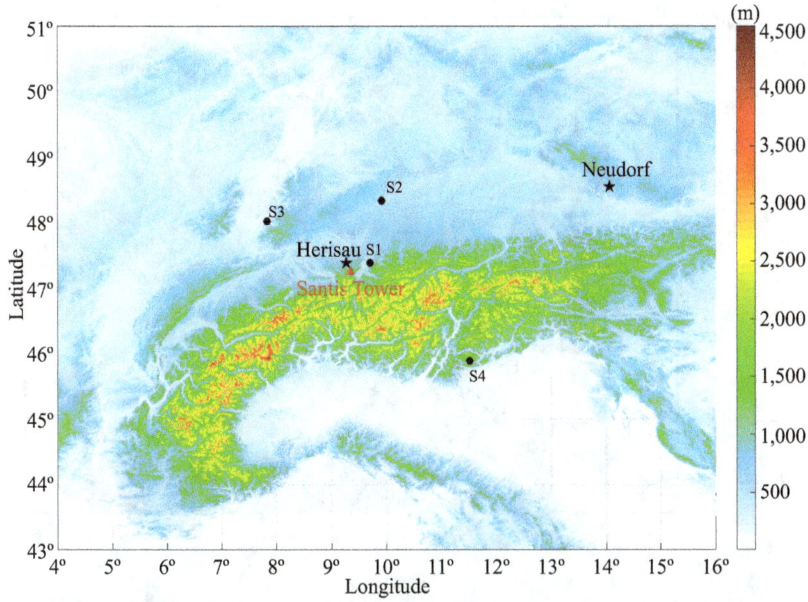

Figure 11.13 The topographic map of the region around the Säntis Tower (red triangle) including the locations of four E field sensors (S1, S2, S3, and S4, black dots) and the Herisau and Neudorf stations (black stars)

topographic map along the direct path between the Säntis Tower and the Herisau station in Switzerland (see Figure 11.14).

For comparison, two terrain profiles are considered in the FDTD model, namely the irregular ground model based on the topographic map shown in Figure 11.14, and a flat ground. Experimental observations consist of simultaneous records of lightning return stroke currents measured at the Säntis Tower and the associated electric fields measured at the Herisau station, about 15 km away from the tower. The measured channel-base current and a fitted version using genetic algorithms are presented in Figure 11.15(a). As shown in Figure 11.15(b), after considering the real irregular terrain between the Säntis Tower and the measurement station, both the waveshape and the amplitude of the simulated electric fields are in excellent agreement with the measured waveform. On the other hand, the assumption of a flat ground results in a significant underestimation of the amplitude of the electric field.

11.2.2.2 3D Cartesian FDTD model

Since a 3D FDTD simulation of the full domain would require prohibitively high computational resources, in this subsection, we consider a reduced 3D FDTD model and we compare it with the 2D symmetric model presented in the previous section. The 3D model utilizes a reduced version of the full 3D space consisting of

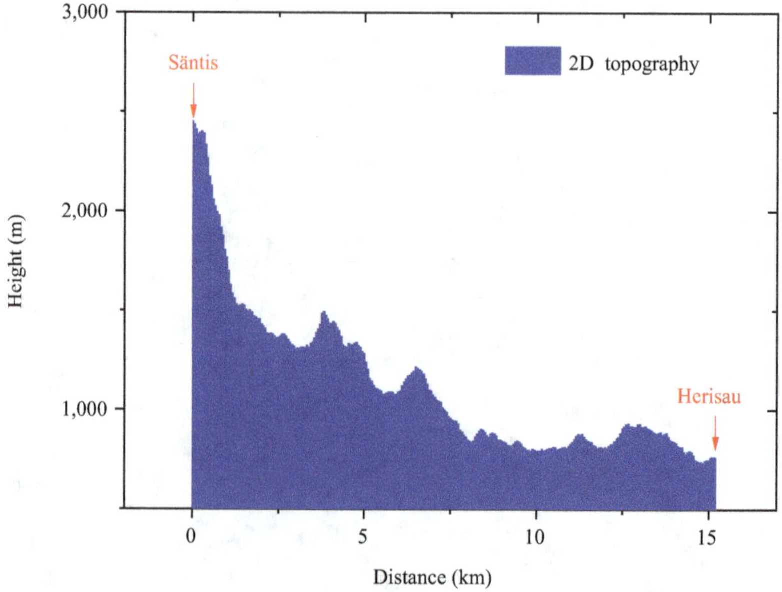

Figure 11.14 Cross-section of the topographic map along the direct path between the Säntis Tower and the Herisau E-field measurement station

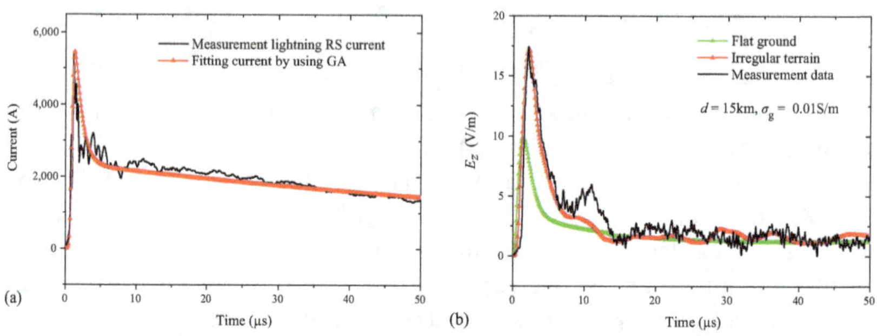

Figure 11.15 (a) The lightning return stroke currents measured at the Säntis Tower (black solid line) and its fitted current based on Heidler's functions (Red symbol line). (b) The associated electric fields measured at the Herisau station (black solid line), the FDTD simulated waveforms assuming a flat ground (green symbol line) and FDTD simulated waveforms taking into account the terrain profile (red symbol line). More details can be found in [26]

a 20-km wide rectangular area between the lightning at the Säntis Tower and the field point (see Figure 11.16).

We use four field points corresponding to lightning location system sensors S1, S2, S3, and S4 belonging to the European Cooperation for Lightning Detection

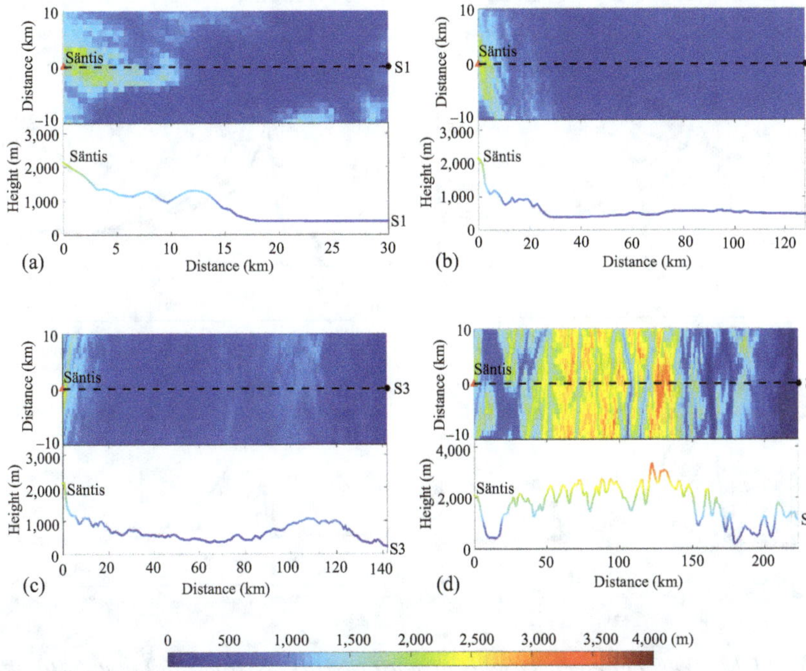

Figure 11.16 The reduced 3D rectangular region (upper) and the 2D cross-section (bottom) of the topographic profiles along the direct path between the Säntis Tower and the observation sensor site S1 (a), S2 (b), S3 (c), and S4 (d)

(EUCLID) network (see Figures 11.13 and 11.16). For comparison, as shown in Figure 11.17, we consider four different conditions for the ground: (i) perfectly conducting flat ground, (ii) finitely conducting flat ground, (iii) finitely conducting ground taking into account the 2D cross-section terrain profile, and (iv) finitely conducting ground taking into account the reduced 3D terrain profile. Note that cases (i), (ii), and (iv) are calculated using the 3D FDTD method, while the results corresponding to case (iii) are computed by way of the 2D FDTD approach presented in Section 11.2.2.1. It is found that the peak and the rise times of vertical electric fields can be strongly affected by the presence of mountainous terrain and the finite ground conductivity. It is interesting to note that the results obtained using the approximate 2D approach are very similar to those obtained using the 3D FDTD model for the cases corresponding to sensors S1, S2, and S3. However, for sensor S4, the results obtained using the 2D FDTD approach are significantly different from those obtained using the 3D approach. This is due, on the one hand, to the fact that the propagation distance to sensor 4 is much longer than for the other sensors and, on the other hand, to the complex terrain profile along the path between the Säntis Tower and the S4 sensor site (see Figure 11.16(d)).

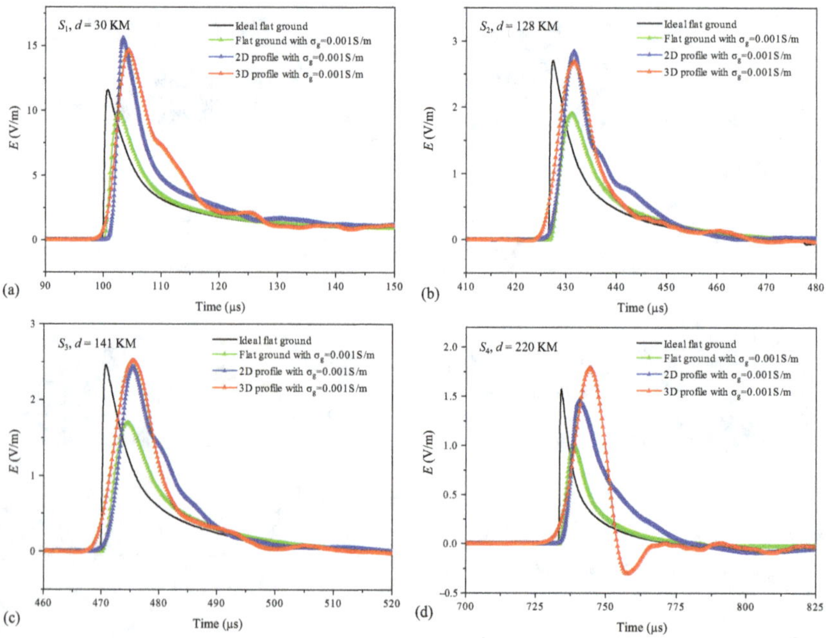

Figure 11.17 Vertical electric fields at sensor S1(a) , S2(b), S3(c), and S4(d) considering four different conditions for the ground corresponding to the cases from (i) to (iv). More details can be found in [5]

A more detailed theoretical analysis of the propagation effects of lightning electromagnetic fields over a mountainous terrain can be found in [5,26,27].

11.2.3 Lightning electromagnetic field propagation in the EIWG and its interaction with the lower D-region ionosphere

Very low-frequency (VLF)/low-frequency (LF) waves generated by lightning discharges are well known to constitute an efficient tool to probe localized perturbations of the ionospheric D region (60–90 km) [28,29]. In order to add the ionospheric characteristics of the D region, the FDTD model, as suggested by Marshall [4], needs to solve three fundamental equations including Maxwell's equations and a modified Ohm equation (sometimes called Langevin equation). In the simulation, we only consider the electron mobility, since ion motion can be neglected in the lower D region ionosphere within our timescale of interest [30]. The resulting equations are

$$\varepsilon_0 \frac{\partial \vec{E}}{\partial t} = \nabla \times \vec{H} - J_e - J_s, \tag{11.39}$$

$$\mu_0 \frac{\partial \vec{H}}{\partial t} = -\nabla \times \vec{E}, \tag{11.40}$$

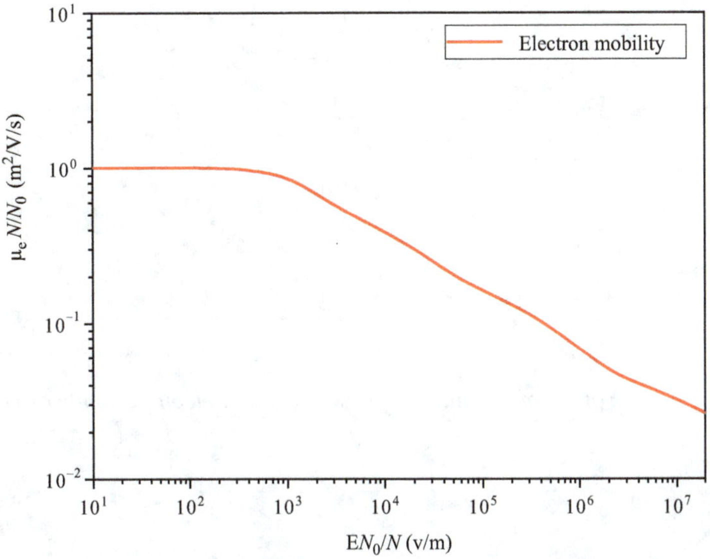

Figure 11.18 Electron mobility as a function of the local reduced electric field, N_0 and N are the air densities at a given altitude and at sea level

$$\frac{\partial \vec{J}_e}{\partial t} + \nu_e \vec{J}_e = \vec{\omega}_{c,e} \times \vec{J}_e + \omega_{p,e}^2 \varepsilon_0 \vec{E}, \tag{11.41}$$

where J_s is the current density of the source, J_e is the self-consistent conduction current driven by the electric field, $\vec{\omega}_{c,e} = e\vec{B}_0/m_e$ is the gyrofrequency vector associated with the geomagnetic field vector $\vec{B}_0$, $\omega_{p,e} = \sqrt{e^2 N_e/m_e \varepsilon_0}$ is the plasma frequency, $\nu_e = e/(\mu_e m_e)$ is the collision frequency between electrons and the neutral air, and e, m_e, and N_e are, respectively, the charge, mass, and number density of electrons. The electron mobility μ_e is considered as a function of the local electric field based on the profile proposed by [31] (see Figure 11.18). Since the electric fields scale as the number dentisity N of the gas, the unity Towsend or T_d has been introduced in Figure 11.18 for the local reduced electric field E/N, where $1\mathrm{T_d} = 1 \times 10^{-17}\mathrm{Vcm^2}$ [32]. The background gas number density N_{gas} is taken from the MSIS-E-90 model [33] by assuming a mixed air with 78% N_2 and 22% O_2 (see Figure 11.19(a)).

11.2.3.1 2D symmetric polar FDTD model

In this subsection, we present a 2D symmetric polar FDTD model including the effect of the ionospheric cold plasma characteristics, the effect of the earth curvature, and the propagation effects over a mountainous terrain. The simulation results are validated against simultaneous experimental data obtained from the Säntis Tower in Switzerland and the 380-km distance electric field station at Neudorf in Northern Austria (see Figure 11.16). The 2D cross-section of the topographic

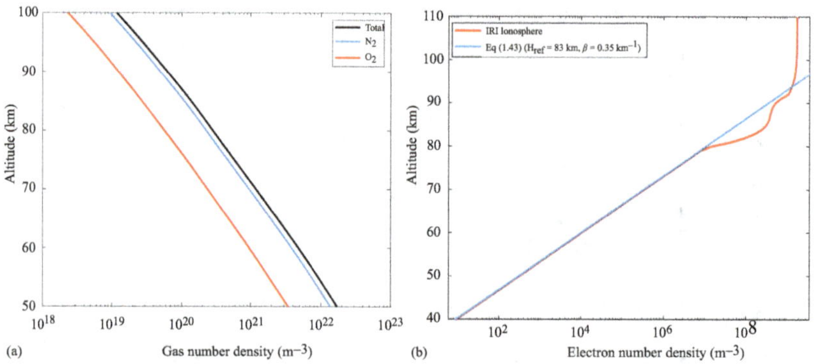

Figure 11.19 Profiles of gas number density (a) and electron number density (b)

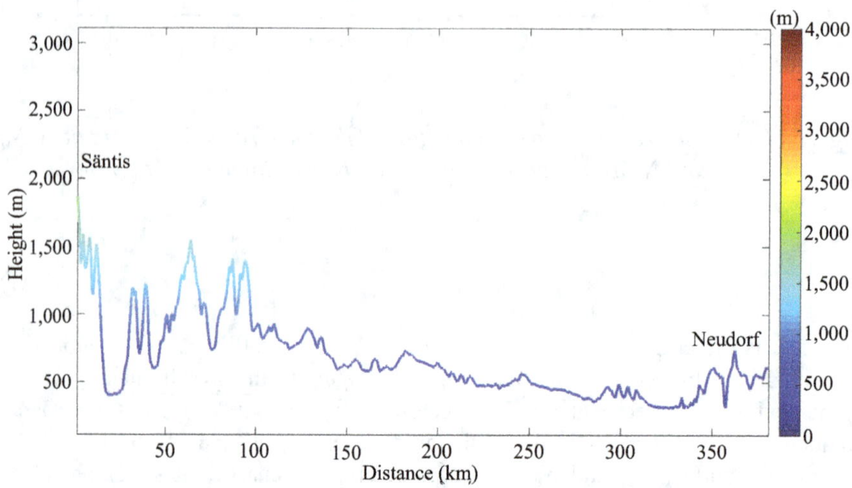

Figure 11.20 2D cross-section of the topographic profile along the direct path between the Säntis Tower and the 380-km distant Neudorf station

profile along the direct path between the Säntis Tower and the 380-km distant Neudorf station is given in Figure 11.20.

The effect of the geomagnetic field is neglected due to the 2D symmetrical conditions in the FDTD model. With this approximation, (11.41) reduces to

$$\frac{\partial \vec{J}_e}{\partial t} + \nu_e \vec{J}_e = \omega_{p,e}^2 \varepsilon_0 \vec{E}. \tag{11.42}$$

The profile of the electron number density is assumed to be exponential given by [34]

$$N_e = n_e^{ref} e^{\beta(z - H_{ref})}, \tag{11.43}$$

where β is the steepness of the profile and $n_e^{ref} = 1 \times 10^9 \mathrm{m}^{-3}$ is the electron density where most of the VLF energy is reflected at the effective reflected height H_{ref}. As shown in Figure 11.19(b), the exponential profile of electron number density proposed by Wait and Spies [34] gives a reasonable approximation for the electron density profile below 90 km compared to the International Reference Ionosphere (IRI) model [35].

Figure 11.21 (a,c,e) shows three measured lightning return stroke currents obtained from the Säntis Tower. Note that the return stroke and an M component† in the current are marked as RS and M, respectively. Figure 11.21(b,d,f) further shows the comparison between the measurement and the FDTD model. It can be seen that the vertical electric fields calculated using the FDTD model are in good agreement with the measurements obtained from the 380-km sensor in Neudorf for all the considered cases. It is further shown that the time delays and amplitudes of the lightning-radiated electromagnetic fields could be affected by the medium parameters in the EIWG including the effect of the ionospheric cold plasma characteristics, the effect of the Earth curvature, and the propagation effects over mountainous terrain. More details about the propagation effects of lightning electromagnetic fields over mountainous terrain in the EIWG can be found in [37].

11.2.3.2 3D spherical FDTD model

In this subsection, we extend the previous 2D polar FDTD model to a 3D spherical FDTD model to investigate the nonlinear effect of the ionospheric cold plasma characteristics, including the ionization, attachment and detachment, and optical emissions. Comparing to the 2D FDTD model, the 3D FDTD model allows to quantify the effect of the geomagnetic field in the lower ionosphere and to investigate the propagation of the EMP pulse of the lightning source at any position with any inclination in the EIWG. The nonlinear effect of the ionosphere is based on the rate coefficients used for ionization, attachment, and detachment shown in Figure 11.22, which are calculated based on the Boltzmann solver BOLSIG+‡ [38].

Equations (11.39)–(11.41) must be coupled with the balance equations for the electron and negative ion densities, N_e and N_{O^-}, which are affected by ionization, attachment and detachment rates [4,39]:

$$\frac{\partial N_e}{\partial t} = (\nu_i - \nu_a)N_e + \nu_d N_{O^-}, \tag{11.44}$$

$$\frac{\partial N_{O^-}}{\partial t} = \nu_a N_e - \nu_d N_{O^-}, \tag{11.45}$$

†An M component refers to a temporary increase of luminosity in a faintly luminous lightning channel, it may occur within microseconds or up to a few milliseconds of a return stroke. More details can be found in [36].

‡BOLSIG+ uses a two-term expansion of the electron distribution function in spherical harmonics to solve the steady-state Boltzmann equation for electrons in weakly ionized gases under uniform electric fields in the bulk of collisional low-temperature plasmas. More detail for BOLSIG+ can be found in the website https://nl.lxcat.net/solvers/BOLSIG+/.

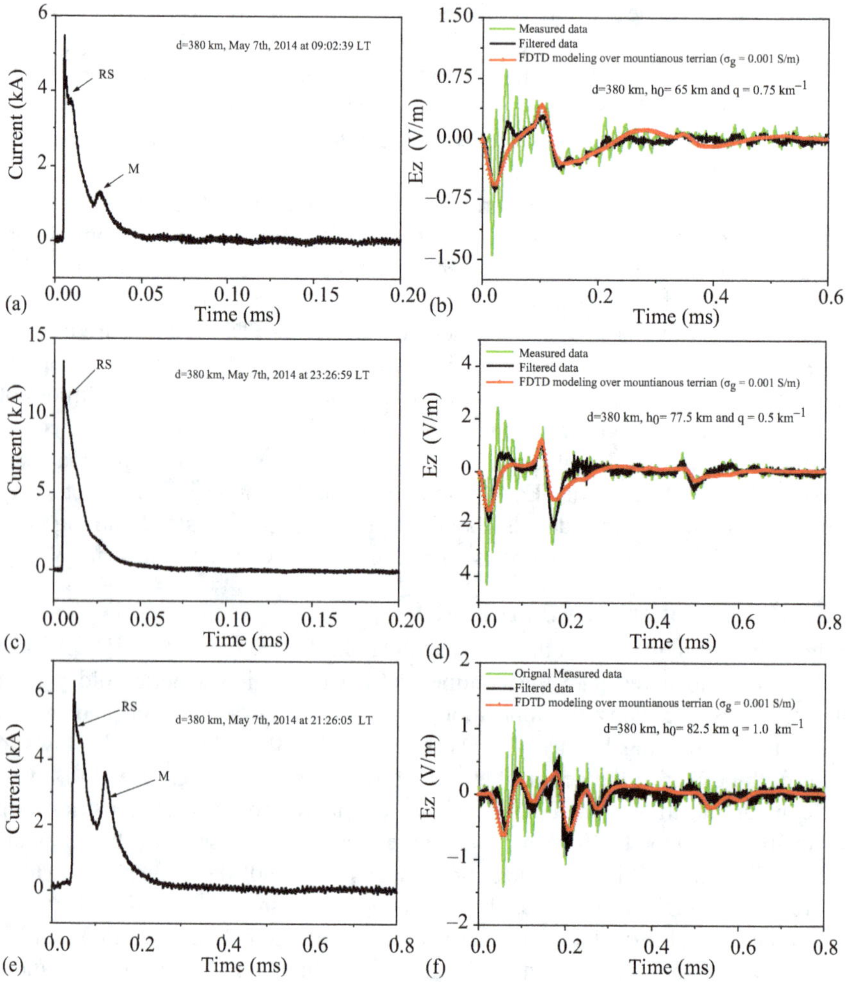

Figure 11.21 Measured lightning return stroke current from the Säntis Tower (a,c,e) and the comparison between the simultaneous measurement and the FDTD modeling results at the 380-km Neudorf station (b,d,f). More details can be found in [37]

where ν_i, ν_a, and ν_d are the ionization, attachment, and detachment rates, respectively. As shown in Figure 11.22, the rates of the ionization, attachment, and detachment will be updated by following the values of the local electric fields during the FDTD calculation.

11.2.3.3 Optical emission model of elves

In this subsection, the 3D spherical FDTD model is used to simulate one of the most common Transient Luminous Events (TLEs) – elves ("Emission of Light and VLF perturbations due to Electromagnetic Pulse Sources"). Elves are caused by the

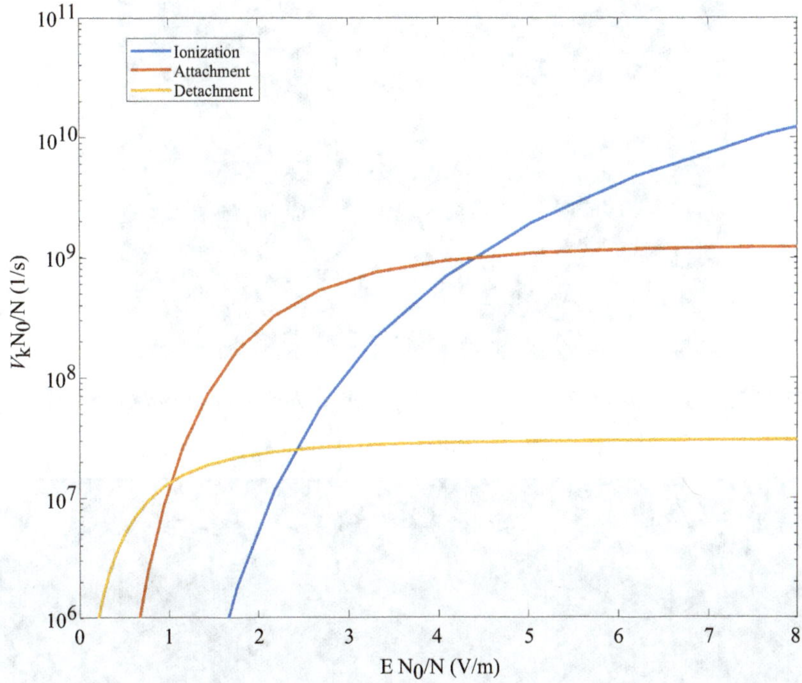

Figure 11.22 The ionization, attachment, and detachment rates as a function of the local reduced electric field, N_0 and N are the air densities at a given altitude and at sea level

propagation of lightning EMPs (LEMP) and their interaction with the lower ionosphere. When a LEMP reaches the lower ionosphere, the electrons there will be energized and produce molecular excitation and optical emissions, which are known as elves. Elves normally occur over thunderstorms at around 80–90 km above the ground and typically present a ring shape expanding to around 200 km in the radial direction and lasting about 1 ms. Figure 11.23 gives an example of ring-shaped optical emissions of an elve and its parent lightning observed by the Imager for Sprites and Upper Atmospheric Lightning (ISUAL) on the Taiwanese FORMOSAT-2 spacecraft [40]. Figure 11.24 shows another example of elve observation in the southeast of the Bay of Biscay, captured by using a ground-based camera.

In the optical emission model, we consider four main emission band systems of elves. They are the first positive ($1PN_2$) and second positive ($2PN_2$) band systems of N_2, the Lyman–Birge–Hopfield (LBH) band system of N_2 and the first negative band system of N_2^+ ($1NN_2^+$). The optical excitation rate corresponding to each emission band system is shown in Figure 11.25, which can be obtained by using the previously mentioned Boltzmann solver BOLSIG+. In order to determine the optical emissions, we need to calculate the number of molecules n_k in the excited

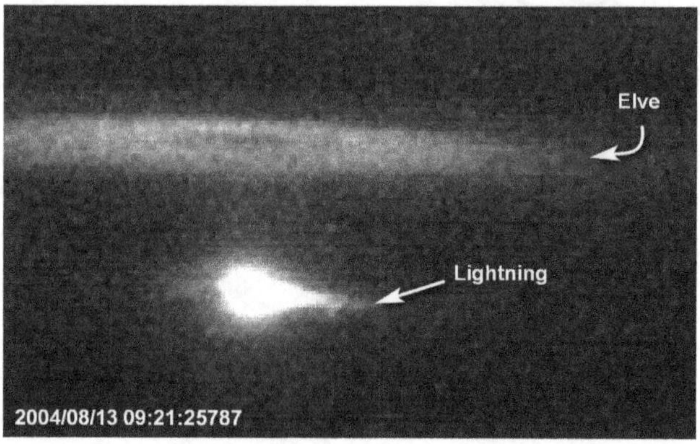

Figure 11.23 An example of an elve and its parent lightning observed by ISUAL [41]

Figure 11.24 An example of an elve captured by Oscar van der Velde on January 16, 2015 at southeastern Bay of Biscay (more TLE photos can be found at http://lightningwizard.com/)

state k corresponding to each emission band system [42]:

$$\frac{\partial n_k}{\partial t} = \nu_k N_e - \frac{n_k}{\tau_k} + \sum_m n_m A_m, \tag{11.46}$$

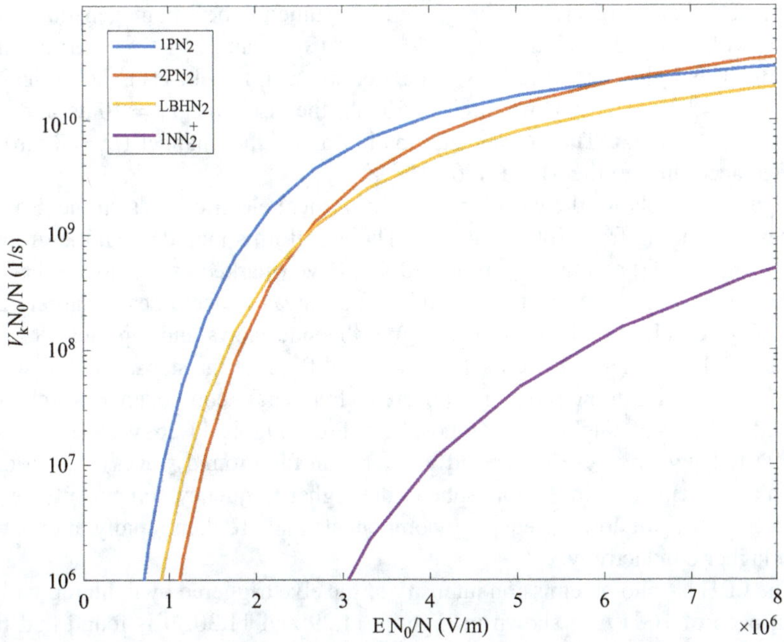

Figure 11.25 The optical excitation rates as a function of the local electric fields for the main four emission band systems of elves

Table 11.2 The values of A_k, α_1, and α_2 used in the optical emission band systems

Emission band system	Excited states	A_k (1/s)	α_1 (m^3/s)	α_2 (m^3/s)
$1PN_2$	$N_2(B^3\Pi_g)$	1.7×10^4	1.6×10^{-17}	1.5×10^{-16}
$2PN_2$	$N_2(C^3\Pi_u)$	2.0×10^7	1.12×10^{-17}	2.85×10^{-16}
$LBHN_2$	$N_2(a^1\Pi_g)$	1.8×10^4	2.2×10^{-17}	4.3×10^{-16}
$1NN_2^+$	$N_2^+(B^2\Pi_u^+)$	1.6×10^7	4.53×10^{-16}	7.36×10^{-16}

where $\tau_k = (A_k + \alpha_1 N_{N_2} + \alpha_2 N_{O_2})^{-1}$ is the total lifetime of excited state k, α_1, and α_2 are the quenching rates due to the collisions between N_2 and O_2 molecules, respectively. N_{N_2} and N_{O_2} are the gas number densities of N_2 and O_2 molecules. The sum term $\sum_m n_m A_m$ represents the increase of n_k due to cascading from higher states m with Einstein coefficients A_m. Table 11.2 shows the values of A_k, α_1, and α_2 used in the four considered emission band systems [41].

The photon emission intensity I_k, with units of photons per unit volume per unit time (e.g., ph/cm^3/s) is:

$$I_k = A_k n_k. \tag{11.47}$$

The initial profile of electron number density N_e is taken from the IRI 2007 model (see Figure 11.19). The ground is assumed to be a homogeneous lossy ground with ground conductivity $\sigma_g = 0.001$ S/m and relative permittivity $\varepsilon_{rg} = 10$. The lightning current is simulated by using a bi-Gaussian function $I(t) = I_0(e^{-t^2/\tau_1^2} - e^{-t^2/\tau_2^2})$, where $I_0 = 50$ kA, the rise time $\tau_1 = 10$ μs and the fall time $\tau_2 = 100$ μs. The current distribution along the channel ($H = 8$ km) is specified according to the MTLE model.

Figure 11.26 shows the waveforms of the vertical electric fields on the ground at distances ranging from 100 to 400 km. The waveforms include both the ground wave (marked as G) and the first reflected sky wave (marked as S). Notice that, as expected, the delay between the ground and sky waves decreases at larger distances. Figures 11.27 and 11.28 further give the side views and top views of the normalized electric fields in the FDTD model at different time steps. The full-wave EMP from the lightning contains different frequency components which are reflected at different heights in the ionosphere. From Figure 11.26, we also see that the reflected skywaves are wider and weaker than the ground waves. Due to dispersion and attenuation in the ionosphere, the higher frequency components penetrate deeper than the lower frequency components and are then rapidly attenuated by the higher conductivity.

The LBH N_2 photon emission intensity of the elve triggered by lightning with a current peak of 100 kA is shown in Figures 11.29 and 11.30. It is found that the

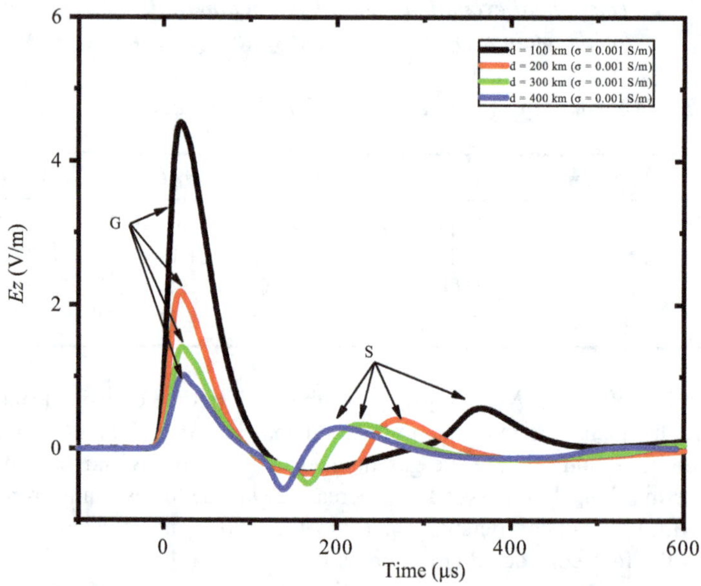

Figure 11.26 The vertical electric fields on the ground at distances ranging from 100 to 400 km. Ground waves and reflected sky waves are marked as G and S, respectively

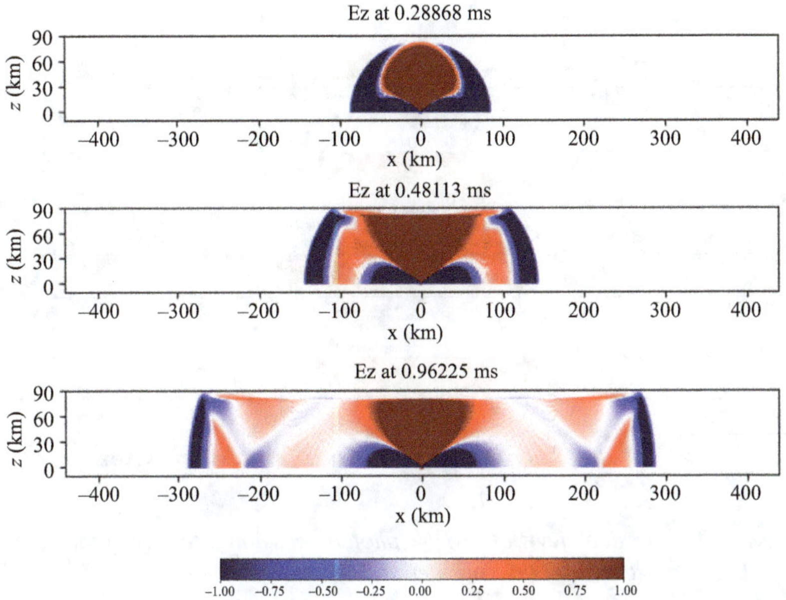

Figure 11.27 The side views of the normalized electric fields for FDTD model at different time steps

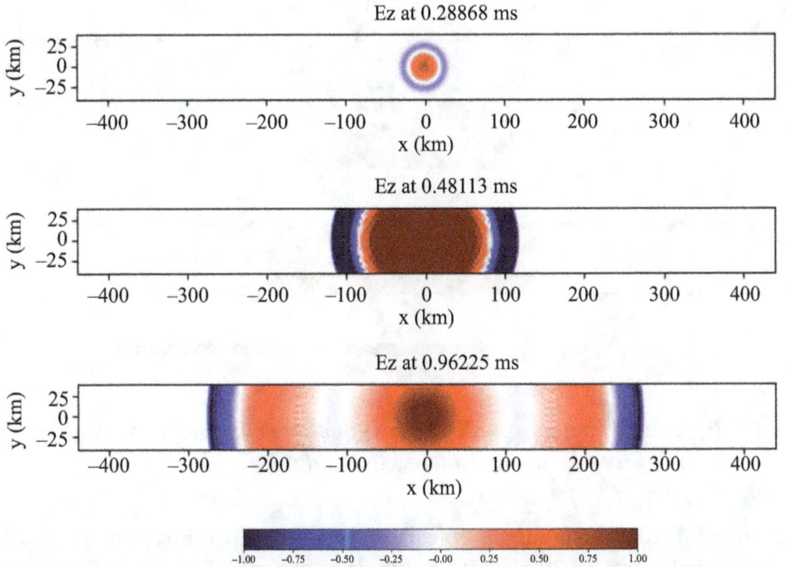

Figure 11.28 The 80 km top views of the normalized electric fields for FDTD model at different time steps

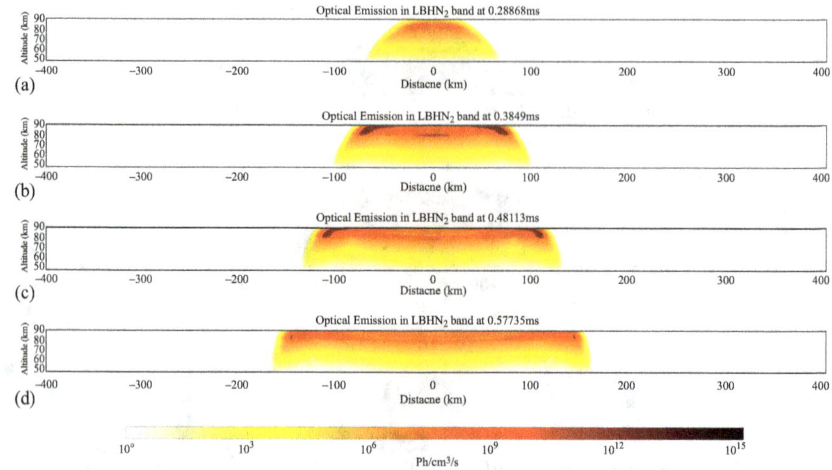

Figure 11.29 The side views of LBHN$_2$ photon emission intensity of the elve triggered by the considered lightning

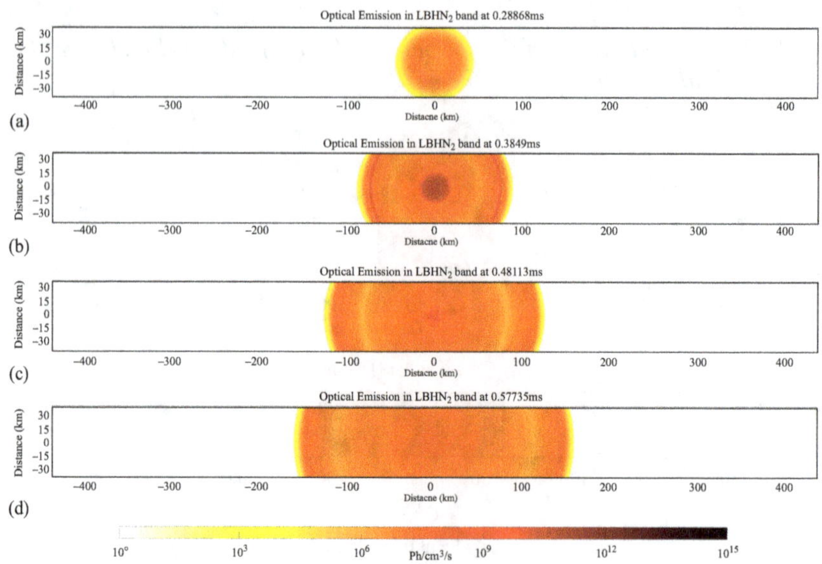

Figure 11.30 The 80 km top views of LBHN$_2$ photon emission intensity of the elve triggered by the considered lightning

optical emission of the simulated elve is consistent with the typical ring shape expanding radially around 200 km at an altitude of about 80 km. Based on previous studies, the brightness of the optical emission of elves strongly depends on the ambient electron number density and the peak current of the parent lightning

[4,41,43]. Furthermore, the dip angle of geomagnetic field B_0, a tilted lightning discharge channel and the local chemical impact may also create significant density perturbations in the overlying ionosphere and further affect the shape of the optical emission of elves [43,44].

11.3 Conclusion and summary

We have seen in this chapter that the FDTD method for solving Maxwell's equations is accurate and versatile in a very wide variety of applications related to lightning. One can analyze the lightning electromagnetic field propagation over distances ranging from a few meters to hundreds of kilometers by just adjusting the smallest wavelength governed by the highest frequency of the band being simulated with the FDTD method. One can also take into account complex media in the EIWG by adding the media parameters through the simple discretization process. Moreover, a number of powerful software packages exist for FDTD modeling, including free, open-source packages and commercial program suites. One can also modify the basic codes to adapt them to any problem of interest. With the continued growth of computing resources, FDTD method will continue to serve as a helpful tool in research and engineering applications requiring lightning electromagnetic pulse (LEMP) simulations.

References

[1] Yee K. Numerical solution of initial boundary value problems involving Maxwell's equations in isotropic media. *IEEE Transactions on Antennas and Propagation.* 1966;14(3):302–307.

[2] Baba Y, Rakov VA. Applications of the FDTD method to lightning electromagnetic pulse and surge simulations. *IEEE Transactions on Electromagnetic Compatibility.* 2014;56(6):1506–1521.

[3] Hu W, Cummer SA. An FDTD model for low and high altitude lightning-generated EM fields. *IEEE Transactions on Antennas and Propagation.* 2006;54(5):1513–1522.

[4] Marshall RA. An improved model of the lightning electromagnetic field interaction with the D-region ionosphere. *Journal of Geophysical Research: Space Physics.* 2012;117(A3).

[5] Li D, Azadifar M, Rachidi F, *et al.* Analysis of lightning electromagnetic field propagation in mountainous terrain and its effects on ToA-based lightning location systems. *Journal of Geophysical Research: Atmospheres.* 2016;121(2):895–911.

[6] Taflove A, Hagness SC. *Computational Electrodynamics: The Finite-Difference Time-Domain Method*. Norwood, MA: Artech House; 2005.

[7] Inan US, Marshall RA. *Numerical Electromagnetics: The FDTD Method.* Cambridge: Cambridge University Press; 2011.

[8] Holland R. THREDS: A finite-difference time-domain EMP code in 3D spherical coordinates. *IEEE Transactions on Nuclear Science.* 1983;30 (6):4592–4595.
[9] Otsuyama T, Sakuma D, Hayakawa M. FDTD analysis of ELF wave propagation and Schumann resonances for a subionospheric waveguide model. *Radio Science.* 2003;38(6):11–1.
[10] Uusitupa T, Ilvonen S, Laakso I, *et al.* The effect of finite-difference time-domain resolution and power-loss computation method on SAR values in plane-wave exposure of Zubal phantom. *Physics in Medicine & Biology.* 2007;53(2):445.
[11] Uman MA, McLain DK. Magnetic field of lightning return stroke. *Journal of Geophysical Research.* 1969;74(28):6899–6910.
[12] Rakov VA, Dulzon A. A modified transmission line model for lightning return stroke field calculations. In: *Proceedings of 9th International Symposium on Electromagnetic Interference and Compatibility*; 1991. pp. 229–235.
[13] Nucci CA. On lightning return stroke models for LEMP calculations. In: *Proceedings of 19th International Conference on Lightning Protection*, Graz, Austria; 1988.
[14] Rachidi F, Nucci C. On the Master, Uman, Lin, Standler and the modified transmission line lightning return stroke current models. *Journal of Geophysical Research: Atmospheres.* 1990;95(D12):20389–20393.
[15] Baba Y, Rakov VA. On the transmission line model for lightning return stroke representation. *Geophysical Research Letters.* 2003;30(24):1301–1304.
[16] Uman MA, McLain DK, Krider EP. The electromagnetic radiation from a finite antenna. *American Journal of Physics.* 1975;43(1):33–38.
[17] Roden JA, Gedney SD. Convolution PML (CPML): an efficient FDTD implementation of the CFS–PML for arbitrary media. *Microwave and Optical Technology Letters.* 2000;27(5):334–339.
[18] Rachidi F, Janischewskyj W, Hussein AM, *et al.* Current and electromagnetic field associated with lightning-return strokes to tall towers. *IEEE Transactions on Electromagnetic Compatibility.* 2001;43(3):356–367.
[19] Shoory A, Mimouni A, Rachidi F, *et al.* Lightning horizontal electric fields above a two-layer ground. In: 2010 30th International Conference on Lightning Protection (ICLP). IEEE; 2010. pp. 1–5.
[20] Zhang Q, Li D, Fan Y, *et al.* Examination of the Cooray-Rubinstein (C-R) formula for a mixed propagation path by using FDTD. *Journal of Geophysical Research: Atmospheres.* 2012;117(D15).
[21] Zhang Q, Li D, Zhang Y, *et al.* On the accuracy of Wait's formula along a mixed propagation path within 1 km from the lightning channel. *IEEE Transactions on Electromagnetic Compatibility.* 2012;54(5):1042–1047.
[22] Zhang Q, Li D, Tang X, *et al.* Lightning-radiated horizontal electric field over a rough-and ocean-land mixed propagation path. *IEEE Transactions on Electromagnetic Compatibility.* 2013;55(4):733–738.

[23] Romero C, Paolone M, Rubinstein M, *et al.* A system for the measurements of lightning currents at the Säntis Tower. *Electric Power Systems Research.* 2012;82(1):34–43.

[24] Romero C, Rachidi F, Paolone M, *et al.* Statistical distributions of lightning currents associated with upward negative flashes based on the data collected at the Säntis (EMC) tower in 2010 and 2011. *IEEE Transactions on Power Delivery*. 2013;28(3):1804–1812.

[25] Azadifar M, Paolone M, Pavanello D, *et al.* An update on the instrumentation of the Säntis Tower in Switzerland for lightning current measurements and obtained results. In: *CIGRE International Colloquium on Lightning and Power Systems, Lyon, France*; 2014.

[26] Li D, Azadifar M, Rachidi F, *et al.* On lightning electromagnetic field propagation along an irregular terrain. *IEEE Transactions on Electromagnetic Compatibility.* 2015;58(1):161–171.

[27] Li D, Zhang Q, Wang Z, *et al.* Computation of lightning horizontal field over the two-dimensional rough ground by using the three-dimensional FDTD. *IEEE Transactions on Electromagnetic Compatibility.* 2013;56(1):143–148.

[28] Inan US, Cummer SA, Marshall RA. A survey of ELF and VLF research on lightning-ionosphere interactions and causative discharges. *Journal of Geophysical Research: Space Physics.* 2010;115(A6).

[29] Shao XM, Lay EH, Jacobson AR. Reduction of electron density in the nighttime lower ionosphere in response to a thunderstorm. *Nature Geoscience.* 2013;6(1):29.

[30] Han F, Cummer SA. Midlatitude daytime D region ionosphere variations measured from radio atmospherics. *Journal of Geophysical Research: Space Physics.* 2010;115(A10).

[31] Pasko VP, Inan US, Bell TF, *et al.* Sprites produced by quasi-electrostatic heating and ionization in the lower ionosphere. *Journal of Geophysical Research: Space Physics.* 1997;102(A3):4529–4561.

[32] Ebert U, Nijdam S, Li C, *et al.* Review of recent results on streamer discharges and discussion of their relevance for sprites and lightning. *Journal of Geophysical Research: Space Physics.* 2010;115(A7).

[33] Hedin AE. Extension of the MSIS Thermosphere Model into the middle and lower atmosphere. *Journal of Geophysical Research: Space Physics.* 1991;96(A2):1159–1172.

[34] Wait JR, Spies KP. Characteristics of the Earth-ionosphere waveguide for VLF radio waves. US Dept. of Commerce, National Bureau of Standards: For sale by the Supt. of Doc., US Govt. Print. Off.; 1964.

[35] Bilitza D, Reinisch BW. International reference ionosphere 2007: improvements and new parameters. *Advances in Space Research.* 2008;42(4): 599–609.

[36] Rakov VA, Thottappillil R, Uman MA, *et al.* Mechanism of the lightning M component. *Journal of Geophysical Research: Atmospheres.* 1995;100 (D12):25701–25710.

[37] Li D, Luque A, Rachidi F, *et al.* The propagation effects of lightning electromagnetic fields over mountainous terrain in the earth-ionosphere waveguide. *Journal of Geophysical Research: Atmospheres.* 2019;124 (24):14198–14219.

[38] Hagelaar G, Pitchford L. Solving the Boltzmann equation to obtain electron transport coefficients and rate coefficients for fluid models. *Plasma Sources Science and Technology.* 2005;14(4):722.

[39] Luque A, Gordillo-Vázquez F. Mesospheric electric breakdown and delayed sprite ignition caused by electron detachment. *Nature Geoscience.* 2012;5 (1):22–25.

[40] Chern J, Hsu RR, Su HT, *et al.* Global survey of upper atmospheric transient luminous events on the ROCSAT-2 satellite. *Journal of Atmospheric and Solar-Terrestrial Physics.* 2003;65(5):647–659.

[41] Kuo CL, Chen A, Lee Y, *et al.* Modeling elves observed by FORMOSAT-2 satellite. *Journal of Geophysical Research: Space Physics.* 2007;112 (A11).

[42] Sipler DP, Biondi MA. Measurements of $O(^1D)$ quenching rates in the F region. *Journal of Geophysical Research.* 1972;77(31):6202–6212.

[43] Marshall RA, Inan US, Glukhov VS. Elves and associated electron density changes due to cloud-to-ground and in-cloud lightning discharges. *Journal of Geophysical Research: Space Physics.* 2010;115(A4):A00E17.

[44] Pérez-Invernón FJ, Luque A, Gordillo-Vázquez FJ. Modeling the chemical impact and the optical emissions produced by lightning-induced electromagnetic fields in the upper atmosphere: the case of halos and elves triggered by different lightning discharges. *Journal of Geophysical Research: Atmospheres.* 2018;123(14):7615–7641.

Chapter 12

Modeling of lightning and its interaction with overhead conductors

Amedeo Andreotti[1], Vladimir Rakov[2] and Luigi Verolino[1]

Lightning is one of the most powerful and dangerous natural phenomena on the Earth. It has had a profound impact on human society. We will focus here on the so-called "cloud-to-ground" lightning discharges. These discharges can cause damage not only when they strike the structure directly but also when they hit ground nearby. In the case of overhead conductors, whether they are power or tele-communication lines, due to overvoltages produced, these discharges can cause outages, disturbances on the network, or failure of electronic components and/or electrical equipment. In order to mitigate lightning effects via effective technological solutions and protective measures, we need to know lightning parameters and have models of both lightning itself and its interaction with the strike object or system. We will focus on the so-called indirect or nearby lightning and its effects. The following topics will be covered: (1) specification of channel-base current, (2) return-stroke models, (3) calculation of lightning electromagnetic field, and (4) electromagnetic coupling between the lightning channel and overhead conductors. Both analytical and distributed-circuit model (based on generalized telegrapher's equations) approaches will be considered.

12.1 Specification of channel-base current

Direct effects of lightning current injection and induced effects from nearby lightning are often evaluated by numerical simulation involving assumed lightning current waveforms at the channel base. A variety of equations are used to approximate such waveworms.

Although it is well known that the return-stroke current waveform is a pulse, the step-function lightning current representation remains in use, particularly in analytical solutions (e.g., [1–5]). The Rusk formula for computing the maximum overvoltage magnitude, based on the step-function current waveform is found in

[1]Department of Electrical Engineering and Information Technology, University of Naples Federico II, Italy
[2]Department of Electrical and Computer Engineering, University of Florida, USA

IEEE Std 1410-2010. Also, applying Duhamel's integral, the unit-step-function response of the system can be used for finding system's response to an arbitrarily-time-varying excitation (e.g., [6]).

A linearly-rising current with linearly falling tail (triangular pulse) is a reasonable representation of lightning return-stroke current. Current waveforms of this type were used, for example, in [7–11]. The channel-base current was represented by a single linearly-rising ramp for computing the initial part (first few microseconds) of the induced-voltage waveform in [12,13]. IEEE Std 1410-2010 allows the use of "equivalent" linear front times (1.28 μs for first strokes and 0.31 μs for subsequent strokes) for induced voltage calculations.

A generalized formula for lightning channel-base current was introduced in [14]

$$I(0,t) = \begin{cases} I_p(\lambda^{t/t_p} - 1)/(\lambda - 1) & t < t_p \\ I_p \exp[-\beta(t - t_p)] & t > t_p \end{cases} \tag{12.1}$$

where I_p is the current peak, t_p is the time of current peak (current risetime), β is the rate of exponential current decay with time when $t > t_p$, and λ is the ratio of the steepness of the rising part of current pulse at $t = t_p$ to that at $t = 0$. For $\lambda = 1$, the rising part is a linear ramp, for $\lambda < 1$ it is convex and for $\lambda > 1$ concave, the latter being the most realistic case. For $\lambda = 1$ and $\beta = 0$, current given by equation (12.1) linearly rises from 0 to t_p and then remains constant at I_p (trapezoidal waveform; e.g., [15]).

The typical subsequent-stroke current waveform at the channel base is often approximated by the Heidler function [16]

$$I(0,t) = \frac{I_0}{\eta}\frac{(t/\tau_1)^n}{(t/\tau_1)^n + 1}\exp(-t/\tau_2) \tag{12.2}$$

where

$$\eta = \exp[-(\tau_1/\tau_2)(n\tau_1/\tau_2)^{(1/n)}] \tag{12.3}$$

where I_0, τ_1, n, and τ_2 are constants. This function allows one to change conveniently the current peak, maximum current derivative, and associated electrical charge transfer nearly independently by changing I_0, τ_1, and τ_2, respectively. Equation (12.2) reproduces the observed concave rising portion of a typical current waveform, as opposed to the once more commonly used double-exponential function, introduced independently in [17,18], which is characterized by an unrealistic convex wavefront with a maximum current derivative at $t = 0$. The IEC Lightning Protection Standard (IEC 62305-1, 2010) recommends the 10/350 μs current waveform for first positive strokes, 1/200 μs for first negative strokes, and 0.25/100 μs for subsequent strokes, which are approximated by Heidler functions with the following parameters: for first positive strokes, $I_0 = 200$ kA, $n = 10$, $\eta = 0.93$, $\tau_1 = 19$ μs, and $\tau_2 = 485$ μs; for first negative strokes, $I_0 = 100$ kA, $n = 10$, $\eta = 0.986$, $\tau_1 = 1.82$ μs, and $\tau_2 = 285$ μs; for subsequent strokes, $I_0 = 50$ kA, $n = 10$, $\eta = 0.993$, $\tau_1 = 0.454$ μs, and $\tau_2 = 143$ μs. Sometimes the sum of two Heidler functions with different parameters is used to approximate the desired current waveshape (e.g., [6,19–21]).

One popular channel-base current waveform, which was used for the calculation of subsequent return-stroke electric and magnetic fields, is represented by the sum of a Heidler function and a double-exponential function (e.g., [22,23])

$$I(0,t) = \frac{I_{01}}{\eta} \frac{(t/\tau_1)^n}{(t/\tau_1)^n + 1} \exp(-t/\tau_2) + I_{02}[\exp(-t/\tau_3) - \exp(-t/\tau_4)] \tag{12.4}$$

where $n = 2$, $I_{01} = 9.9\,\text{kA}$, $\eta = 0.845$, $\tau_1 = 0.072$ µs, $\tau_2 = 5.0$ µs, $I_{02} = 7.5\,\text{kA}$, $\tau_3 = 100.0$ µs, $\tau_4 = 6.0$ µs. For this latter waveform, the current peak is 11 kA, and the maximum current rate-of-rise is 105 kA/µs. It is worth reiterating that a double-exponential function alone (the last two terms in (12.4)) does not allow representation of concave current front, because the maximum time-derivative of this function occurs at $t = 0$. Multiple Heidler functions were employed in [24] to reproduce with higher precision the concave front in first- and subsequent-stroke current waveforms, as well as the secondary peak typically seen in first-stroke current waveforms recorded on instrumented towers.

The basic idea of the Heidler function is factorization of the current waveform into three separate parts: the first one relevant to the peak value, the second one relevant to the front portion, and the third one relevant to the decay portion. As a result, the peak value, the rise time, and the fall time become decoupled, which is usually a desired property:

$$I_0 x(t) y(t) \tag{12.5}$$

where I_0 is the peak current, $x(t)$ and $y(t)$ are the current-front and current-decay functions, respectively, which are given by

$$x(t) = \frac{(t/\tau_1)^n}{1 + (t/\tau_1)^n} \qquad y(t) = e^{-(t/\tau_2)} \tag{12.6}$$

$x(t)$ is a function which asymptotically ($n \to \infty$) represents a step function, while $y(t)$ is a classical exponential decay function. These functions have been chosen so as during the front time $y(t) \cong 1$, and during the decay time $x(t) \cong 1$. Since, as pointed out in [25], for $t > 0, x(t) < 1$, the maximum of the current is less than I_0. For this reason an "amplitude correction factor" η given by (12.3) is used in (12.2).

Nowadays the Heidler function is widely considered standard for describing the lightning return-stroke current at the channel base.

12.2 Return-stroke models

Depending on the type of governing equations, in [26], the return stroke models were classified into four categories:

1. The first category is the gas-dynamic models, also called "physical" models, which describe the radial evolution of a short segment of the lightning channel. This type of models generally involves the solution of three gas-dynamic

equations which account for conservation of mass, momentum, and energy, coupled to two equations of state. The input parameter is the channel current expressed as a function of time. The outputs are usually pressure, temperature, and mass density, each expressed as a function of radial coordinate and time.
2. The second category of models is the electromagnetic models. In these models, the current distribution along the lightning channel is obtained by numerical solution of Maxwell's equations.
3. The third category is the distributed-circuit models that can be viewed as an approximation to the electromagnetic models. They represent the lightning discharge as a transient process on a transmission line (TL) characterized by resistance (R), inductance (L), and capacitance (C), all per unit length. The current distribution along the channel is obtained by numerical solution of telegrapher's equations.
4. The fourth category is the "engineering" models, in which the channel current (or the line charge density) is specified based on such observed lightning return-stroke characteristics, as the current at the channel base, the speed of the upward propagating front, and the channel luminosity profile.

The gas-dynamic models are primarily used to reproduce physical parameters of the return-stroke process. All types of models, except for the gas-dynamic ones, were used for computing lightning electromagnetic fields. In the following, we focus on engineering models.

An engineering model of the lightning return stroke is just a mathematical description of the spatial and temporal distribution of return-stroke current $i(z', t)$, or line charge density $\rho(z', t)$, along the lightning channel. Such a mathematical specification generally includes the channel-base current waveform, return-stroke front speed, and dependence of current magnitude on height. The resultant distribution of current along the channel can be directly used for computation of electromagnetic fields. Note that in these models, the lightning channel is generally assumed to be straight and vertical (perpendicular to the ground plane), as shown in Figure 12.1. From an engineering point of view, the models of main interest are those in which the return-stroke current at any height z' , $i(z', t)$can be related to the channel-base current $i(0, t)$, which can be measured directly. Most engineering models found in the literature can be described by the following equation [26]:

$$i(z', t) = P(z')i\left(0,\ t - \frac{z'}{v}\right)u\left(t - \frac{z'}{v_f}\right), \tag{12.7}$$

where $u(\cdot)$ is the Heaviside function, v_f is the return-stroke front propagation speed, v is the propagation speed of return-stroke current wave, and $P(z')$ is the generalized current attenuation function of the return-stroke current along the channel, which was proposed in [23].

The return-stroke models most commonly adopted for calculation of lightning-induced voltages are:

- Bruce–Golde model [17].
- Traveling current source (TCS) model, proposed by Heidler [27].

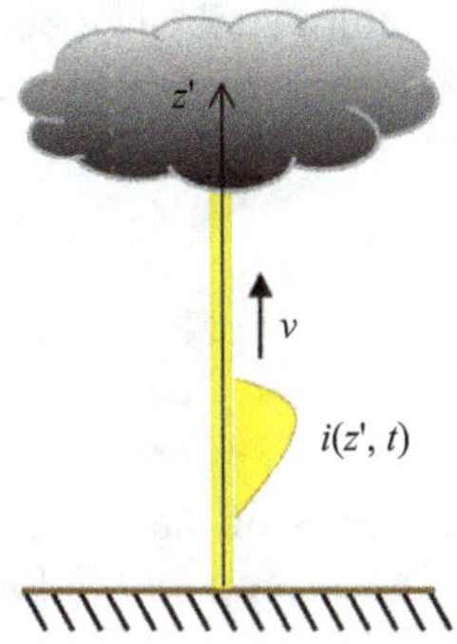

Figure 12.1 Schematic representation of return-stroke current wave i(z′,t)*, propagating at speed* v *along the lightning channel*

Table 12.1 A summary of return-stroke models described by equation (12.4)

Model	***P*(*z*′)**	***v***
BG (Bruce and Golde [17])	1	∞
TCS (Heidler [27])	1	$-c$
TL (Uman and McLain [28])	1	v_f
MTLL (Rakov and Dulzon [23])	$1 - z'/h_c$	v_f
MTLE (Nucci *et al.* [29])	$e^{(-z'/\lambda)}$	v_f

- TL model, proposed by Uman and McLain [28];
- Modified TL model with linear current decay with height (MTLL), proposed by Rakov and Dulzon [23].
- Modified TL model with exponential current decay with height (MTLE), proposed by Nucci *et al.* [29].

These models are summarized in Table 12.1, where the current attenuation function $P(z')$ and current wave speed v are given for each model. The parameter h_c in the MTLL model is the total channel length, λ in the MTLE model is the current decay height constant (assumed in [29] to be 2,000 m), and c is the speed of light in free space.

In the following, each of these models will be formulated in terms of both current and line charge density and briefly discussed.

12.2.1 Bruce–Golde (BG) model

Bruce and Golde proposed a simple return-stroke model in which a current discontinuity (front) moves up along the channel at speed v_f, and current behind the front is uniform and equal to the front current at any instant of time. The variation of front current is determined by the time variation of current at the channel base.

Thus, at any instant of time, the return stroke can be viewed as a vertical Hertzian dipole whose length is equal to $v_f t$ and whose current is $i(0, t)$. There is no current propagation behind the front ($v = \infty$). The model can be described mathematically by the following expression:

$$i(z', t) = \begin{cases} i(0,\ t) & z' \le v_f t, \\ 0 & z' > v_f t. \end{cases} \tag{12.8}$$

An equivalent expression in terms of the line charge density on the channel, based on the continuity equation, was proposed in [30]:

$$\rho(z', t) = \frac{i(0, z'/v_f)}{v_f}. \tag{12.9}$$

According to the BG model, the return stroke instantaneously compensates the leader charge stored at the position of the front.

12.2.2 TCS model

In the TCS model, which has been proposed by Heidler [27], the return-stroke current may be viewed as generated at the upward-moving return-stroke front (by a single moving source or via activation of multiple shunt sources distributed along the channel) and propagating downward. The channel is implicitly matched at $z' = 0$. Similar to the BG model, there is current discontinuity at the front, but, in contrast with the BG model, there is a traveling current wave ($v = -c$) behind the front. The channel current expression reads as follows:

$$i(z', t) = \begin{cases} i(0,\ t + z'/c) & z' \le v_f t, \\ 0 & z' > v_f t \end{cases}. \tag{12.10}$$

The corresponding line charge density equation is

$$\rho(z', t) = -\frac{i(0, z'/v)}{c} + \frac{i(0, z'/v*)}{v*}, \tag{12.11}$$

where $v* = v_f/(1 + v_f/c)$

The TCS model reduces to the BG model if $v = -c$ is replaced with $v = \infty$. Both TCS and BG models have an inherent current discontinuity at the front, treating of which requires an additional term in electric and magnetic field equations.

12.2.3 TL model

This model (also referred to as the original TL model) has been proposed by Uman and McLain [28]. The return-stroke current is assumed to travel without distortion or attenuation from an implied source at ground level up along the lightning channel at a constant speed v. There is no inherent current discontinuity at the front.

The expression for the current at any height z' along the lightning channel is given by:

$$i(z',t) = \begin{cases} i(0,\ t - z'/v) & z' \le v_f t, \\ 0 & z' > v_f t \end{cases}. \tag{12.12}$$

In the TL model, no charge is deposited to (or removed from) the channel. Instead, all the charge is transferred to the channel top, causing an unrealistic "mirror image" in radiation field waveforms. Nevertheless, earlytime field predictions of the TL model are very similar to those of the more physically reasonable models. The current is a classical travelling wave, as is the corresponding charge density $\rho(z',t) = i(0,t - z'/v)/v$. Finally, it is worth noting that the TL model reduces to the BG model when v is set to ∞, while v_f remains finite.

12.2.4 Modified TL with linear current decay with height (MTLL) model

The modified TL model with linear current decay with height was proposed by Rakov and Dulzon [23]. This model can be viewed as incorporating a current source at the channel base, which injects a specified current wave into the channel, with that wave propagating upward without distortion but with specified linear attenuation, as seen from the corresponding current expression:

$$i(z',t) = \begin{cases} (1 - z'/H)i(0,\ t - z'/v) & z' \le v_f t, \\ 0 & z' > v_f t \end{cases}, \tag{12.13}$$

where H is the total channel length (labeled h_c in Table 12.1). The equivalent formulation of this model in terms of line charge density is:

$$\rho(z',t) = \frac{(1 - z'/H)i(0,t - z'/v)}{v} + \frac{1}{H}Q(z',t), \tag{12.14}$$

where $Q(z',t)$ is the total charge deposited on the channel at height z' at time t, to neutralize the leader charge previously stored there. It is given by:

$$Q(z',t) = \int_{z'/v}^{t} i(0,\tau - z'/v)d\tau. \tag{12.15}$$

Clearly, as $t \to \infty$, Q is the same at all heights; that is, the MTLL model implies a uniformly charged leader channel. It is the only engineering model in which the boundary conditions at the channel top $(i(H,t) = 0)$ is specified explicitly.

12.2.5 Modified TL with exponential current decay with height (MTLE) model

This model was proposed by Nucci *et al.* [29]. It is similar to the MTLL model, except for the current decay with height being exponential. The current equation reads:

$$i(z',t) = \begin{cases} e^{-z'/\lambda} i(0,\, t - z'/v) & z' \le v_f t, \\ 0 & z' > v_f t, \end{cases} \tag{12.16}$$

where λ is the current decay height constant, that is usually assumed to be equal to 2,000 m. The equivalent formulation of this model in terms of the line charge density is:

$$\rho(z',t) = e^{-z'/\lambda} \frac{i(0,t-z'/v)}{v} + \frac{e^{-z'/\lambda}}{\lambda} Q(z',t), \tag{12.17}$$

In contrast with the MTLL model, (a) the current in the MTLE model does not vanish at the channel top, although it is very small for $\lambda = 2{,}000$, and (b) the implied leader charge distribution is highly nonuniform (has the highest value at ground level and exponentially decays with height).

At early times (a few microseconds or so), electric and magnetic fields predicted by all the TL-type models (TL, MTLL, and MTLE) are similar. For this reason, peak overvoltages computed using those three models are also similar.

12.3 Electromagnetic fields generated by lightning

Two main methods have been used to calculate the electric and magnetic fields radiated from a known distribution of currents and charges. One is the *monopole technique* (also called continuity equation technique (e.g., [31,32])) which is used, primarily in the power systems literature, for lightning fields calculation. This technique requires a knowledge of both the current and the charge density as a function of time and space [33]. The *dipole technique* (also called Lorentz condition technique) is widely used in studies of antennas. These two techniques are equivalent, even though their analytical expressions are different, as demonstrated, for example, in [33]. In the following, we shall briefly describe both the monopole and dipole techniques.

12.3.1 General solution

Maxwell's equations for a linear, homogeneous, isotropic, time-invariant medium can be written as follows:

$$\nabla \cdot (\varepsilon_m \boldsymbol{E}) = \rho, \tag{12.18}$$

$$\nabla \cdot (\mu_m \boldsymbol{H}) = 0, \tag{12.19}$$

$$\nabla \times \boldsymbol{E} = -\frac{\partial(\mu_m \boldsymbol{H})}{\partial t}, \tag{12.20}$$

$$\nabla \times \boldsymbol{H} = \boldsymbol{J} + \frac{\partial(\varepsilon_m \boldsymbol{E})}{\partial t}. \tag{12.21}$$

where $\boldsymbol{E}$ is the electric field intensity, $\boldsymbol{H}$ is the magnetic field intensity, ε_m is the electric permittivity of the medium, μ_m is the magnetic permeability of the medium, and $\boldsymbol{J}$ is the conduction current density.

Since (12.19) can be derived from (12.20) by taking the divergence of the latter and integrating over time and (12.20) and (12.21) are generally each a set of three scalar equations, we have seven independent scalar equations containing the following ten scalar source and field variables:

- three components of the electric field intensity $\boldsymbol{E}$;
- three components of the magnetic field intensity $\boldsymbol{H}$;
- three components of the conduction current density $\boldsymbol{J}$;
- the volume charge density ρ.

Thus, we have seven scalar equations with ten unknowns, three of which are to be specified. To facilitate the solution of Maxwell's equations for $\boldsymbol{E}$ and $\boldsymbol{H}$, the so-called vector $\boldsymbol{A}$ and scalar ϕ potentials are commonly used (e.g., [34–36]):

$$\boldsymbol{E} = -\nabla\phi - \frac{\partial \boldsymbol{A}}{\partial t}, \tag{12.22}$$

and

$$\mu_m \boldsymbol{H} = \nabla \times \boldsymbol{A}, \tag{12.23}$$

where, for geometry shown in Figure 12.2

$$\boldsymbol{A}(\boldsymbol{r}_s,\ t) = \frac{\mu_m}{4\pi}\int_{V'} \frac{\boldsymbol{J}\left(\boldsymbol{r}_s',t - \left|\boldsymbol{r}_s - \boldsymbol{r}_s'\right|/c\right)}{\left|\boldsymbol{r}_s - \boldsymbol{r}_s'\right|} dV', \tag{12.24}$$

$$\phi(\boldsymbol{r}_s,\ t) = \frac{1}{4\pi\varepsilon_m}\int_{V'} \frac{\rho\left(\boldsymbol{r}_s',t - \left|\boldsymbol{r}_s - \boldsymbol{r}_s'\right|/c\right)}{\left|\boldsymbol{r}_s - \boldsymbol{r}_s'\right|} dV', \tag{12.25}$$

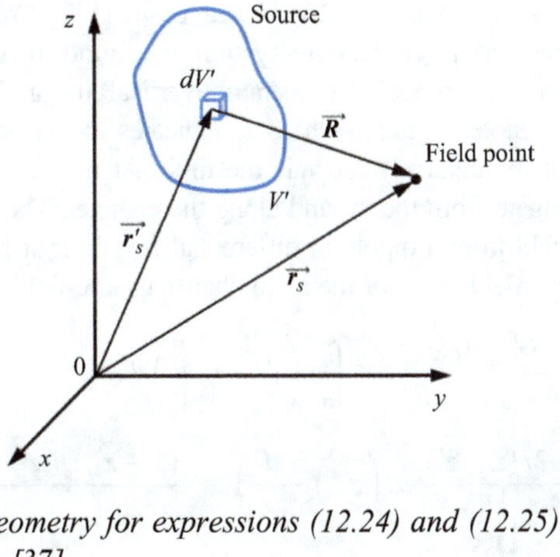

Figure 12.2 The geometry for expressions (12.24) and (12.25). Adapted from Uman [37].

A and ϕ are related by the so-called Lorentz condition:

$$\nabla \cdot A = -\mu_m \varepsilon_m \frac{\partial \phi}{\partial t} \tag{12.26}$$

In the dipole technique, (12.24) is used to find A and (12.26) to find ϕ ((12.25) is not needed). Then A and ϕ are used in (12.22) and (12.23) to find E and H, respectively. In the monopole technique, (12.24) is used to find A and (12.25) to find ϕ, with ρ found from the continuity equation.

$$\nabla \cdot J = -\frac{\partial \rho}{\partial t}, \tag{12.27}$$

which follows from taking the divergence of (12.21) and combining the resultant equation with (12.18). In the monopole technique, the source is described in terms of both charge density and current density, while in the dipole technique only current density is needed. The Lorentz condition (12.26) is not used in the monopole technique ((12.27) and (12.26) are equivalent to each other).

In the more specific case of a vertical lightning channel over an infinitely conducting plane, we can use the above equations along with the method of images to calculate the electric and magnetic fields radiated by lightning return stroke represented by one of the models described in Section 12.2.

12.3.2 Solutions for a vertical lightning channel

In the context of lightning-induced voltages calculation, the most commonly used field equations adopted in the time domain have been derived in the cylindrical coordinate system by Master and Uman [38]. The equations were derived for the electric and magnetic fields produced by a vertical dipole of infinitesimal length over perfectly conducting ground by solving Maxwell's equations (12.18)–(12.21) in terms of the retarded scalar and vector potentials (12.25) and (12.24). The geometry of the problem is shown in Figure 12.3. Similar derivations in the spherical coordinate system were performed by Uman *et al.* [37] Additionally, detailed derivations in the cylindrical coordinate system are found in Appendix 3 of [39].

The lightning channel is modeled as a straight vertical line, and the current along the lightning channel is denoted $i(z',t)$, where z' indicates the position along the z axis (origin at the base of the channel) and t is the time. At time $t = 0$, the return stroke current starts to propagate from the ground along the channel. The elementary contributions to the overall field from a dipole of differential length dz' at height z', for the vertical and radial electric fields and for the azimuthal magnetic field are found as follows:

$$de_r(r,z,t) = \frac{dz'}{4\pi\varepsilon_0}\left[\frac{3r(z-z')}{R^5}\int_0^t i\left(z',\tau-\frac{R}{c}\right)d\tau + \frac{3r(z-z')}{cR^4}\cdot i\left(z',t-\frac{R}{c}\right) + \frac{r(z-z')}{c^2R^3}\frac{\partial i(z',t-R/c)}{\partial t}\right], \tag{12.28}$$

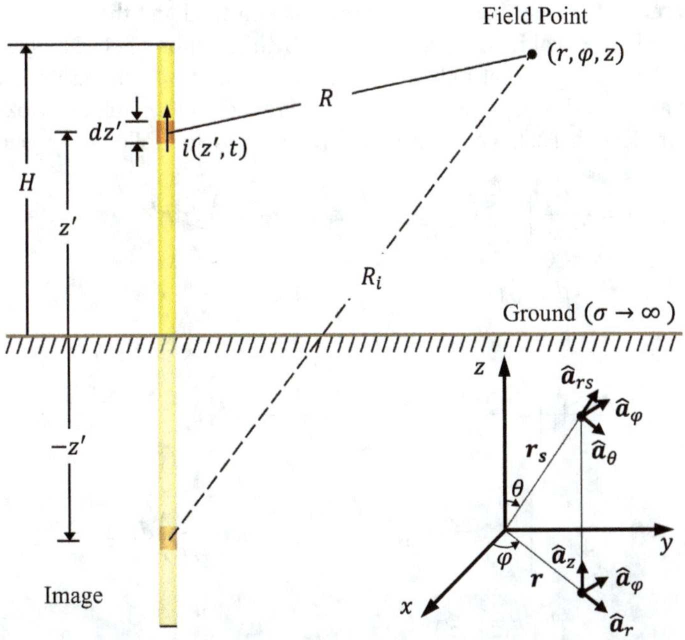

Figure 12.3 Geometry for the return stroke field evaluation: dipole technique

$$de_z(r,z,t) = \frac{dz'}{4\pi\varepsilon_0}\left[\frac{2(z-z')^2 - r^2}{R^5}\int_0^t i\left(z',\, \tau - \frac{R}{c}\right)d\tau + \frac{2(z-z')^2 - r^2}{cR^4}\cdot i\left(z',\, t-\frac{R}{c}\right) - \frac{r^2}{c^2R^3}\frac{\partial i(z',\, t-R/c)}{\partial t}\right] \tag{12.29}$$

$$dh_\varphi(r,z,t) = \frac{dz'}{4\pi}\left[\frac{r}{R^3}i\left(z',\, t-\frac{R}{c}\right) + \frac{r}{cR^2}\frac{\partial i(z',\, t-R/c)}{\partial t}\right], \tag{12.30}$$

where

- $i(z', t)$ is the current along the lightning channel;
- $R = \sqrt{(z-z')^2 + r^2}$ is the inclined distance between the source (idz') and the observation point;
- r is the projection of R on the xy-plane (see Figure 12.3);
- c is the speed of light.

In (12.28) and (12.29), the three terms on the right-hand side are called, respectively, electrostatic, induction, and radiation field components. In (12.30), the first term on the right-hand side is the induction field and the second one is the radiation field. The total fields can be obtained by integration of (12.28)–(12.30) over the entire channel (note that radiating channel length is $\leq H$) and its image. The current as a function of time and height is given by one of the return-stroke models presented in Section 12.2. The total fields are given by the following equations:

$$e_r(r,z,t) = \frac{1}{4\pi\varepsilon_0}\left[\int_{-H}^{H} \frac{3r(z-z')}{R^5}\int_0^t i\left(z',\tau-\frac{R}{c}\right)d\tau dz' + \int_{-H}^{H} \frac{3r(z-z')}{cR^4}\cdot i\left(z',t-\frac{R}{c}\right)dz' + \int_{-H}^{H} \frac{r(z-z')}{c^2R^3}\frac{\partial i(z',t-R/c)}{\partial t}dz'\right], \tag{12.31}$$

$$e_z(r,z,t) = \frac{1}{4\pi\varepsilon_0}\left[\int_{-H}^{H} \frac{2(z-z')^2-r^2}{R^5}\int_0^t i\left(z',\tau-\frac{R}{c}\right)d\tau dz' + \int_{-H}^{H} \frac{2(z-z')^2-r^2}{cR^4} i\left(z',t-\frac{R}{c}\right)dz' - \int_{-H}^{H} \frac{r^2}{c^2R^3}\frac{\partial i(z',t-R/c)}{\partial t}dz'\right] \tag{12.32}$$

$$h_\varphi(r,z,t) = \frac{1}{4\pi}\left[\int_{-H}^{H} \frac{r}{R^3} i\left(z',t-\frac{R}{c}\right)dz' + \int_{-H}^{H} \frac{r}{cR^2}\frac{\partial i(z',t-R/c)}{\partial t}dz'\right]. \tag{12.33}$$

It should be noted that the above equations are derived using the dipole technique so that the source was described only in terms of current and Lorentz condition was employed to find the scalar potential. Field equations derived using the monopole technique are found in [40].

12.3.3 The effect of finite ground conductivity

If the observation point $P(r,\varphi,z)$ is located on the ground surface ($z=0$), and the ground is assumed to be perfectly conducting, only two field components, the vertical electric field and the azimuthal magnetic field, are present. Indeed, the boundary condition on the surface of a perfect conductor, requires that the horizontal electric field component is zero. If the observation point elevated above a perfectly conducting ground, a nonzero horizontal electric field component exists. In the case of a finitely conducting ground, a horizontal electric field exists on the ground surface, and also above and below the ground surface. It has been demonstrated that for distances not exceeding a few kilometers, the perfectly conducting ground assumption is reasonable for the vertical component of the electric field and for the horizontal component of the magnetic field (e.g., [41,42]). Indeed, considering for example the dipole technique, the contributions to these fields from the source dipole and its image add constructively and, consequently, small variations in the image field due to the finite ground conductivity will have small effect. The horizontal component of the electric field is appreciably affected by finite ground conductivity: the effects of the two contributions in this case subtract, and small changes in the image field may lead to appreciable changes in the total horizontal field. It should be noted that, although the intensity of the horizontal field component is generally much smaller than that of the vertical one, it plays an important role in the context of certain coupling models [43–47], in which case an accurate calculation will be essential. The most rigorous approach to the evaluation of the horizontal field is based on Sommerfeld integrals [48], but this approach is impractical because of the long computational time and because the integrands are highly oscillating. There are two approximate solutions of the Sommerfeld integrals, both originally proposed in the frequency domain, which are commonly used for the computation of the horizontal electric field in the case of finitely conducting ground. The first one is termed the "wave-tilt formula" proposed by Zenneck [49], and the second one is the so-called "Cooray-Rubinstein formula", presented by Cooray [50] and by Rubinstein [42].

12.3.3.1 Wave-tilt formula

The term "wave tilt" originates from the fact that when a plane electromagnetic wave propagates over a finitely conducting ground, the total electric field at the ground surface is tilted from the vertical because of the presence of a nonzero horizontal (radial) electric field component. The tilt is in the direction of propagation if the vertical electric field component is directed upward, or in the opposite direction if the vertical electric field component is directed downward, with the vertical component of the Poynting vector being directed into the ground in both cases. The magnitude of this tilt, and hence the amplitude of the horizontal electric field, depends on the conductivity and the dielectric constant (relative permittivity) of the soil.

The wave-tilt formula sates that, for a plane wave, the ratio between the horizontal electric field E_r and the vertical electric field E_z in the frequency domain, is equal to the ratio of the propagation constants in the air and in the ground [49]. Therefore, the horizontal electric field component is given by:

$$E_r(r,z,\omega) = E_z(r,z,\omega)\frac{1}{\sqrt{\varepsilon_r + \sigma/(j\omega\varepsilon_0)}}, \tag{12.34}$$

where σ and ε_r are the conductivity and relative permittivity of the ground, respectively, and ω is the angular frequency. This formula is a special case (valid for grazing incidence) of the theory of the reflection of electromagnetic waves off a conducting surface and, hence, is a reasonable approximation only for relatively distant lightning strikes or at early times of close lightning when the return-stroke front is near ground. Typically, the vertical electric field, $E_z(r, z, \omega)$, in (12.34) is computed assuming that ground is a perfect conductor, or is measured.

Thomson *et al.* [51] presented an expression for the magnitude of the horizontal electric field component as a function of the magnitude of the vertical electric field component by using the time domain approach described by Master [52]. By using this technique, the vertical electric field can be approximated as a sequence of superimposed delay ramps, since a ramp has an analytical inverse transform expression like (12.34). The horizontal electric field is then determined in the time domain as the superposition of the responses to the ramps.

12.3.3.2 Cooray–Rubinstein formula

The Cooray–Rubinstein formula is expressed as follows [42,50]:

$$E_r(r, z, \omega) = E_{rp}(r, z, \omega) - H_{\varphi p}(r, 0, \omega) \frac{c\mu_0}{\sqrt{\varepsilon_r + \sigma/(j\omega\varepsilon_0)}}, \tag{12.35}$$

where μ_0 is the free space permeability, $E_{rp}(r, z, \omega)$ is the Fourier transform of the horizontal electric field at height z above ground and $H_{\varphi p}(r, 0, \omega)$ is the azimuthal magnetic field component at ground level, both computed as if the ground were a perfect conductor. The second term on the right-hand side of (12.35) is equal to zero for $\sigma \to \infty$ and becomes increasingly important as σ decreases.

This approach has been shown to produce satisfactory approximation of the horizontal electric field in a number of practically important cases. In particular, it reproduces the positive, bipolar, and negative waveforms at close (one hundred meters), intermediate (some kilometers), and far (tens of kilometers) distances, respectively, and at all these ranges it yields results that are close to those predicted by more accurate expressions [42,53,54].

12.4 Field-to-line coupling models

Once the electromagnetic field is calculated making use of a return-stroke model, it is used to calculate the voltages and currents induced on the conducting system. The most general and rigorous approach to these calculations is based on Maxwell's equations (e.g. [55]) However, due to the considerable length of power distribution lines, the use of such approach for the evaluation of lightning induced voltages is limited by long computation times. Further, the simplest approach known as "the quasi-static approximation" [56], according to which the propagation of the field can be neglected and coupling between incident fields and the line conductors can be described using lumped circuit elements is not appropriate. Such an approach requires that the dimensions of the line conductors be smaller than about one-tenth of the shortest significant wavelength of the electromagnetic field. This assumption is not valid for power lines excited by lightning electromagnetic field transients (characterized by above 1 MHz frequency, that is, below 300 m wavelength). Another possible approach, that is the most suitable for the problem of interest, is the distributed-circuit theory or "transmission line approximation" [57]. The basic assumptions of this theory are that the transverse dimension of the line is much smaller than the minimum significant wavelength and that the response of the line to the lightning fields is quasi-transverse electromagnetic (quasi-TEM), i.e., the electromagnetic field due to the electric charges and currents along the line is confined to the plane perpendicular (transverse) to the line axis. In this case, the line can be represented by a series of elementary sections, such as the quasi-static approximation applies to each of them. Each section is excited progressively by the incident electromagnetic field so that longitudinal propagation effects are taken into account.

In the literature, the most used coupling models adopted for lighting-induced voltage evaluation are based on the TL approximation. In this section, we shall briefly present three of them, which are the most popular ones:

- Taylor, Satterwhite, and Harrison model [57].
- Agrawal, Price, and Gurbaxani model [58].
- Rachidi model [59].

We will refer to a lossless TL composed of a single conductor (oriented along the x-axis) located at height h above a perfectly conducting ground and terminated in two impedances, Z_0 and Z_L (see Figure 12.4). The line is excited by an *incident* (external) electromagnetic field (E^i, B^i), shown in Figure 12.4, which is the sum of the field radiated by the lightning return stroke and its reflection from the ground, computed at the position of the conductor in its absence. The total field (E or B) is the sum of the incident field and the so-called *scattered* field which represents the reaction of the conductor to the incident field.

As we shall see, all the above-mentioned coupling models lead to a pair of equations involving time and space derivatives of induced voltage and current, as well as source terms, each expressed as a function of the incident electromagnetic field components. These equations are obtained using Maxwell's equations and geometry shown in Figure 12.4, as described in detail, for example, in [60].

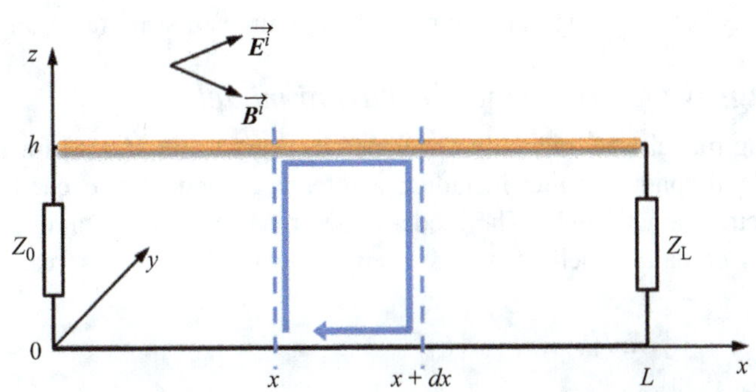

Figure 12.4 Geometry of the field-to-line coupling problem

It has been demonstrated in [61] that the three models considered here are fully equivalent, i.e., they yield identical results in terms of total induced voltage and total induced current.

It is also worth noting that an earlier field-to-line coupling model was developed by Rusck [31]. It has been shown by Nucci *et al.* [62] and by Cooray [53] that Rusck's equations, expressed in terms of scalar potential, are equivalent to the above-mentioned models, provided that the lightning channel is straight and vertical.

12.4.1 Taylor, Satterwhite, and Harrison model

This coupling model was originally developed by Taylor *et al.* [57] for a two-wire TL excited by a general external electromagnetic field. Later (e.g., [60]), it was also formulated for a single wire above ground. The model is described by two coupling equations which in the frequency domain read as follows:

$$\begin{cases} \dfrac{dV(x)}{dx} + j\omega L' I(x) = -j\omega \displaystyle\int_0^h B_y^i(x,z)dz, \\ \dfrac{dI(x)}{dx} + j\omega C' V(x) = -j\omega C' \displaystyle\int_0^h E_z^i(x,z)dz, \end{cases} \tag{12.36}$$

where $V(x)$ is the total induced voltage along the line, $I(x)$ is the total induced current along the line, C' and L' are the per-unit-length capacitance and inductance of the line, respectively; $E_x^i(x,z)$ is the horizontal component of the incident electric field along the line, and $B_y^i(x,z)$ is the azimuthal component of the incident magnetic field (magnetic flux density).

The boundary conditions are

$$V(0) = -Z_0 I(0), \tag{12.37}$$

$$V(L) = Z_L I(L). \tag{12.38}$$

The terms on the right-hand side of (12.36) are the source (excitation) terms, which are expressed in terms of the transverse (y-directed) magnetic and vertical (z-directed) electric field components of the incident field. The equivalent circuit for this coupling model is shown in Figure 12.5. The source terms in (12.36) are represented by series voltage and shunt current sources distributed along the line.

12.4.2 Agrawal, Price, and Gurbaxani model

This coupling model was originally developed by Agrawal *et al.* [58] for the general case of a multi-conductor line including a reference conductor, excited by a non-uniform electromagnetic field. The model is described by two coupling equations in the frequency domain, which, for the geometry shown in Figure 12.4, read as follows:

$$\begin{cases} \dfrac{dV^s(x)}{dx} + j\omega L' I(x) = E_x^i(x,h), \\ \dfrac{dI(x)}{dx} + j\omega C' V^s(x) = 0, \end{cases} \tag{12.39}$$

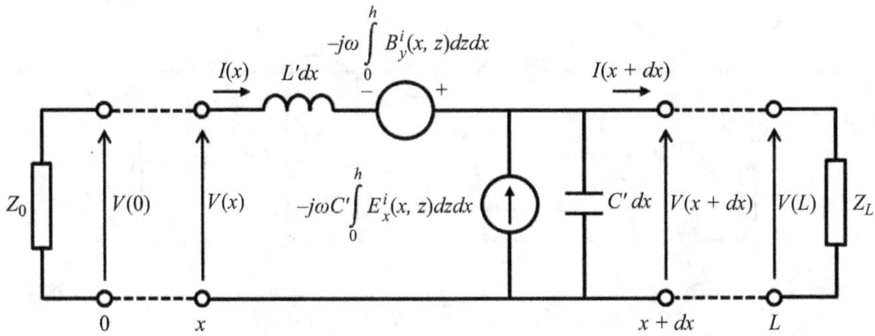

Figure 12.5 Taylor et al. model

where $E_x^i(x,h)$ is the horizontal component of the incident electric field along the line, $C^{'}$ and $L^{'}$ are the per-unit-length capacitance and inductance of the line, respectively, $I(x)$ is the induced current and $V^s(x)$ is the scattered voltage, related to the total voltage $V(x)$ by the following equation:

$$V^s(x) = V(x) - V^i(x) = V(x) + \int_0^h E_z^i(x,z)dz. \tag{12.40}$$

$V^i(x)$ is called the incident (or exciting) voltage, and $E_z^i(x,z)$ is the vertical component of the incident electric field.

The boundary conditions, written in terms of the scattered voltages and the total current are given by

$$V^s(0) = -Z_0 I(0) + \int_0^h E_z^i\,(0,\, z)dz, \tag{12.41}$$

$$V^s(L) = Z_L I(L) + \int_0^h E_z^i\,(L,\, z)dz. \tag{12.42}$$

The equivalent circuit for this coupling model is shown in Figure 12.6. For this model, the source term (forcing function) is the incident electric field tangential to the line conductor, which is represented by series voltage sources distributed along the line. According to boundary conditions (12.41) and (12.42), two lumped voltage sources are inserted at the line terminations.

Note that in the formulation of Agrawal *et al.*, the source term is a function of the horizontal component of the incident electric field only, whereas in the formulation of Taylor *et al.*, the source terms are functions of both electric and magnetic field components.

12.4.3 Rachidi model

The model proposed by Rachidi [59] is sometimes referred to as dual to that of Agrawal *et al.*, because each of them contains a single source terms and in Rachidi

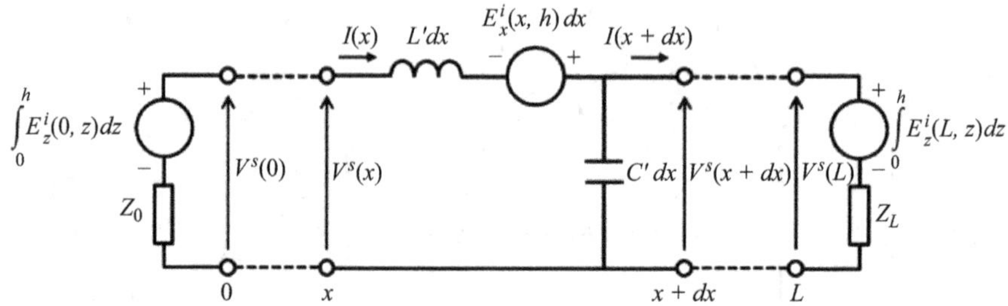

Figure 12.6 Agrawal et al. model

model (12.43) it appears in the second equation, while it is present in the first equation in the Agrawal *et al.* model (12.39). According to Rachidi model,

$$\begin{cases} \dfrac{dV(x)}{dx} + j\omega L' I^s(x) = 0, \\ \dfrac{dI^s(x)}{dx} + j\omega C' V(x) = \dfrac{1}{L'} \displaystyle\int_0^h \dfrac{\partial B^i_x(x,z)}{\partial y} dz, \end{cases} \tag{12.43}$$

where $B^i_x(x,z)$ is the x-directed horizontal component of the incident magnetic field along the line, C' and L' are the per-unit-length capacitance and inductance of the line, and $I^s(x)$ is the scattered current, related to the total current by the following equation:

$$I^s(x) = I(x) - I^i(x) = I(x) + \frac{1}{L'} \int_0^h B^i_y(x,z)dz. \tag{12.44}$$

$I^i(x)$ is called the incident (or exciting) current, and $B^i_y(x,h)$ is the y-directed horizontal component of the incident magnetic field normal to the line.

The boundary conditions, written in terms of the scattered current are:

$$I^s(0) = -\frac{V(0)}{Z_0} + \frac{1}{L'} \int_0^h B^i_y\,(0,\ z)dz, \tag{12.45}$$

$$I^s(L) = \frac{V(L)}{Z_L} + \frac{1}{L'} \int_0^h B^i_y\,(L,\ z)dz. \tag{12.46}$$

In this formulation, the source terms are expressed only in terms of magnetic field components. This model is particularly useful when the exciting field is determined experimentally, since only the measurement of magnetic field (generally much easier than that of electric field) is needed.

The equivalent circuit of this coupling model is shown in Figure 12.7. The source term is represented by shunt current sources distributed along the line. To satisfy boundary conditions (12.45) and (12.46), two lumped current sources are connected at line terminations.

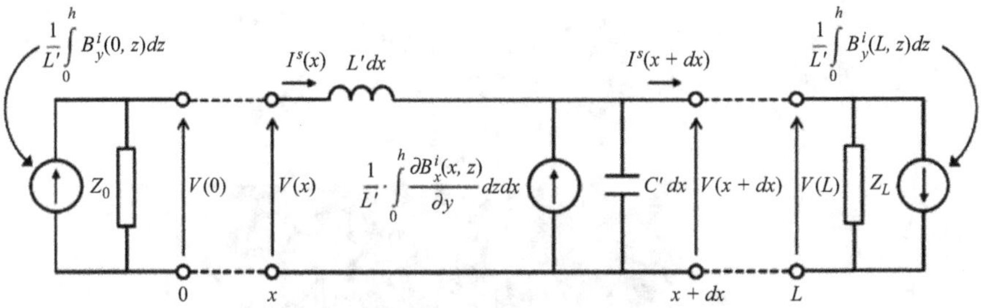

Figure 12.7 Rachidi model

12.5 Analytical solutions for lightning-induced voltage calculations

As noted above, to determine the induced voltages on overhead lines the following three steps are to be followed. First, a channel-base current is to be specified and a return stroke model is to be adopted (the most popular options have been reviewed in Sections 12.1 and 12.2). Then, the resultant electromagnetic field must be evaluated. Finally, this field and a coupling model are to be used for calculating induced voltages. Both numerical (e.g., [63–67]) and analytical (e.g., [4,12,31,68–70]) approaches have been proposed for these calculations. Numerical approaches can treat power network geometries of virtually any complexity and provide a better approximation to the reality. For example, they allow a better approximation of the actual return-stroke current waveform and an easier analysis of the effects of nonlinearities, such as the influence of surge arresters. Analytical solutions are much more limited in those respects: a number of simplifying assumptions, e.g., representation of channel-base current by a step or trapezoidal function, have to be made. However, analytical solutions are much more practical than numerical ones at the design stage, since compact and straightforward solutions are needed at this stage. They are also fundamental in carrying out sensitivity and parametric assessments (e.g., [71]). Furthermore, they give an insight into the physical mechanisms, which are often obscured in numerical solutions. We will now review the analytical solutions found in the literature.

12.5.1 Step-function channel-base current

The simplest configuration is represented by a single lossless conductor horizontally extending above a perfectly conducting ground plane, which is illuminated by the electromagnetic field generated by a step-function channel-base current wave. The wave propagates at constant velocity, without attenuation or distortion, along a straight vertical lightning channel (as per the TL model [28]). Figure 12.8(a) shows this basic configuration.

Despite the extreme simplification of the problem, many attempts have been made in the literature to obtain the solution, with some of those solutions being

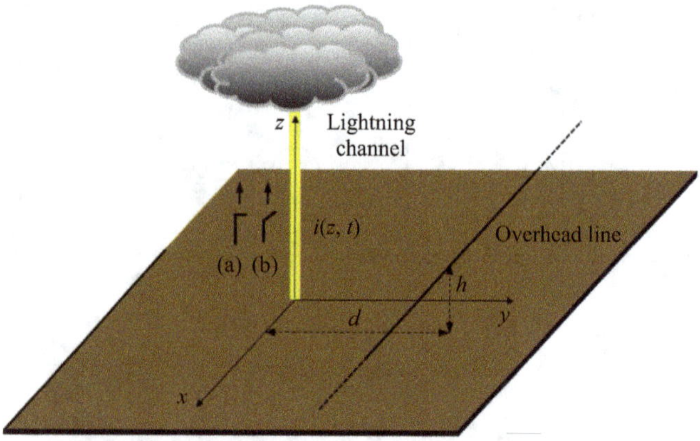

Figure 12.8 Lightning channel and infinitely long overhead conductor above ground: (a) step-function current and (b) linearly rising current

Table 12.2 Solutions for step-function current

Solution (reference)	Electromagnetic field model	Coupling model	Type of solution	Notes
1. Andreotti *et al.* [4]	Monopole technique [1,40]	Taylor *et al.* [57]	Exact	
2. Rusck [1]	Monopole technique [1,40]	Rusck [1]	Approximate[a]	
3. Høidalen [12]	Dipole technique [28,40]	Agrawal *et al.* [58]	Approximate[a]	
4. Chowdhuri and Gross [32]	Monopole technique [1,40]	Chowdhuri–Gross [32]	Approximate (via *s*-domain)	Incorrect coupling model/ Incorrect procedure
5. Liew and Mar [72]	Monopole technique [1,40]	Chowdhuri–Gross/ Jakubowski correction [72]	Approximate (via *s*-domain)	Incorrect procedure
6. Haldar and Liew [73]	Monopole technique [1,40]	Chowdhuri–Gross/ Jakubowski correction [72]	Approximate (via *s*-domain)	Incorrect procedure

[a]First term of Taylor expansion about $h = 0$.

incorrect. The exact solution to this problem has been obtained relatively recently. In Table 12.2, the solutions are summarized, along with the models employed.

It can be seen that there are three solutions which have been obtained directly in the time domain and three solutions obtained via Laplace transforms. The first three are all related, in the sense that two of them (numbered 2 and 3) are first order approximations to the exact solution (numbered 1). Unfortunately, all three solutions derived in the Laplace domain and numbered 4 to 6 are incorrect. We now analyze the solutions in more detail.

12.5.1.1 Rusck solution

The first analytical solution for the calculation of the lightning-induced voltage assuming a step function for the channel-base current was proposed by Rusck in 1958 [1, Eqs. (104) and (105)]. This solution, derived directly in the time domain, has been obtained by using as a coupling model a model developed by Rusck himself, which for this reason is named the "Rusck coupling model" [1]. This model was found to be incorrect since it is missing one of the source terms, which makes the model "source-dependent": it provides correct results only if the source of the electromagnetic field is spatially distributed so that the contribution of the vector potential to the horizontal electric field is zero, as in the case of a vertical lightning channel. This limitation in the coupling equations was identified by Cooray [53] through a comparison with the Agrawal *et al.* model [58].

12.5.1.2 Chowdhuri–Gross solution

Another approximate solution has been proposed by Chowdhuri and Gross in 1965 [33, Eqs. (20)–(25)]. This equation has been obtained in the Laplace domain by using as coupling model a model developed by the authors themselves, known as "Chowdhuri-Gross coupling model" [4]. In this case too, one of the source terms of the electromagnetic field is missing and, as a result, the coupling with the line is modeled incorrectly. This inaccuracy was identified by Nucci *et al.* [61], comparing the Chowdhuri–Gross model with the one proposed by Agrawal *et al.* [58]. Due to this error, the solution proposed by Chowdhuri and Gross can provide very inaccurate results for some problem geometries, even for the vertical channel case.

12.5.1.3 Liew–Mar solution

Liew and Mar have published their solution in 1986 [76, Eq. (27)]. Although they used the Rusck coupling model, and applied the same mathematical and geometrical approximations as those used by Rusck [1], their expression is completely different in analytical terms than the one proposed by Rusck. It yields incorrect results, as shown by Andreotti *et al.* [4].

12.5.1.4 Høidalen solution

Høidalen derived his solution in 1990 [75, Eq. (13)]. This equation is based on the Agrawal *et al.* coupling model [57], and it is identical to the Rusck solution, although it has been obtained by considering a different coupling model. As shown by Andreotti *et al.* [4], Høidalen's expression gives correct results.

12.5.1.5 Andreotti *et al.* solution

As pointed out before, Andreotti *et al.*, in 2009, derived the exact analytical solution for the considered case [4, Eqs. (51)–(55)]. The expression is exact because it has been obtained without mathematical approximations; it is based on the Taylor *et al.* coupling model [57]. The authors demonstrated in [4] that the Rusck solution corresponds to the first order approximation of this exact solution.

In [78, Eqs. (17)–(19)], the same solution has been presented in a simpler form:

$$\begin{aligned} v(x,t) &= \zeta_0' I[-v_1(h) + v_1(-h) + v_2(\beta ct, h) \\ &\quad + v_2(-\beta ct, h) - v_2(\beta ct, -h) - v_2(-\beta ct, -h) \\ &\quad + v_3(\beta ct) + v_3(-\beta ct) - v_4] \end{aligned} \tag{12.47}$$

where

$$v_1(h) = \ln\left\{[-(\beta ct - h) + \beta x]^2 + d^2\right\},$$

$$v_2(\beta ct, h) = \ln\left[\beta(\beta ct - h) + x/\gamma^2 + \sqrt{(\beta ct - h)^2 + (d^2 + x^2)/\gamma^2}\right],$$

$$v_3(\beta ct) = 2\beta \ln\left[-\beta ct + h + \sqrt{(h - \beta ct)^2 + (d^2 + x^2)/\gamma^2}\right],$$

$$v_4 = 2\beta \ln\left[(d^2 + x^2)/\gamma^2\right].$$

Symbols meaning can be found in [74]. As an example, a 3D plot of the induced voltage as a function of time t and position along the line x is shown in Figure 12.9.

12.5.2 Linearly rising channel-base current

The linearly rising case is depicted in Figure 12.8(b). The linearly rising channel-base current represents an important improvement compared to the step function, because one can account for the current "rise time," which is a fundamental

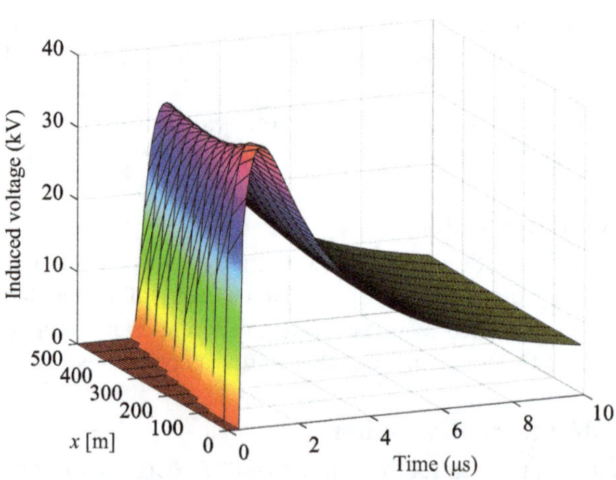

Figure 12.9 3D plot of the induced voltage computed using (12.47) (x is the point along the line where the induced voltage is evaluated, h = 10 m, d = 100 m, I = 10 kA, β = 0.4)

Table 12.3. Solutions for linearly rising current

Solution (reference)	Electromagnetic field model	Coupling model	Type of solution	Notes
1. Andreotti *et al.* [8]	Monopole technique [1,40]	Taylor *et al.* [57]	Exact	Available only for $x = 0$
2. Andreotti *et al.* [8]	Monopole technique [1,40]	Taylor *et al.* [57]	Approximate	
3. Rusck [1]	Monopole technique [1,40]	Rusck [1]	Approximate[a]	
4. Sekioka [13]	Monopole technique [1,40]	Agrawal *et al.* [57]	Approximate[a]	
5. Chowdhuri and Gross [32]	Monopole technique [1,40]	Chowdhuri–Gross [32]	Approximate (via s-domain)	Incorrect coupling model/ Incorrect procedure
6. Liew and Mar [72]	Monopole technique [1,40]	Chowdhuri–Gross/ Jakubowski correction [72]	Approximate (via s-domain)	Incorrect procedure
7. Haldar and Liew [73]	Monopole technique [1,40]	Chowdhuri–Gross/ Jakubowski correction [72]	Approximate (via s-domain)	Incorrect procedure

parameter in assessing the induced voltage. Analytical solutions for this case are summarized in Table 12.3.

12.5.2.1 Rusck solution

Rusck was the first to give an analytical solution for the case of a linearly rising channel-base current [1, Eq. (132)]. Based on a well-known property of linear systems, Rusck derived this equation by performing a convolution with the solution obtained for the step-function current case [1, Eqs. (104) and (105)], i.e., the step response of the system, and a ramp (linearly-rising) function.

12.5.2.2 Sekioka solution

Sekioka, in 2005, directly derived a time-domain solution, without any convolution with step-function channel base current. This solution can be found in [13, Eq. (12)], and has been obtained by using the field-to-line coupling model proposed by Rusck [1]. As demonstrated in [8], the Sekioka solution corresponds to the first order approximation of the Andreotti *et al.* exact solution, that will be discussed below.

12.5.2.3 Chowdhuri–Gross, Liew–Mar, and Haldar–Liew solutions

By adopting the same approach as Rusck, Chowdhuri, and Gross [33, Eqs.(32) and (33)], Liew and Mar [76, Eq. (28), and Haldar and Liew [77, Eq. (18)], derived their solutions for the linearly rising current case.

12.5.2.4 Andreotti *et al.* solution

Andreotti *et al.* have derived an exact analytical solution [8, Eq. (31)],] by, evaluating, as done by Rusck, the convolution between the step-current solution [4, Eqs. (51)–(55)] and a ramp function. This solution reads:

$$v_r(0,t) = v'_{r1}(0,t) - v''_{r1}(0,t) - v'_{r2}(0,t) + v''_{r2}(0,t), \tag{12.48}$$

where

$$\begin{aligned} v'_{r1}(0,t) = -\frac{\mu_0}{4\pi}\alpha\Bigg\{\lambda\Bigg\{\Bigg\{\ln\left(\lambda+\sqrt{\lambda^2+\delta^2}\right)+\frac{1}{\beta}\left[\ln\left(-\beta\lambda+\sqrt{\lambda^2+\delta^2}\right)-1\right]\Bigg\} \\ +\frac{d}{\beta}\left[\operatorname{atan}\left(\frac{\lambda}{d}\right)-\operatorname{atan}\left(\frac{\beta d}{\sqrt{\lambda^2+\delta^2}}\right)\right]-\sqrt{\lambda^2+\delta^2}\Bigg\}u(t-t_0). \end{aligned} \tag{12.49}$$

$v''_{r1}(0,t)$ can be obtained by replacing λ with λ_0 in $v'_{r1}(0,t)$, whereas $v'_{r2}(0,t)$ and $v''_{r2}(0,t)$ can be obtained by reversing the sign of h in $v'_{r1}(0,t)$ and $v''_{r1}(0,t)$, respectively. Symbol meanings can be found in [8]. We note that this exact solution can be applied only for the evaluation of induced voltages at the point of the overhead line closest to the lightning channel. An approximate solution for voltages induced at any point along the line is found in [8].

References

[1] S. Rusck, "Induced lightning overvoltages on power transmission lines with special reference to the overvoltage protection of low-voltage networks," *Trans. R. Inst. Technol.*, no. 120, pp. 1–118, 1958.

[2] P. P. Barker, T. A. Short, A. Eybert-Berard, and J. B. Berlandis, "Induced voltage measurements on an experimental distribution line during nearby rocket triggered lightning flashes", *IEEE Trans. Power Del.*, vol. 11, no. 2, pp. 980–995, 1996.

[3] M. Darveniza, "A practical extension of Rusck's formula for maximum lightning induced voltages that accounts for ground resistivity," *IEEE Trans. Power Del.*, vol. 22, no. 1, pp. 605–612, 2007.

[4] A. Andreotti, D. Assante, F. Mottola, and L. Verolino, "An exact closed-form solution for lightning-induced overvoltages calculations," *IEEE Trans. Power Del.*, vol. 24, no. 3, pp. 1328–1343, 2009.

[5] J. O. S. Paulino, C. F. Barbosa, I. J. S. Lopes, and W. C. Boaventura, "An approximate formula for the peak value of lightning-induced voltages in overhead lines," *IEEE Trans. Power Del.*, vol. 25, no. 2, pp. 843–851, 2010.

[6] A. Piantini and J. M. Janiszewski, "Lightning-induced voltages on overhead lines—application of the extended Rusck model," *IEEE Trans. Electromagn. Compat.*, vol. 51, no. 3, pp. 548–558, 2009.

[7] A. Piantini, "Extension of the Rusck model for calculating lightning-induced voltages on overhead lines considering the soil electrical parameters," *IEEE Trans. Electromagn. Compat.*, vol. 59, no. 1, pp. 154–162, 2017.

[8] A. Andreotti, A. Pierno, V. A. Rakov, and L. Verolino, "Analytical formulations for lightning-induced voltage calculations," *IEEE Trans. Electromagn. Compat.*, vol. 55, no. 1, pp. 109–123, 2013.

[9] A. Andreotti, A. Pierno, and V. A. Rakov, "An analytical approach to calculation of lightning induced voltages on overhead lines in case of lossy ground – Part I: model development," *IEEE Trans. Power Del.*, vol. 28, no. 2, pp. 1213–1223, 2013.

[10] A. Andreotti, A. Pierno, and V. A. Rakov, "An analytical approach to calculation of lightning induced voltages on overhead lines in case of lossy ground – Part II: comparison with other models," *IEEE Trans. Power Del.*, vol. 28, no. 2, pp. 1224–1230, 2013.

[11] A. R. Rodrigues, G. C. Guimarães, M. L. R. Chaves, W. do Couto Boaventura, D. A. Caixeta, and M. A. Tamashiro, "Lightning performance of transmission lines based upon real return-stroke current waveforms and statistical variation of characteristic parameters", *Electric Power Syst. Res.*, vol. 153, pp. 46–59, 2017.

[12] H. K. Høidalen, "Analytical formulation of lightning-induced voltages on multiconductor overhead lines above lossy ground," *IEEE Trans. Electromagn. Compat.*, vol. 45, no. 1, pp. 92–100, 2003.

[13] S. Sekioka, "An equivalent circuit for analysis of lightning-induced voltages on multiconductor system using an analytical expression," in *Proc. Int. Conf. Power Syst. Trans.*, Montreal, Canada, pp. 1–6, June 2005.

[14] V.A. Rakov and A.A. Dulzon, "Results of calculation of the electromagnetic fields of lightning discharges," *Tekh. Elektrodinam.*, vol. 1, pp. 87–89, 1987.

[15] J. O. S. Paulino, C. F. Barbosa, I. J. S. Lopes, and W. C. Bonaventura, "The peak value of lightning-induced voltages in overhead lines considering the ground resistivity and typical return stroke parameters," *IEEE Trans. Power Del.*, vol. 26, no. 2, pp. 920–926, 2011.

[16] F. Heidler, "Travelling current source model for LEMP calculation," in *Proc. 6th Symp. Electromagn. Compat.*, Zurich, Switzerland, pp. 157–162, 1985.

[17] C. E. R. Bruce and R. H. Golde, "The lightning discharge," *J. Inst. Electr. Eng.*, vol. 88, Part II, no. 6, pp. 487–505, 1941.

[18] I. S. Stekolnikov, "Lightning stroke parameters and computations of the current waveform of the direct stroke," *Electricity*, vol. 2, pp. 217–229, 1941.

[19] G. Diendorfer and M. A. Uman, "An improved return stroke model with specified channel-base current," *J. Geophys. Res.*, vol. 95, no. D9, pp. 13621–13644, 1990.

[20] A. Piantini, "Lightning protection of overhead power distribution lines," in *Proc. 29th Intl Conf. Lightning Protection*, pp. 1–29, Uppsala, Sweden, 2008.

[21] A. Piantini, J. M. Janiszewski, and C. P. Braz, "Utilization of reduced models for the analysis of lightning induced overvoltages on overhead lines," in *Proc. 2010 Intl. Conf. High Voltage Engineering and Application*, pp. 160–163, 2010.

[22] C. A. Nucci, G. Diendorfer, M. A. Uman, F. Rachidi, M. Ianoz, and C. Mazzetti, "Lightning return stroke current models with specified channel-base current: a review and comparison," *J. Geophys. Res.*, vol. 95, no. D12, pp. 20395–20408, 1990.

[23] V. A. Rakov and A. A. Dulzon, "A modified transmission line model for lightning return stroke field calculations," in *Proc. 9th Int. Zurich. Symp. Electromagn. Compat.*, Zurich, Switzerland, pp. 229–235, 1991.

[24] A. De Conti and S. Visacro, "Analytical representation of single- and double-peaked lightning current waveforms," *IEEE Trans. Electromagn. Compat*, vol. 49, no. 2, pp. 448–451, 2007

[25] A. Andreotti, A. Pierno and L. Verolino, "A new channel-base current model for lightning-induced voltage calculations," *IEEE Trans. Electromagn. Compat*, vol. 61, no. 3, pp. 617–622, 2019.

[26] V. A. Rakov and M. A. Uman, "Review and evaluation of lightning return stroke models including some aspects of their application," *IEEE Trans. Electromagn. Compat.*, vol. 40, no. 4, pp. 403–426, 1998.

[27] F. Heidler, "Travelling current source model for LEMP calculation," in *Proc. 6th Symp. on Electromagn. Compat.*, Zurich, Switzerland, pp. 157–162, 1985.

[28] M. A. Uman, D. K. McLain, "Magnetic field of lightning return stroke," *J. Geophys. Res.*, vol. 74, no. 28, pp. 6899–6910, 1969.

[29] C. A. Nucci, C. Mazzetti, F. Rachidi, and M. Ianoz, "On lightning return stroke models for LEMP calculations," in *Proc. 19th Int. Conf. Lightning Protection*, Graz, Austria, April 25–29, pp. 463–470, 1988.

[30] R. Thottappillil, V. A. Rakov, and M. A. Uman, "Distribution of charge along the lightning channel: relation to remote electric and magnetic fields and to return stroke models," *J. Geophys. Res.*, vol. 102, no. D6, pp. 6987–7006, 1997.

[31] S. Rusck, "Induced lightning over-voltages on power transmission lines with special reference to the overvoltage protection of low-voltage networks," *Trans. Royal Inst. Tech.*, no. 120, pp. 1–118, 1958.

[32] P. Chowdhuri and E. T. B. Gross, "Voltage surges induced on overhead lines by lightning strokes," in *Proc. Inst. Elect. Eng.*, vol. 114, no. 12, pp. 1899–1907, 1967.

[33] M. Rubinstein and M. A. Uman, "Methods for calculating the electromagnetic fields from a known source distribution: application to lightning," *IEEE Trans. Electromagn. Compat.*, vol. 31, no. 2, pp. 183–189, 1989.

[34] J. A. Stratton, *Electromagnetic Theory*, New York, NY: McGraw-Hill, 1941.

[35] J. D. Jackson, *Classical Electrodynamics*, New York, NY: Wiley, 1962.

[36] J. R. Reitz, F. J. Milford, and R. W. Christy, *Foundations of Electromagnetic Theory*, 3rd ed., Reading, MA: Addison-Wesley, 1980.

[37] M. A. Uman, D. K. McLain, and E. P. Krider. "The electromagnetic radiation from a finite antenna", *Am. J. Phys.*, vol. 43, 1975.

[38] M. J. Master and M. A. Uman, "Transient electric and magnetic fields associated with establishing a finite electrostatic dipole," *American J. Phys.*, vol. 51, no. 2, pp. 118–126, 1983.

[39] V. A. Rakov, *Fundamentals of Lightning*, Cambridge: Cambridge University Press, 2016.

[40] R. Thottappillil and V. A. Rakov, "On different approaches to calculating lightning electric fields", *J. Geophys. Res. Atmosp.* vol. 106, n. D13, pp. 14191–14205, 2001.

[41] B. Djebari, J. Hamelin, C. Leteinturier, and J. Fontaine, "Comparison between experimental measurements of the electromagnetic field emitted by lightning and different theoretical models. Influence of the upward velocity of the return stroke," in *Proc. 4th Int. Symp. Electromagn. Compat.*, Zurich, Switzerland, March 1981.

[42] M. Rubinstein, "An approximate formula for the calculation of the horizontal electric field from lightning at close, intermediate, and long range," *IEEE Trans. Electromagn. Compat.*, vol. 38, no. 3, pp. 531–535, 1996.

[43] M. J. Master and M. A. Uman, "Lightning induced voltages on power lines: theory," *IEEE Trans. Power Appar. Syst.*, vol. 103, no. 9, pp. 2502–2518, 1984.

[44] M. Rubinstein, A. Y. Tzeng, M. A. Uman, P. J. Medelius, and E. M. Thomson, "An experimental test of a theory of lightning-induced voltages on an overhead wire," *IEEE Trans. Electromagn. Compat.*, vol. 31, no. 4, pp. 376–383, 1989.

[45] G. Diendorfer, "Induced voltage on an overhead line to nearby lightning," *IEEE Trans. Electromagn. Compat.*, vol. 32, no. 4, pp. 292–299, 1990.

[46] C. A. Nucci, F. Rachidi, M. Ianoz, and C. Mazzetti, "Lightning-induced overvoltages on overhead lines," *IEEE Trans. Electromagn. Compat.*, vol. 35, no. 1, pp. 75–86, 1993.

[47] M. Ishii, K. Michishita, Y. Hongo, and S. Ogume, "Lightning-induced voltage on an overhead wire dependent on ground conductivity," *IEEE Trans. Power Del.*, vol. 9, no. 1, pp. 109–118, 1994.

[48] A. Sommerfeld, "On the propagation of waves in wireless telegraphy," *Ann. Phys.*, vol. 81, no. 25, pp. 1135–1153, 1926.

[49] J. Zenneck, *Wireless Telegraphy*, New York, NY: McGraw-Hill, 1915 (English translation, A. E. Seelig).

[50] V. Cooray, "Horizontal fields generated by return strokes," *Radio Sci.*, vol. 27, no. 4, pp. 529–537, 1992.

[51] E. M. Thomson, P. J. Medelius, M. Rubinstein, M. A. Uman, J. Johnson, and J. W. Stone, "Horizontal electric fields from lightning strokes," *J. Geophys. Res.*, vol. 93, no. D3, pp. 2429–2441, 1998.

[52] M. J. Master, "Lightning induced voltages on power lines: theory and experiment," *Ph.D. dissertation*, University of Florida, Gainesville, 1982.

[53] V. Cooray, "Calculating lightning-induced overvoltages in power lines: a comparison of two coupling models," *IEEE Trans. Electromagn. Compat.*, vol. 36, pp. 179–182, 1994.

[54] F. Rachidi, C. A. Nucci, M. Ianoz, and C. Mazzetti, "Influence of a lossy ground on lightning-induced voltages on overhead lines," *IEEE Trans. Electromagn. Compat.*, vol. 38, no. 3, pp. 250–264, 1996.

[55] Y. Baba and V. A. Rakov, "The finite-difference time domain method for solving Maxwell's equations," in *Electromagnetic Computation Methods for Lightning Surge Protection Studies*, IEEE, 2016, pp. 43–72, 2019.

[56] C. T. A. Johnk, *Engineering Electromagnetic Field and Waves,* New York, NY: John Wiley and Sons, 1975.

[57] C. D. Taylor, R. S. Satterwhite, and C. W. Harrison, "The response of a terminated two-wire transmission line excited by a non-uniform electromagnetic field," *IEEE Trans. Antennas Propagat.*, vol. 13, no. 6, pp. 987–989, 1965.

[58] A. K. Agrawal, H. J. Price, and S. H. Gurbaxani, "Transient response of multiconductor transmission lines excited by a nonuniform electromagnetic field," *IEEE Trans. Electromagn. Compat.*, vol. 22, no. 2, pp. 119–129, 1980.

[59] F. Rachidi, "Formulation of the field-to-transmission line coupling equations in terms of magnetic excitation field," *IEEE Trans. Electromagn. Compat.*, vol. 35, no. 3, pp. 404–407, 1993.

[60] Y. Baba and V.A. Rakov "Lightning-induced Effects in Electrical and Telecommunication Systems", IET, 2020.

[61] C. A. Nucci, and F. Rachidi, "On the contribution of the electromagnetic field components in field-to-transmission line interaction," *IEEE Trans. Electromagn. Compat.*, vol. 37, no. 4, pp. 505–508, 1995.

[62] C. A. Nucci, F. Rachidi, M. Ianoz, and C. Mazzetti, "Comparison of two coupling models for lightning induced overvoltage calculations," *IEEE Trans. Power Del*, vol. 10, pp. 330–338, 1995.

[63] F. Rachidi, C. A. Nucci, and M. Ianoz, "Transient analysis of multiconductor lines above a lossy ground," *IEEE Trans. Power Del.*, vol. 14, no. 1, pp. 294–302, 1999.

[64] G. Diendorfer, "Induced voltage on an overhead line due to nearby lightning", *IEEE Trans. Electromagn. Compat.*, vol. 32, no. 4, pp. 292–299, 1990.

[65] H. K. Høidalen, J. Slebtak, and T. Henriksen, "Ground effects on induced voltages from nearby lightning," *IEEE Trans. Electromagn. Compat.*, vol. 39, no. 4, pp. 269–278, 1997.

[66] A. Andreotti, U. De Martinis, A. Maffucci, and G. Miano, "A mixed frequency and time domain approach for accurate evaluation of the dynamics of LEMP excited lossy multiconductor power lines," *Electrical Engineering*, vol. 83, no. 3, pp. 147–155, 2001.

[67] A. Andreotti, U. De Martinis, and L. Verolino, "A spice simulation of lightning induced overvoltages on power transmission lines", *Adv. Eng. Softw.*, vol. 31, no. 10, pp. 757–762, 2000.

[68] J. O. S. Paulino, C. F. Barbosa, and W. C. Boaventura, "Effect of the surface impedance on the induced voltages in overhead lines from nearby lightning," *IEEE Trans. Electromagn. Compat.*, vol. 53, no. 3, pp. 749–754, 2011.
[69] F. Napolitano, "An analytical formulation of the electromagnetic field generated by lightning return strokes," *IEEE Trans. Electromagn. Compat.*, vol. 53, no. 1, pp. 108–113, 2011.
[70] C. F. Barbosa and J. O. S. Paulino, "A time-domain formula for the horizontal electric field at the earth surface in the vicinity of lightning," *IEEE Trans. Electromagn. Compat.*, vol. 52, no. 3, pp. 640–645, 2010.
[71] P. Chowdhuri, "Parametric effects on the induced voltages on overhead lines by lightning strokes to nearby ground," *IEEE Trans. Power Del.*, vol. 4, no. 2, pp. 1185–1194, 1989.
[72] A. C. Liew and S. C. Mar, "Extension of the Chowdhuri-Gross model for lightning induced voltage on overhead lines," *IEEE Trans. Power Syst.*, vol. 1, no. 2, pp. 240–247, 1986.
[73] M. K. Haldar and A. C. Liew, "Alternative solution for the Chowdhuri-Gross model of lightning-induced voltages on power lines," *IEE Proc. C: Gener. Transmiss. Distrib.*, vol. 135, no. 4, pp. 324–329, 1988.
[74] A. Andreotti, V. A. Rakov, and L. Verolino "Exact and approximate analytical solutions for lightning-induced voltage calculations," *IEEE Trans. Electromagn. Compat.*, vol. 60, no. 6, pp. 1850–1856, 2018.

Chapter 13
Transient behaviour of grounding systems

Bo Zhang[1] *and Jinliang He*[1]

Grounding is a hot topic in electrical engineering, especially when it is necessary to deal with its transient behaviour under a lightning surge where inductive and conductive couplings among elements and the soil ionisation are not negligible. The chapter will introduce to the nonlinear and dynamic phenomenon caused by soil ionisation and the frequency variation of soil parameters first, and then the numerical methods applied to simulate the transient behaviour of grounding grids.

13.1 Introduction of the transient behaviour of grounding systems

The impulse performance of the grounding system, which is used to release the lightning current to the earth, plays an important role in lightning protection. The grounding system is required to have "sufficiently low impedance and current-carrying capacity to prevent the buildup of voltages that may result in undue hazard to connected equipment and to persons" [1]. Unlike that at power frequency, the grounding system under a lightning surge is not only nonlinear and dynamic, but also frequency dependent.

First, the performance of the grounding system under a lightning surge is nonlinear and dynamic. Due to the large lightning current, the current density in the soil near the grounding electrode may be very large, and hence the corresponding electric field may be strong. When it reaches a threshold, the soil discharge will occur [2,3]. The discharge develops with the development of lightning current, which shows dynamic and nonlinear characteristics. Because the lightning current contains a lot of high-frequency components and the longitudinal impedance of the grounding electrode under high frequency is very large, the grounding system can no longer be regarded as equipotential under lightning strike. The longitudinal impedance of the grounding electrode prevents current from flowing far away through it. Therefore, under a lightning strike, the current leaks into the earth mainly from the grounding conductor near the current injection point, and there is little current dissipating through the remote grounding electrode if the electrode is

[1]Department of Electrical Engineering, Tsinghua University, China

long. Thus, the current discharge area of the grounding system is limited. Due to this reason, under the same peak of the injection current the current density of the soil near the current injection point during lightning strikes is higher than that at power frequency, which intensifies the discharge in the soil.

Second, the performance of the grounding system under a lightning surge is also frequency dependent. This is not only because of the skin effect in the grounding electrode, a large number of studies have also found that the soil parameters vary with frequency [4,5]. The resistivity and permittivity are large at low frequencies, but gradually decrease as the frequency increases. Therefore, the performance of the grounding system under a lightning surge is frequency dependent.

In order to obtain the accurate transient behaviour of grounding systems, the effect of the nonlinear and dynamic phenomenon caused by soil ionisation and the effect caused by the frequency variation of soil parameters should be considered.

13.2 Discharge characteristics of soil under a lightning surge

13.2.1 Soil ionisation process

When a lightning impulse current is injected into a grounding electrode, the soil neighbouring the grounding electrode may be ionized. The discharge makes the resistivity of ionised soil decrease to a low value. Jinliang reviewed the progress in study on soil ionisation under lightning [6]. Figure 13.1 shows that a soil resistivity profile decays as the applied impulse current rises [7]. The soil ionisation process can be classified into four zones, which is the further modified model from the four-zone model in [7] and the three-zone model proposed by Liew and Darveniza [8]; the model of Liew is presented by the gravy region in Figure 13.1.

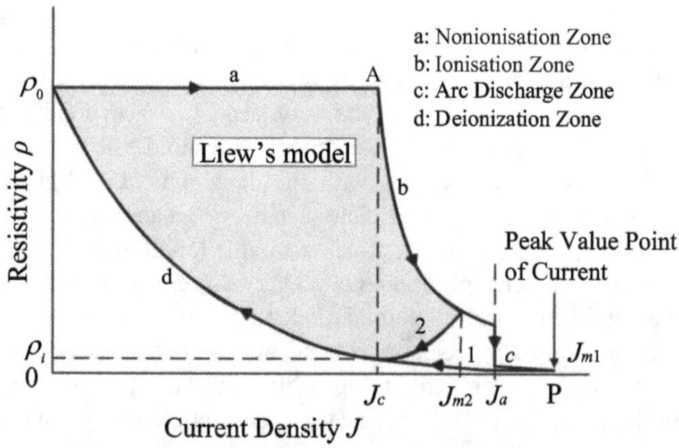

Figure 13.1 Modified transient resistivity profile of the soil ionisation zone (this figure is modified from [6,7])

Nonionisation zone: Soil resistivity remains constant at ρ_0 until the current density exceeds the critical current density value J_c, i.e., $\rho = \rho_0$. It is presented by line a (path a) [7,8].

Ionisation zone: When the current density is greater than a critical ionisation value J_c, soil ionisation happens. With the current density increases, the ionisation initializes from the soil neighbouring the grounding conductors and continues outward up to the points at which the electric field drops to below the critical value [7,8]. At the same time, ionized tracking appears along the irregular surface of soil grains. The tracking grows as current density increases [9].

The soil resistivity has a time-dependent decay from the normal value of ρ_0 (point A) along path b. This decay corresponds to breakdown and is assumed to occur by a time-dependent diffuse growth of increasing ionisation [10].

After the current reaches its peak value, the current will still provide energy for soil ionisation channels before the current density is smaller than J_c, and the residual resistivity of ionised soil continues to decrease to the lowest value ρ_i along path 2.

The residual resistivity of the ionized soil was estimated in a large range from 1% to 5% of the original soil resistivity [11]. The geometric mean of those data is 6.77%, the respective medium value is 6.6%, and 7% was adopted in simulation. For simplification, the resistivity of the ionised soil is assumed zero in many simulations, and the soil ionisation region was considered as an enlarged grounding conductor [7].

This zone is also called a nonlinear zone due to the nonlinear variation of soil resistivity. The relationship between electrical field E and current density J behaves nonlinearly [7], i.e.,

$$E = AJ^{\beta} \tag{13.1}$$

where A and β are coefficients related to soil and applied current, i.e., $0 < \beta < 1$. It is difficult to measure the critical current density to generate soil ionisation and spark discharge. E_c is defined as the critical breakdown electric field of soil, i.e., the electric field when spark discharge takes place in the soil contacting with the grounding electrode, which can be presented by the data in Table 13.1. Another electrical field E_e is defined as the electric field on the edge of the spark region. When the electric field in the soil reaches E_c, the spark discharge in the soil begins; then, if the electric field on the edge of the spark region can reach E_e, the spark discharge can be maintained and developed outside [7].

Usually, E_e is assumed to be equal to E_c. However, research shows that E_e is much smaller than E_c; this means the electric field maintaining and developing the spark discharge is much low, and both satisfy the following relationship [7]:

$$E_e = E_c\sqrt{r/r_i} \tag{13.2}$$

where r is the radius of the grounding conductor, and r_i is the equivalent radius of the spark zone.

Table 13.1 Critical breakdown field E_c suggested in the literature [6,7]

Soil type or soil resistivity (Ωm)	E_c (kV/cm)	References
	1.6–5.2	Towne [13]
100	3	Bellaschi [14,15]
100	2.7	
85	1.27	
75	2.2	
300	4.25	
Sand	6.6 (negative impulse)	Nor *et al.* [16]
	5.5 (positive impulse)	(hemispherical test cell)
Sand	9.0 (negative impulse)	Nor *et al.* [16]
	7.9 (positive impulse)	(parallel plate test cell)
50	1–3	Liew and Darveniza [8]
60	0.5	
150	2	
300	0.5	
Humous soil	3.41	He *et al.* [10]
Clay	5.04	
Sand-clay	6.02	
Sand	9.91	
Wet soil	6	Oettle[12] and
Dry soil	18.5	Chisholm [17]
290	8.3	Petropoulos [18]
100	8	Korsuntcev [19]
180	10	
470	12	
	3	Mousa [3]
	4	CIGRE [20]
	5.6–9	Kosztaluk *et al.* [21] and Loboda *et al.* [22]
Various soils	$E_c = 241\rho^{0215}$	Oettle [23]

Deionisation zone: When the current density is smaller than J_c, as the current decreases, the soil resistivity recovers to its original value ρ_0 from ρ_i along path d.

Arc zone: When current density exceeds J_a, the arc discharge happens on the roots of sparking channels in the soil neighbouring the grounding conductor. The residual soil resistivity of the arc zone drops to a very small value along path c, which can be assumed to be zero.

According to the results of the impulse experiment, in the arc region, the critical breakdown electric field satisfies [7]

$$E_c = N/\sqrt{a_i} \tag{13.3}$$

where a_i is the equivalent radius of the arc region. According to different experimental results [12], the maximum, mean, and minimum values of N are 1.78×10^5, 7.35×10^4, and 2.19×10^4 V/m$^{1/2}$.

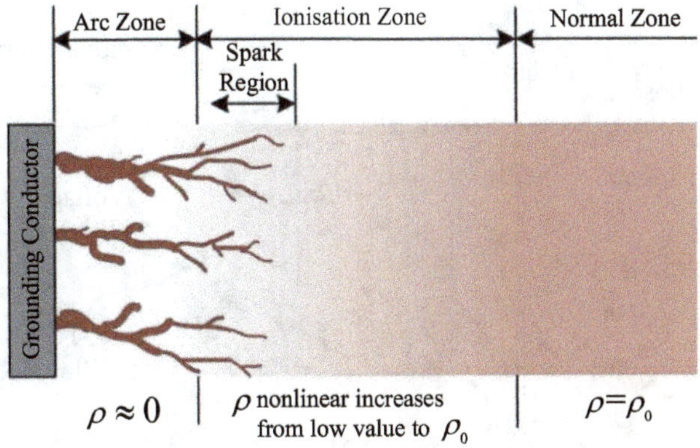

Figure 13.2 Breakdown region in the soil neighbouring the grounding electrode under lightning [6]

When an impulse current is applied, as the impulse current rises from zero to the peak value, if the maximum current density J_{m2} is smaller than J_a, then only the nonionisation zone and the ionisation zone happen. As the current decays from the peak value to zero, the soil resistivity will reduce to the lowest value along path 2 and then recover to its initial value along path d [7]. The soil resistivity changes along paths a, b, 2, and d.

If the maximum current density J_{m1} is larger than J_a, then the nonionisation zone, the ionisation zone, and the arc zone happen. As the current decays from the peak value to zero, the soil resistivity will recover from 0 (point P) to its initial value along path d through path 1 [7]. The soil resistivity changes along paths a, b, c, 1, and d.

According to the analysis above, the breakdown process and the residual resistivity of soil neighbouring a grounding conductor are shown in Figure 13.2. When lightning current exceeds the critical value J_c, an ionisation region will surround the grounding conductor; outside this ionisation zone, the soil remains as the normal resistivity. If the applied current is high enough, an arc discharge zone will neighbour the conductor; outside this arc zone will be the nonlinear ionisation zone. Usually, the residual resistivity of the arc zone is very low and assumed to be zero. In the nonlinear ionisation zone, the soil resistivity nonlinearly increases from a low value to the normal value ρ_0.

13.2.2 Method for observing soil discharge

A large number of literatures have carried out research on soil discharge characteristics by using different electrodes, such as spherical electrode, rod-to-rod and rod-to-plane [24,25], which cannot reflect the influence of the shape of the grounding electrode. In order to simulate the electric field distribution around the grounding rod, a coaxial cylindrical structure can be set up as shown in Figure 13.3 [26]. A copper cylinder with both ends open is put on an insulation board. The

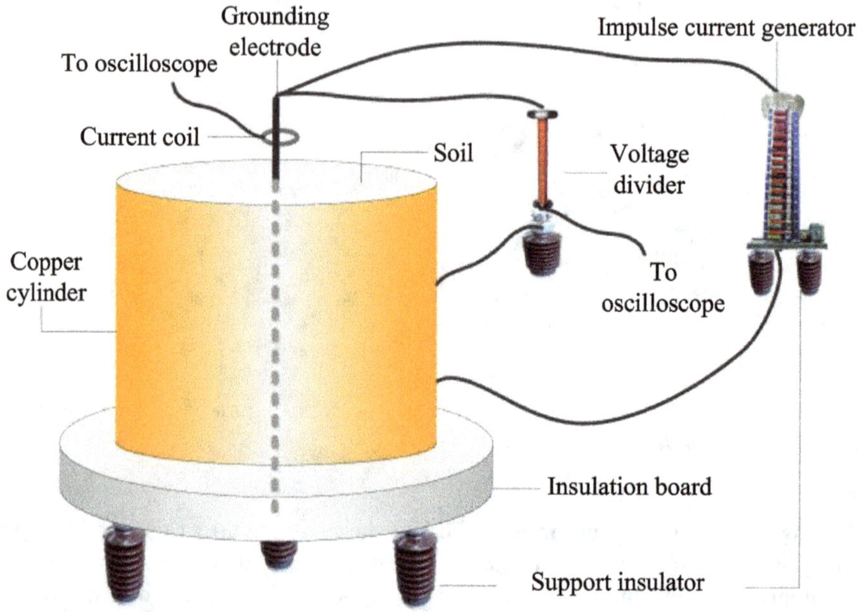

Figure 13.3 Introduction of the experiment [26]

cylinder is filled with soil. The grounding rod under test is at the axis of the cylinder, whose two ends are all exposed to air (there is a hole in the insulation board to let the electrode pass though). All the test facilities are insulated from the ground. The oscilloscopes are powered by batteries.

An impulse current is applied between the grounding rod and the copper cylinder. As the rod is much shorter compared with the wavelength of the main frequency of the impulse, the leakage current is almost evenly distributed along the rod, and the current in the soil is in radial direction. This means that the electric field in the soil is also in radial direction. When the current is large enough, the strong electric field will result in soil ionisation near the electrode, which will decrease the equivalent soil resistivity and change the waveforms of the impulse current and voltage between the electrode and the cylinder. Thus, theoretically, by observing the variation of the waveforms of the measured impulse voltage and current, the onset voltage of the soil ionisation may be obtained. Then, the corresponding critical breakdown electric field can be derived by

$$E_c = V_c / \left(r \ln \frac{R}{r} \right) \tag{13.4}$$

where, V_c is the onset voltage of the soil ionisation, r is the radius of the electrode, and R is the radius of the cylinder.

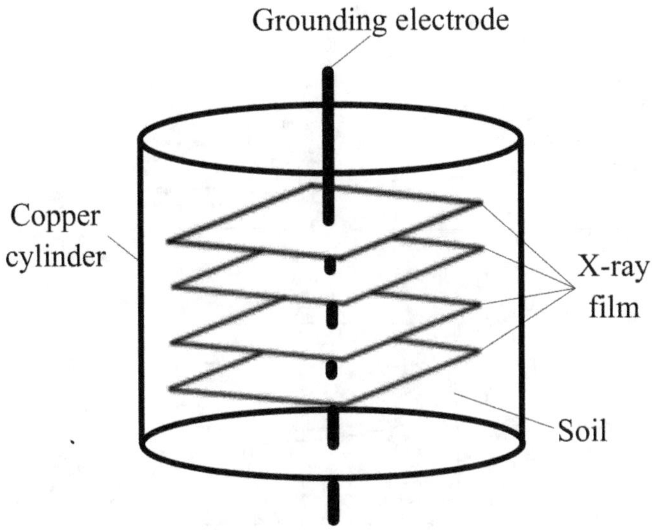

Figure 13.4 Arrangement of the X-ray films

In order to make sure that the soil ionisation takes place. X-ray imaging technology can be used [25]. In [26], taking into account the dispersion of the discharge position, several layers of X-ray films crossed by the electrode were buried in the soil as shown in Figure 13.4. Because both the films and the path of the current are perpendicular to the electrode, the influence of the films on the current path as well as the electric field around the electrode is very small. The onset voltages of the soil ionisation with and without the films were measured, and the results showed that the films had almost no influence on the onset voltage.

13.2.3 Discharge characteristics of soil under lightning

By increasing the charging voltage of the impulse current generator gradually, the waveforms of the impulse current and voltage can be measured. Figure 13.5 shows the typical measured waveforms. The transient resistance which is the ratio of voltage to current at each moment is also calculated as shown in Figure 13.5(c). It can be seen that there are three kinds of transient curves.

First, when the impulse voltage or current are small, the waveforms of the current and voltage are similar. There is an apparent current peak with a very short rise time. The transient resistance is a constant when the current begins to decrease. Meanwhile, the curves of the transient resistances at different low voltages coincide. Therefore, it seems that there is little ionisation in the soil.

Then, with the increase of the impulse voltage, the current peak time lags the voltage peak time apparently, which means that the current continues to increase as the voltage begins to decrease. This can be explained by the ionisation in the soil. When the voltage increases to a certain extent, soil ionisation appears near the

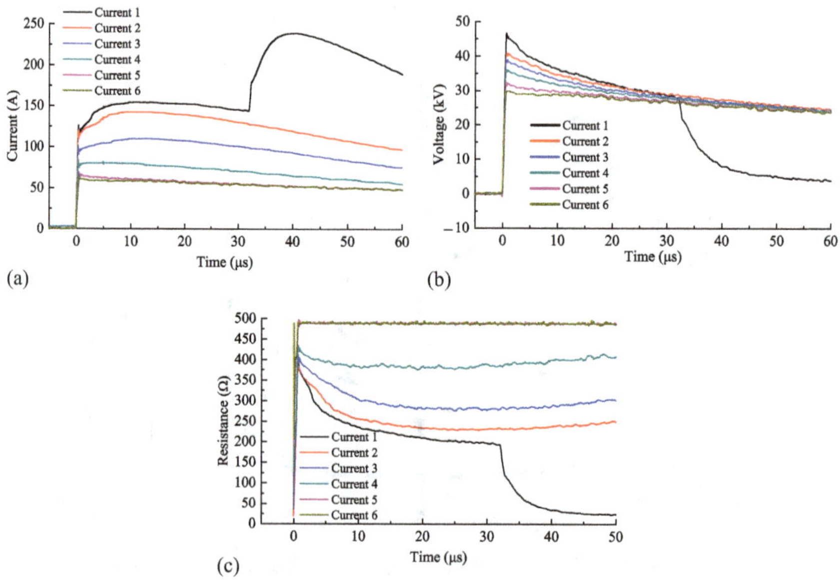

Figure 13.5 Waveforms of the impulse current and voltage and transient resistance in different stages [26]. (a) Current waveforms; (b) voltage waveforms; (c) transient resistance

electrode and the soil resistivity decreases, which results in the increase of the current. Then, the ionisation continues to develop and the current continues to increase even when the voltage begins to decrease. Finally, the soil ionisation decreases and disappears as the voltage decreases. In this case, the transient resistance varies with time apparently. For each impulse voltage, the transient resistance decreases first and then increases and finally approaches its initial value. The trend is opposite to the impulse current.

Finally, when the impulse current is very high, a second peak of the current appears, and there are apparent voltage drop and transient resistance drop at the moment. It seems that the soil resistivity becomes very low which results in the increase of the current and the decrease of the voltage. Maybe there is another kind of discharge different from the ionisation in the soil.

The above analysis can be verified by the X-ray photos in Figure 13.6. Figure 13.6(a) shows that when the impedance just begins to decrease, there has already been ionisation near the grounding electrode. Then, with the increase of the applied voltage, discharge branches appear as shown in Figure 13.6(b), which become longer and longer until they touch the cylinder. At this time, there are obvious discharge channels as shown in Figure 13.6(c), which results in the sudden decrease of the transient resistance and a sudden increase of the current as shown in Figure 13.5.

It is also interesting that at the beginning, there is a sudden drop for the transient resistance in Figure 13.5(c). This is due to the capacitance between the

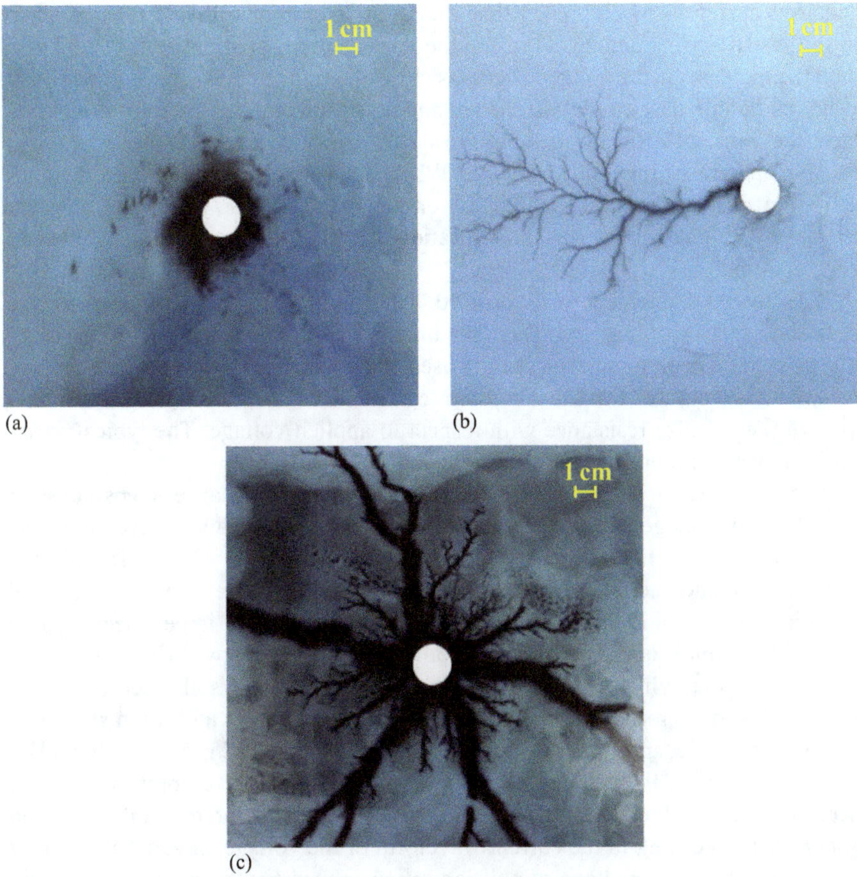

Figure 13.6 X-ray photos (the electrode passes through the centre of the film. The dark area is the discharge area) [26]. (a) Ionisation; (b) discharge branches; (c) discharge channels

electrode and the cylinder since the current and voltage have following relationship in time domain:

$$I(t) = \frac{U(t)}{R_s} + C_s \frac{\mathrm{d}U(t)}{\mathrm{d}t} \tag{13.5}$$

where I and U are the transient current and the voltage respectively, R_s and C_s are the resistance and capacitance between the electrode and the cylinder respectively. It should be noted that the frequency variations of R_s and C_s are neglected in (13.5). The capacitive current is dependent on the variation of the voltage with time. If the variation is great, the capacitive current is high. Therefore, when the voltage suddenly starts to increase, the capacitance between the electrode and the cylinder plays an important role, and the transient resistance drops greatly. For example, the capacitive current is much greater than the resistive current at the first 0.1 μs when the soil resistivity is hundreds Ω·m and the relative permittivity is up to tens. Then, when the increase of the voltage becomes slow, the transient resistance grows back toward its initial value quickly. If the voltage is so high that the soil ionisation takes place, the transient resistance cannot grow back to its initial value and varies with the development of the soil ionisation. When the voltage varies slowly, the contribution of the second term in (13.5) is very small. Therefore, finally the transient resistance with low voltage (no soil ionisation occurs) returns to its initial value and keeps constant.

To draw the transient resistance to determine whether soil ionisation has occurred or not is troublesome. Then, the impulse resistance which is defined as the ratio of voltage peak to current peak is used to depict the characteristic of the soil ionisation. The start of the soil ionisation can be determined according to the variation of the impulse resistance with respect to applied voltage. The typical variation is shown in Figure 13.7.

It can be seen from Figure 13.7 that the impulse impedance keeps constant when the peak voltage is smaller than 32 kV. Then, the impedance decreases as the voltage increases until there is a sudden impedance drop at 46 kV. In fact, the appearance of the second peak of the current and the sudden voltage drop and resistance drop in Figure 13.5 corresponds to the sudden impedance drop in Figure 13.7, which is due to the breakdown of the soil between the grounding electrode and the cylinder. Before that, there is ionisation in soil when the impedance begins to decrease. Thus, the onset electric field of soil ionisation should be that when the impedance just begins to decrease. Since the soil is not totally broken down at this stage, there is neither apparent increase of the current or apparent decrease of the voltage. It is difficult to determine whether the soil ionisation happens just according to the current waveform or the voltage waveform. The best way to find the onset voltage of soil ionisation is to draw the impulse resistance curve varying with the peak voltage to find the inflection point. For example, the ionisation onset voltage in Figure 13.7 is 32 kV not 46 kV.

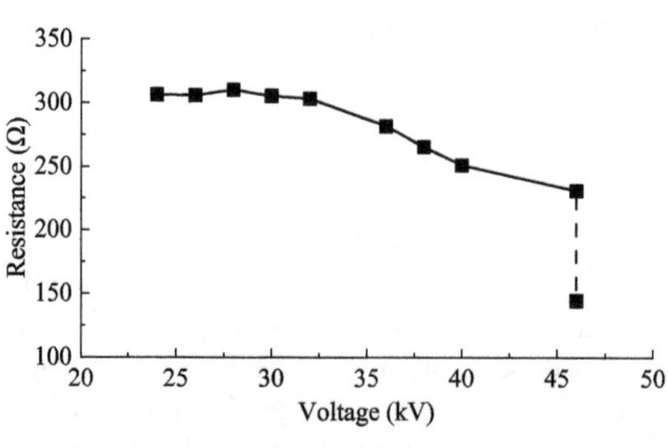

Figure 13.7 Typical impulse resistance with respect to applied voltage [26]

Based on the above analysis, if the impulse resistance varying with peak voltage is obtained, the onset voltage and electric field of soil ionisation can be derived. There are many factors affecting the soil ionisation. That may be the reason that the value of E_c reported in the literatures varies so great. In [26], the onset value with respect to the soil resistivity and the radius of the grounding electrode was provided:

$$E_c(r,\ \rho) = (19.916 + 30.861/\sqrt{r})\rho^{0.17} \tag{13.6}$$

where, E_c is in kV/m, ρ is the soil resistivity in Ω·m, and r is the electrode radius in m. It can be seen that the ionisation is related to the shape of the electrode. The larger the radius, the lower the onset electric field will be. This is because the smaller the radius of the conductor, the more uneven the electric field around it and the faster the attenuation in the radial direction. Generally, the self-sustainment of the discharge requires not only a relatively high electric field on the electrode surface, but also a relatively high electric field along the electron movement path to obtain sufficient energy. Therefore, to maintain the breakdown condition, the field strength around the conductor with small radius needs to be high.

13.2.4 Discussion of discharge characteristics of soil under lightning

Soil is composed of particles and pores. The former constitutes the skeleton, and the latter is filled with air. For moist soil, in addition to the above two components, there is water which covers the surface of solid particles and fills the pores. The parameters of soil such as the soil moisture are the important factors impacting on E_c [27].

At present, there are two main mechanisms for explaining the discharge of soil under lightning. One is that the soil breakdown is mainly an electrical process [28], the other is that the soil breakdown is a thermal processes [29]. Some studies also propose that the above two mechanisms exist simultaneously, but under different humidity, the decisive mechanism is different [30]. When the humidity is low, electrical breakdown plays a major role. In this case, due to the large amount of air in the pores of the soil, the irregularity of each pore structure makes the local electric field distorted. Once the electric field in the pore reaches the condition of breakdown, discharge occurs. When the humidity is high, thermal breakdown plays a major role. In this case, large current flows through the water covering the surface of the solid particles. Due to the current heating effect, the temperature of the channel begins to rise, and the resistivity of the water channel decreases accordingly. Due to the unevenness of local heating, thermal instability occurs in the soil. And then the current converges to areas with higher temperature and lower resistivity, and eventually several low-resistivity channels will be formed, which will conduct most of the current. At this time, the evaporation of water in the channel is intensified, and the air gap left by the evaporation of water carries most of the voltage, and then discharge occurs.

Usually the relationship curve between discharge voltage and soil moisture is U-shaped as shown in Figure 13.8. As the soil moisture increases, the discharge gradually transforms from the air gap discharge between the particles to the surface

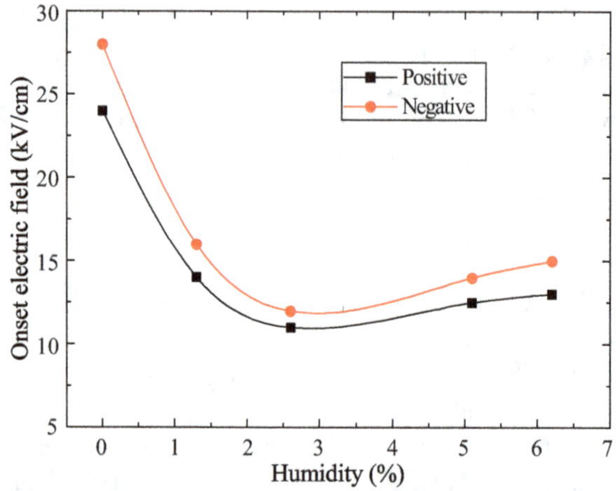

Figure 13.8 Onset electric field of arc discharge of sandy soil around grounding electrode with a radius of 0.015 m [27]

discharge of the particles, and then to the liquid discharge. Because the size of the soil pores is very small, generally between tens of μs and a few mm, for dry soil, the discharge in the pores is mainly composed of air gap breakdown. However, compared with dry soil, the surface of particles in wet soil is covered with water. According to the law of surface flashover, wet flash is much easier than dry flash and air gap discharge. Therefore, the breakdown of wet soil is mainly composed of the surface wet flashover. For wet soils with different humidity, the water mainly covers the surface of the particles, which means that the more water content on the surface of the particles, the thicker the water layer will be, which squeezes air in the pores. Then, the resistance of the water channel decreases. Under the same current, the voltage on the pore will be low. Therefore, the very wet soil has a higher onset electric field of the arc discharge compared with the slightly wet soil. When the water content of the soil is saturated, since there is no air, the discharge of the soil is only the discharge of the liquid in the gap of the particles. Therefore, the discharge field strength is higher than that of the gas-liquid mixed medium when the humidity is low.

For the effect of the lightning polarity on the arc discharge in soil, the onset electric field under negative lightning is higher than that under the positive lightning. This is because the streamer is more likely to be generated under positive impulse according to the theory of discharge.

Meanwhile, the discharge in soil does not always occur during the period of current rise. If the current peak is low, the arc discharge usually occurs when the current has passed the peak value and started to decrease, and there is a time delay which is defined as the time from the application of the current impulse to the occurrence of arc discharge in the soil, as the T_d shown in Figure 13.9. This kind of discharge is mainly due to thermal breakdown. The lower the discharge voltage, the

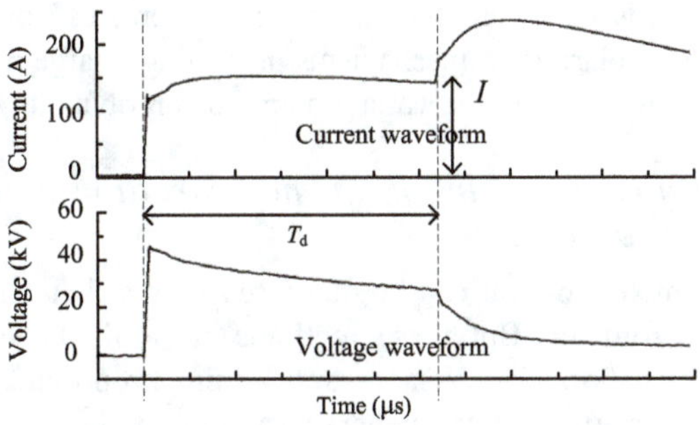

Figure 13.9 Time delay of the discharge in the soil [27]

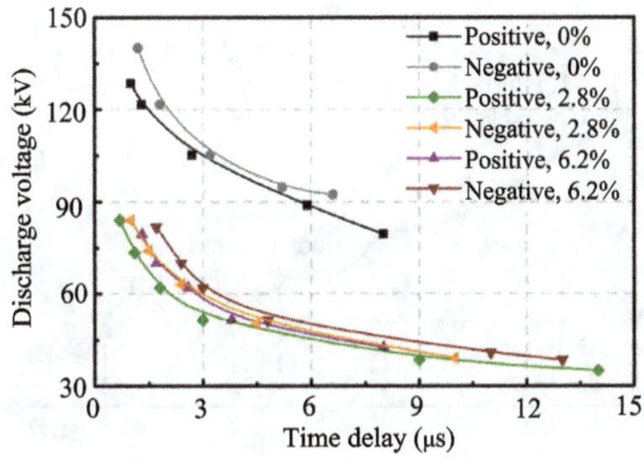

Figure 13.10 Relationship curve between discharge voltage and discharge delay of sandy soils with different humidity [27] (the radius of the grounding electrode is 0.015 m)

longer the time delay will be, as Figure 13.10 shows. At the same discharge voltage, T_d first decreases fast and then increases slowly with the increase of soil humidity. The dry soil has the longest time delay, and T_d of soils with high humidity is only a little longer than that of soils with low humidity. The one under negative polarity is longer than that under positive polarity.

13.3 Frequency variation of soil electrical parameters

The electrical parameters of most materials such as the resistivity and permittivity vary with frequency, so do soil parameters. These parameters are closely related to the polarisation characteristics of the materials. There are various polarisations mechanisms, such as electronic, ionic, dipolar and interfacial polarisation. The curve of Figure 13.11 depicts the approximate behaviour of such parameters along a range of frequencies. At the low frequency range, all of these mechanisms are able to contribute to this polarisation while the electric field alternates slowly. However, as frequency rises, those polarisation mechanisms presenting a large moment of inertia are no longer able to follow the field variation and to contribute to polarisation.

13.3.1 Method for measuring frequency variation of soil electrical parameters

The electrical parameters of soil can be measured with a dielectric spectrometer or impedance spectrometer. But these methods can only be used for sample measurement in the laboratory. Visacro and Alipio proposed an experimental methodology to determine the frequency variation of the soil resistivity and permittivity based on measurements carried out directly in the field [31]. As indicated in Figure 13.12, according to this methodology, impulsive low-intensity

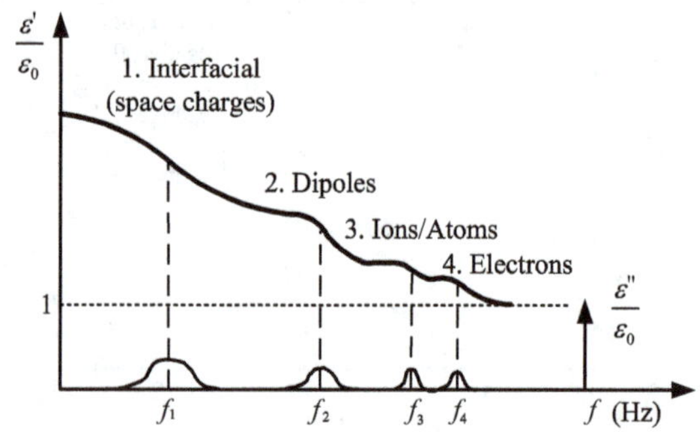

Figure 13.11 Typical profile of the complex permittivity along frequency spectrum relaxation-frequency: f_1, ...f_4. ε′ and ε″ correspond to the real and imaginary components of permittivity, respectively

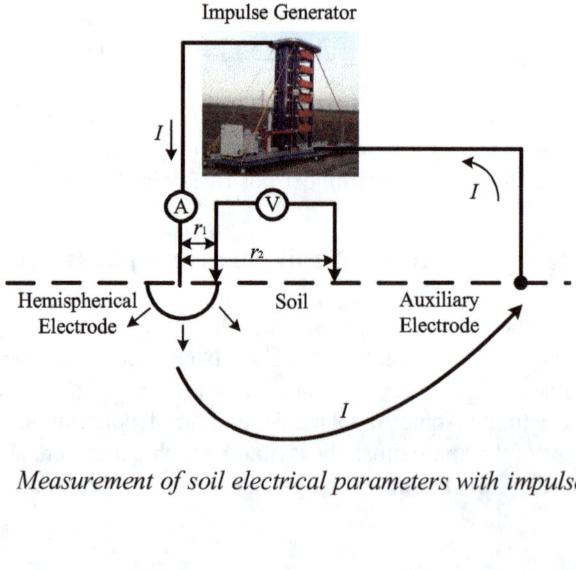

Figure 13.12 Measurement of soil electrical parameters with impulse current [31]

currents with very short front times are impressed on a metallic hemispherical electrode buried in the soil, from an impulse generator whose output is connected to the electrode and to auxiliary current-return electrode. Because the distance between the two electrodes is long, the current around the hemispherical electrode can be regarded as evenly distributed and the skin effect of ground current is not obvious. Therefore, near the hemispherical electrode the equipotential surfaces in the ground can be considered as a series of concentric hemispheres. Then by measuring the impressed current and the voltage developed between two points at the soil surface simultaneously, the resistivity and permittivity of the soil can be obtained from

$$Z = \frac{V(\omega)}{I(\omega)} = \frac{\left(\frac{1}{r_1} - \frac{1}{r_2}\right)}{2\pi\left(\frac{1}{\rho} + j\omega\varepsilon\right)}$$

where $V(\omega)$ and $I(\omega)$ are the voltage and current phasors measured in Figure 13.12, and ω is the angular frequency, r_1 and r_2 are shown in Figure 13.12 also, ρ and ε are the resistivity and the permittivity of the soil, respectively.

Both the impressed current and the voltage can be recorded using a digital oscilloscope. After each data acquisition, the fast Fourier transform (FFT) was applied to both the current and the voltage waveforms and the harmonic impedance $Z(\omega)$ of the soil (which is equal to the ratio of the Fourier transform of the voltage ($V(\omega)$) and the Fourier transform of the current ($I(\omega)$), namely $Z(\omega) = V(\omega)/I(\omega)$ was calculated for each frequency. Then, the values of the soil resistivity and permittivity can be derived at each frequency.

In addition, some literatures proposed the use of Wenner's method to measure soil resistivity at different frequencies and then the soil resistivity and permittivity could be derived. But until now, there have been no published measurement results.

13.3.2 *Frequency variation of soil resistivity and permittivity*

In [4], based on the plenty of experimental results, Visacro and Alipio developed the empirical expressions that are able to predict the approximate variation of both soil resistivity and permittivity in the broad 100 Hz–4 MHz frequency range:

$$\varepsilon_r(f) = \begin{cases} 192.2 & f < 10\ \text{kHz} \\ 7.6 \cdot 10^3 f^{-0.4} + 1.3 & f \geq 10\ \text{kHz} \end{cases} \tag{13.7a}$$

$$\rho(f) = \rho_0\left\{1 + \left[1.2 \cdot 10^6 \rho_0^{0.73}\right] \cdot \left[(f - 100)^{0.65}\right]\right\}^{-1} \tag{13.7b}$$

where f represents the frequency and ρ_0 represents the soil resistivity at 100 Hz.

13.4 Time-domain methods to simulate the transient behaviour of grounding grids

Grounding is very important for lightning protection. Sunde developed a large number of theoretical formulas and approximate expressions for calculating the performance of simple grounding electrode, some of which are still of great value to the development of numerical methods. He also preliminarily analysed the transient of grounding grids [32].

With the development of the numerical calculations of electromagnetic field, calculations for grounding systems have been gradually developed. At first, the method was mainly for low frequency and Giao and Dawalibi made a great contribution [33,34]. Since the 1980s, researchers had begun to use the transmission line method to simulate the transient characteristics of the grounding grid [35–37]. After that, Dawalibi and Grcev developed the electromagnetic methods for high frequency and transient [38,39]. Subsequently, circuit method, partial element equivalent circuit method (PEEC), finite element method (FEM), finite-difference time-domain (FDTD) method, hybrid electromagnetic method (HEM) and other methods have been used to analyse the transient characteristics of grounding grids [40–45]. Now, there have been many numerical methods to obtain the transient performance of the grounding system under impulse condition. In the beginning, these methods did not consider the influence of soil ionisation. With the gradual understanding of soil discharge phenomenon, this nonlinear problem was taken into account by the simulation method. Bo Zhang developed a method based on the approach in the frequency domain, which used the Fourier forward transform and inverse transform repeatedly to judge the degree of soil ionisation in the time domain and calculated the current and voltage distribution in the frequency domain [46]. After that, Baba *et al.* provided a FDTD approach, Habjanic and Jingli Li provided FEMs respectively, and so on [47–52].

Usually circuit models based on electromagnetic field theory are more intuitive. Because of the nature of the studied phenomenon, a time domain approach seems more suitable to analyse the grounding system. Jinpeng Wu and Bo Zhang developed a time-domain circuit approach to obtain the transient performance of the grounding system with consideration of the frequency-dependence soil parameters and the soil ionisation altogether, which will be introduced in detail below [53].

13.4.1 Equivalent circuit

When an impulse current is injected into a conductor buried under ground, the currents flow both along and out of the conductor. Because the current distribution along the conductor is uneven, for convenience, the grounding conductor is firstly cut into K short segments and N nodes. The conductor segment can be regarded as a thin-wire structure. And some hypotheses are used as follows:

Regarding a segment, the current along it, or so-called longitudinal current I_l, remains constant and concentrates on the axial line. Besides, the current leaks from the segment's two terminals evenly and intensively.

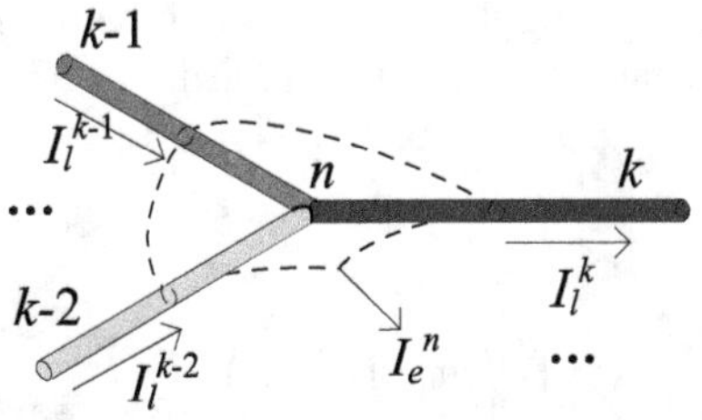

Figure 13.13 Current arrangements [53]

Regarding a node, the current out of it, or the so-called *leakage current* I_e, is the half of the sum of all the currents out of the segments connected to the node.

Figure 13.13 shows the current arrangements based on the above hypotheses. I_l^k is the current along segment k, while I_e^n is the current leaking out of node n, which is the half of the sum of all the leakage currents of segments k, k-1 and k-2.

For each point on the surface of segment k, the interface condition must be satisfied as:

$$E_{a,i}^k - E_{a,e}^k = 0 \tag{13.8}$$

where $E_{a,i}{}^k$ is the axial component of the internal electric field caused by the longitudinal current of the conductor itself, and $E_{a,e}{}^k$ is the axial component of the external electric field impressed on the conductor. $E_{a,i}{}^k$ can be expressed as:

$$E_{a,i}^k = Z_c^k I_l^k \tag{13.9}$$

where $Z_c{}^k$ is the internal self-impedance. $E_{a,e}{}^k$ can be expressed as:

$$E_{a,e}^k = -j\omega \sum_{j=1}^{K} A_a^{j,k} - \sum_{i=1}^{N} \nabla\varphi^{i,k} \tag{13.10}$$

where $A_a{}^{j,k}$ is the axial component of the vector magnetic potential impressed on the surface of segment k caused by the longitudinal current of segment j; $\varphi^{i,k}$ is the scalar electric potential, impressed on the surface of segment k, caused by the leaking current of segment i; ω is the angular frequency. $A_a{}^{j,k}$ can be expressed as

$$A_a^{j,k} = \frac{\mu_0 I_l^k}{4\pi} \int_{l^j} \left(\frac{1}{r} \mathrm{d}\boldsymbol{l}^j \right) \cdot \frac{\boldsymbol{l}^k}{l^k} \tag{13.11}$$

where $\boldsymbol{l}_j$ and $\boldsymbol{l}_k$ are the vector lengths of segments j and k, respectively, and l_j and l_k are their scalar quantity, r is the distance between $\mathrm{d}\boldsymbol{l}^j$ and $\boldsymbol{l}^k$.

By substituting (13.9), (13.10) and (13.11) into (13.8), the following equation can be achieved:

$$Z_c^k I_l^k + j\omega \sum_{j=1}^{K} A_a^{j,k} + \sum_{i=1}^{N} \nabla\varphi^{i,k} = 0 \tag{13.12}$$

By integrating (13.12) along the axial direction between the two ends 1 and 2 of segment k, the following expression is obtained:

$$\int_1^2 E_{a,i}^k \mathrm{d}l = Z_c^k I_l^k l^k$$

$$\int_1^2 j\omega \sum_{j=1}^{K} A_a^{j,k} \mathrm{d}l = j\omega \sum_{j=1}^{K} \int_1^2 \left(\frac{\mu_0 I_l^j}{4\pi} \int_{l^j} \frac{1}{r} \mathrm{d}l^j \right) \cdot \frac{l^k}{l^k} dl = j\omega \sum_{j=1}^{K} L^{j,k} I_l^j \tag{13.13a}$$

$$\int_1^2 \left(\sum_{i=1}^{N} \nabla\varphi^{i,k} \right) dl = \sum_{i=1}^{N} \left(\varphi^{i,k_2} - \varphi^{i,k_1}\right) = \sum_{i=1}^{N} Z_e^{i,k_2} I_e^i - \sum_{i=1}^{N} Z_e^{i,k_1} I_e^i \tag{13.13b}$$

where

$$L^{j,k} = \frac{\mu_0}{4\pi} \int_{l^k} \int_{l^j} \left(\frac{1}{r} \mathrm{d}l^j \right) \mathrm{d}l^k \tag{13.14}$$

and $Z_e^{i,k1}$ and $Z_e^{i,k2}$ are the leakage impedance, which represents the potential value of node k_1 and k_2, respectively, caused by a unit current leaking out of node i. Thus, (13.12) can be derived as:

$$Z_c^k l_k I_l^k + j\omega \sum_{j=1}^{K} L^{j,k} I_l^j + \sum_{i=1}^{N} \left(Z_e^{i,k_2} - Z_e^{i,k_1}\right) I_e^i = 0 \tag{13.15}$$

Based on the derivation above, the equivalent circuit of the grounding conductor is established as Figure 13.14 shows. There are four electromagnetic mechanisms, which are named as leakage self-impedance Z_{es}, leakage mutual-impedance Z_{em},

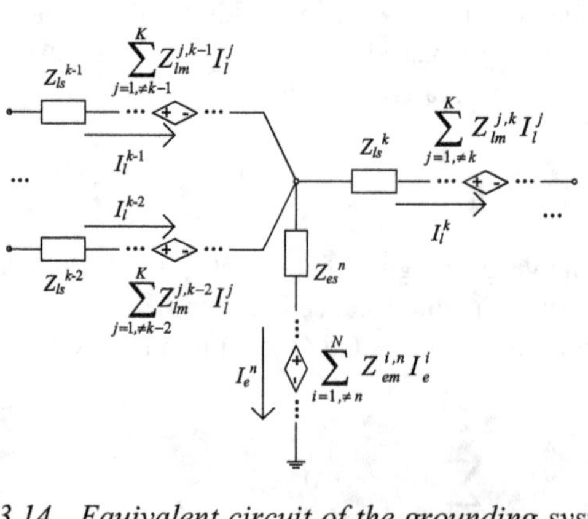

Figure 13.14 Equivalent circuit of the grounding system [53]

longitudinal self-impedance Z_{ls} (including internal one $Z_c{}^k l^k$ and external one $j\omega L^k$), and longitudinal mutual-impedance $Z_{lm}(j\omega L^{j,k})$. The leakage parameters can be computed by using the method in the next section. The expression of Z_c of a unit length of cylindrical conductor was derived in [54,55].

$$Z_c = \frac{j\omega\mu_c}{2\pi r\sqrt{j\omega\sigma_c\mu_c}}\frac{J_0\left(r\sqrt{j\omega\sigma_c\mu_c}\right)}{J_1\left(r\sqrt{j\omega\sigma_c\mu_c}\right)} \tag{13.16}$$

where r, σ_c and μ_c are the radius, conductivity and permeability of the conductor, respectively, J_0 is the zero order Bessel's function and J_1 is the first-order Bessel's function.

The equivalent circuit of the grounding system is in frequency domain. If each frequency-dependent element in the equivalent circuit can be transferred as a group of frequency-independent elements, the circuit can be solved in the time-domain.

13.4.2 Scalar electric potential φ in multilayer soil

Although the frequency of lightning can reach several MHz, its main frequency is below hundreds of kHz. At this frequency, the theory of quasi-electrostatic field is still suitable. Since usually earth can be approximated by a horizontally stratified multi-layer structure, to calculate the scalar electric potential $\varphi^{i,k}$ in (13.10), the potential produced by a current source should be obtained [56]:

$$\varphi = \frac{\rho I}{4\pi}\int_0^\infty [\varphi(\lambda)e^{-\lambda z} + \psi(\lambda)e^{\lambda z}]J_0(\lambda r)\mathrm{d}\lambda \tag{13.17}$$

where ρ is the resistivity of the earth layer in which the current source I locates, z and r are the vertical and the horizontal distances from the current source respectively, $J_0(\lambda h)$ is the zero-order Bessel's function of the first kind, $\varphi(\lambda)$ and $\psi(\lambda)$ which have been derived in [56] are determined by the boundary conditions on the interface of the layers. If the observation point and the source point are in the same earth layer, φ_0 should be also added to (13.17), where

$$\varphi_0 = \frac{\rho I}{4\pi}\int_0^\infty e^{-\lambda|z|}J_0(\lambda r)\mathrm{d}\lambda = \frac{\rho I}{4\pi\sqrt{r^2+z^2}} \tag{13.18}$$

Because there exists an improper integral with a Bessel's function and an infinite limit of integration, it is troublesome to calculate (13.17) by numerical integration. In this chapter, the complex image method is employed which can calculate (13.17) quickly and efficiently [57]. That is, $\varphi(\lambda)$ as well as $\psi(\lambda)$ can be approximated by a sum of complex exponential series [58]:

$$\begin{cases} \varphi(\lambda) \approx \sum\limits_{t=1}^{n} b_t e^{a_t\lambda} \\ \psi(\lambda) \approx \sum\limits_{t=1}^{n} d_t e^{c_t\lambda} \end{cases} \tag{13.19}$$

where a_t, b_t, c_t and d_t are complex values, n can be set to two times of the earth layers' number according to the experience. Then, according to (13.18), by using (13.19), (13.17) can be well approximated as

$$\varphi \approx \frac{\rho I}{4\pi} \sum_{t=1}^{n} \left[\frac{b_t}{\sqrt{r^2 + (z - a_t)^2}} + \frac{d_t}{\sqrt{r^2 + (z + c_t)^2}} \right] \tag{13.20}$$

The merit of this method is that no matter how many earth layers are involved, equation (13.20) has only few terms. There are many ways to obtain (13.19), such as Prony's method.

If the current leaks from the segment evenly, φ can be obtained by integrating (13.20) along the segment. Also, the potential at any point in the earth can be calculated by summing up the integrals along all segments. Thus, the potential rise at any point in the soil can be obtained.

In the frequency domain, both the resistivity and the permittivity of soil affect the distribution of electric fields in the soil. The electric field strength $\boldsymbol{E}$ and the current density $\boldsymbol{J}$ in the soil have the following relationship:

$$\boldsymbol{J} = (\rho^{-1} + j\omega\varepsilon)\boldsymbol{E} \tag{13.21}$$

where ρ and ε are the resistivity and the permittivity of soil, respectively. Then, a complex resistivity can be defined as:

$$\rho' = \frac{1}{\rho^{-1} + j\omega\varepsilon} = \frac{\rho}{1 + j\omega\varepsilon\rho} \tag{13.22}$$

By replacing the resistivity in (13.17) with the complex resistivity, both the resistivity and permittivity can be taken into account when calculating the scalar electric potential in soil, even they are frequency dependent.

13.4.3 Frequency-independent model

In order to use time domain method to consider the soil ionisation, all the frequency-dependent element such as Z_c can be approximated by a group of frequency-independent elements [59]. For example, Z_c can be fitted by a series of rational expressions in the frequency range of the lightning current:

$$Z_c(s) \approx hs + b + \sum_{p=1}^{P} \frac{c^p}{s - a^p} \tag{13.23}$$

where s represents the complex frequency jr, P is the number of poles and h, d, a^p and c^p are constant coefficients. There are several reasonable approximation methods to obtain (13.23). Here vector fitting method is chosen because of its reliability and efficiency [60]. Then, the group of the frequency-independent elements can be obtained as Figure 13.15 shows, where the elements have following

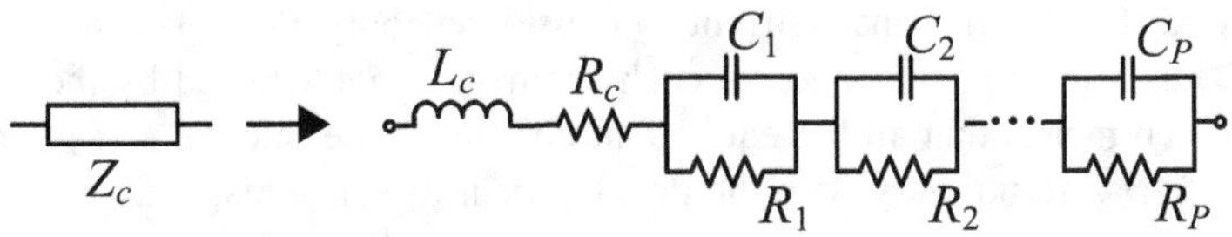

Figure 13.15 Equivalent circuit of the internal self-impedance [59]

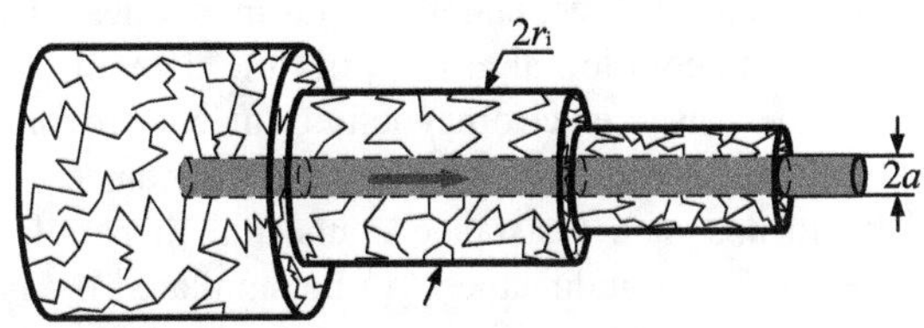

Figure 13.16 Shape of the ionized zone around an electrode

relationship with the parameters in (13.23):

$$L_c = h, \ R_c = b,$$

$$C_p = 1/c^p, \ \ R_p = c^p/a^p, \ \ p = 1, \ 2, \ \cdots, \ P$$

Then, by vector fitting, each frequency-dependent element in the equivalent circuit is transferred as a group of frequency-independent elements. Thus, the circuit can be solved in time-domain.

13.4.4 Soil ionisation effect

When a high magnitude current is injected into the grounding system, soil ionisation will occur if the electric fields surrounding the segments exceed the critical breakdown value E_c. The affected portion of the soil will become a good conductor. In order to take account of the effect of the soil ionisation, some assumptions are adopted.

First, in the ionized region, voltage drop is considered negligible. The ionized zone is assumed concentric with the conductor, and its radius extends up to a distance where the electric field has decreased to E_c as Figure 13.16 shows [46]. This is equivalent to changing the radius of the conductor by

$$r = \rho I_e/(2\pi E_c \Delta l) \tag{13.24}$$

where I_e and l are the leakage current and the length of the segment, respectively, and ρ is the soil resistivity.

Second, the longitudinal current is considered only flowing inside the conductor. Thus, the self-impedance of each segment is not affected by the soil ionisation, which means that all the equivalent circuits of the internal self-impedances in Figure 13.14 do not vary with the development of soil ionisation.

Then, it can be seen that the soil ionisation only affects the self-impedance Z_{es}, and the mutual-impedance Z_{em} in Figure 13.14. At this time, the voltage source is not the potential on the surface of corresponding segment, but the potential on the boundary of the ionized zone around the segment. The boundary is dynamic according to (13.24) and the voltage source is time-varying accordingly.

The time-varying frequency-independent circuit model described above can be solved by the trapezoidal integration algorithm in time domain which has already been adopted by the widely used electromagnetic transient program (EMTP) [61]. According to the algorithm, at each time step, all the inductances and capacitances can be simulated by means of a resistance connected in parallel with an ideal current source. For example, an inductance can be simulated by means of a resistance R_L connected in parallel with an ideal current source $I_L(t-\Delta t)$ as Figure 13.17 shows [59]. In the figure, $i_{km}(t)$ is the current flowing in the inductance from node k to m, $u_k(t)$ and $u_m(t)$ are the potentials on the nodes respectively. R_L and $I_L(t-\Delta t)$ can be obtained with $i_{\mathrm{km}}(t-\Delta t)$, $u_{\mathrm{k}}(t-\Delta t)$ and $u_{\mathrm{m}}(t-\Delta t)$:

$$R_L = 2L/\Delta t \tag{13.25}$$

$$I_L(t-\Delta t) = i_{km}(t-\Delta t) + [u_k(t-\Delta t) - u_m(t-\Delta t)]/R_L \tag{13.26}$$

Thus, if $i_{km}(t-\Delta t)$, $u_k(t-\Delta t)$ and $u_m(t-\Delta t)$ are known, the equivalent circuit of L at time t can be set up. For a capacitance, an equivalent circuit similar to Figure 13.17 can also be obtained based on the same idea. Then, the total equivalent circuit at time t can be set up which can be solved by node potential method.

In order to consider the effect of soil ionisation accurately, an iterative scheme is developed at each time step:

First, the current distribution on the grounding grid with new lightning current at this time step is calculated based on the radii of the segments at the last time step.

Then, the electric field on the surface of each segment is obtained by the leakage current

$$E(t) = \rho I_e(t)/(2\pi a \Delta l) \tag{13.27}$$

where a is the original radius of the segment. If the electric field on the surface of a segment is higher than E_c, soil ionisation occurs. The equivalent radius of corresponding segment during the ionisation is calculated by (13.24).

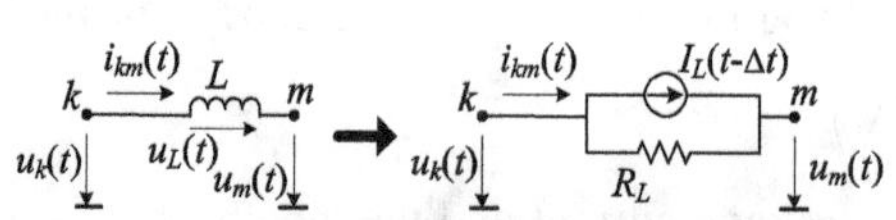

Figure 13.17 Equivalent circuit of an inductance [59]

With the new radius, current distribution on the grounding grid is calculated again and another group of new equivalent radius of the ionised segments is obtained.

If there is almost no difference between the two groups of equivalent radii, for example, the relative errors are smaller than 5% for all segments, the final results at this time are obtained. Else, the analysis goes back to step 3.

In summary, the whole procedure is displayed in Figure 13.18. The nonlinear and dynamic soil ionisation, the frequency-dependent parameters are considered all together in the model.

13.4.5 Determination of initial state for each time-step

It is worth noting that there are many non-resistant elements such as the inductances and capacitance in the equivalent circuit. Once the soil is ionised, the elements in the equivalent circuit will change over time. This encounters the question of how to determine the initial state for each time-step in the iterating process.

Take one specified element $Z_{es}{}^{n}$ for example. By vector fitting, the frequency-dependent element can be equivalent to a network as Figure 13.19(a) shows, where the network has a convergent state for the time-step t. Considering soil ionizing, $Z_{es}{}^{n}$ may change at the next time-step $t+\Delta t$, and the equivalent network becomes as Figure 13.19(b) shows. Herein lays a problem to be settled regarding how to determine the initial state of $t+\Delta t$, based on the convergent state of t.

In order to solve the problem, this chapter proposes the following rule: It is assumed that the topology of the network keeps invariable at each time-step, which can be achieved by applying the same number of vector fitting poles P. Further, the

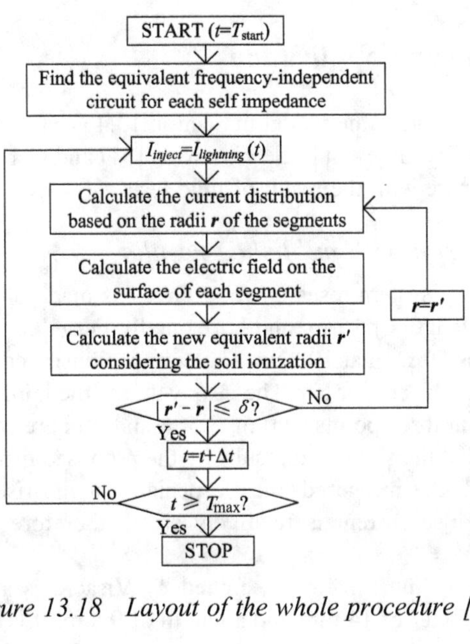

Figure 13.18 Layout of the whole procedure [59]

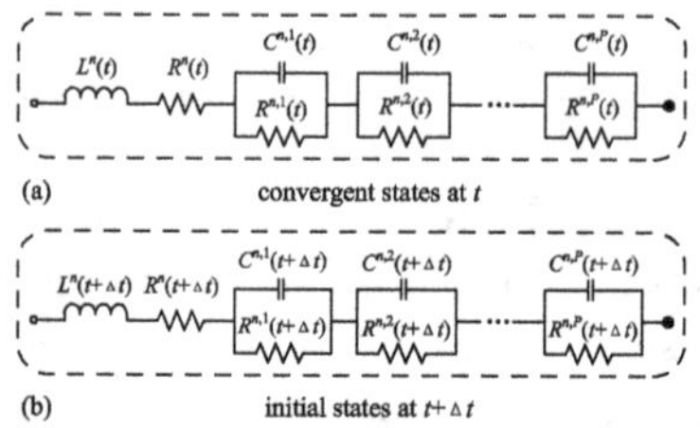

Figure 13.19 Equivalent circuits of one specified element at two time-steps [53]

initial states are determined by the following principle: the flux linkage of each inductance, as well as the electric charge of each capacitance, remains constant, correspondingly and respectively.

In Figure 13.19, the circuit topologies of the two time-steps must be kept the same. At t, assume that the convergent current of the inductance $L^n(t)$ is $I_e^n([t]_1)$, and the convergent voltage of capacitance $C^{n,i}(t)$ is $U_c^{n,i}([t]_1)$. While soil ionizing, the parameters in the network change, from $L^n(t)$ and $C^{n,i}(t)$ to $L^n(t+\Delta t)$ and $C^{n,i}(t+\Delta t)$, respectively. Based on the rule, the initial state of the elements at $t+\Delta t$ can be obtained as:

$$\begin{aligned} I_e^n([t+\Delta t]_0) &= \frac{I_e^n([t]_1)L^n(t)}{L^n(t+\Delta t)} \\ U_c^{n,i}([t+\Delta t]_0) &= \frac{U_c^{n,i}([t]_1)C^{n,i}(t)}{C^{n,i}(t+\Delta t)} \quad (i=1,2,\cdots,P) \end{aligned} \tag{13.28}$$

13.5 Validation and application

In this part, the validation of the model in Section 13.4 is carried out. The model is compared with those by Grcev in [62], Visacro in [31] and Geri in [64]. Moreover, it is applied to compare with the result of field test.

13.5.1 Comparison with published results

In [62], the grounding system researched by Grecv is made of copper with a diameter of 12 mm, a length of 15 m, and horizontally buried at the depth of 0.6 m. The soil resistivity is 70 Ω·m at 100 Hz. The current, illustrated in Figure 13.20, is injected at one end of the copper bar. The peak value of the impulse current is about 35 A. The voltages at three points as 0 m, 3.5 m and 7 m are concerned.

It can be seen that the simulated results by the proposed method in this chapter correlate better with the measured ones, especially at the rise time. It is worth noting that the impulse current is relatively small, therefore the soil ionisation effect is ignored.

In [31], the grounding system researched by Visacro is a copper horizontal electrode with a diameter of 14 mm and a length of 9.6 m. It is buried at the depth

0.5 m. The soil resistivity is 1400 Ω·m at 100 Hz and is frequency dependent as defined by (13.7), so is the relative permittivity. The peak value of the impulse current is about 0.3 A, which is injected at one end of the electrode.

From Figure 13.21, it can be seen that the simulated result by the proposed method fits the measured one by Visacro well. Moreover, the result with frequency-dependence ignored is also illustrated in the figure. The comparison shows that the frequency-dependence should be considered while the grounding device is buried in the soil with high resistivity. This conclusion can be validated and explained by [63].

In [64], the grounding system is made of copper and buried at the depth of 0.6 m. The copper bar is with a diameter of 12 mm and a length of 5 m. The soil

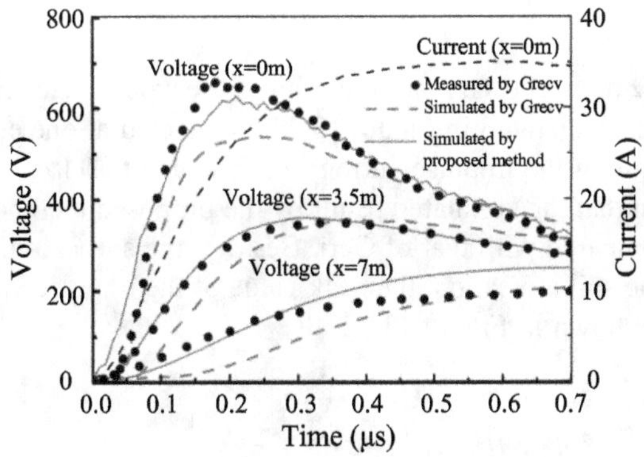

Figure 13.20 Comparison with Grecv for validating [53]

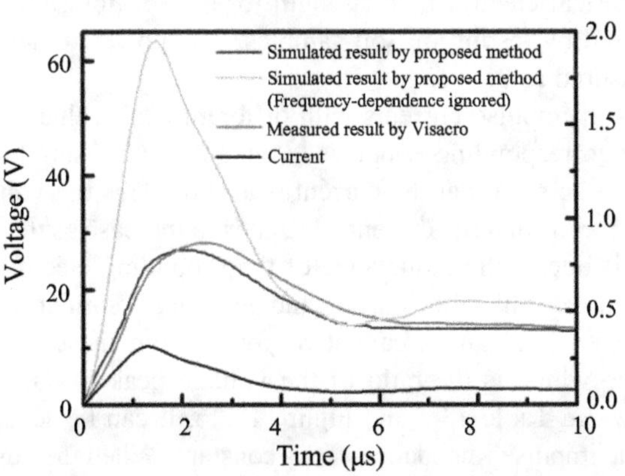

Figure 13.21 Comparison with Visacro for validating [53]

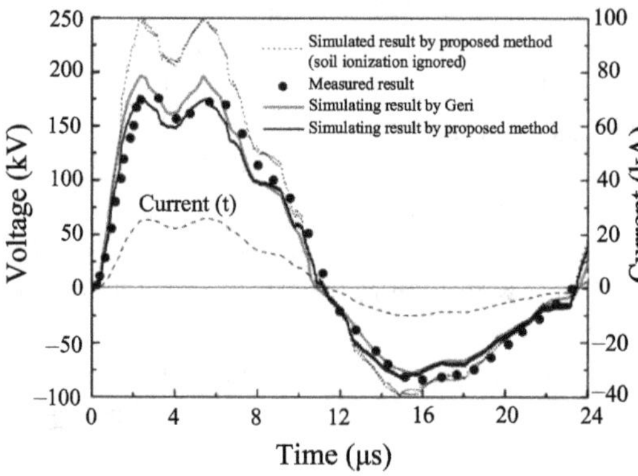

Figure 13.22 Comparison with Geri for validating [53]

resistivity is 42 Ω·m at 100 Hz, and the critical breakdown electrical field is 300 kV/m. The current, illustrated in Figure 13.22, is injected at one end of the copper bar. The peak value of the impulse current reaches about 30 kA.

It can be seen that the simulated result by the proposed method is closer to the measured one, compared with that of Geri. Besides, the soil is ionised at this time. While ignoring the soil ionisation, the peak value of the impulse voltage will reach about 250 kV, as shown in Figure 13.22.

13.5.2 Comparison with field test

The approach is also validated by the test field. In the test, a cross horizontal conductor made of steel with a diameter of 12 mm is buried in the soil with a depth of 1 m, as shown in Figure 13.23. The soil has two horizontal layers; the resistivity of the top layer with a thickness of 6.2 m is 15.8 Ω·m, and that of the bottom layer is 6.6 Ω·m. The critical electric field strength for the soil ionisation is 270 kV/m, which was measured by testing on soil samples. The power frequency grounding impedance is measured as 3.14 Ω.

Inject a series of impulse currents with different peak values as Figure 13.24 shows, and the corresponding impulse voltage is measured as shown in Figure 13.25. It can be seen that the current peak time lags the voltage peak time apparently, which means that the current continues to increase as the voltage begins to decrease. This is due to the inductance of the grounding electrode. Beside this phenomenon, the waveforms of the current and voltage are similar. It seems that the contribution of the soil ionisation cannot be found from these curves. Then, the impulse impedance which is the ratio of the voltage peak to the current peak is calculated as shown in Table 13.2 and Figure 13.26. It can be seen that when the current is small, the impulse impedance keeps constant. When the current is 3.3 kA, the impulse impedance begins to decrease with the increase of the current. After that, the greater the current, the faster the decrease. This means that the soil

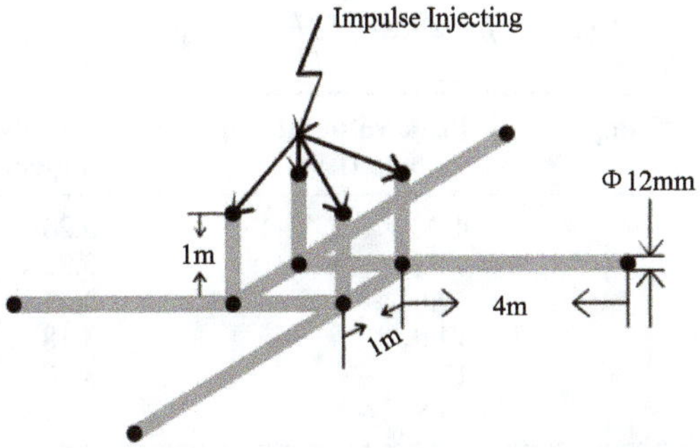

Figure 13.23 The grounding model for validating in the field test [53]

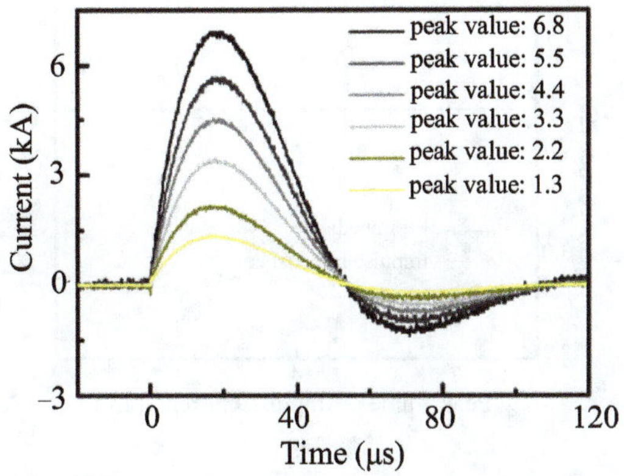

Figure 13.24 The injection impulse current in the field test [53]

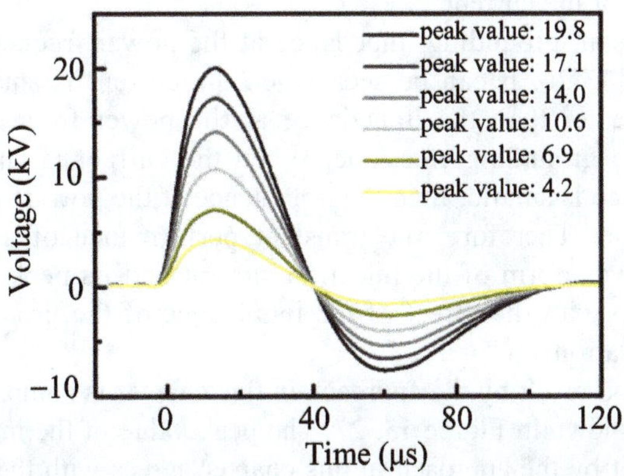

Figure 13.25 The corresponding impulse voltage in the field test [53]

Table 13.2 Results of field test [53]

Peak value of impulse current (kA)	Peak value of impulse voltage (kV)	Impulse impedance (Ω)
1.3	4.2	3.23
2.2	6.9	3.22
3.3	10.6	3.21
4.4	14.0	3.18
5.5	17.1	3.10
6.8	19.8	2.91

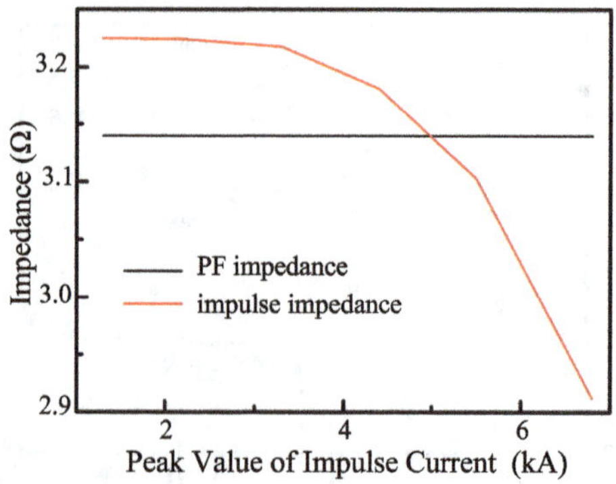

Figure 13.26 The impulse impedance in the field test [53]

ionisation appears when the current is 3.3 kA and becomes stronger and stronger with the increase of the current.

For comparison, grounding impedance at the power frequency is also presented in Figure 13.26. It can be seen when the current is small, the impulse impedance is greater than the impedance at the power frequency due to the inductance of the grounding electrode. When the current is large enough, the impulse impedance is smaller than the impedance at the power frequency due to the soil ionisation. Therefore, the transient performance of a ground device depends on the waveform of the injection current and its peak which have the close relationship with the effect of the inductance of the grounding electrode and the soil ionisation.

The simulation result by the approach in this chapter is compared with that of the field test, as shown in Figure 13. 27. The peak value of the impulse current is 6.8 kA. The result by the approach in this chapter agrees with the test result very well (in the figure, the simulated and measured lines almost overlap together), which shows the validity of the proposed approach in this paper.

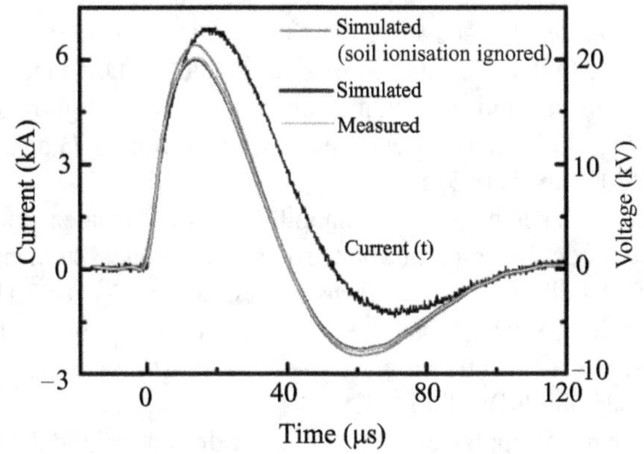

Figure 13.27 Comparison with the field test for validating [53]

References

[1] IEEE Standard Dictionary of Electrical and Electronics Terms. IEEE Standard; 1993.

[2] A. Geri. 'Behavior of grounding systems excited by high impulse currents: the model and its validation'. *IEEE Transactions on Power Delivery*. 1999; 14(3): 1008–1017.

[3] A. M. Mousa. 'The soil ionisation gradient associated with discharge of high currents into concentrated electrodes'. *IEEE Transactions on Power Delivery*. 1994; 9(3): 1669–1677.

[4] S. Visacro and R. Alipio. 'Frequency dependence of soil parameters: experimental results, predicting formula and influence on the lightning response of grounding electrodes'. *IEEE Transactions on Power Delivery*. 2012; 27(2): 927–935.

[5] CIGRE TB 781 – Impact of Soil-Parameter Frequency Dependence on the Response of Grounding Electrodes and on the Lightning Performance of Electrical Systems; 2019.

[6] J. L. He. 'Progress in lightning impulse characteristics of grounding electrodes with soil ionisation'. *IEEE Transactions on Industry Applications*. 2015; 51(6): 4924–4933.

[7] J. L. He, R. Zeng, and B. Zhang. *Methodology and Technology for Power System Grounding*. Hoboken, NJ: Wiley; 2012.

[8] A. C. Liew and M. Darveniza. 'Dynamic model of impulse characteristics of concentrated earths'. *Proceedings Institution of Electrical Engineers*. 1974; 121(2): 123–135.

[9] J. P. Wang, A. C. Liew, and M. Darveniza. 'Extension of dynamic model of impulse behavior of concentrated grounds at high currents'. *IEEE Transactions on Power Delivery*. 2005; 20(3): 2160–2165.

[10] J. L. He, Y. Q. Gao, R. Zeng, B. Zhang, and X. D. Liang. 'Soil ionization phenomenon around grounding electrode under lightning impulse'. In *Proceedings VIII International Symposium Lightning Protection*; São Paulo, Brazil, 2005, pp. 520–525.

[11] Y. Q. Liu, N. Theethayi, R. Thottappillil, R. M. Gonzalez, and M. Zitnik. 'An improved model for soil ionisation around grounding system and its application to stratified soil'. *Journal of Electrostatics*. 2004; 60(2/4): 203–209.

[12] E. E. Oettle. 'A new general estimation curve for predicting the impulse impedance of concentrated earth electrodes'. *IEEE Transactions on Power Delivery*. 1988; 3(4): 2020–2029.

[13] H. M. Towne. 'Impulse characteristics of driven grounds'. *General Electric Review*. 1929; 31(11): 605–609.

[14] P. L. Bellaschi. 'Impulse and 60-cycle characteristics of driven ground—II'. *AIEE Transactions*. 1942; 61(3): 349–363.

[15] P. L. Bellaschi. 'Impulse and 60-cycle characteristics of driven ground—III'. *AIEE Transactions*. 1943, 62(3): 333–345.

[16] N. M. Nor, A. Haddad, and H. Griffiths. 'Determination of threshold electric field Ec of soil under high impulse currents'. *IEEE Transactions on Power Delivery*. 2005; 20(3): 2108–2113.

[17] W. A. Chisholm and W. Janischewskyj. 'Lightning surge response of ground electrodes'. *IEEE Transactions on Power Delivery*. 1989; 4(2): 1329–1337.

[18] G. M. Petropoulos. 'The high-voltage characteristics of earth resistances'. *Journal of the Institution of Electrical Engineers – Part II: Power Engineering*. 1948; 95(2): 59–70.

[19] A. V. Korsuntcev. 'Application of the theory of similitude to the calculation of concentrated earth electrodes'. *Elektrichestvo*. 1958; 5: 31–35.

[20] CiGRE Working Group on Lightning. Guide to procedures for estimating the lightning performance of transmission lines. CIGRE, Paris, France; 1991.

[21] R. Kosztaluk, M. Loboda, and D. Mukhedkar. 'Experimental study of transient ground impedances'. *IEEE Transactions on Power Apparatus and Systems*. 1981; 100(11): 4653–4660.

[22] M. Loboda and V. Scuka. 'On the transient characteristics of electrical discharges and ionisation processes in soil'. In *Proceedings of 23rd International Conference Lightning Protection*; Florence, Italy, Sep 1996. pp. 539–544.

[23] E. E. Oettle. 'A new general estimation curve for predicting the impulse impedance of concentrated earth electrodes'. Presented at the IEEE PES Summer Meeting; San Francisco, CA, USA, July 1987. pp. 2020–2029.

[24] M. Hayashi. 'Observation of streamer in the soil by surge current' (in Japanese). *Journal of IEE of Japan*. 1967; 87(1): 133–141.

[25] Y. Gao, J. He, J. Zou, R. Zeng, and X. Liang. 'Fractal simulation of soil breakdown under lightning current'. *Journal of Electrostatics*. 2004; 61 (3–4): 197–207.

[26] B. Zhang, Z. Li, and S. Wang. 'Onset electric field of soil ionisation around grounding electrode under lightning'. *High Voltage.* 2020; 5(5): 614–619.

[27] B. Zhang, S. Wang, and J. Wu. 'Effect of soil moisture on arc discharge in sandy soils under lightning struck'. *IEEE Letters on Electromagnetic Compatibility Practice and Applications*. 2020; 2(4): 111–114.

[28] T. M. Flanagan, C. E. Mallon, and R. Denson. 'Electrical breakdown characteristics of soil'. *IEEE Transactions Nuclear Science*. 1982; 29(6): 1887–1890.

[29] M. Narayanan, S. Parameswaran, and D. Mukhedkar. 'Transient performance of grounding grids'. *IEEE Transactions on Power Delivery*. 1989; 4(4): 2053–2059.

[30] D. P. Snowdon and J. W. Erler. 'Initiation of electrical breakdown of soil by water vaporisation'. *IEEE Transactions Nuclear Science*. 1983; 30(6): 4568–4571.

[31] S. Visacro, R. Alipio, and M. H. Murta Vale. 'The response of grounding electrodes to lightning currents: the effect of frequency-dependent soil resistivity and permittivity'. *IEEE Transactions on Electromagnetic Compatibility*. 2011; 53(2): 401–406.

[32] E. D. Sunde. *Earth Conduction Effects in Transmission Systems*. New York, NY: D. Van Nostrand Company Inc.; 1949.

[33] T. N. Giao and M. P. Sarma. 'Effect of two-layer earth on the electric field near HVDC ground electrodes'. *IEEE Transactions Power Apparatus and Systems*. 1972; 91(6): 2356–2365.

[34] D. Mukhedkar, Y. Gervais, and F. Dawalibi. 'Modelling of potential distribution around a grounding electrode'. *IEEE Transactions Power Apparatus and Systems*. 1973; 92(5): 1455–1459.

[35] B. R. Gupta and B. Thapar. 'Impulse impedance of grounding grids'. *IEEE Transactions Power Apparatus and Systems*. 1980; 99(6): 2357–2362.

[36] A. P. Meliopoulos and M. G. Moharam. 'Transient analysis of grounding systems', *IEEE Transactions Power Apparatus and Systems*. 1983; 102(2): 389–397.

[37] R. Verma and D. Mukhedkar. 'Impulse impedance of buried ground wire'. *IEEE Transactions Power Apparatus and Systems*. 1980; 99(5): 2003–2007.

[38] W. Xiong and F. P. Dawalibi. 'Transient performance of substation grounding systems subjected to lighting and similar surge current'. *IEEE Transactions on Power Delivery*. 1994; 9(3): 1412–1420.

[39] L. Grcev and F. Dawalibi. 'An electromagnetic model for transients in grounding systems'. *IEEE Transactions on Power Delivery*. 1990; 5(4): 1773–1781.

[40] A. F. Otero, J. Cidras, and J. L. del Alamo. 'Frequency-dependent grounding system calculation by means of a conventional nodal analysis technique'. *IEEE Transactions on Power Delivery*. 1999; 14(3): 873–878.

[41] P. Yutthagowith, A. Ametani, N. Nagaoka, and Y. Baba. 'Application of the partial element equivalent circuit method to analysis of transient potential rises in grounding systems'. *IEEE Transactions on Electromagnetic Compatibility*. 2011; 53(3): 726–736.

[42] M. Trlep, M. Jesenik, and A. Hamler. 'Transient calculation of electromagnetic field for grounding system based on consideration of displacement current'. *IEEE Transactions on Magnetics*. 2012; 48(2): 207–210.

[43] W. Sima, B. Zhu, T. Yuan, Q. Yang, and J. Wang. 'Finite-element model of the grounding electrode impulse characteristics in a complex soil structure based on geometric coordinate transformation'. *IEEE Transactions on Power Delivery*. 2016; 31(1): 96–102.

[44] M. Tsumura, Y. Baba, N. Nagaoka, and A. Ametani. 'FDTD simulation of a horizontal grounding electrode and modeling of its equivalent circuit'. *IEEE Transactions on Electromagnetic Compatibility*. 2006; 48(4): 817–825.

[45] S. Visacro and A. Soares. 'HEM: a model for simulation of lightning-related engineering problems', *IEEE Transactions on Power Delivery*. 2005; 20(2): 1206–1208.

[46] B. Zhang, J. He, R. Zeng, *et al*. 'Numerical analysis of transient performance of grounding systems considering soil ionisation by coupling moment method with circuit theory'. *IEEE Transactions on Magnetics*. 2005; 41(5): 1440–1443.

[47] Z. Fan, H. Tanaka, Y. Baba, and N. Nagaoka. 'FDTD surge simulation of a vertical grounding rod considering soil ionisation'. In *Proceedings 2016 Asia–Pacific International Symposium Electromagnetic Compatibility*; Shenzhen, China. May 2016. 1. pp. 260–262.

[48] Z. Feng, X. Wen, X. Tong, H. Lu, L. Lan, and P. Xing. 'Impulse characteristics of tower grounding devices considering soil ionisation by the time-domain difference method'. *IEEE Transactions on Power Delivery*. 2015; 30(4): 1906–1913.

[49] A. Habjanic and M. Trlep. 'The simulation of the soil ionisation phenomenon around the grounding system by the finite element method'. *IEEE Transactions on Magnetics*. 2006; 42(4): 867–870.

[50] J. L. Li, T. Yuan, Q. Yang, W. X. Sima, C. X. Sun, and M. Zahn. 'Numerical and experimental investigation of grounding electrode impulse-current dispersal regularity considering the transient ionisation phenomenon'. *IEEE Transactions on Power Delivery*. 2011; 26(4): 2647–2658.

[51] H. Chen and Y. Du. 'Lightning grounding grid model considering both the frequency-dependent behavior and ionisation phenomenon'. *IEEE Transactions on Electromagnetic Compatibility*. 2019; 61(1): 157–165.

[52] S. Sekioka. 'Frequency and current-dependent grounding resistance model for lightning surge analysis'. *IEEE Transactions on Electromagnetic Compatibility*. 2019; 61(2): 419–425.

[53] J. Wu, B. Zhang, J. He, and R. Zeng. 'A comprehensive approach for transient performance of grounding system in the time domain', *IEEE Transactions on Electromagnetic Compatibility*. 2015; 57(2): 250–256.

[54] R. L. Stoll. *The Analysis of Eddy Currents*. London: Oxford University Press; 1974.

[55] J. R. Wait. 'Electromagnetic wave propagation along a buried insulated wire'. *Canadian Journal of Physics*. 1972; 50: 2402–2409.

[56] T. Takahashi and T. Kawase. 'Calculation of earth resistance for a deep-driven rod in a multi-layer earth structure'. *IEEE Transactions on Power Delivery*. 1991; 6: 608–614.

[57] Y. L. Chow, J. J. Yang, and K. D. Srivastava. 'Complex images of a ground electrode in layered soils'. *Journal of Applied Physics*. 1992; 71: 569–574.

[58] R. W. Hamming. *Numerical Methods for Scientists and Engineers*, 2nd ed. New York, NY: McGraw-Hill Book Company; 1973.

[59] B. Zhang, J. Wu, J. He, and R. Zeng. 'Analysis of transient performance of grounding system considering soil ionisation by time domain method'. *IEEE Transactions on Magnetics*. 2013; 49(5): 1837–1840.

[60] B. Gustavsen. 'Improving the pole relocating properties of vector fitting'. *IEEE Transactions on Power Delivery*. 2006; 21(3): 1587–1592.

[61] H. W. Dommel. 'Digital computer solution of electromagnetic transient in single and multiple networks'. *IEEE Transactions on Power Apparatus and Systems*. 1969; 88(4): 388–399.

[62] L. D. Grcev and F. E. Menter. 'Transient electromagnetic fields near large earthing systems'. *IEEE Transactions on Magnetics*. 1996; 12(3): 1525–1528.

[63] R. Alipio and S. Visacro. 'Impulse efficiency of grounding electrodes: effect of frequency-dependent soil parameters'. *IEEE Transactions on Power Delivery*. 2014; 29(2): 716–723.

[64] A. Geri, G. M. Veca, E. Garbagnati, and G. Sartorio. 'Non-linear behaviour of ground electrodes under lightning surge currents: computer modelling and comparison with experimental results'. *IEEE Transactions on Magnetic*. 1992; 28(2): 1442–1445.

Chapter 14

Statistics of electromagnetic reverberation chambers and their simulation through time domain modeling

Philippe Besnier[1]

This chapter discusses the fundamental properties of electromagnetic reverberation chambers (RCs) and the simulation of their behavior through a simplified time-domain numerical approach. Electromagnetic RCs are used as a test environment with growing range of applications. Their unique properties allow for powerful radiated immunity and radiated emissivity tests for electromagnetic compatibility assessment of various pieces of electronic equipment in industry. Moreover, their application is also spread to antenna characterization, performance assessment of wireless communication systems, exposure of animals to electromagnetic fields, and radar cross-section estimations. This list may be further extended and will be in the future, researchers around the world being very active in this matter.

Though electromagnetic RCs behavior is now well established and documented [1–3], the accurate prediction of field cartography remains a challenge. As far as an empty chamber (i.e. with only antennas and probes in a rectangular cavity equipped with a metallic stirrer and free of any other object) is concerned, the solution of Maxwell's equations does not predict experimental measurements, due to the extreme difficulty to provide an exact description of boundary conditions in the broad sense. Adding a device under test (DUT) in interaction with the EM field (generated by an external source or by the DUT itself) yields an even worse situation due to the lack of description of all its electric and geometric parameters.

This does not obliterate the possibility of using analytic solutions derived from the wave equation or performing numerical simulations of EM wave equations but from a different perspective. However, the statistical theory of RC encompasses in a quiet efficient way the statistical (spatial or spectral) experimental observations. Statistical models, based on some underlying hypotheses, are able to describe the statistics of the field in the chamber from a limited knowledge of the RC geometry.

The theoretical models of RCs may be roughly decomposed into two main strategies. The first and the most popular modeling avenue is purely statistical and based on the plane-wave integral representation of the EM field. This model was

[1]INSA Rennes, CNRS, IETR – UMR 6164, F-35000 Rennes, France

synthesized by D. Hill [4], following Kostas and Boveris [5] early proposals. It is an asymptotic model in which the modal density of RCs plays a key role. It is indeed applicable for highly overmoded cavities, assuming a continuous spectrum of plane waves coming from all directions and arbitrary polarization. A second type of model consists of a hybridization involving modal theory and statistical description. Such a modal description of RCs doe not only rely on the supposed modal density but also on their modal bandwidth (or modal width). In other words, it accounts for losses in RC. The role of losses appears to be indeed fundamental to reach the significant excitation of several modes within the RC, due to modal overlap, which is only possible beyond some low-frequency limit. The statistical part of these models may come from attributing random weights to the modes, according to Cozza's model [6]. On the other hand, waves chaos also provides a scope to describe the field statistics based on the statistics of eigenmodes of the RCs retrieved from random matrix theory. The statistics of eigenmodes solve the problem of retrieving a more realistic distribution of neighboring modes than that of the Weyl distribution [7,8].

We could therefore conclude that numerical investigations are useless for RCs. This is not the case at all, reminding that the two key parameters of RCs are the modal density (and more precisely its modal distribution) and the modal bandwidth. Therefore, a numerical simulation is of interest since it may provide a realistic modal distribution and could mimic the modal overlap if all losses mechanisms are properly accounted for. However, these deterministic-like simulations do not play the usual role of simulations. They cannot indeed be compared to totally equivalent experimental conditions. The description of exact boundary conditions is out of reach and that would be required due to extreme sensitivity of the RC response to any tiny variations. Therefore, these simulations are also used to check for statistical properties according to some parameters. Such numerical simulations start by imposing metallic boundary conditions representing the boundaries of the RC under investigation. These boundaries determine both modal density and distribution. In case the simulated cavity is rectangular, the modal distribution is that of a regular cavity and not that of a chaotic cavity. This is a huge difference regarding individual modes since none of them, individually, is spatially ergodic due to the regular pattern of field cartography. This is further discussed in Section 14.3 where we will introduce image theory for RC simulations. However, adding a metallic stirrer (if large enough) probably transforms a regular cavity into a more or less chaotic cavity, depending on the geometry and location of the stirrer. Therefore, statistics may be extracted from a set of simulations at different stirrer positions using brute force simulations. However, one needs a Maxwell solver and large computing resources, depending on the frequency spectrum to be analyzed. If available, these simulations offer various interesting viewpoints of analysis as developed, for instance in [9,10]. In this chapter, we describe a simulation technique that, despite some simplifying hypothesis, provides simulations of RC over a very large bandwidth within reasonable computation time. It is based on the time propagation of the wave emitted from a point source according to image theory. Image sources are readily obtained if the cavity is considered empty and rectangular. This is a limitation since it does not account for any chaoticity but provides a clear indication about the deviation of such parallelepiped-shaped cavities with regard to Hill's model, in the worst case. It is easy to compute (a few tens of lines of code) and gives approximate descriptions of chamber statistics confirming both non-ideal and ideal regimes of RCs.

Before describing this simulating approach, we provide in a first section an overview of the main features of RCs and introduce the main theoretical approaches. For the sake of brevity, we only underline the main aspects of these approaches and provide some references for further in-depth analysis. The second section of this chapter briefly discusses numerical modeling of RCs. We describe a selection of different techniques used in the literature to highlight the behavior of RCs with regard to statistical theory predictions. The third section describes a method of time domain simulation using a simple ray-tracing technique able to calculate the response of parallelepiped-shaped RCs to any point-source excitation. Possible applications of this method are then illustrated in a fourth section.

14.1 Introduction to electromagnetic RCs

Fundamental theoretical and experimental aspects of RCs may be found in some books entirely dedicated to this subject [1,2]. Moreover, advanced description or applications are provided in two more recent books [3,11]. This section highlights the key aspects of the theory of RCs and describe the parameters that govern their properties. At a first glance, a RC is made of a Faraday cage whose dimensions are large enough with regard to the wavelength. The Faraday cage imposes the electromagnetic field is confined in the chamber, the boundary conditions at the walls being close to those provided by perfect electric conductor boundaries. The role of losses is fundamental but let first suppose that no losses are accounted for and that the walls of the Faraday cage are indeed made of perfect electric conductors.

The electromagnetic field is therefore strictly confined in this cavity, obeys Maxwell's equations, and is a solution of the wave equation which is given below in the harmonic steady state:

$$\Delta\vec{E} + k^2\vec{E} = 0. \tag{14.1}$$

This is the Helmholtz equation for electromagnetic waves propagation in terms of the electric field vector $\vec{E}$ and the wave number $k = \omega/v_m$, with ω being the angular frequency and v_m being the propagation speed of the wave in the considered medium. Since continuous wave signals are considered using the convention $exp(j\omega t)$ where j is the unit complex number, this term should appear as common factor in (14.1) but was removed for convenience. This equation is that of an eigenmode problem, since no source of electromagnetic field excitation is considered, the right-hand side of (14.1) being zero. It therefore admits non trivial solutions for discrete values of k or ω. We suppose a rectangular cavity of size $a \times b \times d$ with the coordinate system $Oxyz$ according to Figure 14.1. Stating that the tangential electric field strength and the normal magnetic field vanishes at

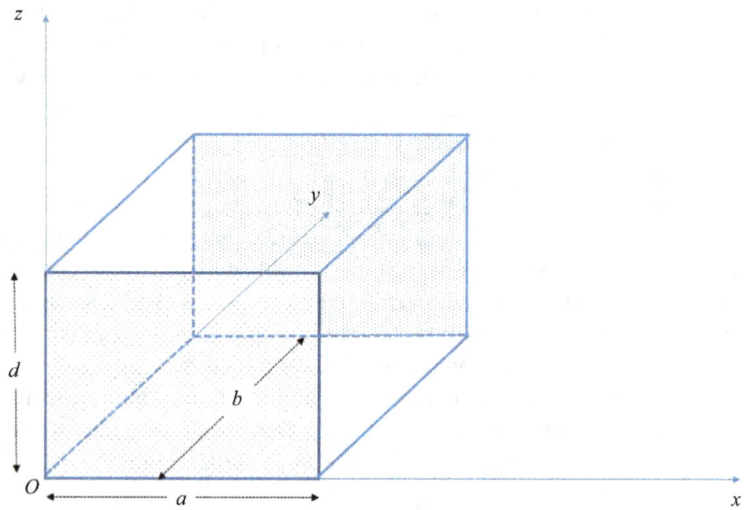

Figure 14.1 Geometrical parameters of a rectangular Faraday cage

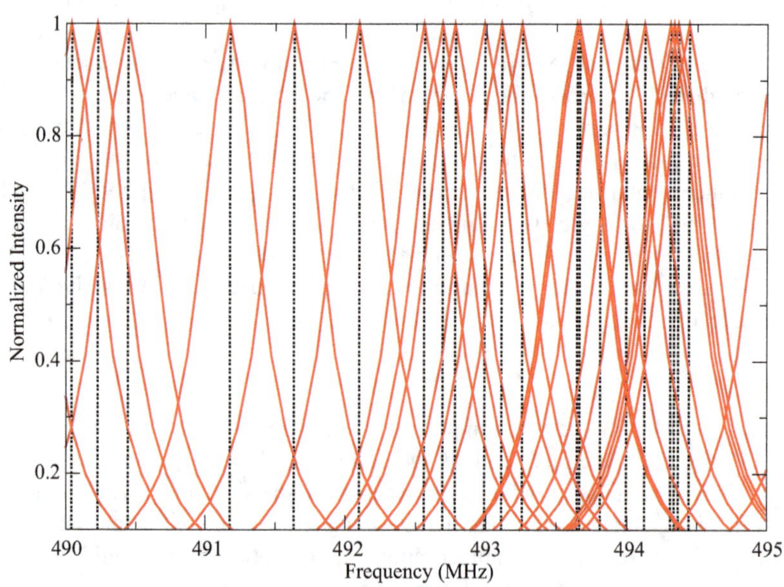

Figure 14.2 Illustration of modal density and modal width (here constant with $\Delta_f = 0.5$ MHz) in a rectangular and empty RC. Dotted black lines are the real eigenvalues of a lossless cavity according to (14.2). Red curves represent the width of the modes in a lossy cavity according to (14.6)

boundaries ($x = 0, a$; $y = 0, b$; $z = 0, d$), an infinite set of non-trivial solutions appears for specific values of the wave number $k_{m,n,p}$ such that:

$$k_{m,n,p} = \pi\sqrt{\left(\frac{m}{a}\right)^2 + \left(\frac{n}{b}\right)^2 + \left(\frac{p}{d}\right)^2}, \tag{14.2}$$

where m, n, p are positive integer numbers with at least two of them being different from zero. In other words, at any angular frequency $\omega_{m,n,p} = k_{m,n,p}v_m$, the cavity exhibits an eigenmode with an undamped resonance with a specific field spatial cartography $\vec{E}_{m,n,p}$. Excitation at a different angular frequency, in this lossless case, corresponds to a non-significant excitation of an infinite set of modes. On the contrary, a well-behaved RC relies on its ability to supply multiple eigenmodes with significant excitation (depending on coupling conditions regarding the source excitation) in order to exhibit the properties of a perfect random or diffuse field.

Two main factors mutually contribute to that end. First, the modal density (i.e. the number of modes per Hertz) increases with the square of the frequency. It may be inferred from (14.2) but it is confirmed asymptotically from Weyl's law. The modal density D_{f_0} (in Hz^{-1}) at a high enough frequency f_0 is given by:

$$D_{f_0} = 8\pi V\frac{f_0^2}{v_m^3}, \tag{14.3}$$

where V is the volume of the cavity. Second, and in combination with the increase of modal density, thermal losses in walls contribute to introduce some width to the natural modes identified in (14.2). From a theoretical point of view, if losses are small enough, the eigenmodes follow the same spectral distribution but eigenmode wave numbers are no longer given by real value and become:

$$k'_{m,n,p} = k'_{m,n,p}\left(1 - \frac{j}{2Q_{m,n,p}}\right), \tag{14.4}$$

where $Q_{m,n,p}$ is the quality factor accounting for the losses in the wall associated with the modal field m, n, p. Note that losses are fundamentally involved to reach steady state. The steady state is reached once the dissipated power in the RC balances the transmitted power in the chamber (through an antenna). The quality factor may be defined as:

$$Q_{m,n,p} = \frac{f_{m,n,p}}{\Delta f}, \tag{14.5}$$

where Δf may be called the mode width. In the lossless case, the mode width is null and the intensity of the field is not properly defined. In our practical case of a lossy cavity, the shape of the resonance follows a Lorentzian function (here normalized to a peak intensity of 1):

$$I_{norm}(f) = \frac{1}{1 + Q^2_{m,n,p}\left(\frac{f}{f_{m,n,p}} - \frac{f_{m,n,p}}{f}\right)^2}. \tag{14.6}$$

We illustrate this through a simple example. Let us suppose that we dispose of a rectangular cavity of size $3.0 \times 3.25 \times 3.40$ m^3. The distribution of its modes in the lossless and lossy cases, assuming an hypothetical constant mode width $\Delta_f = 0.5$ MHz are shown in Figure 14.2 In the considered frequency range (490 – 495 MHz), the modal density is already high (about 5 modes per MHz). Losses enable a significant excitation at an arbitrary frequency of at least one mode or several modes at a time. The latter case defines the modal overlapping situation. Overlapping of several modes enables to reach a diffuse field with ideal statistical properties. However, these ideal properties may be reached in different ways according to the operating mode of the RC.

Looking at the spatial distribution of a single mode in a rectangular cavity, if one moves a receiver along any rectilinear path in the chamber, one may easily find spatial periodic pattern of the field amplitude. Such a mode is described by a simple product of sinusoidal functions. Coming to ray dynamics, such mode corresponds to a sum of 8 non-independent plane waves or equivalently to a simple pattern of ray dynamics with absence of field along some of these rectilinear paths. The mode is not itself ergodic and the field does not follow any identified statistical distribution along any path.

The story would be different in case of a chaotic cavity that can be introduced by curving the panels of a rectangular cavity or by the presence of a large mechanical stirrer or other wave diffractors [12]. In such a situation, the field cartography is no longer described by close-form equations, each mode becoming ergodic and even follows an ideal Gaussian distribution [7].

However, a standard mode of operation of electromagnetic RCs for EMC purpose is to modify the boundary conditions in different ways to reach different independent states of the cavity. The statistical behavior is then observed as ensemble statistics, through the observation of the field at a single or at a few spatial locations in the chamber. These modifications may involve one or several mechanisms such as the use of a mechanical stirrer, walls vibration [13], source or excitation stirring using different antennas or with slight modifications of the frequency of excitation. In any case, it contributes to stir the electromagnetic field in such a way that different modes overlap with varying weights.

From Figure 14.2, we may see that at some frequencies, typically around 494.5 MHz, the distance between neighboring modes becomes reasonably low with regard to their width. It is likely that any tiny perturbation of boundary conditions would provide, for any state, various cartographies of electromagnetic field, each of them being composed of one or multiple weighted modes. In such conditions, ergodicity may be retrieved through the combination of multiple modes using several independent states of the cavity. On the contrary, the situation seems less favorable at around 491 MHz. In this case, a stronger perturbation is required (e.g. a more important change of the excitation frequency in case of frequency stirring or an important modal shift from mechanical stirring). Note that the frequency spacing between modes looks particularly irregular with a set of modes concentrated around 494.5 MHz. This is a typical signature of non-chaotic cavities for which the distance between neighboring modes follows a Poisson distribution. Not that in case an efficient mechanical stirrer is present, the cavity is no longer strictly rectangular and one may change of view with regard to whether it becomes chaotic or not.

14.1.1 Stochastic description versus deterministic solution of the electromagnetic wave equation

The above description yields to the conclusion that a well-operated RC is an oversized cavity where multiple modes come into play and their combination varies substantially with adequate changes of the boundary conditions. The whole picture may be formally predictable from the solution of the wave equation associated to Maxwell's equations. Once supplementary equations describing boundary conditions are established, it would be found that the spectral and spatial field distribution in a RC fits with the output of the theoretical model, confirming that it obeys Maxwell's equations.

Unfortunately, in practical situation, the ability to describe in a sufficiently exact manner the fine geometrical and electrical details of a RC prevents for such an attempt. This does not mean at all that analytic or numerical analysis based on Maxwell's equation are useless. However, their purpose is either to assess statistical properties and their relation to the fundamental parameters of a RC. Models are also useful to evaluate or compare stirrer performances according to their shape or position [10].

Models of RC rather describe the electromagnetic field as a stochastic quantity although it is not a random process at all. Rather, it behaves as a random process, and statistical models describe adequately the experimental observations. This is particularly true in the asymptotic range, beyond a certain frequency limit for which ideal conditions are met.

Dealing with purely statistical models refers to the random plane wave model briefly described below. This model was made popular at the end of the last century within the electromagnetic compatibility (EMC) community of experts. The second type of model is hybrid in nature. It starts from a modal analysis and stick to the properties of the solutions of Maxwell's equation in cavities. Then, the statistical part of the model consists in randomly weighting these solutions to retrieve the statistical behavior. The fundamental difference between those two families of model is that the latter one is able to provide a parametric analysis in terms of the fundamental relationship between modal density and modal width. We briefly describe these in the following two sections. We then discuss about the relevance of full-wave numerical simulation in that context.

14.1.2 Random plane wave spectrum

A complete statistical description in terms of probability density functions of the EM fields was provided in [5] and then theoretically consolidated in [4]. The main assumption of this model is related to the only mode density. It is considered to be high enough (regardless to losses which are not taken into account) to provide the excitation of many modes both simultaneously and through the change of boundary conditions. This leads to the concept of a continuous spectrum of eigen wavenumbers ($k_{m,n,p}$) and equipartition of corresponding plane waves from all directions.

Using Fourier transform in the wave number domain (see pp. 64–67 in [2]), we indeed establish that the field cartography of a mode is equivalent to that obtained from a combination of 8 plane waves coming from three pairs of opposite directions. Having many modes with various *m*, *n*, *p* values is therefore equivalent to populate all direction of space with sets of 8 plane waves in a uniform manner with arbitrary polarization. In the asymptotic limit this becomes a continuous spectrum of plane waves with arbitrary angle of arrival, polarization, and random phase.

This description corresponds to the ensemble statistics obtained through changes of boundary conditions (mainly through mechanical stirring). The electric field may be decomposed in its real and imaginary part in the volume of the chamber (at a sufficient distance from any walls or object) such as:

$$\vec{E} = (Ex_{\Re} + jEx_{\Im})\vec{x} + (Ey_{\Re} + jEy_{\Im})\vec{y} + (Ez_{\Re} + jEz_{\Im})\vec{z}, \tag{14.7}$$

where *Ex*, *Ey*, *Ez* are the rectangular components of the electric field in the x, y, z directions, respectively and $\Re$, $\Im$ denote the real and imaginary part of each of these components. These six variables exhibit random fluctuations. Due to the arbitrary angle of arrivals and phase of all plane waves the ensemble average of the electric field equals zero for each of them:

$$\mathrm{E}[\vec{E}] = \mathrm{E}[Ex_{\Re}] = \mathrm{E}[Ex_{\Im}] = \mathrm{E}[Ey_{\Re}] = \mathrm{E}[Ey_{\Im}] = \mathrm{E}[Ez_{\Re}] = \mathrm{E}[Ez_{\Im}] = 0. \tag{14.8}$$

For the same reasons, the fluctuations of each of these variables are identically distributed and furthermore, supposed to be independent. In particular, their distributions share the same moments and in particular their variance:

$$\mathrm{Var}[Ex_{\Re}] = \mathrm{Var}[Ex_{\Im}] = \mathrm{Var}[Ey_{\Re}] = \mathrm{Var}[Ey_{\Im}] = \mathrm{Var}[Ez_{\Re}] = \mathrm{Var}[Ez_{\Im}] = \sigma^2 \tag{14.9}$$

It turns out that, according to the general properties of a distribution, and to the principle of maximum entropy, the most suitable distribution is the Gaussian distribution. Using the dummy variable Er_p in substitution to the set of the six variables, we have:

$$p(Er_p) = \frac{1}{\sqrt{(2\pi)}\sigma} \exp\left[-\frac{1}{2}\frac{(Er_p)^2}{\sigma^2}\right]. \tag{14.10}$$

This is the most suitable distribution but not the best one since, its support is infinite and the field is not due to losses. As above mentioned, losses are not directly accounted for in this model. However, it is implicitly in the backyard since the superposition of modes implies that each mode has a certain width. However, it depicts correctly the situation for reasonable size sample as those collected in most experiments in much oversized cavity with large modal overlap.

Therefore, this plane wave spectrum model provides a picture of the ideal random field in a much oversized cavity but does not afford to depict less ideal situations. Those non-ideal situations are encountered using such cavities in intermediate frequency ranges. The modal theory allows analyzing the evolution of RC properties from intermediate frequency ranges up to the ideal regime.

14.1.3 Modal theory

The ideal random field model suffers from two major limitations. First, the plane wave continuous spectrum gives no indication about the number of modes (through any stirring process) that should be involved to render such a theoretical view close enough to reality. Second, losses play a fundamental role in that matter and they are not taken into account at all.

There is therefore a lack in describing the different regimes of RC that take place below the frequency limit above which the ideal random field (i.e. perfect diffuse field) conditions are met. Several works have highlighted, from empirical means, that the RC behavior below the perfect diffuse field regime could be described by the more general Weibull distribution function. This was proposed in [14,15] and confirmed several times since then [16,17]. Furthermore, numerical simulations have also shown that the Weibull parameters evolve (at least their average trend) with frequency according to two parameters only, the volume of the chamber and the Q-factor of the RC. This was extracted from numerical simulations in the time domain regime and will be further illustrated in Section 14.4 of this chapter.

On a theoretical point of view, the intermediate regime has been deeply described in several contributions based on both modal theory and statistical description of modal spectrum and weights. Modal theory states that the EM field involves a superposition of modes that are only significant according to their bandwidth. The resulting significant number of modes is equivalent to the degrees of freedom. In other words, each mode provide a discrete contribution to the spectrum of plane waves in Hill's theory. Assuming an independently and identically distributed magnitude for each mode excitation, it is shown that the variance of the field get closer to that of ideal random field as the number of significant mode increases [18]. It turns out that about 10 or 20 modes are enough to reach a non-detectable difference from a statistical inference point of view. The key feature is therefore the modal overlap, dictated by the Q-factor and the average spectral spacing between modes. The modal overlap d at frequency f_0, in the asymptotic case, is the product of the modal density $D(f_0)$ and the mode width:

$$d = 8\pi V \frac{f^3}{c^3 Q}. \tag{14.11}$$

We have assumed in (14.11) and in the rest of this chapter that waves propagate at the speed of light c. However, the behavior of the modal overlap is not monotonous with regard to the increasing frequency, since the spectral spacing between modes also exhibits some statistical fluctuations. Another, but related theoretical point of view comes from the notion of universal fluctuations of electromagnetic fields in chaotic cavities. In that context, the EM field also follows a modal description and losses are accounted for thanks to an effective Hamiltonian.

The problem is then solved using random matrix theory: the wave equation is replaced or modeled by a linear matrix equation where the elements of the matrix are random variables. According to Wignerâ€™s hypothesis, eigenvalues for chaotic cavity have the same statistics as those of matrices drawn from a suitable ensemble. Therefore, the random matrix theory provides the universal random fluctuations of the modes. Their spectral distance obeys a Wigner distribution, for which spacing between modes is much less irregular due to the unlikelihood of too closed spacing between them. This effect was also highlighted through the numerical simulation of the eigenmode equation (14.1) using finite element method [19]. Furthermore, the electromagnetic response is also predictable, in a statistical sense [20,21].

14.1.4 About the relevance of full-wave simulations for RC

According to the above section, full-wave simulations of 3-D cavities could be thought as useless in view of the appropriate predictions of either fully statistical models or modal theory. Furthermore, a deterministic simulation is unable to predict the exact response of the chamber at a specific location and frequency due to the practical issue of assessing exact boundary conditions. It is indeed an ill-posed problem in the sense that the cavity response is very sensitive to tiny variations of these boundary conditions.

However, deterministic full-wave simulations remain useful for several reasons. They possibly provide a set of data whose statistics can be closed enough to experimental ones if some conditions are fulfilled. In particular, the modal width (or modal overlap) must be well controlled. In other words, the simulated boundary conditions must provide a realistic Q-factor, close enough to that of the real RC under investigation. It is then possible to check for statistical properties of RCs that are confirmed through statistical models or with experiments. Statistical models discard geometry but full-wave simulations take them into account with fine details. As a consequence, one may investigate stirrer performance as a function of its size and geometry. Other stirring procedures may be investigated as well and the description of real source antennas may be available. The spatial properties of EM field are another possible outcome of simulations (including at boundaries) as well as spatial correlation. The impact of object size or losses is another important case study among many others.

14.2 Numerical modeling of RCs

14.2.1 Numerical modeling using 3-D solvers of Maxwell's equations

Several research groups have extensively used full-wave simulation for RC studies. For example, some authors of the above-mentioned references [9,10] have published many different investigations using the conventional finite difference time domain (FDTD) method to solve Maxwell's equations. These investigations require important computational resources. Numerical solutions in the time domain could indeed provide some insight about RC behavior over a wide spectrum providing that the simulated Q-factor is realistic over that spectrum. On a practical point of view, the relatively slow variation of the time constant of real RCs with respect to frequency eases this possibility.

The main cost of calculation is related to the mechanical stirrer when its effect is accounted for in the simulation. Fine details must be represented which implies possible meshing issues. Most importantly, as many simulations as stirrer positions are required over a significant time window. The use of over types of time-domain or frequency domain simulation methods was also considered by some authors. We could mention a set of analysis about TEM excitations of RCs using a set of wires at vicinity of walls, for which different methods such as transmission line matrix (TLM) and finite integration technique (FIT) [22] were involved. Frequency domain methods based on the electric field integral equation (EFIE) method of moments (MoM) solvers were also considered using acceleration techniques [23,24]. More precisely, the MoM is used to model the stirrer and hybridized with a high-order discrete singular convolution method with a coarse mesh to handle the rectangular cavity. A Huygen's box surrounding the stirrer delimits the two calculation methods. Although the inconvenience lies in the repetition of calculus at each frequency of investigation, it offers much faster techniques than using MoM only.

In summary, numerical investigations may certainly be used to investigate the behavior of RCs. This could be thought as the only way of investigating stirrer performance, according to its shape, size, and position. However, it was recently proposed to study the two first items, using Doppler spectrum analysis of a stand-alone stirrer [25].

This section dealt only with a very brief glance about possibilities offered by 3-D simulation techniques using possibly on-the-shelf solvers or more advanced techniques. If those are available together with adequate resources, these enable to extract useful information to interpret measurement data. In any other case, geometrical optics approximation may offer an interesting alternative.

14.2.2 Geometrical optics approximation

Numerical investigations using asymptotic representation of Maxwell's equations should provide simpler models to play with, while keeping the essential properties of RCs. In these conditions, the electromagnetic field induced at any location is approximated as the contribution of bouncing rays, representing the wavefront propagation according to geometrical optics. A particularity of RCs is that an actual source gives birth to equivalent multiple secondary sources whose contributions are summed at any location where the electromagnetic response is calculated. Such a proposal was illustrated in [26] where the authors simulated the effect of a moving wall in a rectangular 2-D cavity and in [27] for 3-D RCs. We further illustrate this method in the following of this chapter where we present the practical concepts for designing and using such tools. Following the initial contribution in [27], this chapter details the principle of RC simulations based on ray-tracing and theory of images. The contents of this chapter are much inspired from the contributions of the Ph.D. manuscript of E. Amador (in French) and related publications in [28,29].

14.3 Time-domain simulations using theory of images and ray-tracing

14.3.1 Introduction

Time-domain simulations are particularly suitable for broadband electromagnetic simulations for 3D configurations. Solving Maxwell's equations through the conventional time–space differentiation, i.e. using FDTD, is certainly a possible option. Nevertheless, it requires many degrees of freedom, since RCs are oversized with regard to the wavelength. In the following, we rather introduce a simple simulation concept based on theory of images and ray dynamics of waves according to the asymptotic description of Maxwell's equations. This may be not considered as a totally comprehensive method but rather provides a reasonable approximation of RC statistics as further illustrated in the present section of this chapter. Also, it assumes simplifying geometrical hypothesis, the simulated RCs having a strict cuboid shape. This involves possible algorithmic simplifications based on symmetry. This concept involves the description of wave propagation from elementary current sources of radiation.

14.3.2 Image theory of elementary currents

The image of an elementary current over en infinite perfect electrical conductor (PEC) plane may be resumed according to Figure 14.3. The electromagnetic field radiated by the current source $\vec{i}$ above the infinite PEC plane is strictly equivalent to the electromagnetic field generated by both the source and its image $\vec{i'}$ in the absence of the plane. As a consequence, the field may be described by the Green's function of elementary sources that embed the boundary conditions of the PEC plane without its explicit description.

The image theory allows for the description of reflection of waves, described as rays, in a simple way. Figure 14.4 illustrates the direction of the reflected wave R incoming from a transmitting source at point E as a consequence of the position of the image source E' according to Snell-Descartes law for specular reflection.

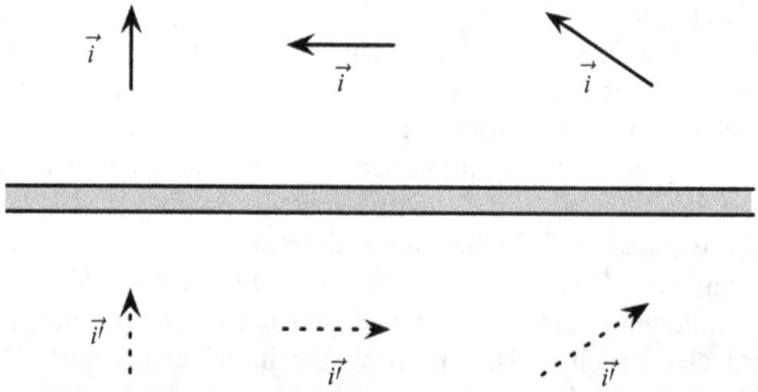

Figure 14.3 Currents orientation above a PEC plane (grey slab) and their corresponding images (dotted arrows) below that plane

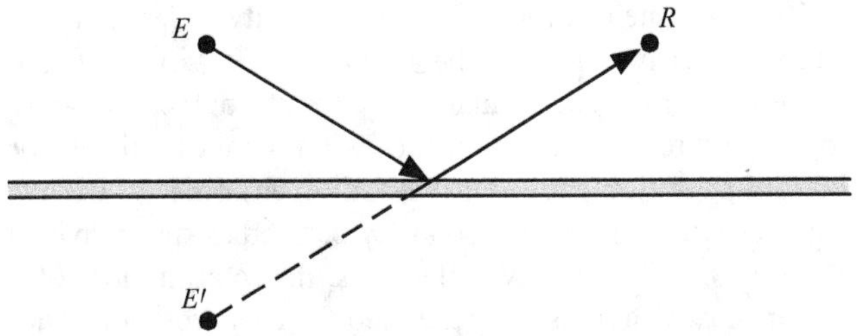

Figure 14.4 *Reflected wave direction R deduced from the transmitter E and its image E'*

14.3.3 Application to cavities

14.3.3.1 Image order of a cavity and introduction of losses

Using image theory inside a cavity involves multiple images with respect to the different PEC planes surrounding the elementary current source(s). Moreover, it implies multiple reflections according to time delay. The energy being confined in the cavity, if losses are not accounted for, these multiple reflections occur indefinitely. This is described by multiple images of the cavity itself, going further in the virtual space according to the number of involved reflections. This number of successive reflections defines the order of the corresponding image cavities. The original cavity is the zeroth-order cavity and is the only one where the electromagnetic field is calculated. It gives birth after an appropriate time delay to first-order cavities and themselves generate second order cavities and so forth. Consecutive images of the zeroth-order source current may be called the *n*th order image currents in the *n*th order cavities after a total of *n* reflections.

However, as early mentioned in Section 14.1 of this chapter, losses play a key role to account for mode overlapping in a RC. Since these losses are mainly due to the finite conductivity of the walls, each reflection may be endowed with a fraction of energy loss. The current magnitude after each reflection is reduced by a $(1 - \xi)$ quantity with $\xi \leq 1$. Thus, after *n*th reflections, we obtain:

$$\|\vec{i}_n\| = (1 - \xi)^n \|\vec{i}_0\|, \text{ with } 0 \leq \xi \leq 1, \tag{14.12}$$

where ξ is a scalar representing average losses over all incident angles and $\vec{i}_n$ is the vector of an image current of *n*th order with $\vec{i}_0$ the initial vector of the current source.

14.3.3.2 1-D cavity

The most simple application case is that of a uni-dimensional (1-D) cavity. Such a model may describe multiple reflections of waves at normal incidence with respect to the boundaries of the cavity associated with two parallel and infinite walls. For instance, this may represent a Fabry-Perot cavity at normal incidence where ξ accounts for wall losses.

Figure 14.5 features the example of an initial cavity of length l (i.e. the zeroth-order one) and its successive images. The zeroth-order cavity in Figure 14.5 lies in between solid vertical lines and contains a current source whose vector points towards the top page direction. Through successive images, this vector is oriented towards the bottom page (top page) for the odd (even) order of cavities. The anti-symmetry of images with respect to $x = 0$ may be used to speed up calculations. In order to reach the Nth order after N reflections, the total number of 1-D cavities equals $2N + 1$, since two cavities are generated for each iteration. The positions x_i of image currents in the nth order cavities along the x-axis may be found from the following expressions where $i = \pm n$:

$$x_i = \begin{cases} il + x_0 & \text{if } i \text{ is even,} \\ (i+1)l - x_0 & \text{if } i \text{ is odd.} \end{cases} \tag{14.13}$$

Without accounting for the precise source location, the successive images of the cavity are generated for discrete time intervals t_i such as:

$$t_i = n \times \frac{l}{c}, \tag{14.14}$$

where c is the speed of light. In order to get some preliminary results of simulation, we must attribute a propagation law for the radiated wave from the source. In the far-field, the electric field strength is proportional to the current source and inversely proportional to the distance. According to (14.12), we may define a loss coefficient (say $\xi = 0.01$ e.g.). We first assign a Dirac excitation for the initial current $I_0 = \delta(0)$ where δ stands for the Dirac function in order to calculate the impulse response. This impulse response is calculated from the set of all images according to their distance from the selected receiving point in the cavity at abscissa x_r, with $0 < x_r < l$. The electric field strength response in the cavity (at any arbitrary location other than the current source position) is given by:

$$E(t) \propto I(t) = \sum_{i=-n}^{i=n} \frac{I_{|i|}\delta(t - D_i/c)}{D_i}. \tag{14.15}$$

The term D_i corresponds to the distance between the two images of each $n = |i|$ order cavity and the observation point. Therefore the impulse response is composed of all images accounting for successive attenuated reflections according to (14.12) and free space propagation.

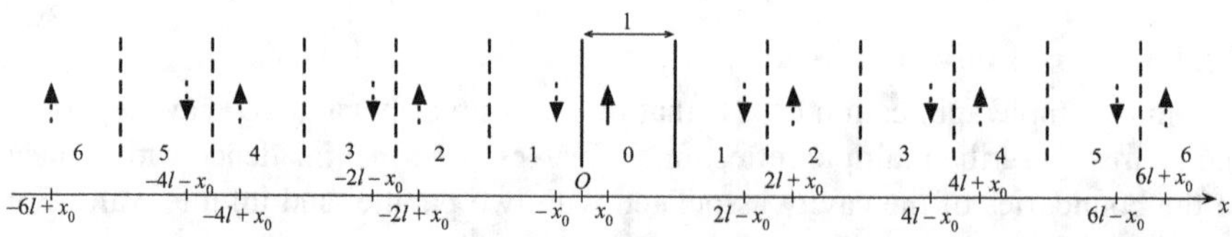

Figure 14.5 Successive images of a 1-D cavity of length l

An example of the corresponding impulse response and its Fourier transform are given in Figure 14.6. The time-domain response exhibits an exponential decay related to $\xi = 0.01$. This decay is related to the decrease of electromagnetic power density in the cavity as a consequence of ξ being non-zero. The electromagnetic power density P_{dens} at any time t may be written as:

$$P_{dens}(t) \propto E^2 \propto \sum_{i=-n}^{i=n} \frac{I_{|i|}^2 \delta(t - D_i/c)}{D_i^2}. \tag{14.16}$$

According to (14.14), at the time t_n, the power density is related to the only images of the nth-order image sources and considering (14.12) the remaining power density writes:

$$P_{dens}(t_n) \propto (1 - \xi)^{2t_n c/l} \|\vec{i_0}\|. \tag{14.17}$$

It means that the power density indeed follows an exponential decay of type:

$$P_{dens}(t_n) = \exp(-t_n/\tau), \text{ with } \tau = -\frac{1}{2cln(1-\xi)}. \tag{14.18}$$

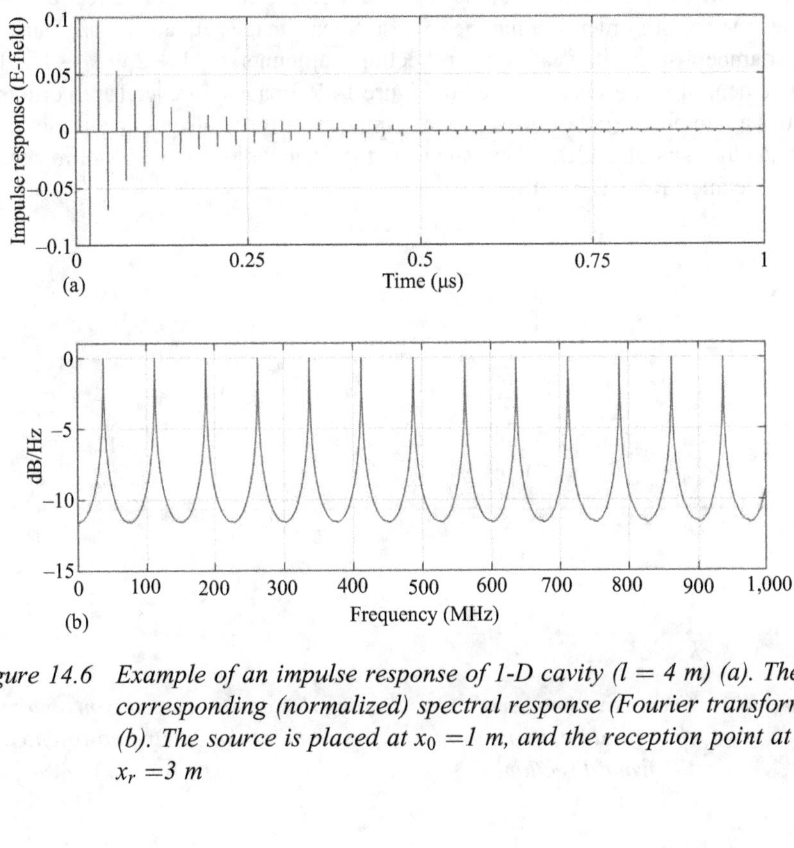

Figure 14.6 Example of an impulse response of 1-D cavity ($l = 4$ m) (a). The corresponding (normalized) spectral response (Fourier transform) (b). The source is placed at $x_0 = 1$ m, and the reception point at $x_r = 3$ m

In this example ($l = 4$ m and $\xi = 0.01$), the decay time constant τ is equal to ≈ 0.66 μS. The response consists of two interleaved trains of positive and negative pulses at regular time intervals (equals to $2l/c$, i.e. ≈ 26.7 nS) and associated to odd and even images. Those different echoes are only in phase for discrete wavelengths. Due to phase inversion with successive cavity orders, this happens for the only following frequencies:

$$f_m = (2m + 1)\frac{c}{2l}, \qquad \text{with } m \in \mathbb{N}, \tag{14.19}$$

Figure 14.6(b) confirms that the spectral response is composed of equally spaced cavity modes. Naturally, this does not offer favorable conditions to transform such a 1D-cavity in a reverberating one. The successive resonant frequencies are 37.5 MHz, 112.5 MHz, 187.5 MHz, and so on. This is consistent with (14.19).

14.3.3.3 2-D cavity

We briefly discuss the case of a 2-D cavity. The principle is exactly the same but the number of generated cavities increases much rapidly and is now proportional to n^2. The initial cavity is now a rectangle surrounded by four metallic walls. Images are then created in 4 directions instead of two, leading to 4 first-order image cavities. Then, each of these four cavities give birth to a new set of second order cavities. Then the cavities continue to extend as the number of reflections progresses. At the nth order, the number of additional images equals $4 \times n$. Thus, the total number of cavities after N reflections amounts to $1 + 2N(N + 1)$. The arrangement of images is illustrated in Figure 14.7. Images for even (odd) orders of cavities are in phase (out of phase). The symmetry of these images according to the four quadrants is also clear. This suggests that calculation may be optimized, as further detailed for 3-D cavities.

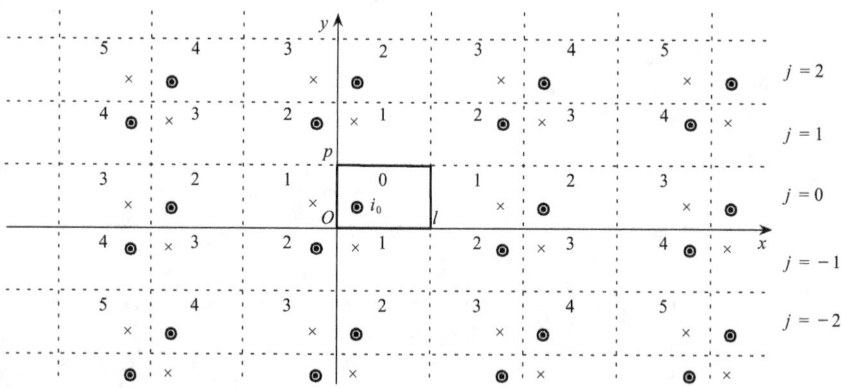

Figure 14.7 Successive images of a 2-D rectangular cavity. The current source in the initial cavity is normal to the plane of the cavity and oriented in the front direction

14.3.3.4 3-D cavity

The appearance of images with successive order of cavities is depicted in Figure 14.8. This figure only describes the generation of cavities along the two *Oxy* and *Oxz* cut planes in the Cartesian coordinate system. At the *n*th order, the number of images M_n may be deduced from the number of additional images for a 2-D cavity and is given from:

$$M_n = 4n^2 + 2, \qquad \mathrm{n} \geq 1. \tag{14.20}$$

It means that the total number of cavities/images to reach N reflections is:

$$M_N = 1 + \sum_{n=1}^{N} M_n = 1 + 2N + \frac{2N(N+1)(2N+1)}{3}. \tag{14.21}$$

Therefore, the time domain simulations of a 3-D cavity requires important memory resources to store all images, since M_N goes with N^3. The total simulation time window, T_s, of a $l \times p \times h$ cavity may be related to N using some approximation. Since the cavity considered for RC simulations is not cubic, time intervals ΔT between consecutive reflections can only be considered on average. This problem was tackled in acoustic through the definition of the average distance L between two reflections such as [30]:

$$\Delta t \approx \frac{L}{c}, \tag{14.22}$$

where L is called the reverberation distance in acoustics approximated as:

$$L \approx \frac{V}{4S}. \tag{14.23}$$

14.3.4 Preliminary result example

As in the case of a 1-D cavity, we provide some preliminary results for a 3-D cavity in Figure 14.9, accounting for electromagnetic propagation through (14.15). The

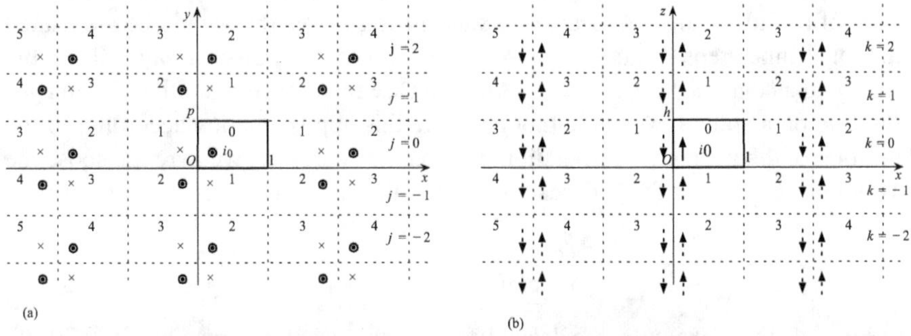

Figure 14.8 Successive images in a 3-D cavity in the Cartesian coordinate system, in the Oxy plane (a) and 0xz plane (b). The origin O is at a corner of the initial cavity

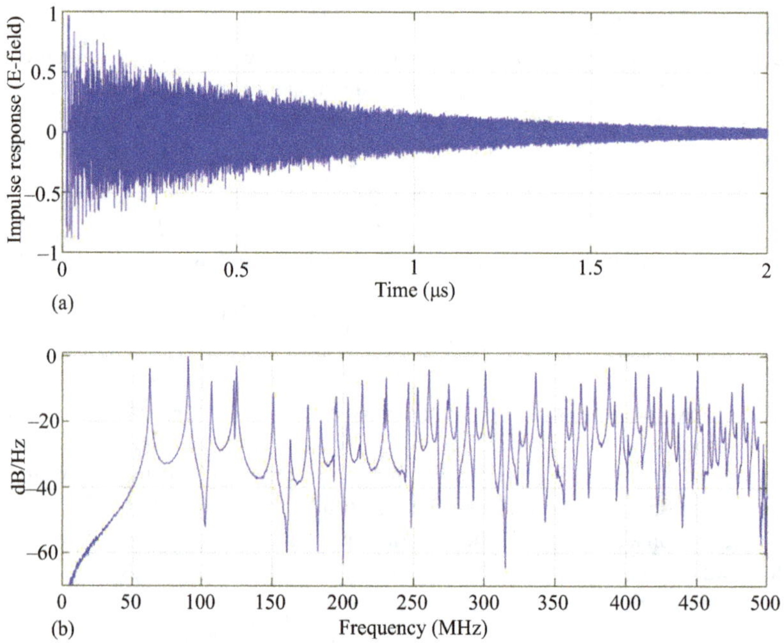

Figure 14.9 Example of an impulse response of a $4 \times 3 \times 2$ m^3 3-D cavity (a). The corresponding (normalized) spectral response (Fourier transform) (b). The source is placed at $x_0 = 1$ m, $y_0 = 2$ m, $z_0 = 1$ m and the reception point at $x = 3$ m, $y = 1$ m, $z = 2$ m

impulse response of a 4 m $\times$ 3 m $\times$ 2 m cavity is simulated with the same arbitrary attenuation factor $\xi = 0.01$.

The temporal impulse response for a 3-D cavity appears much more irregular than for a 1-D cavity. This is due to the non-symmetric nature of all images in a parallelepiped but non cubic cavity. Images at the nth order (even for a centered source of radiation) are not located on concentric spheres that would be associated with equal time steps δt given by (14.22). However, the power decay follows the same trend as that of the 1-D cavity. Although we mentioned that nth order images do not lie on a sphere, we can although make this approximation according to the definition of the average reverberation distance L. Therefore, we may transpose the 1-D definition of τ for the 3-D case for which the approximate time constant is:

$$\tau \approx -\frac{L}{2cln(1-\xi)} = -\frac{4V}{S}\frac{1}{2cln(1-\xi)}. \tag{14.24}$$

For this given example, the decay time constant is approximately $\tau \approx 0.30\ \mu S$. We may indeed check in Figure 14.9 that the impulse response in term of electromagnetic field (E-field) strength at $t = \tau$ corresponds approximately to 37% of the maximum E-field strength. The Fourier transform exhibits the modal response of the cavity. More precisely, the peak values of this oscillating curve correspond to the modes of the cavity that are significantly coupled with the current source, given its position and polarization. According to (14.2), the first seven modes of a $4 \times 3 \times 2$ m^3 cavity appear theoretically at $f_{011} = 62.5$ MHz, $f_{101} = 83.9$ MHz, $f_{012} = 90.1$ MHz, $f_{110} = 90.1$ MHz, $f_{111} = 97.6$ MHz, $f_{102} = 106.1$ MHz, and $f_{021} = 106.9$ MHz. The two first modes are significantly excited as well as one of the two last ones. That considered, the frequency response is much more irregular, as expected. Given the size of the cavity, modes do not overlap and are clearly distinguishable from each other up to 500 MHz and obtaining a diffuse field may only be expected in the upper frequency range.

14.4 Simulation of RCs through image theory

The previous section provided the fundamental rules regarding image theory for cavities. However, there are a number of additional recipes needed for a realistic and possible simulation of RCs. We list here three major aspects to deal with:

- Evaluation of loss coefficient according to the composite Q-factor of the RC under investigation.
- Generation of current source with arbitrary orientation and computation of the vectorial electric field.
- Introduction of a stirring method.

14.4.1 Assessment of the loss coefficient ξ

The loss coefficient ξ was previously set to an arbitrary value without knowledge of the physical system under investigation. Practically, this coefficient may be related to the composite Q-factor of the cavity [2,3]. The composite Q-factor is an ensemble quantity involving the simultaneous or successive excitation of several modes through a stirring mechanism. It is a specific generalization of a single mode Q-factor $Q_{m,n,p}$ in (14.5) and is defined, in steady state, as the ratio of the average stored energy per unit of time and the dissipated power. In fact, the composite Q-factor Q_g links the ensemble average energy over different states of the chamber and the overall dissipated power as follows:

$$Q_g = 2\pi f_0 \frac{\mathbb{E}[W_E]}{\mathbb{E}[P_d]}. \tag{14.25}$$

This definition of the quality factor Q_g applies to steady state harmonics fields at frequency f_0. The notation $\mathbb{E}$ stands for the expected value. The term W_E, stands for the electromagnetic energy in the chamber and P_d represents the active power dissipated in the RC.

The composite Q-factor is related to the decay time constant of the chamber. Its relationship with the Q-factor is linked to the charge or the discharge time of resonators to reach equilibrium. Let suppose that the source of energy is switched off once the steady state is reached. The elementary rate of change of the dissipated power P_d over a small time interval dt is compensated by the elementary rate of change of the stored energy (dW_E) in the chamber:

$$dW_E = -P_d dt. \tag{14.26}$$

Therefore, according to the definition in (14.25), it implies that $dW_E(t)$ decays exponentially with the time constant τ such as:

$$dW_E(t) = \exp\frac{-t}{\tau} = \exp\frac{-t}{Q_g/(2\pi f_0)}, \tag{14.27}$$

and, therefore:

$$Q_g = 2\pi f_0 \tau. \tag{14.28}$$

Combining (14.28) and (14.24), we find an approximate relation between the loss coefficient and the composite quality factor:

$$ln(1-\xi) \approx -\frac{\pi f_0 L}{cQ_g}. \tag{14.29}$$

For $\xi \ll 1$:

$$\xi \approx \frac{\pi f_0 L}{cQ_g}. \tag{14.30}$$

There are various ways of determining the intrinsic Q-factor both in the frequency domain and in the time domain. The estimation of τ is a popular method using time-domain instruments or inverse Fourier transform from frequency domain measurements. It turns out that, although τ is not constant over a very large frequency bandwidth, an average value provides reasonable estimation of most existing RCs.

14.4.2 Current source in arbitrary orientation and rectangular components of electric field

The current source in the real cavity may be oriented in any arbitrary direction and hence all image currents. The response at an observation point P in the cavity is the superposition of all elementary currents of all cavities, including the real elementary current source.

Any arbitrary nth order image current $\vec{i_a}$ is created after a total of n reflections including i reflections along the Ox axis, j reflections along the Oy axis, and k reflections along the Oz axis, with $n = i + j + k$. Introducing $R = 1 - \xi$, we thus obtain for the reflection coefficient R_a of $\vec{i_a}$:

$$R_a = R^i R^j R^k, \text{with } i + \mathrm{j} + \mathrm{k} = \mathrm{n}. \tag{14.31}$$

In the above equation, the decomposition along the three Cartesian axis is useless if the same reflection coefficient R is used for all RC walls. Attributing different reflection coefficients for each wall is therefore straightforward.

Then, we define a local Cartesian coordinate system $\vec{u}_a, \vec{v}_a, \vec{w}_a$ for this arbitrary nth order image current $\vec{i}_a$, where $\vec{w}$ corresponds to its orientation. The latter is given by a tilt angle α, defined by the angle between $\vec{w}_a$ and $\vec{e}_z$ in the global coordinate system, and an azimuthal angle β defined by the angle between $\vec{w}_a - (\vec{w}_a.\vec{e}_z)\vec{w}_a$ and $\vec{e}_x$ as illustrated in Figure 14.10. Furthermore, we attribute a dipole radiation pattern ($-\sin\theta_a$). As a result, the impulse response at the reception point P, in term of the electric field vector $\vec{E}_a(t)$, writes as follows:

$$\vec{E}_a(t) = -\omega\mu\frac{dhI_0R_a\delta(t-t_a)}{4\pi d_a}\sin\theta_a\begin{cases}\cos\theta_a\cos\phi_a\ \vec{u}_a\\ \cos\theta_a\sin\phi_a\ \vec{v}_a,\\ -\sin\theta_a\ \vec{w}_a\end{cases} \tag{14.32}$$

where dh is the elementary length of the dipole. The angle θ_a is the elevation angle under which the reception point P is seen from the considered image current $\vec{I}_a$ in its local spherical coordinate system. The angle ϕ_a corresponds to the azimuth angle in the same local coordinate system. The term d_a is the distance between the reception point P and the current image $\vec{I}_a$. As a result, (14.32) provides the electric field vector response at point P, but in the local coordinate system of the image currents. By using $\omega\mu(dhI_0)/4\pi = 1$ (in Volt) for the simulations, the power radiated from the source remains constant over the considered bandwidth.

The expression of this response in the global coordinate system requires the definition of a transformation matrix from the first to the second coordinate system such that:

$$\begin{aligned}\vec{E}_{a\left(\vec{e_x},\vec{e_y},\vec{e_z}\right)} &= \mathcal{R}_{\alpha,\beta}^{-1}\vec{E}_{a\left(\vec{u_a},\vec{v_a},\vec{w_a}\right)}\\ &= \mathcal{R}_{-\alpha,-\beta}\vec{E}_{a\left(\vec{u_a},\vec{v_a},\vec{w_a}\right)}\end{aligned}. \tag{14.33}$$

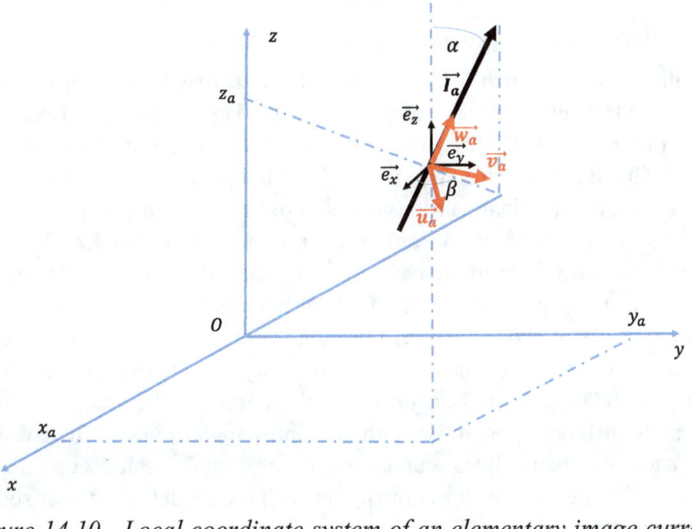

Figure 14.10 Local coordinate system of an elementary image current

The matrix $\mathcal{R}_{\alpha,\beta}$ defines the rotation matrix of angle α around the unit vector: $\vec{e_\beta} = -\cos\beta\vec{e_x} + \sin\beta\vec{e_y}$. This matrix $R_{\alpha,\beta}$ writes:

$$\mathcal{R}_{\alpha,\beta} = \begin{pmatrix} \cos^2\beta + (1-\cos^2\beta)\cos\alpha & -\cos\beta\sin\beta(1-\cos\alpha) & \sin\beta\sin\alpha \\ -\cos\beta\sin\beta(1-\cos\alpha) & \sin^2\beta + \left(1-\sin^2\beta\right)\cos\alpha & \cos\beta\sin\alpha \\ -\sin\beta\sin\alpha & -\cos\beta\sin\alpha & \cos\alpha \end{pmatrix} \tag{14.34}$$

So far, we considered that all image sources position and orientation $(\vec{i_a}, \theta_a, \phi_a)$ were predefined and stored in memory. This process may be deeply investigated to save memory location and time of calculation. There are a number of duplicable patterns and symmetry simplifications that may be exploited. For example, due to symmetry of the parallelepiped cavity only one eighth of the images may be stored and their position and orientation may be retrieved from simple equations for the rest of the images. These simplifications are detailed in [28].

14.4.3 Stirring process

There are three types of stirring process available to generate multiple states of RCs. The most popular one in electromagnetic compatibility applications is to change the boundary conditions through a mechanical stirrer. We may think of a virtual representation of a stirrer movement with a set of moving current sources. This is rather straightforward for image theory simulation but does not handle the boundary conditions of the real stirrer.

On the contrary, frequency stirring is easily handled with this method using the Fourier transform of the impulse response. Source stirring is also possible using a set of simulations at different positions of the source in the cavity. Above all, this model takes advantage of being able to evaluate the response for a set of reception points giving access to spatial distribution of the field over a large bandwidth.

14.4.4 Time domain results

We first illustrate how such 3-D simulations using image theory may succeed in simulating the transient response of a loaded RC. The measurement configuration appears in Figure 14.11. The RC size is 8.7 m × 3.7 m × 2.9 m and is the larger one at the IETR (Institut d'Electronique et des Technologies du numerique) laboratory. A few pieces of pyramidal carbon-loaded urethane absorbers are lying on the metallic (steel coated with zinc) floor. Two discone antennas (ARA CMA-118/A), operating in the 1–18 GHz frequency range, are placed in the chamber in arbitrary position. They are co-polarized in the vertical polarization. The one in the front of the photography is at fixed position throughout the experiment. It is used as the transmitting antenna and connected to an arbitrary wave generator (AWG), more precisely a AWG7052 from Tektronix. On the contrary, the receiving antenna is moved in 50 different positions, with a minimum spacing of half-wavelength between each of them. It is connected to a digital oscilloscope (Tektronix TDS6124C). For the sake of fair comparison with the model, the stirrer seen on the picture remains in a fixed position.

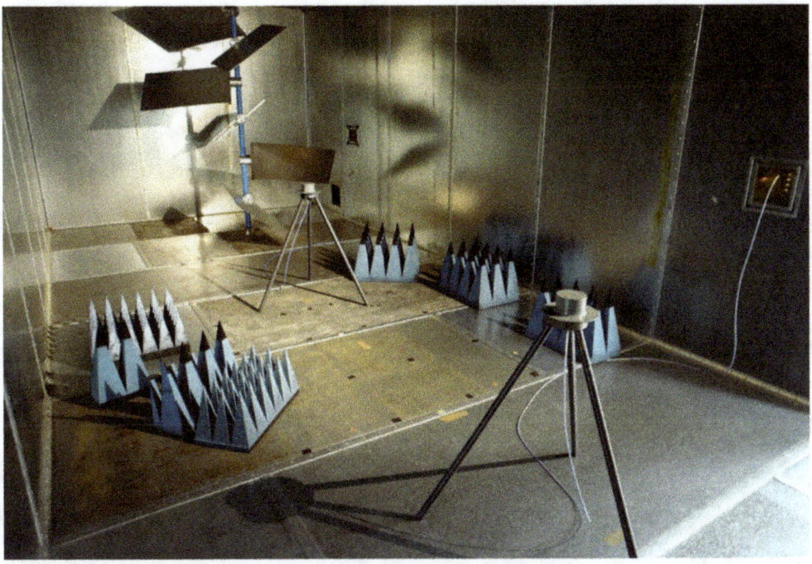

Figure 14.11 Experimental configuration of a loaded RC with some pyramidal absorbers. The RC size is 8.7 m × 3.7 m × 2.9 m. The estimated corresponding loss factor was $\xi \approx 0.025$ ($R = 0.975$)

The first input parameter of the model consists of the size of the RC. Then, the loss coefficient ξ is determined from (14.30), using the estimation of the average distance between two reflections in (14.23) and an experimental evaluation of the quality factor of the loaded RC. This evaluation may be carried out in many different ways. Here, we use the impulse response of the channel using the AWG. Due to the antenna response and the bandwidth of the AWG this corresponds to a filtered impulse response in the 1–2.5 GHz frequency band. The decay time of the response provides an estimation of τ. Here, $\tau \approx 180$ ns, and therefore $\xi \approx 0.025$. The discone antenna is assimilated to a vertical dipole and therefore, the E-field is calculated at any reception point P using (14.32). Simulations of the z-component of the electric field are performed in 50 random reception points P in the model (that do not have to correspond to the 50 arbitrary locations of the receiving antenna in the RC).

The transient regime of the RC is observed using a source of excitation consisting of a sine wave signal (here at 1 GHz) modulated by a unique rectangular pulse of 300 ns duration. The received simulated or measured power are calculated from the square envelope or recorded signals. The mean received power in then averaged over the 50 random or arbitrary locations of the receiver.

The average measured and simulated transient responses are provided in Figure 14.12. The impulse duration being less than two times the reverberation time constant, both barely reach a plateau before the input signal is switch off. The simulation is able to reproduce the rising and falling edge of the measured signal.

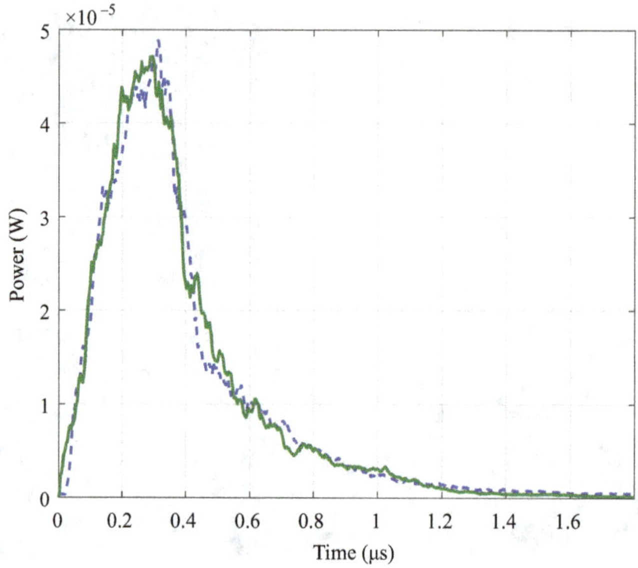

Figure 14.12 Average power measured or simulated at every moment over 50 positions of the receiving antenna. The source signal is a continuous wave signal at 1 GHz modulated by a 300 ns pulse

The mean statistic of the signal at any time is therefore well reproduced using a very limited set of information about the RC, i.e. its size and Q-factor only. This is the case for such an oversized cavity for which we obtain a perfect diffuse field, despite the cavity is perfectly parallelepiped. Observation of simulation data in the frequency domain confirms this statement as shown in the next section.

14.4.5 Frequency domain results

A classical issue regarding RCs is the determination of the required conditions to obtain an ideal random field, i.e. a Gaussian field according to (14.10). Such a condition is reached if the modal overlap is high enough. This modal overlap is defined as the average distance between modes divided by the width of the mode and was given in (14.11) in the asymptotic case. In case the modal overlap is much less than 1, the modes are isolated from each other, and ideal random field conditions are not met. There is no strict threshold that determines the ideal diffuse field condition. It is a matter of tolerance with regard to the Gaussian statistics. As a rule of thumb, we are close to Gaussian field conditions if the modal overlap is about 10 or 20.

We use the theory of image simulation to examine this issue. Since the 3-D ray-tracing simulation is performed with a purely parallelepiped cavity, no chaotic effects can be accounted for, such as those associated with the mechanical stirrer that is used for stirring fields in experiments. The role of these simulations is therefore dedicated to the analysis of the RC size and Q-factor only.

We present the results of E-field statistics obtained in the empty chamber as a function of frequency. The same RC without absorbers exhibits a much higher Q-factor or time constant. The time constant of the empty RC chamber is roughly estimated as a constant $\tau \approx 2.75$ μs over a 200–2,000 MHz frequency range. The time window for simulation is fixed to 12 μs, allowing the RC to reach steady state (99% of energy is stored according to the exponential rise time). The discrete time response over the entire time windows is obtained with an elementary time step of 200 pS. A fast Fourier transform is then executed to obtain the response in the frequency domain. The simulated data consists of the magnitude of x, y, z rectangular components of the electric field collected in 150 random positions of the reception point P for a single position of the transmitter. The transmitter is once again a dipole antenna, but oriented along the $\vec{e}_x + \vec{e}_y + \vec{e}_z$ direction to avoid any preferential polarization. According to the plane wave spectrum model, a rectangular component of electric field corresponds to the square root of the sum of two independent and identical Gaussian distributions of the real and imaginary parts. As a consequence, it follows a two degree of freedom χ distribution, i.e. a Rayleigh distribution. For the subsequent analysis, the data are first normalized to their mean. The probability density function of a Rayleigh distribution with data normalized to their mean writes

$$f_r(x) = \frac{\pi}{2} x \exp\left(-\frac{\pi}{4}x^2\right), \tag{14.35}$$

and the standard deviation of this normalized distribution is $\sigma_r = \sqrt{(4/\pi - 1)} \approx 0.52$.

Since the above simulation based on ray-tracing and theory of images provides samples of 150 realizations for the three rectangular components, it is possible to test if a Rayleigh distribution can fit them as a function of frequency. This is done using a goodness-of-fit (GoF) test. A GoF test consists in evaluating the likelihood of a hypothetical distribution regarding the sample collected from a population. These tests are generally based on a distance evaluation between the empirical cumulated distribution function and the hypothetical distribution function. Such a distance is tabulated according to the sample size and some significance level (e.g. 5%). In our case, we use the Anderson-Darling (AD) goodness-of-fit-test [15]. The distance evaluated from an AD test a weighted distance that gives more importance to the tails of the distribution. It is therefore less conservative than a chi-square or a Kolmogorov–Smirnov test.

Since we collected many samples according to frequency, the results of AD tests can be translated into rejection rate or acceptance rate. The fast Fourier transform of the 12 μs window from the time domain simulation provides a spectral response with a frequency step of around 83.3 kHz. However, in order to select uncorrelated frequency sample, we pick up one sample per MHz and calculate the rejection rate on 50 MHz bandwidths between 0 and 2.5 GHz.

The results are shown in Figure 14.13. It highlights that the rejection of the Rayleigh distribution tends to decrease with frequency and that the probability to reject it becomes lower than the probability to accept it above around 900 MHz. At

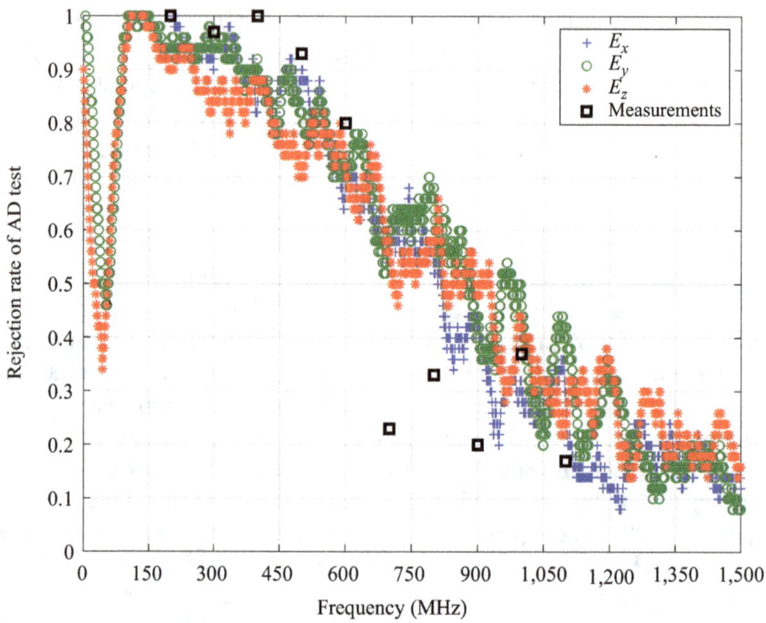

Figure 14.13 Rejection rate from Anderson Darling goodness-of-fit test applied to the Rayleigh distribution according to the three rectangular components of the simulated electric-field. Comparison with the same test from experimental data. The sample size is N = 150 and risk threshold is 5%

900 MHz, the Q-factor reaches $15,550$ according to (14.28). Thus, the model overlap d may be approximated from (14.11) and is found to be $d \approx 4$. These results are confronted with experimental ones represented by square symbols in Figure 14.13. Therefore these simulations confirmed the trend of experimental data. However, the rejection or acceptance rate of Rayleigh distribution gives a pass or fail indication only. As any GoF test, one can only conclude that the hypothetical hypothesis remains acceptable if the rejection rate is low enough reaching the significance level. If the test fails, then we have no clear information about how far it fails. The Weibull distribution, already identified as a candidate distribution for non-ideal random field in RC provides this feature. The Weibull probability density function writes:

$$f_W(x) = \alpha\beta x^{\beta-1} e^{-\alpha x^{\beta}/\beta}, \text{with } x \in \mathbb{R}^+, \alpha > 0 \text{et} \beta > 0. \qquad (14.36)$$

The Weibull distribution consists of a two-parameter distribution with α its scale parameter and β its shape parameter. This parametric distribution encompasses different distributions including the exponential or the Rayleigh distribution. In case of normalized data to the mean, the Weibull distribution is equivalent to a Rayleigh distribution with $\alpha = \pi/4$ and $\beta = 2$. It is therefore interesting to check is a Weibull distribution is able to fit the simulated data and to observe the trend of α and β according to frequency.

Figure 14.14 is obtained in the same conditions as introduced for Figure 14.13. The GoF is now applied to test for a hypothetical Weibull distribution instead. The acceptance rate is higher than the rejection rate at frequencies above 150 MHz only. The rejection rate tends to decrease with frequency and reach a plateau above 150 MHz. We may conclude that the Weibull law is an acceptable hypothesis over a large-frequency range including the lower frequency range for which the ideal random field conditions are not met. These results are also confronted with experimental ones represented by square symbols in Figure 14.14. Therefore, these simulations confirmed the trend of experimental data at the exception of the lowest test frequency (200 MHz).

A better view is offered looking at the corresponding estimated α and β parameters plotted in Figure 14.15. It shows that the scale parameter α reaches its corresponding Rayleigh value as far as the frequency exceeds 600 MHz. However, looking at the β parameter, one may conclude that convergence to a Rayleigh distribution is observed above 1,000 GHz. These simulations are also in agreement with experimental data (square symbols).

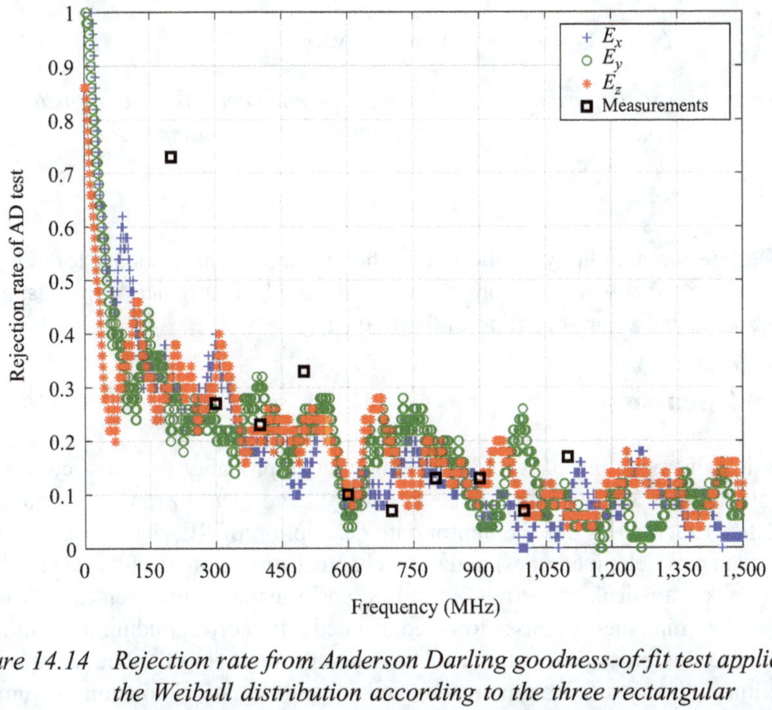

Figure 14.14 Rejection rate from Anderson Darling goodness-of-fit test applied to the Weibull distribution according to the three rectangular components of the simulated electric-field. Comparison with the same test from experimental data. The sample size is $N = 150$ *and risk threshold is 5%*

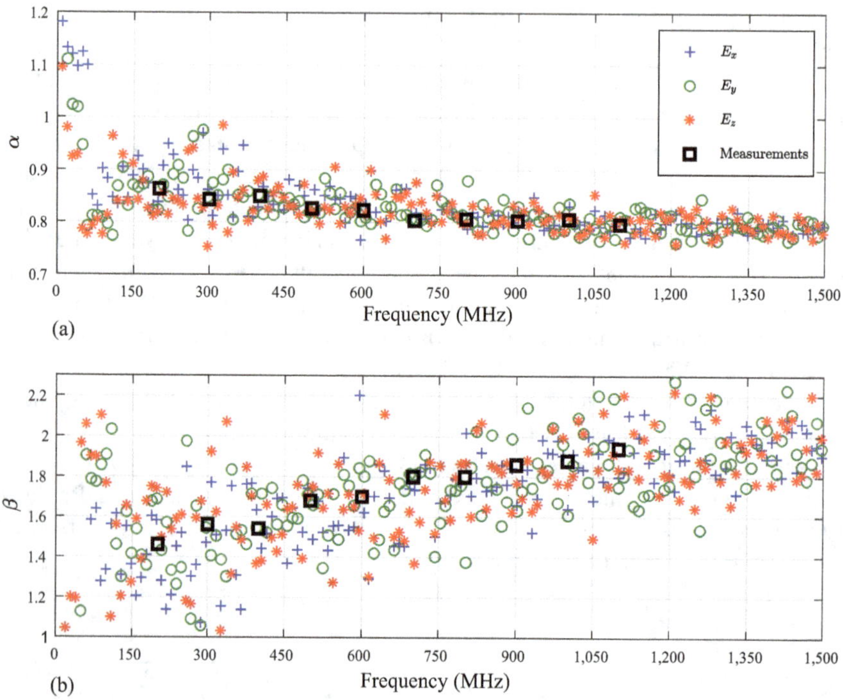

Figure 14.15 Scale parameter α (a) and shape parameter β (b) of adjusted Weibull distribution from simulated data. Comparison with measurements

Despite the simplicity of the image theory model that accounts for size and losses of a perfectly parallelepiped cavity, it is able to predict with reasonable success a RC behavior according to the frequency.

14.5 Discussion

Time domain modeling of Maxwell's equations is probably the most convenient way of studying RC over their wide frequency band of utilization. Deterministic simulations accounting for the appropriate description of RC chamber geometries (other than parallelepiped ones) and of mechanical stirrers are useful tools to check for various statistical properties or stirrer performance. However, they cannot provide deterministic responses to be confronted with corresponding deterministic data due to the extreme sensitivity of the response regarding the boundary conditions. Image theory offers an interesting and complementary simulation instrument.

Using image theory and ray-tracing allows for a rather quick retrieval of a RC behavior, given a minimum amount of information, i.e. its size and its Q-factor. This simulation principle is simple and requires a limited number of code lines. We limited its presentation to results obtained with a single current source but the core of the method offers many degrees of freedom. Manipulating antenna patterns, various reflection coefficients for each wall, simulating the movement of a virtual stirrer, studying the field correlation [31] are among many possible opportunities for image theory application. Despite its intrinsic limitations, this is therefore an interesting tool for testing ideas.

14.6 Acknowledgments

The author wish to thank Emmanuel Amador who conducted his Ph.D. work at IETR, Rennes, under his supervision. He is the author of the software used to produce the results shown in this chapter. This software is available at: https://projets.ietr.fr/imagetheorymodel/.

References

[1] Hill D. Electromagnetic Fields in Cavities: Deterministic and Statistical Theories. New York, NY: IEEE Press Wiley; 2009.

[2] Besnier P, Demoulin B. Electromagnetic Reverberation Chambers. New York, NY: ISTE Wiley & Sons; 2011.

[3] Andrieu G. (ed.) *Electromagnetic Reverberation Chambers: Recent Advances and Innovative Applications*. Stevenage: SciTech Publishing Incorporated; 2020.

[4] Hill D. Plane wave integral equations for fields in reverberation chambers. *IEEE Transactions on Electromagnetic Compatibility*. 1998;40(Pt 3): 209–217.

[5] Kostas JG, Boverie B. Statistical model for a mode-stirred chamber. *IEEE Transactions on Electromagnetic Compatibility*. 1991;33(4):366–370.

[6] Cozza A. The role of losses in the definition of the overmoded condition for reverberation chambers and their statistics. *IEEE Transactions on Electromagnetic Compatibility*. 2011;53(Pt 1):296–307.

[7] Gros JB, Legrand O, Mortessagne F, *et al.* Universal behaviour of a wave chaos based electromagnetic reverberation chamber. *Wave Motion*. 2014; 51(Pt 4):664–672.

[8] Gros JB, Kuhl U, Legrand O, *et al.* Lossy chaotic electromagnetic reverberation chambers: universal statistical behavior of the vectorial field. *Physical Review E*. 2016;93:032108.

[9] Moglie F. Convergence of the reverberation chambers to the equilibrium analyzed with the finite-difference time-domain algorithm. *IEEE Transactions on Electromagnetic Compatibility*. 2004;46(3):469–476.

[10] Moglie F, Primiani VM. Analysis of the independent positions of reverberation chamber stirrers as a function of their operating conditions. *IEEE Transactions on Electromagnetic Compatibility*. 2011;53(2):288–295.

[11] Xu Q, Huang Y. Anechoic and reverberation chambers: theory, design and measurements. *Wiley – IEEE series*; 2019.

[12] Arnaut LR. Operation of electromagnetic reverberation chambers with wave diffractors at relatively low frequencies. *IEEE Transactions on Electromagnetic Compatibility*. 2001;43(4):637–653.

[13] Leferink F, Boudenot J, van Etten W. Experimental results obtained in the vibrating intrinsic reverberation chamber. In: *IEEE International Symposium on Electromagnetic Compatibility*. Symposium Record (Cat. No.00CH37016). vol. 2; 2000. pp. 639–644.

[14] Orjubin G, Richalot E, Mengue S, *et al.* Statistical model of an undermoded reverberation chamber. *IEEE Transactions on Electromagnetic Compatibility*. 2006;48(1):248–251.

[15] Lemoine C, Besnier P, Drissi M. Investigation of reverberation chamber through high-power goodness-of-fit tests. *IEEE Transactions on Electromagnetic Compatibility* 2007;49(Pt 4):745–755.

[16] Chen X. Using Akaike information criterion for selecting the field distribution in a reverberation chamber. *IEEE Transactions on Electromagnetic Compatibility*. 2013;55(4):664–670.

[17] Xue W, Chen X, Zhang A. Investigation of automotive testing in a reverberation chamber. IET Microwaves, *Antennas Propagation*. 2019;13 (15):2605–2609.

[18] Monsef F, Cozza A. Variability and confidence intervals of the power measured in a reverberation chamber. *IEEE Transactions on Electromagnetic Compatibility*. 2014;56(5):1238–1241.

[19] Orjubin G, Richalot E, Picon O, *et al.* Chaoticity of a reverberation chamber assessed from the analysis of modal distributions obtained by FEM. *IEEE Transactions on Electromagnetic Compatibility*. 2007;49(4):762–771.

[20] Gros JB, Kuhl U, Legrand O, *et al.* Statistics of the Electromagnetic Response of a Chaotic Reverberation Chamber; 2014.

[21] Gros JB, Kuhl U, Legrand O, *et al.* Lossy chaotic electromagnetic reverberation chambers: Universal statistical behavior of the vectorial field. *Physical Review E*. 2016;93:032108. Available from: https://link.aps.org/doi/10.1103/PhysRevE.93.032108.

[22] Weinzierl D, Sartori CAF, Perotoni MB, *et al.* Numerical evaluation of noncanonical reverberation chamber configurations. *IEEE Transactions on Magnetics*. 2008;44(6):1458–1461.

[23] Zhao H, Shen Z. Efficient modeling of three-dimensional reverberation chambers using hybrid discrete singular convolution-method of moments. *IEEE Transactions on Antennas and Propagation*. 2011;59(8):2943–2953.

[24] Zhao H, Shen Z. Memory-efficient modeling of reverberation chambers using hybrid recursive update discrete singular convolution-method of moments. *IEEE Transactions on Antennas and Propagation*. 2012;60 (6):2781–2789.

[25] Soltane A, Andrieu G, Reineix A. Doppler spectrum analysis for the prediction of rotating mode stirrer performances in reverberation

chamber. *IEEE Transactions on Electromagnetic Compatibility*. 2019;61 (5):1408–1413.

[26] Cappetta L, Feo M, Fiumara V, *et al.* Electromagnetic chaos in mode-stirred reverberation enclosures. *IEEE Transactions on Electromagnetic Compatibility*. 1998;40(3):185–192.

[27] Do-Hoon Kwon, Burkholder RJ, Pathak PH. Ray analysis of electromagnetic field build-up and quality factor of electrically large shielded enclosures. *IEEE Transactions on Electromagnetic Compatibility*. 1998;40(1):19–26.

[28] Amador E, Lemoine C, Besnier P, *et al.* Reverberation chamber modeling based on image theory: investigation in the pulse regime. *IEEE Transactions on Electromagnetic Compatibility*. 2010;52(4):778–789.

[29] Amador E, Andries MI, Lemoine C, *et al.* Absorbing material characterization in a reverberation chamber. In: *International Symposium on Electromagnetic Compatibility, EMC EUROPE*; September 2011; York, United Kingdom. IEEE; 2011. pp. 117–122.

[30] Kosten CW. The mean free path in room acoustics. *Acustica* 10. 1960;10: 245–250.

[31] Amador E, Lemoine C, Besnier P. Numerical study of spatial correlation in reverberation chamber. *Electronics Letters*. 2011;47(24):1319–1320.

Chapter 15

Analysis of a class of dynamical systems with applications to power conversion circuits

A. El Aroudi[1]*, L. Benadero*[2]*, F. Torres*[3] *and E. Ponce*[3]

15.1 Introduction

In the past, remarkable developments in the field of nonlinear smooth dynamical systems and bifurcation theory have taken place [1–3]. Recently, nonlinear dynamics and bifurcation behavior of *nonsmooth* switched systems have attracted the interest of many research groups all over the world. Outstanding works have been published in different magazines and presented in specialized conferences [4–11]. During the last years, there has been also published several textbooks on this subject [12–15].

Power electronics circuits are examples of switching systems. They are designed so that certain designated signals such as voltages and currents, tracking errors and actuator inputs, do not exceed pre-specified levels. Despite the seemingly obvious requirement of knowing explicitly the system behavior in terms of circuit parameters, it was only in the recent years that researchers begin to talk about using results from bifurcation analysis to design purpose based on explicit mathematical expressions [16–20]. Sure, using simplified models such as averaged models has allowed to obtain such expressions for stability boundaries and control design years ago [21,22] but this was done at the expense of losing some details in the actual behavior of the switching system or ignoring some interesting phenomena.

Resonant power converters can operate efficiently with high switching frequencies allowing the use of small storage components which improve the power density. Besides their traditional use in inductive heating, resonant converters are becoming increasingly popular in many applications such as efficient lighting [23], battery chargers in electrical vehicles [24] and wireless power transfer [25] among others [26,27]. However, their dynamic behavior analysis is more involved due to their increased harmonic content. The dependence of their dynamics, both local and global, with the load is a well known problem of these converters.

[1]Departament d' Enginyeria Electrònica, Elèctrica i Automàtica, Universitat Rovira i Virgili, Spain
[2]Departament de Física, Universitat Poliecnica de Catalunya, Spain
[3]Departamento de Matemática Aplicada II, Universidad de Sevilla, Spain

Indeed, the assumption that the switching frequency is much higher than all the natural frequencies of the system, widely used in classical pulse width modulated converters, is no longer valid in resonant converters because their operating switching frequency is close to the natural frequency of the resonant circuit [28]. Consequently, more advanced modeling tools such as the frequency-domain-based describing function [29–31] or the equivalent time-domain-based Hamel locus [32] are needed to describe the dynamics of the system. Discrete-time modeling is another accurate tool which were used in [25] to determine the possible steady-state operating points of an LCL wireless power transfer resonant converter. A similar approach was also used in [33,34] to study the stability of the desired limit cycles in a series resonant power converter and in a parallel LC resonant converter respectively. All the previous approaches assume a beforehand known switching pattern ignoring any peculiar phenomena that can take place in switched systems such as sliding-mode behavior. Therefore, these analysis techniques fail short in revealing accurately the real global dynamics of a switched system. Namely, under the existence of sliding conditions, sliding motion can take place in this kind of systems [35,36]. In particular, a trajectory of the switched system can partly remain on a sliding-mode region associated with an infinite number of switching between two different subsystems and a part of a limit cycle may emerge on this region. Sliding bifurcations take place when a crossing limit cycle interacts with the sliding-mode region [5]. Such dynamical behavior results in nonsmooth complex behaviors that cannot be described by the approaches detailed in [28–33].

This chapter first presents some methods to analyze the dynamics of nonlinear systems and to predict their bifurcation behavior. The chapter aims to provide the readers with both a basis in dynamical systems theory and the necessary understanding of the approaches, methods, results, and terminology used in the modern applied mathematics literature with special focus on switching systems. The results presented here are hardly covered in standard graduate-level textbooks but seem to be important in applications. It is also aimed to provide the reader with application of explicit mathematical procedures to some particular research problems in the field of power and energy conversion circuits. First, several examples are used for illustrating basic concepts from smooth nonlinear dynamics field. An example from the nonsmooth field is used as an application. A prior course on state-space representation of dynamical systems would be advantageous but is not necessary. Using the illustrative examples, a brief review of dynamical systems analysis is provided introducing the basic terminology related to this field. The notions of Filippov systems, their invariant sets and their stability are introduced.

A system is said to be of Filippov type if the differential equations that govern its dynamics have a discontinuity with respect to the state vector [37]. This discontinuity represents existence of physical switches that change their state according to threshold conditions. In recent years this kind of systems has received a great interest from both theoreticians and researchers seeking for applications in different engineering fields [12–15]. Mathematically, systems of this type can be described by involving in their models discontinuous functions such as Heaveside step function, sign function or similar. They are called therefore systems with *discontinuous right hand sides* [37]. A particular class of these systems are linear systems with discontinuous control signal. This is the case of many power electronics circuits [21,22] and relay feedback systems [5]. The resonant power converter that will be studied is an example that belongs to this class of systems. Other more general cases such as bilinear systems for which a product of the state and the control variables appear in their models can be treated following a similar approach.

Before addressing Filippov systems, methods for analyzing two-dimensional linear autonomous (time-invariant) dynamical systems are reviewed and simple phase portraits are constructed in the plane. For nonlinear systems, linearization technique is used for both equilibrium points and limit cycles. Phase portraits are constructed and stability analysis tools for both the equilibrium points and limit cycles are presented.

The rest of this chapter is organized as follows. After this introduction Section 15.2 lays the foundations and basic principles for the development in the rest of the chapter. The intention of the discussion offered is to provide a brief review of fundamentals including some mathematical background, to make the treatment self-contained. Readers with an introductory background on dynamical systems may omit some sections. Sections 15.3 and 15.4 are sequentially structured to introduce some concepts for analyzing nonlinear dynamical systems. Emphasis here is on the importance of limit cycles in nonlinear systems and their stability analysis and coexistence of multiple attractors. Moreover, Floquet theory, as a tool for analyzing the stability of limit cycles in smooth nonlinear systems and in nonsmoooth piecewise affine systems, is presented. The concept of sliding mode in switched Filippov systems is introduced in Section 15.5 and a simple example is used to illustrate the features of this prominent behavior in this kind of systems. Section 15.6 presents the mathematical model of the Parallel Resonant Converter (PRC). The equations describing the dynamics at the repulsive or escaping sliding-mode region are derived. Simulations from the switched model are presented for different values of the load resistance thus revealing different dynamical behaviors and inferring coexistence of steady-state solutions. A combined analytical-numerical approach is used to obtain the steady-state crossing limit cycles and to perform their stability analysis using the Floquet theory combined with Filippov technique. The boundary of saddle-node bifurcation is located in the same section. The conditions for nonsmooth crossing-sliding limit cycle bifurcations are also derived and a summary of the bifurcation scenario in terms of the load resistance is presented.

15.2 An overview on dynamical systems: basic principles

15.2.1 Modeling: vector fields

The main goal in dynamical system analysis is to understand its behavior, i.e, to reveal how its state variables evolve. In electrical circuits, the state variables are currents in the inductors and voltages in the capacitors. Given the state variables of a dynamical system, one needs a rule describing how the system changes with time. Obtaining the rule refers to modeling of the system. The starting point to derive a model is the fundamental principles and laws that the system must fulfill. The models of electric circuits can be derived by using energy balance or Kirchoff laws. Often, simplifications and approximations based on realistic assumptions are used to obtain simplified models that allow a profound understanding of the system behavior. Good matching between the model and the actual system can be obtained even under doing these approximations. In this chapter, for simplicity of the presentation, the focus is in general on planar dynamical systems. These systems have two state variables and their motion can be plotted in the state plane or phase plane. Some concepts also hold for higher dimensions. First, smooth systems will be considered. Then, systems with discontinuous hand sides known as Filippov systems will be addressed. Some illustrative examples will be used when presenting the concepts.

Generally, a dynamical system can be described by an ordinary differential equation that can be expressed in compact form as follows

$$\dot{\mathbf{x}} = \mathbf{f}(\mathbf{x}, t, \mu) \tag{15.1}$$

where $\mathbf{x}$ is the vector of state variables and μ is a set of system parameters and the overdot stands for differenciating with respect to time t. Let us consider two-dimensional vector fields and hence $\mathbf{x} \in \mathbb{R}^2$ and $\mathbf{f} \in \mathbb{R}^2$. Hereinafter systems with function $\mathbf{f}(\mathbf{x}, \mu)$ not depending explicitly on time are considered. These are called *autonomous* systems. For this kind of systems, the vector function $\mathbf{f}$ is called *vector field*. It is a vector operator that assigns to each vector $\mathbf{x}$ on a domain another vector $\mathbf{f}(\mathbf{x}, \mu)$ in the same domain over a period of time Δt. This vector has the specific amplitude and direction of $\mathbf{f}(\mathbf{x}, \mu)$ and is represented based at the point $\mathbf{x}$, with a directed line segment from $\mathbf{x}$ to $\mathbf{x} + \Delta t\mathbf{f}(\mathbf{x}, \mu)$. Repeating this operation at every point in the plane, we obtain a vector field diagram. A vector field diagram is a way of visualizing functions that have the same number of dimensions in their domain as in their range.

In control systems, the model given by (15.1) is replaced by

$$\dot{\mathbf{x}} = \mathbf{f}(\mathbf{x}, \mu, u), \quad u = \sigma(\mathbf{x}) \tag{15.2}$$

where, as before, $\mathbf{x}$ is the vector of the state variables but u now represents a suitable control input signal. Such a control system is called linear if the function $\mathbf{f}$ is linear in both $\mathbf{x}$ and u and the function σ is linear in $\mathbf{x}$. A linear control system can be represented by the following state-space model

$$\dot{\mathbf{x}} = \mathbf{A}\mathbf{x} + \mathbf{B}u, \quad u = U_0 - \mathbf{K}^{\top}\mathbf{x} \tag{15.3}$$

where $\mathbf{A}$ is a 2×2 matrix, $\mathbf{B}$ and $\mathbf{K}$ are 2×1 vectors and U_0 is a scalar constant.

15.2.2 Phase portraits

The phase portrait of a dynamical system is a geometric representation of its trajectories in the phase space. It consists of a plot of typical trajectories in the phase space starting at different initial conditions. Each initial condition gives a unique trajectory or point in the phase space. The phase portrait reveals about the existence of equilibrium points and limit cycles, both attractors and repellors, in the system for the chosen parameter values. Therefore, plotting the phase portrait of a dynamical system in the phase plane allows to understand its dynamical behavior. For that, the vector field describing the velocity state vector is needed.

15.2.3 Equilibrium points and stability of planar linear systems

A point $\mathbf{x}_q$ is said to be an equilibrium point of the autonomous system $\dot{\mathbf{x}} = \mathbf{f}(\mathbf{x}, \mu)$ if it has the property that whenever the state of the system starts at $\mathbf{x}_q$, it remains at $\mathbf{x}_q$ for all future time, i.e $\mathbf{x}(t_0) = \mathbf{x}_q \Rightarrow \mathbf{x}(t) = \mathbf{x}_q \; \forall t \geq t_0$, which means that $\dot{\mathbf{x}}(t) = 0$. Therefore, equilibrium points can be obtained by zeroing the vector field and solving the resulting equation for the vector $\mathbf{x}$. They are therefore the roots of the equation $\mathbf{f}(\mathbf{x}, \mu) = \mathbf{0}$. For the linear system $\dot{\mathbf{x}} = \mathbf{A}\mathbf{x} + \mathbf{B}$, if $\mathbf{A}$ is a non singular matrix, the equilibrium point is unique and is given by $\mathbf{x}_q = -\mathbf{A}^{-1}\mathbf{B}$. If $\mathbf{B} = \mathbf{0}$, the equilibrium point is the origin. By a change of variable $\mathbf{y} = \mathbf{x} - \mathbf{x}_q$ it is always possible to move the equilibrium point to the origin without changing the qualitative behavior of the system. The stability of an equilibrium point is the property that determines whether or not trajectories close to this point remain nearby, get closer or move further away. A solution $\phi(\mathbf{x}(t_0), t)$ that starts at the initial condition $\mathbf{x}(t_0)$ at an initial time t_0 can be represented by a curve in the two-dimensional plane. These curves are called trajectories or orbits. Without loss of generality let us consider $\mathbf{B} = \mathbf{0}$. The trajectories near the equilibrium points of linear systems can be expressed as follows

$$\mathbf{x}(t) := \phi(\mathbf{x}(t_0), t) = e^{\mathbf{A}(t-t_0)}\mathbf{x}(t_0) \tag{15.4}$$

The equilibrium points may be classified depending upon the type of the roots of the characteristic equation $\det(s\mathbf{I} - \mathbf{A}) = 0$. These roots are the eigenvalues of the matrix $\mathbf{A}$. The equilibrium points are called hyperbolic if the real part of their corresponding eigenvalues of the Jacobian matrix are nonzero. If the real part of either of the eigenvalues of the matrix $\mathbf{A}$ is equal to zero, then the critical point is called *nonhyperbolic*. If the eigenvalues of the matrix $\mathbf{A}$ are real, the trajectories $\mathbf{x}(t)$ will form hyperbolic paths in the vicinity of the equilibrium point $\mathbf{x}_q$. If the eigenvalues are imaginary, the trajectories will spiral or rotate about $\mathbf{x}_q$. The matrix $e^{\mathbf{A}(t-t_0)}$ is also called the *evolution operator* since it maps $\mathbf{x}(t_0)$ at an initial time t_0 to any future instant t. If the real parts of the eigenvalues are negative, those trajectories will evolve exponentially towards the equilibrium $\mathbf{x}_q$ and this is called an *attractor*. They will evolve away from it if the real parts of the eigenvalues are positive and the equilibrium point is called in this case a *repellor*. The neutral case is when the real part are zero. In this case, the trajectories nor they are attracted nor they are repelled by the equilibrium point.

Example 1. *As a first example, suppose that we have a system that incorporates two integrators* [38,39]. *In this case, it follows that the system differential equation in the time domain is $\ddot{x} = u$, where u is the system input and x is its output. It is clear that the open loop system (u constant) is unstable since its trajectories* *are unbounded. Let $y = \dot{x}$ and $\mathbf{x} = (x,\ y)^\top$. A way to stabilize the system is by using state feedback, i.e, making u dependent on the state $\mathbf{x}$. The feedback is used to change the behavior of dynamical systems so that desired features are met according to the specifications. For instance, systems that are unstable without feedback can be stabilized by appropriate feedback, systems that are slow can be made fast and oscillatory systems can be damped as will be shown below.*

A state space representation of the open loop system (without feedback) can be expressed as follows $\dot{\mathbf{x}} = \mathbf{A}\mathbf{x} + \mathbf{B}u$, where $\mathbf{A}$ and $\mathbf{B}$ are given by

$$\mathbf{A} = \begin{pmatrix} 0 & 1 \\ 0 & 0 \end{pmatrix}, \qquad \mathbf{B} = \begin{pmatrix} 0 \\ 1 \end{pmatrix} \tag{15.5}$$

Let us consider the control law $u = U_0 - \mathbf{K}^\top \mathbf{x} := U_0 - k_1 x - k_2 y$, where k_1 and k_2 and U_0 are suitable scalar constants. The state space representation becomes $\dot{\mathbf{x}} = (\mathbf{A} - \mathbf{B}\mathbf{K}^\top)\mathbf{x} + \mathbf{B}U_0$. The equilibrium point is $\mathbf{x}_q = (U_0/k_1, 0)^\top$. Without loss of generality let $U_0 = 0$. The closed loop system can be described by $\dot{\mathbf{x}} = \mathbf{A}\mathbf{x}$, where, with some abuse of notation, now $\mathbf{A}$ is given by

$$\mathbf{A} = \begin{pmatrix} 0 & 1 \\ -k_1 & -k_2 \end{pmatrix} \tag{15.6}$$

The equilibrium point is the origin, i.e, $\mathbf{x}_q = (0 \quad 0)^\top$. The characteristic equation of the closed loop system is $\det(s\mathbf{I} - \mathbf{A}) = s^2 + k_2 s + k_1 = 0$. For the roots of characteristic equation to lie in the left half side of the complex plane and equivalently for the system to be stable, both k_1 and k_2 must be positive. Let $\Delta = k_2^2 - 4k_1$. The eigenvalues of the state matrix are $\lambda_1 = (-k_2 + \sqrt{\Delta})/2$ and $\lambda_2 = (-k_2 - \sqrt{\Delta})/2$. If $\Delta > 0$, the roots are real. If $\Delta < 0$, they are complex conjugate. In this case, it can be seen that the feedback coefficient k_2 is related to the damping factor. The oscillation angular frequency is $\omega_0 = \sqrt{|\Delta|}/2$. The system is asymptotically stable and it converges to the equilibrium if $k_2 > 0$. If $k_2 = 0$, the system is marginally stable and the origin is a *center*. In this case, the trajectories on the phase plane are invariant closed curves (periodic solutions) whose amplitude depends on the initial conditions.

Computer simulations are used below to illustrate the previous cases.

Figure 15.1 shows the phase portrait near the equilibrium point (origin) and the vector field corresponding to the system with the parameter values depicted in the figure caption. The type of phase portrait depends on the eigenvalues λ_1 and λ_2, as summarized below:

1. If λ_1 and λ_2 are complex conjugate with negative real part, then the equilibrium point is an stable *focus*.
2. If λ_1 and λ_2 are complex conjugate with positive real part, then the equilibrium point is an unstable *focus*.
3. If λ_1 and λ_2 are distinct, real, and positive, then the equilibrium point is an unstable *node*.
4. If λ_1 and λ_2 are distinct, real, and negative, then the critical point is called a stable node.

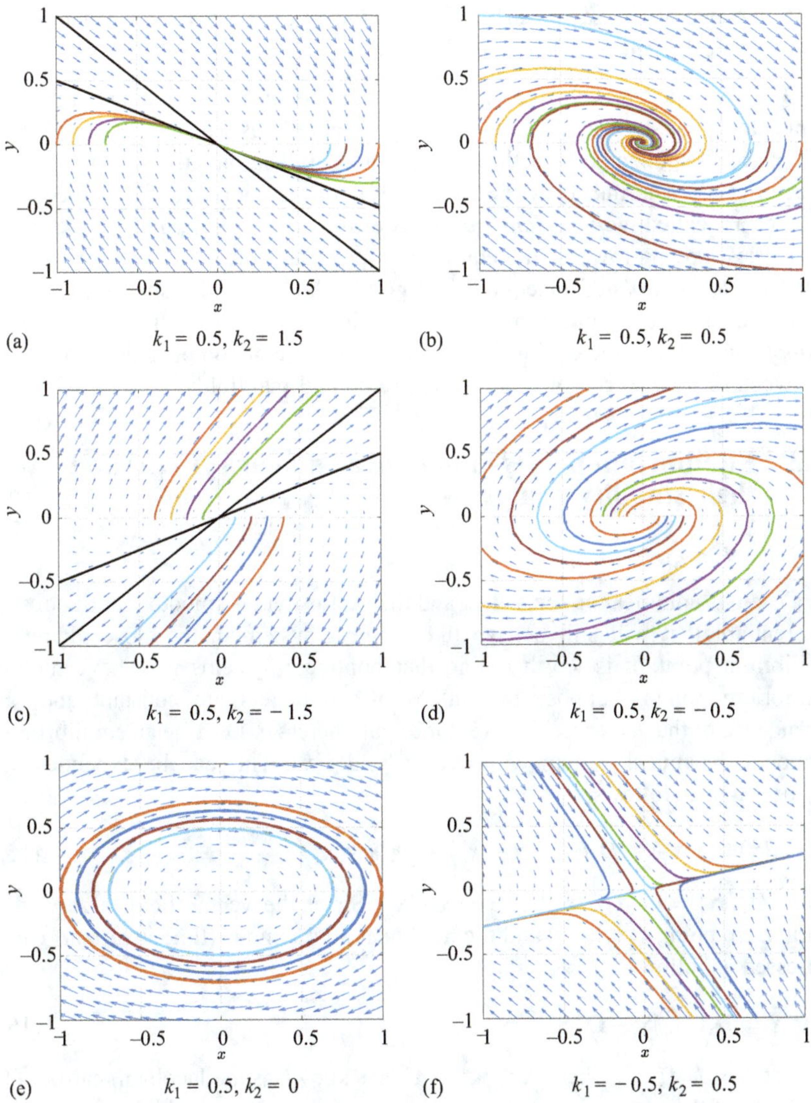

Figure 15.1 *Phase portraits of the double integrator system with state feedback for different values of feedback coefficients k_1 and k_2. The equilibrium is a (a) stable node: $\lambda_1 = -1$, $\lambda_2 = -0.5$, (b) stable focus: $\lambda_1 = -0.25 + 0.66j$, $\lambda_2 = -0.25 - 0.66j$, (c) unstable node $\lambda_1 = 1$, $\lambda_2 = 0.5$, (d) unstable focus $\lambda_1 = 0.25 + 0.66j$, $\lambda_2 = 0.25 - 0.66j$, (e) center $\lambda_1 = 1/\sqrt{2}j$, $\lambda_2 = -1/\sqrt{2}j$ and (f) saddle $\lambda_1 = -1$, $\lambda_2 = 0.5$*

5. If one eigenvalue, say λ_1, is positive and the other one, say λ_2, is negative, then the critical point is called a saddle point.

The phase portraits corresponding to real eigenvalues all contain straight lines. A trajectory that starts at a point belonging to these lines will remain on it forever and exhibits exponential growth or decay along it. The direction of those special trajectories are determined by the eigenvectors of the matrix $\mathbf{A}$. These special trajectories are called the *invariant manifolds* of the equilibrium point. Those special trajectories moving towards the equilibrium point as t increases, constitute the so called *stable manifold* of the equilibrium point, while all of this special trajectories moving away the equilibrium point as t increases, constitute the so called *unstable manifold* of the equilibrium point. In the case of complex eigenvalues, the eigenvectors are also complex and there are no real manifolds.

15.3 Equilibrium points in nonlinear planar systems and their stability analysis

15.3.1 Linearization

If a system is nonlinear on the state $\mathbf{x}$ and the vector field is smooth (continuous and differentiable), it will still behave like a linear system in the close vicinity of equilibrium point. It is worth noting that nonlinear systems may have multiple equilibrium points. Suppose the values of parameters are constant and omit dependence of the vector field on vector of parameters μ. Let $\mathbf{x}$ be an equilibrium of a smooth dynamical system with vector field $\mathbf{f}(\mathbf{x})$. This can be expanded as a Taylor series as follows

$$\mathbf{f}(\mathbf{x}) \approx \mathbf{f}(\mathbf{x}_q) + D_{\mathbf{x}}\mathbf{f}(\mathbf{x}_q)(\mathbf{x} - \mathbf{x}_q) + O(|\mathbf{x} - \mathbf{x}_q|^2) \tag{15.7}$$

where $D_{\mathbf{x}}\mathbf{f}(\mathbf{x}) = d\mathbf{f}(\mathbf{x})/d\mathbf{x}$. Let $\hat{\mathbf{x}} = \mathbf{x} - \mathbf{x}_q$. Since the vector field is zero at an equilibrium point, $\mathbf{f}(\mathbf{x}_q) = 0$ and the nonlinear autonomous (time invariant) can be expressed as follows

$$\dot{\hat{\mathbf{x}}} = \mathbf{D}_{\mathbf{x}}\mathbf{f}(\mathbf{x}_q)\hat{\mathbf{x}} + \mathbf{O}(|\hat{\mathbf{x}}|^2) \tag{15.8}$$

Let $\mathbf{A} = D_{\mathbf{x}}\mathbf{f}(\mathbf{x}_q)$, where $\mathbf{A} \in \mathbb{R}^{2\times 2}$. This is known as the Jacobian matrix of the system or of the vector field. Close to the expansion point, the higher-order terms can be neglected if the Jacobian matrix is non-singular. Therefore in the very close vicinity of the equilibrium point $\mathbf{x}_q$, the system can be represented by

$$\dot{\hat{\mathbf{x}}} \approx \mathbf{A}\hat{\mathbf{x}} \tag{15.9}$$

Higher order terms are needed if the Jacobian matrix is singular. Close to the equilibrium point the trajectories can be approximated as follows

$$\hat{\mathbf{x}}(t) \approx e^{\mathbf{A}(t-t_0)}\hat{\mathbf{x}}(t_0) \tag{15.10}$$

where $\mathbf{x}(t_0)$ is a suitable initial condition at the initial time t_0. According to the theorem of Hartman-Grobman [40], the behavior of the equilibrium point depends on the eigenvalues of the Jacobian matrix of the vector field $\mathbf{f}$ whenever the real part of the quoted eigenvalues are different from zero.

Example 2. *Consider the system represented by the following two-dimensional model*

$$\dot{x} = a\tanh(mx) + b\tanh(my) \tag{15.11}$$

$$\dot{y} = c\tanh(mx) + d\tanh(my) \tag{15.12}$$

where $a \in \mathbb{R}$, $b \in \mathbb{R}$, $c \in \mathbb{R}$, $d \in \mathbb{R}$ and $m \in \mathbb{R}^+$ are the system parameters. The equilibrium point is the origin and the Jacobian matrix is

$$\mathbf{J}(x, y) = \begin{pmatrix} am(1 - \tanh^2(mx)) & bm(1 - \tanh^2(my)) \\ cm(1 - \tanh^2(mx)) & dm(1 - \tanh^2(my)) \end{pmatrix} \tag{15.13}$$

At the origin this matrix becomes

$$\mathbf{J}(0, 0) = m\begin{pmatrix} a & b \\ c & d \end{pmatrix} \tag{15.14}$$

The characteristic equation is $\det(s\mathbf{I} - \mathbf{J}(0, 0)) = 0$ and can be expressed as follows

$$s^2 - m(a + d)s + m^2(ad - bc) = 0 \tag{15.15}$$

The eigenvalues of the Jacoban matrix $\mathbf{J}(0, 0))$ are given by

$$\lambda_1 = \frac{m}{2}\left(a + d + \sqrt{(a - d)^2 + 4bc}\right) \tag{15.16}$$

$$\lambda_2 = \frac{m}{2}\left(a + d - \sqrt{(a - d)^2 + 4bc}\right) \tag{15.17}$$

Close to the origin, similar behavior as in **Example 1** can be obtained depending on the values of parameters. A phase portrait of the system is depicted in Figure 15.2 for the set of parameter values shown in the figure caption. For this set of parameter values, it is worth to note that trajectories are spirals converging to the origin. However, the effect of non linearity makes these spirals quite different from their counterparts in a linear system. Neverthless in a close vicinity of the equilibrium point, only linear terms are dominant and the spirals are very similar to the linear case. Indeed, Hartman-Grobman or linearization theorem [40] shows that there is a neighborhood of the equilibrium point on which the phase portrait for the nonlinear system resembles that of the linearized system whenever the real part of the eigenvalues of the Jacobian matrix are non null. Other dynamical behavior are also possible depending on the parameter values. Note that for the origin to be stable in this example, $a + d$ must be negative and ad-bc must be positive.

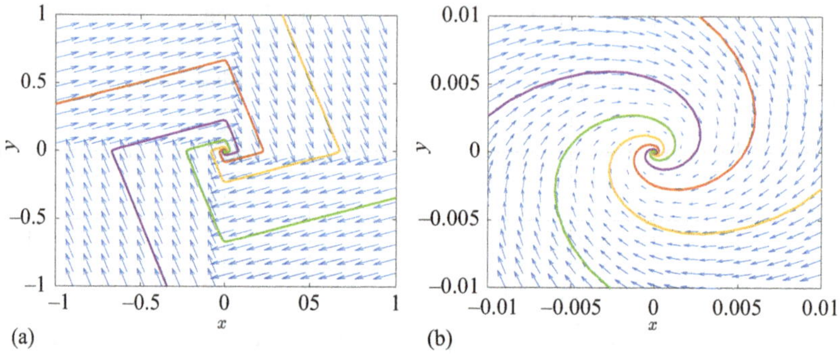

Figure 15.2 Phase portrait of the system of Example 2. Parameter values: $a = -1$, $b = 2$, $c = -2$, $d = -1$ and $m = 100$. The eigenvalues are $\lambda_1 = -100 + 200j$ and $\lambda_2 = -100 - 200j$ and the origin is a stable focus. (a) general view, (b) zoom near the origin

15.3.2 Basic concepts for bifurcation analysis

In a dynamical system, a bifurcation takes place when a small change made to a suitable parameter causes a sudden qualitative or topological change in its phase portrait. This kind of bifurcations can be analyzed through studying the local stability properties as parameters changes in the vicinity of the critical points.

Example 3. *Consider the two dimensional system described by the following model*

$$\dot{x} = ax + y \tag{15.18}$$

$$\dot{y} = -x + b\tanh(my) \tag{15.19}$$

where $a \in \mathbb{R}$, $b \in \mathbb{R}$ and $m \in \mathbb{R}^+$ are suitable parameters.

Remark 1. *Let consider the vector field* $\mathbf{F} = (ax + y, -x + b\tanh(my))$. *One has that* $\mathbf{F}$ *is odd-symmetric, i.e.,* $\mathbf{F}(-ax - y, x - b\tanh(my)) = -\mathbf{F}(ax + y, -x + b\tanh(my))$. *Therefore, the following statements hold*

- if $\mathbf{x}_q$ is a an equilibrium point, then $-\mathbf{x}_q$ is also an equilibrium point.
- if $\mathbf{x}_1(t)$ is a limit cycle with period T, then, either $\mathbf{x}_1(t)$ is asymmetric and there exists another limit cycle with period T, $\mathbf{x}_2(t)$ given by $\mathbf{x}_2(t) = -\mathbf{x}_1(t)$ or $\mathbf{x}_1(t)$ is a unique symmetric solution (limit cycle) with half period odd symmetry, *i.e.,* $\mathbf{x}_1(t) = -\mathbf{x}_1\left(t \pm \frac{T}{2}\right)$.

The expression of the Jacobian matrix of the previous model is as follows

$$\mathbf{J}(x, y) = \begin{pmatrix} a & 1 \\ -1 & -bm(\tanh^2(my) - 1) \end{pmatrix} \tag{15.20}$$

Let $\mathbf{x}_q = (x^*, y^*)$ be an equilibrium point for the system. From (15.18)–(15.19), y^* is the solution of the equation $y = -ab\tanh(my)$ and $x^* = -y^*/a$. The number of solutions mainly depends on the sign of ab. The characteristic equation is as follows

$$s^2 + (bm\tanh(my^*)^2 - (a + bm))s + abm\left(1 - \tanh(my^*)^2\right) + 1 = 0 \tag{15.21}$$

Depending on the values of the parameters, a pair of two different symmetric equilibrium points may coexist. If $ab > 0$, only the origin is an equilibrium point of the system. Figure 15.3 shows the plots of the functions $f(y) = -ab\tanh(my)$ together with the function y for $ab > 0$ and $ab < 0$. The intersections between the two plots determine the y-coordinates of the possible equilibrium points. In Figure 15.3(d), the dots indicate the approximated analytical equilibrium points for high enough m (see below) and the squares stand for the numerically computed equilibrium points. Note that in this particular case, an error between the exact equilibrium points and the approximated ones can be clearly noticed. In the case of

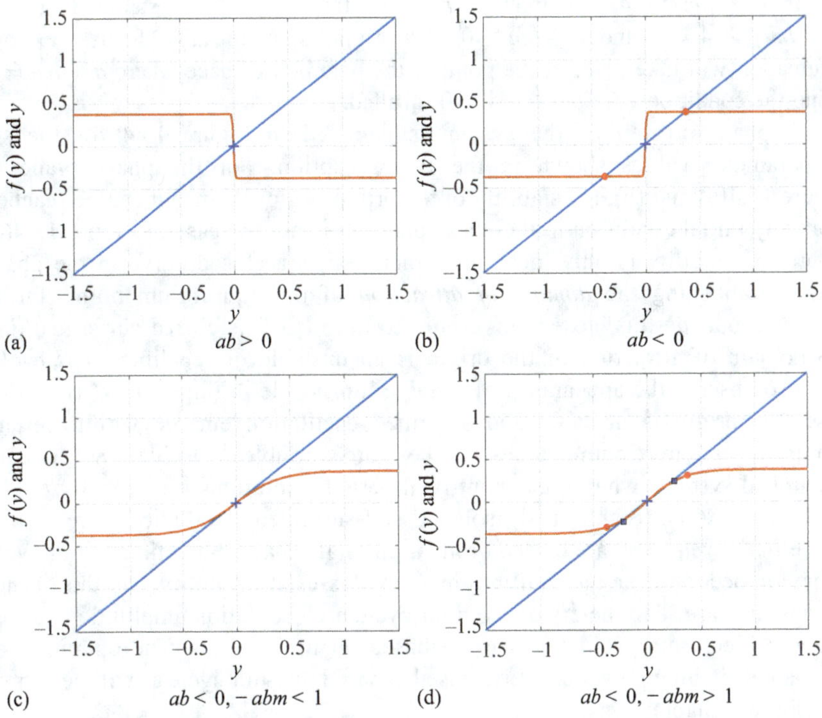

Figure 15.3 The plots of the functions $f(y) = -ab\tanh(my)$ together with the function y for (a) $ab > 0$ and (b)–(d) $ab < 0$. (c) $-abm < 1$ and (d) $-abm > 1$. In (d), the dots indicate the approximated analytical equilibrium points for high enough m and the squares stand for the numerically computed equilibrium points

$ab < 0$, another condition is necessary for the existence of two equilibrium points apart from the origin. The slope of the function $-ab\tanh(my)$ at the origin must be greater than the unity, i,e. $df(y)/dt|_{y=0} > 1$, which results in $-abm > 1$. For high enough values of m ($m > -1/(ab)$) this condition is always fulfilled. If $-abm < 1$, only the origin is an equilibrium point for the system.

Stability analysis for the origin: The origin is always an equilibrium point for the system. Its stability depends on the characteristic equation evaluated at $y^* = 0$ which is $s^2 - (a + bm)s + ab + 1 = 0$. Its corresponding eigenvalues are

$$\lambda_1 = \frac{1}{2}\left(a + bm + \sqrt{(a - bm)^2 - 4}\right) \tag{15.22}$$

$$\lambda_2 = \frac{1}{2}\left(a + bm - \sqrt{(a - bm)^2 - 4}\right) \tag{15.23}$$

The dynamical behavior close to the origin depends on the sign of the terms $a + bm$ and $(a - bm)^2 - 4$.

If $(a - bm)^2 - 4 < 0$ and $a + bm < 0$, the origin is a stable focus. If $(a - bm)^2 - 4 < 0$ and $a + bm > 0$, the origin is an unstable focus. A Hopf bifurcation will take place at the point in the parametric space where $a + bm = 0$ being the condition $(a - bm)^2 - 4 < 0$ fulfilled.

A phase portrait of the system is depicted in Figure 15.4 for the set of parameter values shown in the figure caption. For the phase plane in Figure 15.4(a) and (b), the stability of the origin is global since it can be reached from any initial condition in the phase plane. It is not the case of Figure 15.4(c) where the stability is only local since there exists a closed curve in the phase plane establishing the *domain of attraction* of the equilibrium point. Initial condition outside this closed curve goes to infinity. This closed curve marking the domain of attraction of the origin is an unstable or repelling *limit cycle*. We also observe the appearance of a stable limit cycle in Figure 15.4(d). In this case, the origin is unstable and sustained oscillation emerge within certain minimum and maximum values of the state variables. Oscillations arise in dynamical systems when it has nontrivial periodic solutions $\mathbf{x}^p$: $\exists T > 0\ \forall t \geq T$; $\mathbf{x}^p(t + T) = \mathbf{x}^p(t)$. Stable and isolated self-sustained oscillations are called stable *limit cycles* and are closed orbits in the state space. Sometimes limit cycle behavior occurs after an equilibrium point losses its stability but this is not always the case. Parameters of the limit cycle are oscillation amplitude and frequency. Electronic oscillators can be based on limit cycles. The operation of resonant switching power converters is also based on limit cycle as will be shown later in this chapter.

Stability of the symmetric equilibrium points: If $ab < 0$ two other equilibrium points may exist if the additional condition $-abm > 1$ is fulfilled. These two equilibrium points are $(-y^*/a, y^*)$ and $(y^*/a, -y^*)$. For m high enough ($-abm \gg 1$), $y^* \approx \pm ab$ and therefore the two equilibrium points coexisting with the origin can be approximated by $\mathbf{x}_q^- = (-b, ab)$ and $\mathbf{x}_q^+ = (b, -ab)$. For both

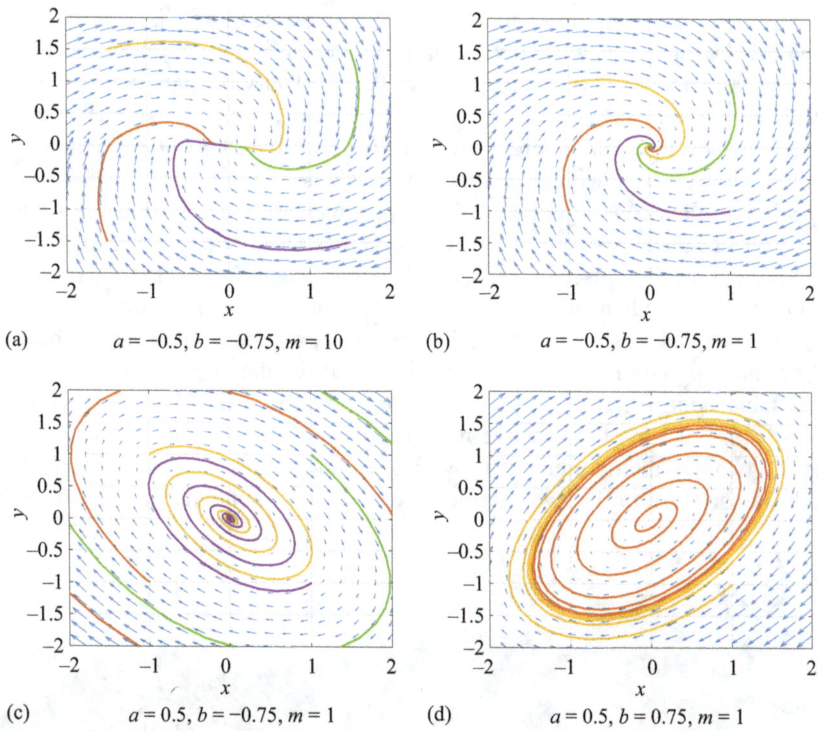

Figure 15.4 Phase portrait of the system of Example 2. (a) $\lambda_{01} = -7.35$, $\lambda_{02} = -0.64$. *(b)* $\lambda_{01} = -0.62 + 0.99j$, $\lambda_{02} = -0.62 - 0.99j$. *(c)* $\lambda_{01} = -0.12 + 0.78j$, $\lambda_{02} = -0.12 - 0.78j$. *(d)* $\lambda_{01} = 0.12 + 0.78j$, $\lambda_{02} = 0.12 + 0.78j$

equilibrium points, the characteristic equation is the same and can be expressed as follows

$$s^2 + (bm\tanh^2(abm) - (a + bm)s + abm(1 - \tanh^2(abm)) + 1 = 0 \tag{15.24}$$

Solving the previous equation for s yields the following expressions for the two roots in terms of the system parameters

$$\lambda_1 = \frac{1}{2}\left(a + bm(1 - \tanh^2(abm)) + \sqrt{(bm(\tanh^2(abm) - 1) + a)^2 - 4}\right) \tag{15.25}$$

$$\lambda_2 = \frac{1}{2}\left(a + bm(1 - \tanh^2(abm)) - \sqrt{(bm(\tanh^2(abm) - 1) + a)^2 - 4}\right) \tag{15.26}$$

The existence condition for the two symmetric equilibria is $-abm > 1$. As stated before, for $-abm < 1$, only the origin is an equilibrium point for the system. Let us fix all parameters $a = -0.5$ and $b = 0.75$ and vary the parameter m. As this parameter is increased beyond $m > -1/(ab) \approx 2.67$, the two symmetric equilibrium points emerge. But before that occurs, other changes take place in the system behavior. Let us start with relatively small values of $m = 0.5$. In this case, the origin is a stable focus (Figure 15.5(a)). As m is increased, the real part $(a + bm)/2$ of the roots of the characteristic equation becomes positive and the focus loses its stability and as a consequence a stable limit cycle emerges in its vicinity (Figure 15.5(b)). When this occurs, it is said that the system undergoes a *Hopf bifurcation*. By increasing m further, the roots of the characteristic equation become real because $(a - bm)^2 - 4$ becomes positive and the the unstable focus becomes an unstable node (Figure 15.5(c)). As m is increased beyond the critical value $m > -1/(ab) \approx 2.67$, the unstable node becomes a saddle point and at the same time the two symmetric equilibrium points $\mathbf{x}_q^-$ and $\mathbf{x}_q^+$ are born (Figure 15.5(d)). The behavior close to the two symmetric equilibrium points depends on the term

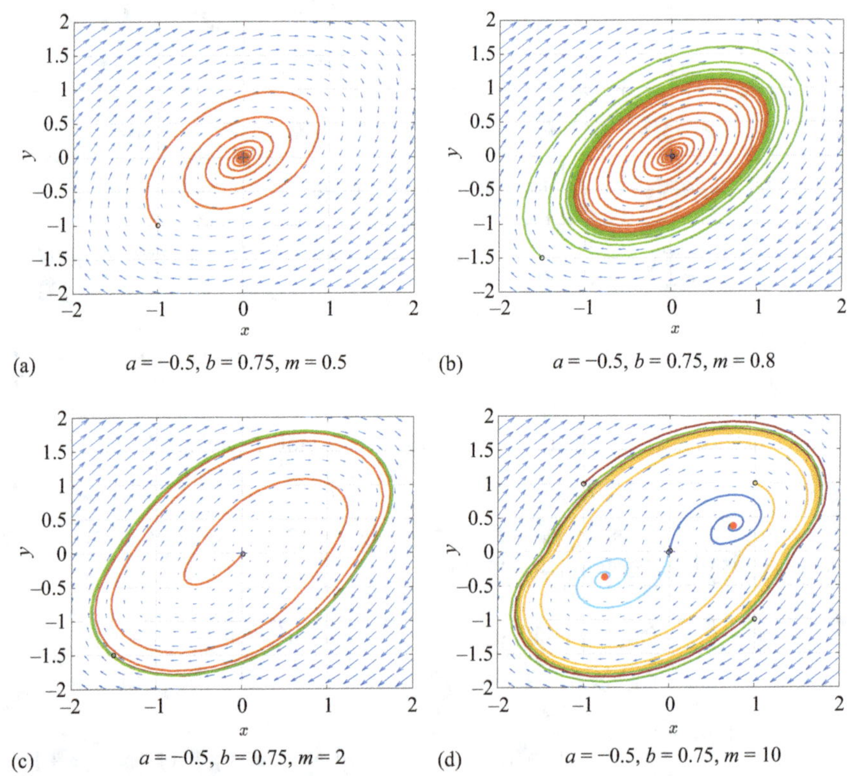

Figure 15.5 Evolution of the dynamics of the system of Example 3 as the parameter m varies

$(bm(\tanh^2(abm) - 1) + a)^2 - 4$ which is negative. These two equilibrium points are therefore stable foci. Since the two equilibrium points are stable these are also two attractors for the system. They coexist with the stable limit cycle. Therefore, depending on the initial conditions, three different steady-state behavior can be reached. The system can reach either the stable limit cycle or any one of the two equilibrium points. Each attractor has its basin of attraction. The basin of attraction of the two equilibrium points are two unstable symmetric limit cycles whose size is very small when the two equilibrium points are born (Figure 15.5(e)). The size of these unstable limit cycles get bigger by increasing m. At a critical value of $m \approx 10$, the unstable symmetric limit cycles touch each other and a so called double homoclinic connection takes place (Figure 15.5(f)). In such a connection, the unstable limit cycles are connected with unstable manifold of the saddle point at the origin. Further increasing of m leads to an increase of the size of a unique unstable limit cycle hence making the basin of attraction of the equilibrium points larger. Such unstable limit cycles plotted in Figure 15.6 have been obtained by backward time simulation. This is basically performed reversing time in the system model, i.e, when time t is replaced by its reverse value $-t$. By reversing time, stable solutions become unstable ones and *vice-versa*.

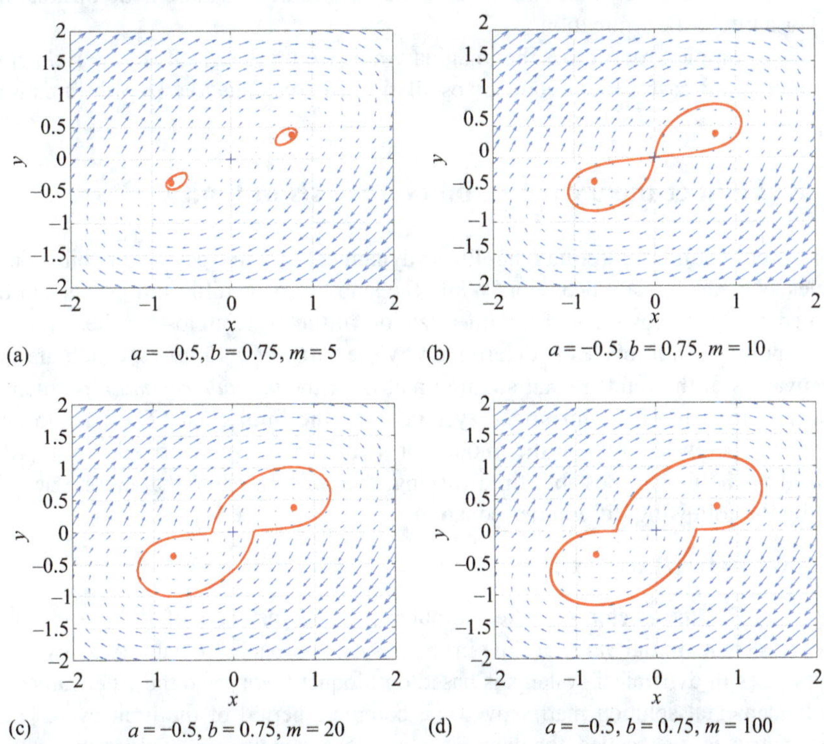

Figure 15.6 Evolution of the basin of attraction of the symmetric equilibrium point when the parameter m varies

In forward time, trajectories with initial conditions selected outside the unstable limit cycle will converge to the stable limit cycle. This is also the case of trajectories initiated outside the stable limit cycle.

From the previous example it can be observed that the number of equilibrium points in a nonlinear dynamical system and their stability depends on the system parameters. By varying a suitable parameter, the number of coexisting equilibrium points and/or their stability may change. If this happens, we say the system has undergone a bifurcation. In dynamical systems, a bifurcation occurs when a small smooth change made to the parameter values (the bifurcation parameters) causes a sudden qualitative or topological change in its behavior. Generally, at a bifurcation point, the local stability properties of equilibria, periodic orbits or other invariant sets changes. During the last three decades bifurcation theory has been well developed. Currently, it is still playing a vital role in understanding the nonlinear behavior of dynamical systems in both nonlinear sciences and many applied fields.

Recall that for an equilibrium point to be stable all roots of the characteristic equation must have negative real part. Three different scenarios for equilibrium point to become unstable exist depending on *from where* roots of the characteristic equation leave the left hand side of the complex plane

- One root becomes 0 and in this case we have either a saddle-node bifurcation or a pitchfork bifurcation.
- Two complex roots cross the imaginary axis from the left side to the right side and in this case a low frequency oscillation emerges due to a Hopf bifurcation.

15.4 Floquet theory for stability analysis of limit cycles

Apart from equilibrium points, nonlinear dynamical systems can also exhibit limit cycles as a steady-state behavior. While the stability of equilibrium points can be studied by the eigenvalues of the linearization of the system close to these points, the stability of limit cycles is determined by the *Floquet multipliers* which are the eigenvalues of the fundamental solution matrix or the *monodromy matrix* which is also the Jacobian of the linearized system along the limit cycle. The background presented below to obtain the monodromy matrix is summarized from [12]. Consider a nonlinear, possibly time varying, dynamical system that can be described by the following differential equation

$$\dot{\mathbf{x}}(t) = \mathbf{f}(\mathbf{x}(t), \mu), \quad \mathbf{x}(t_0) = \mathbf{x}_0 \tag{15.27}$$

where $\mathbf{x}(t)$ is the vector of state variables at time instant t of the system and $\mathbf{x}(t_0) = \mathbf{x}_0$ is its initial value at time instant t_0. An approach for stability analysis of limit cycles in dynamical systems is based on Floquet theory and the eigenvalues of the fundamental solution matrix over one complete period of the limit cycle [12]. This matrix is also called the *monodromy matrix* and its eigenvalues are called *Floquet multipliers*. The stability of a limit cycle $\mathbf{x}^p(t)$ can be analyzed by monitoring

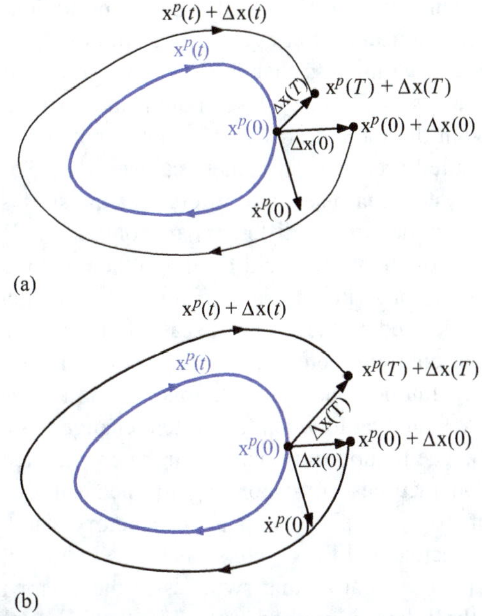

Figure 15.7 Perturbing a limit cycle in a nonlinear dynamical system for performing stability analysis. (a) stable limit cycle, (b) unstable limit cycle

the evolution of a small perturbation in its vicinity. The limit cycle is stable if the size of the perturbation goes to 0 as t tends to ∞ according to a certain norm.

Figure 15.7 illustrates a stable and an unstable limit cycle. Let $\mathbf{x}^p(t)$ be a T−periodic orbit or a limit cycle for the system and let $\mathbf{x}^p(0)$ be its value at an initial time that can be selected to be zero. Let $\Delta\mathbf{x}(t) = \mathbf{x}(t) - \mathbf{x}^p(0)$. Let $\Delta\mathbf{x}(t+T)$ the value of $\Delta\mathbf{x}(t)$ one period later. According to [12], it can be demonstrated that

$$\Delta\mathbf{x}(t+T) \approx \mathbf{\Phi}(T)\Delta\mathbf{x}(t) + O(|\Delta\mathbf{x}|^2) \tag{15.28}$$

where the matrix $\mathbf{\Phi}(T) := \mathbf{M}$ is the state transition matrix evaluated at the period T and it is called *monodromy matrix*. Therefore, from (15.28) one has

$$\Delta\mathbf{x}(T) \approx \mathbf{M}\Delta\mathbf{x}(0) \tag{15.29}$$

$$\Delta\mathbf{x}(2T) \approx \mathbf{M}\Delta\mathbf{x}(T) \approx \mathbf{M}^2\Delta\mathbf{x}(0) \tag{15.30}$$

$$\vdots \qquad \vdots$$

$$\Delta\mathbf{x}(nT) \approx \mathbf{M}^n\Delta\mathbf{x}(0) \text{ for } n = 1, 2, 3, \ldots \tag{15.31}$$

Therefore the eigenvalues of $\mathbf{M}$, called Floquet multipliers, gives the rate of expansion or compression near a limit cycle. For a limit cycle to be stable all these multipliers must have modulus less than 1. The monodromy matrix plays a vital role on the stability analysis of limit cycles in nonlinear dynamical systems. Below, the procedure to obtain the monodromy matrix for a switched system is summarized. For a more detailed presentation readers can see [12]. For simplicity. let us consider a switched system that uses two different affine subsystems. In this case the state space can be divided in two different partitions Σ_1 and Σ_2. In the region Σ_1 the system is described by the vector field $\mathbf{F}_1$ and in the region Σ_2 by $\mathbf{F}_2$. The set Σ is the switching manifold or switching surface at which switching takes place. The regions Σ_1 and Σ_2 are defined by the sign of a scalar function $\sigma(\mathbf{x}, t) = \mathbf{K}^\top \mathbf{x} + r(t)$, where $\mathbf{K}$ is a suitable column vector of feedback coefficients and $r(t)$ is a time varying function. For autonomous systems, this function is zero. Let t_s be the switching instant corresponding to when the system changes its dynamics from the vector field $\mathbf{F}_1$ to $\mathbf{F}_2$. The monodromy matrix can be constructed from the product of the state transition matrices corresponding to each sub-cycle and the corresponding saltation matrix [9,12,37]. Suppose a trajectory starts at time instant t_0 and described by the vector field $\mathbf{F}_1(\mathbf{x})$ intersects the switching boundary described by the equation $\sigma(\mathbf{x}(t), t) = 0$ at t_1, and switches to the vector field $\mathbf{F}_2(\mathbf{x})$. It has been shown, using the Filippov method [9,12,37], that when there is a *crossing* behavior, the state transition matrix across the switching boundary, called also the saltation matrix $\mathbf{S}_1$, is given by

$$\mathbf{S}_1 = \mathbf{I} + \frac{(\mathbf{F}_2(\mathbf{x}(t_1)) - \mathbf{F}_1(\mathbf{x}(t_1)))\mathbf{K}^\top}{\mathbf{K}^\top \mathbf{F}_1(\mathbf{x}(t_1)) + \frac{\partial \sigma}{\partial t}|_{t=t_1}} \tag{15.32}$$

Similarly, suppose the same trajectory starts at time instant t_2 and described by the vector field $\mathbf{F}_2(\mathbf{x})$ intersects the switching boundary described by the equation $\sigma(\mathbf{x}(t), t) = 0$ at t_2, and switches to the vector field $\mathbf{F}_1(\mathbf{x})$. Then, the new state saltation matrix $\mathbf{S}_2$, is given by

$$\mathbf{S}_2 = \mathbf{I} + \frac{(\mathbf{F}_1(\mathbf{x}(t_2)) - \mathbf{F}_2(\mathbf{x}(t_2)))\mathbf{K}^\top}{\mathbf{K}^\top \mathbf{F}_2(\mathbf{x}(t_2)) + \frac{\partial \sigma}{\partial t}|_{t=t_2}} \tag{15.33}$$

Finally, the monodromy matrix $\mathbf{M}$ for the switched system can be composed during a complete period as follows

$$\mathbf{M}(\mathbf{x}(0)) = \mathbf{S}_2 \mathbf{\Phi}_2 \mathbf{S}_1 \mathbf{\Phi}_1 \tag{15.34}$$

Unfortunately, the state transition matrices like $\mathbf{\Phi}_1$ and $\mathbf{\Phi}_2$ cannot be obtained in closed-form in most of the cases and numerical computations must be used to obtain them. However, these matrices can be analytically obtained for switched systems that uses linear or affine subsystems described by $\mathbf{F}_i = \mathbf{A}_i \mathbf{x} + \mathbf{B}_i$ and they are the exponential matrices $\mathbf{\Phi}_i(\Delta t_i) = e^{\mathbf{A}_i \Delta t_i}$ evaluated at the corresponding time interval Δt_i. Later, this matrix will be analytically obtained for a resonant power converter and the result will be used to analyze the different kinds of instability that the system can undergo. It is worth noting that for limit cycles with small enough amplitude, this approach can be also applied to nonlinear switched system by *piecewise linearizing* the vector field in the vicinity of the limit cycle [44].

Limit cycles can undergo three different types of bifurcations. Their classification is beyond the scope of this chapter. Interested readers may see related text books cited in the list of references. Three different scenarios for a limit cycle to become unstable exist depending on *from where* the eigenvalues leave the unit disk in the complex plane. These are

- One eigenvalue becomes $+1$ and in this case we have a saddle-node or a cyclic fold bifurcation. At this singularity, two limit cycles of the system collapse and disappear.
- One eigenvalue becomes -1 and in this case we have a period doubling bifurcation which leads to subharmonic oscillation.
- A pair of complex conjugate eigenvalues have modulus equal to 1 and in this case a low frequency oscillation emerges due to a Neimark-Sacker bifurcation.

15.5 Dynamic behavior of Filippov control systems and concept of sliding mode motion

Control systems with discontinuous control action lead to Flilipov systems. These are systems with discontinuous vector field. These are also called variable structure systems (VSS, for short) or switching systems. Due to the non smooth nature of this kind of systems, they can exhibit richer variety of dynamical behaviors than smooth systems being the most prominent one *sliding mode* motion [35,36,39]. In control systems, this behavior is exploited to design sliding mode controllers whose most known and significant property is robustness. This is the most known application of sliding motion in control theory.

The theory related to this control technique were first proposed in the 1950s resulting also to the notion of variable structure systems. At that time, the resulting sliding-mode controller and VSS describe a system in which the structure is changed with different inputs while the plant does not switch. Later the concept was also applied to systems containing switches that making them to switch between different subsystems. This is the case, for instance, of power electronic converters that naturally belong to variable structure systems. Therefore, sliding mode control is a typical kind of variable structure control method. But variable structure control does not always lead to sliding motion. The basic idea in sliding mode control is to force the system state to slide along the preset surface or manifold, until the controlled variable reaches its desired reference (Figure 15.8). Unlike other control techniques, which also can be used to make a suitable controlled variable to reach its desired reference, precisely sliding mode control does that by making the trajectory of the system to be constrained in the sliding manifold. This control technique was widely used in power and energy systems [41,42]. Sliding mode motion is characterized by an order reduction and infinite switching frequency. In practice, this frequency is finite due to the limited

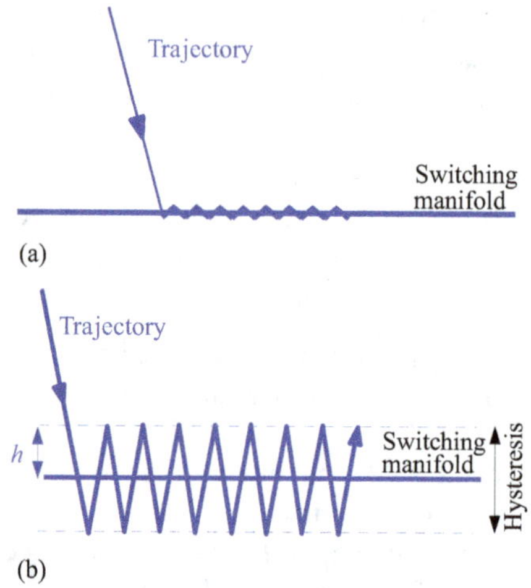

Figure 15.8 Trajectories close to a sliding manifold (a) without hysteresis and (b) with hysteresis

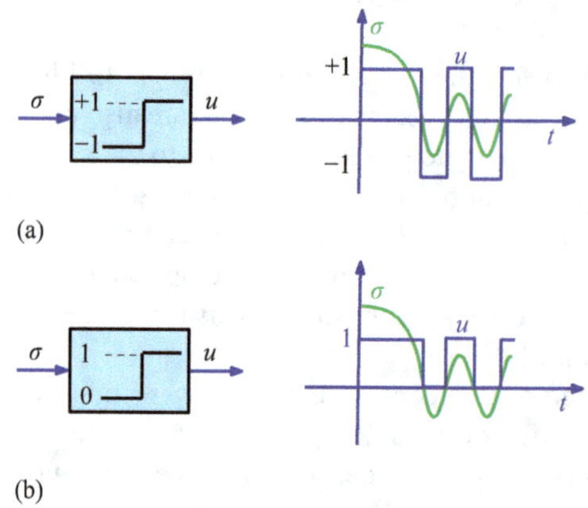

Figure 15.9 Block diagram of a discontinuous control system. (a) Relay and (b) comparator

speed of switching devices leading to uncontrolled and undesired chattering phenomenon. It appears as oscillations of the system variables, whose magnitude is related to the influence of the limited speed of switching devices. To limit the switching frequency and induce control of chattering oscillations, hysteresis or other impurities such as filtering or delay are added. Although switching frequency is limited with such impurities, the motion is also considered sliding mode. Most Filippov systems contain in their vector field the signum function representing an ideal relay or the Heaviside step function representing a comparator. The two cases

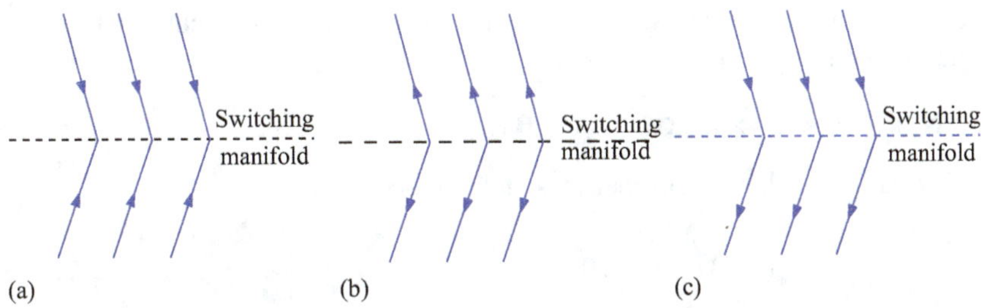

Figure 15.10 Possible behaviors of trajectory of a switched system close to the switching manifold. (a) Attractive sliding, (b) repulsive sliding and (c) crossing

are illustrated in Figure 15.9. The well known type of sliding-mode motion used in control theory is attractive sliding which is illustrated in Figure 15.10(a). However, sliding motion can also be repulsive or escaping which is illustrated in Figure 15.10 (b). The other type of switching behavior in Filippov systems is the simple crossing which is Figure 15.10(c). In some switched systems, it may appear that they do not exhibit sliding mode motion, they indeed exhibit this kind of behavior but in backward time. Such a behavior in forward time manifests itself as a repulsive sliding.

15.5.1 A simple example of Filippov system: the double integrator with relay feedback

Let us consider the double integrator of **Example 15.1**. As before, let $y = \dot{x}$ and $\mathbf{x} = (x, y)^{\top}$. A state space representation of the open loop system (without feedback) can be expressed as follows $\dot{\mathbf{x}} = \mathbf{A}\mathbf{x} + \mathbf{B}u$, where $\mathbf{A}$ and $\mathbf{B}$ are given by

$$\mathbf{A} = \begin{pmatrix} 0 & 1 \\ 0 & 0 \end{pmatrix}, \quad \mathbf{B} = \begin{pmatrix} 0 \\ 1 \end{pmatrix} \tag{15.35}$$

Instead of a linear state feedback, let us consider a variable structure discontinuous control based on a relay feedback that can be expressed as follows [39]

$$u = \begin{cases} -1 \text{ if } \sigma > 0, \\ +1 \text{ if } \sigma < 0. \end{cases} = -\mathrm{sign}(\sigma) \tag{15.36}$$

where $\sigma = \tau y + x := \mathbf{K}^{\top}\mathbf{x}$ with $\tau > 0$ and $\mathbf{K}^{\top} = (1\ \tau)$ is the function used to decide the value of the control variable and it is therefore called switching function [39]. The discontinuous control law defines two different vector fields at both sides of the switching line defined by $\sigma = \tau y + x = 0$ in the phase plane, one vector field is $\mathbf{F}^{+} = \mathbf{A}\mathbf{x} + \mathbf{B}$ and it corresponds to $u = 1$ and the other one is $\mathbf{F}^{-} = \mathbf{A}\mathbf{x} - \mathbf{B}$ and it corresponds to $u = -1$. The sliding mode condition is $\sigma\dot{\sigma} < 0$ or more precisely [35]

$$\lim_{\sigma \to 0^{+}} \dot{\sigma} < 0 \text{ and } \lim_{\sigma \to 0^{-}} \dot{\sigma} > 0 \tag{15.37}$$

which is called the *reachability condition* and it means that in the two sides of the sliding region within the switching line, the system trajectories and two vector

vector fields point toward the switching line. The sliding mode condition $\sigma\dot{\sigma} < 0$ can be expressed as follows

$$\mathbf{K}^\top \mathbf{x} \mathbf{K}^\top \dot{\mathbf{x}} < 0 \Rightarrow \mathbf{K}^\top \mathbf{x} \mathbf{K}^\top (\mathbf{A}\mathbf{x} + \mathbf{B}u) < 0 \tag{15.38}$$

Evaluation of the system matrices and vectors yields

$$(1 \quad \tau)\begin{pmatrix} x \\ y \end{pmatrix}(1 \quad \tau)\begin{pmatrix} y \\ u \end{pmatrix} < 0 \tag{15.39}$$

which implies

$$(x + \tau y)(y + \tau u) < 0 \tag{15.40}$$

or equivalently,

$$\sigma(y - \tau \operatorname{sign}(\sigma)) < 0 \tag{15.41}$$

This condition can further be simplified as follows

$$|y| < \tau \tag{15.42}$$

Under this condition, the system trajectories will be confined on the switching line and therefore one has $\tau y + x = 0$ or equivalently, $\dot{x} = -x/\tau$. This is a first order differential equation corresponding to ideal sliding mode dynamics. Its solution is $x(t) = x(t_s)e^{-(t-t_s)/\tau}$, where t_s is the reaching time of the sliding region. Note that τ establishes the convergence speed to zero for the variable x.

15.5.2 Equivalent control within sliding mode region

In Filippov systems there is an ambiguity in the definition of the control signal within the sliding region. This ambiguity is overcome by using the concept of the equivalent control [35,36]. With this technique, the time derivative of switching function $\sigma(x)$ along the system trajectory is equaled to zero, and the resulting equation is solved for the control input. The result is a continuous-time signal that describes the averaged dynamics of the controlled system on the sliding region. Let t_s be the time instant at which a certain trajectory reaches the sliding region. Starting from this time instant one has $\sigma = 0$ for all future time $t > t_s$ if sliding mode takes place. Therefore for $t > t_s$, one also has $\dot{\sigma} = 0$ or equivalently $\tau\dot{x} + u = 0$ which implies that

$$u = -\frac{y}{\tau} := u_{eq} \tag{15.43}$$

It is worth to note that u_{eq} is not the control applied to the system. However, starting from the instant t_s when the system trajectory is confined in the sliding region, it can be considered as the control signal that would lead to the same motion as with the switched signal u.

Because $u = -\operatorname{sign}(\sigma)$, sliding motion will exist if the condition $-1 < u_{eq} < 1$ is fulfilled, i.e, $-\tau < y < \tau$ or equivalently $|y| < \tau$ which coincides with (15.42).

Therefore with $\tau = 0$, sliding mode does not exist. The sliding mode region in the phase space gets wider by increasing τ. For values of y not fulfilling $|y| < \tau$, the system trajectories will cross the manifold without sliding. Note also that τ establishes the convergence speed to zero for the variable x since within the sliding manifold one has $\dot{x} = -x/\tau$ which is a first order differential equation whose solution is $x(t) = x(t_0)e^{-t/\tau}$. The same result can be obtained if convex method of Filippov ([35,36]) is used. According to this method, the resulting vector field in the sliding region can be expressed as a convex combination of the vector fields at both sides of the switching boundary, *i.e.*,

$$\mathbf{F}^S(\mathbf{x}) = \lambda\mathbf{F}^+(\mathbf{x}) + (1-\lambda)\mathbf{F}^-(\mathbf{x}), \text{ with the condition } \tau y + x = 0 \tag{15.44}$$

Therefore, trajectories on the sliding region are solutions of the Filippov vector field $\mathbf{F}^S$, which can be obtained by substituting λ with $\lambda_{eq} \in [0, 1]$ which is the expression of λ that would make the sliding region a local invariant manifold of the system, that is the solution of the equation $\dot{\sigma}(\mathbf{x}) := \mathbf{K}^\top\mathbf{F}^S(\mathbf{x}) = 0$. Therefore, if $\lambda_{eq}(\mathbf{x})$ exists, its expression is given by

$$\lambda_{eq}(\mathbf{x}) = \frac{\mathbf{K}^\top\mathbf{F}^-(\mathbf{x})}{\mathbf{K}^\top(\mathbf{F}^-(\mathbf{x}) - \mathbf{F}^+(\mathbf{x}))}, \tag{15.45}$$

requiring that $\mathbf{K}^\top(\mathbf{F}^- - \mathbf{F}^+) \neq 0$. For the double integrator, one has that

$$\mathbf{K}^\top(\mathbf{F}^-(\mathbf{x}) - \mathbf{F}^+(\mathbf{x})) = 2\tau \neq 0. \tag{15.46}$$

Hence, A necessary condition for sliding motion to exist is $\tau \neq 0$. The equivalent control (15.45) becomes

$$\lambda_{eq} = \frac{1}{2}\left(1 + \frac{y}{\tau}\right). \tag{15.47}$$

By substituting λ_{eq} for λ in (15.44), the Filippov reduced-order model $\dot{x} = \mathbf{F}^S(x)$ on the sliding region is obtained. The sliding vector field $\mathbf{F}^S$ is given by the following expression

$$\mathbf{F}^S(x) = -\frac{x}{\tau}. \tag{15.48}$$

It is worth to note that the same results would have been obtained if one substitute in the model of the system $\dot{\mathbf{x}} = \mathbf{A}\mathbf{x} + \mathbf{B}u$ the switched signal u by the equivalent control $u_{eq} = 2\lambda_{eq} - 1$. Note also that u_{eq} is the smooth control law that would make the sliding region a local invariant manifold of the switched system. It can be observed that the origin is an attractive equilibrium point for the ideal sliding dynamics and also for the switched system, since the sliding-mode region is attracting. Note also that, as already revealed before, τ determines the rate of convergence to the equilibrium point. It is worth to note that both the Filippov convexation method and the equivalent control method lead to the same final results in terms of ideal sliding mode dynamics equations. The only difference is the interval to which belongs each corresponding variable. Namely, the convexation parameter $\lambda_{eq} \in (0, 1)$ while the equivalent control u_{eq} is bounded between the lower and the upper limits of the switched signal u. For instance, in the case of the double integrator with relay feedback example, one has $u_{eq} \in (-1, 1)$.

Computer simulations are used to illustrate the previous ideas and to get a good insight about the behavior of the system. Figure 15.11 shows the phase portraits and the time domain waveforms of the double integrator under the discontinuous control defined in (15.36) for two different values of τ ($\tau = 0.3$ and $\tau = 3$) for the same initial condition $(1.5, 1.5)$. Note that for small values of τ the sliding-mode regime is also small. The system exhibits several crossing behavior before entering in sliding-mode operation. The switching frequency corresponding to the crossing dynamics is finite as can be observed in time domain waveforms of Figure 15.11(b) and Figure 15.11(d). It becomes infinite when the system enters the sliding mode region. This is a subset of the switching line defined by $\sigma := x + \tau y = 0$ in the phase space. This is represented in dashed line in the same figure for the different

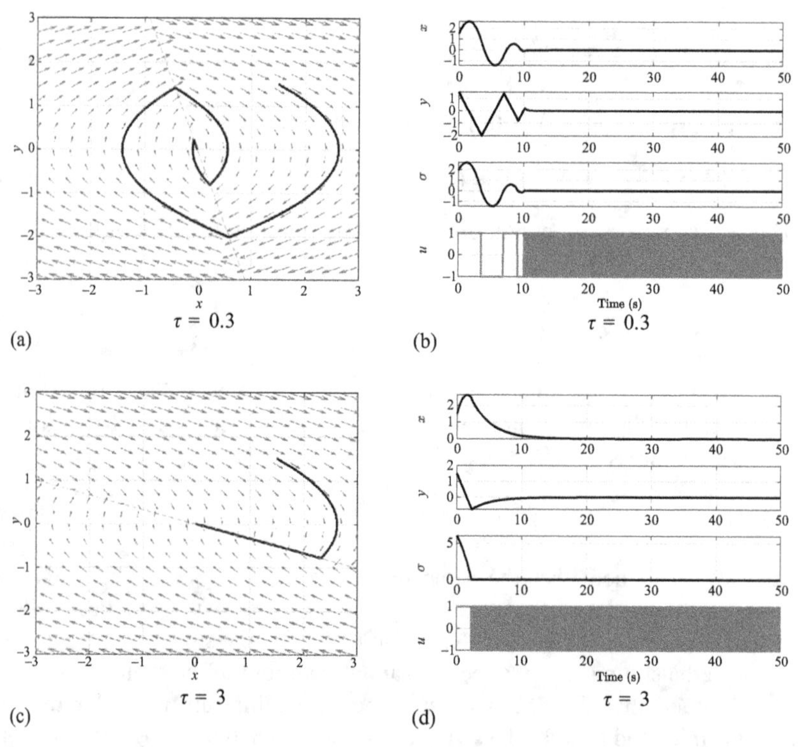

Figure 15.11 Phase portraits and time domain waveforms of the double integrator under the discontinuous control action defined in (15.51c). Starting from (1.5,1.5), trajectories are plotted in (a), (c) and time domain waveforms are depicted in (b), (d)

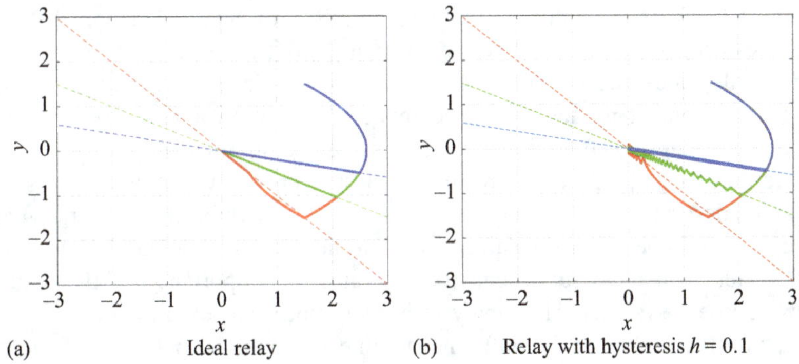

Figure 15.12 Phase portraits and time domain waveforms of the double integrator under the discontinuous control action defined in (15.51c) for $\tau = 1$, 3 and 5. The trajectories are started from the point (1.5,1.5). The ideal switching manifolds without hysteresis are also plotted for the three cases

values of the parameter τ. The value of $-1/\tau$ represents the slope of the switching line in the phase plane.

As stated before the sliding region is defined by $|y| < \tau$. For larger value of τ, the sliding region increases and the sliding reaching time is shorter as can be observed in the time domain waveforms in Figure 15.11(d). In all the cases, it can be observed that after a certain time period depending on the value of τ during which the system exhibits crossing dynamics and the switching frequency is finite, this becomes infinite once the system trajectory reaches the sliding region within the switching line.

To further show the effect of the parameter τ on the sliding mode region, Figure 15.12(a) shows the system trajectories for three different values of τ and for the same initial condition $(1.5, 1.5)$. It is worth to note that once the system enters in sliding-mode operation, its behavior depends only on τ and therefore on the switching function σ but not on the system parameters. In all the cases, the two phases of the dynamics can be clearly identified. Namely, the initial phase (parabolic trajectories) toward the switching line and the sliding phase for which the trajectories are confined on the switching line while converging to the origin can be observed. In Figure 15.12(b) the same results are reproduced by using a small hysteresis $h = 0.1$ in the switching decision. By playing with the hysteresis width, the switching frequency can be controlled at desired values.

15.6 Application to a resonant power converter

In this section, the approach presented in previous sections will be applied to a DC-AC PRC (Parallel Resonant Converter) under ZCS (Zero Current Switching) control [30,31]. Most of the material presented in this section is summarized from a previous work of the authors published in [43]. An analysis of both PRC and Series Resonant Converter (SRC) is given in [45] by means of a unified non-dimensional model. Moreover, in [46], a voltage feedback is added, thus giving a simple and compact model with an extended bifurcation analysis regarding the existence and stability of equilibria and crossing and sliding limit cycles.

The PRC is particularly interesting since it is only a two-dimensional system that can be described by a piecewise linear model, yet exhibiting different types of limit cycles and coexistence of attractors. The fact that such a simple low dimensional system can have such behavior indicates the importance of the study of piecewise linear systems. This analysis is then motivated by an attempt to accurately study the global nonsmooth and nonlinear dynamical behavior of the converter in terms of loading conditions. The generation of the different types of limit cycles that the system may have will be explained in the light of piecewise smooth dynamical systems and Filippov convex method. In particular, it will be shown that at the switching boundary an escaping sliding set exists that plays a key role in some limit cycle bifurcations and that under certain initial conditions, the circuit could not reach the desired crossing limit cycle. The bifurcation patterns, which includes a cyclic-fold bifurcation (saddle node bifurcation of two limit cycles), a crossing-sliding bifurcation and a double homoclinic connection are analyzed, getting a valuable information for the correct operation and design of the converter.

15.6.1 System description and modeling

Figure 15.13 shows the circuit diagram of the system considered in this study which consists of an LC PRC under ZCS control [30,31,43].

An analysis of both PRC and SRC (Series Resonant Converter) is given in [45] by means of a unified nondimensional model. Moreover, in [46], a voltage feedback is added, thus giving a simple and compact model with an extended bifurcation analysis regarding the existence and stability of equilibria and crossing and sliding limit cycles.

Generally speaking, the PRC is a switched system that includes a resonant tank circuit and a switching network that actively participates in determining power

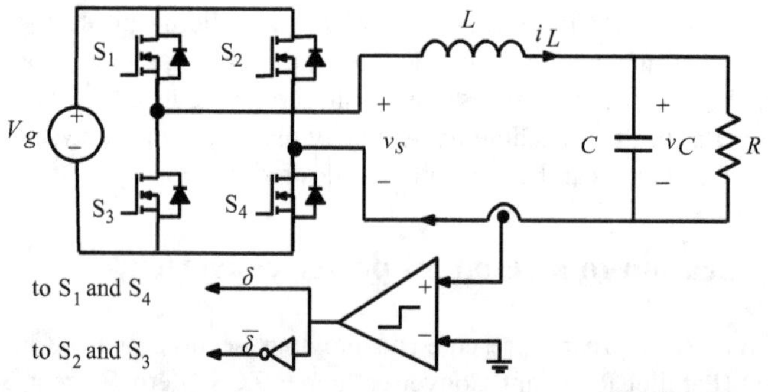

Figure 15.13 Schematic circuit diagram of the LC PRC under ZCS control

flow. The switches S_1 and S_4 are on (therefore $\delta = 1$, $\overline{\delta} = 0$ and $v_s = V_g$) when $i_L > 0$ and they are turned off (therefore $\delta = 0$, $\overline{\delta} = 1$ and $v_s = -V_g$) when $i_L < 0$. Note that the switches S_2 and S_3 are driven in a complementary way to S_1 and S_4. The voltage source at the input together with the bridge of switches generate a bipolar square wave signal v_s at the LC resonant tank input. Therefore, a simplified equivalent circuit diagram of the system is represented in Figure 15.14 where the square wave signal v_s is equivalently generated by a controlled voltage source. The control signal for this source is the discontinuous control signal $u = \text{sign}(i_L)$.

Let v_C be the voltage of the output capacitor and i_L the inductor current. By applying KVL, the following dynamical model of the system is obtained

$$\frac{\mathrm{d}v_C}{\mathrm{d}t} = -\frac{v_C}{RC} + \frac{i_L}{C}, \tag{15.49a}$$

$$\frac{\mathrm{d}i_L}{\mathrm{d}t} = -\frac{v_C}{L} + \frac{V_g}{L}u, \tag{15.49b}$$

where C is the capacitance of the output capacitor, L is the inductance of the inductor, R is the load resistance and V_g is the input source voltage.

All these parameters can be identified in the schematic circuit diagram of Figures 15.13 and 15.14. Let $\mathbf{x} = (v_C,\ i_L)^\top$ be the vector of the state variables of the power stage, then the state-space model of the converter can be expressed as follows

$$\dot{\mathbf{x}} = \mathbf{A}\mathbf{x} + \mathbf{B}u, \tag{15.50a}$$

$$\sigma(\mathbf{x}) = \mathbf{K}^\top\mathbf{x}, \tag{15.50b}$$

$$u = \text{sign}(\sigma(\mathbf{x})), \tag{15.50c}$$

where the matrix $\mathbf{A}$ and the vector $\mathbf{B}$ are given by

$$\mathbf{A} = \begin{pmatrix} -\frac{1}{RC} & \frac{1}{C} \\ -\frac{1}{L} & 0 \end{pmatrix}, \quad \mathbf{B} = \begin{pmatrix} 0 \\ \frac{V_g}{L} \end{pmatrix},$$

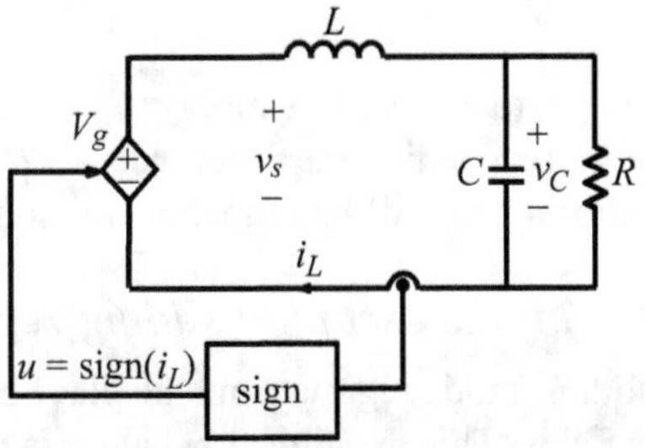

Figure 15.14 Simplified schematic circuit diagram of the LC PRC under ZCS control

and $\mathbf{K}^\top = (0,\ 1)$, so that $\sigma(\mathbf{x}) = i_L$. One has the following partitions in the state plane

$$\Sigma^+ = \{\mathbf{x} : \sigma(\mathbf{x}) = \mathbf{K}^\top \mathbf{x} > 0\}, \quad u = 1, \tag{15.51a}$$

$$\Sigma^- = \{\mathbf{x} : \sigma(\mathbf{x}) = \mathbf{K}^\top \mathbf{x} < 0\}, \quad u = -1. \tag{15.51b}$$

Note that at the switching boundary Σ, the driving signal u jumps between its two possible values $+1$ and -1. Hence, in accordance with the ZCS control, the system (15.50a)–(15.50c) can be rewritten as $\dot{\mathbf{x}} = \mathbf{F}(\mathbf{x})$, where $\mathbf{F}(\mathbf{x})$ is given by

$$\mathbf{F}(\mathbf{x}) = \begin{cases} \mathbf{F}^+(\mathbf{x}) = \mathbf{A}\mathbf{x} + \mathbf{B} \text{ in } \Sigma^+, \\ \mathbf{F}^-(\mathbf{x}) = \mathbf{A}\mathbf{x} - \mathbf{B} \text{ in } \Sigma^-. \end{cases} \tag{15.52}$$

Remark 2. *As in Remark 1, the vector field* $\mathbf{F}$ *is odd-symmetric, i.e.,* $\mathbf{F}(-\mathbf{x}) = -\mathbf{F}(\mathbf{x})$. *Therefore, if* $\mathbf{x}_1(t)$ *is a limit cycle, then, either* $\mathbf{x}_1(t)$ *is asymmetric and there exists another solution* $\mathbf{x}_2(t)$ *given by* $\mathbf{x}_2(t) = -\mathbf{x}_1(t)$ *or* $\mathbf{x}_1(t)$ *is a unique symmetric solution with half period odd symmetry, i.e.,* $\mathbf{x}_1(t) = -\mathbf{x}_1\left(t \pm \frac{T}{2}\right)$.

Each vector field is linear and time-invariant and the corresponding system of state equations can be solved in closed-form. Provided that the matrix $\mathbf{A}$ is invertible, the trajectory $\mathbf{x}(t)$ of the system at time t, starting from an initial condition $\mathbf{x}(t_c)$ at time instant t_c, can be expressed as follows

$$\mathbf{x}(t) = \mathbf{\Phi}(t - t_c)\mathbf{x}(t_c) \pm \mathbf{\Psi}(t - t_c) \tag{15.53}$$

where $\mathbf{\Phi}(t) = e^{\mathbf{A}t}$, $\mathbf{\Psi}(t) = \mathbf{A}^{-1}(\mathbf{\Phi}(t) - \mathbf{I})\mathbf{B}$, and $\mathbf{I}$ is the identity matrix with appropriate dimensions. The sign $+$ in (15.53) applies in the region Σ^+ and the sign $-$ applies in the region Σ^-.

The equilibria of both vector fields $\mathbf{F}^\pm$ can be obtained from the null field condition $\mathbf{F}^\pm(\mathbf{x}) = \mathbf{0}$ or, equivalently, since the matrix $\mathbf{A}$ is Hurwitz, by taking the limit $\lim_{t\to\infty}\mathbf{x}(t)$ in (15.53), noting that $\lim_{t\to\infty}\mathbf{\Phi}(t) = \mathbf{0}$ and therefore $\lim_{t\to\infty}\mathbf{\Psi}(t) = -\mathbf{A}^{-1}\mathbf{B}$. The equilibrium point corresponding to the vector field $\mathbf{F}^+(\mathbf{x})$ is given by

$$\mathbf{x}_q^+ = -\mathbf{A}^{-1}\mathbf{B} = \begin{pmatrix} V_q \\ I_q \end{pmatrix} = \begin{pmatrix} V_g \\ \dfrac{V_g}{R} \end{pmatrix}, \tag{15.54}$$

while the one corresponding to $\mathbf{F}^-(\mathbf{x})$ is simply $\mathbf{x}_q^- = -\mathbf{x}_q^+$. Both equilibria are stable and located in the corresponding partition that is $\mathbf{x}_q^+ \in \Sigma^+$ and $x_q^- \in \Sigma^-$, so they are natural real equilibria and will be attracting for some initial conditions.

15.6.2 The dynamics on the escaping sliding region

As commented before, sliding-mode regime can take place in systems described by piecewise smooth vector fields. This dynamical behavior is produced in a subset of the switching manifold, at which the normal component of the two involved vector fields are of opposite directions. Thus, if both vector fields point inwards or outwards the switching surface, an attracting or an escaping sliding set takes place respectively. Considering (15.52), the sliding manifold can be expressed as

$$\Sigma^S = \{\mathbf{x} \in \Sigma : \mathbf{K}^\top \mathbf{F}^-(\mathbf{x})\mathbf{K}^\top \mathbf{F}^+(\mathbf{x}) \leq 0\}, \text{ and } i_L = 0 \tag{15.55}$$

After simple algebra, one gets

$$\Sigma^S = \{\mathbf{x} : -V_g \leq v_C \leq V_g, i_L = 0\}. \tag{15.56}$$

Remark 3. *Let* $\mathbf{x}_b^+ = (V_g,\ 0)^\top$ *and* $\mathbf{x}_b^- = (-V_g,\ 0)^\top$ *the right and the left extreme points of the sliding set* Σ^S*. Hence, one has that* $\mathbf{K}^\top \mathbf{F}^+(\mathbf{x}_b^+) = 0$ *and* $\mathbf{K}^\top \mathbf{F}^-(\mathbf{x}_b^-) = 0$.

The sliding subset in the system is escaping because

$$\frac{di_L}{dt} > 0 \text{ for } i_L > 0 \text{ and } \frac{di_L}{dt} < 0 \text{ for } i_L < 0$$

This is in a clear contrast with sliding-mode controlled power converters for which the switching decision is dictated in such a way that when i_L is positive, its derivative is negative and viceversa, so that the fields $\mathbf{F}^+$ and $\mathbf{F}^-$ points inwards Σ^S. Escaping sliding motion has received less attention in the literature [7,8], but plays a key role in organizing the state plane of the system under study as will be shown later.

According to the Filippov method [35,36], the field in the sliding region can be expressed as a convex combination of the two piecewise vector fields, *i.e.*,

$$\mathbf{F}^S(\mathbf{x}) = \lambda \mathbf{F}^+(\mathbf{x}) + (1-\lambda)\mathbf{F}^-(\mathbf{x}), \quad \text{with} \quad \mathbf{x} \in \Sigma^S \text{ or } i_L = 0. \tag{15.57}$$

Therefore, trajectories on the sliding set located between its extreme points are solutions of the Filippov vector field $\mathbf{F}^S$, which can be obtained by substituting λ with $\lambda_{eq} \in [0, 1]$. Let $u_{eq} = 2\lambda_{eq} - 1$. This is the smooth control law that would make Σ^S a local invariant manifold of the switched system, that is the solution of the equation $\dot{\sigma}(\mathbf{x}) := \mathbf{K}^\top \mathbf{F}^S(\mathbf{x}) = 0$ in (15.57). Therefore, if $\lambda_{eq}(\mathbf{x})$ exists, its expression is given by

$$\lambda_{eq}(\mathbf{x}) = \frac{\mathbf{K}^\top \mathbf{F}^-(\mathbf{x})}{\mathbf{K}^\top(\mathbf{F}^-(\mathbf{x}) - \mathbf{F}^+(\mathbf{x}))}, \tag{15.58}$$

requiring that $\mathbf{K}^\top(\mathbf{F}^- - \mathbf{F}^+) \neq 0$. For the PRC considered in this study, one has that

$$\mathbf{K}^\top(\mathbf{F}^-(\mathbf{x}) - \mathbf{F}^+(\mathbf{x})) = -\frac{2V_g}{L} \neq 0. \tag{15.59}$$

Hence, theoretically, sliding motion may exist and the equivalent control (15.58) becomes

$$\lambda_{eq} = \frac{1}{2}\left(1 + \frac{v_C}{V_g}\right). \tag{15.60}$$

By substituting λ_{eq} for λ in (15.57), the Filippov reduced-order model $\dot{\mathbf{x}} = \mathbf{F}^S(\mathbf{x})$ is obtained, where $\mathbf{F}^S(\mathbf{x})$ is given by the following expression

$$\mathbf{F}^S(\mathbf{x}) = \mathbf{A}\mathbf{x} + \frac{v_C}{V_g}\mathbf{B}\Big|_{\mathbf{x}\in\Sigma^S} = \begin{pmatrix} -\dfrac{v_C}{RC} \\ 0 \end{pmatrix}. \tag{15.61}$$

It is worth to note that the reduced-order ideal sliding dynamics is 1-dimensional and the vector field in (15.61) can be simplified as

$$\mathbf{f}^S(v_C) = -\frac{v_C}{RC}, \qquad i_L = 0 \tag{15.62}$$

It can be observed that the origin is an attractive equilibrium point for the ideal sliding dynamics and it is a pseudo-saddle for the switched system, since the sliding-mode region is escaping *i.e.*, trajectories evolve on Σ^S according to (15.62), but also escape into the regions Σ^+ and Σ^-.

15.6.3 The crossing switching dynamics: simulations from the switched model

The construction of the state plane trajectories for the switched vector field greatly facilitates the study of the global dynamics and it will therefore be used in the sequel. The trajectories corresponding to different values of the load resistance R will reveal different dynamical behaviors. Therefore, the dynamics will be explored by varying this parameter while maintaining the rest of the parameters fixed. The values of these parameters are depicted in Table. The resulting oscillation frequency of the PRC is in the range 500–570 kHz depending on the load resistance.

In Figure 15.15, two pairs of symmetric orbits evolving around the equilibria (dots) in the plane (v_C, i_L) are depicted. The starting points for all the orbits have been selected close to the unstable sliding set Σ^S. For one pair of symmetric orbits, the initial conditions have been chosen such that they do not reach the switching manifold Σ, hence tending to the corresponding equilibria. Conversely, for the other pair, the initial conditions have been selected in such a way that the trajectories cross the switching manifold Σ so that they will evolve to a limit cycle, which is made up of two symmetric pieces, one in each half state plane. Regarding the two kind of regular orbits plotted in Figure 15.15, either flowing to an equilibrium or to a limit cycle, three different attractors may exist, hence, there must exist at least a boundary of attraction between the different limit sets. Trajectories starting outside the stable limit cycle will spiral towards this limit cycle, hence, every point outside this limit cycle belongs to the basin of attraction of the same cycle. Trajectories in the interior of the stable limit cycle might converge to the same cycle, but they could spiral to any one of the stable equilibria either after hitting the switching boundary or without reaching it.

Table 15.1 The used parameter values for the self-oscillating parallel resonant converter

V_g	L	C
20 V	7.3 μH	10.7 nF

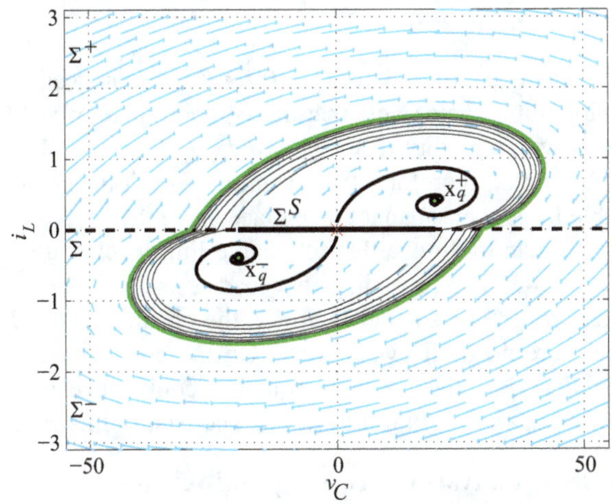

Figure 15.15 Two pairs of symmetric orbits, which start near the unstable sliding-mode region Σ^S, evolving around the equilibria in the state plane (v_C, i_L). Observe that one pair tend to a corresponding equilibrium and the other pair cross the switching manifold Σ, so tending to a limit cycle. $R = 50\ \Omega$

As will be shown later, the boundary between the basins of attraction of the different attractors might be an unstable crossing limit cycle, a sliding limit cycle interacting with the escaping sliding set Σ^S or two unstable sliding limit cycles.

15.6.4 Determination of crossing limit cycles and their stability analysis

15.6.4.1 Computation of the crossing limit cycles

Let T be the period of the crossing limit cycle and let $\mathbf{x}_{ss}(0) = \mathbf{x}_{ss}(T)$ be the steady-state value of the state variables at the beginning/end of the switching cycle.

Let $\mathbf{x}_{ss}(T/2)$ be the steady-state value of the state variables at time instant $T/2$. According to (15.53), $\mathbf{x}_{ss}(T)$ is given by

$$\mathbf{x}_{ss}(T) = \mathbf{\Phi}\left(\frac{T}{2}\right)\mathbf{x}_{ss}\left(\frac{T}{2}\right) - \mathbf{\Psi}\left(\frac{T}{2}\right), \tag{15.63}$$

and due to the odd symmetry of the vector field $\mathbf{F}$

$$\mathbf{x}_{ss}\left(\frac{T}{2}\right) = -\mathbf{x}_{ss}(T). \tag{15.64}$$

Putting (15.64) into (15.63), solving for $\mathbf{x}_{ss}(T)$ and imposing the switching condition $\sigma(\mathbf{x}_{ss}(T)) := \mathbf{K}^\top \mathbf{x}_{ss}(T) = 0$ one obtains

$$\mathbf{x}_{ss}(T) = -\left(\mathbf{I} + \mathbf{\Phi}\left(\frac{T}{2}\right)\right)^{-1}\mathbf{\Psi}\left(\frac{T}{2}\right), \tag{15.65a}$$

$$\sigma(\mathbf{x}_{ss}(T)) = \mathbf{K}^\top \mathbf{x}_{ss}(T) = 0. \tag{15.65b}$$

A solution of the equation $\sigma(\mathbf{x}_{ss}(T)) = 0$ for T gives a steady-state switching period of a possible limit cycle. Note that the v_C-coordinate of $\mathbf{x}_{ss}(T)$ is negative.

It is clear that with a quality factor $Q := R/\sqrt{L/C} < 1/2$, the eigenvalues of the matrix $\mathbf{A}$ will be real, therefore, the response of each linear topology is over-damped and stable limit cycles cannot take place since the trajectories will at most cross the switching manifold once. In fact, stable limit cycles are born only for a quality factor $Q \approx 1.855$ corresponding to a load resistance $R = R_{\mathrm{sn}} \approx 48.459\ \Omega$. For values of load resistance smaller than R_{sn}, the equation $\sigma(\mathbf{x}_{ss}(T)) = 0$ has no solution and therefore no limit cycle exists. On the contrary, for values of load resistance larger than R_{sn}, the previous equation could have two solutions and therefore two different crossing limit cycles can coexist in this case.

15.6.4.2 Stability analysis of crossing limit cycles

Since the PRC is a piecewise affine, the corresponding monodromy matrix can be constructed analytically and can be expressed as the product of the state transition matrices corresponding to each sub-cycle and the corresponding saltation matrices as detailed in Section 15.4. The switching condition $\sigma(\mathbf{x}) = 0$ does not explicitly depends on time and one has $\partial\sigma/\partial t = 0$. The resulting saltation matrix $\mathbf{S}$ is given by

$$\mathbf{S} = \mathbf{I} + \frac{(\mathbf{F}^{\mathrm{after}}(\mathbf{x}(t_c)) - \mathbf{F}^{\mathrm{before}}(\mathbf{x}(t_c)))\mathbf{K}^\top}{\mathbf{K}^\top \mathbf{F}^{\mathrm{before}}(\mathbf{x}(t_c))}, \tag{15.66}$$

where $\mathbf{F}^{\mathrm{before}}$ and $\mathbf{F}^{\mathrm{after}}$ are, respectively, the vector fields before and after a switching taking place at time instant t_c. Within a switching cycle, two different switching take place. In steady-state these occur at the beginning/end of the period and its half. Then, one has two different saltation matrices which are given by

$$\mathbf{S}^+ = \mathbf{I} + \frac{(\mathbf{F}^+(\mathbf{x}_{ss}(T)) - \mathbf{F}^-(\mathbf{x}_{ss}(T)))\mathbf{K}^\top}{\mathbf{K}^\top \mathbf{F}^-(\mathbf{x}(T))}, \tag{15.67a}$$

$$\mathbf{S}^- = \mathbf{I} + \frac{(\mathbf{F}^-(\mathbf{x}_{ss}(\frac{T}{2})) - \mathbf{F}^+(\mathbf{x}_{ss}(\frac{T}{2})))\mathbf{K}^\top}{\mathbf{K}^\top \mathbf{F}^+(\mathbf{x}_{ss}(\frac{T}{2}))}. \tag{15.67b}$$

By using the expressions of the different vector fields, and taking into account that in steady-state one has that $\mathbf{x}_{ss}(T/2) = -\mathbf{x}_{ss}(T)$, the saltation matrices for the crossing limit cycles of the PRC become as follows

$$\mathbf{S}^- = \mathbf{S}^+ = \mathbf{S} = \mathbf{I} + \frac{2\mathbf{B}\mathbf{K}^\top}{\mathbf{K}^\top \mathbf{F}^-(\mathbf{x}_{ss}(T))} \tag{15.68}$$

Then, the monodromy matrix $\mathbf{M}$ corresponding to these limit cycles can be expressed as follows

$$\mathbf{M} = \mathbf{S}^+ \mathbf{\Phi}\left(\frac{T}{2}\right) \mathbf{S}^- \mathbf{\Phi}\left(\frac{T}{2}\right) = \left(\mathbf{S}\mathbf{\Phi}\left(\frac{T}{2}\right)\right)^2 \tag{15.69}$$

All the terms needed for calculating the monodromy matrix are therefore available in closed-form except the switching period T which must be computed numerically. For that, the value of $T_0 = \sqrt{LC}$ can be used as an initial guess for solving the equation $\mathbf{K}^\top \mathbf{x}_{ss}(T) = 0$.

Remark 4. *Being the system two-dimensional, it has two Floquet multipliers of which one is expected to be always equal to one because the system is autonomous, which, as will be shown, implies that the remaining Floquet multiplier for each crossing limit cycle must be real and therefore Neimark-Sacker bifurcation cannot occur. Also, according to (15.69), negative Floquet multipliers are not possible implying that period doubling cannot take place and that only saddle-node (or cyclic-fold) or symmetry breaking bifurcation can occur. The analysis of the limit cycles before and after the bifurcation point shows that it is a cyclic fold bifurcation.*

Remark 5. *The saltation matrices become singular if* $\mathbf{K}^\top \mathbf{F}^+(\mathbf{x}) = 0$ *or symmetrically* $\mathbf{K}^\top \mathbf{F}^-(\mathbf{x}) = 0$. *For crossing limit cycles, generally, this will not happen. However, when the sliding set is approached by these limit cycles, this term becomes very small and consequently one of the Floquet multipliers will become very large.*

Figure 15.16 shows the computed logarithm of the absolute values of the Floquet multipliers $\ln|\lambda|$ corresponding to the two computed crossing limit cycles when the load resistance is varied. This plot shows that one of the multipliers is always located at 1 independently of the value of the load resistance since the system is autonomous. As the value of this parameter approaches the critical value,

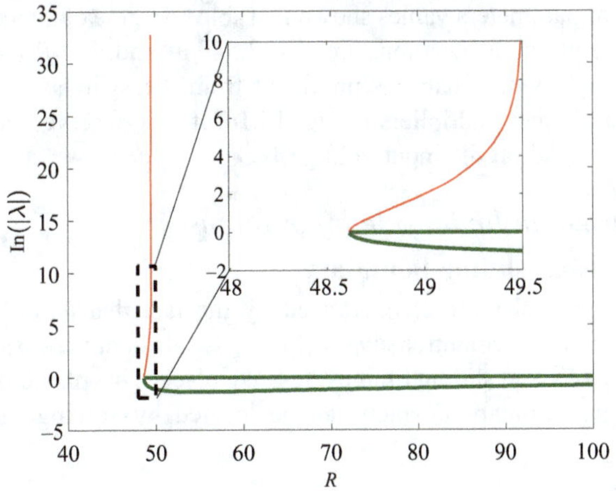

Figure 15.16 *Computed logarithm for absolute values of the Floquet multipliers $|\lambda|$ of both the stable and the unstable limit cycles versus the load resistance of the PRC and a close view near the cyclic-fold bifurcation. For the stable limit cycle all the Floquet multipliers are within the unit circle one of them being always equal to one. For the unstable limit cycle, one of the Floquet multipliers becomes infinite when the sliding set is approached by this limit cycle*

the remaining eigenvalue for both limit cycles also approaches one indicating a cyclic-fold bifurcation. According to Remarks 2 and 5, one of the Floquet multipliers becomes infinite at $R = R_{\text{cs}} \approx 49.353\ \Omega$ because close to this value, $\mathbf{x}_{ss}(T) \approx \mathbf{x}_b^+$ and equivalently $\mathbf{x}_{ss}(T/2) \approx \mathbf{x}_b^-$.

15.6.4.3 Derivation of cyclic-fold bifurcation boundary of the crossing limit cycles

Smooth cyclic-fold bifurcation is a local phenomenon which takes place when two limit cycles collide and annihilate each other. For the system considered in this study, at the boundary of this bifurcation, there is a tangency between the plot of $\sigma(\mathbf{x}_{ss}(T))$ and the T-axis in such a way that two solutions of the equation $\sigma(\mathbf{x}_{ss}(T)) = 0$ coalesce and disappear. Therefore, the following two equalities hold simultaneously at this critical point

$$\mathbf{K}^{\top}\frac{\partial \mathbf{x}_{ss}(T)}{\partial T} = 0, \tag{15.70a}$$

$$\mathbf{K}^{\top}\mathbf{x}_{ss}(T) = 0. \tag{15.70b}$$

The partial derivative in (15.70a) can be obtained by using (15.65a) and differentiating the involved matrix functions, hence the equation (15.70a) becomes

$$\mathbf{K}^{\top}\left(\mathbf{I}+\mathbf{\Phi}\left(\frac{T}{2}\right)\right)^{-1}\mathbf{\Phi}\left(\frac{T}{2}\right)\mathbf{F}^{+}(\mathbf{x}_{ss}(T)) = 0. \tag{15.71}$$

For the set of parameters values shown in Table 15.1, the obtained solutions for T and R from the previous equations are $T \approx 2.549\ \mu\text{s}$ and $R = R_{\text{sn}} \approx 48.459\ \Omega$ in a perfect agreement with what was predicted from the switched model with the evolution of the Floquet multipliers in Fig. 15.16. It was observed that this critical value does not depend on the input voltage V_g.

15.6.5 Nonsmooth limit cycle bifurcations

15.6.5.1 Crossing-sliding boundary

The crossing limit cycles are characterized by the fact that $|v_C(T)| > V_g$. If, by varying a parameter, the conditions $|v_C(T)| = V_g$ is fulfilled, the system limit cycle hits the sliding set Σ^S and sliding limit cycles take place. This phenomenon is called a crossing-sliding bifurcation which can be located by solving its initial value problem.

$$\mathbf{x}_{ss}(T) = \mathbf{x}_b^- \quad \text{or} \quad \mathbf{x}_{ss}\left(\frac{T}{2}\right) = \mathbf{x}_b^+. \tag{15.72}$$

Once the period T is computed using (15.72), the v_C-coordinate for the corresponding limit cycle can be obtained by simply evaluating the trajectory at this time instant. With the set of parameter values depicted in Table 15.1, the solution for the previous equation gives $R_{\text{cs}} \approx 49.352\ \Omega$.

15.6.5.2 Sliding limit cycles

Let us define the particular orbit in $\Sigma^+ \cup \Sigma$ for the linear field $\mathbf{F}^+$ that, starting at the switching manifold, reaches the extreme of the sliding set, $\mathbf{x}_b^+ = (V_g, 0)^\top$ in a time t_{sb}. Due to the contractive character of $\mathbf{F}^+$, this orbit always exists, together with an equivalent symmetric orbit in $\Sigma^- \cup \Sigma$ that starts at Σ and addresses to $\mathbf{x}_b^- = (-V_g, 0)^\top$. Let $\mathbf{x}_{sb} \in \Sigma$ and its symmetric $-\mathbf{x}_{sb}$ be the starting points of these orbits. If $\mathbf{x}_{sb}$ is located on the left of $\mathbf{x}_b^-$, that is outside Σ^S, then sliding cycles do not exist. If $\mathbf{x}_{sb}$ is located between $\mathbf{x}_b^-$ and the origin, then this orbit can be continued backwards in time on Σ^S to $\mathbf{x}_b^-$ and linked with its symmetric counterpart, thus defining a unique sliding limit cycle, encircling the origin and the two equilibrium points. Actually, this unstable limit cycle is the boundary between the basins of attraction of the two equilibrium points. Finally, if $\mathbf{x}_{sb}$ is located between the origin and $\mathbf{x}_b^+$, then this orbit can be continued backwards in time on Σ^S to $\mathbf{x}_b^+$ itself, thus defining a sliding limit cycle which encircles the equilibrium point $\mathbf{x}_q^+$, thus defining the boundary of its basin of attraction. Also, in this case, a symmetric sliding limit cycle, which encircles $\mathbf{x}_q^-$ and is its boundary of attraction, exists.

Taking into account the expression (15.53), the time t_{sb} and its corresponding point $\mathbf{x}_{sb}$ can be computed from the following condition

$$\mathbf{K}^\top \mathbf{x}_{sb} = \mathbf{K}^\top \mathbf{\Phi}^{-1}(t_{sb})(\mathbf{x}_b^+ - \mathbf{\Psi}(t_{sb})) = 0. \tag{15.73}$$

Note that the critical case $\mathbf{x}_{sb} = \mathbf{x}_b^-$ corresponds to the crossing sliding bifurcation obtained from (15.72), in which $t_{sb} = T/2$. Therefore, unstable sliding limit cycles exist only for $R > R_{\text{cs}}$.

15.6.5.3 Double homoclinic connection of sliding limit cycles

This phenomenon takes place when a single symmetric limit cycle bifurcates into two different asymmetric limit cycles [8], being the origin $\mathbf{0} = (0, 0)^\top$ the only common point between them. Therefore, this boundary can be determined by solving the following equation

$$\mathbf{\Phi}^{-1}(t_{sb})(\mathbf{x}_b^+ - \mathbf{\Psi}(t_{sb})) = \mathbf{0}. \tag{15.74}$$

With the set of parameter values depicted in Table 15.1, the solution for the previous equation gives $R_{\text{hc}} \approx 68.108\ \Omega$. From a practical point of view, this is the minimum value of load resistance that will guarantee the convergence of the system to the desired crossing limit cycle starting from zero initial conditions.

15.6.6 Summary of the bifurcation scenario in the PRC under ZCS control

Because the sliding region and the sliding limit cycles themselves are unstable, it is not possible to compute them in forward time and calculation in backward time is needed. The escaping sliding region in the forward time becomes an attracting sliding region in backward time.

To get a clear view of the behavior of the switched system, a bifurcation diagram is computed by using the descriptive equations of all the bifurcations described in the previous sections. The corresponding steady-state vector was computed and the switching condition was imposed. The bifurcation diagram when varying the parameter R is depicted in Figure 15.17, which shows four different branches with periodic solutions. In that figure the switching v_C-coordinate is plotted against the bifurcation parameter R.

Branch Π_1 (thick) corresponds to the desired stable symmetric crossing limit cycle. For relatively small load resistance values, no periodic solution exits under ZCS control and are only created at the turning point A at which $R = R_{sn} \approx 48.459$ Ω. For sufficiently large load resistance values, the periodic solutions in Branch Π_1 are close to the ones corresponding to harmonic resonance in the linear LC tank with almost sinusoidal symmetric crossing limit cycles with output capacitor voltage amplitude approximately equal to $4QV_g/\pi$ being the switching frequency close to the resonant frequency in this case [30].

Branch Π_2 contains different types of limit cycles. Close to the turning point A, an unstable crossing limit cycle (thin) exists in a small range of the load resistance $R \in (R_{sn}, R_{cs})$, $R_{cs} \approx 49.352$ Ω. If the load resistance increases, this limit cycle hits the sliding set at the critical point B where $R = R_{cs}$ and sliding limit cycles take place (dotted). At this critical point, the crossing Branch Π_2 becomes infinitely unstable in the sense that one of the characteristic multipliers becomes infinite.

Unstable symmetric limit cycles on Branch Π_2 for values of the load resistance in the range $R_{sn} < R < R_{cs}$ between Point A and Point B do not contain sliding-mode regime. Unstable symmetric limit cycles on Branch Π_2 for $R_{cs} < R < R_{hc}$, $R_{hc} \approx 68.108$ Ω, between Point B and Point C contain sliding-mode regime whose corresponding time during which the inductor current is zero interval increases with R. At $R \approx R_{hc}$ the unstable limit cycle on Branch Π_2 is destroyed at the point C and two asymmetric sliding limit cycles are created forming the unstable branches Π_3 and Π_4.

To sum up, we have three different bifurcation points when the load resistance R is varied. These points correspond to three different values of R: $R_{sn} \approx 48.459$ Ω corresponding to a smooth cyclic-fold bifurcation, $R_{cs} \approx 49.353$ Ω to a border collision between a crossing and a sliding limit cycle and $R_{hc} \approx 68.108$ Ω which corresponds to a double homoclinic connection. Consequently, four regions having different limit sets can be clearly identified as shown in Figure 15.17. Below, a description of the different cases is given and one example of phase plane for some representative cases are represented in Figures 15.18–15.20.

1. For $R < R_{sn}$, no limit cycles exist and the only steady-state stable solutions are the equilibrium points corresponding to the linear configurations of the switched system.
2. At $R = R_{sn}$, two crossing limit cycles are created.
3. For $R_{sn} < R < R_{cs}$, two crossing limit cycles coexist. Orbits starting outside the inner unstable cycle converge to the outer stable limit cycle. The basin of attraction of each equilibrium point is defined by backward time orbits from the boundary of the sliding set.

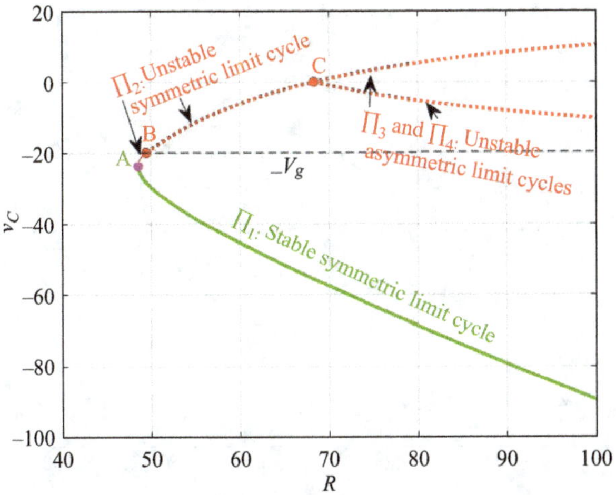

Figure 15.17 Summary of the bifurcations as the load resistance is varied. For the crossing limit cycles, the evolution of the switching v_C-coordinate of the state vector $\mathbf{x}$ in terms of the load resistance R is shown. The birth of two different solutions can be appreciated at $R = R_{\text{sn}} \approx 48.459\ \Omega$ ($Q = Q_{\text{sn}} \approx 1.855$). For the unstable sliding limit cycles, the evolution of the v_C-coordinate of the steady-state vector $\mathbf{x}_{ss}(t_{sb})$ is shown. A crossing-sliding bifurcation takes place at $R = R_{\text{cs}} \approx 49.352\ \Omega$ ($Q = Q_{\text{cs}} \approx 1.889$) when $v_C(T) = -V_g$ and equivalently $v_C(T/2) = V_g$. At Point C, $R = R_{\text{hc}} \approx 68.108\ \Omega$ ($Q = Q_{\text{hc}} \approx 2.608$), a double homoclinic connection takes place

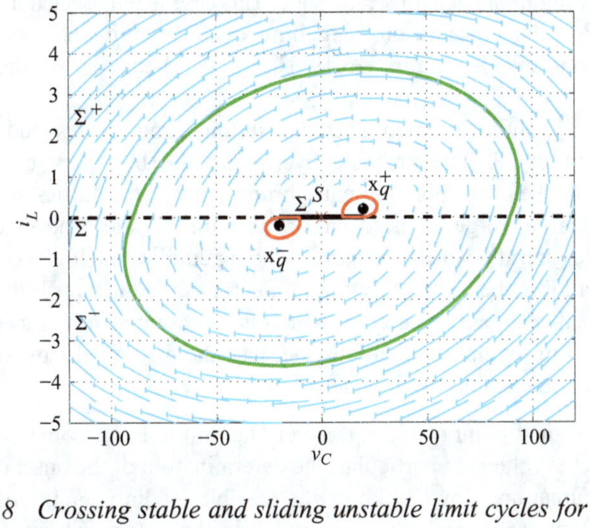

Figure 15.18 Crossing stable and sliding unstable limit cycles for $R = 100\ \Omega$

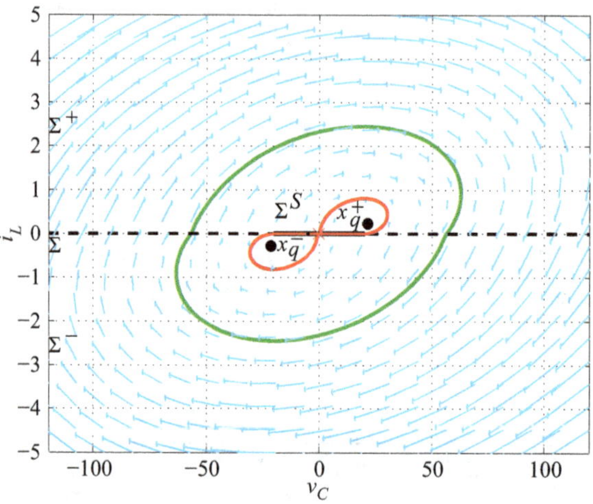

Figure 15.19 Crossing stable and sliding unstable limit cycles for $R = 69\ \Omega$

4. At $R = R_{cs}$, the unstable crossing limit cycle hits the sliding set and unstable sliding limit cycle is about to be created.
5. For $R_{cs} < R < R_{hc}$, apart from the outer stable crossing limit cycle, the unstable sliding limit cycle is unique and of sliding type, with a non-zero sliding time interval. The sliding-mode region inside the unstable limit cycle complete the boundary of the basin of attraction of the two equilibrium points. The origin does not belong to the basin of attraction of the desired crossing limit cycle. Initial conditions must be outside this basin for the system to reach the desired stable crossing limit cycle in steady-state. This case is represented in Figure 15.20.
6. At $R = R_{hc}$, the unstable crossing limit cycle is split onto two asymmetric sliding limit cycles. Figure 15.19 shows the behavior of the system for $R \simeq R_{hc}$.
7. For $R > R_{hc}$, there are two unstable sliding limit cycles and one crosing stable limit cycle. The unstable cycles are the boundary of the basin of attraction of the corresponding equilibrium point. Trajectories starting outside these cycles converge to the desired stable limit cycle. Those starting inside these cycles spiral towards the corresponding equilibria. However, these cycles are so small that the probability for a trajectory to tend towards these equilibria is very small. Actually this case, which is represented in Figure 15.18, is the one desired in regular applications, because the origin belongs to the basin of attraction of the stable limit cycle.

From a practical point of view, it is very helpful to know some of the previous critical parameter values. In particular, the determination of the onset of the smooth cyclic-fold bifurcation allows to know the feasible loading conditions for the system to exhibit desired stable oscillations while the determination of the double homoclinc connection corresponds to the minimum load resistance values

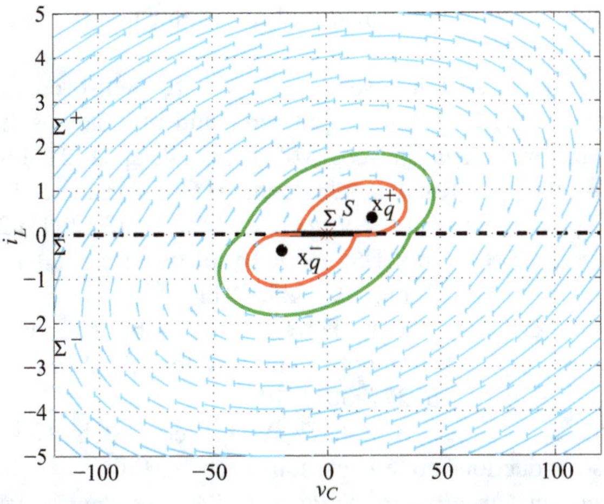

Figure 15.20 Crossing stable and sliding unstable limit cycles for R = 54 Ω

guaranteeing the convergence to the desired crossing limit cycle from zero initial conditions.

It is worth mentioning here that the same phenomena are obtained regardless the values of the remaining fixed parameters such as the inductance L, the capacitance C and the input voltage V_g. Moreover, the variation of V_g does not alter the critical values of the load resistance where bifurcations take place.

15.7 Acknowledgments

This work has been sponsored by the Spanish Agencia Estatal de Investigación (AEI) and the Fondo Europeo de Desarrollo Regional (FEDER) under grant PID2020-12015R-I00 and also by the Spanish *Ministerio de Economía y Competitividad*, in the frame of project PGC2018-096265-B-I00 and by the *Consejería de Economia y Conocimiento de la Junta de Andalucía* under grants P20_01160 and US-1380740.

References

[1] J. Hale and H. Kocak, *Dynamics and Bifurcations*, Springer-Verlag, 1991.

[2] R. Hilborn, Chaos and Nonlinear Dynamics: An Introduction for Scientists and Engineers 2nd Edition, Oxford University Press, 2000.

[3] S. H. Strogatz, Nonlinear Dynamics and Chaos With Applications to Physics, Biology, *Chemistry, and Engineering*, second edition, CRC Press, 2015.

[4] M. di Bernardo, K. Johanson, and F. Vasca, "Sliding orbits and their bifurcations in relay feedback systems," in Proc. 38th Circuits and Devices Conf., 1999, pp. 708–713.
[5] M. di Bernardo, K. H. Johansson, and F. Vasca, "Self-oscillations and sliding in relay feedback systems: Symmetry and bifurcations," *International Journal of Bifurcation and Chaos*, vol. 11, no. 4, pp. 1121–1140, 2001.
[6] Yu. A. Kuznetsov, S. Rinaldi, A. Gragnani, One-Parameter Bifurcations in Planar Filippov Systems, *Int. J. Bifurc. and Chaos* 13 (2003) 2157–2188.
[7] M. R. Jeffrey, S.J. Hogan, "The geometry of generic sliding bifurcations," *SIAM Rev.*, vol. 53, no. 3, pp. 505–525, 2011.
[8] D. J. Pagano, E. Ponce and F. Torres, "On Double boundary equilibrium bifurcations in piecewise smooth planar systems," *Qualitative Theory of Dyn. Systems*, vol. 10, pp. 277–301, 2011.
[9] D. Giaouris, S. Banerjee, B. Zahawi, V. Pickert, "Stability analysis of the continuous-conduction-mode buck converter via Filippov's method," *IEEE Transactions on Circuits and Systems I: Regular Papers*, vol. 55, no. 4, pp. 1084–1096, 2008.
[10] E. Freire, E. Ponce, F. Torres, A general mechanism to generate three limit cycles in planar Filippov systems with two zones. *Nonlinear Dynamics* 78 (2014) 251–263.
[11] E. Freire, E. Ponce, F. Torres, On the critical crossing cycle bifurcation in planar Filippov systems, *J. Differ. Equations* 259 (2015) 7086–7107.
[12] R. L. Leine and H. Nijemeijer, *Dynamics and bifurcations of non-nmooth mechanical systems,* Lecture Notes in Applied and Computational Mechanics, Vol. 18, *Springer*, 2004.
[13] M. di Bernardo, C. Budd, A. Champneys and P. Kowalczy *Piecewise-smooth Dynamical Systems: Theory and Applications*, Springer, 2008.
[14] D. J. W. Simpson, Bifurcations In Piecewise-smooth Continuous Systems, World Scientific Series, *Series A*, vol. 70, 2010.
[15] M. R. Jeffrey, *Modeling with Nonsmooth Dynamics, Part of the Frontiers in Applied Dynamical Systems: Reviews and Tutorials book series (FIADS, volume 7)*, Springer, 2020.
[16] Y. Chen, C. K. Tse, S. S. Qiu, L. Lindenmuller and W. Schwarz, "Coexisting fast-scale and slow-scale instability in current-mode controlled DC/DC converters: Analysis, simulation and experimental results," *IEEE Transactions on Circuits and Systems I: Regular Papers*, vol. 55, no. 10, pp. 3335–3348, 2008.
[17] E. Rodriguez, A. El Aroudi, F. Guinjoan and E. Alarcon, "A Ripple-Based Design-Oriented Approach for Predicting Fast-Scale Instability in DC–DC Switching Power Supplies," *IEEE Transactions on Circuits and Systems I: Regular Papers*, vol. 59, no. 1, pp. 215–227, Jan. 2012.
[18] C. Fang, "Closed-Form Critical Conditions of Subharmonic Oscillations for Buck Converters," *IEEE Transactions on Circuits and Systems I: Regular Papers*, vol. 60, no. 7, pp. 1967–1974, July 2013.

[19] A. El Aroudi, "A new approach for accurate prediction of subharmonic oscillation in switching regulators–Part I: mathematical derivations," *IEEE Transactions on Power Electronics*, vol. 32, no. 7, pp. 5651–5651, 2017.

[20] A. El Aroudi, "A new approach for accurate prediction of subharmonic oscillation in switching regulators–part II: case studies," *IEEE Transactions on Power Electronics*, vol. 32, no. 7, pp. 5835–5849, 2017.

[21] N. Mohan, T. M. Undeland, W. P. Robbins, *Power Electronics, Converters, Applications, and Design*, John Wiley & Sons, Inc., 1995.

[22] R. W. Erickson, D. Maksimovic, Fundamentals of power electronics, *Lluwer*, Springer, 2001.

[23] P. D. Teodosescu, M. Bojan, and R. Marschalko, "Resonant LED driver with inherent constant current and power factor correction," *Electronics Letters*, vol. 50, no. 15, pp. 1086–1088, 2014.

[24] J. Park, M. Kim, S. Choi, "Zero-current switching series loaded resonant converter insensitive to resonant component tolerance for battery charger", *IET Power Electronics*, vol. 7, no. 10, pp. 2517–2524, 2014.

[25] C. S. Tang, Y. Sun, Y. G. Su, S. K. Nguang, A. P. Hu, Determining multiple steady-state ZCS operating points of a switch-mode contactless power transfer system, *IEEE Trans. on Power Electron.* 24 (2) (2009) 416–425.

[26] T. S. Chan and C. L. Chen, "A primary side control method for wireless energy transmission system," *IEEE Transactions on Circuits Systems I, Reg. Papers*, vol. 59, no. 8, pp. 1805–1814, 2012.

[27] T. Qian, "A converter combination scheme for efficiency improvement of PV systems," *IEEE Transactions on Circuits Systems II, Exp. Briefs, early access*, 2017, DOI:10.1109/TCSII.2017.2764027.

[28] V. Volpérian, "Approximate small-signal analysis of the series and the parallel resonant converters," *IEEE Transactions on Power Electronics*, vol. 4, no. 1, pp. 15–24, 1989.

[29] H. Pinheiro, P. K. Jain, G. Joos, Self-sustained oscillating resonant converters operating above the resonant frequency, *IEEE Trans. on Power Electron.* 14 (5) (1999) 803–815.

[30] R. Bonache-Samaniego, C. Olalla, L. Martínez-Salamero, Design of self-oscillating resonant converters based on a variable structure systems approach, *IET Power Electron.* 57, (1) (2015) 111–119.

[31] R. Bonache-Samaniego, C. Olalla, L. Martínez-Salamero, Dynamic Modeling and Control of Self-Oscillating Parallel Resonant Converters Based on a Variable Structure Systems Approach, *IEEE Trans. on Power Electron.* 32 (2) (2017) 1469–1480.

[32] D. Williams, C. Bingham, M. Foster, D. Stone, Hamel locus design of self-oscillating DC-DC resonant converters, *IET Power Electron.* 3 (1) (2010) 86–94.

[33] V. M. Hernandez, R. Silva, H. Sira-Ramirez, "On the stability of limit cycles in resonant DC-to-DC power converters," *Proc. of the 42nd IEEE Conf. on Decision and Control*, pp. 1141–1146, 2003.

[34] O. Dranga, B. Buti, and I. Nagy, "Stability analysis of a feedback-controlled resonant DC-DC converter," *IEEE Transactions on Industrial Electronics*, vol. 50, no. 1, pp. 141–152, 2003.

[35] V. I. Utkin, Variable structure systems with sliding modes, *IEEE Transactions on Automatic Control*, vol. 22, no.2, pp. 212–222, 1977.

[36] V. I. Utkin, Sliding Modes and their Application in Variable Structure Systems. Moscow, *U.S.S.R.: MIR*, 1978.

[37] A. F Filippov, Differential Equations with discontinuous righthand sides, *Kluwer Academic Publishers, Dortrecht*, 1998.

[38] J. Liu, and X. Wang, *Advanced sliding mode control for mechanical systems*, Springer, 2012.

[39] C. Edwards, S. Spurgeon, *Sliding Mode Control: Theory And Applications*, CRC Press, 1998.

[40] P. Hartman. *Ordinary differential equations.* Wiley: New York, 1964.

[41] L. Martinez-Salamero, A. Cid-Pastor, A. El Aroudi, R. Giral, J. Calvente, G. Ruiz-Magaz, Sliding-Mode Control of DC-DC Switching Converters, *IFAC Proceedings Volumes*, vol. 44, no. 1, 2011, pp. 1910–1916.

[42] V. Utkin, Sliding mode control of DC/DC converters, *Journal of the Franklin Institute*, vol. 350, no. 8, 2013, pp. 2146–2165.

[43] A. El Aroudi, L. Benadero, E. Ponce, C. Olalla, F. Torres and L. Martínez-Salamero, "Nonlinear Dynamic Modeling and Analysis of Self-Oscillating H-Bridge Parallel Resonant Converter Under Zero Current Switching Control: Unveiling Coexistence of Attractors," *IEEE Transactions on Circuits and Systems I: Regular Papers*, vol. 66, no. 4, pp. 1657–1667, April 2019.

[44] A. El Aroudi, R. Haroun, M. S. Al-Numay, J. Calvente and R. Giral, "Fast-Scale Stability Analysis of a DC–DC Boost Converter With a Constant Power Load," *IEEE Journal of Emerging and Selected Topics in Power Electronics*, vol. 9, no. 1, pp. 549–558, Feb. 2021.

[45] L. Benadero, E. Ponce, A. El Aroudi, and F. Torres, Limit cycle bifurcations in resonant LC power inverters under zero current switching strategy, *Nonlinear Dynamics*, 91, 1145–1161, 2018.

[46] L. Benadero, F. Torres, E. Ponce, A. El Aroudi, Dynamic analysis of self-oscillating H-bridge inverters with state feedback, *Journal of the Franklin Institute*, vol. 357, no. 1, 2020, pp. 494–521.

Chapter 16

Pure time domain multiconductor transmission line formalism

J. A. Brandão Faria[1]

An important electrical engineering subject, the classical theme of multiconductor transmission lines (MTL) has been receiving continued attention for quite a long time—for almost a century indeed [1–3]. Eclectic, the theme is of interest for a wide variety of areas, ranging from power systems engineering ("slow" time-varying phenomena) to microwave engineering ("rapid" time-varying phenomena). Overhead lines, power cables, crosstalk, electromagnetic compatibility, printed-circuit boards, ribbon cables, high-speed interconnects, very large-scale integration, monolithic microwave integrated circuits, coplanar waveguides, couplers, filters, etc., are a few items in the list of topics where MTL is involved.

Considering the various electrical engineering subfields concerned with MTL problems, considering the time-domain and frequency-domain approaches utilized, considering the analytical and numerical methods developed, the counting of significant contributions published on MTL theory and applications could easily reach a few thousand. This work is not the place for that; in any case, key contributions by Wedepohl and coworkers, mainly published in *IEE Proceedings*, related to matrix algebra procedures in MTL analysis, are worthy of recognition [4–9]. Also, a few books published before 2000 deserve mentioning, namely those by Frankel [10], Djordjevic *et al.* [11], Faria [12], and Paul [13]. The last book is specially recommended not only because of its quality and coverage, but also because of the amount of bibliographic references there offered.

This chapter builds around the issue if MTL theory can or cannot be formulated and solved purely in the time-domain when losses are considered, excluding any detours or shortcuts into the frequency domain. Closely related to the above question a pure time-domain approach aimed to include conductors' skin effect in MTL equations is offered. Also, based directly on Maxwell's equations, time-domain conservation laws regarding multiconductor line voltages and currents are presented.

[1]Instituto de Telecomunicações, Instituto Superior Técnico, Universidade de Lisboa, Av. Rovisco Pais, Lisboa, Portugal

16.1 Time domain electromagnetic field equations

According to the fundamental theorem of vector analysis or Helmholtz theorem [14], any continuous field vector regular at infinity can be uniquely determined from its *curl* and *div* operators; electromagnetic field vectors are no exception.

Electromagnetic fields are originated by electric charges either at rest or in motion—electric currents. These field sources are, respectively, characterized by charge density ρ and current density **J**.

The link between field sources and field vectors* is given by,

$$\begin{cases} curl\ \mathbf{H} = \mathbf{J} + \dfrac{\partial \mathbf{D}}{\partial t} \\ div\ \mathbf{D} = \rho \end{cases} \tag{16.1}$$

where **D** is the electric displacement vector and **H** is the magnetic field vector.

It should be noted that the two equations in (16.1) are not fully independent. In fact, if the charge conservation law†

$$0 = div\ \mathbf{J} + \frac{\partial \rho}{\partial t} \tag{16.2}$$

is taken, *ab initio*, as a physical postulate, it is easy to conclude that the application of the *div* operator to the *curl* equation in (1) yields $div\ \mathbf{D} = \rho$, employing the identity $div\ curl \equiv 0$.

The field pair {**D**, **H**} is insufficient to describe the electromagnetic phenomena. A more fundamental source-independent field pair, is required: the electromagnetic pair {**E**, **B**}, where **E** is the electric field vector and **B** is the magnetic induction vector, which obey to

$$\begin{cases} curl\ \mathbf{E} = -\dfrac{\partial \mathbf{B}}{\partial t} \\ div\ \mathbf{B} = 0 \end{cases} \tag{16.3}$$

It should be noted that the *div* equation in (16.3), stating that the field lines of **B** are closed, is a corollary of the companion *curl* equation. Actually, the application of the *div* operator to the first equation in (16.3), using again the identity $div\ curl \equiv 0$, leads to the conclusion that *div* **B** must be time-invariant. Since at some instant in time the **B** field must have been zero, one will find $div\ \mathbf{B} = 0$.

The set of linear equations in (16.1) and (16.3) define the Maxwell equations [15–19] in the differential form. Those four equations cannot be deduced, are axiomatic, they derive from experimental and theoretical work on electricity, magnetism and optics carried out in the eighteenth and nineteenth centuries and, up until today, have never been contradicted. They remain the pillars of electrical engineering.

*Throughout the chapter, boldface letters will be employed to denote vectors (and matrices as well).
†If a net current is flowing into a given region, the contained charge will be increasing at the same rate.

The time-domain Maxwell's equations are linear differential equations that, themselves, are not very hard to solve. But there is a problem, external to them that can cause serious difficulties. The field pair {**D**, **H**} in (16.1) and the field pair {**E**, **B**} in (16.3) are deeply interrelated. The relationship among them is intermediated by the physical properties of the material media where sources are placed in and where fields are impressed on and propagate.

The description of the electromagnetic behavior of materials is not part of Maxwell's equations; it is another subject that belongs mostly with atomic physics. But one thing is sure; Maxwell's equations cannot be solved without the help of the so-called constitutive equations:

$$\begin{cases} \mathbf{D} = D(\mathbf{E}, \mathbf{H}) \\ \mathbf{B} = B(\mathbf{E}, \mathbf{H}) \end{cases} \tag{16.4}$$

For a wide class of material media, the electric displacement vector **D** only depends on the electric field vector **E**; likewise, the magnetic induction vector **B** only depends on the magnetic field vector **H**. Moreover, if the medium is exceptionally well-behaved, i.e., isotropic, linear, time-invariant, instantaneously and locally reacting [20,21], the constitutive equations in (16.4) greatly simplify yielding

$$\mathbf{D} = \varepsilon \mathbf{E}, \quad \mathbf{B} = \mu \mathbf{H} \tag{16.5}$$

where the constitutive parameters ε and μ are, respectively, the electric permittivity and magnetic permeability of the material medium.

In addition to (16.5), the constitutive equation $\mathbf{J} = \sigma \mathbf{E}$ is also of frequent use to describe the conduction properties of material media, where σ is the conductivity, the inverse of resistivity.

For such an exceptionally simple medium, the *curl* equations from (16.1) and (16.3) allow us to characterize the dynamical evolution of the electromagnetic field pair {**E**, **B**} through [22],

$$\begin{cases} \dfrac{\partial \mathbf{E}}{\partial t} = \partial_{\mathbf{E}} = -\dfrac{\mathbf{J}}{\varepsilon} + \dfrac{1}{\varepsilon} curl\left(\dfrac{\mathbf{B}}{\mu}\right) \\ \dfrac{\partial \mathbf{B}}{\partial t} = \partial_{\mathbf{B}} = -curl\ \mathbf{E} \end{cases} \tag{16.6}$$

and, upon numerical time integration, with step Δt, calculate **E** and **B**

$$\begin{cases} \mathbf{E}(t+\Delta t) = \mathbf{E}(t) + \partial_{\mathbf{E}}(t)\ \Delta t \\ \mathbf{B}(t+\Delta t) = \mathbf{B}(t) + \partial_{\mathbf{B}}(t)\ \Delta t \end{cases} \tag{16.7}$$

at every point in space.

But even for the simplest case of linear, time-invariant, instantaneously and locally reacting media, the computation of the electric derivative $\partial_{\mathbf{E}}$ in (16.6) can turn complicated, and inaccurate too, because of inevitable space inhomogeneity owing to the presence of many different materials, which, are frequently described by imprecisely known constitutive parameters $\varepsilon\ (x, y, z)$, $\mu\ (x, y, z)$, and $\sigma\ (x, y, z)$.

In the next section, we will focus attention on geometries containing multiple wires, good conductors, made of metal, parallel among them, aligned with the z-axis, immersed in dielectric media made of good insulating materials. A system like this is called a MTL.

If the system geometry or the media properties vary along z, the MTL is said to be nonuniform. However, if media properties vary transversally, the MTL is said to be inhomogeneous. In any case, throughout this chapter, the MTL system will always be considered linear and time-invariant.

16.2 Time-domain MTL equations

Figure 16.1 shows, schematically, longitudinal and cross-sectional views of a MTL with $N + 1$ parallel wires, the wire labeled 0 denoting the ground or reference conductor (where the scalar potential is set to zero). A network view is also shown, including MTL terminations, i.e., sources and loads.

The shaded strips in Figure 16.1(a), numbered from 0 to N, represent wires *lato sensus*, that is, conductors whose cross-section's shape is arbitrary, e.g., solid or hollow cylinders, solid or hollow parallelpipeds, etc.

The quantities $v_k\ (z,t)$ and $i_k\ (z,t)$ are, respectively, the voltage and the current of the kth wire, at time t, defined at the transverse plane z = constant. Reference directions for line voltages are from wires to ground; the reference direction for line currents is the positive z-axis, except the return current i_0 along negative z.

Quasi-static approaches based on transverse electromagnetic (TEM) parameters are ordinarily used in MTL analysis, yielding very good results, provided that two basic assumptions are fulfilled [12,13]:

- The transverse distance between MTL wires must be smaller than a quarter of the wavelength corresponding to the higher frequency of importance included in the spectral content of the z-propagating signals guided by the MTL.
- The z-components of the electromagnetic field must be negligibly small compared to the transverse field components.

These assumptions are employed throughout this chapter, enabling the consistent use of such concepts as voltages, currents, resistance, conductance, capacitance, and inductance. Violation of those assumptions leads to increased inaccuracy since contributions of non-TEM higher-order modes of propagation should no longer be ignored.

16.2.1 MTL currents and voltages

We assume that MTL wires carry purely longitudinal currents ($\mathbf{J}_k = J_k\widehat{\mathbf{z}}$), hence assuring the magnetic field is purely transverse, $\mathbf{H} \times \widehat{\mathbf{z}} = \mathbf{0}$. As a consequence, one may unambiguously define wire currents and wire voltages at any transverse plane (z constant)—see Figure 16.2.

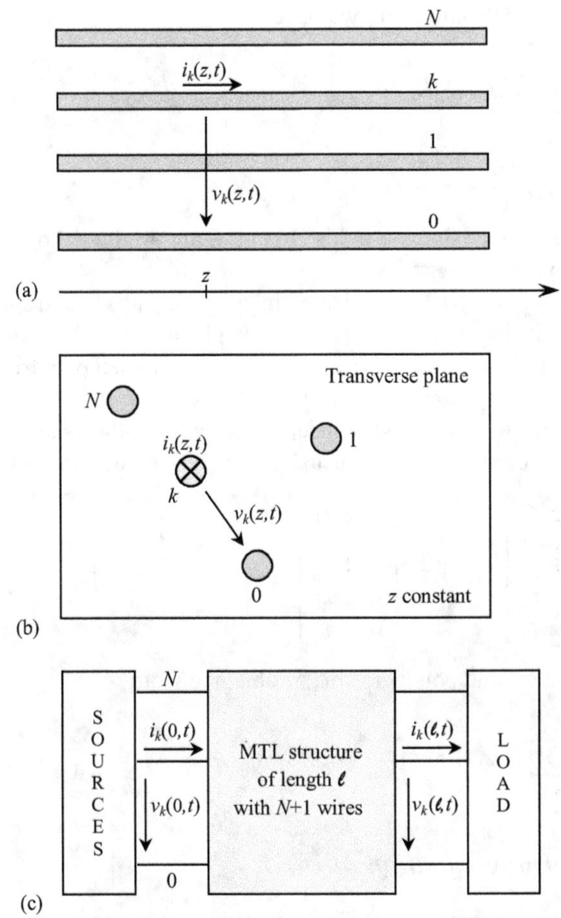

Figure 16.1 Representation of a MTL with N + 1 wires. (a) Longitudinal view. (b) Cross section. (c) Network view

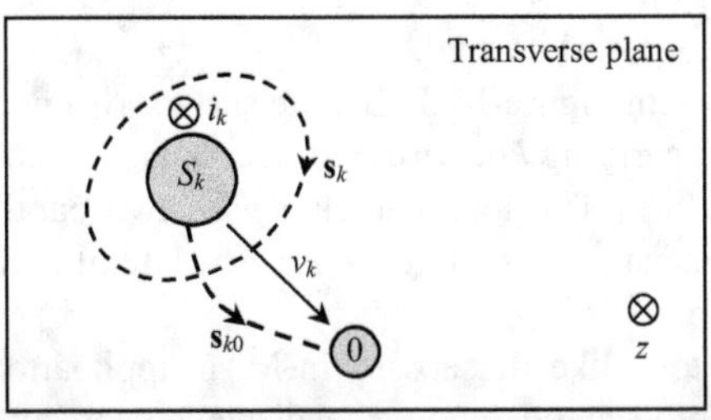

Figure 16.2 Definition of line voltage and line current using integration paths belonging to the transverse plane

For example for the kth wire, we have,

$$i_k(z,t) = \int_{S_k} J_k(x,y,z,t)\ dS = \oint_{\mathbf{s}_k} \mathbf{H}(x,y,z,t)\cdot d\mathbf{s} \tag{16.8}$$

$$v_k(z,t) = \int_{\mathbf{s}_{k0}} \mathbf{E}(x,y,z,t)\cdot d\mathbf{s} \tag{16.9}$$

where it must be stressed that the fields $\mathbf{H}$ and $\mathbf{E}$ are originated by the entire set of wires, not just by the kth wire.

As shown in Figure 16.2, the oriented integration paths $\mathbf{s}_k$ and $\mathbf{s}_{k0}$, in (16.8) and (16.9), reside exclusively on the x,y transverse plane ($d\mathbf{s} = dx\,\widehat{\mathbf{x}}+dy\ \widehat{\mathbf{y}}$). The first path is closed and around the kth wire, the second is an open path from the kth wire to ground.

In a MTL with $N+1$ wires, N independent wire voltages and N independent wire currents are defined; these are usually gathered in column vectors:

$$\mathbf{v} = \begin{bmatrix} v_1(z,t) \\ \vdots \\ v_k(z,t) \\ v_N(z,t) \end{bmatrix}, \quad \mathbf{i} = \begin{bmatrix} i_1(z,t) \\ \vdots \\ i_k(z,t) \\ i_N(z,t) \end{bmatrix} \tag{16.10}$$

the current in the ground conductor being obtained from

$$i_0(z,t) = \sum_{k=1}^{N} i_k(z,t) = [1 \quad 1\cdots \quad 1]\ \mathbf{i} \tag{16.11}$$

16.2.2 Governing equation of $\partial\mathbf{i}/\partial z$

Maxwell's *curl* equation in (16.1) imply that the total current density vector has zero divergence, that is, the flux of conduction currents plus displacement currents across a closed surface S_V must be zero,

$$\int_{S_V} \left(\mathbf{J}\cdot\mathbf{n}_o + \frac{\partial \mathbf{D}\cdot\mathbf{n}_o}{\partial t} \right) dS = 0 \tag{16.12}$$

where $\mathbf{n}_o$ denotes the outward normal.

Consider, as shown in Figure 16.3, that the surface S_V encloses an infinitesimal segment of the current carrying kth wire. Surface S_V is divided into three parts, two tops and a lateral surface. The tops coincide with two parallel transverse cross-sections of the wire, one at z, another at $z+\Delta z$. The lateral part of S_V is the external surface of the kth wire.

For good conductors, like those used in MTL applications, where relaxation times are in the fempto-second range, the displacement current can always[‡] be

[‡]Up to optical frequencies, terahertz band included.

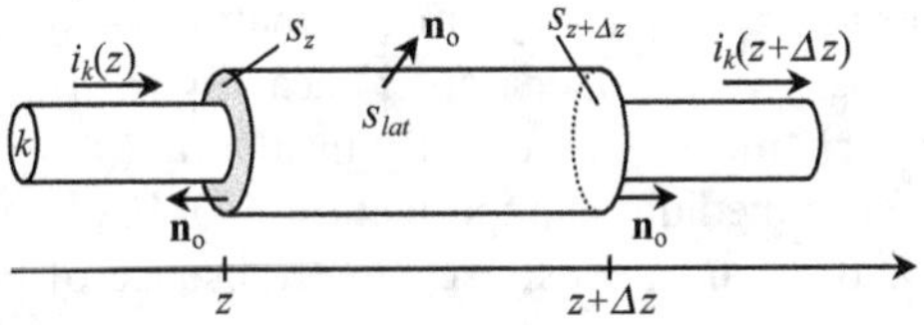

Figure 16.3 Finding $\partial i_k/\partial z$ by application of curl $H = J + \partial D/\partial t$ to the kth wire

disregarded compared to conduction currents, hence, from (16.12),

$$\int_{S_{z+\Delta z}} J_k(z+\Delta z, t)dS - \int_{S_z} J_k(z,t)dS = -\int_{S_{lat}} \left(\mathbf{J}\cdot\mathbf{n}_o + \frac{\partial \mathbf{D}\cdot\mathbf{n}_o}{\partial t}\right)dS \quad (16.13a)$$

or, more compactly,

$$i_k(z+\Delta z, t) - i_k(z,t) = -(i_k(z,t))_T \Delta z \quad (16.13b)$$

In the right hand side of (16.13), the per-unit-length *T*ransverse current $(i_k)_T$ across S_{lat} has two distinct components. The conduction part is associated with leakage currents due to dielectric imperfection ($\sigma \neq 0$). The displacement current part is related with the time-varying electric charge distributed on and along the *k*th wire's lateral surface,

$$\int \mathbf{D}\cdot\mathbf{n}_o dS_k = \int D_n dS_k = q_k(z,t)\ \Delta z \quad (16.14)$$

where q_k denotes the per-unit-length surface charge of the *k*th wire.

The many open paths of the transversally flowing current begin at the *k*th wire and end at every other of the remaining wires, determined by the system voltages.

Dividing (16.13b) by Δz, taking the limit as $\Delta z \to 0$, yields

$$\frac{\partial}{\partial z} i_k(z,t) = -(i_k(z,t))_T \quad (16.15a)$$

or, in the matrix form,

$$-\frac{\partial}{\partial z}\mathbf{i} = \mathbf{i}_T \quad (16.15b)$$

For linear time-invariant systems, the general relationship between time-varying voltages and time-varying currents is defined by ordinary linear constant-coefficients differential equations on *t*. Since the transverse currents collected in $\mathbf{i}_T$depend on all the MTL voltages, the result in (16.15b) can be written in the general form:

$$-\frac{\partial}{\partial z}\mathbf{i}(z,t) = \sum_{n=0}^{\infty} \mathbf{A}_n \frac{\partial^n}{\partial t^n}\mathbf{v}(z,t) \quad (16.16)$$

where the $\mathbf{A}_n$ matrices are real square symmetric matrices—symmetry arising from reciprocity principles; $\mathbf{A}_n = \mathbf{A}_n^T$, superscript T meaning matrix transposition.

If the MTL dielectric media are perfectly insulating ($\sigma = 0$), then $\mathbf{A}_0 = \mathbf{0}$.

If the MTL dielectric medium is imperfect ($\sigma \neq 0$) but homogeneous (σ and ε independent of x,y) and instantaneously reacting (absence of polarization losses), then we have

$$\mathbf{A}_n = \mathbf{0} \quad \text{for } n \geq 2, \quad \text{but}: \quad \mathbf{A}_1 \; = \; \mathbf{C}, \quad \mathbf{A}_0 = \mathbf{G} = \frac{\sigma}{\varepsilon}\mathbf{C} \tag{16.17}$$

that is,

$$-\frac{\partial}{\partial z}\mathbf{i}(z,t) = \mathbf{G}\mathbf{v}(z,t) + \mathbf{C}\frac{\partial}{\partial t}\mathbf{v}(z,t) \tag{16.18}$$

where $\mathbf{C}$ and $\mathbf{G}$, respectively, denote the per-unit-length capacitance matrix and per-unit-length conductance matrix. The product $\mathbf{Cv}$ represents the per-unit-length electric charge $\mathbf{Q}$ of the MTL wires:

$$\mathbf{Q} = \begin{bmatrix} q_1(z,t) \\ \vdots \\ q_k(z,t) \\ \vdots \\ q_N(z,t) \end{bmatrix} = \mathbf{Cv} \tag{16.19}$$

If the MTL dielectric media are inhomogeneous or have polarization losses (asynchronous **D** and **E** fields), then the standard result shown in (16.18), usually seen in the literature, is strictly incorrect [23]; it may only be used for sinusoidal single-frequency regimes, where $\mathbf{G}(\omega)$ and $\mathbf{C}(\omega)$ are calculated for the specific frequency.

However, fortunately, for very many MTL applications, the assumption of negligible dielectric losses is a realistic good approximation and the utilization of

$$-\frac{\partial}{\partial z}\mathbf{i}(z,t) = \mathbf{C}\frac{\partial}{\partial t}\mathbf{v}(z,t) \tag{16.20}$$

is acceptable and is not a critical issue.

We wish to emphasize that the results presented here, regarding $\partial\mathbf{i}/\partial z$, hold true no matter the MTL wires are considered lossless or lossy.

16.2.3 Governing equation of $\partial\mathbf{v}/\partial z$

Maxwell's *curl* equation in (16.3) is the base of Faraday's induction law [18], which will be used here to find $\partial\mathbf{v}/\partial z$.

$$\oint_{s_k} \mathbf{E}\cdot d\mathbf{s} = -\frac{\partial\psi_k}{\partial t}, \qquad \psi_k = \int_{S_k} \mathbf{B}\cdot\mathbf{n}dS \tag{16.21}$$

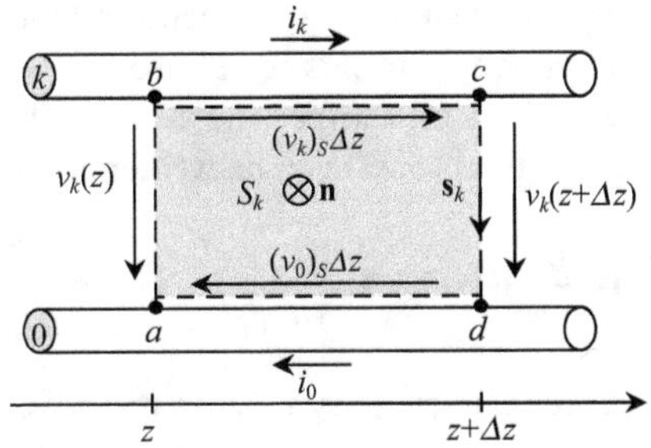

Figure 16.4 Finding $\partial v_k/\partial z$ by application of curl $E = -\partial B/\partial t$ to the kth wire

where $\mathbf{s}_k$ is an oriented closed path, S_k is an open surface bounded by $\mathbf{s}_k$, and $\mathbf{n}$ is the Stokes unit normal to S_k. The scalar quantity ψ_k in (16.21) is the magnetic flux linked with the $\mathbf{s}_k$ path.

As depicted in Figure 16.4, the $\mathbf{s}_k$ path is a rectangle (*abcda*). Two of its opposite sides belong to the transverse planes z and $z + \Delta z$ where kth wire voltages are defined. The other two sides of infinitesimal length are along z, on the surfaces of the kth and 0th wires. The orientation of the integration path coincides with the direction of i_k.

The magnetic induction field $\mathbf{B}$ in (16.21) fills the space external to the wires (the dielectric medium) and is originated by the currents of all the MTL wires, not only by i_k. Since the dielectric medium is nonmagnetic, $\mathbf{B} = \mu_0\mathbf{H}$, magnetization losses are absent; hence, the relationship between $\mathbf{B}$ and the MTL currents is the simplest possible—instantaneously proportional.

From Figure 16.4 and (16.21), we have

$$\underbrace{\int_{ab} \mathbf{E}\cdot d\mathbf{s}_k}_{-v_k(z,t)} + \overbrace{\int_{bc} \mathbf{E}\cdot d\mathbf{s}_k}^{(v_k)_S\Delta z} + \underbrace{\int_{cd} \mathbf{E}\cdot d\mathbf{s}_k}_{v_k(z+\Delta z,t)} + \overbrace{\int_{da} \mathbf{E}\cdot d\mathbf{s}_k}^{(v_0)_S\Delta z} = -\frac{\partial}{\partial t}\underbrace{\phi_k\Delta z}_{\psi_k} \tag{16.22}$$

where ϕ_k is the per-unit-length magnetic flux linked with the $\mathbf{s}_k$ path, which can be expressed as a linear combination of the MTL currents with weighting factors equal to the so-called per-unit-length external inductance coefficients of the MTL.

In (16.22), the *bc* and *da* contributions arise from imperfectly conducting wires ($\sigma \neq \infty$). The quantities $(v_k)_S$ and $(v_0)_S$ are the per-unit-length. Surface voltage drops along the kth and 0th wires caused by the respective currents i_k and i_0.

Dividing (16.22) by Δz, taking the limit as $\Delta z \to 0$, and rearranging terms yields

$$-\frac{\partial}{\partial z}v_k(z,t) = (v_k(i_k) + v_0(i_0))_S + \frac{\partial}{\partial t}\phi_k \tag{16.23}$$

The second term $(v_0)_S$ in the right-hand side of (16.23) involves the return current, which, from (16.11), is given by $i_0 = \Sigma i_k$. This means that this second term (like the third) depends on all the MTL currents.

For linear time-invariant systems, the general relationship between time-varying voltages and time-varying currents is defined by ordinary linear constant-coefficients differential equations on t. Since the surface voltage drops depend on all the MTL currents, the result in (16.23) can be written in the general matrix form:

$$-\frac{\partial}{\partial z}\mathbf{v}(z,t) = \sum_{n=0}^{\infty} \mathbf{B}_n \frac{\partial^n}{\partial t^n}\mathbf{i}(z,t) + \mathbf{L}_e \frac{\partial}{\partial t}\mathbf{i}(z,t) \tag{16.24}$$

where $\mathbf{L}_e$ and $\mathbf{B}_n$ matrices are real square symmetric matrices. $\mathbf{L}_e$ is the familiar per-unit-length external inductance matrix. All matrices $\mathbf{B}_n$ have the same type of structure:

$$\mathbf{B}_n = \mathbf{B}_n^T = \begin{bmatrix} b_{n1} & & & \\ & b_{n2} & & \\ & & \ddots & \\ & & & bnN \end{bmatrix} + b_{n0}\begin{bmatrix} 1 & 1 & \cdots & 1 \\ 1 & 1 & \cdots & 1 \\ \vdots & \vdots & \ddots & \vdots \\ 1 & 1 & \cdots & 1 \end{bmatrix} \tag{16.25}$$

Matrix entries $b_{00}, b_{01}, b_{02} \cdots b_{0N}$ are the per-unit-length DC resistance of the 0th, 1st, 2nd,… Nth wires. Matrix entries $b_{10}, b_{11}, b_{12}, \ldots, b_{1N}$ are the per-unit-length DC internal inductance of the 0th, 1st, 2nd … Nth wires.

If MTL wires are identical and indiscernible, then

$$\mathbf{B}_n = b_n \begin{bmatrix} 2 & 1 & \cdots & 1 \\ 1 & 2 & \cdots & 1 \\ \vdots & \vdots & \ddots & \vdots \\ 1 & 1 & \cdots & 2 \end{bmatrix} \tag{16.26}$$

If the ground wire is a perfect conductor, then $\mathbf{B}_n$ is a diagonal matrix ($b_{n0} = 0$). If all MTL wires are perfect conductors then $\mathbf{B}_n = \mathbf{0}$, and one finds

$$-\frac{\partial}{\partial z}\mathbf{v}(z,t) = \mathbf{L}_e \frac{\partial}{\partial t}\mathbf{i}(z,t) \tag{16.27}$$

It is rather frequent to see a simplified version of (16.24) of the type

$$-\frac{\partial}{\partial z}\mathbf{v}(z,t) = \mathbf{R}\mathbf{i}(z,t) + \mathbf{L}\frac{\partial}{\partial t}\mathbf{i}(z,t) \tag{16.28}$$

where $\mathbf{R}$ and $\mathbf{L}$ denote, respectively, the per-unit-length resistance and inductance matrices of the MTL, with $\mathbf{L} = \mathbf{L}_e + \mathbf{L}_i$, where $\mathbf{L}_i$ describes the contribution of the wires' internal inductance, both matrices $\mathbf{R}$ and $\mathbf{L}_i$ being associated with skin effect phenomena. But the result in (16.28) is strictly incorrect [23]—as can be readily understood as follows.

The skin effect is a well-known linear phenomenon, easy to deal with in the frequency domain. The per-unit-length skin effect impedance operator $\bar{Z}_S(\omega)$, [13,18,19], is employed to relate the phasors of two sinusoidal quantities, the per-unit-length surface voltage along a wire $\bar{V}_S$ and the wire's current intensity $\bar{I}$. The complex character of the impedance is a consequence of two concurring mechanisms, losses due to wire's finite conductivity and internal magnetic energy storage, that is,

$$\bar{V}_S = \bar{Z}_S \bar{I}, \quad \bar{Z}_S = R(\omega) + j\omega L_i(\omega) = Z(\omega)\ e^{j\phi(\omega)} \tag{16.29}$$

where the per-unit-length wire's resistance R and absolute impedance Z increase with ω, the per-unit-length wire's internal inductance L_i decreases with ω, and the angle ϕ increases with ω, from 0 to $\pi/4$.

In the same way $\bar{D}$ lags behind $\bar{E}$ in a lossy dielectric medium, also, $\bar{I}$ lags behind $\bar{V}_S$ in a lossy conductor, phase lag depending on the working frequency.

So, the question is about the real meaning of writing $v_S(t) = Ri(t) + L_i di(t)/dt$ for a unit-length wire, knowing that both parameters R and L_i depend on the time-dynamics of $i(t)$. Such a time-domain equation can only apply to single-frequency regimes or to stationary regimes, in no other case [23].

In fact, (16.28) can be derived from the general result in (16.24) by truncating the summation at its first-order term,

$$\sum_{n=0}^{1} \mathbf{B}_n \frac{\partial^n}{\partial t^n} \mathbf{i}(z,t) = \mathbf{B}_0 \mathbf{i}(z,t) + \mathbf{B}_1 \frac{\partial}{\partial t} \mathbf{i}(z,t), \quad \begin{cases} \mathbf{B}_0 = (\mathbf{R})_{DC} \\ \mathbf{B}_1 = (\mathbf{L}_i)_{DC} \end{cases} \tag{16.30}$$

We wish to emphasize that the results presented here, regarding $\partial\mathbf{v}/\partial z$, hold true no matter the MTL insulation media are considered lossless or lossy.

16.2.4 MTL conservation laws

In the preceding sections, we made use of Maxwell's *curl* equations to establish the rate of evolution in space ($\partial/\partial z$) of MTL currents and voltages. We did that by applying Gauss and Stokes theorems to an MTL of infinitesimal length $\Delta z \to 0$.

Here we utilize the same tools, but apply them to the entire length of the MTL for the purpose of relating the time-integrals ($\int dt$) of MTL currents and voltages observed at both ends of the structure, at $z = 0$ and $z = \ell$.

We exemplify the problem with a simple lossless but inhomogeneous MTL defined by a coaxial cable running above a ground plane at height h—see Figure 16.5. The inhomogeneous character results from the presence of dielectric media with different permittivity, ε inside the cable, ε_0 outside. The MTL system comprises three conductors: the cable inner conductor (1) of radius r_1, the cable sheath (2) of radius r_2, and the ground plane (0).

At the far end of the line $z = \ell$, the cable is terminated by its matched load R_w and the cable sheath is bonded to ground. The cable inner wire is driven at $z = 0$ by a voltage generator that produces a single rectangular pulse v_{10} of duration T and magnitude V. Hence, boundary conditions are defined by:

$$v_1(0,t) = v_{10}$$

$$= \begin{cases} V, \text{ for } t \in [0,\ T] \\ 0, \text{ for } t < 0 \cup t > T \end{cases}, \begin{cases} i_2(0,t) = i_{20} = 0 \\ v_2(\ell,t) = v_{2\ell} = 0 \end{cases}, \begin{cases} i_1(\ell,t) = i_{1\ell} \\ v_1(\ell,t) = v_{1\ell} = R_w i_{1\ell} \end{cases} \tag{16.31}$$

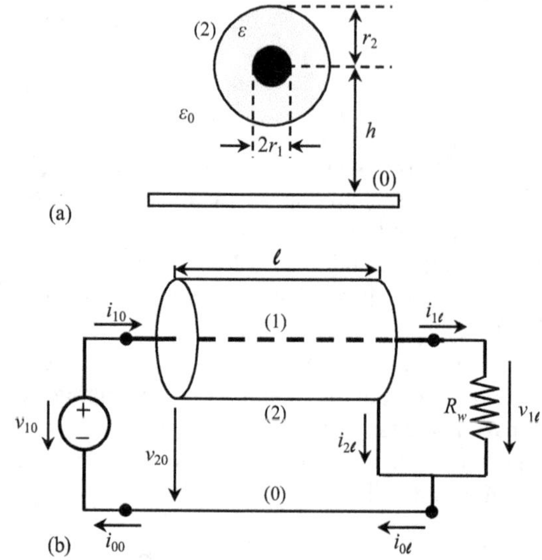

Figure 16.5 MTL system used for illustrating the conservation laws regarding the time-integrals of MTL voltages and currents. (a) Cross section. (b) Longitudinal view

Time-domain voltages and currents of the lossless MTL structure in Figure 16.5 can be easily found using ordinary methods [13], like those due to Bergeron [24], and Branin [25].

For the geometries being considered, coaxial and cylinder/plane, simple closed form expressions exist for the per-unit-length parameters (inductance, capacitance) and wave parameters (velocity, characteristic resistance) [16–19,26],

$$\begin{matrix} \text{Coaxial} & & \text{Cylinder/plane} \\ \left\{ \begin{array}{l} L = \mu_0 g_c, \;\; C = \varepsilon / g_c \\ g_c = \frac{1}{2\pi} \ln(r_2/r_1) \\ v = (\varepsilon \mu_0)^{-1/2} \\ R_w = g_c (\mu_0/\varepsilon)^{1/2} \end{array} \right\} & , & \left\{ \begin{array}{l} L_0 = \mu_0 g_0, \;\; C_0 = \varepsilon_0 / g_0 \\ g_0 = \frac{1}{2\pi} \text{arccosh}\,(h/r_2) \\ v_0 = c = (\varepsilon_0 \mu_0)^{-1/2} \\ R_{w0} = g_0 (\mu_0/\varepsilon_0)^{1/2} \end{array} \right\} \end{matrix} \tag{16.32}$$

from which the 2×2 symmetric **L** and **C** matrices are obtained

$$\mathbf{L} = \begin{bmatrix} L_0 + L & L_0 \\ L_0 & L_0 \end{bmatrix}, \quad \mathbf{C} = \begin{bmatrix} C & -C \\ -C & C + C_0 \end{bmatrix} \tag{16.33}$$

The product matrices **LC** and **CL** share the same pair of eigenvalues equal to $(\mu_0\varepsilon)$ and $(\mu_0\varepsilon_0)$, i.e.

$$\mathbf{T}^{-1}\mathbf{LCT} = \mathbf{W}^{-1}\mathbf{CLW} = \mu_0\begin{bmatrix} \varepsilon & \\ & \varepsilon_0 \end{bmatrix}, \quad \mathbf{T} = \begin{bmatrix} 1 & 1 \\ 0 & 1 \end{bmatrix}, \quad \mathbf{W} = \begin{bmatrix} 1 & 0 \\ -1 & 1 \end{bmatrix} \tag{16.34}$$

where **T** and **W** (with $\mathbf{T}^{-1} = \mathbf{W}^T$) are the similarity transformation matrices that, respectively, bring **LC** and **CL** into diagonal form; the columns of **T** and **W** (the eigenvectors) determining, respectively, the distribution of voltages and currents among the MTL wires at any transverse plane, for each of the propagation modes [12,13].

The result in (16.34) shows that the analysis of the structure can be conducted using two very simple independent modes: the cable internal mode, with $v_1 \neq 0,\ v_2 = 0,\ i_1 = -i_2$, and with wave velocity $v_w = c(\varepsilon_0/\varepsilon)^{1/2}$; and the external aerial mode, with $v_1 = v_2,\ i_1 = 0,\ i_2 \neq 0$, and with wave velocity $v_{w0} = c$. This mode decomposition should be no surprise; it is a consequence of the screening action of the cable sheath.

Because of MTL boundary conditions in (16.31) the internal mode is load matched and, therefore, cable waves reaching $z = \ell$ are absorbed. However, in general, the external mode is mismatched at both ends of the line and the aerial-wave will suffer successive reflections; the forward propagating voltage wave arriving at $z = \ell$ observes a short-circuit and is reflected back multiplied by a reflection factor of -1; the backward propagating voltage wave arriving at $z = 0$ observes the cable entrance and is reflected back, multiplied by a reflection factor Γ given by

$$\Gamma = \frac{\xi - 1}{\xi + 1}, \quad \xi = \frac{R_w}{R_{w0}} = \frac{g_c}{g_0}\sqrt{\frac{\varepsilon_0}{\varepsilon}} \tag{16.35}$$

For illustration purposes, the following numerical values are employed: $\varepsilon = 4\varepsilon_0$ and $g_c = 2g_0$. With this peculiar set of data, we get $v = c/2$ and $R_{w0} = R_w$, meaning that the backward aerial-wave is matched[§] at $z = 0$. Further, we assume that the pulse duration T is twice the aerial-wave travel time $T = 2\tau_0$, with $\tau_0 = \ell/c$.

For the above data, wire voltages and wire currents at the input port $z = 0$, and at the output port $z = \ell$, are calculated in (16.36), where $I = V/R_w$. Results are graphically illustrated in Figure 16.6:

$$\text{Wire 1}\begin{cases} \text{Input}\begin{cases} v_{10}(t) = \begin{cases} V, & 0 < t < 2\tau_0 \\ 0, & \text{otherwise} \end{cases} \\ i_{10}(t) = \begin{cases} \frac{1}{2}I, & 0 < t < 4\tau_0 \\ 0, & \text{otherwise} \end{cases} \end{cases} \\ \text{Output}\begin{cases} v_{1\ell}(t) = \begin{cases} \frac{1}{2}V, & 2\tau_0 < t < 6\tau_0 \\ 0, & \text{otherwise} \end{cases} \\ i_{1\ell}(t) = \begin{cases} \frac{1}{2}I, & 2\tau_0 < t < 6\tau_0 \\ 0, & \text{otherwise} \end{cases} \end{cases} \end{cases} \tag{16.36a}$$

[§]The interested reader can find in [26] other examples where $\Gamma \neq 0$.

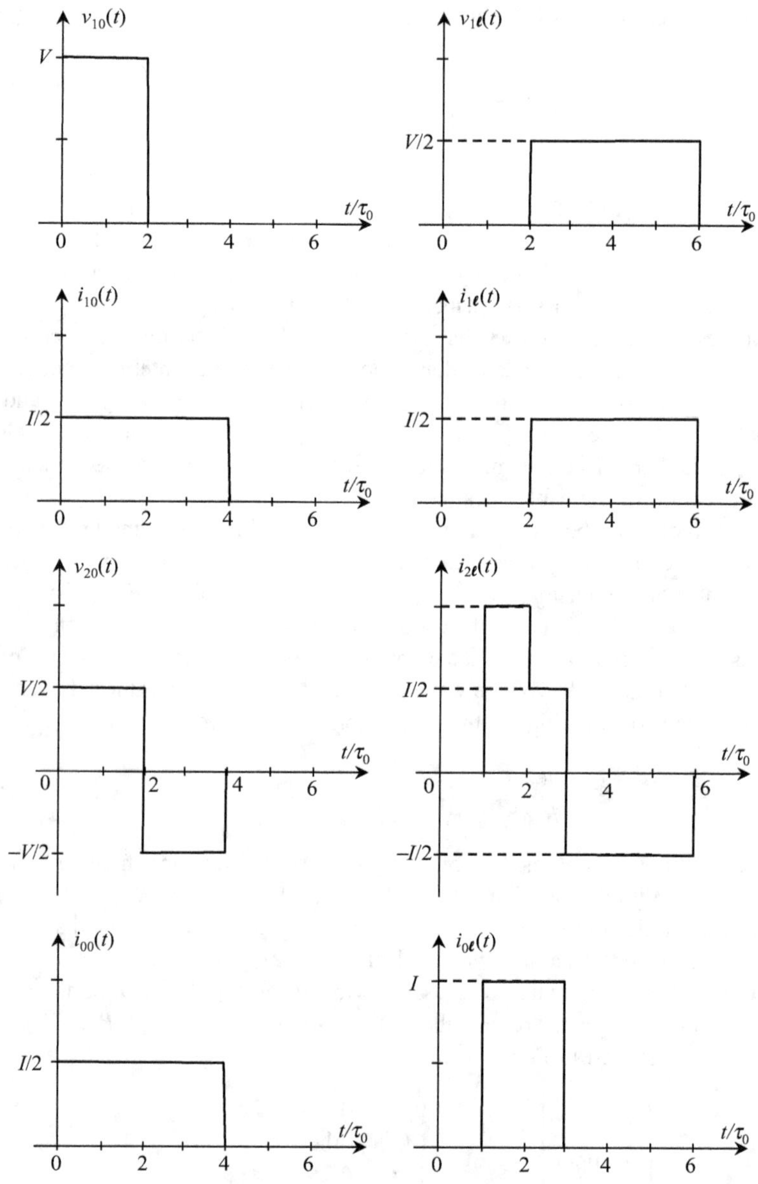

Figure 16.6 Time-domain wire voltages and wire currents regarding the MTL structure in Figure 16.5. Note that the areas on the left (for $z = 0$) are equal to the areas on the right (for $z = \ell$).

$$\text{Wire 2}\begin{cases}\text{Input}\begin{cases}v_{20}(t)=\begin{cases}+\frac{1}{2}V, & 0<t<2\tau_0\\ -\frac{1}{2}V, & 2\tau_0<t<4\tau_0\\ 0, & \text{otherwise}\end{cases}\\ i_{20}(t)=0\end{cases}\\ \text{Output}\begin{cases}v_{2\ell}(t)=0\\ i_{2\ell}(t)=\begin{cases}+I, & \tau_0<t<2\tau_0\\ +\frac{1}{2}I, & 2\tau_0<t<3\tau_0\\ -\frac{1}{2}I, & 3\tau_0<t<6\tau_0\\ 0, & \text{otherwise}\end{cases}\end{cases}\end{cases} \tag{16.36b}$$

$$\text{Wire 0}\begin{cases}\text{Input}: \; i_{00}(t)=\begin{cases}\frac{1}{2}I, & 0<t<4\tau_0\\ 0, & \text{otherwise}\end{cases}\\ \text{Output}: i_{0\ell}(t)=\begin{cases}I, & \tau_0<t<3\tau_0\\ 0, & \text{otherwise}\end{cases}\end{cases} \tag{16.36c}$$

Now, from (16.36), if we compute the time-integrals of the MTL voltages and MTL currents at the distinct ports located in $z = 0$ and $z = \ell$, we find that the integrals remain invariant:

$$\text{Wire 1}\begin{cases}\text{Voltages}\begin{cases}\int_{-\infty}^{\infty} v_{10}(t)dt=\int_{0}^{2\tau_0} v_{10}(t)dt=VT\\ \int_{-\infty}^{\infty} v_{1\ell}(t)dt=\int_{2\tau_0}^{6\tau_0} v_{1\ell}(t)dt=VT\end{cases}\\ \text{Currents}\begin{cases}\int_{-\infty}^{\infty} i_{10}(t)dt=\int_{0}^{4\tau_0} i_{10}(t)dt=IT\\ \int_{-\infty}^{\infty} i_{1\ell}(t)dt=\int_{2\tau_0}^{6\tau_0} i_{1\ell}(t)dt=IT\end{cases}\end{cases} \tag{16.37a}$$

$$\text{Wire 2}\begin{cases}\text{Voltages}: \; \int_{-\infty}^{\infty} v_{20}(t)dt=\int_{0}^{4\tau_0} v_{20}(t)dt=\int_{-\infty}^{\infty} \overbrace{v_{2\ell}(t)}^{0}\,dt=0\\ \text{Currents}: \; \int_{-\infty}^{\infty} \underbrace{i_{20}(t)}_{0}\,dt=\int_{-\infty}^{\infty} i_{2\ell}(t)dt=\int_{\tau_0}^{6\tau_0} i_{2\ell}(t)dt=0\end{cases} \tag{16.37b}$$

$$\text{Wire 0}: \quad \text{Currents}\begin{cases}\int_{-\infty}^{\infty} i_{00}(t)dt=\int_{0}^{4\tau_0} i_{00}(t)dt=IT\\ \int_{-\infty}^{\infty} i_{0\ell}(t)dt=\int_{\tau_0}^{3\tau_0} i_{0\ell}(t)dt=IT\end{cases} \tag{16.37c}$$

The preceding results are summarized and generalized to MTLs with $N+1$ wires in (16.38)

$$\text{For } k = 1 \text{ to } N: \quad \begin{cases} \int_{-\infty}^{\infty} i_{k0}(t)dt = \int_{-\infty}^{\infty} i_{k\ell}(t)dt \\ \int_{-\infty}^{\infty} v_{k0}(t)dt = \int_{-\infty}^{\infty} v_{k\ell}(t)dt \end{cases} \tag{16.38}$$

The results in (16.36)–(16.38) were found by considering the example of a completely lossless MTL structure with perfect conductors and perfect dielectrics. But the validity of (16.38) is not so strict.

The identity for currents in (16.38) is a consequence of Maxwell's *curl* **H** equation and can be derived the same way we did with the governing equation of $\partial \mathbf{i}/\partial z$, but considering not an infinitesimal line section of length Δz, but the whole line length from $z = 0$ to ℓ. Like in (16.13), and Figure 16.3, we write

$$i_k(z = \ell, t) - i_k(z = 0, t) = -\int_{S_{lat}} \left(\mathbf{J} \cdot \mathbf{n}_o + \frac{\partial \mathbf{D} \cdot \mathbf{n}_o}{\partial t} \right) dS \tag{16.39}$$

where the right-hand side refers to the flux of the transverse total current density through the lateral surface of the kth wire of length ℓ. If the MTL dielectric media are lossless, then, from (16.39),

$$i_{k0}(t) - i_{k\ell}(t) = \frac{d}{dt} q_k(t), \qquad q_k = q_k(v_1, \quad \cdots, \quad v_N) \tag{16.40}$$

where q_k is the total electric charge distributed along the length ℓ of the kth wire due to all of the MTL voltages, these voltages being, naturally, considered to be time-limited, they start and end in time, that is $v_k(z, t = \pm\infty) = 0$.

Note that (16.40) remains valid even if the distributed charge q_k also includes any discrete contributions due to the possible existence of lumped capacitors shunt-connected between MTL wires.

Finally, time integration of (16.40) gives

$$\int_{-\infty}^{\infty} i_{k0}(t)dt - \int_{-\infty}^{\infty} i_{k\ell}(t)dt = (q_k)_{t=\infty} - (q_k)_{t=-\infty} = 0$$

$$\rightarrow \int_{-\infty}^{\infty} i_{k0}(t)dt = \int_{-\infty}^{\infty} i_{k\ell}(t)dt \tag{16.41}$$

thereby proving the conservation law of the time-integral of MTL currents in (16.38). As a corollary to (16.41) it follows that if the kth wire is interrupted by an open-circuit at a given place in the MTL ($i_k = 0$), then the time-integral of the wire's current at any other place will also be zero.

At this stage we underline that the sole simplifying assumption behind (16.41) is the absence of transversal losses (perfect dielectric media, and ideal shunt capacitors if these are added to the MTL structure). Longitudinal losses are not

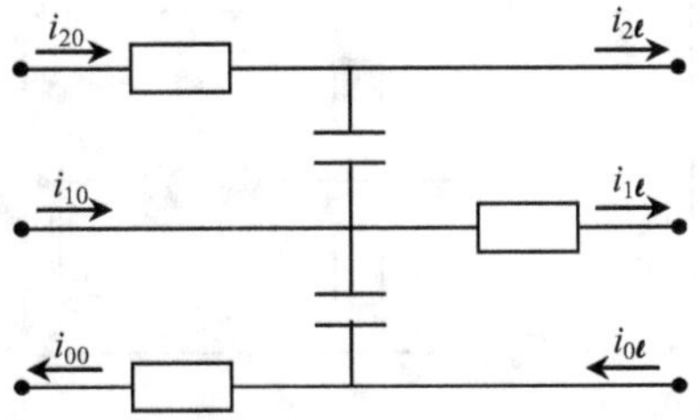

Figure 16.7 Example of a MTL structure with longitudinal losses (wire losses included) where the conservation law for currents in (16.38) still holds true

precluded; losses in MTL wires are not a limiting factor of the conservation law in (16.41). Also, lumped devices, linear or not, can be series-connected with any wire at discrete places along the MTL [27]—see Figure 16.7.

The identity for voltages in (16.38) is a consequence of Maxwell's *curl* **E** equation and can be derived the same way we did with the governing equation of $\partial \mathbf{v}/\partial z$, but considering not an infinitesimal line section of length Δz, but the whole line length from $z = 0$ to ℓ. Like in (16.21) and (16.22), and Figure 16.4, we write

$$\underbrace{\int_{ab} \mathbf{E} \cdot d\mathbf{s}_k}_{-v_k(z=0,t)} + \int_{bc} \mathbf{E} \cdot d\mathbf{s}_k + \underbrace{\int_{cd} \mathbf{E} \cdot d\mathbf{s}_k}_{v_k(z=\ell,t)} + \int_{da} \mathbf{E} \cdot d\mathbf{s}_k = -\frac{d}{dt}\psi_k(t) \qquad (16.42)$$

where the right-hand side refers to the time-derivative of the total magnetic flux linked with the *abcda* closed path of length ℓ that includes the kth and 0th wires.

If the MTL wires are lossless, then, in the left-hand side, the *bc* and *da* voltages are zero, and we have

$$v_{k0}(t) - v_{k\ell}(t) = \frac{d}{dt}\psi_k(t), \qquad \psi_k = \psi_k(i_1, \ \cdots, \ i_N) \qquad (16.43)$$

where the magnetic flux Ψ_k is due to all of the MTL currents, which, naturally, are considered to be time-limited, they start and end in time, that is $i_k(z, t = \pm\infty) = 0$.

Note that (16.43) remains valid even if the total distributed magnetic flux Ψ_k also includes discrete contributions due to the possible existence of lumped inductors series-connected with any MTL wire, along the MTL.

Finally, time integration of (16.43) gives

$$\int_{-\infty}^{\infty} v_{k0}(t)dt - \int_{-\infty}^{\infty} v_{k\ell}(t)dt = (\psi_k)_{t=\infty} - (\psi_k)_{t=-\infty} = 0$$

$$\rightarrow \int_{-\infty}^{\infty} v_{k0}(t)dt = \int_{-\infty}^{\infty} v_{k\ell}(t)dt \qquad (16.44)$$

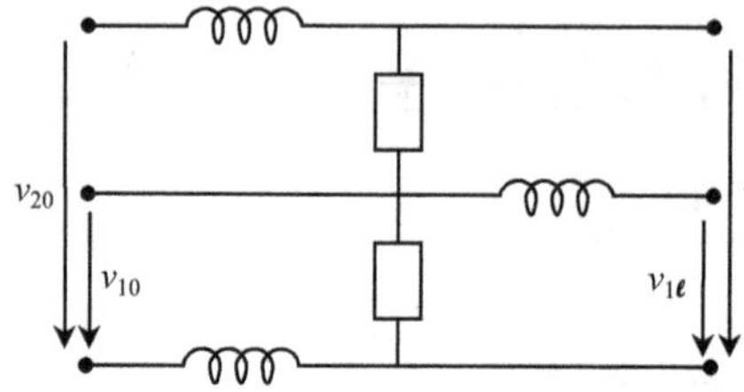

Figure 16.8 Example of a MTL structure with transversal losses (dielectric losses included) where the conservation law for voltages in (16.38) still holds true

thereby proving the conservation law of the time-integral of MTL voltages in (16.38). As a corollary to (16.44) it follows that if the kth wire is short-circuited to ground at a given place in the MTL ($v_k = 0$), then the time-integral of the wire's voltage at any other place will also be zero.

At this stage, we underline that the sole simplifying assumption behind (16.44) is the absence of longitudinal losses (perfect MTL wires, and ideal series inductors if these are added to the MTL structure). Transversal losses are not precluded; losses in the MTL dielectric media are not a limiting factor of the conservation law in (16.44). Also, lumped devices, linear or not, can be shunt-connected between wires at discrete places along the MTL [27]—see Figure 16.8.

16.3 Skin effect in the time domain

In this section, we look back into (16.22), paying attention to the contributing terms due to wires' finite conductivity, namely, the per-unit-length surface voltage v_S appearing in the calculation of the longitudinal voltage drop ($v_S \Delta z$) in the kth wire of the MTL, as accounted by the line integral bc in (16.22). Our aim is to show how a full time-domain relationship between $v_S(t)$ and $i(t)$ can be established.

Skin effect phenomena in current-carrying wires is one of the most important issues in MTL analysis, they are the main responsible for the attenuation of travelling waves. The analysis of skin effect in the frequency domain poses no special difficulties, but pure time domain approaches are very rare; contributions to time-domain skin effect theory are due to Maxwell [13], Giacoleto [28], Faria and Raven [29,30], for example.

A full time-domain analysis of solid circular cylindrical wires is presented, excluding proximity effects. We keep considering that the wire is a very good conductor (displacement current disregarded) and that it carries a longitudinal z-directed current density $\mathbf{J}$. The wire, with radius a, is characterized by conductivity σ and magnetic permeability μ. To ensure that axial field variations are negligibly small, all the calculations refer to a wire's section with infinitesimal unit length. The longitudinal **E**-field and the azimuthal **H**-field are assumed to vary radially,

$$\begin{cases} \mathbf{J}(r,t) = \sigma\mathbf{E}(r,t) = \sigma E(r,t)\hat{\mathbf{z}} \\ \mathbf{B}(r,t) = \mu\mathbf{H}(r,t) = \mu H(r,t)\hat{\boldsymbol{\theta}} \end{cases} \tag{16.45}$$

therefore, in a cylindrical reference frame, from Maxwell's *curl* equations in (16.1) and (16.3), we write

$$\text{For } 0 \le r \le a : \begin{cases} \dfrac{\partial H(r,t)}{\partial r} + \dfrac{H(r,t)}{r} = \sigma E(r,t) \\ \dfrac{\partial E(r,t)}{\partial r} = \mu \dfrac{\partial H(r,t)}{\partial t} \end{cases} \tag{16.46}$$

In the analysis that follows we assume that, for $r \le a$, both fields H and E can be described by convergent radial expansions with time-dependent coefficients $h_n(t)$ and $e_n(t)$, respectively.

Since $H(r) = -H(-r)$, the radial expansion of H includes only odd powers of r; however, because $E(r) = E(-r)$, the radial expansion of E only includes even powers of r, that is,

$$\begin{cases} E(r,t) = \displaystyle\sum_{n=0}^{m} e_{2n}(t) \left(\frac{r}{a}\right)^{2n} \\ H(r,t) = \displaystyle\sum_{n=0}^{m} h_{2n+1}(t) \left(\frac{r}{a}\right)^{2n+1} \end{cases} \tag{16.47}$$

where m is the truncation order of the summations.

Denoting by $i(t)$ the time-varying wire's current intensity, and denoting by $v_S(t)$ the per-unit-length time-varying voltage along the surface of the wire, we see, from (16.47), making $r = a$, that the time-dependent coefficients $h_n(t)$ and $e_n(t)$ are such that

$$\begin{cases} \displaystyle\sum_{n=0}^{m} e_{2n}(t) = E(a,t) = v_S(t) \\ \displaystyle\sum_{n=0}^{m} h_{2n+1}(t) = H(a,t) = \frac{i(t)}{2\pi a} \end{cases} \tag{16.48}$$

Substituting (16.47) into (16.46) yields

$$e_{2n} = \frac{\mu a}{2n} h'_{2n-1} = \frac{2(n+1)}{\sigma a} h_{2n+1} \tag{16.49}$$

where $h'_k = dh_k/dt$.

The time-dependent coefficients e_{2n} and h_{2n+1} are determined recursively with the help of (16.50)

$$e_0 \to h_1 \to h'_1 \to e_2 \to h_3 \quad \cdots \quad e_{2n} \to h_{2n+1} \to h'_{2n+1} \to e_{2n+2} \to h_{2n+3} \quad \cdots \tag{16.50}$$

The set of functions e_{2n} and h_{2n+1} can be expressed in terms of e_0 and all its time-derivatives $e_0^{(n)} = d^n e_0/dt^n$. For example, for $n = 0, 1, 2$, we find

$$n = 0 \begin{cases} e_0 \\ h_1 = \dfrac{\sigma a e_0}{2} \rightarrow h_1' = \dfrac{\sigma a e_0^{(1)}}{2} \end{cases} \tag{16.51a}$$

$$n = 1 \begin{cases} e_2 = \dfrac{\mu a h_1'}{2} = \dfrac{(a^2\sigma\mu)}{2^2} e_0^{(1)} \\ h_3 = \dfrac{\sigma a e_2}{4} = \dfrac{\sigma}{4}\dfrac{\sigma\mu}{2^2} e_0^{(1)} \rightarrow h_3' = \dfrac{\sigma a}{4}\dfrac{(a^2\sigma\mu)}{2^2} e_0^{(2)} \end{cases} \tag{16.51b}$$

$$n = 2 \begin{cases} e_4 = \dfrac{\mu a h_3'}{4} = \dfrac{(a^2\sigma\mu)^2}{4^2 2^2} e_0^{(2)} \\ h_5 = \dfrac{\sigma a e_4}{6} = \dfrac{\sigma a}{6}\dfrac{(a^2\sigma\mu)^2}{4^2 2^2} e_0^{(2)} \rightarrow h_5' = \dfrac{\sigma a}{6}\dfrac{(a^2\sigma\mu)^2}{4^2 2^2} e_0^{(3)} \end{cases} \tag{16.51c}$$

In general, for every n, the following results apply

$$\begin{cases} e_{2n} = \left(\dfrac{a^2\sigma\mu}{4}\right)^n \dfrac{e_0^{(n)}}{n!n!} = (n+1) f_n \tau^n e_0^{(n)} \\ h_{2n+1} = \dfrac{\sigma a}{2}\left(\dfrac{a^2\sigma\mu}{4}\right)^n \dfrac{e_0^{(n)}}{n!(n+1)!} = \dfrac{\sigma a}{2} f_n \tau^n e_0^{(n)} \end{cases} \tag{16.52}$$

where

$$\begin{cases} f_n = \dfrac{1}{n!(n+1)!} \\ \tau = \sigma\mu\left(\dfrac{a}{2}\right)^2 = \dfrac{2L_{dc}}{R_{dc}}, \quad \text{with} \begin{cases} L_{dc} = \mu/(8\pi) \\ R_{dc} = 1/(\pi\sigma a^2) \end{cases} \end{cases} \tag{16.53}$$

where τ is a characteristic time constant of the wire, L_{dc} and R_{dc} denoting, respectively, the per-unit-length internal inductance and resistance of the wire for time-invariant regimes.

For compactness, the normalized nth time-derivative of $e_0(t)$ is introduced, identified by a tilde mark,

$$\tilde{e}_0^{(n)} = \tau^n e_0^{(n)} = \tau^n \frac{d^n e_0(t)}{dt^n} \tag{16.54}$$

Substituting (16.52)–(16.54) into (16.47), gives the time-domain field solutions

$$\begin{cases} E(r,t) = \displaystyle\sum_{n=0}^{m} (n+1) f_n \, \tilde{e}_0^{(n)} \left(\frac{r}{a}\right)^{2n} \\ H(r,t) = \dfrac{\sigma r}{2} \displaystyle\sum_{n=0}^{m} f_n \, \tilde{e}_0^{(n)} \left(\frac{r}{a}\right)^{2n} \end{cases} \tag{16.55}$$

These field solutions require that the function $e_0(t)$ and all of its derivatives be known. To find $e_0(t)$, we start by enforcing the boundary condition of the magnetic field at $r = a$,

$$i(t) = \oint_{r=a} H ds = 2\pi a H(a,t) \tag{16.56}$$

then, from (16.55) and (16.56),

$$i(t) = \sigma\pi a^2 \sum_{n=0}^{m} f_n\, \widetilde{e}_0^{(n)} \quad \rightarrow \quad \sum_{n=0}^{m} f_n\, \widetilde{e}_0^{(n)} = R_{dc} i(t) \tag{16.57}$$

Successive time differentiation of (16.57), from $k = 0$ to m, yields

$$\sum_{n=0}^{m-k} f_n\ \widetilde{e}_0^{(n+k)} = R_{dc}\ \widetilde{i}^{(k)} \tag{16.58}$$

where, from (16.54), account has been taken of $d^k \widetilde{e}_0^{(n)}/dt^k = \widetilde{e}_0^{(n+k)}/\tau^k$.

The quantity $\widetilde{i}^{(k)}$ in (16.58) is the normalized kth time-derivative of $i(t)$,

$$\widetilde{i}^{(k)} = \tau^k \frac{d^k i(t)}{dt^k} = \tau^k i^{(k)} \tag{16.59}$$

The result in (16.58) describes a set of $m+1$ algebraic equations, which can be written in matrix notation:

$$\begin{bmatrix} \widetilde{e}_0^{(0)} \\ \widetilde{e}_0^{(1)} \\ \widetilde{e}_0^{(2)} \\ \vdots \\ \widetilde{e}_0^{(m)} \end{bmatrix} = R_{dc} \begin{bmatrix} 1 & f_1 & f_2 & \cdots & f_m \\ & 1 & f_1 & \ddots & \vdots \\ & & 1 & \ddots & f_2 \\ & & & \ddots & f_1 \\ & & & & 1 \end{bmatrix}^{-1} \begin{bmatrix} \widetilde{i}^{(0)} \\ \widetilde{i}^{(1)} \\ \widetilde{i}^{(2)} \\ \vdots \\ \widetilde{i}^{(m)} \end{bmatrix} = R_{dc} \begin{bmatrix} 1 & a_1 & a_2 & \cdots & a_m \\ & 1 & a_1 & \ddots & \vdots \\ & & 1 & \ddots & a_2 \\ & & & \ddots & a_1 \\ & & & & 1 \end{bmatrix} \begin{bmatrix} \widetilde{i}^{(0)} \\ \widetilde{i}^{(1)} \\ \widetilde{i}^{(2)} \\ \vdots \\ \widetilde{i}^{(m)} \end{bmatrix} \tag{16.60}$$

The determinant of the upper-triangular matrix with entries f_k is equal to 1. The inverse matrix exists and is also upper-triangular, whose entries a_k can be determined recursively through

$$a_k = -\sum_{n=1}^{k} f_n\ a_{k-n} = -\frac{a_{k-1}}{1!2!} - \frac{a_{k-2}}{2!3!} - \cdots - \frac{a_1}{(k-1)!k!} - \frac{1}{k!(k+1)!} \tag{16.61}$$

where the first a_k coefficients are $a_0 = 1,\ a_1 = -1/2,\ a_2 = 1/6,\ a_3 = -7/144$.

From (16.60), the generating function $\widetilde{e}_0(t)$ and all its derivatives can be evaluated

$$\widetilde{e}_0^{(n)} = R_{dc} \sum_{k=0}^{m-n} a_k \, \widetilde{i}^{(n+k)} \tag{16.62}$$

Substituting the result for $\widetilde{e}_0^{(n)}$ back into (16.55) allows a complete description of the time dynamics of the electromagnetic field inside the wire, based on the knowledge of $i(t)$.

The step that follows is the evaluation of the per-unit-length voltage drop $v_S(t)$ measured longitudinally along the surface of the wire, ultimately showing how it dependents on $i(t)$. From (16.55), (16.62), and (16.59), we obtain

$$v_S(t) = E(a,t) = \sum_{n=0}^{m} (n+1) f_n \, \widetilde{e}_0^{(n)} = \sum_{n=0}^{m} \frac{\widetilde{e}_0^{(n)}}{n!n!} = R_{dc} \sum_{n=0}^{m} u_n \, \widetilde{i}^{(n)} \tag{16.63}$$

where the dimensionless u_n coefficients are evaluated through

$$u_n = \sum_{k=0}^{n} \frac{a_k}{((n-k)!)^2} = a_n + \frac{a_{n-1}}{1!1!} + \frac{a_{n-2}}{2!2!} + \cdots + \frac{a_1}{(n-1)!(n-1)!} + \frac{1}{n!n!} \tag{16.64}$$

The first u_n coefficients are: $u_0 = 1,\ u_1 = 1/2,\ u_2 = -1/12,\ u_3 = 1/48$. Values of u_n, up to $n = 12$, are presented in Table 16.1.

Finally, noting that $L_{dc} = R_{dc} u_1 \tau$, the result in (16.63) can be put in form

$$v_S(t) = R_{dc} i(t) + L_{dc} \frac{di(t)}{dt} + \left(R_{dc} \sum_{n=2}^{m} u_n \tau^n \frac{d^n i(t)}{dt^n} \right) \tag{16.65}$$

showing how the per-unit-length time-domain voltage along the wire depends on the time-domain current in the wire.

Table 16.1 Coefficients u_n for $n = 1$ to 12

n	u_n	n	u_n
1	$+5.0000 \times 10^{-1}$	7	$+1.1143 \times 10^{-4}$
2	-8.3333×10^{-2}	8	-3.0355×10^{-5}
3	$+2.0833 \times 10^{-2}$	9	$+8.2697 \times 10^{-6}$
4	-5.5556×10^{-3}	10	-2.2530×10^{-6}
5	$+1.5046 \times 10^{-3}$	11	$+6.1381 \times 10^{-7}$
6	-4.0923×10^{-4}	12	-1.6723×10^{-7}

Strangely the interesting result in (16.65) is rarely seen or utilized in skin effect studies. There are two main reasons for that. On the one hand, the function $i(t)$ must be infinitely differentiable and, on the other hand, the summation on n must be convergent. We address the question considering the simple case of a sinusoidal regime, where

$$\begin{cases} i(t) = I\cos(\omega t) = \mathrm{Re}\{Ie^{j\omega t}\} \\ v_S(t) = Ri + L\dfrac{di}{dt} = \mathrm{Re}\{(R + j\omega L)Ie^{j\omega t}\} = R(\omega)I\cos(\omega t) - \omega L(\omega)I\sin(\omega t) \end{cases} \tag{16.66}$$

where $R = R(\omega)$ and $L = L(\omega)$ represent the per-unit-length skin effect resistance and internal inductance of the wire at frequency ω.

The expression for $v_S(t)$ in (16.65) requires the nth derivative of $i(t)$ be known. For the sinusoidal case, with $i = I\cos(\omega t)$, we have

$$\frac{d^n i(t)}{dt^n} = (-\omega^2)^k \times \begin{cases} I\cos(\omega t) = i(t), & \text{for } n = 2k \text{ (even)} \\ -\omega I\sin(\omega t) = \dfrac{di(t)}{dt}, & \text{for } n = 2k+1 \text{ (odd)} \end{cases} \tag{16.67}$$

Substituting (16.67) into (16.65), yields the following convenient result:

$$v_S(t) = \overbrace{\left(\sum_{k=0}^{m} R_k\right)}^{R(\omega)} i(t) + \overbrace{\left(\sum_{k=0}^{m} L_k\right)}^{L(\omega)} \frac{di(t)}{dt} \tag{16.68}$$

with

$$R_k = R_{dc}u_{2k}(-(\omega\tau)^2)^k, \quad L_k = 2L_{dc}u_{2k+1}(-(\omega\tau)^2)^k \tag{16.69}$$

Successful convergence of the R and L summations in (16.68) is achieved if

$$\lim_{k\to\infty}\left|\frac{R_{k+1}}{R_k}\right| < 1, \quad \lim_{k\to\infty}\left|\frac{L_{k+1}}{L_k}\right| < 1 \quad \rightarrow \quad \omega\tau < \lim_{n\to\infty}\sqrt{\frac{u_n}{u_{n+2}}} \tag{16.70a}$$

but, from (16.64), or Table 16.1, it can be shown that:

$$\lim_{n\to\infty}\sqrt{\frac{u_n}{u_{n+2}}} \approx 3.67 \tag{16.70b}$$

Bearing in mind that the skin effect penetration depth δ [18] is related to the characteristic time constant τ in (16.53) through $(a/\delta)^2 = 2\omega\tau$, one may see that the convergence condition in (16.70) can be rewritten as $a/\delta < 2.71 \approx e$.

In short, the R and L summations involved in the computation of $v_S(t)$ in (16.68) fail to converge even when skin effect is a phenomenon of modest importance. For strong skin effect regimes—as it is the case for many MTL applications—the constraint $\delta > a/e$ is a severe limitation to the applicability of (16.65).

16.4 Frequency-domain MTL equations

In view of the preceding analysis concerning time-domain skin effect phenomena, our conclusion is that the conditions required for a full time-domain theory of MTL structures, comprising lossy wires, are very restrictive and, therefore, we believe that, at present, the standard frequency-domain approach remains the best tool for dealing with general MTL problems.

Enforced MTL excitations in the time-domain are Fourier converted to the frequency-domain. Frequency-domain MTL complex equations, accounting for wire and dielectric losses, are solved in the frequency-domain yielding the spectral solution for all the MTL voltages and currents. The complex spectral results are inverse Fourier converted to the time-domain, producing the sought MTL time-domain response to the time-domain excitation. The unique constraint to this procedure is that the MTL structure cannot include component parts behaving non-linearly.

Here, we briefly review the frequency-domain MTL complex equations and its matrix modal solution [12,13,31].

The time-domain sinusoidal components of the MTL currents and voltages in (16.10), $\mathbf{i}(t)$ and $\mathbf{v}(t)$, are substituted by the corresponding time-independent complex phasors $\mathbf{I}(\omega)$ and $\mathbf{V}(\omega)$ using

$$\mathbf{i}(z,t) = \mathrm{Re}\left(\mathbf{I}(z)e^{j\omega t}\right), \quad \mathbf{v}(z,t) = \mathrm{Re}\left(\mathbf{V}(z)e^{j\omega t}\right) \tag{16.71}$$

The complex frequency-domain governing equations for $d\mathbf{I}/dz$ and $d\mathbf{V}/dz$ are obtained from (16.18) and (16.28), yielding

$$\frac{d}{dz}\mathbf{I}(z) = -\mathbf{Y}(\omega)\mathbf{V}(z), \quad \frac{d}{dz}\mathbf{V}(z) = -\mathbf{Z}(\omega)\mathbf{I}(z) \tag{16.72}$$

where $\mathbf{Y}$ and $\mathbf{Z}$ are complex square symmetric matrices whose real and imaginary parts are positive definite [12],

$$\mathbf{Y}(\omega) = \mathbf{G}(\omega) + j\omega\mathbf{C}(\omega), \quad \mathbf{Z}(\omega) = \mathbf{R}(\omega) + j\omega\mathbf{L}(\omega) \tag{16.73}$$

Matrix $\mathbf{Y}$ is the per-unit-length transverse admittance matrix of the MTL. It can account for the possible existence of dielectric losses (polarization and conduction). If losses are absent or negligible, we have $\mathbf{G} = \mathbf{0}$ and $\mathbf{C}$ is the frequency-independent electrostatic capacitance matrix.

Matrix $\mathbf{Z}$ is the per-unit-length longitudinal impedance matrix of the MTL. It can account for the possible existence of wire losses (skin and proximity effects). If losses are absent or negligible, we have $\mathbf{R} = \mathbf{0}$ and $\mathbf{L}$ is the frequency-independent external inductance matrix $\mathbf{L}_e$.

The coupled differential equations in (16.72) can be decoupled by introducing an appropriate change of variables

$$\mathbf{I} = \mathbf{W}\mathbf{I}_{\mathrm{m}} \text{ and } \mathbf{V} = \mathbf{T}\mathbf{V}_{\mathrm{m}} \tag{16.74}$$

where $\mathbf{I}_{\mathrm{m}}$ and $\mathbf{V}_{\mathrm{m}}$ are arrays for the so-called modal currents and modal voltages, respectively, [12].

The $N \times N$ matrices $\mathbf{W}$ and $\mathbf{T}$ in (16.74) are the modal transformations matrices, whose entries are, ordinarily, frequency dependent. Note that these transformation matrices have already appeared to us in (16.34) when the lossless MTL structure in Figure 16.5 was studied.

By substituting (16.74) into (16.72), we get

$$\frac{d}{dz}\mathbf{I}_{\mathrm{m}}(z) = -\underbrace{\left(\mathbf{W}^{-1}\mathbf{Y}\mathbf{T}\right)}_{\mathbf{Y}_{\mathrm{m}}}\ \mathbf{V}_{\mathrm{m}}(z),\quad \frac{d}{dz}\mathbf{V}_{\mathrm{m}}(z) = -\underbrace{\left(\mathbf{T}^{-1}\mathbf{Z}\mathbf{W}\right)}_{\mathbf{Z}_{\mathrm{m}}}\ \mathbf{V}_{\mathrm{m}}(z) \tag{16.75}$$

where $\mathbf{Y}_{\mathrm{m}}$ and $\mathbf{Z}_{\mathrm{m}}$ are the per-unit-length transverse admittance and longitudinal impedance matrices in modal coordinates, respectively.

To ensure that the equations in (16.75) are indeed decoupled, both $\mathbf{Y}_{\mathrm{m}}$ and $\mathbf{Z}_{\mathrm{m}}$ must be diagonal matrices, that is

$$\frac{d}{dz}\begin{bmatrix} I_{1\mathrm{m}} \\ \vdots \\ I_{k\mathrm{m}} \\ \vdots \\ I_{N\mathrm{m}} \end{bmatrix} = \begin{bmatrix} Y_{1\mathrm{m}} & & & & \\ & \ddots & & & \\ & & Y_{k\mathrm{m}} & & \\ & & & \ddots & \\ & & & & Y_{N\mathrm{m}} \end{bmatrix}\begin{bmatrix} V_{1\mathrm{m}} \\ \vdots \\ V_{k\mathrm{m}} \\ \vdots \\ V_{N\mathrm{m}} \end{bmatrix} \tag{16.76a}$$

$$\frac{d}{dz}\begin{bmatrix} V_{1\mathrm{m}} \\ \vdots \\ V_{k\mathrm{m}} \\ \vdots \\ V_{N\mathrm{m}} \end{bmatrix} = \begin{bmatrix} Z_{1\mathrm{m}} & & & & \\ & \ddots & & & \\ & & Z_{k\mathrm{m}} & & \\ & & & \ddots & \\ & & & & Z_{N\mathrm{m}} \end{bmatrix}\begin{bmatrix} I_{1\mathrm{m}} \\ \vdots \\ I_{k\mathrm{m}} \\ \vdots \\ I_{N\mathrm{m}} \end{bmatrix} \tag{16.76b}$$

or, which is the same,

$$\text{For } k = 1 \text{ to } N:\ \begin{cases} \dfrac{d}{dz} I_{k\mathrm{m}} = -Y_{k\mathrm{m}}\ V_{k\mathrm{m}} \\ \dfrac{d}{dz} V_{k\mathrm{m}} = -Z_{k\mathrm{m}}\ I_{k\mathrm{m}} \end{cases} \tag{16.76c}$$

The results in (16.76c), for mode-k, are the familiar two-wire transmission line equations, whose well-known solution can be written in the form of a superposition of incident and reflected waves:

$$\begin{cases} V_{k\mathrm{m}}(z) = (V_i)_k\ e^{-\gamma_k z} + (V_r)_k\ e^{+\gamma_k z} \\ I_{k\mathrm{m}}(z) = Z_{wk}^{-1}\left((V_i)_k\ e^{-\gamma_k z} - (V_r)_k\ e^{+\gamma_k z}\right) \end{cases} \text{ where } \begin{cases} \gamma_k = \sqrt{Z_{k\mathrm{m}} Y_{k\mathrm{m}}} \\ Z_{wk} = \sqrt{Z_{k\mathrm{m}}/Y_{k\mathrm{m}}} \end{cases} \tag{16.77}$$

or in the form of a transfer matrix (or ABCD matrix):

$$\begin{bmatrix} V_{km}(z) \\ I_{km}(z) \end{bmatrix} = \begin{bmatrix} A & B \\ C & D \end{bmatrix} \begin{bmatrix} V_{km}(0) \\ I_{km}(0) \end{bmatrix} \quad \text{where} \quad \begin{cases} A = \cosh(\gamma_k z) \\ B = -\sinh(\gamma_k z) Z_{wk} \\ C = -Z_{wk}^{-1} \sinh(\gamma_k z) \\ D = A \end{cases} \tag{16.78}$$

where γ_k and Z_{wk} denote the propagation constant and the characteristic wave impedance of the kth mode, respectively.

So far, the procedure described above is straightforward. However, we still have to deal with the problem of the determination of the modal transformation matrices **T** and W.

Taking into account the definition of the diagonal matrices $\mathbf{Y}_\mathrm{m}$ and $\mathbf{Z}_\mathrm{m}$ in (16.75), and bearing in mind that diagonal matrices commute, we obtain

$$\begin{cases} \mathbf{Z}_\mathrm{m}\mathbf{Y}_\mathrm{m} = \boldsymbol{\gamma}^2 = \mathbf{T}^{-1}\mathbf{ZYT} & \rightarrow \quad (\mathbf{ZY} - \mathbf{1}\gamma_k^2)\ \mathbf{t}_k = \mathbf{0} \\ \mathbf{Y}_\mathrm{m}\mathbf{Z}_\mathrm{m} = \boldsymbol{\gamma}^2 = \mathbf{W}^{-1}\mathbf{YZW} & \rightarrow \quad (\mathbf{YZ} - \mathbf{1}\gamma_k^2)\mathbf{w}_k = \mathbf{0} \end{cases} \tag{16.79}$$

where $\boldsymbol{\gamma}^2$ is a diagonal matrix that gathers the squared propagation constants of the N independent propagation modes.

The kth column of matrix **T** denoted by $\mathbf{t}_\mathrm{k}$ (i.e., the kth eigenvector of the **ZY** matrix product) defines the distribution of voltages among the set of MTL wires for mode-k. Likewise, the kth column of matrix **W** denoted by $\mathbf{w}_\mathrm{k}$ (i.e., the kth eigenvector of the **YZ** matrix product) defines the distribution of currents among the set of MTL wires for mode-k.

The modal transformation matrices **T** and **W** in (16.79) are interrelated. In fact, keeping in mind that both **Z** and **Y** are symmetrical, by transposing the second equation in (16.79) leads to $\boldsymbol{\gamma}^2 = \mathbf{W}^T\mathbf{ZY}(\mathbf{W}^{-1})^T$, and comparing the latter with the first equation in (16.79) we come to the conclusion that $\mathbf{T}^{-1} = \mathbf{W}^T$.

The transfer matrix solution for mode-k in (16.78), extended to all the modes, can be brought back to natural coordinates via (16.74) yielding [12],

$$\begin{bmatrix} \mathbf{V}(z) \\ \mathbf{I}(z) \end{bmatrix} = \begin{bmatrix} \mathbf{A} & \mathbf{B} \\ \mathbf{C} & \mathbf{D} \end{bmatrix} \begin{bmatrix} \mathbf{V}(0) \\ \mathbf{I}(0) \end{bmatrix} \quad \text{where} \quad \begin{cases} \mathbf{A} = \mathbf{T}(\cosh \boldsymbol{\gamma} z)\mathbf{T}^{-1} \\ \mathbf{B} = -\mathbf{T}(\sinh \boldsymbol{\gamma} z)\mathbf{T}^{-1}\mathbf{Z}_w \\ \mathbf{C} = -\mathbf{Z}_w^{-1}\mathbf{T}(\sinh \boldsymbol{\gamma} z)\mathbf{T}^{-1} \\ \mathbf{D} = \mathbf{A}^T \end{cases} \tag{16.80}$$

where the characteristic wave impedance matrix $\mathbf{Z}_w$ in natural coordinates is a full square symmetric matrix, obtained from its modal counterpart (diagonal $\mathbf{Z}_{w\mathrm{m}}$) through [12],

$$\mathbf{Z}_w\mathbf{W} = \mathbf{T}\mathbf{Z}_{w\mathrm{m}} = \mathbf{T}(\mathbf{Z}_\mathrm{m}/\mathbf{Y}_\mathrm{m})^{1/2} \tag{16.81}$$

Let us, finally, illustrate the procedure with a typical example.

Suppose that the time-domain input voltages $\mathbf{v}(0,t)$ are known. Suppose that all MTL parameters are known, as well as the MTL termination at $z = \ell$ described by a certain load impedance matrix $\mathbf{Z}_L(\omega)$. We wish to find $\mathbf{i}(0,t)$, $\mathbf{i}(z,t)$ and $\mathbf{v}(z,t)$.

First, the spectral representation of the input voltages is defined

$$\widetilde{\mathbf{V}}(0,\omega) = \frac{1}{\sqrt{2\pi}}\int_{-\infty}^{\infty} \mathbf{v}(0,t)\ e^{-j\omega t}dt$$

Second, in the frequency-domain, using the transfer matrix formalism, the spectral representations of the unknowns is evaluated,

$$\begin{cases} \widetilde{\mathbf{I}}(0,\omega) = (\mathbf{B} - \mathbf{Z}_L\mathbf{D})^{-1}(\mathbf{Z}_L\mathbf{C} - \mathbf{A})\widetilde{\mathbf{V}}(0,\omega) \\ \widetilde{\mathbf{I}}(z,\omega) = \mathbf{C}\widetilde{\mathbf{V}}(0,\omega) + \mathbf{D}\widetilde{\mathbf{I}}(0,\omega) \\ \widetilde{\mathbf{V}}(z,\omega) = \mathbf{Z}_L\widetilde{\mathbf{I}}(z,\omega) \end{cases}$$

Finally, inverse Fourier transformations produce the sought time-domain voltages and currents,

$$\begin{Bmatrix} \mathbf{i}(0,t) \\ \mathbf{i}(z,t) \\ \mathbf{v}(z,t) \end{Bmatrix} = \frac{1}{\sqrt{2\pi}}\int_{-\infty}^{\infty} \begin{Bmatrix} \widetilde{\mathbf{I}}(0,\omega) \\ \widetilde{\mathbf{I}}(z,\omega) \\ \widetilde{\mathbf{V}}(z,\omega) \end{Bmatrix} e^{j\omega t}d\omega.$$

References

[1] Fallou J.A. 'Propagation des courants de haute fréquence polyphase le long des lignes aériennes de transport d'energie affectées de court-circuits au de defaults d'isolement'. *Bulletin Société Française des Électriciens. Series 5*, 1932; 2: 787.

[2] Pipes L.A. 'Matrix theory of multiconductor transmission lines'. *Philosophical Magazine*, 1937; 24: 97–113.

[3] Rice S.O. 'Steady state solutions of transmission line equations'. *Bell System Technical Journal*, 1941; 20: 131–178.

[4] Wedepohl L.M. 'Application of matrix methods to the solution of traveling-wave phenomena in polyphase systems'. *IEE Proceedings*, 1963; 110: 2200–2212.

[5] Wedepohl L.M. 'Electrical characteristics of polyphase transmission systems with special reference to boundary-value calculations at power-line carrier frequencies'. *IEE Proceedings*, 1965; 112: 2103–2112.

[6] Wedepohl L.M. 'Wave propagation in nonhomogeneous multiconductor systems using the concept of natural modes'. *IEE Proceedings*, 1966; 113: 622–626.

[7] Wedepohl L.M., Mohamed S.E. 'Multiconductor transmission lines: theory of natural modes and Fourier integral applied to transient analysis'. *IEE Proceedings*, 1969; 116: 1553–1563.

[8] Wedepohl L.M., Mohamed S.E. 'Transient analysis of multiconductor transmission lines with special reference to non-linear problems'. *IEE Proceedings*, 1970; 117: 979–988.

[9] Wedepohl L.M., Ephtymiadis A.E. 'Wave propagation in transmission lines over lossy ground: a new complete field solution'. *IEE Proceedings*, 1978; 125: 505–510.

[10] Frankel S. *Multiconductor Transmission Line Analysis*. Norwood: Artech House; 1977.

[11] Djordjevic A.R., Harrington R.F., Sarkar T.K., Bazdar M. *Matrix Parameters for Multiconductor Transmission Lines: Software and User's Manual*. Norwood: Artech House; 1989.

[12] Faria J.A.B. *Multiconductor Transmission-Line Structures: Modal Analysis Techniques*. New York, NY: John Wiley & Sons; 1993.

[13] Paul C.R. *Analysis of Multiconductor Transmission Lines*. New York, NY: John Wiley & Sons; 1994.

[14] Bladel J. *On Helmholtz's Theorem in Finite Regions*. MURA Report no. 440, Wisconsin University, Madison, WI, 1958.

[15] Maxwell J.C. *A Treatise on Electricity and Magnetism*. 2 vols. New York, NY: Dover; 1954.

[16] Simonyi K. *Foundations of Electrical Engineering*. Oxford: Pergamon Press; 1963.

[17] Ramo S., Whinnery J., Duzer T. *Fields and Waves in Communication Electronics*. New York, NY: John Wiley & Sons; 1965.

[18] Faria J.A.B. *Electromagnetic Foundations of Electrical Engineering*. Chichester: John Wiley & Sons; 2008.

[19] Orfanidis S.J. *Electromagnetic Waves and Antennas*. New Brunswick-Piscataway, NJ: Rutgers University; 2013. Available from www.ece.rutgers.edu/~orfanidi/ewa

[20] DeHoop A. 'The electromagnetic constitutive relations', in *Handbook of Radiation and Scattering of Waves*. London: Academic Press, 1995 (Chapter 19).

[21] Karlsson A., Kristensson G. Constitutive Relations, Dissipation and Reciprocity for the Maxwell Equations in the Time Domain. Tech. Report LUTEDX/TEAT-7005, Department of Electroscience, Lund Institute of Technology, 1989.

[22] Soodak H., Tiersten M. S. 'Dynamic interpretation of Maxwell's equations'. *American Journal of Physics*, 1994; 62(10): 907–913.

[23] Faria J.A.B. 'On the time-domain transmission-line equations'. *Microwave and Optical Technology Letters*, 1999; 22(3): 194–197.

[24] Bergeron L. *Water Hammer in Hydraulics and Wave Surges in Electricity*. New York, NY: John Wiley & Sons; 1961.

[25] Branin F.H. 'Transient analysis of lossless transmission lines'. *Proceedings of the IEEE*, 1967; 55(11): 2012–2013.

[26] Faria J.A.B. 'A pedagogical example of the application of transmission-line theory to study conductor grounding effects'. *IEEE Transactions on Education*, 2004; 47(1): 141–145.

[27] Faria J.A.B. 'Conservation laws for time-domain signals propagating along multiconductor transmission-line structures'. *IEE Proceedings – Microwaves, Antennas and Propagation*, 2003; 150(5): 375–378.

[28] Giacoletto L.J. 'Frequency- and time-domain analysis of skin effects'. *IEEE Transactions on Magnetics*, 1996; 32(1): 220–229.

[29] Faria J.A.B., Raven M.S. 'On the success of electromagnetic analytical approaches to full time-domain formulation of skin effect phenomena'. *Progress in Electromagnetics Research M*, 2013; 31: 29–43.

[30] Raven M.S. 'Skin effect in the time and frequency domain: comparison of power series and Bessel function solutions'. *Journal of Physics Communications*, 2018; 2: 035028.

[31] Faria J.A.B. 'Multimodal propagation in multiconductor transmission lines'. *Journal of Electromagnetic Waves and Applications*, 2014; 28(14): 1677–1702.

Chapter 17
Shielding in time domain

Rodolfo Araneo[1] *and Salvatore Celozzi*[1]

Analysis and design of shielding enclosures [1–4] are often complex tasks with several conflicting constraints coming from the need of assuring an adequate protection level at reasonable cost and weight and with the maximum possible accessibility, ventilation, and so forth [5]. The protection level must be guaranteed in the presence of either intentional or unintentional electromagnetic (EM) sources [6], and the emissions from the enclosure must be kept below the thresholds fixed by standards.

When the concern is on the emission, the source inside the enclosure is well known and may emit under transient or continuous wave (CW) conditions. In this scenario, a frequency-domain (FD) approach based on the IEEE Standard 299 [7,8] is generally advisable and immediately applicable, since standards generally prescribe limits in terms of FD masks.

On the contrary, when the design of the enclosure is aimed at protecting its content against external transient threats, a direct time-domain (TD) approach appears more reliable and closer to the intrinsic nature of the involved phenomena. Recently, some attempts to define the shielding effectiveness (SE) in the TD have been presented, e.g., [9,10], but widespread standards have not yet included the direct TD approach in the general framework of testing procedures.

As a matter of fact, the definition of any direct TD parameter should stem from the key roles that a shielding configuration plays in the protection from transient EM threats. Then, considering that most electric and electronic devices are sensitive to one or more of the following physical quantities:

1. the maximum value of the electric (magnetic) field,
2. the maximum value of induced effects caused by time variations of magnetic (electric) flux density, and/or
3. the total (unintentional plus intentional) energy delivered to the susceptible content (victim);

[1]University of Rome—La Sapienza, Department of Astronautical, Electrical and Energy Engineering—Via Eudossiana 18, Italy

shielding parameters accounting for

- a reduction of the peak value(s) of the EM transient field in the shielded region,
- a reduction of the maximum time derivative of the EM transient field in the shielded region, and
- a reduction of the energy density flowing through the shielded region

should be introduced to characterize the actual TD behavior of any shielding structure [11,12]. They complement each other and offer an educated guess about the expected performance in dealing with transient EM threats. Unfortunately, as in the case of CW classical shielding parameters, also the previous parameters give the value of the shielding performance at a given position and, thus, are local quantities: no information may be inferred regarding the performance in the whole shielded volume or in other positions. Different strategies may be envisaged to introduce global shielding parameters at the same time synthetic and meaningful of the overall performance [13–15].

17.1 Rationale: when a direct TD approach is mandatory, when convenient

A direct time-domain approach may represent the only feasible way to address shielding configurations exhibiting a non-linear behavior (e.g., when saturable ferromagnetic materials have been used to realize the shielding enclosure) or when exists the possibility that a non-linear behavior is excited in the protected volume, e.g., because of a partial or complete discharge or in case of triggering a content reaction to the external field above a specific threshold. In both cases, a direct TD approach cannot be avoided, unless some rude approximations are introduced.

Frequently, the resort to a direct TD analysis may be suggested by the transient nature of the actual threat whose typical waveform would require (direct and inverse) Fourier Transforms eventually misleading in catching the essential features and performance of the shielding configuration under analysis. Typical transient sources suggesting a direct approach are those ensuing from discharges (e.g., lightning or electrostatic discharges, ESDs) or coming from pulsed intentional EM interferences. In fact, in these cases, the transient waveforms are generally characterized by a fast front steepness and a long-lasting duration, involving very different time constants (and a wide frequency spectrum).

17.2 Reference standards and commonly used FD approaches

A parameter that is commonly used to describe the reduction in electric or magnetic field intensity due to a shield configuration is the so-called *shielding effectiveness* (SE); the SE in the frequency domain is defined by the IEEE dictionary and by the military standard MIL-STD 285, replaced by IEEE Std. 299 [7,16] as the ratio of electric- or magnetic-field intensities at a point $\mathbf{P}_0 \equiv (x_0, y_0, z_0)$ of a Cartesian coordinate system before ($|\mathbf{E}^{\mathrm{i}}|$ or $|\mathbf{H}^{\mathrm{i}}|$) and after ($|\mathbf{E}|$ or $|\mathbf{H}|$) the placement of the shield between a given external source and the considered observation point, i.e.,

$$\mathrm{SE}_E(\mathbf{P}_0,\omega) = \left|\frac{\mathbf{E}^{\mathrm{i}}(\mathbf{P}_0,\omega)}{\mathbf{E}(\mathbf{P}_0,\omega)}\right| \tag{17.1}$$

$$\mathrm{SE}_H(\mathbf{P}_0,\omega) = \left|\frac{\mathbf{H}^{\mathrm{i}}(\mathbf{P}_0,\omega)}{\mathbf{H}(\mathbf{P}_0,\omega)}\right|, \tag{17.2}$$

where $\omega = 2\pi f$ is the radian frequency of the source (assumed monochromatic with a time-harmonic dependence $e^{j\omega t}$).

These two figures of merit are usually expressed in decibels (dB), as:

$$\mathrm{SE}_{E,\mathrm{d}B}(\mathbf{P}_0,\omega) = 20\log\left|\frac{\mathbf{E}^{\mathrm{i}}(\mathbf{P}_0,\omega)}{\mathbf{E}(\mathbf{P}_0,\omega)}\right| \tag{17.3}$$

$$\mathrm{SE}_{H,\mathrm{d}B}(\mathbf{P}_0,\omega) = 20\log\left|\frac{\mathbf{H}^{\mathrm{i}}(\mathbf{P}_0,\omega)}{\mathbf{H}(\mathbf{P}_0,\omega)}\right|. \tag{17.4}$$

The performance of a shield can also be characterized by the *shielding factor* (SF) which is defined as the ratio (in dB and in the frequency domain) of the total electric- or magnetic-field intensities on two surfaces of the shield (the inner and the outer surface for a closed three-dimensional shield):

$$\mathrm{SF}_{E,\mathrm{d}B}(\omega) = 20\log\left|\frac{\mathbf{E}^{\mathrm{i}}(\mathbf{P}_1,\omega)}{\mathbf{E}(\mathbf{P}_2,\omega)}\right| \tag{17.5}$$

$$\mathrm{SF}_{H,\mathrm{d}B}(\omega) = 20\log\left|\frac{\mathbf{H}^{\mathrm{i}}(\mathbf{P}_1,\omega)}{\mathbf{H}(\mathbf{P}_2,\omega)}\right|. \tag{17.6}$$

The point $\mathbf{P}_1 \equiv (x_1, y_1, z_1)$ is assumed to be located at the surface of the shield on the side of the source, while the point $\mathbf{P}_2 \equiv (x_1, y_1, z_1)$ is assumed on the opposite surface.

Figure 17.1 shows the configurations leading to the evaluation of SE and SF, respectively. Both these quantities, SE and SF, raised some criticism, basically because of the problems occurring during their experimental determination. In fact, as regards the SE, when the shield is in the near field of the source, it affects the field produced by the source itself, acting as a load for the antenna. On the other hand, also the SF cannot be easily measured due to the practical difficulty to place field sensors on the surface of the shield and being the sensors themselves of finite dimensions and perturbing the field distributions. The clear and unambiguous definition of significant and measurable quantities is of paramount importance in the shielding practice and these issues have been addressed in some recent works (e.g., [5,6,8]), aimed at making possible the experimental verification of predictions, the design of effective shielding structures, and the promulgation of standards.

Shielding theory has been developed in order to obtain guidance in practical problems and to clearly identify the key-factors upon which the whole EM problem

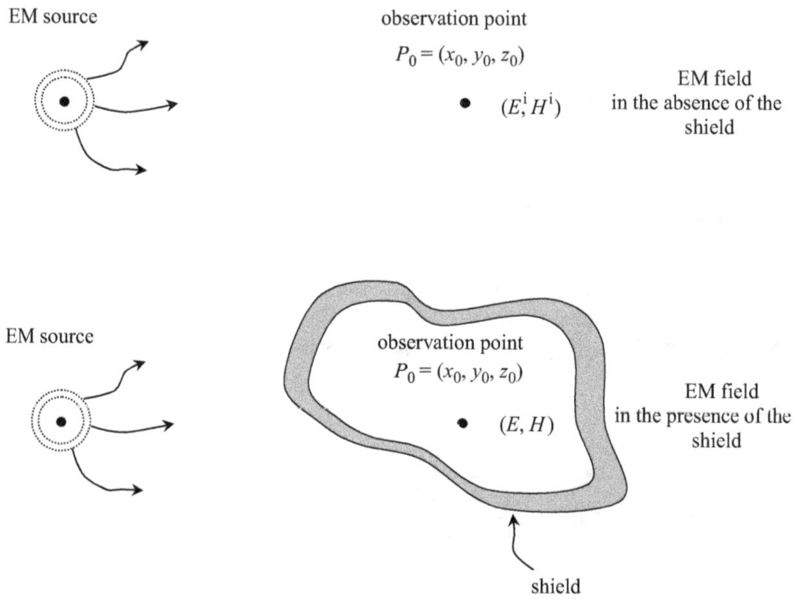

(a) Shielding effectiveness.

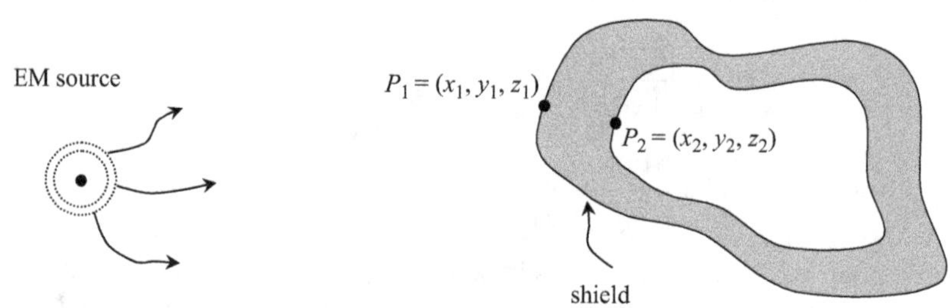

(b) Shielding factor.

Figure 17.1 System configurations for the evaluation of the SE and SF

depends. The evaluation of the SE can be performed analytically only for shields of simple shape and under idealized assumptions. The analysis of realistic problems often needs the use of numerical techniques (both in the frequency and in the time domain), which usually require high computational efforts, depending on the geometrical configuration, the EM characteristics of the shield material, and on the source type.

The most common documents describing the test procedures recommended to assess the shielding characteristics of enclosures are:

- IEEE 299: Standard method of measuring the effectiveness of electromagnetic shielded enclosures [7,8].

- National Security Agency (NSA) Specification 94-106 [17] (formerly 65-6): General specifications for RF shielded enclosures for communications equipment.

These standards describe antenna geometries and configurations, and delineate some measurement practices, classifying the source as magnetic-, electric-, or plane-wave-type. The data obtained considering any of the above-described set-ups cannot be applied to source configurations different from those used in the experiments; hence, they serve only as reference value [18].

Figure 17.2(a) and (b) shows the configurations prescribed by NSA 94-106 and IEEE Std. 299 for the measurement of the SE against a magnetic field source [19]. The NSA 94-106 is known as the parallel setup, where both transmitting and receiving antennas are in parallel planes, and parallel with the plane of the sample under test. The IEEE Std. 299 (deriving from MIL STD 285) is known as the coplanar setup, where both antennas are located in the same plane, but perpendicular to the sample.

When performing SE measurements using both methodologies, but identical samples and equipment, different results are obtained for the resulting SE values.

Figure 17.3(a) and (b) shows the configurations requested in the measurements of the SE against an electric field source and a plane wave source, respectively.

Theoretical analysis has been done in the past by different authors, as Bannister, Dahlberg and others [6,20–22]. This resulted in some general integral formulations of the SE values, and also in some closed form approximations under certain conditions. A thorough analysis of both setups has been presented in [23] and the most important results of that work are summarized in the next paragraphs.

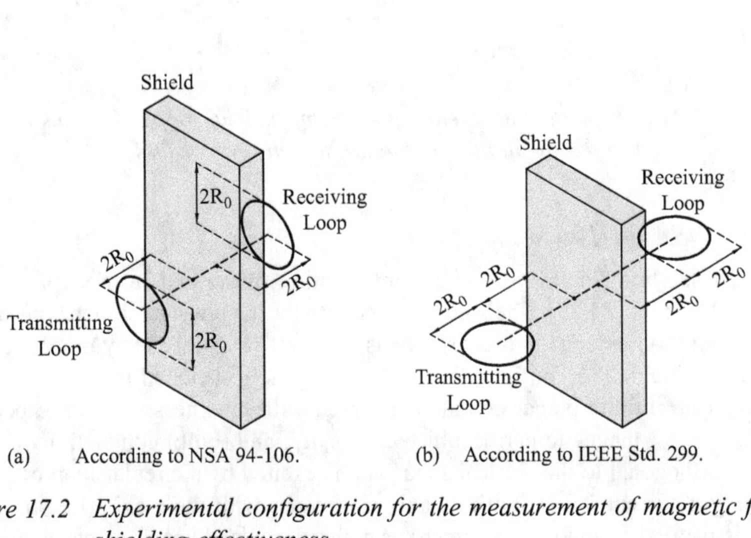

(a) According to NSA 94-106. (b) According to IEEE Std. 299.

Figure 17.2 Experimental configuration for the measurement of magnetic field shielding effectiveness

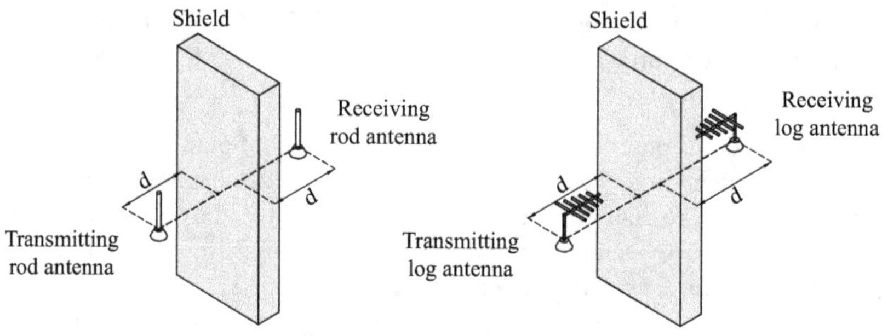

(a) High-impedance source configuration. (b) Plane-wave source configuration.

Figure 17.3 Typical setup for SE measurements

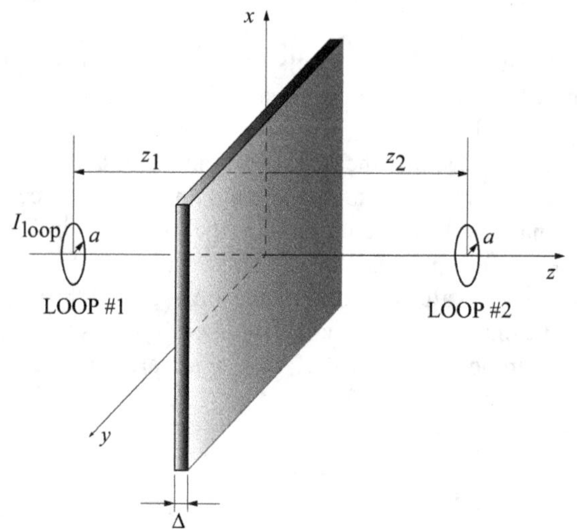

Figure 17.4 Electromagnetic shielding problem: coaxial loops of radius a separated by an infinite planar screen of thickness Δ

17.2.1 NSA 94-106 setup

The exact solution for the SE setup configuration prescribed by NSA 94-106 is given by Moser in [24,25]. The solution accounts for the physical dimensions of the transmitting loop antenna and considers as observation point the symmetrical one beyond the shield. The configuration under analysis is sketched in Figure 17.4. It consists of an infinite planar conductive screen with thickness Δ, electric permittivity $\varepsilon_s = \varepsilon_0\varepsilon_{sr}$, magnetic permeability $\mu_s = \mu_0\mu_{sr}$, and (bulk) conductivity σ. The screen is orthogonal to the vertical z axis and is excited by a circular loop of radius a carrying a current $i_{loop}(t)$ with its axis coincident with the z axis (i.e., the loop source is parallel to the planar screen). Another circular loop of same radius a is located in the halfspace $z > 0$ parallel to the planar screen at $z = d$ and coaxial with the loop in $z < 0$.

The classical formulation of Moser for the solution of the problem in the frequency domain is here briefly resumed. In particular, by assuming a frequency-domain source

$$\mathbf{j}_{\rm i}(\mathbf{r},\omega) = I_{\rm loop}(\omega)\frac{\delta(\rho-a)}{\rho}\delta(z+z_1)\mathbf{u}_\phi, \tag{17.7}$$

where $I_{\rm loop}(\omega)$ indicates the Fourier transform of $i_{\rm loop}(t)$ and $\delta(\cdot)$ indicates the Dirac delta function, the z-component of the magnetic induction is obtained at $\rho = 0$ and for $z > 0$ as

$$B_z(z,\omega) = 2\mu_{\rm s} a I_{\rm loop}\cdot\int_0^{+\infty}\frac{\tau(\lambda)C(\lambda)}{\tau_0^2(\lambda)}J_1(\lambda a\cdot \\ \cdot {\rm e}^{-\tau_0(\lambda)(z_1+z)-\Delta[\tau(\lambda)-\tau_0(\lambda)]}\lambda^2\,d\lambda \tag{17.8}$$

where

$$\tau(\lambda) = \sqrt{\lambda^2-k^2},\quad \tau_0(\lambda) = \sqrt{\lambda^2-k_0^2} \tag{17.9}$$

with $k = \sqrt{j\omega\mu_{\rm s}\sigma}$ and $k_0 = \frac{\omega}{c_0}$ (c_0 is the speed of light in free space), and

$$C(\lambda) = \frac{1}{[K(\lambda)+\mu_{\rm sr}]^2-[K(\lambda)-\mu_{\rm sr}]^2 e^{-2\tau(\lambda)d}} \tag{17.10}$$

with

$$K(\lambda) = \frac{\tau(\lambda)}{\tau_0(\lambda)}. \tag{17.11}$$

The incident z-component of the magnetic induction (i.e., the z-component of the magnetic induction in the absence of the screen) is given by

$$B_z^{\rm inc}(\rho=0,z) = \frac{\mu_0 a}{2}I_{\rm loop}\cdot\int_0^{+\infty}\frac{1}{\tau_0(\lambda)}J_1(\lambda a){\rm e}^{-\tau_0(\lambda)(z_1+z)}\lambda^2\,d\lambda. \tag{17.12}$$

The magnetic shielding effectiveness ${\rm SE}_H$, accounting for the physical dimensions of the transmitting loop antenna and considering as observation point the symmetrical one beyond the shield, can be obtained as

$${\rm SE}_H = \frac{1}{4\mu_{\rm sr}}\frac{\int_0^{+\infty}\frac{\lambda^2}{\tau_0(\lambda)}J_1(\lambda a){\rm e}^{-\tau_0(\lambda)(z_1+z_2)}\,d\lambda}{\int_0^{+\infty}\lambda^2\frac{\tau(\lambda)C(\lambda)}{\tau_0^2(\lambda)}J_1(\lambda a){\rm e}^{-\tau_0(\lambda)(z_1+z_2)-\Delta[\tau(\lambda)-\tau_0(\lambda)]}\,d\lambda} \tag{17.13}$$

Important in this configuration is the dependency of the resulting SE from only the total distance $z_1 + z_2$ between both antennas, the radius a of the transmitting antenna and the thickness Δ of the sample. It is worth noting that the integrals in

(17.8) and (17.12) are in general, oscillating, slowly converging, and many times their numerical evaluation suffers of loss of numerical accuracy. For this reason, many efforts have been dedicated in the past to address such calculations [24–26].

To obtain the transient magnetic induction, an inverse Fourier transform has to be performed, i.e.,

$$b_z(z,t) = \frac{1}{2\pi}\int_{-\infty}^{+\infty} B_z(z;\omega)e^{j\omega t}\,d\omega = \frac{1}{\pi}\int_{0}^{+\infty} B_z(z;\omega)e^{j\omega t}\,d\omega \tag{17.14}$$

having exploited the fact that b_z is a real function. To obtain the transient shielding effectiveness, the same has to be done for the incident field.

In order to simplify the exact Moser formulation, in [20] a quasi-near field approximation has been derived, leading to the following formula for SE_H:

$$\mathrm{SE}_H = 8.686\alpha\Delta + 20\log\left[\frac{\alpha D_\Delta}{8.485\mu_{\mathrm{rs}}}\frac{D_\Delta}{(2d-\Delta)}\left(\frac{D_\Delta}{D}\right)^3\right] \tag{17.15}$$

where

$$D = \sqrt{a^2 + (z_1 + z_2)^2} \tag{17.16}$$

$$D_\Delta = \sqrt{a^2 + (z_1 + z_2 - \Delta)^2} \tag{17.17}$$

and

$$\alpha = \sqrt{\pi f \mu_s \sigma} \tag{17.18}$$

Formula (1.15) is valid for $p > 10$, being

$$p = \frac{\alpha D_\Delta}{\mu_{\mathrm{rs}}} \tag{17.19}$$

When $z_1 + z_2 \gg \Delta$ and $z_1 + z_2 \gg a$, (1.15) can be approximated as

$$\mathrm{SE}_H = 8.686\alpha\Delta + 20\log\left(\frac{p}{8.485}\right) \tag{17.20}$$

On the other hand, when $p < 0.1$ the following formula is obtained:

$$\mathrm{SE}_H = 8.686\alpha\Delta + 20\log\left[\frac{\mu_{rs} D_\Delta (D_\Delta + z_1 + z_2 - \Delta)}{5.657\alpha D^3}\right] \tag{17.21}$$

which, for $z_1 + z_2 \gg \Delta$ and $z_1 + z_2 \gg a$, can be approximated as

$$\mathrm{SE}_H = 8.686\alpha\Delta + 20\log\left(\frac{1}{2.828p}\right). \tag{17.22}$$

Formulas (1.15) and (1.21) are particular cases of the following formula, valid also at high frequencies [20]:

$$\mathrm{SE}_H = 8.686\alpha\Delta + 20\log\left|\frac{aD_\Delta}{2.828\mu_r}\frac{D_\Delta}{z_1+z_2-\Delta}\cdot\left(\frac{D_\Delta}{D}\right)^3\frac{1+jk_0D}{\left[3+3jk_0D_\Delta-k_0^2D_\Delta^2\right]}\right| \tag{17.23}$$

which, for $z_1 + z_2 \gg \Delta$ and $z_1 + z_2 \gg a$, can be approximated as

$$\mathrm{SE}_H = 8.686\alpha\Delta + 20\log\left|\frac{p}{2.828}\frac{[1+jk_0(z_1+z_2)]}{\left[3+jk_0(z_1+z_2)-(z_1+z_2)^2k_0^2\right]}\right|. \tag{17.24}$$

However, formula (17.23) is valid for screen thicknesses Δ larger than twice the skin depth $\delta = 1/\alpha$ [20].

In the Schelkunoff transmission-line theory, the following simple formula is obtained [22]:

$$\mathrm{SE}_H = A + R + B \tag{17.25}$$

where

$$A = 8.686\alpha\Delta, \quad R = 20\log\frac{1}{|p|}, \quad B = 20\log\left|1-qe^{-2\alpha\Delta}\right| \tag{17.26}$$

with

$$p = \frac{4\psi}{(\psi+1)^2}, \quad q = \frac{(\psi-1)^2}{(\psi+1)^2}, \tag{17.27}$$

where ψ is the ratio between the wave impedances in air and in the shield, i.e., $\psi = Z_{\mathrm{air}}/Z_{\mathrm{shield}}$, with $Z_{\mathrm{shield}} = \eta = \sqrt{\frac{\mu_s}{\varepsilon_s}}$. The correct definition of the wave impedances is crucial to determine the accuracy of the formulation; it is related to a small dipole source parallel to an conductive plane [27]. An accurate approximate formula, based on a small dipole assumption, is

$$Z_{\mathrm{air}} = \eta_0\frac{\left(\frac{j\omega z}{c_0}\right)^2+\left(\frac{j\omega z}{c_0}\right)}{\left(\frac{j\omega z}{c_0}\right)^2+3\left(\frac{j\omega z}{c_0}\right)+3} \tag{17.28}$$

where $\eta_0 = \sqrt{\frac{\mu_0}{\varepsilon_0}}$ is the free-space impedance and z is the distance from the source point (the transmitting loop antenna). In order to apply the TL equivalent, $z = z_1$ at the position of the shield plane. Although both the tangential electric and magnetic field components tend to vanish as the observation point tend to the loop axis, their ratio i.e. the wave impedance, is constant. Its LF approximation is valid when

$\frac{j\omega z}{c_0} \ll 1$ which implies, considering a loop radius a and its distance from the shield as requested by IEEE Std. 299 ($2a = z_1 = 305$ mm), a maximum frequency of 16 MHz. As further tests were performed up to 20 MHz, the small dipole LF approximation proposed by Whitehouse [22,28] may readily be used:

$$Z_{\text{air}} = \eta_0 \left(\frac{j\omega z}{c_0} \right) = j\omega\mu_0 \frac{z}{3} \tag{17.29}$$

Regarding a practical measurement setup, the source will have some finite dimensions. Therefore, a correction, ensuing from the exact integral expression, may be introduced to account for these finite dimensions of the transmitting source:

$$Z_{\text{air}} = j\omega\mu_0 \frac{z^2 + a^2}{3} \tag{17.30}$$

17.2.2 IEEE Std 299 setup

The exact solution for the SE setup configuration prescribed by IEEE Std 299 is given by Celozzi in [16,29]. The configuration under analysis is sketched in Figure 17.5.

The incident x-component of the magnetic vector potential of the second-order U (i.e., the z-component of the magnetic induction in the absence of the screen) is given by

$$U_x^{inc}(x, y, z) = \int_0^\infty \int_0^\infty G_0(\alpha, \beta) \cos(\alpha x) \cos(\beta y) e^{-D_1|z+z_1|} d\alpha d\beta \tag{17.31}$$

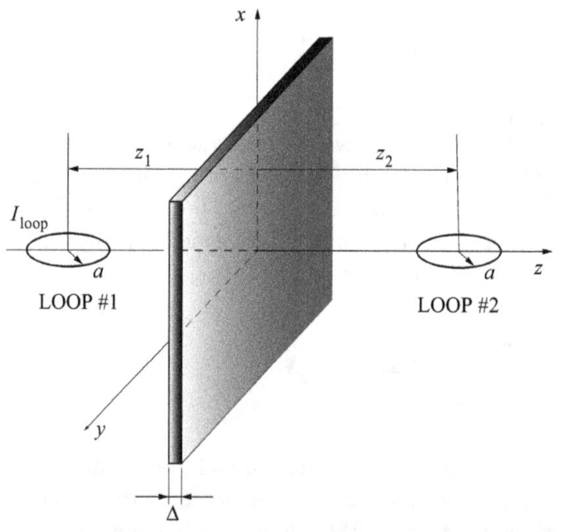

Figure 17.5 Electromagnetic shielding problem: coaxial loops of radius a separated by an infinite planar screen of thickness Δ

where

$$G_0 = -\frac{\mu_0 I_{\text{loop}} a}{\pi} \frac{1}{\beta D_1} I_1(\beta a) \quad D_1 = \sqrt{\alpha^2 + \beta^2}, \tag{17.32}$$

with $I_1(\cdot)$ the modified Bessel function of the first kind of order one.

The x-component of the magnetic vector potential of the second order when the shield is present, reads:

$$U_x(x, y, z) = \int_0^\infty \int_0^\infty G_1(\alpha, \beta) \cos(\alpha x) \cos(\beta y) e^{-D_1|z+z_1|} d\alpha d\beta \tag{17.33}$$

where

$$G_1 = -\frac{4\mu_0 I_{\text{loop}} a}{\pi} \frac{D_3}{D_4} \frac{e^{-(D_2\Delta - D_1\Delta)}}{\beta} I_1(\beta a) \tag{17.34}$$

with

$$D_2 = \sqrt{\alpha^2 + \beta^2 + j\omega\mu_s\sigma} \tag{17.35}$$

$$D_3 = \frac{D_2}{\mu_{rs}} \tag{17.36}$$

$$D_4 = (D_3 + D_1)^2 - (D_3 - D_1)^2 \ e^{-2D_2\Delta}. \tag{17.37}$$

The SE_H can be computed as the ratio between (17.31) and (17.33). From the formulas, the direct dependency of the resulting SE from the distance $z_1 + z_2$ (from the transmitting antenna to the sample), the radius a of the transmitting antenna and the thickness Δ of the sample is observed for this configuration.

In the Schelkunoff transmission-line theory, the following approximate incident wave impedance, obtained by means of a small dipole approximation, must be used:

$$Z_{\text{air}} = \eta_0 \frac{\left(\frac{j\omega z}{c_0}\right)^2 + \left(\frac{j\omega z}{c_0}\right)}{\left(\frac{j\omega z}{c_0}\right)^2 + \left(\frac{j\omega z}{c_0}\right) + 1}. \tag{17.38}$$

17.3 Local figures of merit in TD: when they are sufficient

Considering that victim sensitivity characteristics are often largely unknown, without loss of generality it can be simplified the notation and avoided a lengthy formalism accounting for direction- and polarization-dependence of the source characteristics. The expressions useful to quantify the performance of a shielding configuration with respect to the above coupling mechanisms are: *Shielding Effectiveness as to the reduction of the peak value of the electric (or magnetic) field*

$$\mathrm{SE}_E^{\mathrm{PR}} = 20\log\frac{\left|e_{\mathrm{MAX}}^{\mathrm{inc}}(t,x,y,z)\right|}{\left|e_{\mathrm{MAX}}(t,x,y,z)\right|}, \tag{17.39}$$

$$\mathrm{SE}_H^{\mathrm{PR}} = 20\log\frac{\left|h_{\mathrm{MAX}}^{\mathrm{inc}}(t,x,y,z)\right|}{\left|h_{\mathrm{MAX}}(t,x,y,z)\right|}, \tag{17.40}$$

where $e_{\mathrm{MAX}}^{\mathrm{inc}}(\cdot)$ or $h_{\mathrm{MAX}}^{\mathrm{inc}}(\cdot)$ represents the maximum value of the electric (magnetic) field (modulus of the vector) at the observation point, in the absence of the enclosure (incident field). When the enclosure is present, the maximum electric (magnetic) field is denoted as $e_{\mathrm{MAX}}(\cdot)$ or $h_{\mathrm{MAX}}(\cdot)$. The parameter introduced is based on the peak value reduction in the electric or magnetic field waveforms. The numerator is the peak value of the incident field at the observation point in the absence of the shield, while the denominator represents the peak value of the field in the shielded region, at the same observation point. Of course, the electric (magnetic) field is a vector quantity and a modulus should be considered, unless a specific direction is of special and well-known interest.

The second parameter represents the time derivative reduction (as usual, a dot above the field component denotes the time derivative) and accounts for the limitation of the induced effects associated with the time derivatives of magnetic or electric flux density, $\mathbf{b}(t,x,y,z)$ or $\mathbf{d}(t,x,y,z)$, respectively, being proportional to induced emf's or currents: the *Shielding Effectiveness as to the reduction of the peak value of the time derivative of the electric (magnetic) field* is given by

$$\mathrm{SE}_H^{\mathrm{DR}} = 20\log\frac{\left|\dot{h}_{\mathrm{MAX}}^{\mathrm{inc}}(t,x,y,z)\right|}{\left|\dot{h}_{\mathrm{MAX}}(t,x,y,z)\right|} \tag{17.41}$$

$$\mathrm{SE}_E^{\mathrm{DR}} = 20\log\frac{\left|\dot{e}_{\mathrm{MAX}}^{\mathrm{inc}}(t,x,y,z)\right|}{\left|\dot{e}_{\mathrm{MAX}}(t,x,y,z)\right|}, \tag{17.42}$$

where the effect is evaluated considering the reduction of the maximum values of the time derivatives, i.e. the reduction of the maximum induced effect under analysis. In (17.41) and (17.42), the following notation and hypotheses have been adopted:

$$\dot{b}_{\mathrm{MAX}}^{inc}(\cdot) = \mu\cdot\dot{h}_{\mathrm{MAX}}^{\mathrm{inc}}(\cdot) = \mu\cdot\ \max\left[\frac{\partial h^{\mathrm{inc}}(\cdot)}{\partial t}\right] \tag{17.43}$$

$$\dot{d}_{\mathrm{MAX}}^{inc}(\cdot) = \varepsilon\cdot\dot{e}_{\mathrm{MAX}}^{\mathrm{inc}}(\cdot) = \varepsilon\cdot\ \max\left[\frac{\partial e^{\mathrm{inc}}(\cdot)}{\partial t}\right]. \tag{17.44}$$

The first equality holds if and only if the magnetic permeability (permittivity) may be considered constant over time, otherwise it cannot be extracted by the time-derivative. It should be noted that the use of ferromagnetic materials for the enclosures is not so frequent and, as well, non-linear dielectric materials have not yet been considered for shielding applications, to date. Thus, the previous assumptions are generally legitimate, especially considering that, from an experimental point of view, commercially available sensors measure directly the time derivative of the magnetic flux density or that of the electric induction.

The third parameter accounts for the reduction of the energy passing through the observation point where the victim is located. Such an energy may be evaluated integrating the flow of the power density over the period of time under consideration (theoretically infinite) and is expressed in [J/m^2]: the *Shielding Effectiveness as to the reduction of the energy passing through a given position* is evaluated as

$$\mathrm{W} = \int_0^{\infty} |e(t,x,y,z) \times \mathbf{h}(t,x,y,z)| \mathrm{d}t. \tag{17.45}$$

Of course, the upper limit in the time integral is finite in transients of actual interest: the improper integral will reduce to a definite one. However, the duration of the transient may be different in the presence and in the absence of the shielding configuration and attention must be paid to its actual duration.

The energy density reduction shielding effectiveness is:

$$\mathrm{SE}_W = 10 \log \frac{\mathrm{W}^{\mathrm{inc}}(x,y,z)}{\mathrm{W}(x,y,z)}. \tag{17.46}$$

The advantages of the proposed parameters are:

- immediate link with victim susceptibilities,
- easy computation of protection and effects of a shielding structure.

The disadvantages are mainly due to habitual practices in terms of FD quantities and to the need of re-thinking measurement procedures and sensors. Moreover, the proposed parameters are local and, thus, suffer from the same lack of information typical of the classical SE defined in IEEE-Std. 299: their values are valid at a specific point in the shielded region and nothing can be argued about the performance achieved at other points.

Moreover, it should be noted that the energy in (17.45) and (17.46) is due only to the external electromagnetic field threat and any contribution deriving, for instance, from indirect effects (e.g., conducted through the cable or line connections) or from normal operative conditions is not included.

An additional comment applies: the sensitivity of a potential victim may be not isotropic and in general anisotropy is the most frequent occurrence, in the sense that the immunity levels corresponding to EM threats with different polarizations and angle of incidence are different. In most cases, however, either the accurate orientation of the internal layout of the victims and their immunity levels with respect to threats having different characteristics are not known a priori. Thus, a general investigation about the single spatial components is not particularly useful, especially when a synthetic performance parameter is looked for. Of course, when mission critical components or systems should be installed a deeper analysis should be conducted and these details are worthy of further investigation.

Finally, it should be noted that only a fraction of the energy density will be effectively delivered to the victim located at a given position: in order to measure such a quantity (i.e., the deliverable energy) some further discussion is in order, requiring the knowledge of the victim geometrical, physical and functional characteristics. The approach proposed in [1] could be extended to the time domain to overcome some of the above limitations, but of course requires a preliminary consensus about the most representative content(s) to be adopted.

Last, but not least, comment is that the above performance parameters are expressed by just a number and neither masks or trend need to be analyzed, making easy any comparison between susceptibility and protection levels or between different shielding configurations.

17.4 Global figures of merit in TD: when they are recommended

It is known that classical shielding effectiveness parameters are intrinsically local and has been already highlighted mentioning the drawbacks; some attempts there have been in the FD to overcome the lack of information about the EM shielding behavior of an enclosure at positions different from that (or those, but very often just one point is observed) where the SE is evaluated or measured. It is worth noting that, even though measurements are conducted ideally at a specific point, finite dimensions of most receiving antennas imply that generally somewhat "unwanted" integral responses are obtained as the result of practical tests.

As a matter of fact, the spatial dependence of the three local shielding parameters in closed shielding enclosures is generally such that their values may strongly depend on the specific observation point, where the potential victim is or could be located. This occurs, in particular, when the frequency spectrum of the source waveform exhibits a content such that a resonant behavior, eventually exciting multiple modes, is expected. A simple FD check will put in evidence the eventual presence of resonances and, thus, the usefulness of a global evaluation. Even in the case of several apertures in the enclosure it is expected that the EM field is non-uniform inside and, thus, a global evaluation may be recommended.

In order to get more general information, two sets of new, global parameters may be introduced, obtained by scanning the whole shielded volume and looking for

- the average level of shielding effectiveness in the protected region,
- the minimum performance.

The first set of quantities include three parameters similar to those described in the previous Section; the first, the *Average Peak-Value Reduction Shielding Effectiveness*, is introduced to quantify the volume-averaged peak value reduction appearing in the electric or magnetic field waveforms when the shield is applied. Numerator is the integral of the peak value of the incident field over the shielded volume U, in the absence of the shield, while the denominator represents the integral of the same quantity over the same region, in the presence of the considered shield. Hence:

$$\mathrm{ASE}_E^{\mathrm{PR}} = 20\ \log \frac{\int_U \left|E_{\mathrm{MAX}}^{\mathrm{in}}(t,x,y,z)\right| \mathrm{d}U}{\int_U \left|E_{\mathrm{MAX}}(t,x,y,z)\right| \mathrm{d}U} \tag{17.47}$$

$$\mathrm{ASE}_H^{\mathrm{PR}} = 20\ \log \frac{\int_U \left|H_{\mathrm{MAX}}^{\mathrm{in}}(t,x,y,z)\right| \mathrm{d}U}{\int_U \left|H_{\mathrm{MAX}}(t,x,y,z)\right| \mathrm{d}U}, \tag{17.48}$$

where $E_{\mathrm{MAX}}^{\mathrm{in}}(\cdot)$ or $H_{\mathrm{MAX}}^{\mathrm{in}}(\cdot)$ represents the maximum (peak) value of the electric (magnetic) field in each point, in the absence of the enclosure. When the enclosure is present, the electric (magnetic) field at each point inside the shielded region is denoted as $E_{\mathrm{MAX}}(\cdot)$ or $H_{\mathrm{MAX}}(\cdot)$. It should be noted that the time instant in which the maximum is achieved is not relevant, is generally position-dependent and at each position may vary after the shield is applied.

The second parameter, denoted as *Average Derivative Reduction Shielding Effectiveness*, is introduced to consider the reduction of the maximum values of the time derivatives, averaged over the shielded volume:

$$\mathrm{ASE}_H^{\mathrm{DR}} = 20\ \log \frac{\int_U \left|\dot{H}_{\mathrm{MAX}}^{\mathrm{in}}(t,x,y,z)\right| \mathrm{d}U}{\int_U \left|\dot{H}_{\mathrm{MAX}}(t,x,y,z)\right| \mathrm{d}U} \tag{17.49}$$

$$\mathrm{ASE}_E^{\mathrm{DR}} = 20\ \log \frac{\int_U \left|\dot{E}_{\mathrm{MAX}}^{\mathrm{in}}(t,x,y,z)\right| \mathrm{d}U}{\int_U \left|\dot{E}_{\mathrm{MAX}}(t,x,y,z)\right| \mathrm{d}U}. \tag{17.50}$$

The third parameter, denoted as *Average Incident Energy Shielding Effectiveness*, accounting for the energy transiting through each position inside the enclosure and quantifying its average reduction in the considered volume, reads:

$$\mathrm{ASE}_W = 10\ \log \frac{\int_U W^{\mathrm{in}}(x,y,z)\mathrm{d}U}{\int_U W(x,y,z)\mathrm{d}U}. \tag{17.51}$$

When the protection levels of mission-critical devices or systems must be determined, the minimum reduction level of specific field features guaranteed by the shielding configuration are generally looked for. To this end, three quantities may be introduced:

(i) *the minimum Peak-Value Reduction Shielding Effectiveness*, $\min\left(\mathrm{SE}_{E,H}^{PR}\right)$,

(ii) *the minimum Derivative Reduction Shielding Effectiveness*, $\min\left(\mathrm{SE}_{E,H}^{DR}\right)$,

(iii) *the minimum Incident Energy Shielding Effectiveness*, $\min(\mathrm{SE}_W)$.

As in resonant cavities in the frequency domain, also in the time domain, it may happen that an enhancement of some observed electromagnetic quantities occurs in some positions inside a shielded volume when a shielding enclosure is applied.

In [30], the authors have conducted an interesting comparison between the proposed figures of merits on an enclosure with an aperture considering an impinging plane wave. The study was conducted on simulated and measured results. The comparison between the five TD figures of merit showed that:

- the reliability of the electric field indicators is better, since discrepancies for the electric field are lower than for the magnetic field;
- among the five figures of merit, the ASE_W indicator is the most restrictive and is recommended against the others;
- for the practical cases, it is interesting to use the ASE_W parameter, as it combines both field distributions;
- the proposed TD ASE_W figure of merit can be used alternatively to the proposed FD SE parameter from IEEE Std. 299.1-2013 valid for physically small and electrically large enclosures.

17.5 Statistical approach

17.5.1 When it is useful

In the frequency domain (FD), a statistical approach with reference to different positions in a shielded volume of enclosures has been considered in the past [31]. Moreover, it has been put in evidence that, in overmoded cavities, the choice of only one observation point may be misleading [32], because the electromagnetic field may be extremely varying point to point. However, it can be immediately recognized that the exact analysis of the field distribution in loaded overmoded cavities may be lengthy and even useless, since little characteristic variations in the loading can produce different modes excitation, reducing the general validity of any conclusion drawn [33]. Statistical studies on the electromagnetic field distribution inside metallic cavities date back several decades [34], but perhaps the most important impulse to this approach was given by the diffusion of reverberating chambers [35,36] for EMC/EMI testing. Different perspective angles were also considered to overcome the misleading information provided by the classical SE parameter [1,37,38].

In the time domain (TD), the situation is not too much different: It should be noted that any transient excitation may be thought of as a superposition of infinite monochromatic sinusoidal waveforms, thus resulting in a field distribution which is generally complex and different at any given time instant, at any position.

If a simple statistical investigation about the characteristics of the EM field distribution in loaded and unloaded cavities is pursued, only the *probability density function* (PDF) $P(x)$ and the *cumulative distribution function* (CDF) $F_X(x)$ may be investigated for the above introduced TD shielding parameters. They are evaluated in the highest number of observation points inside the protected volume compatible with the computational burden, and considered either relevant and significant. A deeper investigation on the statistical features of the EM field distribution and of the shielding parameters should be in order and useful in a limited number of cases.

It should be observed that, although the PDF's are characterized by trends that are difficult to interpret and use for shielding design or comparisons, their corresponding CDF's are much easier to handle and offer very useful information about the EM behavior of the enclosures under the considered specific excitation [13].

17.5.2 Statistical analyses

The first statistical check we can conduct concerns the *skewness* γ and the *kurtosis* β. We recall that the skewness is a measure of the asymmetry of the pdf with respect to the median value, while the kurtosis measures how heavily the tails of a distribution differ from the tails of a normal distribution. Kurtosis should not be not confused with peakedness or flatness. In other words, it identifies whether the tails of a given distribution contain extreme values, i.e., whether outliers are present in the distribution.

Biased skewness is computed according to the Fisher–Pearson formula, as

$$\gamma = \frac{\frac{1}{N}\sum_{i=1}^{N}(y_i - \bar{x})^3}{\sigma^3}, \tag{17.52}$$

where $\bar{x}$ is the mean

$$\bar{x} = \frac{1}{N}\sum_{i=1}^{N} x_i \tag{17.53}$$

and σ is the standard deviation of the distribution

$$\sigma = \sqrt{\frac{1}{N}\sum_{i=1}^{N}(x_i - \bar{x})^2}. \tag{17.54}$$

The skewness is zero for a normal distribution and any symmetric population should have a skewness near zero. Negative values indicate that the population is skewed left (i.e., the left tail is longer than the right tail) and positive values indicate a population skewed right (i.e., the right tail is longer than the left tail). If the population are multi-modal, this may affect the sign of the skewness.

Often, the adjusted or bias-corrected Fisher–Pearson skewness is computed as

$$\gamma_0 = \sqrt{\frac{N(N-1)}{N-2}}\gamma. \tag{17.55}$$

The adjustment approaches 1 as N gets large.

For univariate data $\{y_1, y_2, y_3 \ldots y_N\}$, the kurtosis is computed as

$$\beta = \frac{\frac{1}{N}\sum_{i=1}^{N}(y_i - \bar{y})^4}{\sigma^4}. \tag{17.56}$$

The kurtosis for a standard normal distribution is three. Kurtosis values greater than three indicates a *heavy-tailed* distribution, indicating large outliers. Values lower than three indicates a distribution with *flat tails*, due to the small outliers in a distribution.

A second series of statistical checks we can perform are the tests of Kolmogorov–Smirnov, Lilliefors, Anderson–Darling and Chi-square goodness-of-fit, to verify, with some degree of significance, whether the shielding computed data may be associated with a population of continuous normal distribution or with one different from normal, e.g., exponential, lognormal, Weibull or extreme value distributions.

The main PDFs used in our computations are defined by the following functions:

Exponential PDF:

$$P(x;\alpha) = \begin{cases} \alpha e^{-\alpha(x-\mu)} & x > 0 \\ 0 & \text{otherwise}, \end{cases} \tag{17.57}$$

where μ is the location parameter and α is the scale parameter. The case where $\mu = 0$ and $\lambda = 1$ is called the standard exponential distribution.

Lognormal PDF:

$$P(x;\gamma;\mu;\alpha) = \frac{1}{(x-\mu)\gamma\sqrt{2\pi}} \exp\left\{\frac{1}{2\gamma^2}\left[\frac{\ln(x-\mu)}{\alpha}\right]^2\right\}$$

$$x > \mu;\ \ \alpha, \gamma > 0 \tag{17.58}$$

where γ is the shape parameter (it is the standard deviation of the log of the distribution), μ is the location parameter and α is the scale parameter (it is also the median of the distribution). The case where $\mu = 0$ and $\alpha = 1$ is called the standard lognormal distribution.

Weibull PDF:

$$P(x;\gamma;\mu;\alpha) = \frac{\gamma}{\alpha}\left(\frac{x-\mu}{\alpha}\right)^{\gamma-1} \exp\left[-\left(\frac{x-\mu}{\alpha}\right)^{\gamma}\right] \quad x \geq \mu;\gamma, \quad \alpha > 0. \tag{17.59}$$

When $\mu = 0$ and $\alpha = 1$, the PDF is called the standard Weibull distribution. The case where $\mu = 0$ is called the 2-parameter Weibull distribution.

Extreme Value PDF:

$$P(x;\mu;\alpha) = \frac{1}{\alpha} e^{\frac{x-\mu}{\alpha}} \cdot \exp\left[-e^{\frac{x-\mu}{\alpha}}\right] \quad -\infty < x < \infty, \quad \alpha > 0. \tag{17.60}$$

For the definition of the Kolmogorov–Smirnov, Lilliefors, Anderson–Darling and Chi-square goodness-of-fit, the interested reader may find several statistical textbooks, e.g. [39].

The third statistical analysis concerns the fitting of the CDFs of the computed data using several distributions as possible candidates: beta, binomial, Birnbaum-Saunders, Burr, exponential, extreme value, gamma, generalized extreme value, generalized Pareto, inverse Gaussian, kernel, logistic, log-logistic, log-normal, Nakagami, negative binomial, normal, Poisson, Rayleigh, Rician, t-location scale, and Weibull. We have found that, according to the previous step, the GEV distribution provides the best approximation for the distribution of the TD shielding parameters CDF inside an enclosure. The GEV PDF is defined as:

$$y = \frac{1}{\sigma} e^{-t} t^{k+1}, \tag{17.61}$$

where

$$t = \left(1 + k\frac{x - \mu}{\sigma}\right)^{-\frac{1}{k}}, \tag{17.62}$$

and μ, σ, and k are, respectively, the location parameter, the scale parameter, and the shape parameter (we assume $k \neq 0$). The CDF is defined as

$$y = e^{-t}. \tag{17.63}$$

17.6 Results

As in [13], we have considered an electrostatic discharge (ESD) [40–42] as typical transient source of interference. We have considered a metallic enclosure of a computer case whose dimensions along the three axes x, y, z are 470 × 180 × 400 mm, as shown in Figure 17.6; a real 3D model of the enclosure has been developed in Autocad. The ESD has been simulated by means of a known impressed current

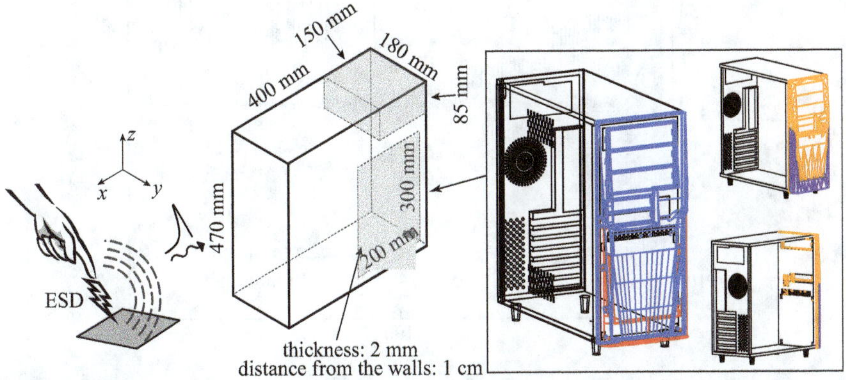

Figure 17.6 Shielding enclosure under transient excitation (dimensions are 470 × 180 × 400 mm) with a typical content represented by a printed circuit board and a power supply unit

flowing through a conducting path [43]:

$$i(t) = I_1(1 - \exp[-t/\tau_1])^p \exp[-t/\tau_2] + I_2(1 - \exp[-t/\tau_3])^q \exp[-t/\tau_4] + \\ + I_3(1 - \exp[-t/\tau_5])^r \exp[-t/\tau_6],$$

where the values of the parameters I_k and τ_k are reported in [43], along with the exponents p, q, r.

The ESD channel is parallel to the z-axis and located at 162 mm from the closest side of the victim, centered with respect to the yz face.

The integrals in the formulae for the computation of the global figures of merit have been computed using a grid of 195 probes positioned according to a Gaussian integration rule along the three x-, y- and z-coordinate axes. We have used the commercial software CST Microwave Studio to perform the simulations, using a mesh grid with 1,523,312 cells. To perform the volumetric integration of the field quantities, we have used a Gaussian quadrature rule over an elementary cube with equal sides h that provides a fourth order accuracy $\mathcal{O}(h^4)$ [44]. Thus, we have subdivided the shielded enclosure in elementary cubic cells where the integration is performed. We have chosen the side h of the elementary cell in order to ensure an accurate sampling at the highest frequency of interest, i.e., at least 10 samples per wavelength ($h \leq \lambda/10$). Different approaches are also available [4,42,45].

In Figure 17.7, we show the time trend of the z-component of the electric field E_z without and with the shielding enclosure, at the center position of the protected volume, in the absence of the load.

In Figure 17.8, we show the maps of z-component of the electric field in the planes yz, xz and xy, respectively, with reference to the time instant in which the peak

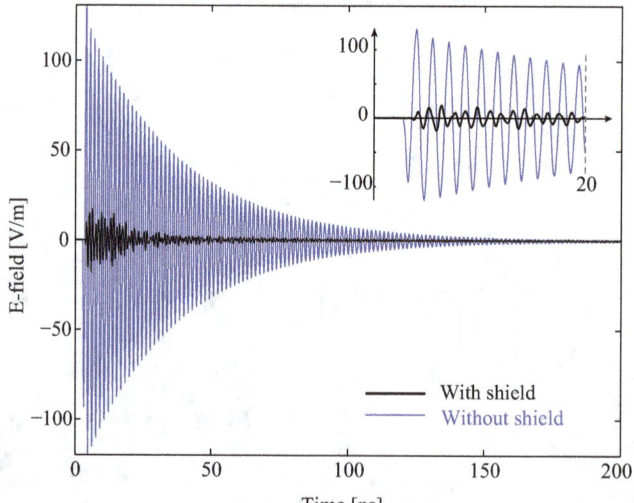

Figure 17.7 Electric field z-component without and with the shielding enclosure, assuming as observation point the center of the protected volume: in the inset the first 20 ns are shown

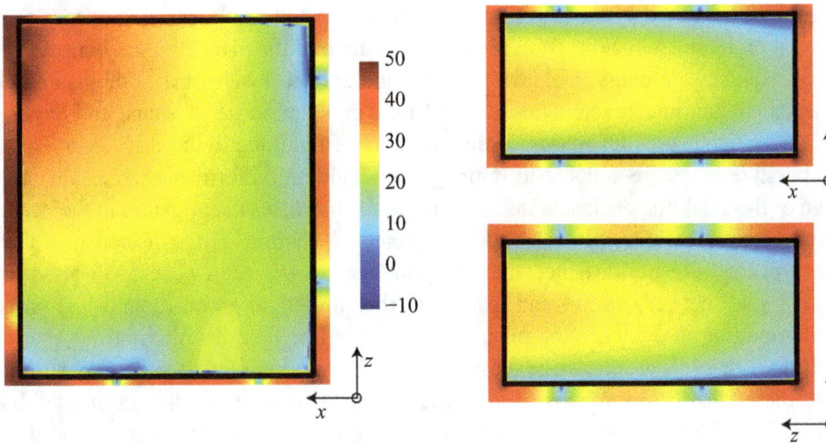

Figure 17.8 Maps of the electric field z-component, in the time instant in which the peak value is reached at the center position of the shielding enclosure

Table 17.1 Values of the figures of merit

(a) Global shielding parameters		
Figure of merit	**Empty (dB)**	**Loaded (dB)**
ASE_E^{PR}	12.4	9.3
ASE_H^{DR}	10.8	7.2
ASE_W	21.8	14.3
$\min\left(SE_E^{PR}\right)$	−3.9	−4.8
$\min\left(SE_H^{DR}\right)$	−8.4	−8.9
$\min(SE_W)$	−4.7	−4.3
(b) Local (center) shielding parameters		
Figure of merit	**Empty (dB)**	**Loaded (dB)**
SE_E^{PR}	14.6	9.2
SE_H^{PR}	40.1	20.3
SE_E^{DR}	28.2	18.3
SE_H^{DR}	21.1	10.0
SE_W	38.0	26.2

value is reached at the center of the enclosure. To investigate the influence of the content on the proposed parameters, the calculations have been also conducted in the presence of a printed circuit board and a power supply unit. Even though positions and time instants were different with respect to the unloaded empty configuration, the observed variations in the proposed figures of merit were limited, as shown in Table 17.1a, where we report also the minimum values of the global figures of merit.

In the considered system configuration, the computed results put in evidence that the worst problems may arise from the time derivative of the magnetic flux density. As in the frequency domain, shielding performance are the result of a combination of elements: excitation characteristics and EM field penetration paths, shape and dimensions of the enclosure, absorption of the interior EM field due to the content or walls. Two possible ways are suitable to improve the shielding performance: to modify the design of the shielding enclosure and/or to place the most sensitive devices in the better (i.e., the most protected) places inside the protected volume. The consciousness that some types of TD performance may be poor (i.e., lower than necessary) at some locations may be very useful and lead to further investigation on a statistical basis aimed at putting in evidence the main aspects of the EM field distribution.

In Figure 17.9, we show the classical SE values, computed at several positions inside the enclosure by means of a Fourier Transform. The SE data computed by means of CST Microwave Studio in the time domain are compared with the data computed by means of the commercial code FEKO (based on the Method of Moments) in the frequency domain for validation purposes. In Figure 17.9, the dashed vertical lines mark the resonant frequencies of a rectangular lossless cavity without apertures, having the same dimensions of the considered enclosure.

It is evident that no easy way exists for the comparison of the actual TD shielding performance with the predicted FD spectra of the SE. Since the spectrum of the source appear to be relevant up to 1 GHz, one could argue that at least 10 dB of protection are achieved, but results in Table 17.1a reveal that, at some positions, the actual shielding is much lower than that predicted.

In Table 17.1b, the values of the local TD parameters are reported with reference to the center of the enclosure.

Some relevant comments can be inferred by the values reported in Table 17.1a and b:

1. there exist positions where the effectiveness of the shielding enclosure is extremely poor;

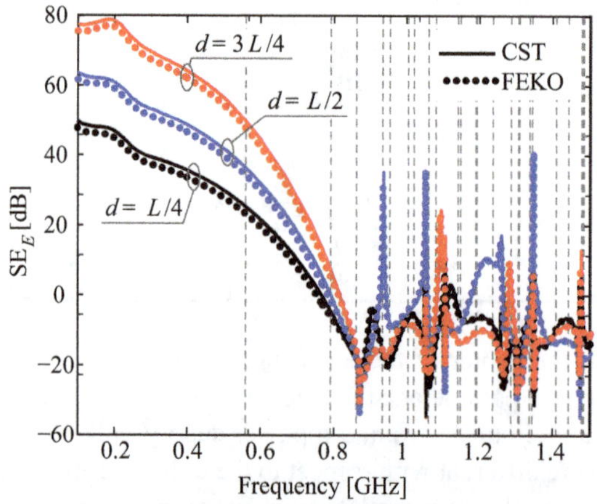

Figure 17.9 Frequency spectrum of the classical SE at three different positions

2. the volume averaged performance may range of several dB depending on which figure of merit is looked for;
3. the volume averaged performances are much less sensitive to the presence of the considered enclosure content with respect to the local parameters; and
4. the presence of the simple considered loads (a printed circuit board and a PEC body) may worsen local shielding parameters up to several dB.

The first observed behavior is corresponding to field (or time-derivative) enhancement due to field reflections at boundary PEC walls and is analogous to the field enhancement occurring in the frequency domain because of resonances. From the last two observations, it may be inferred that the presence of the considered loads changes the most critical points within the protected volume but the values of the average and worst performance are not considerably affected.

As concern the statistical investigation, the first statistical check concerns the kurtosis β and the skewness γ of the computed performances: the results are summarized in Table 17.2 and confirm that, since β is always less than 3, the distribution is flat-topped (with respect to a normal Gaussian distribution) and that the left tail is more pronounced than the right tail, because of the negative values of γ. These values mean that, in the considered example, the performance values observed in the various locations are more "spread" with respect to a normal Gaussian distribution and that the locations where the performance is lower than the average value are more than those where the average value is exceeded.

A second series of statistical checks have been the tests of Kolmogorov–Smirnov, Lilliefors, Anderson–Darling, and Chi-square goodness-of-fit, to verify whether, with some degree of significance, the computed data may be associated with a population of continuous normal distribution or with one different from normal, e.g., extreme value, exponential, lognormal or Weibull distributions. All the tests failed, that is to say they rejected the null hypothesis with a 5% significance level, except the Kolmogorov–Smirnov test [46] when using the GEV CDF. This means that the difference between the sampled CDF (either of peak values of the electric field, magnetic field and energy) and the continuous GEV CDF is not significant enough to say that they have different distributions. Anyway the resulting *p*-values are always below 0.5 (the *p*-value ranges from 0 to 1) – 0.060 for the peak electric field CDF, 0.25 for the peak time derivatives

Table 17.2 Values of kurtosis and skewness in the distributions of the local (center) shielding parameters

	Empty enclosure		Loaded enclosure	
Figure of merit	β_{Empty}	γ_{Empty}	β_{Loaded}	γ_{Loaded}
SE_E^{PR}	2.19	−0.48	2.29	−0.51
SE_H^{PR}	2.24	−0.31	2.38	−0.30
SE_E^{DR}	2.14	−0.45	2.39	−0.51
SE_H^{DR}	2.35	−0.42	2.55	−0.42
SE_W	2.16	−0.42	2.24	−0.45

of the electric field CDF, 0.34 for the peak magnetic field CDF, 0.37 for the peak time derivatives of the magnetic field CDF, and 0.22 for energy CDF – and show somehow that the probability that the sampled values really come from a GEV distribution is low, even if the test is passed, that is to say that the null hypothesis is accepted. These tests are useful to decide whether a fitting of the performance values distribution is reasonable and feasible or not.

The third statistical analysis has concerned the fitting of the CDFs of the computed data using the previously mentioned distributions as candidates. We have found that, according to the previous step, the GEV distribution provides the best approximation. The CDF of the computed data compared with the fitted continuous distributions are shown in Figures 17.10 and 17.11. It is interesting to note that the fitted CDFs show a very good agreement with the computed data even if any attempt to fit the PDFs did not succeed. This means that the statistical approach fails to predict how many samples are within a specified interval, but succeeds to predict, with reasonable accuracy, how many samples are below a prescribed

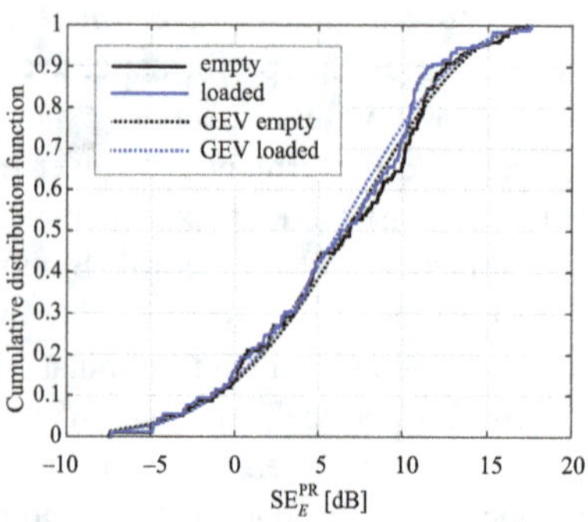

(a) SE_E^{PR} (empty enclosure: $k = -0.449$, σ = 5.482, $\mu = 5.132$)

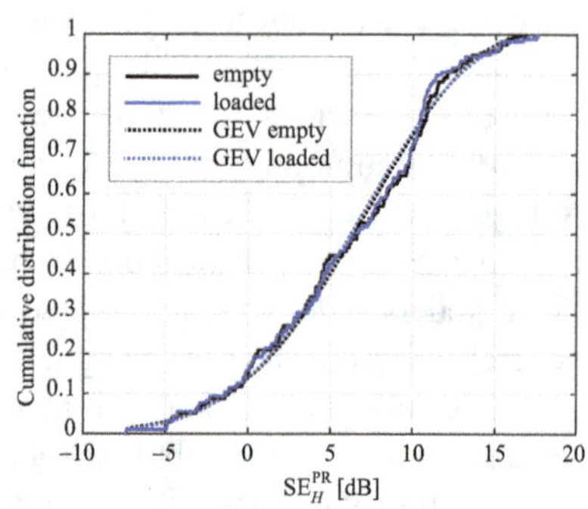

(b) SE_H^{PR} (empty enclosure $k = -0.428$, σ = 5.835, $\mu = 4.733$)

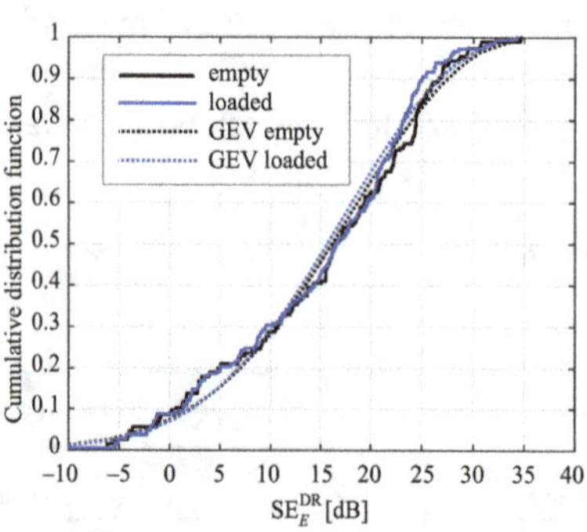

(c) SE_E^{DR} (empty enclosure $k = -0.453$, σ = 10.537, $\mu = 12.577$)

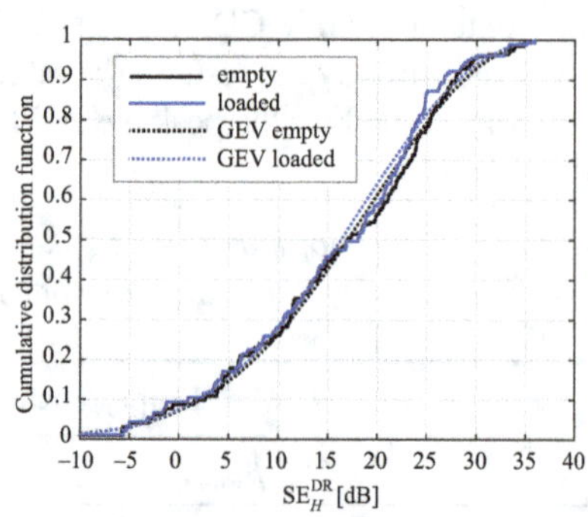

(d) SE_H^{DR} (empty enclosure $k = -0.455$, σ = 10.909, $\mu = 13.355$)

Figure 17.10 Cumulative distribution functions for the peak and derivative reductions

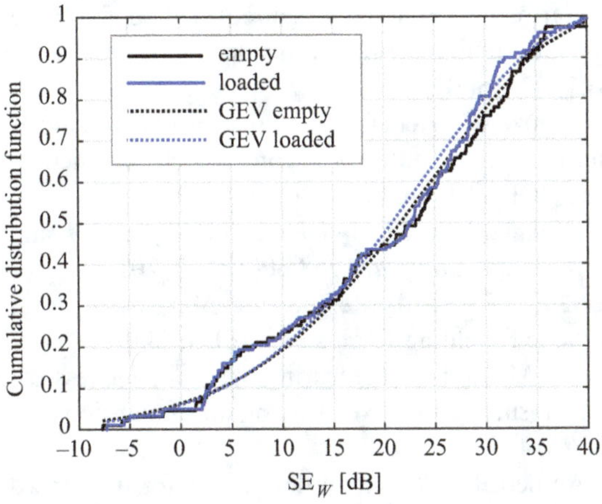

Figure 17.11 Cumulative distribution functions for the SE^W *(empty enclosure: $k = -0.515$, $\sigma = 12.918$, $\mu = 17.315$)*

threshold, which is of great practical interest in emission and susceptibility tests where it is necessary to comply with upper limits or when the goal is the prediction of the adequacy of the achievable protection levels.

17.7 A word of caution: the influence of load on the shielding performance

It should be observed that although the field distribution may strongly depend on the content [47], it is generally useful to assess the performance achievable by means of an empty enclosure, since the load may be variable or even unknown a priori. Of course, the availability of a content representative of actual loading could be of guidance in the prediction of the EM shielding behavior in practical applications and would imply generally an additional computational effort (with respect to the empty case) which depends on the discretization constraints ensuing from the representative content characteristics.

References

[1] Marvin AC, Dawson JF, Ward S, *et al.* A proposed new definition and measurement of the shielding effect of equipment enclosures. *IEEE Trans. Electromagn. Compat.* 2004;46(3):459–468.

[2] Araneo R, Celozzi S. Actual performance of shielding enclosures. In: *Proc. IEEE Electromagn. Compat. Symp.*, vol. 2; 2004. pp. 539–544.

[3] Li M, Nuebel J, Drewniak JL, *et al.* EMI from cavity modes of shielding enclosures – FDTD modeling and measurements. *IEEE Trans. Electromagn. Compat.* 2000;42(1):29–38.

[4] Celozzi S, Sarto MS. Equivalent source method for the evaluation of the electromagnetic field penetration inside enclosures. *IEEE Trans. Magn.* 1996;32(3):1497–1500.

[5] Araneo R, Celozzi S. Actual challenges in electromagnetic shielding. In: 2015 International Conference on Electromagnetics in Advanced Applications (ICEAA); pp. 867–870.

[6] Schulz RB, Plantz VC, Brush DR. Shielding theory and practice. *IEEE Trans. Electromagn. Compat.* 1988;30(3):187–201.

[7] IEEE Std. 299: Standard Method for Measuring the Effectiveness of Electromagnetic Shielding Enclosures. IEEE; 2007.

[8] IEEE Standard Method for Measuring the Effectiveness of Electromagnetic Shielding Enclosures and Boxes Having all Dimensions between 0.1 m and 2 m. IEEE; 2014.

[9] Celozzi S, Araneo R. Alternative definitions for the time-domain shielding effectiveness of enclosures. *IEEE Trans. Electromagn. Compat.* 2014;56(2): 482–485.

[10] Araneo R, Celozzi S. Suitability of new definitions of shielding effectiveness for enclosures. In: Proc. International Symposium EMC EUROPE, Barcelona, Spain; 2006. pp. 6–10.

[11] Yan Z, Qin F, Cai J, Zhong S, Lin J. Shielding performance of materials under the excitation of high-intensity transient electromagnetic pulse. *IEEE Access* 2021;9:49697–49704.

[12] Yan Z, Qin F, Cai J. Shielding effectiveness of materials under the excitation of high-power microwave. *IEEE Trans. Electromagn. Compat.* 2020;62 (5):2317–2320.

[13] Araneo R, Attolini G, Celozzi S, *et al.* Time-domain shielding performance of enclosures: a comparison of different global approaches. *IEEE Trans. Electromagn. Compat.* 2016;58(2):434–441.

[14] Cai J, Sun X, Zhao X. A total energy reduction shielding effectiveness description for enclosures. *Int. J. Appl. Electromagn. Mech.* 2016;50(2):283–295.

[15] Chen P, Mao C, Wu G. Shielding effectiveness assessment of enclosure by means of EMP excitation. In: 2016 Asia-Pacific International Symposium on Electromagnetic Compatibility (APEMC). vol. 01; pp. 273–275.

[16] Celozzi S. Shielding effectiveness prediction in MIL-STD 285 loop source configuration. In: 1998 IEEE EMC Symposium. International Symposium on Electromagnetic Compatibility. Symposium Record (Cat. no. 98CH36253). vol. 2. IEEE; 1998; pp. 1125–1130.

[17] NSA 94-106: National Security Agency Specification for Shielded Enclosures. NSA; 1994.

[18] Bridges JE. Proposed recommended practices for the measurement of shielding effectiveness of high-performance shielding enclosures. *IEEE Trans. Electromagn. Compat.* 1968;10(1):82–94.

[19] Catrysse J, Vanhee F, Pissoort D, *et al.* Differences between NSA 94-106 and IEEE 299 LF magnetic shielding measurements. In: IEEE International

Symposium on Electromagnetic Compatibility (EMC), Dresden, Germany; 2015. pp. 13–16.

[20] Bannister PR. New theoretical expressions for predicting shielding effectiveness for the plane shield case. *IEEE Trans. Electromagn. Compat.* 1968;(1):2–7.

[21] Bannister PR. Further notes for predicting shielding effectiveness for the plane shield case. *IEEE Trans. Electromagn. Compat.* 1969;(2):50–53.

[22] Whitehouse ACD. Screening: new wave impedance for the transmission line analogy. *IEE Proc.* 1969;116(7):1159–1164.

[23] Celozzi S, Araneo R, Lovat G. *Electromagnetic Shielding*. Hoboken: Wiley-IEEE; 2008.

[24] Moser JR. Low-frequency low-impedance electromagnetic shielding. *IEEE Trans. Electromagn. Compat.* 1988;30(3):202–210.

[25] Moser JR. Low-frequency shielding of a circular loop electromagnetic field source. *IEEE Trans. Electromagn. Compat.* 1967;9(1):6–18.

[26] Ryan CM. Computer expression for predicting shielding effectiveness for the low-frequency plane shield case. *IEEE Trans. Electromagn. Compat.* 1967;9(2):83–94.

[27] Celozzi S, D'Amore M. Magnetic field attenuation of nonlinear shields. *IEEE Trans. Electromagn. Compat.* 1996;38(3):318–326.

[28] Hansen RC. Shielding formulas for near fields. *Microwave Journal.* 2000;43 (11):154–167.

[29] Araneo R, Celozzi S. Exact solution of the low-frequency coplanar loops shielding configuration. *IEE Proc. Sci. Meas. Tech.* 2002;149(1):37–44.

[30] Garcìa-Pèrez LG, Lozano-Guerrero AJ, Blázquez-Ruiz JM, *et al.* Time-domain shielding effectiveness of enclosures against a plane wave excitation. *IEEE Trans. Electromagn. Compat.* 2017;59(3):789–796.

[31] Bunting CF, Yu SP. Field penetration in a rectangular box using numerical techniques: an effort to obtain statistical shielding effectiveness. *IEEE Trans. Electromagn. Compat.* 2004;46(2):160–168.

[32] Khan ZA, Bunting CF, Deshpande MD. Shielding effectiveness of metallic enclosures at oblique and arbitrary polarizations. *IEEE Trans. Electromagn. Compat.* 2005;47(1):112–122.

[33] Price RH, Davis HT, Wenaas EP. Determination of the statistical distribution of electromagnetic-field amplitudes in complex cavities. *Phys. Rev. E.* 1993;48(6):4716.

[34] Donaldson E, Free WR, Robertson DW, *et al.* Field measurements made in an enclosure. *Proc. IEEE.* 1978;66(4):464–472.

[35] Freyer GJ, Hatfield MO, Johnson DM, *et al.* Comparison of measured and theoretical statistical parameters of complex cavities. In: IEEE International Symposium on Electromagnetic Compatibility (EMC), Santa Clara, USA; 1996. pp. 250–253.

[36] Hill DA, Ladbury JM. Spatial–correlation functions of fields and energy density in a reverberation chamber. *IEEE Trans. Electromagn. Compat.* 2002;44(1):95–101.

[37] Celozzi S. New figures of merit for the characterization of the performance of shielding enclosures. *IEEE Trans. Electromagn. Compat.* 2004;46(1):142–142.

[38] Klinkenbusch L. On the shielding effectiveness of enclosures. *IEEE Trans. Electromagn. Compat.* 2005;47(3):589–601.

[39] Ross SM. *Introduction to Probability and Statistics for Engineers and Scientists*, 6th ed. London: Academic Press; 2021.

[40] Maheshwari P, Li T, Lee JS, *et al.* Software-based analysis of the effects of electrostatic discharge on embedded systems. In: IEEE 35th Annual Computer Software and Applications Conference; 2011; pp. 436–441.

[41] Muchaidze G, Koo J, Cai Q, *et al.* Susceptibility scanning as a failure analysis tool for system-level electrostatic discharge (ESD) problems. *IEEE Trans. Electromagn. Compat.* 2008;50(2):268–276.

[42] Cerri G, de Leo R, de Rentiis R, Primiani M. ESD field penetration through slots into shielded enclosures: a time domain approach. *IEEE Trans. Electromagn. Compat.* 1997;39(4):377–386.

[43] Yuan Z, Li T, He J, *et al.* New mathematical descriptions of ESD current waveform based on the polynomial of pulse function. *IEEE Trans. Electromagn. Compat.* 2006;48(3):589–591.

[44] Abramowitz M, Stegun LA. *Handbook of Mathematical Functions*. New York, NY: Dover Publications Inc.; 1972.

[45] Araneo R, Lovat G. Fast MoM analysis of the shielding effectiveness of rectangular enclosures with apertures, metal plates, and conducting objects. *IEEE Trans. Electromagn. Compat.* 2009;51(2):274–283.

[46] Kolmogorov AN. *Foundations of the Theory of Probability*. Eastford, CT: Martino Fine Books.

[47] Hill DA. *Electromagnetic Fields in Cavities: Deterministic and Statistical Theories, vol. 35*. New York, NY: John Wiley & Sons; 2009.

Index

www.ingramcontent.com/pod-product-compliance
Lightning Source LLC
LaVergne TN
LVHW011538200926
848484LV00001B/280